Reversibility of Chronic Degenerative Disease and Hypersensitivity

Regulating Mechanisms of Chemical Sensitivity

VOLUME 1

Reversibility of Chronic Degenerative Disease and Hypersensitivity

Regulating Mechanisms of Chemical Sensitivity

VOLUME 1

William J. Rea, M.D.
Kalpana Patel, M.D.

CRC Press is an imprint of the
Taylor & Francis Group, an **informa** business

CRC Press
Taylor & Francis Group
6000 Broken Sound Parkway NW, Suite 300
Boca Raton, FL 33487-2742

© 2010 by Taylor and Francis Group, LLC
CRC Press is an imprint of Taylor & Francis Group, an Informa business

No claim to original U.S. Government works

Printed in the United States of America on acid-free paper
10 9 8 7 6 5 4 3 2 1

International Standard Book Number: 978-1-4398-1342-3 (Hardback)

This book contains information obtained from authentic and highly regarded sources. Reasonable efforts have been made to publish reliable data and information, but the author and publisher cannot assume responsibility for the validity of all materials or the consequences of their use. The authors and publishers have attempted to trace the copyright holders of all material reproduced in this publication and apologize to copyright holders if permission to publish in this form has not been obtained. If any copyright material has not been acknowledged please write and let us know so we may rectify in any future reprint.

Except as permitted under U.S. Copyright Law, no part of this book may be reprinted, reproduced, transmitted, or utilized in any form by any electronic, mechanical, or other means, now known or hereafter invented, including photocopying, microfilming, and recording, or in any information storage or retrieval system, without written permission from the publishers.

For permission to photocopy or use material electronically from this work, please access www.copyright.com (http://www.copyright.com/) or contact the Copyright Clearance Center, Inc. (CCC), 222 Rosewood Drive, Danvers, MA 01923, 978-750-8400. CCC is a not-for-profit organization that provides licenses and registration for a variety of users. For organizations that have been granted a photocopy license by the CCC, a separate system of payment has been arranged.

Trademark Notice: Product or corporate names may be trademarks or registered trademarks, and are used only for identification and explanation without intent to infringe.

Library of Congress Cataloging-in-Publication Data

Rea, William J.
   Reversibility of chronic degenerative disease and hypersensitivity / William J. Rea.
      p. ; cm.
   Includes bibliographical references and index.
   ISBN 978-1-4398-1342-3 (hardcover : alk. paper)
   1. Chronic diseases. 2. Multiple chemical sensitivity. I. Title.
   [DNLM: 1. Chronic Disease--prevention & control. 2. Hypersensitivity--prevention & control. 3. Environmental Pollution--adverse effects. 4. Homeostasis--physiology. WD 300 R281r 2010]

RB156.R43 2010
616'.044--dc22
                                                                                       2010003795

Visit the Taylor & Francis Web site at
http://www.taylorandfrancis.com

and the CRC Press Web site at
http://www.crcpress.com

This book is dedicated, first, to Hartmut Heine, PhD, Professor of Anatomy Emeritus at the University of Witten-Herdecke Witten, North Rhine-Westphalia, Germany. He helped coalesce Professor Alfred Pischinger's data and concepts about the holistic approach to health and pushed further the basic science of homeostasis. Second, this book is dedicated to Theron Randolph, MD, who is the father of environmental medicine and who developed the concept of individual susceptibility to ambient chemicals in air, food, and water. He also defined the clinical adaptation syndrome for humans that had previously been worked out in animals by Hans Selye. He conceived, designed, and contracted the first environmental control unit. Third, we dedicate this book to Carlton Lee, MD, who developed the intradermal provocative neutralization (IDPN) technique for foods and chemicals, which was derived from Dr. Herbert Rinkle's intradermal serial dilution techniques. In addition to the Environmental Control Unit, the IDPN technique has been the cornerstone of diagnosis and treatment for the patient with food and chemical sensitivity.

Contents

Prologue xiii
Acknowledgments xv
Overall Perspective xvii
About the Authors xix

Chapter 1   The Physiologic Basis of Homeostasis 1

Introduction 1
Overview of the Dynamics of Noxious Incitant Entry and Fate 3
    Nature of the Incitant Stimulus 4
    Nature of the Body's Homeostatic Response 6
    Nature of Information Intake and Distribution 7
    Automaticity of the Body—Energy 8
    Clinical Signs and Symptoms after Chronic Noxious Incitant Entry 9
The Body's Communication System—Ground Regulation System 10
    Anatomical Categories Used in Communication for
    Regulating the Homeostatic Mechanism 12
    Skin and Mucous Membranes 12
    Connective Tissue Matrix 12
        Mechanical Support System (Support Structure) 12
        Communication System Properties 13
        Connective Tissue and Electromagnetic Energy 25
        Molecular Sieve 29
        Latent Free Radical Response 32
    Vascular Function—Homeostasis 48
        Diffusion 48
        Vascular Sieve 49
        Microcirculation 52
        Vascular Tone 55
        Lymphatics 56
    Intravascular Content 62
        Cells 67
        Plasma Proteins 82
        Plasma Lipids 83
        Neurological Aspects of Homeostasis 85
Pollutant Entry and the Body's Homeostatic Response to and Fate
of the Noxious Stimuli 85
    Total Environmental Load 85
    Total Body Load 98
    Local Receptors 99
        Local Homeostatic Responses and Information Reception 99
        Matrix Receptor 103
    Adjustment Responses 107
        Local Cell and Matrix Reactions for the Adjustment Response 107
        Regional Homeostatic Adjustment 110

Central Homeostasic Adjustment 112
Principles and Facts about Adjustment Responses 117
Defense System 132
Redox System Latency 132
The Early Defense Mechanism 133
Tissue Response to Local Entry (Initially Subinflammatory) 135
Immune Response 137
Nonimmune Detoxification in Relation to Homeostasis and the Defense
Mechanism 139
Repair Mechanism—Healing 143
Nonspecific Mesenchyme Reaction 144
Rate of Enzyme and Nutrient Deficiency 155
Homeostatic Dysfunction Leading to a Disordered Homeostatic Response 156
Periodic Response 156
Aperiodic Response 158
Specific Mechanisms of Inflammation and Function of Neutrophils
and Macrophages 159
Proinflammatory Cytokines 165
Interleukin-6 165
Neurological System Connection with the Connective Tissue Matrix
and Inflammation 166
Summary 167
References 167

SECTION   Amplification Systems: Neurological, Immune, and Endocrine

Chapter 2   Nervous System 185

Introduction 185
Linear and Biphasic Effects of Noxious Stimuli Entry 187
Receptors 188
Autonomic Nervous System 190
Anatomy and Physiology of the Peripheral Autonomic Nervous System 190
Central Connection of the Autonomic Nervous System with the
Hypothalamus 198
Reflex Pathways 202
Dysfunction of the Autonomic Nervous System 205
Denervation Supersensitivity 212
Law of Denervation 215
Denervation and Spondylosis: Neuropathic Pain 219
Neurally Mediated Hypotension and Tachycardia 220
Neuromuscular Hypotension and Chronic Fatigue 223
Reflex Sympathetic Dystrophy (Causalgia) 227
Pollutant Injury to the Eye 228
Pollutant Injury to the Nervous System of the Eye 228
Cranial Nerves 228
Autonomic Nerves 234
Neurogenic Vascular Responses to Pollutant Stimuli 237

Contents

Pathogenesis of Neuroimmunological Mediators (Neuropeptide Triggering by Noxious Stimuli) 243
The Acupuncture Energy Flow System (AES) 251
    NAET 260
Voluntary Central Nervous System 260
    Noxious Injury to the Blood–Brain Barrier 260
    Noxious Injury after Penetration of the Blood–Brain Barrier 264
    Unique Biochemistry 266
    General Principles of Neuronal Physiology 270
    Principles of Response after Toxic Exposure 278
    Reversibility vs. Irreversibility of Cell Damage after Pollutant Exposure 278
        Mechanism of Acute Central Nervous System Injury 279
    Neurological Effects of Toxic Chemicals 286
    Fixed-Named Disease 288
    Toxic Neuropathy 288
Summary 317
References 317

Chapter 3   Immune System 337

Introduction 337
Properties of Entering Noxious Excitants 337
    Stem Cells 338
Lymph Nodes, Lymphatic Organs, and the Lymphatic System 342
    Development of Lymphatic Channels and Lymphocytes 342
Regionalization of the Immune Response by Lymphatic Tissue Nodes
and Channels after Mucosal Entry 344
    Mucosa Associated Lymphoid Tissue (MALT); Nasal Associated Lymphoid
    Tissue (NALT); Bronchial Associated Lymphoid Tissue (BALT); Gut
    Associated Lymphoid Tissue (GALT) 344
Neuroimmune Regulation 350
Clinical Implications of Food and Chemical Sensitivity in Relation to the
Autonomic Nervous System and Immune System 351
Immunity 353
    Innate Immunity 353
    Transition to Innate and Humoral Immunity in the Newborn and Infant 355
    Acquired Immunity 357
Lymphocytes 360
    T-lymphocytes 362
    General Intrinsic Mechanisms 364
B-Lymphocytes 368
Onset of Antibodies 372
    Classes of Antibodies 374
    Inflammatory Effects 376
    Clinical Regulation of the Immune System 377
    Cell Cycle 389
    Abnormal Cell Cycle Progression in Patients with Chronic Fatigue
    Syndrome (CFS) 392
    Cell Cycle and Cancer 393
    Autoimmunity 395
Failure of the Tolerance Mechanism Causes Autoimmune Diseases 395

	Autogenous Lymphocytic Factor (ALF)	396
	Cell Death	396
	Allergy and Hypersensitivity	401
	Aging and the Immune System	402
	Summary	406
	References	406
Chapter 4	Endocrine System	419
	Introduction	419
	Integrated Physiology	419
	Pituitary Gland	428
	Pollutant Effects on the Nervous System of the Pituitary Gland	429
	Pollutant Effects on the Physiology of the Pituitary	429
	Growth Hormone	433
	Pineal Gland	434
	Neuroendocrine System (Paraganglia, Paraneuron)	436
	Neuroendocrine Phenotype	437
	Biosynthetic Profile	438
	Secretory Granules	440
	Specialization of Neuroendocrine Cells	440
	Topography of Neuroendocrine Cell Subsets	442
	Branchiomeric Group	442
	Intravagal Neuroendocrine Cells	444
	Visceral-Autonomic Paraganglion Cells	444
	Genitourinary Paraganglion Cells	445
	Paragangliomas	446
	Clinical Manifestations of Neuroendocrine Stimulation	446
	Adrenal Glands	447
	Physiology of the Dynamics of Homeostasis and Dyshomeostasis in the Adrenal Glands	447
	Clinical Picture of Pollutant Injury and Adrenal Dysfunction	454
	Hyperadrenalism	454
	Adrenal Insufficiency/Hypoadrenalism	454
	Parathyroid Gland	456
	Ovary	457
	Organization/Activation Hypothesis	460
	Male Homosexuality	461
	Sexual Dysfunction	468
	Uterus and Tubes	469
	Premenstrual Syndrome	469
	Vasculitis of the Reproductive System	472
	Endometriosis	473
	Spontaneous Abortion	478
	Hormones	479
	Influence of Toxics on Hormonal Homeostasis	479
	PCBs	479
	Pesticides/Herbicides	479
	Menopause Dysfunction	479
	Vagina and Vulva	480
	Testes and Seminal Vesicles	480

Mechanisms of Action and Fetal Vulnerability	484
Hormones and Neurobehavioral Effect	484
Beyond Endocrine Disruption	486
Implications and Ongoing Activities	487
Exogenous Estrogens	487
Phytoestrogens	488
Estrogens in Milk	491
Estrogenic Chemicals	492
Sertoli Cell Number and Sperm Output	492
Thyroid	493
Physiology and Pathophysiology	493
Agents Acting Directly on the Thyroid	495
Class I	495
Class II	496
Class III	496
Class IV	496
Class V	496
Agents Acting Indirectly on the Thyroid	497
General Properties, Distribution, and Epidemiology	498
Sulfurated Organics Thiocyanate (SCN), Isothiocyanates, and Thioglycosides (Goitrin)	498
Aliphatic Disulfides	499
Polyphenols	499
Phenolic and Phenolic-Carboxylic Derivatives	500
Resorcinol (1,3 Dihydroxybenzene)	500
2,4-DNP	501
Phthalate Esters and Phthalic Acid Derivatives: DHBAs	501
PCBs and PBBs	502
PAH, 3,4,-BaP, MCA, and 7,12-DMBA	504
Summary	505
References	505
Index	527

Prologue

One may think it strange for a surgeon and a pediatrician to team up and attempt to describe the reversibility of the mechanisms and anatomy of hypersensitivity and chronic degenerative disease. However, this unique combination has allowed us to have the broadest scope of thought on why people become ill and what we can do to help patients get better.

Chronic disease, whether it be hypersensitivity, or chronic degenerative disease, or a combination, prevails and should now be the basis of thinking for all physicians, scientists, and lay people in society. We hope this book will aid the clinician to trace inciting noxious substances in our environment to their initial disturbance of the metabolic processes, to the triggering of the adjustment response, followed by the induction of the defense and repair mechanism. These incitants trigger the homeostatic mechanism to overcome adversity, thus allowing the patient to obtain and maintain robust health without taking medications. Also, we emphasize the fact that most hypersensitivity and many chronic degenerative diseases if diagnosed early can be eliminated if the environmental triggers are removed or neutralized.

William J. Rea
Kalpana Patel

Acknowledgments

Thanks to all the great anatomists and physiologists of the ages, especially Arthur Guyton, MD; Alfred Pischinger, MD; Jonathan Brostoff, MD; Joe Miller, MD; Manfred von Ardenne, PhD; Hans Selye, MD; Walter Cannon, MD; Chan Gunn, MD; J. Alexander Bralley, PhD; Richard S. Lord, PhD; Jon Pangborn, PhD; Jean Monto, MD; Sherry Rogers, MD; and many others whose ideas and facts we used liberally to solidify the concepts of hypersensitivity and chronic degenerative disease. Special thanks to Martha Stark, MD, for her challenging inquiries and important insights into the specific details of the ground regulation system and homeostasis.

Thanks also to the members of the American Academy of Environmental Medicine, the Pan American Medical Society, the American Academy of Otolarynic Allergy, the American College of Nutrition, the Society of Thoracic Surgeons, and the American College of Surgeons for contributing their ideas, case reports, and clinical facets of hypersensitivity and chronic degenerative disease.

Thanks to Ya Qin Pan, MD, for her work with references and tabulation of data; Irwin Fenyves, PhD, for his statistical analysis; Bertie Griffiths, PhD, for his bacterial and immune studies, and development of autogenous lymphocytic factor; Ron Overberg, PhD, Jon Pangborn, PhD, Jeff Bland, PhD, and Jonathan Wright, MD, for their nutritional concepts and analysis; and Patricia Smith, MD, for collection of data and analysis. Also, thanks to Joel Butler, PhD, and Nancy Didrikson, PhD, for their consultations in behavioral medicine; Dr. David Hickey for his studies in thermography; Dr. Theodore Simon for his SPECT scan studies; Dr. Daniel Martinez and Dr. Kaye Kilburn for the balance studies; and Dr. Satoshi Isikawa for the development of the ANS measurements by pupillography. We also wish to thank deeply Deborah Singleton and her team at the Arasini Foundation for their development and practice of energy balancing work that helps the patient with chemical sensitivity and/or chronic degenerative disease, get well. Thanks to Judith Lyle for her ability to computerize this whole book and to Professor Chris Bishop and Dr. Stephanie McCarter, who corrected the English and helped make sense out of our ideas. Last but not least, we thank our families—Bill's wife, Vera, Kalpana's husband, Dilup, and her mother, Sharda Desai, who have endured this laborious work. We also thank Jeanette Plusnik for her generous support of these works.

Overall Perspective

There is much information about subunit function of the body in relation to chronic degenerative disease and hypersensitivity (chemical sensitivity) but little unified information on the holistic response of the integrated function of the body's homeostatic mechanism. It is the authors' contention that chronic overload of this mechanism causes homeostatic dysfunction from pollutant entry, information distribution, reception, and alteration in the adjustment mechanism, to metabolic changes (hypoxia, pH changes), tissue changes (sols to gels, gels to sols), nonspecific mesenchyme reaction and, eventually, to fixed end-stage disease. The clinician must understand, as well as possible, the known noxious stimuli kinetics (half-life, biotransformation toxicity and fate) and the integrated physiology of how the total environment, both external and internal, affects the environmental receptors in the body and triggers the near automatic adjustment mechanisms to stabilize the organism and to prevent chronic degenerative disease and chemical sensitivity. It is hoped that a better understanding of the dynamic mechanism of homeostasis, and the integration of its thousands and perhaps millions of subunits, will enable the clinician to prevent disease before it is fixed and, thus, becomes irreversible. It was Professors Alfred Pischinger and Hartmut Heine who developed the concept of the integrated relationship between physiology and the dynamics of the ground regulation system, which is the basis for the body's communication system.

In addition, this book emphasizes how the total environmental load, as well as the specific environmental load, affects, alters, and triggers the nongenetic, epigenetic and genetic responses of the homeostatic mechanism in generating chronic degenerative disease and chemical sensitivity. The authors have drawn information from their experiences and those of other environmental physicians especially Theron Randolph, MD, and surgeons, nutritionally oriented physicians, environmental scientists, biochemists, immunologists, physiologists, pathologists, and other sources. Over the last twenty years, the experience of environmentally and nutritionally oriented physicians and surgeons who have studied and treated patients with chronic degenerative disease and/or chemical sensitivity who are in distress has become vast, encompassing well over a million people worldwide. The authors of this book have studied and treated a combined total of 40,000 patients who were environmentally wounded and who developed either chemical sensitivity and/or chronic degenerative disease. Integrating basic scientific facts with this vast clinical experience has provided us with a holistic perspective for the prevention and treatment of body malfunction before end-organ failure occurs. The development of the environmental control unit by Dr. Theron Randolph for the study and treatment of the chronically ill patient has been a significant advance in diagnosing and treating the environmental aspects of chronic disease. The principles and facts about construction of these units have allowed new knowledge to be gleaned and used for building construction and remodeling of polluted homes, clinics, and hospital units.

Dr. Carlton Lee's development of the intradermal injection provocative-neutralization technique (which appears to be based on the concept of hormesis) has given us another wonderful tool that can be used for the treatment of adverse reactions to food, mold, chemical, bacterial, viral, and other substances. These injection techniques allow the clinician modalities that can push the dyshomeostasis induced by the environment back into normal function without the complications of long-term medication treatment. It is hoped that students of medicine in every specialty will incorporate this holistic perspective into their practice.

We would like to emphasize that the term *homeostasis* could be considered by some to be outdated since all body functions are dynamic and always fluctuating, but the modern meaning must be taken into account that normal body function even though fluctuating within a range when working properly, is a stable situation. We will continue to use homeostasis as generic in this book to mean normal stable dynamic function. However, we will often use *homeodynamics* and the *dynamics of homeostasis* to emphasize that homeostasis connotes dynamic equilibrium. Professor Hartmut Heine has crystallized this dynamic function with his concepts of information transfer through the ground regulation system, emphasizing the dynamic holistic homeostatic response.

About the Authors

William J. Rea, MD, is a thoracic, cardiovascular, and general surgeon, with an added interest in the environmental aspects of health and disease. Founder of the Environmental Health Center–Dallas (EHC–D) in 1974, Dr. Rea is currently director of this highly specialized Dallas-based medical facility.

Dr. Rea was awarded the Jonathan Forman Gold Medal Award in 1987 for outstanding research in environmental medicine, the Herbert J. Rinkle Award in 1993 for outstanding teaching, and the 1998 Service Award, all by the American Academy of Environmental Medicine. He was named Outstanding Alumnus by Otterbein College in 1991. Other recognition includes the Mountain Valley Water Hall of Fame in 1987 for research in water and health, the Special Achievement Award by Otterbein College in 1991, the Distinguished Pioneers in Alternative Medicine Award by the Foundation for the Advancement of Innovative Medicine Education Fund in 1994, the Gold Star Award by the International Biographical Center in 1997, a Five Hundred Leaders of Influence Award in 1997, listing in *Who's Who in the South and Southwest* in 1997, the Twentieth Century Award for Achievement in 1997, the Dor W. Brown, Jr., MD, Lectureship Award by the Pan American Allergy Society in 2002, and the O. Spurgeon English Humanitarian Award by Temple University in 2002. Author of four medical textbooks, *Chemical Sensitivity,* and coauthor of *Your Home, Your Health and Well-Being,* he also published the popular how-to book on building less-polluted homes *Optimum Environments for Optimum Health and Creativity.* Dr. Rea has published more than 150 peer-reviewed research papers related to the topic of thoracic and cardiovascular surgery as well as that of environmental medicine.

Dr. Rea currently serves on the board and is president of the American Environmental Health Foundation, is vice president of the American Environmental Board of Medicine, and previously served on the board of the American Academy of Environmental Medicine. He previously held the positions of chief of surgery at Brookhaven Medical Center and chief of cardiovascular surgery at Dallas Veteran's Hospital, and he is a past president of the American Academy of Environmental Medicine and the Pan American Allergy Society. He has also served on the Science Advisory Board for the U.S. Environmental Protection Agency, the Research Committee for the American Academy of Otolaryngic Allergy, the Committee on Aspects of Cardiovascular, Endocrine and Autoimmune Diseases of the American College of Allergists, the Committee on Immunotoxicology for the Office of Technology Assessment, and the panel on Chemical Sensitivity of the National Academy of Sciences. He was previously adjunct professor with the University of Oklahoma Health Science Center College of Public Health. Dr. Rea is a fellow of the American College of Surgeons, the American Academy of Environmental Medicine, the American College of Allergists, the American College of Preventive Medicine, the American College of Nutrition, and the Royal Society of Medicine. He is on the editorial board of the *Journal of Implant Complications*, *Journal of Environmental Biology*, and *Management of Environmental Quality.* Born in Jefferson, Ohio, and raised in Woodville, Ohio, Dr. Rea graduated from Otterbein College in Westerville, Ohio, and Ohio State University College of Medicine in Columbus. He then completed a rotating internship at Parkland Memorial Hospital in Dallas, Texas. He held a general surgery residency from 1963 to 1967 and a cardiovascular surgery fellowship and residency from 1967 to 1969 with the University of Texas Southwestern Medical School system, which includes Parkland Memorial Hospital, Baylor Medical Center, Veteran's Hospital, and Children's Medical Center. He was also part of the team that treated Governor Connelly when President Kennedy was assassinated.

From 1969 to 1972, Dr. Rea was assistant professor of cardiovascular surgery at the University of Texas Southwestern Medical School; from 1984 to 1985, Dr. Rea held the position of adjunct professor of environmental sciences and mathematics at the University of Texas, and from 1972 to 1982, he acted as clinical associate professor of thoracic surgery at the University of Texas Southwestern Medical School. Dr. Rea held the First World Professorial Chair of Environmental Medicine at the University of Surrey, Guildford, England, from 1988 to 1998. He also served as adjunct professor of psychology and guest lecturer at North Texas State University.

Kalpana Patel, MD, is a board-certified pediatrician, with an added interest in the environmental aspects of health and disease. She is a diplomat of the American Board of Environmental medicine. Founder of the Environmental Health Center–Buffalo (EHC–Buffalo) in 1995, Dr. Patel is currently a director of this highly specialized medical facility.

Dr. Patel was awarded the Jonathan Forman Gold Medal Award in 2006 for her contribution and research in the subject of environmental medicine and the Herbert J. Rinkle Award in 2008 for excellence in teaching the techniques of environmental medicine by the American Academy of Environmental Medicine. She also received the 2003 Physician of the Year Award and 2004 Pioneer of Healthcare Reform award. She received a 2006 B. J. Medical College Award for the Outstanding Alumnus by B. J. Medical College Alumni Association. Other awards include the Hind Ratan Award by the Non Resident Indians (NRI) Society.

Dr. Patel serves on the board and is president of the Environmental Health Foundation of Western New York (WNY), is president of the American Board of Environmental Medicine, and previously has served on the board of the American Academy of Environmental Medicine. She previously held the position of chief of pediatrics at Deaconess Hospital in Buffalo, New York. Dr. Patel is a fellow of the American Academy of Pediatrics and the American Academy of Environmental Medicine.

Born in Poona, India, and raised in Ahmedabad, Gujarat, India, Dr. Patel graduated from St. Xavier's College in Ahmedabad, attended B. J. Medical College in Ahmedabad, and received a medical degree from Gujarat University. She then completed a rotating internship at Civil Hospital in Ahmedabad. She held a pediatric internship and residency from 1969 to 1971 at the University of Texas Medical School at San Antonio, which includes Bexar County Hospital and Santa Rosa Medical Center. She finished her senior residency at Albany Medical Center Hospital in 1972.

From 1972 to 1973, Dr. Patel was assistant clinical professor of pediatrics at the West Virginia University Medical Center, Morgantown. From 1973 to 1976, she was a director of child health at the Erie County Health Department in Buffalo, New York. From 1976 to 1982, Dr. Patel held the position of clinical assistant professor of family practice at the State University of New York at Buffalo. She has been a clinical assistant professor of pediatrics at the State University of New York at Buffalo since 1976. Currently, she is in private practice of pediatrics and environmental medicine. She is actively involved in national and international education, teaching the subject of environmental medicine.

1 The Physiologic Basis of Homeostasis

INTRODUCTION

While studying health and disease, it has become clear to the authors that a switch occurs where the normal body function changes to dysfunction. If the dysfunction occurs over a period of time, chronic disease develops. **Chronic degenerative disease and both acute and chronic hypersensitivity are the conditions that occur during and after health deteriorates and before fixed-named end-stage disease occurs.** Understanding this process is the cornerstone of evaluating health and chronic disease. Much information exists about subunit function of the body in relationship to chronic degenerative disease and chemical sensitivity. Little is discussed about the switch from health to body dysfunction, which when chronic, leads to hypersensitivity and chronic degenerative disease.

In addition, little unified information has been recorded on the holistic response of the integrated function of the body's homeostatic mechanism to chronic noxious incitant entry (e.g., chemicals, electromagnetics, bacteria, viruses, pollens, molds, parasites, etc.). **This unified information for the understanding of chronic noxious incitant entry and the body's subsequent homeostatic and dyshomeostatic responses helps the clinician to prevent or reverse early degenerative disease and hypersensitivity by a series of metabolic, immune, neurological and biochemical manipulations.** The authors' contention is that chronic pollutant overload of this normal homeostatic mechanism **causes homeostatic dysfunction ranging from constant noxious stimuli entry to information reception and then to the triggering of the adjustment and the defense mechanism and eventually, to the repair mechanism.** We see this sequence of events in the body's function and dysfunction often as the originators of disease but when not overloaded with pollutants as the guarantors of health. It appears that environmental triggers for the genetic (time bomb) or epigenetic makeup such as the body's total environmental nonspecific and specific noxious incitant load induce metabolic and physiologic dysfunction, spiraling down, continually deteriorating, and if not corrected eventually causes end-stage disease. **Absence of proper nutrition will create greater susceptibility to the normal level of noxious stimuli, strongly contributing to this downward spiral to end-stage fixed disease (heart, kidney, lung failure, etc.).**

Once triggered by single or multiple noxious environmental stimuli, the body's homeostatic function (both the process of self-righting and the end result of such a process) leads to metabolic changes (pH changes, hypoxia, molecular change, up and down regulation of receptors, enzymes, and immune cells) that signal the activation or inhibition of the immune system, the sensory and the autonomic nervous system, and other detoxification processes. These processes bring about the body's normal adjustment physiology, tissue changes (sols to gels, gels to sol), and the activation of the nonspecific mesenchyme reaction, which signals the activation of the repair mechanism. **Eventually, if the noxious stimuli are not eliminated, the normal righting mechanism is chronically overloaded and leads to aberrant physiological function and to fixed end-stage disease, e.g., renal, heart, kidney, brain failure.**

In order to prevent, reverse, and control disease, the clinician must understand as much as possible about the kinetics (half-life, biotransformation, toxicity, and virulence) of the known noxious stimuli that enter the body.[1] This is not always easy to do since the pollutant(s) entering information is often complex, compounded, and hazy due to the entry of many simultaneous noxious stimuli and their linear and biphasic (U- or J-like) dose effects.[2] The clinician does not always know these facts, which at times makes it extremely difficult to plot the homeostatic course or reduce adverse reactions.

In addition to the linear toxic dose effect of pollutants, Calebrese's[2] voluminous review and incorporation of the knowledge of the hormetic (biphasic) effects of any incitant whether chemical, biological, or physical have crystallized the potential individual and combined noxious stimuli effects of the total body nonspecific and specific pollutant load. **This hormetic concept complements the linear dose response curve eliminating the observed gaps in the dose-response observation.** A better understanding of the dynamic mechanism of homeostasis and the integration of function of its thousands and perhaps millions of subunits and their triggers will enable the clinician to recognize homeostatic dysfunction and to prevent disease before it is fixed thus, becoming irreversible.

Also of importance is how the integrated interactions of the total environment, both external and internal, affect the environmental receptors in the body. The action of these receptors, triggers the near automatic adjustment process of the normal homeostatic mechanism to stabilize the organism in order to prevent chronic degenerative disease and/or chemical sensitivity.

Prevention, reversibility, and control of disease also have to do with the state of nutrition of the body. The nutrient-deficient individual will have a lower threshold for similar strengths of noxious stimuli triggering, which sets off a cascade of reactions resulting in dyshomeostasis and/or fixed-named disease.

Professors Pischinger and Hartmut Heine developed the concept of the integrated relationship between physiology and the dynamics of the ground regulation system and **its information collection and dissemination,**[3] **which is the basis for the body's communication system**. This concept gives physicians and surgeons the groundwork for understanding basic pollutant entry physiology.

In addition to Pischinger[3] and Heine's[4] concept, this book emphasizes how the total nonspecific environmental load (composite weight of pollens, dust, molds, toxics, bacteria, viruses, etc.), as well as the "specific" environmental load (e.g., Streptococcus hemolyticus, chlorodane, etc.) affects, alters, and triggers nongenetic, epigenetic, and genetic responses of the homeostatic mechanism where, if the triggers are not adequately neutralized, results in the generation of chronic degenerative disease and/or hypersensitivity (i.e., chemical sensitivity). Inclusion of the use of these principles (the incitant hormetics, the body's homeostasis, the total nonspecific and specific environmental load) in daily practice allows for reversibility in the treatment of early chronic disease as well as the prevention of disease.

The authors have drawn information from their experiences and those of other environmental physicians and surgeons, nutritionally oriented physicians, environmental scientists, biochemists, immunologists, physiologists, pathologists, and other sources to allow integration of information about the holistic clinical overview. Over the last forty years, the experience of environmentally and nutritionally oriented physicians who have studied and treated patients with chronic degenerative disease and/or hypersensitivity (chemical sensitivity) who are in distress has become vast, encompassing well over a million people worldwide. The authors of this book have studied and treated a combined total of over 40,000 patients who were environmentally wounded and who developed either a hypersensitivity state (e.g., chemical sensitivity and/or sound sensitivity), chronic degenerative disease, or both. Integrating basic scientific facts that are now legion with this vast clinical experience provided us with a holistic perspective for the prevention and treatment of early body malfunction before end-organ failure occurs. In our experience, **the hypersensitivity phase with its total nonspecific and specific noxious incitant load of chronic disease is the most often mishandled and even ignored**. The poor handling of this hypersensitivity phase causes many mishaps and alterations of otherwise potentially successful treatment.

Several concepts and tools have been advocated and integrated for dissecting complex homeostatic dysfunction. **The development of the less-polluted environmental control unit by Dr. Theron Randolph**[5] **for the study and treatment of the chronically ill patient has been the major significant advance in diagnosing and treating the environmental aspects of chronic disease**. The principles and facts about construction of these units has allowed new knowledge to be gleaned and used for new building construction and remodeling of once-polluted homes, clinics, and hospital units.[6] The practical application of this knowledge and information based on Selye's[7] studies of adaptation in his stressed-animal models has also allowed us to dissect each facet of the generation of chronic disease under controlled conditions without muddying the vision of actual triggering agents versus fellow travelers. For the first time, this knowledge provides adequate explanation of the significance that multiple entering noxious stimuli have on each organ and the induction of their specific responses in the body. Randolph[5] elaborated on Selye's[7] concept to humanize and individualize the practical aspects of dissecting out the triggering agents of homeostasis and dyshomeostasis. He developed and described the entity of chemical sensitivity that had been observed by Pischinger and Heine but was not elaborated on by them.

Dr. Carlton Lee's[8] development of the intradermal injection provocation-neutralization technique (which appears to be partially based on the concept of hormesis) and evolved from Dr. Herbert Rinkle's[9] studies of endpoint titration of antigens has given us another wonderful tool that can be used for the diagnosis and treatment of adverse reactions to food, biological inhalants (i.e., molds, pollens, dust), chemical, bacterial, viral, and other substances. These treatments, once a safe endpoint is found, result in intradermal and subcutaneous injections or sublingual therapy. For treatment of environmental sensitivities these sublingual[10–26] and intradermal treatment[27–41] routes of the administration of specific dose antigens are coupled with avoidance techniques and optimum nutrition. These techniques allow the clinician modalities that can push the dyshomeostasis induced by the total nonspecific and specific load of the environmental pollutants back into normal function without the complications of long-term medication treatment. The authors hope that students of medicine and surgery in every specialty will incorporate this holistic perspective into their practice integrating it with other newfound long-term knowledge of their specialty.

The noxious stimuli affects of the specific environmental incitants(s) (low-dose stimulation and high-dose inhibition) are independent of the body's homeostatic response, adding another layer of physiologic response before the body's own alterations enter the picture. This hormetic effect[2] of the entering incitant complicates clinical observations of dose-response because it can be confused with the body's homeostatic response. These hormetic effects have radically changed the old concept of only a linear-dose relationship of noxious substance entry and response because now explanations of varied responses can be seen for the same substance that may be inhibitory or stimulatory at various doses.

We would like to emphasize that the term *homeostasis*[43]—both the process and the end result—could be considered by some to be outdated since all body functions are dynamic and always fluctuating. However, the modern meaning takes into account that normal body function even though fluctuating within a range when working properly is a stable situation. We will continue to use homeostasis as generic in this book to mean "normal stable **dynamic** function." We will often use "homeodynamics" and the "dynamics of homeostasis" to emphasize that homeostasis connotes dynamic equilibrium. *Dyshomeostasis* will mean that the body's fluctuations are outside of the physiologic range. Professor Pischinger[3] and Heine[4] have crystallized this dynamic function with their concepts of information transfer through the ground regulation system, emphasizing the dynamic holistic homeostatic response in the body. Their meticulous work on the anatomy, physiology, and electrodynamics of the communication system has been paramount in understanding chronic degenerative disease and hypersensitivity. In this book, ***hypersensitivity,* originally described by Randolph[5], means an exaggerated response of the body's normal fluctuating physiology, which is periodic in nature. This type of response is usually above the response line. Chronic degenerative disease is an eventful deterioration below the response line and is, generally, aperiodic (both are discussed later in the chapter).**

OVERVIEW OF THE DYNAMICS OF NOXIOUS INCITANT ENTRY AND FATE

The body has the ability to sense through sensory nerves, receptors, autonomic nervous system and connective tissue matrix entry of noxious stimuli and to communicate this entry to the body's homeostatic mechanism, including reception, adjustment, defense, and repair in order to neutralize or eliminate the substance(s). When a noxious substance enters the body, several things happen. First after neurological & receptor perception there is a lowering of the pH due to the cleavage of sulfates, acetates, etc., in the connective tissue matrix (CTM). The local calcium increases and the magnesium decreases. The local fixed tissue macrophages are released to surround the noxious stimuli (e.g., bacteria, viruses, and/or chemicals). The local microcirculation dilates with fluid leaking from the vessels to give an influx of gamma globulin, dilute the noxious incitant, and supply extra oxygen for metabolism of the noxious incitant and supply leukocytes to combat the local incitant. This microcirculation then contracts in order to delay the spread of offending noxious substances. After this sequence of events occurs in 5–7 days, the macrophages move in coming from the bone marrow and other tissues, e.g., lymph nodes, liver, etc. The macrophages release the proteolytic enzymes as well as engulf the debris, the noxious substance, and the damaged polymorpho-nuclear leukocytes. Next, fibroblasts put out collagen and PG/GAGs in order to heal the wound. If the injury is large or chronic enough, inflammation sets in;, if not, healing occurs. If the injury is large a scar forms and/or if the injury is chronic, inflammation sets in. If the wound does not heal lymphocytes move in signaling chronic disease. If the injury is small, no scar forms, and healing occurs.

If the local matrix and cells cannot control the noxious incitant, regionalization occurs and the central alarm system in the brain is activated. The regionalization occurs in the spinal cord and at the place where the regional injury happened or an adjacent muscle mass is present. The central homeostatic mechanism involves the reticular activating system, the hypothalamus, the limbic system, the area postrema of the fourth ventricle, and perhaps, the pineal gland, all of which activate the rest of the body by an alarm reaction through the autonomic nerves and the adrenal glands. Part of this ability to sense and act upon a noxious incitant depends on the dynamic nutritional state of the body since its completeness is necessary for a proper response.

Understanding the basis of the dynamics of the body's homeostatic mechanism, as well as the linear and hormetic (biphasic) effect of the pollutant entry agent(s) (high-dose inhibits, low dose stimulates or vice versa) will help the clinician understand the early changes that occur in the patient before the advent of chronic degenerative disease and/or hypersensitivity, especially chemical sensitivity. If the deviations from the body's holistic normal homeostasis, which are driven by entering noxious stimuli and modulated by the state of nutrition are found early the clinician can often institute therapy, avoiding costly fixed-named end-stage disease (pulmonary, renal, heart, brain failure) treatment, revert the dysfunction to normal homeodynamics, and offer the patient a way back to health without the use of long-term medication or surgery. However, on some occasions, precise surgery after a precise diagnosis will often revert the body's physiology back to normal (e.g., appendectomy, cholecystectomy, tooth removal, etc.). **This understanding of the total nonspecific and specific total environmental pollutant load depends upon the kinetics (half-life, biotransformation capabilities, toxicity or virulence) of the entering noxious stimuli, the weight of the total combined entering pollutant load (total body load), the body's state of nutrition, and the capability of the body's holistic and specific homeostatic mechanism to deal with the entering pollutant(s) in an energy efficient manner.**

Classically, the noxious stimulus if it is chemical, biological, or physical depends on the quantity and quality of the dose (physical and chemical properties) or its metabolites or conversion products reaching appropriate stable sensory and autonomic nerves and receptors in the body at a concentration and for a length of time sufficient to initiate a change. The coalescence of new information now shows that in addition to the linear dose effect, there are biphasic effects for many chemicals. These facts complicate the reception and adjustment response because of symptoms and information overloading of the body's response mechanism.

Nature of the Incitant Stimulus

The susceptibility of the person (genetic, acquired, epigenetic, environmental, and nutritional) and the particular biologic system involved are also crucial to a proper homeostatic response. The

The Physiologic Basis of Homeostasis

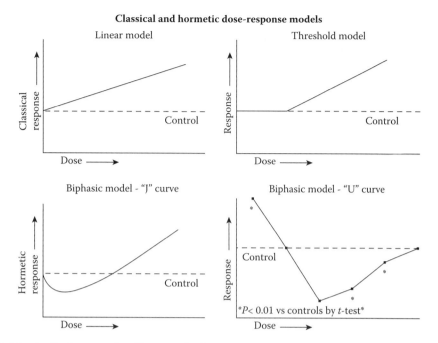

FIGURE 1.1 Stylized curves that illustrate the linear, threshold, and hormetic dose-response model, the J and U curve. (Courtesy of Calabrese, E.J., *The Scientist.* 22, Fig.7, 2005.)

noxious stimuli portal or avenue of entry is also extremely important, as is quantity of the dose on which a higher one will at times give more toxicity or change function from stimulation to inhibition. There is a linear dose response relationship for some chemicals, incitants, but a biphasic effect from others where there is low-dose stimulation and high-dose inhibition or even vice versa.

However, newer evidence shows that with many chemicals and biologics there is a biphasic effect that will complicate the perception of response. **Hormesis is the property of the noxious stimulus that can have the individual properties of low-dose stimulation and high-dose inhibition,**[2] or another type of the biphasic response would be the opposite effect.[44] See Figure 1.1.[45]

The biphasic dose response challenges such long-standing toxicological linear dose-response model mindsets but clearly has a place in clinical medicine. Hormetic and other nonmonotonic dose responses are biphasic, displaying either an inverted U- or J-shape depending on the endpoint measured as shown in Figure 1.1. It is generally recognized, for example, that adults who consume a glass of wine most days have reduced risk of cardiovascular disease compared to nondrinkers, while excessive consumption increases such risks. This type of J-shaped dose response is now known to be quite common in toxicology and pharmacology, being seen with many dozens of chemicals, and for hundreds of important endpoints such as cancer risks, longevity, growth, performance on various types of intelligence test, and more.[45]

Comprehensive assessments of the literature especially by Calabrese have shown the hormetic model to be biologically as fundamental as either major dose-response rival, i.e., the linear and threshold response. The biphasic response is more common and valid in head-to-head comparison, in a generalized biological model (i.e., where a plant is stimulated with a growth enhancer at a certain dose, but with a higher dose its growth is inhibited), where the endpoint is measured, and the chemical class or physical agents are studied. Based on these conclusions, the biphasic dose-response model should be equal to both the threshold and linear models as the model in risk assessments for noncarcinogens and carcinogen. Further, according to Calebrese, the biphasic dose response should be considered not just the dominant model in toxicology but also in the broader domain of the biomedical sciences including immunology, cancer cell biology, neuroscience, and all other fields that rely upon dose-response relationship.[45] In addition to the linear model this biphasic model applies to the onset and propagation of chemical sensitivity and chronic degenerative disease. **It also suggests effects that can be applied to their reversibility**.

In order to understand under what conditions linear effects and biphasic effects of the entering pollutant(s) (noxious stimuli) may be expected, it is critical to identify the rate-determining process or the rate-limiting processes and their half-lives and the kinetics of the compound(s) upon the particular pollutant(s) entry. This information enables experimentalists to choose appropriate times for sample collection and clinicians opportune times to manipulate adverse conditions to the patient's favor. Attention to this hormetic fact allows the clinician a broader tool to reverse early disease.

If the kinetics of the offending entering stimuli drives the dynamics of the effect upon homeostasis, then a rate determining effect may originate either from the absorption (slow or fast) or from elimination of an agent (slow or fast). Most frequently, elimination of the entering substances is slower than absorption and hence represents the crucial step in evaluation of the hormetic effect. For example, if a low-dose effect results from a distinct stimulation, then ongoing absorption may provide an explanation for the low-dose stimulating effect (gradual increase in entering substance concentration) and an inhibitory effect at a high-dose (on its way to a neutralizing equilibrium) that will eventually equilibrate as the entering offender is neutralized (see section describing hormesis).[46]

One can see that the biphasic response of many chemicals entering the body simultaneously or nearly at the same time might cause a strong influence on the receptors (sensitivity or numbers), adjustment, defense, and repair mechanism thus disturbing homeostasis. These elements might also change receptor adjustment, define, and repair sensitivity or numbers. If chronic stimuli occur over a period of time one will see changes in physiology resulting in chronic degenerative disease and/or hypersisitivity (chemical sensitivity).

NATURE OF THE BODY'S HOMEOSTATIC RESPONSE

In addition to the entry of noxious stimuli with their linear and hormetic effects into the body, the dynamics of the body's normal homeostasis are the most important and most complex processes in human physiology. Homeostasis has both linear and hormetic responses. The principle of these dynamics of homeostasis is often ignored or never even verbalized in some aspects of clinical medicine and surgery because their effects are assumed and thus, overlooked. However, **use of the data on physiologic homeostatic processes and principles by the clinician are commonplace when understood because of their great help to patient management**. This complex process of holistic homeostasis must be understood as much as possible, in order for the clinician to help the individual patient obtain and maintain optimal health and prevent end-organ failure.

The dynamics of homeostasis are maintained by constant fluctuating and compensatory micro adjustments, which consist of the dynamic, harmonic, energy efficient reactions in the internal environment within physiologic boundaries.[4] These dynamics are controlled not only by the property of the noxious incitants, whether stimulating or inhibiting, and their half-life but also by the integrated local, regional, and central physiologic homeostatic responses (stimulating or inhibiting) that lead to automatic adjustments, in the body's physiology, which keep the individual in optimal health. Homeostasis is always dynamic and not static. There are always fluctuations of all the components of the homeostatic mechanism within a range of normal variations. **Maintaining the dynamics of the energy-efficient holistic, homeodynamic response is essential for optimum physical and mental energy as well as creativity**. To a point, this efficient response can be accomplished even with minor subsystem malfunction. However, longer and chronic changes in this process can lead to chronic degenerative disease and/or hypersenitivity (chemical sensitivity).

Nature of Information Intake and Distribution

The holistic homeostatic response system has much redundancy of reserve in order to maintain proper function over a life time. Vast nutrient reserves enable the individual to maintain wellness even at times of an acute increase in the total toxic environmental load.

For survival, the body has the innate ability to respond to external stimuli. There are several criteria that are met in order to have normal homeostasis. These involve information collection, by the sensory nerves, autonomic nervous system and receptors of the skin connective tissue matrix electromagnetics, and automaticity as well as the proper functioning of the redox process. According to Heine, the most suitable energy form that gives the body the ability to self-correct (homeostasis) is information input and processing.[4] **Information from the entering stimuli triggers the highly intermeshed open receptor systems of the body (the skin and mucosa membranes and particularly the CTM) to initiate its own latent ability not only to form structure and organization (i.e., matrix, cells, organs, systems) but also to maintain these for optimum function.** The major significance of information from the entering stimuli as a nonchaotic energy form is that it is not tied to any particular energy carrier (i.e., sound waves, vision, tactile connection, gastrointestinal function and nutritional absorption, etc.)[4] and thus, can enter via multiple vehicles of transportation to multiple areas of the body often simultaneously. Every type of sensor in the body can perceive some information input of an entering substance both locally and centrally, which allows the body to have many early warning systems with or without integrated responses. Therefore, information entering into a living system is thus, the most suitable energy carrier for setting off both local and distant extracellular and intracellular reciprocal feedback via the body's homeostatic mechanism including the adjustment, the defense, and the repair mechanisms. **Entry information almost never has a neutral effect.[4] It usually has a linear effect or biphasic effect (e.g., low dose produces stimulation, high-dose produces inhibition) or vice versa and a homeostatic response effect, which either or both effects can cause alterations in the body's function.** The neutral effect of the entering information is either not recorded or is just sequestered, while the positive information appears to affect the holistic response in a positive way of well being and the autonomic nervous system balance is recorded as a positive response.

The basic molecular mechanism that information triggers in the body for initiating homeostasis appears to have electromagnetic features,[4] which become communication vehicles. **The electric and magnetic tone established by isotonia, isoosmia, isionia, and redox reactions reacts to every change the skin and mucous membrane but particularly in the extracellular matrix (ECM) with deviations in the electrical potential.**[4] The entered information of an external stimulus can be encoded in the organized water in the body, and in this way, can inform the cell membrane as a potential deviation for a reciprocal feedback reaction.[4] Thus, these facts emphasize the role of electromagnetics in orderly body function.

Each organ is an aggregate of cells, autonomic or sensory nerves and CTM that communicate with each other by several modalities. These modalities include neurotransmitters, nerve impulses, multiple different cytokine alterations, growth factors, etc., which distribute information locally and eventually throughout the body. These modalities including skin, mucous membrane and CTM receptors interact reciprocally, being part of the integrated homeodynamic adjustment response, including the defense, and repair mechanisms, all of which communicate that a stimulus has entered the body, needs to be responded to, and is controlled. In turn, the aggregate of organ cells and CTM also communicate with other organs and systems and the rest of the individual's environment in order to obtain and maintain general biphasic energy-efficient holistic homeodynamic function. Alterations in this sequence of events can lead to chronic degenerative disease and/or hypersensitivity (chemical sensitivity). See Figure 1.2.

The fact that we remain alive, to a point, is almost beyond our control because automatic homeodynamic function prevails for survival.[47] For example, we are forced by the body to seek water and food by automatic thirst and hunger. Sensations of cold or excess heat force us to provide warmth

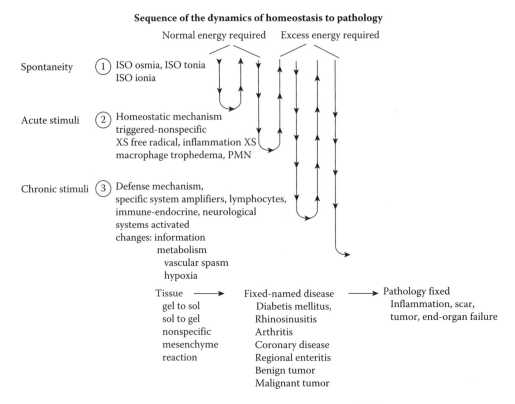

FIGURE 1.2 Sequence of Homeostasis to Pathology. (From EHC-Dallas, 2002.)

or cooling. Other forces cause us to seek fellowship and reproduce. Instinct allows us to nurse and protect our young.

AUTOMATICITY OF THE BODY—ENERGY

This automaticity of the body for survival is manifested by the dynamic holistic homeostatic mechanism, which consists of entering stimulus information reception, neutralization of a noxious stimuli by the adjustment mechanism, the defense, and the repair processes to correct and repair damage caused by noxious stimuli, handling both external and internal environmental stimuli. The neural, humoral, and cellular control circuits interact with each other and with the CTM as an intermeshed holistic homeostatic control system.[4] Adverse changes in this process will lead to chronic degenerative disease and/or chemical sensitivity.

In order to assure the energy metabolism for the homeostatic process with the oxygen-reducing electrical potential (oxidation–reduction, B_1 (TPP) and B_2 (FAD)[4] dependent electron transfer reaction) at its core, the cell and its CTM environment continue to interact as regulators of each other. These interactive processes that maintain the body's ability to function not only for survival but also for optimum function are influenced by both the total sum of the external and internal environmental stimuli, as well as, the state of nutrition, which is either average, optimum, or depleted. These redox processes easily alter the body's function positively or negatively and with technology today can be measured by the evaluation of redox potentials (Eh) using Nehm (redox assays) equations, potentiometrically by appropriate working and reference electrode, and chemoluminescence (light reaction measurements). Registered potentiality by measurement can be modified by introducing redox active compounds that go toward oxidosis (toxic chemicals) or redosis (foods, or herbs and nutrients).

Every group of organ cells and its CTM needs its own fixed oxygen-reducing potential and its own energy potential in order to work efficiently. The normal individual functions on the principles

The Physiologic Basis of Homeostasis

of efficient reception and collection. These functions include the state of entering stimulus information, containment of noxious stimuli, coordination of information reception, adjustment, neutralizing processes, defense, and repair mechanisms. **Energy efficient regulation of the** body as a whole is the goal. The dynamics of homeostasis can be maintained only if the organism is working economically otherwise disease processes such as hypersensitivity and chronic degenerative disease evolve. The task of the regulating and control systems is to adapt all the metabolic processes in accordance with economic principles to the energy demand at any given time, utilizing the shortest route, in the shortest time, using a minimum of energy. Of course, this is not always so easily possible because of the kinetics of some environmental toxics and the poor state of nutrition of many patients with chemical sensitivity and/or chronic degenerative disease. For example, some types of dioxins etc. have a half-life of over one-half to one year, thus making it impossible to have a short neutralization period and if nutrient deficiency is present detoxification will take longer or be indefinite.[48]

In diseased individuals, symptoms (pain, itching weakness, and fatigue) can be regarded as reflections of disturbances in homeodynamic regulation. Many other types of symptoms including transient anxiety, depression, tachy- and bradycardia, extra systoles, palpitations, breathlessness, cough, bloating, cramping, headache, light-headedness, numbness, tingling, hyper- and hypotension, dry mouth, anorexia, nausea, constipation, diarrhea, fibromyalgia, urinary urgency and frequency, etc. are reactions of altered physiology. Many of these symptoms previously have been considered by the medical profession as functional. However, in the light of new details, the process of homeostasis and dyshomeostasis must now be considered clinical reflections of the homeostatic response. These aforementioned symptoms are transient being periodic in nature. Disease is the consequence of a persistent disturbance of the entry substances (information) and the feedback mechanism of the dynamics of homeostasis. These symptoms reflect this altered physiology.

CLINICAL SIGNS AND SYMPTOMS AFTER CHRONIC NOXIOUS INCITANT ENTRY

After entry of a noxious stimulus a time interval occurs between the onset of symptoms such as generalized pain, itching fatigue, or weakness or any of the aforementioned symptoms, which are due to early homeodynamic disturbances and the period when diseases are fixed-named and irreversible such as hypertension, myocardial infarction, and stroke. **To diagnose the disease at the early homeodynamic dysfunction (pain, itching fatigue, weakness) stage rather than waiting for the occurrence of fixed-named disease (myocardial infarction, respiratory, or renal failure) is, usually, the best time for the clinician to reverse the process**. Thus, the clinician must understand the linear and biphasic effects of the entering noxious stimuli and the dynamics of the body's homeostasis in order to prevent severe metabolic variation and eventually end-organ failure. Our impression is that **most diseases**, especially those that are chronic, **have prodromes that last from minutes to hours, to days, to weeks, or years,** depending on the kinetics of the entering stimuli, their rate of absorption and their half-life as well as, the body's ability to respond through its homeostatic neutralization and adjustment responses. Therefore, this stage of initial entry and injury is the time for therapeutic manipulation before the dyshomeostasis becomes fixed.

Although there is a vast amount of information in this book about system and subsystem responses, the reader should keep in mind that the understanding and manipulation of **the total body pollutant load, the kinetics of the entering stimuli, and their potential linear and biphasic effects, as well as, the holistic response of the body's homeostatic mechanism is paramount in obtaining and maintaining optimum health**. A holistic approach is frequently forgotten in modern medicine because we are deluged with so many facts about subsystem function and we are fragmented by subspecialization in almost all fields of medicine and surgery. For example, we are deluged by facts about the kidney, liver, immune, and nonimmune detoxification systems, how the endocrine and neurological systems work, including opinions from medical and surgical therapists in each field but little is said about how the holistic harmonic integrated function of all of these systems occurs (response to entering foreign information, reception, neutralization, adjustment, defense, and repair) and how it is used to maintain health. The understanding of this integrated holistic response is the key to prevention of and possible reversibility of disease. Many new authors think their own single modality such as exercise, nutrition, or immune manipulation will be the solution for health. Of course, at times this is true but usually over the long run, their narrow therapy is only a partial solution. An entire holistic diagnostic and treatment program is needed for maintenance of health and prevention of ill health. **The purpose of this book is to integrate this knowledge for developing a holistic treatment program.**

When the clinician attempts to diagnose early homeostatic dysfunction in order to prevent disease and to reverse early periodic homeostatic disturbances as seen in chemical sensitivity (the adverse reaction to ambient doses of toxic and nontoxic chemicals), (aperiodic homeostatic disturbance) or in chronic degenerative disease (eventful deterioration of the dynamics of homeostasis, which never returns to normal, e.g., spondylosis and spondylitis), it is of paramount importance to understand as many facts as possible about the dynamics of the holistic end-organ homeostatic process.

THE BODY'S COMMUNICATION SYSTEM— GROUND REGULATION SYSTEM

The body's communication system starts with the skin and mucous membranes of the gut, respiratory tree, and urogenital systems, which initially perceive and fend off many environmental toxic stimuli including bacteria, viruses, parasites, toxic chemicals, pollens, dust, molds, electromagnetic radiation, and even toxic foods. The body performs these information gathering perceptions triggering adjustment and defense mechanisms by a series of physiologic and mechanical functions including actual dedicated information reception lines networked to deep areas including the CTM, brain, mechanical (proteins, lipids, cellular barriers), immunological (IGA), enzymes (i.e., superoxide dismutase, glutathione peroxidase, catalase, etc.) and other nutrients (vitamin A, C, E).

The concept and facts about the communication system of which the CTM receptors, as well as other receptors such as the skin, mucous membranes, eyes and ears, play a significant role in the body collecting information and responding holistically. This relationship toward function and environment arose in the Germanic nations. From here the body's information and communication of the environment knowledge was then passed throughout most of central and Eastern Europe. The ground regulation system for communication is beginning to be understood in the United States, where people discuss it as the cellular communication network. Both are a part of the general master reciprocal communication system of the information acquired by perception of the environment for regulating the dynamics of homeostasis in the body. The dynamics of this sequence of events will be incorporated as the ground regulation system in this book. **This communication with noxious environmental stimuli and their fate have great influence on the generation and propagation of chemical sensitivity and chronic degenerative disease**.

The scientific understanding of the functional integration of the entry of foreign information and its reception is fragmented and not well verbalized. A plethora of facts however, are known about the local communication between the skin, mucous membranes, CTM, and the amplification systems,[4] which include the immune system, the autonomic nervous system, and the endocrine system. The extent of the dynamic function of the local connective tissue has not previously been well integrated in the North American thinking and an attempt to increase understanding of this function will be an objective incorporated in this book. Our goal is to delineate the integration of the communication from the skin and mucous membrane and the local CTM and its cellular components with the amplification systems for more complex responses involving the immune, the endocrine, and the autonomic nervous systems, as well as the delineating the dynamics of the regional and central homeostatic control centers in the brain.[49-51] **This communication network has to be integrated between pollutant-nutrient balance and pollutant overload**, both of which can initiate disease. This chapter is dedicated to delineating the facts that make the communication system work, by showing the mechanisms that allow the body to develop dysfunction without producing true fixed-name

The Physiologic Basis of Homeostasis

disease through the creation of inflammation and scarring. Also this book will show how the repair mechanism works versus how the propagation of chronic inflammation occurs.

The ground regulation communication system (GRS) is a local information reception and collection, adjustment, defense, repair, and reciprocal feedback response system with its functional unit consisting of the connective tissue (CT); its cells, including the fibroblast, macrophage, the leukocyte, and mast cell.[4] The CTM makes up 40% of the body's tissue surrounding every cell and organ and is 87% of the dermis and submucosal area. This system is also comprised of the loose areolar tissue containing glucose aminoglycans (GAGs), proteoglycans (PGs), fibrils of collagen, and elastin, which are produced by the fibroblast.

The ground regulation system also includes the end vascular pathway, including lymphatics, the autonomic nerve endings, and, at times, accompanied by the somatic sensory nerve. All of the above make up the ground regulations system's communication attachments for local, regional (spinal reflexes) and central functions, and feedback (reticular formation, hypothalamus, pineal gland, area postrema of the fourth ventricle, and limbic system). The total response of this system results in the generalized homeostatic response.[4] See Figure 1.3.[4]

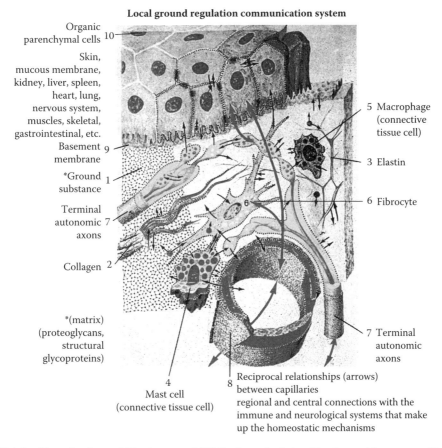

FIGURE 1.3 (See color insert following page 364.) Reciprocal relationships (arrows) between capillaries (8), ground substance (matrix), (proteoglycans and structural glycoproteins 1, collagen 2, elastin 3), connective tissue cells (mast cell 4, defense cell 5) fibrocyte (6), terminal autonomic axons (7), and organic parenchymal cells (10). Occurs Basement membrane (9), the fibrocyte (6) is the regulatory center of the ground substance. Only this type of cell is able to synthesize extracellular component. The main mediators and filters of information are the proteoglycans, structural glycoproteins, and the cell surface sugar film, glycocalyx (dotted line on all the cells, collagen and elastin). (Modified from Pischinger, A., *Matrix and Matrix Regulation: Basis for a Holistic Theory in Medicine*, Ed. H. Heine, Eng. Trans. N. MacLean, Brussels, Belgium: Editions Haug International, 21, 1991.)

Anatomical Categories Used in Communication for Regulating the Homeostatic Mechanism

The dynamics of homeostasis, which are driven by noxious stimuli and are reflected by the communication system, can be divided into four large anatomical categories of bodily function. There is an integrated harmonic orchestration of anatomical categories and their functions through the ground regulation system that generates global responses. The function of these anatomical categories will be discussed as separate entities because of their complexities and vastness. Due to all of their physiologic responses these anatomical categories can be used for communication. **These categories include the connective tissue (ECM), the neurovascular tree, the intravascular contents, and the cell, all of which make up the ground regulation system used as communication for homeostasis.**[4] The harmonic orchestration of the dynamics of total body homeostasis, which are used for communication, as well as the individual subunit function of these complex interactions when extremely disturbed, results in hypersensitivity and especially conditions such as chemical sensitivity and chronic degenerative diseases.[4]

Four general anatomical categories of the communication system for homeostatic regulation are presented with each one discussed separately. These categories include the skin, mucosa, CTM complex, the blood vessels, the intravascular content, and the neurological system.

Skin and Mucous Membranes

The skin and mucosa have a barrier and sieve effect for most major pollutants. They also have a communication physiology that is intimately connected with the CTM for communication of environmental pollutants and entering substances that can act as noxious stimuli to disturb the internal organs and their physiology. The skin because of its toughness and thickness is much more difficult to penetrate by noxious incitants but it also has receptors that can go to the CTM or directly through certain receptors and nerves to the brain. (This is discussed later under receptors.)

The mucous membranes are not as difficult a barrier to penetrate because of their thinness and therefore, are much more easily damaged. In addition, their receptors are much more sensitive allowing for greater and easier triggering, signaling an earlier warning to the internal milieu. Both organs (the skin and the mucosa) have 80–89% of their component parts as the CTM. Therefore, these two organs function physiologically along with the connective tissue matrix and its physiology.

Connective Tissue Matrix

Mechanical Support System (Support Structure)

Connective tissue is a support structure for many areas of the body including the skin, gastrointestinal tract, respiratory tract, urogenital system, liver, spleen, lungs, heart and blood vessels, kidney, bladder, adrenals, arteries and veins, cartilage and bones, lymphatics, some of the bone marrow, and the nervous system. It makes up 40% of the body's tissue and works by supporting organs and systems so they can function properly. The connective tissue acts with a springlike effect that buffers minor and at times even major trauma from doing damage to both the overlying and underlying tissue. Due to this meshlike superstructure, PG/GAGs act as a shock absorbing system that works like a lubricating substance (joint lubricant), which changes to a viscous elastic (gel to sol, sol to gel) substance with severe repeated mechanical demands and in addition, due to the mechanical pressure, triggers electrical responses.

Professional athletes, runners, and people with multiple episodes of skeletal injuries experience this shock-absorbing damage phenomena, frequently succeeding in sensitizing the matrix and in triggering other secondary sensitivities to foods, molds, and chemicals. This type of injury is highly elastically malleable and has a high energy consuming effect. We see this phenomenon, due not only to trauma but also to fascial sequestration of pollutants, in a subset of chemically sensitive as well as chronic degenerative diseased patients who have had multiple recurrent injuries. These patients are often weak and have severe muscle, fascial, and joint pain because they transfer so much energy in trying to obtain harmonic energy flow. In this type of injury homeostasis is difficult to obtain and maintain, frequently requiring physical therapy, cranial manipulation, acupuncture, energy balancing, detoxification, nutritional therapy, and even at times surgery. These patients also often have problems with weather changes flaring the joints, fascia, and muscle pain. Internal joint pathology shows changes, which in turn transforms the colloid chemistry and response thus, affecting the electrical and homeostatic responses. **Recent evidence has now been shown that in vitro gels used in the printing industry, have significant changes in form just before and during weather front changes. Presumably, the same change happens to the colloid in the matrix in injured patients**.[4]

Connective tissue consists of cells, fibroblasts, macrophages, mast cells, plasma cells, extra cellular fibers, and extracellular amorphous ground substance.[4] Soft connective tissue (SCT) is categorized by a relatively small amount of loosely arranged collagen with abundant amorphous ground substance (PGs and GAGs) surrounding cells.[4] All fiber types and all cell types may be present. As stated previously, wandering cell types such as the monocyte-macrophage and the leukocyte are looking for action in this type of tissue where confrontation with antigens is common. Traumatic injury will make this tissue very susceptible to many environmental pollutants causing total body dysfunction.

Harder connective tissue is composed of dense irregular connective tissue (dermis, vaginal wall, periosteal, perichondria, perineuron, and organ capsules) and regular connective tissue (fascia), as well as loose areolar tissue with reticular fibers.[51] This latter type of tissue (loose areolar and reticular) contains an abundance of reticular fibers, which end in proteogylcans and GAGs. These fibers form a loose netlike stroma for such organs as lymph nodes, spleen, and bone marrow. The term stroma refers to mesenchymal derivative of an organ, i.e., connective tissue and vascular. **The parenchyma is the functionally differentiated epithelium appropriate for a particular organ.**

Communication System Properties

The connective tissue is an intermeshed internal mediator between the blood vessel and nerve supply on one side and the end organ on the other. The terrain of the matrix where local disturbances take place has three parts: the parenchyma, the basement membrane, and the amorphous ground substance, including the capillary and the nerve endings. **The ECM is primarily involved in communication of the disturbances of other tissues as well as a support structure for the cells of the skin, gastrointestinal tract, or any organ**.

The connective tissue being a major part of the communication system is dominant when one considers information reception, transmission, adjustment, and healing in the dynamic process of homeostasis.[4] About one sixth of the body consists of spaces between the cells, which are filled by the interstitium and fluid between the spaces.[4] The connective tissue is the largest system in the body penetrating every organ and completely regulates nutrition and wastes. It is also a part of every inflammatory and defense reaction, with which it communicates freely thus, controlling the cell milieu.

Another category of information spread for communication is the three-dimensional architecture of the ECM, which can modify and alter the expression of the genetics of the cell under some certain circumstances.[4] This three-dimensional configuration and receptor information of the matrix appears significant in tumor elimination and thus, may become a powerful tool in the elimination of carcinoma.

All organ cells depend on the connective tissue's intact function for their existence because the connective tissue becomes the cells immediate surrounding environment. The CTM surrounds all the cell segments and organs. **The two types of solid structures of the connective tissue are the** collagen fiber bundles and the loose areolar tissue, consisting of fibrils of type III collagen, PGs and glucoseaminoglycan filaments, interwoven on a hyaluronic acid tree.[4] These structures make up a major part of the communication network whose receptors take information and telegraph it through pathways. The information pathways direct adjustments and changes in the response to pollutant entry of noxious stimuli.

The CTM determines the tissue's physical properties and is the architectural framework of the body where the amounts found vary from organ to organ. For example, the connective tissue is a major component of skin and bone and a minor component of the brain and spinal cord. Variation in the relative amounts of the different types of matrix macromolecules and in the way they are organized in the ECM gives rise to an amazing diversity of forms each adapted to the functional requirements of the particular tissue. These forms then demonstrate a diversity of function. Bissal et al. have shown that CTM's three-dimensional architecture has a communication system that can control and modify gene function most likely through epigenetics; thus, cell function is dependent on proper architectural configuration of the matrix.[4] The matrix plays a far more active and complex role in regulating the behavior of the cells to which it comes in contact by influencing development, migration, proliferation, shape, and function.[3,4] See Figure 1.4.

Collagen is synthesized primarily by fibroblasts but also, by chondroblasts, osteoblasts, smooth muscle cells, endothelial cells, and epithelial cells. Collagen is generally classified in five types: **Type I:** fibrils, bundles of dermis, bones, tendons, gut mucosa, and fibrocartilage; **Type II:** fibrils, includes all types of cartilage; **Type III:** fibrils; blood vessels, dermis, stroma of internal organs, lamina reticularis of basement membranes; **Type IV:** no defined organization—basal lamina; **Type V:** no defined organization—placental and fetal lamina. There are also reticular fibers (0.05–2gM), which consist mainly of Type III collagen but with some associated glycoproteins and

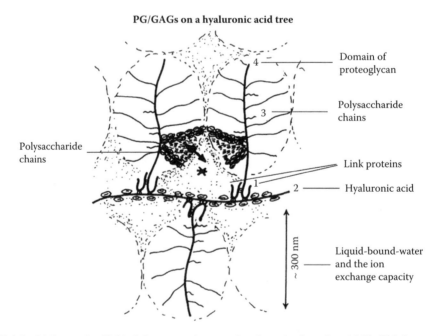

FIGURE 1.4 Link proteins (1) bind the proteoglycan molecules to hyaluronic acid (2). This is stretched due to its negative charge. The same situation exists with the polysaccharide chains (3), which are stretched away from their protein backbone. The interrupted lines give the "domain" of a proteoglycan molecule. The double arrow shows the liquid-crystal-bound water and the ion exchange capacity (arrow) between the polysaccharide chains. (Modified from Pischinger, A., *Matrix and Matrix Regulation: Basis for a Holistic Theory in Medicine*, Ed. H. Heine, Eng. Trans. By Ingeborg Eibl, North Atlantic Books, 25, 1991. With permission.)

PGs. Elastic fibers also occur. The collagen fiber bundles extend long distances in the interstitium providing most of the tensile strength and elasticity of the connective tissue.[4]

According to Linsenmayer[52], collagen has numerous other properties than just a supporting structure. These other functional properties for communication include involvement in cell attachment (cell adhesion) and differentiation, as chemotactic agents, as antigens in immunopathological processes, and as the detective components in central pathological processes.

The amorphous ground substance of the CTM engulfs all cells and fibers and is gel containing much water. GAGs formerly called acid mucopolysaccharides are long negatively charged polysaccharide chains consisting of repeating disaccharide units.[4] **The seven types of GAGs include hyaluronic acid, chondroitin 4-sulfate, chondrotin 6-sulfate, keratin sulfate, heparin sulfate, dermatin sulfate, and heparin**.[4] All GAGs, except hyaluronic acid, link with core proteins to form giant glycoproteins (mucoproteins). These are giant molecules that are 90–95% carbohydrate by weight. They attract water, which adds turgor to the matrix, which not only prevents compression but also **functions as an environmental receptor and communication vehicle**.

Glycoproteins for cell adhesion are also protein carbohydrate compounds, but protein comprises their bulk (90–95%). These include fibronectin and chondronectin.

The PGs and glucoseaminoglycans (PG/GAGs) are extremely thin-coiled molecules composed of 98% hyaluronic acid with glucose molecules and 2% protein. They form a mat of very fine reticular filaments (loose areolar tissue) that are similar to a brush pile.

The fluid in the interstitium, which contains almost all the components of plasma except for a filtered lower size protein content that is even lower than plasma, is supplied by filtration and diffusion from the capillaries. **The interstitial fluid (IF) is mainly entrapped in minute spaces in the matrix of the proteoglycan filaments thus, making the complex a peripheral local regulator of IF and lymphatic flow for regional and central homeostasis**.[4] This phenomenon is another way communication occurs.

The combination of[53] proteoglycan filaments and the fluid entrapped within them has a characteristic of a gel. It is difficult for fluid to flow through this gel; therefore, it diffuses slowly molecule by molecule in kinetic motion. This diffusion of fluid, as it flows freely through the gel is 95–99% as rapid as when it leaves the capillaries. For the short distances between the capillaries and the end-organ tissue and cells this diffusion allows rapid transport through the interstitium not only of water and water soluble substances but also electrolytes, nutrients, cellular wastes, lipid soluble substances, especially oxygen, carbon dioxide, and unfortunately most organic toxics.[54] Because these substances can permeate all areas of the capillary membrane without having to go through the pores only (as do only water soluble substances) **the rates of transport through the capillary membrane are many times faster than the rates for substances such as sodium ions and glucose**. Thus, distribution of toxics can be much faster and partially explains why many chemically sensitive patients react in a few seconds to minutes of an exposure. Distribution is also taken into account as an effective communication system.

The small vessels filter a greater portion of large proteins, which means not only that communication changes but also that the toxics that are bound to that particular protein are kept in the vascular tree. These toxics can then repeatedly be taken by the blood vessels to the liver, kidneys, and lungs for further detoxification, but can also cause damage to the vessel wall due to the reoccurring reactions that trigger inflammation, endothelial injury, and ultimately, environmentally triggered vasculitis.

The extracellular fluid bathes cells uniformly in the body. This allows information to be transmitted by proper signals and responses through growth factors, neurotransmitters, and other hormones traveling to and from the intracellular space. Communication is well established by this process.

Due to this anatomical organization of engulfing every body cell, the clinician can enter the ground regulation system at any place to trigger reactions or enhance therapy. This GRS was initially thought to react uniformly and entirely with each stimulus.[4] However, further study at our center (and by others) shows that with or without pollutant damage there can be exceptions where the information of the initial local insult did not go past the local or regional area and local containment of the entry of the noxious stimuli occurs. An alternative reason for the lack of information transmission could be that dissemination of the noxious stimuli because of the defense strength in this area the pollutant is contained or minimized. It also might be that the body's resistance around the injured area increased, or that the stimuli went only distally to certain regional areas in the body, or perhaps the stimuli went with less intensity than the initial local insult. In fact, with understanding of this concept and facts **one of the strategies used in prevention and containing of disease spread and the reversal of periodic focal homeostatic disturbances (chemical sensitivity) and aperiodic homeostatic disturbance (chronic degenerative disease) is to stop the spread and feedback of altered information.** (See Treatment chapter in Volume 4 - Mechanisms of Cardiovascular Disease and Chemical Sensitivity.) These constant noxious stimuli would cause generalized inflammation if the area of assault were not able to contain the response locally.

The clinician desires to keep the response from noxious stimuli as a pharmacological response (finite) rather than a pathological one (inflammation). This pharmacological response will cease in one to four hours, thus prohibiting a repetitive strong response that would trigger inflammation. Again, this adjustment response corresponds with the alarm stage of Selye[7] for local, regional, and central communication. All too often the total body pollutant load is chronically exceeded, triggering a latent redox reaction with release of free radicals and the patient then develops chronic inflammation with pain and swelling in certain areas.

It has been observed by many people that the resultant adverse body response can be relieved by a variety of therapies, including reduction of the total body load, injection therapy, good nutrition, and precise immune modalities such as injection of gamma globulin or autogenous lymphocytic factor (ALF). Due to the anatomical configuration of the Heine cylinders,[4] and the sequestration of toxics phenomena in muscle and fascia, a variety of different techniques such as massage, cranial manipulation, osteopathic manipulation, application of magnets and ELF fields, acupuncture and other needle modalities, local anesthesia, and laser beams can be used at the acupuncture points to release toxic substances that result in sensitivities. Injection of anesthetics, nutrients, and antiinflammatory substances at the point of maximum spasm can also be used to restore the homeostatic regulation processes in patients with musculoskeletal complaints. We also emphasize the importance of decreasing the total body pollutant load and enhancing nutrition as therapy to prevent the build up of sensitivity.

Loose SCT corresponding to the embryonic mesenchyme is particularly rich in the ECM and also is thus particularly reactive.[55] It has a typical distribution pattern accompanying all capillaries and forms the reticular cell tissue that underlies epithelial groups like the epidermis and mucosa (e.g., the tunica propria of the esophagus and the gastrointestinal tract), the splenic pulp, lymphatic tissue, fat tissue, the uterine mucosa, the ovarian cortex, the tooth pulp, and the Virchow-Robin spaces in the central nervous system (CNS). See Figure 1.5.

As the loose endoneurium, endomysium, and peritonium internum, connective tissue accompanies the vessels between nerve, muscle, and tendon fibers.[55] The SCT forms interstitial connective tissue, thereby enhancing the communication system. This loose connective tissue (LCT), among other functions, subdivides glandular tissues into lobules as vascular and nerve conducting interstitial connective tissue. LCT includes the adventitia of the large blood vessels, the serous membrane tissues, the endocardium, the soft brain membranes, and the inner most layer of the periostium.[56] See Figure 1.6.

The finely structured CTM of the organ capsules forms a transition to stiff and hard connective tissue (HCT) types. Tendons, fascia, ligaments, aponeurosis, dura matter, the stratum texticulare of the skin, the cornea, cartilage, bone and dentin already have the character of organs. These organs are also provided with LCT, accompanying vessels, and nerves (e.g., Havers and Volkmann's) channel in bone, dentin, cannuliculi in the teeth).[57] **Only the cornea, postnasal cartilages, and other joint cartilages are free of blood vessels and thus LCT.**

The Physiologic Basis of Homeostasis 17

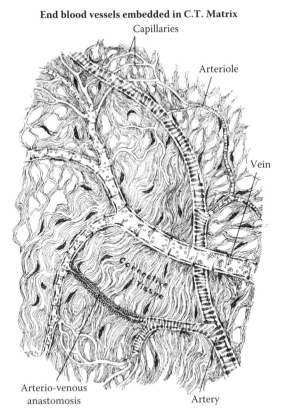

FIGURE 1.5 End blood vessels embedded in the connective tissue matrix, which consist of loose areolar tissue. (Modified from Woodburne, R.T., *Essentials of Human Anatomy*, New York: Oxford University Press, 21, 1957. With permission.)

The connection between the HCT substances penetrated by SCT accompanied by vessels and nerves is a clear illustration of the importance of the loose connective tissue bearing ECM. The reason for this importance, according to Pischinger[3] and Heine,[4] **is that the LCT allows information and energy flow and contains acupuncture centers that make up the basis for intradermal neutralization therapy, which is necessary for the treatment of sensitivities to biological incitants, foods, and some chemicals**.[4] The LCT's purpose appears to include maintenance of the functional capacity of the hard CT substance by providing nourishment and energy. Recently, the closely circumscribed perforations in the superficial body fascia (diameter 3–7 mm) have been shown to be of particular significance as they are involuted in LCT containing the vessels and nerve bundles of the fascia, which penetrate it deeply. They present the morphological correlation of the acupuncture points[58,59] also supports this observation and has shown the subdermal twin papillae containing the autonomic nerve, the vessels, and loose areolar connective tissue to be acupuncture points. See Figure 1.7.

After connection with the vessels and nerves the fascial nerves finally reach the spinal nerves of the CNS; conversely, fine vascular and nervous bundles, involuted in LCT leave the spinal cord, resulting in another anatomical configuration for communication. These nerve connections set up reflex loops for regional homeostatic feed back loops.

The area of the ossified sulci of the dura matter penetrates through the bone of the skull and appears in the skin of the scalp. The exit points in bone also correspond to acupuncture points.[60] Since the "points" (or rather perforations) are always found in the same place, irrespective

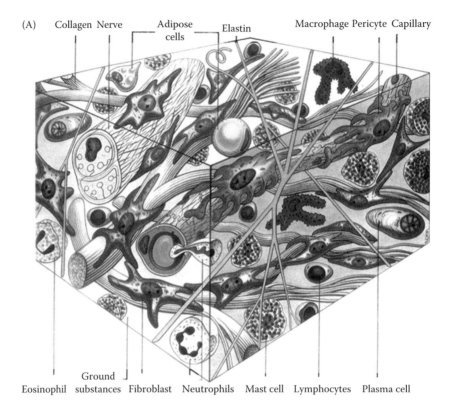

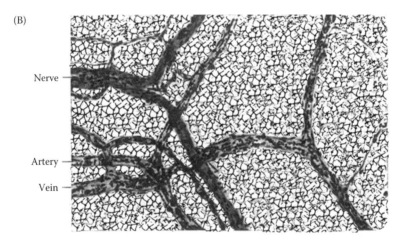

FIGURE 1.6 (A) Diagramatic reconstruction of loose connective tissue showing characteristic cell types, fibres and intercellular spaces. (B) Messentery evidencing small arteries and veins in loose connective tissue. (From EHC-Dallas, 2005.)

of racial differences or the living or dead, they seem to have a genetic origin. However, these points in the normal condition apparently, are normally conduits for energy flow through acupuncture meridians. Obviously, the acupuncture points can also become diseased themselves (perhaps this is a causative explanation for some peripheral neuropathies, Sudett's atrophy, dystrophy syndrome, painful trigger points, etc.)

The Physiologic Basis of Homeostasis

Loose connective tissue with the dermal acupuncture points

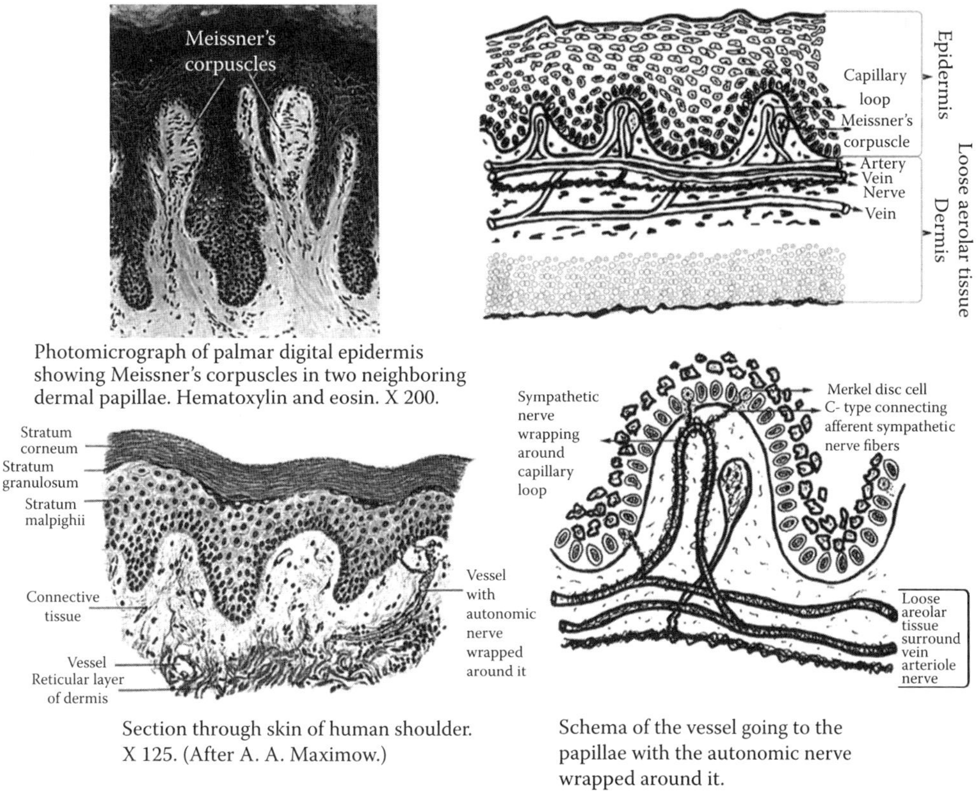

FIGURE 1.7 According to Croley, two dermal papillae, which contain the autonomic nerve wrapped around the vessel, are the acupuncture points. (Reproduced from Croley, T. E. and Carlson M., *American Journal of Acupuncture, 19*(3), 247–253, 1991. With permission.)

One of the signs of neurotoxicity in the chemically sensitive patient, which is demonstrated on triple headed SPECT brain scan, is soft tissue reactivity that may connote damage of the matrix and acupuncture systems. When anatomy is damaged and communication is disturbed often, faulty information transfer occurs (Figure 1.8).

These findings also offer a rational basis for neurotherapy and related therapies.[60] But also, if the organism becomes dysfunctional due to **excess total body load the observed fact is that the acupuncture, neural, homeopathic and electrical frequency therapies usually hold the dynamics of homeostasis for only a few hours to a few days and then symptoms exacerbate again**. If the total body load is not decreased, structural stability is not restored nor maintained. We have observed this phenomenon in our study of over 40,000 chemically sensitive and chronic degenerative diseased patients. Once the total body load of noxious stimuli is decreased to a subthreshold level the benefit from these therapies holds or stays in the body and the patient is markedly improved.

Dynamic Properties and Physiology of the Connective Tissue Matrix

The clinician must understand that the connective tissue is an extremely dynamic and electrically active tissue because of the mesh like PGs, GAGs, and hyaluronic acid, as well as the structural proteins like fibrin, elastin, collagen, and the cell surface glycoproteins like laminin, selectin, fibronectin[61], and immunomodulin[62] of the matrix. These substances make the body ready for excellent communication. This tissue is extremely dynamic having the capability of forming and absorbing

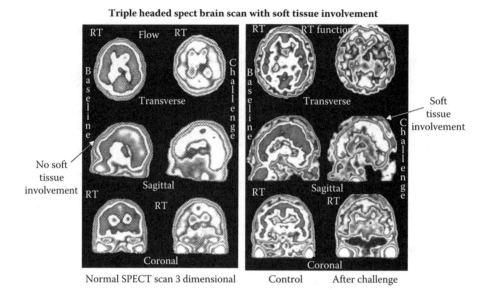

FIGURE 1.8 Abnormal SPECT scan after phenol challenge. Note soft tissue involvement shown as dark areas of increase uptake as dye traveled through the exit channels in the dura, skull, and to the nasal soft tissue. (From EHC-Dallas, Drs. Simon and Hickey).

heat energy generated from redox reactions[63] plus their tissue damaging free radicals),[64] of generating piezoelectric energy by changes of body pressure resulting from body movements,[65] of absorbing and generating energy from glucose breakdown[66] such as ATP, as well as being the environmental receptor and response system.

This capacity for handling the redox generated heat and free radicals is extremely large in a normal individual. However, as the patient is compromised by an increase in total body pollutant load, with resulting nutrient deficiencies, the ability to neutralize a large chronic total body load decreases until inflammation occurs. **Inflammation further changes the orderly dynamics of the homeostatic mechanism often requiring fewer free radicals to trigger a reaction because the quantity of free radicals in the tissue is already in abundance or the tissue threshold has been lowered by previous incitant injury.**

The matrix has strong electrical properties that allow for the reception and dissemination of information and communication throughout the body.[67] Due to its electrical lability characteristics this CTM can also respond to any external environmental energy (molds, food, toxic chemicals, bacteria, viruses, parasites) as well as electromagnetic energy (radio frequencies, transmitter emissions, music), subtle, and other energies from other individuals, and the universe in general making the CTM a giant environmental receptor system (ERS). The CTM has the capability of absorbing,[68] transmitting, and storing bioinformation (memory) due to its structure of GAGs, PGs, and the configuration and ordering of water[68] (see Figure 1.4). **This information transfer and dissemination is what gives biological systems their form (morphology) and maintenance thus, allowing the body to create new tissue while monitoring and repairing older tissue.**

Open-Ended System—Extra Cellular Matrix Instability

Biological environmental information receptor systems like those in the CTM are energetically open, i.e., they assimilate suitable "dissipative" energy (food, O_2, etc.) from their environment and expel or excrete waste substances (feces, CO_2, etc.) of lower energy back into the environment.

The Physiologic Basis of Homeostasis

As the CTM is an open system it receives and reacts to all stimuli no matter how small. This fact again emphasizes the presence of varied local responses that the organism has for dealing with intake of environmental information from any external substance or environment entering the body (e.g., complement change, nerve loop response, serums, change in T-cells and antipollutant enzymes [i.e., glutathione peroxidase, catalase, superoxide dismutase] or mucous flow). **Energetically open systems such as the CTM, swing back and forth like a pendulum due to its stimulation, inhibition, and the spring effect**. Spontaneous molecular changes of the matrix occur in addition to a response to a noxious stimulus accounting for some nonlinear characteristics of response. The open system receives and disseminates information throughout the body but due to its physiology and anatomy is extremely unstable. When a noxious stimulus enters the CTM there is an instant depolarization response that is either contained locally or regionally or spreads throughout the system. Once the information is sent usually by depolarization of the receptor tissue and depending upon the severity or chronicity of the incitant, normally, repolarization occurs immediately making the tissue ready to deal with another incitant.

Organisms that have energetically open information collecting and response systems are characterized by unstable conditions that normally swing back and forth, changing local temperature and even generalized thermic balance (thermal equilibrium) as well as, electromagnetic tone. When pathology occurs the patient with chemical sensitivity or chronic degenerative disease will become more unstable and fragile to any noxious as well as nonnoxious stimuli resulting in further swings in his/her physiology. This instability is particularly seen in the overloaded chemically sensitive patient who suddenly cannot tolerate noxious and nonnoxious exposures, e.g., food, or their neutralizing injections. **Their temperatures will fluctuate from 89–97°F (often being confused with hypothyroidism)** and they will also have electrical changes that fluctuate widely on the skin and CTM that will be similar to a normal individual even though these responses are exaggerated to an abnormal level. These chemically sensitive or chronic degenerative diseased patients who are going through a hypersensitive stage will still have physiologic fluctuation but at lower temperatures. Often, there are rigid (none or poor local temperature responses) or excess local homeostatic thermal responses (high local area temperatures) when the clinician measures a specific area of the body using specific topical thermography as we have done in thousands of individuals with chemical sensitivity and chronic degenerative disease.

One can see how these **multiple end-organ local temperature changes might alter stable efficient dynamic homeostatic physiology and when this alteration occurs on a chronic basis the patient will feel not only hypothermic but also weakness, fatigue, and perhaps pain**.

Thermodynamic fluctuations are observed clinically in changes in electromagnetic tone when measured by a thermogram, skin galvanometer,[69] the pupilograph,[70] or heart rate variability apparatus[71] in these patients (see Cardiovascular chapter in Volume 2 - Mechanisms of Chemical Sensitivity and Chronic Degenerative Disease for more information).

The abnormal thermodynamic response whether it be lowered or elevated emphasizes how excess chronic environmental noxious stimuli will initially change the body's homeostatic response subtly. If left uncatabolized or unneutralized, the entering pollutant will continue to stimulate or inhibit the physiologic adjustment process, which will eventually start and continue to deteriorate due to nutrient depletion until the noxious stimulus half-life is equilibrated. Therefore, the clinician will then observe an abnormal fluctuation of response in both the periodic (chemically sensitive) and aperiodic (chronic degenerative diseased) homeostatically disturbed patient, resulting in a subsequent nutrient drain.

There are frequent electrical changes due to chronic pollutant exposure, usually in the CTM, but also in the end-organ cell in the periodic homeostatic disturbance (chemically sensitive) or aperiodic homeostatic disturbances (chronic degenerative diseased patient) due to this open system. This instability will render the patient increasingly vulnerable to new incitants. **These electrical changes appear to be due to the more rapid fluctuation and oscillation of the PGs and GAGs due to stimulation, inhibition, and the spring effect**.[72] **Due to this open system these characteristics and the poor nutrient integrity make the tissue unstable electrically therefore, allowing** greater numbers of less intense stimuli to trigger a response because of the tissue electrical lability. Since the system is open and lability is increased a less intense total environmental load (nonspecific) or less virulent specific incitant (e.g., *Staphylococcal epidermis*) may then easily trigger the homeostatic adjustment response, where previously these less intense stimuli could not. Since the CTM is always open this labile condition continues to drive the patient with chemical sensitivity and/or chronic degenerative disease down hill. These unstable conditions may spontaneously emerge and change, respectively, and depends not only upon the intensity, chronicity, and type of environmental factors to which he/she is exposed but also on how fast the patient can replace their body's nutrition.

Open systems show that when suitable energy sources (such as nutritional substances, oxygen, etc.) are fed into the system the information can spread suddenly throughout the entire system. This homeostatic information to the subsequential healing process leads to the physical induction of new structures to repair the environmentally induced autocatalytic damaged structures resulting from proteases, hydrolases, etc., released by macrophages and other cells in local tissues.[73]

Due to the open system there is often a nonlinearity of response. This nonlinearity is a necessary characteristic for spontaneous organization, autocatalytic breakdown, and reorganization of the ECM tissue. These changes occur after a noxious stimuli,[74] or electromagnetic entry,[75] or at times spontaneously. This perceived spontaneous reaction may also be due to subtle external forces such as sun and earth electromagnetic changes but can also be due to spontaneous oxidation.

This dynamic open reception of external stimuli of information and subsequent adjustment cycling is accomplished morphologically and histologically in the CTM by the new production of PG/GAGs and collagen from incitant-stimulated fibroblasts.[76] Subsequently, matrix breakdown by release of local proteases, lysozymes, hydrolases, etc. secreted by macrophages, mast cells, leukocytes,[77] etc. occurs. This breakdown is followed again by fibroblastic production of new PG/GAG, collagen, etc., with reformation of a new matrix. Usually, if the lesion is not large enough, there is no scar. This cycle is constantly occurring in the body with normal homeostatic function. The body always pays a price nutritionally and energetically for entering noxious stimuli or moving internal stimuli no matter how small or intense is the noxious stimuli.

If inflammation occurs from excess noxious stimuli because free radicals break through then in severe cases scar formation results. However, in this dynamic spontaneous process or low intensity stimulation, allowed by open systems, new canaliculi or **channels are constantly formed and broken down in the matrix without scar formation because the matrix's latent inflammatory potential was not exceeded.**[74] **This tissue breakdown and healing process is extremely important because it is subinflammatory as long as the stimulus is not too strong or too chronic and proper nutrition is maintained**.

In other words, once the noxious incitant enters the body it is perceived by the matrix and other receptors, which give out information locally to the rest of the body. This information stimulates the homeostatic adjustment mechanism to mount a defense. The adjustment mechanism then integrates the summation of the smallest homeostatic adjustment changes in order to physically neutralize the incitant. This process of adjustment, neutralization, and integration results in an increased demand on the energy system. More energy will be used for the process of altered dynamics of homeostasis than for the normal, orderly process of dynamic homeostasis. **This energy drain if left uncorrected is the origin of disease**. The true cost of efficient energy use of the physiologic processes can be accomplished at this level, if the incitant is withdrawn and the dynamics of homeostasis are corrected. This open informational feedback loop response system makes continuous adjustments to changing internal and external conditions possible at the expense of increased energy demand and utilization. When the dynamics of homeostasis are disturbed (i.e., by toxic substances, infection, nutritional deficiency, and trauma) the feedback information may be lost or altered and the system becomes overloaded. **Clinically, initial weakness, lassitude, inertia, chronic fatigue, subtle brain dysfunction with mental fogginess, and fibromyalgia** can result from this over utilization of energy, due to stress on the mitochondria and its energy dissemination systems in the CTM, the macrophages, and leukocytes. These symptoms are seen early in both periodic homeostatic disturbances (i.e., chemical sensitivity) and the aperiodic homeostatic disturbances (chronic degenerative disease).

There is a sequence of the individual's ability to initially handle a variety of environmental incitants of varied intensity and direction, and then to send multiple types of information all over the body. This ability is therefore, based on the characteristic of nonlinearity in an open informational receptor system (CTM) with its energy-efficient reception and adjustment response. It appears that the energy of the individual informational feedback process involved in the responses of homeostasis is greater than the sum of the energy of the whole-integrated homeostatic mechanism involved in normal body function without triggering.

The triggering of the local homeostatic mechanism and the resultant change in minor adjustment responses caused by diverse environmental stimuli coming from different areas in the body, dictates that in nonlinear systems, like the ECM, it is not possible to predict final outcomes or states of the dynamics of homeostasis. However, excellent communication occurs out of this system.

These varied fluctuations of the open matrix's homeostatic responses are a characteristic of a chaotic system. Energetically open systems may therefore, within their confines, be described as a "determined chaos," which can, in a mathematical sense, eventually develop a focus of energy; and if the incitant is not chronic, allows for the dynamics of homeostasis to occur in order to restore the body to normal function.[78,79] An example of this varied response has been seen when studying patients in the environmental control unit, where the total body load is decreased, and the patient is in the deadapted basal state. A challenge (oral or inhaled) of a specific incitant is given and a reaction and deviation of the response will always occur from the basal state if the patient is sensitive to that substance. However, the patient will often manifest symptoms and signs related to different end organs after challenges with the same amount and intensity of substances at one time, for example, rhinorrhea, hoarseness, and diarrhea will occur, while the next time one performs a challenge, rhinorrhea, myalgia, and pedal edema may result. The consistency is that the hypersensitive or chronic degenerative diseased patient always reacts but at times with a different response, depending on the total body toxic load, state of the body's nutrition, etc., at the time of exposure. In other words, nonlinear responses may occur after the same quality and quantity of noxious stimuli exposure.

This type of environmental information receptor system that occurs in the ECM allows the open-system flexible responses. These responses include the dynamics of homeostasis and its energy system. These near automatic informational feedback and adjustment loops of homeostasis respond to nonphysiological changes and thereby, permit homeostatic adjustment processes to occur before excess chronic noxious stimuli cause chronic disease with its resultant near total energy drain. When excess chronic noxious stimuli are present with a total body overload occurring, there is a dyshomeostatic adjustment, which means that diseases once they occur will always continue to affect homeostasis and its energy consumption. **The symptoms that result, for example, chronic weakness, fatigue, pain, etc., are due to the excess energy demand needed for the difficult attempt at physiologic correction and communication for correction of the dynamics of homeostasis**. It appears that pain is usually due to an accumulation of excess energy in the damaged area and will not be relieved until the homeostatic physiology is restored with the energy dissipation and restoration to normal flow.

Lability of the Connective Tissue Matrix

In healthy people, the PG/GAGs can absorb the heat of the redox reaction, which acts as fuel for driving the system. In addition, the matrix can absorb free radicals thus, preserving tissue integrity and, to a point, preventing inflammation. These properties of the CTM are essential for normal homeostatic adjustment processes and harmonic resonance of information spread and reciprocal response information. See Figure 1.9.

With excess noxious stimuli the CTM and its receptors become damaged by excess free radical generation, become fragile, and then unstable (labile). The electrical insulating properties of amino acids, configuration and communication properties of glucose, including the ordering and

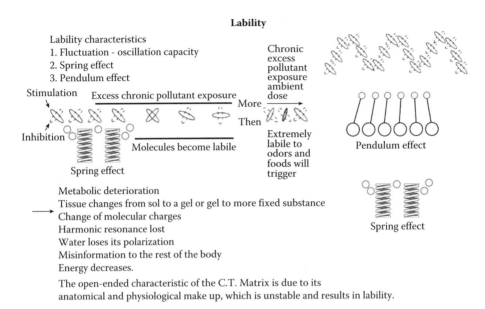

FIGURE 1.9 Instability and lability after pollutant overload of an open-ended connective tissue matrix due to its anatomy and physiology. (From EHC-Dallas, 2002.)

disordering of water, also contribute to this instability state of the environmental receptor of the matrix. Along with the information storage and dissemination function,[80] water can easily be configurationally changed by an external or internal noxious stimulus. The same can be said for the change of PG/GAGs by proteases, lysozymes, and other tissue proteolytic enzymes, which are secreted by leukocytes and macrophages. Their alterations result in not only anatomical change but also changes in the electric and electromagnetic tone of the CTM because they physically and morphologically alter the matrix (tissue changes) and thus, the ability of the matrix to transmit clear information. **Information changes can then alter the dynamics of homeostasis. A vicious downward cycle occurs until fixed-named disease occurs**.

The dynamics of homeostasis at the molecular level in the CTM is thought to be a constantly changing spontaneous movement of molecules ranging from enthalpy (extreme order) to entropy (random distribution of the molecules. These changes may be exaggerated by pollutant overload thus changing communication pathways and healing information. See Figure 1.10.

As the CTM is the ERS it is influenced by electromagnetic frequencies (EMF), external pressure changes, redox electron transfer, piezoelectric anatomical pressure changes, and other changes therefore, it is likely that the molecular changes are quite driven by these forces, rapidly influencing the "spontaneous movements." Certainly, the redox reactions are spontaneous because oxidation can occur without energy expenditure following the way that nature disintegrates biological material to the point of nonexistence. **It appears that the spontaneous oxidation—reduction reaction is also, often one of the sources for matrix instability**.[81]

This spontaneous flux of molecules allows for the rapid countering of the external and internal stimuli on a day-to-day, hour-to-hour, minute-to-minute basis. This ebb and flow of the dynamics of homeostasis, in the connective issue matrix, continues to a point where a large noxious stimulus enters the body and triggers the whole equilibrium mechanism. This dynamic physiology also rapidly neutralizes small amounts of noxious stimuli that are constantly entering the body or that are being self-generated from within the body in a local area.

Several factors emphasize that EM fields emanating from the spontaneous molecular flux are extremely important in understanding the CTM function. It appears that water ordered by the PG/GAGs takes on EMF imprints and the information that is stored in the matrix at times changes

The Physiologic Basis of Homeostasis

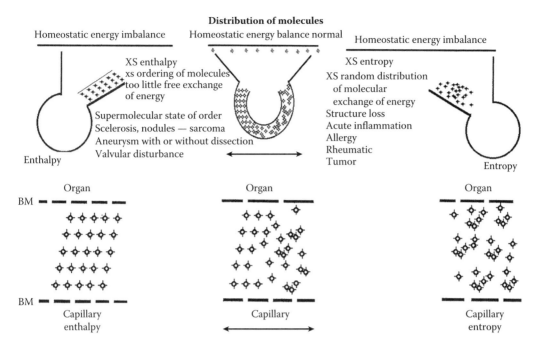

FIGURE 1.10 Random distribution of molecules leads to entropy, and overly ordered molecules leads to enthalpy. (From EHC-Dallas, 2002.)

often, rapidly, and at times, can change slowly. At a certain stage of injury or illness the matrix is imprinted with electrical impulses, which are difficult to erase. Some opinion suggests that a high fever will also erase imprints. However, Trincher[82] has established that the proportion of crystalline water decreases with boiling, which would suggest that the body temperature of 105° F would not be sufficient to erase the EMF imprint. Smith[83] has shown that water looses its EMF imprint with boiling or removal of the electromagnetic field by placing tissue from the CTM in a metal box or in some cases overprinting with another frequency, which may be endogenous or exogenous. Therefore, toxics, bacteria, and various other stimuli in the CTM can cause the tissue to alter or loose its polarization, structure, and order under some environmental receptor conditions. **Due to such damage to this ordering of water from noxious stimuli due to an alteration in the environment in the ground regulation system, the organization factor for homeostasis disappears**.

Connective Tissue and Electromagnetic Energy

According to Pischinger[84] and Heine,[85] **the cell is the internal generator of electromagnetic information**. Electromagnetic discharge and waves are how the cell communicates with the rest of the body. When the electromagnetics becomes disturbed there is a change in the entire communication system. This cell generation of disturbed electromagnetics is emphasized repeatedly in the patient who has become electromagnetically sensitive. This type of patient has a dysfunction of these cell generators for electromagnetics and is more likely to have an extremely low threshold for EMF triggering, resulting in a metabolic reaction. This condition is frequently initiated due to a severe pesticide exposure followed by an electromagnetic insult. For example, a patient has her home sprayed with pesticide and she is also a heavy cell phone user up to eight hours a day. From this combination of pesticide exposure and EMF insult she then becomes supersensitive to external EMF exposure. Physiologically, this sensitivity results from a combination of the disturbance of the CTM and the cell.

> **Case report**: An example of this EMF response was manifested clinically in a 38-year-old white female physician who developed extreme hypersensitivity to her environment (air contaminants, food, water contaminants) along with episodes of headaches in the left temporal area, short-term memory loss, and

an inability to concentrate. For three years she had used the cell phone constantly throughout the day in the left ear. Her home was sprayed with a pyrethrum pesticide monthly and it also became mold and mycotoxin infested. She became incapacitated. Triple headed SPECT brain scan showed a large area of damage in the left temporal area.

From this case report and many others, there is basic scientific data, which helps clarify the electrical reactions of the matrix. According to calculations made by Frohlich,[86] there is a membrane thickness of 6 nanometer and a potential difference of 0.1 volt. Therefore, an electric field strength of 10^7 V/m cm/lc can be expected. These high unstable field strengths cause polarization of dipoles in the membrane. The relationship between the speed of sound and membrane thickness 6nm gives a resonating oscillation in the microwave range (about 1000 GHz or 0.1 TeraHz).[82] This resonance would occur first in the cell then to the CTM, and the PG/GAGs. Information would be sent efficiently resonating throughout the body (Figure 1.11).

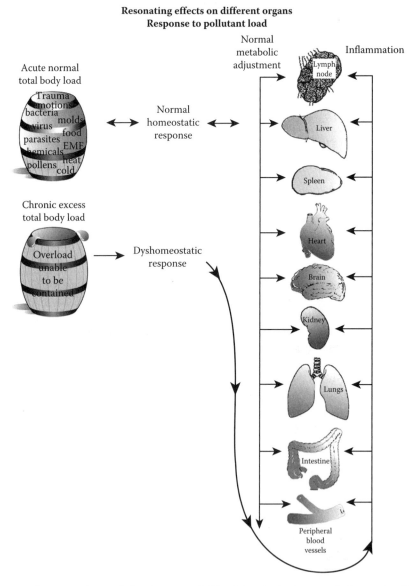

FIGURE 1.11 (See color insert following page 364.) When the body exceeds its total pollutant load, dyshomeostatic responses occur, finally resulting in inflammation. (From EHC-Dallas.)

The Physiologic Basis of Homeostasis

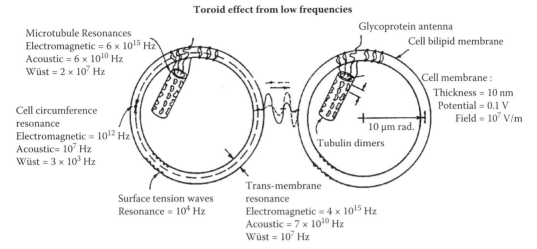

FIGURE 1.12 Toroid effect of Smith. The toroid contains the magnetic filed within itself but radiates the magnetic vector potential into the surrounding space. The magnetic vector potential may appear as a chemical potential term in the wave function. (Courtesy of Dr. Cyril Smith, University of Salford, England. Personal communication.)

Popp (1983)[87] and Klina[88,89] showed that the electromagnetic oscillations of coherent light present an information system for all living organisms and those facts are the model for the communications in the ground regulation system. Here the transition of oxygen from the stimulated single state, which can occur both in the matrix and in the cell to the molecular triple state, was recognized as a laser source with a wavelength of 634 nm. **Thus, the ground regulation system's electromagnetic function appears to be highly significant in the field of metabolic tissue regeneration and (from Bergsman[90] observations) is capable of releasing resonance and dampening effect of extremely low frequencies (ELF).** Externally generated low frequencies are thought to interfere with orderly function of tissue communication because the tissue functions with low frequencies. Smith proposes a toroid effect, which would account for all of the resonating effects. See Figure 1.12. The toroid contains the magnetic field within itself but radiates the magnetic vector potential into the surrounding space. The magnetic vector potential may appear as a chemical potential term in the wave function. In studies at the EHC-Dallas[91] with electromagnetically sensitive patients, specific low frequencies have been shown to trigger a reaction.

Mechanical Coherence

Another characteristic of the CTM (according to Balasz and Gibbs 1929,[92] Buddecke 1921[93]) is that the mesh type macro-molecular three-dimensional superstructure of PG/GAGs plays an important role in the mechanical coherence of tissues, which also aids in the communication process. Through this structure, the terminal axons of the autonomic nerve fibers, for example, are subject to a specific mechanical and electrical tension and can react to these stimuli with the release of neurotransmitters such as acetylcholine, epinephrine, and numerous neuropeptides.

As previously referenced, the basic scientific fact regarding electrical properties is that the PGs and GAGs are charged sugar molecules,[94] which use amino acids as insulators.[94] The PG/GAGs sugar molecules have a high ability for structural diversity. This structural diversity makes sugar polymers more efficient data receptors and carriers for they are able to receive multiple bits of information from the diverse environmental stimuli. These sugars can then, in turn, rapidly transmit the received information locally, regionally, and centrally for the homeostatic adjustment to occur. This information flow occurs with coherent harmonic resonance, which allows for energy efficient communication.

The PG/GAGs order the water thus, allowing for a memory imprint component in the CTM. This dynamic electrical condition will always be changing as new information enters from the environment thus, causing dynamic metabolic and mechanical changes in the connective tissue matrix.

Spring Effect Piezoelectric

In order to keep the proper electrical orientation of the CTM, the spring effect appears to play a crucial role (see Figure 1.13). This spring effect of the matrix, with each bodily movement, will mechanically generate piezoelectric currents, which tends to keep the PG/GAGs polarized with a proper electrical orientation.[95] The effect occurs because of the way the PG/GAGs are set upon the hyaluronic acid tree and also, in the way they are placed in the matrix **with extracellular fluid, creating the brush pile effect**.[96] Since the anatomical makeup is like a spring the matrix cannot totally collapse. Mechanical pressure of each bodily movement creates currents. Just think about this spring motion in the body as we are constantly moving even when we are lying down or sitting still for a long period of time and thus, we are always creating and surrounded by small currents. These piezoelectric currents then allow proper orientation of the electricity of the PG/GAGs with harmonic resonance of information flow throughout the body for instant communication to or from cells and organs.

Also, the matrix will not only act as a mechanical filter excluding molecules (e.g., polyaromatic hydrocarbons) of excessive weight but also aids fluid balance, **due to the spring effect, which never allows the matrix to totally collapse from dehydration**.[97] Thus, again, this spring effect generates proper information flow.

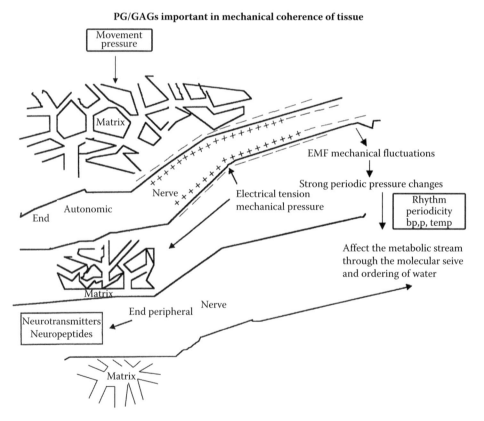

FIGURE 1.13 Piezoelectric Effect of Mechanical Compression allows for electrical reorientation of the connective tissue matrix. (From EHC-Dallas, 2002.)

The piezoelectric phenomena of the PG/GAGs are important because as dipoles they are always oriented toward the direction of growth.[98] Also, they not only produce electrical impulses by pressure, but also alter electromagnetic fields of certain resonance frequency, resulting in mechanical fluctuations. **These mechanical fluctuations lead to strong periodic pressure changes in the surroundings, which have an effect on both the metabolic stream through the molecular sieve of the ECM and on the structure of the water region.** The modulating effects also affect the rhythmic periododicity. Thus, pulse, blood pressure, etc., act in a periodic nature due to the mechanical fluctuations emanating from the piezoelectric effect. Bergsmann[99] and Heine[100] have been able to produce evidence of the modulatory effect of laser frequencies (620–725 nm) on the molecular activity of the matrix. **This fact, again, emphasizes the origin of periodic activity of the body's fluctuating metabolism and reemphasizes that physical exercise will tend to extract pollutants by creating pressure upon the matrix.** We continuously observe this fluctuation in our sauna and physical therapy programs. The pressure from exercise movement will cause piezoelectric phenomena to occur, which helps to reorient the PG/GAGs to normal function of detoxification. The resultant condition of the tissue will be electrically oriented to allow efficient homeostasis.

Molecular Sieve

In addition to being a mechanical structure support, open receptor, and communication system, the matrix acts as a giant filter system with the origin of nutrients coming in from the lungs (O_2) and gastrointestinal tract (food, water) with the waste products going out from the parenchyma to the liver, lungs, kidneys, and gastrointestinal system. **The matrix filters all substances that enter or exit thus, becoming a regulator of all of the body's dynamic physiology. Thus, communication can be modified by a change in filtration and often is changed as the chronic noxious stimuli clog the matrix filters or alter its anatomy and thus, their response.**

Since the connective tissue is also a giant mechanical molecular sieve it can filter out substances (e.g., polyaromatic hydrocarbons, and other toxics) of larger sizes as well as very small substances depending upon their magnitude.[101] **Due to the matrix is dynamic it can change sizes and configuration of the canuliculi and, thus,** filter out substances of other smaller sizes by sequestering them as is seen in toxic metals like mercury, lead, cadmium, and the smaller volatile organics 0–2.5 microns. See Figure 1.14.

The matrix as a molecular sieve is shown in Figure 1.15.[102]

The canuliculi of the matrix function dynamically as the fluids and nutrients pass through the basement membranes of the capillaries to the basement membranes of the end organ and back again. **The matrix, like the movement of the glass pieces in a kaleidoscope (Figure 1.16), is constantly changing as it responds to autocatalysis and proteolysis triggered by the release from macrophage and leukocytes of local peptidases, serine protease's, hydrolases, glucosidases, and lysozymes.**[103] **The interpolarization of matrix molecules followed by lysomal degradation in the cell releases collagenase, which is a metalloprotein that is highly specific, and stromoylsin, which is reactive.** Protein may be targeted by other pathways, such as limited proteolytic cleavage, or by secreted enzymes, proteinases, including collagenase, PMN collagenase, stromoylsin, matrilysain, gelatinase, undolysin (gelatinase 92KOA), plasmin (serine proteases).[104]

These substances help to constantly fashion and reform the matrix area, which is an extremely dynamic tissue that not only acts as a fluid conduit, but also operates as a molecular mechanical sieve and partial regulator of the dynamics of homeostasis due to its electrical generation and heat absorption properties.[104] Communication can be generated instantly and altered very rapidly due to this filtration process. **The molecular sieve is similar to glomeruli and is very dynamic, expanding, and contracting due to environmental stimuli.**

The size of the molecules going through the matrix sieve depends on their charge, the concentrations of PG/GAGs in the tissue areas involved, the molecular weight of the filtered substances (molecular weight), the pH of the matrix, and filtered substances as well as, concentration of electrolytes.

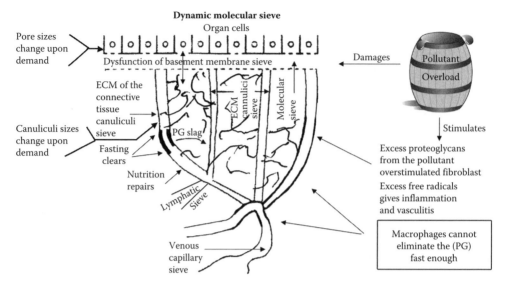

FIGURE 1.14 Dynamic molecular sieve. The basement membrane and the extracellular matrix canuliculi of the connective tissue going to the blood vessels and lymphatics must be kept flowing properly or the vasculitis occurs in the chemically sensitive individual. It is important to note the dynamics of this sieve process can change sizes upon entry of different noxious stimuli which will expand or contract the filters. The filters apparently have a brain like quality to expand or contract in size depending upon the demand of the toxic substance. The ground regulation system plays a key role in healing with a return to homeostasis when it is unloaded. (From EHC-Dallas, 2004.)

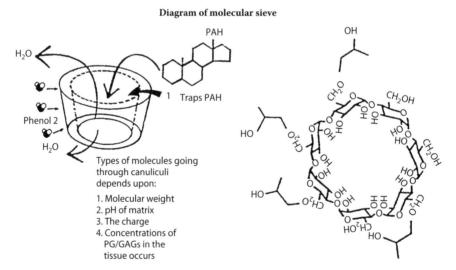

FIGURE 1.15 The dynamic molecular sieve of the connective tissue matrix with the ordered water in the PG/GAGs mechanically trapping polyaromatic hydrocarbons according to Heine. 1. PAH—polyaromatic hydrocarbon. 2. Phenol. This sieve is extremely dynamic due to its size change probably from the different chemical and biological processes. (Modified from Heine, H., and Anastasiadis, P., *Normal Matrix and Pathological Conditions,* Gustar Fischer Verlag, New York, 7, 1992.)

Excess bacteria, virus, or toxics can clog up the matrix causing other alterations in electrical tone thus, giving severe deregulation of homeostasis not only locally but also distally[105] due to misinformation being sent to the central homeostatic control mechanism. Examples of this matrix alteration are the Wing studies where nasal biopsies on chemical and mold sensitive patients have shown the pathologic change of the nasal matrix in literally hundreds of patients.[106] His biopsies

The everchanging canuliculi of the connective tissue matrix

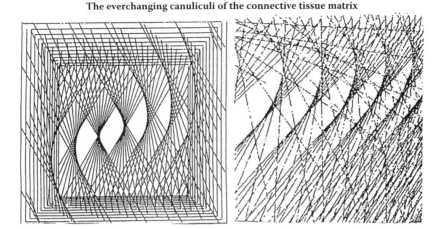

FIGURE 1.16 The dynamic kaleidoscope effect according to Heine. The kaleidoscope effect of autocatalysis generated through hydrolases, proteinases, peptidases, glucocidases, lysozymes secreted by the macrophages and leukocytes with regeneration by proteoglycans and glucosoaminoglycocans secreted from the fibroblast causes the canuliculi to be constantly changing sizes due to the molecular demand. (From Pischinger, A., *Matrix and Matrix Regulation: Basis for a Holistic Biological Medicine*, English translation by Ingeborg Eibl, North Atlantic Books, 1991. With permission.)

Nasal biopsies

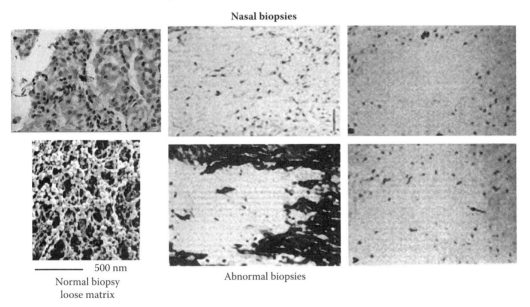

FIGURE 1.17 (See color insert following page 364.) Different patterns of extracellular matrix in mold sensitive and chemically sensitive patients—nasal biopsies. Note: increased density of matrix in each slide due to inflammation. Biopsies show the ECM sieve as clogged. (Courtesy of Dr. Lindsey Wing, Sydney Australia. Personal communication.)

show the clogging of the matrix by both the triggers and their substrates and their reactive responses (Figure 1.17). These changes occur routinely with excess pollutant injury.

Molecular Sieve Control

The dynamic local and generalized molecular sieve of the ECM is also important to obtain and maintain homeostasis in other ways. For example, if the nerve receives a physiological or even a noxious unphysiological stimulus the surface membrane relaxes and becomes more permeable. As a result, potassium, sodium, calcium, and magnesium ions change concentration. Potassium escapes and sodium enters the cell. However, the nerve cell is kept in its electrically charged state by virtue of the differential concentration of ions within and outside of it. An example of this is the loss of potassium from the nerve cell to outside of it, which allows its ion concentration to become equalized. This results in the loss of the electrical potential of the surface membrane, which collapses and the nerve looses its electrical charge. This condition results in depolarization, producing impulses in the nerve, which are transmitted to the spinal cord, the brain stem, the cerebellum, and the cerebrum. The sensation of pain thus, also, registers and controls electrical impulse patterns (depolarization) and signals any unphysiological or noxious stimuli. The higher the initial potential of the nerve cell (membrane resting potential (MRP)) the greater the stimulus needed to produce depolarization and vice versa. **Disturbances in the oxidizing cell metabolism influence the MRP and thus, indirectly affect the depolarization rate and the perception of pain.**

This sieve phenomenon can partially account for both the hypersensitivity (by decreasing the threshold thus, increasing the ability to trigger reactions) and the chronic deteriorating response as seen in chronically ill patients. The molecular sieve of the CTM must be crossed by all the components of the body's metabolism (nutrient and waste products) for extensive communication throughout the body.

In addition, the active cell membrane is like an adjustable sieve. Its permeability to ions of different sizes can be altered electrically. If spreading of noxious stimuli after local entry occurs, bacteria, virus, and other noxious products may try to cross the matrix in order to get to the cell. Thus, a signal of this spreading in the CTM again would trigger the central homeostatic mechanism either locally, regionally, or centrally to adjust the defense mechanism. The result of this signal includes alteration of the connective tissue filter mechanism in order to prevent noxious substances entry into the cell. **Communication is very clear at this point because of this signal as well as other signals. The ground regulation system is the largest system in the body completely penetrating the whole organism collecting and supplying new feedback information to provide nutrition and to remove waste from the cell, receiving and interpreting information, regulating local cellular homeostasis for defense, regulating the response and adjustment mechanism and thus, regulating the dynamics of nutritional homeostasis.**

Experimental Stress and Molecular Sieve

In an experimental stress model, **Heine has shown under sympathetic stimulation for one hour inducing stress increased formation of chondroitin sulfate and dermatan sulfate by myocytes (descendent of fibroblast) occur in the middle of the arterial wall.**[107] Dermatan sulfate augments trapping of lipids and calcium ions, resulting in accelerated atherosclerotic changes. Thus, ground substance synthesis caused by stress, leads to functional disturbances of the ECM and causes changes in the characteristics of the molecular sieve resulting in periodic (chemical sensitivity) homeostatic dysfunction and aperiodic (chronic degenerative disease) including tumor formation, arteriosclerosis, scleroderma, vasculitis, etc. At the EHC-Dallas autonomic nervous system studies on hypersensitivity and chronic degenerative diseased patients have shown that the majority have sympathetic responses (82%) that put out excess norepinephrine thus, stressing the blood vessels causing vascular spasm and thus, tissue hypoxia in addition to the triggering of chondroitin and dermatan sulfate. For every incoming pollutant the CTM has a latent nonspecific mesenchyme reaction (NSMR) the property of which can be triggered producing inflammation.

Latent Free Radical Response

Due to its unstable origination however, the spontaneous reactions caused by the oxidation-reduction process are constantly throwing off excess electrons and protons from the PG/GAG, water, sialic acid, nitrous oxide, carbon dioxide, sugar, and urea. Therefore, the matrix normally must be capable of absorbing free radicals **without causing inflammation** and without upsetting the dynamics of homeostasis and information flow otherwise the communication system can be altered. This adjustment response of the homeostasis of the CTM carries a fine line between normal function and inflammation. Since the matrix must take on and neutralize all pollutants an organ must have lots of reserve nutrients. **Failure to keep up the neutralization of excess free radicals will cause inflammation.**

It should be pointed out that in twenty-first-century America the generation of free radicals is extremely copious once noxious stimuli enter the body. This volume of pollutants is due to the massive amounts of xenobiotics in our contaminated air, food, and water. The contaminants have now been shown to generate a host of free radicals in the body, including $\bar{O}$, O_3, OH (see Table 1.1). With the potential of generating so many types of radicals in different areas of the body it is paramount to reduce the total noxious stimuli load to maintain the dynamics of homeostasis and prevent inflammation.

Macrophages play four major roles in matrix remodeling. They degrade the matrix directly by secreting uroplakins (UPK) and metalloproteinases; they participate in activity of proteinases made by other cells; they produce cytokines (IL-1 and TNFx) that can stimulate fibroblasts and endothelial cells to express metalloproteinases, and they phagocytize. Excess proteinases occur in inflammation and degenerative inflammatory disease including scleroderma, rheumatoid arthritis, emphysema, and tumor invasion.

TABLE 1.1
Radicals Can Be Generated from a Myriad of Noxious Stimuli Including Bacteria, Viruses, Chemicals, and Radiation

Physiological and pathological important radicals or radical progenitors in living systems

Activated Oxygen ("oxidative stress")

Superoxide-Anion-Radical	$O_2^{\cdot-}$	Haber-Weiss-Reaction
Hydroxylradical	$HO^{\cdot}$	$H_2O_2 + O_2 \xrightarrow{Fe\ UN\ CU} HO^{\cdot} + HO^{\cdot} + O_2$
Erhydroxyradical	$^{\cdot}OH_2$	
Singlet Oxygen Molecule	$O_2(1\Delta G)$	Fenton-Reaction $H_2O_2 + Fe^{2+} ® Fe^{3+} + HO^{\cdot} + HO^{-}$
Ozone	O_3	
Hydrogen peroxide	H_2O_2	H_2O_2 Reduction by Semichromes (e.g., Oxygenation, Cytochrome)
		$QH^{\cdot} + H_2O_2 ® Q + OH_2 + HO^{\cdot}$
		↑ ↑

H_2O_2 is not a radical but is able to generate in three ways the very dangerous hydroxyl radical

Lipid peroxidation (chain reaction within biological membranes)		Radicals from Sulfhydrl (SH)-Groups	
Lipidradical	$L^{\cdot}$	SHiyl-Radical (R-S)	
Alkoxyradical	$LO^{\cdot}$	SUlfinyl-Radical (R-SO)	
Erhydroxyradical	$LOO^{\cdot}$	SHiyl-Peroxyl-Radical (R-SOO)	

Radicals from water molecules induced by energy rich radiation ($> 10^{3eV}$) (e.g., gamma radiation)

$H_2O + h\nu \rightarrow H^{\cdot} + HO^{\cdot}\ e^{\cdot}aqua$

Radicals from organic molecules, e.g.:

Methyl-Radical	$CH^{\cdot}$
Phenyl-Radical	C_5H_5
Alkoxy-Radical	$RO^{\cdot}$
Peroxy-Radical	$ROO^{\cdot}$

Non-enzymatic scavengers of radicals can change into radicals; therefore, they must be neutralized by other oxidants: orchestrated action between vitamin E, C and reduced glutathione.

Source: From Levine, S., and Kidd, P., *Antioxidant Adaptation: Its Role in Free Radical Biochemistry,* Biocurrents Press, San Francisco, 1985. From EHC-Dallas.

Extracellular Fluid and Connective Tissue Matrix

All authors agree that the extracellular fluid transport system travels throughout the body to all parts in two stages. The dysfunction of this system in patients with chemical sensitivity and/or chronic degenerative disease allows for further communication and dissemination of misinformation. **The first stage of transport entails movement of blood and plasma around the circulatory system and second involves movement of fluid between the capillaries and the end-organ cells.** All the blood in the circulation traverses the entire circuit of the circulation an average of once each minute when the body is at rest and as many as six times each minute when a person becomes active.[108] This phenomenon allows for rapid ingress and egress of both noxious substances and nutrients. When overloaded with noxious stimuli as in the chemically sensitive and/or chronic degenerative diseased patient, one can at times change the dynamics of the individual's homeostatic response resulting in dyshomeostasis and inappropriate information dissemination. An example of this rapid distribution phenomenon is seen in the chemically sensitive patient who is resting near a busy street. He breathes, absorbs, and redistributes X amount of car exhaust just by breathing normally. He decides to run along the street and immediately absorbs and redistributes 6(X) times the amounts of car exhaust, which causes him to feel extremely weak and fatigued. This rapid excess of a noxious exposure actually overloads him and puts a strain on his immune and enzyme detoxification systems. These systems may not be able to generate the superoxide dismutase, the glutathione peroxidase, aryl hydrocarbon hydralase, and the catalase rapidly enough to immediately combat the excess pollutant exposure. He can now become nutrient depleted because of the nutrition drain, which leads to a vicious downward cycle involving the positive feedback characteristic of homeostasis. An extreme example of internal toxic substance overload would be an individual with tachycardia who has renal failure and cannot keep up removal of the waste. In the normal matrix function the faster the toxic substances cross the matrix filter the more rapidly the matrix fills with individual waste. Since the capillaries are porous diffusion occurs throughout the capillary bed everywhere in the body.

With the constant mixture of diffused fluid almost complete homogenecity of fluid in the connective tissue matrix surrounding the cells is assured throughout the body. Therefore, the noxious incitant is rapidly distributed everywhere unleashing the potential for unparalleled communication. However, in the patient with chemical sensitivity an/or chronic degenerative disease unequal deposition usually occurs due to unequal blood shunting in the microcirculation, due to hypoxia, autoimmune, and electrolyte disturbances, areas of previous injury (secondary homeostatic foci, disturbance fields) nutritional deficits in specific areas, or due to extreme attraction by the type of cell (i.e., lipophilic attraction of fat-soluble organochlorine pesticides) of any toxic substance. Due to these variables individuals with chemical sensitivity and/or chronic degenerative disease will have either different symptoms or early symptoms in different areas. Thus, clinical entities may be difficult for each individual.

Few cells are more than 50 micrometers from a capillary, which ensures diffusion of almost any substance from the capillary across the matrix sieve to the cell within a few seconds. This anatomical make up can be very productive for rapid communication. However, it can also be detrimental for the patient with chemical sensitivity whose sieves are intermittently or periodically clogged. For example, these damaged areas (interference fields) become secondary, nonenvironmental noxious triggers for the cells, which can with increased damage become foci for less intense stimuli triggering of the homeostatic mechanism. **These focal homeostatic disturbances (interference fields) are one reason why the chemically sensitive patients have rapid adverse reactions even to the minutest amount of a substance.** Internal homeostatic disturbance foci (secondary, tertiary end foci in the patient with chemical sensitivity and/or chronic degenerative disease) explain why dyshomeostasis may occur because the foci are closer to the area of disturbed homeostasis and are triggered more easily. They become so easy to trigger, developing to the point that normal movement can trigger the disturbance foci (interference fields) creating more disturbed homeostasis until the process becomes autonomous. The uniform mixing makes up the nonspecific total body noxious incitant load, which then can exhibit a strong adverse effect on all detoxification systems and, especially, the adaptation mechanism, which is so intent on the maintenance of normal efficient homeostatic function. In other words, the total body load that is increased disturbs homeostasis and if the load is reduced, restores homeostasis.

Biochemical Components of Connective Tissue Matrix, Especially Carbohydrates as an Information Communication System

The basic scientific fact about the dynamics of homeostasis is that the informational communication units, functioning between the entry of external and the reentering of sequestered internal environmental incitants and their responses, including repair of tissue, depend upon carbohydrates rather than amino acids.[109] Both carbohydrates and amino acids are important with carbohydrates having a high potential for structural diversity, which allows for gathering and dissemination of information while amino acids have a low potential, which apparently makes them insulators. This structural diversity makes sugar polymers more efficient data receptors and carriers in the CTM, as they are able to record information and function rapidly in response to a multitude of environmental stimuli. For example, four different sugars may in theory form as many as 35,560 different tetrasaccharides for holding information, while four different amino acids can form only 24 tetrapeptides that can hold on to information;[110] thus, amino acids are rendered as more limited interconnecting responders.

The regulatory capacity through its communication ability of the ECM thus, has major significance in the provocation or prevention of disease processes. Not only do the PGs and GAGs have a sugar component but they are also bound to cell membranes by glycoproteins and glycolipids, which further supports the anatomical basis for continued communication ability. All structural proteins such as collagen, elastin, and fibonectin experience glycosilation.

Sugars in the form of nucleotides are involved in most of the enzymatic reactions in the ECM, as well as in the cells, as part of coenzymes. Nucleotides are built up from a base that is a monosaccharide, usually ribose combined with phosphoric acid. Precisely because coenzymes mediate between various enzymes, they have a connecting link in metabolism, which is only possible because of the sugar connection.

Specific monitoring by the genes and their amino acid sequences are available for inherited protein synthesis. **However, the epigenes upon the genetic helix are the environmental switches that can change gene function. The opposite modality of control applies to the environmentally receptive carbohydrates, which according to Heine,[78] are available to receive multiple environmental stimuli and respond to them without specific genetic dominance**. See Figure 1.18.

This local response though genetically programed to respond can generally have a multitude of local nongenetic environmental responses. Genes that have a long-term memory are necessary for the production of macroorganic structures like the skeleton, tendons, nerves, muscles, and processes such as protein synthesis, etc. On the other hand the opposite applies to the immediate environmental receptors and homeostatic regulators of the CTM and the sugars on the cell surface. **The maintenance of the dynamics of homeostasis requires that the CTM be able to adapt very rapidly to complex information inputs and physiologic response changes**. Thus, a "short-term memory" is necessary for a rapid response. This rapid homeostatic response with communication sequences is due to the high diversity, rapid turnover and the coupling characteristics of extracellular sugars.[111] An overlapping of multiple responses is thus possible, which, in itself, permits an individually renewable response for reordering the matrix, which has been disturbed by noxious stimuli, even when it is far removed from homeostatic balance.[112] **The incorporation and ordering of water, sugar, uric acid, salts, and electrolytes by the PG/GAGs are the other components that allow for memory storage and communication with the rest of the body**. The components will be discussed later in the section on CTM.

Monosaccharides and their polymers as communicators

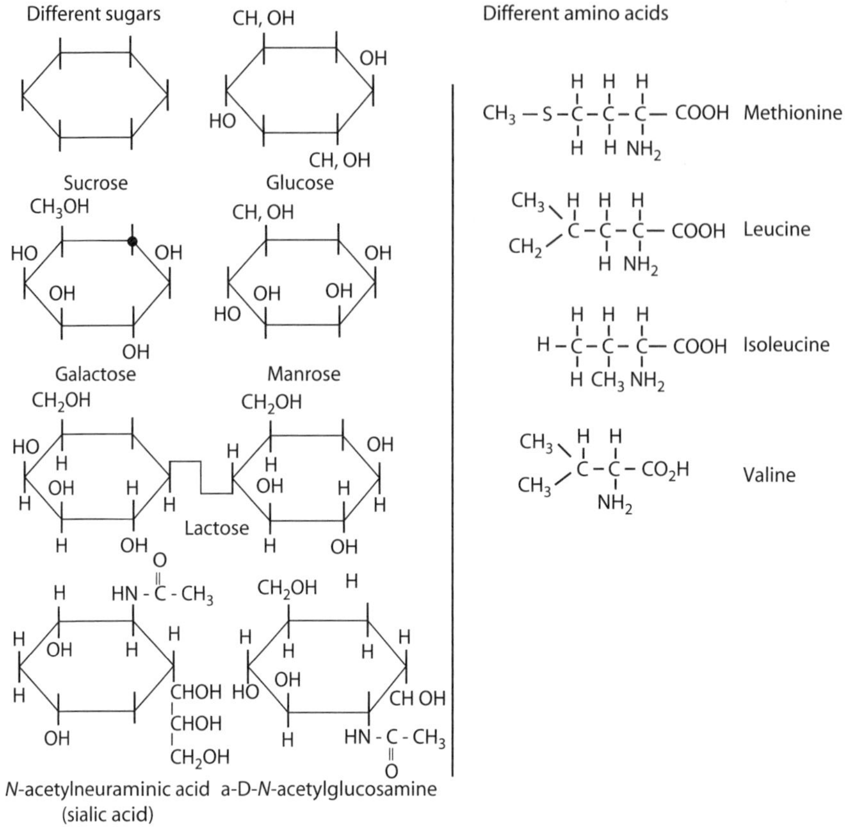

Different sugars / Glucose / Sucrose / Galactose / Mannose / Lactose / N-acetylneuraminic acid (sialic acid) / α-D-N-acetylglucosamine

Different amino acids: Methionine, Leucine, Isoleucine, Valine

Four different sugars may form as many as 35,560 different tetrasaccharides for holding information

Four different amino acids can only form 24 tetrapeptides which can hold information

FIGURE 1.18 Four (4) different monosaccharides in theory may form 35,560 different tetrasaccharides, while four (4) different amino acids can form only 24 tetrapeptides. Carbohydrates have a tremendous potential for extracting information. (From Pischinger, A., *Matrix and Matrix Regulation: Basis for a Holistic Biological Medicine*, English translation by Ingeborg Eibl, North Atlantic Books, 1991. With permission.)

Connective Tissue Matrix Control of the Individual Local Homeostatic Response

Environmentally induced local nutrient depletion or disease triggers in the individual with chemical sensitivity and/or chronic degenerative disease can significantly alter the production of mediators or intracellular communication agents. The production of mediators varies from individual to individual. These factors fashion the make up of the varied homeostatic responses seen in each individual. The patient with chemical sensitivity and/or chronic degenerative disease will have altered homeostatic responses resulting in pain, weakness, and fatigue.

The linear concept of gene expression thought by many in medicine to be the "internal cause of illness" has superimposed upon its function not only the dynamic homeostatic response but also the epigenetic switches and biphasic responses that lend to various non linear responses. This response is initiated by entering noxious stimuli received through the ERS of the ECM as shown in the previous section. **The "external triggers of illness" are multifactorial eliciting nonlinear multiple and varied responses within the body. The delineation of environmental etiology, their resultant varied local receptors and reception, and the body's response to these stimuli are, at times, more difficult to determine without the use and understanding of specific principles and techniques that will be elicited in this book.** This interconnection between external, internal triggers,

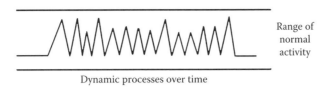

FIGURE 1.19 Everchanging dynamic process. (From EHC-Dallas, 2002.)

and internal homeostatic responses in the body behaves in nonlinear patterns and is executed through informational feedback loops, locally, regionally, and centrally. In other words, reciprocal information occurs usually, triggering an automatic adjustment response, which can be different locally and is chronic centrally. **Homeostasis is a dynamic process that is forever changing, occurring within a range of normalcy, but always fluctuating** (Figure 1.19).

The dynamic homeostatic response to noxious environmental stimuli entry depends on information that is sent to the nonprogramed environmental receptors and response systems, as well as to the programed polymorphic genetic material and its epigenetic switches that can be changed with environmental manipulation. These response systems are capable of causing mutifactorial, environmentally induced alterations of the local physiology and even, at times, the anatomy (by enzyme breakdown and nonspecific mesenchyme repair) of the ECM. **Clearly, the genetic material (genome) is a permanent program necessary for the maintenance and evolution of a biological species (e.g., bones, muscle structure, reproductive organs). However, the genome is not sufficient to predict the individual's lifetime maintenance of daily health or his/her minute-to-minute responses to rapidly changing information received by an individual, as well as, multiple noxious, environmental triggering agents**. Thus, the dynamics of homeostasis are genetically programed to occur; however, the initial information collection and homeostatic responses to environmental stimuli are programed to be flexible, in the sense that the body's receptor systems are supposed to react to every stimulus and elicit a response when dealing with environmental incitants. In other words, **the general genetic program allows for a vast intake of often simultaneous variable information. The epigenetic switch phenomenon results in varied adaptable homeodynamic responses, rather than fixed linear reactions to incoming incitants. Of course this fact allows the body's various response systems to be activated, accounting for different responses to similar entering noxious stimuli in different individuals**. Due to this normal variability, these facts make the etiology of dyshomeostasis difficult for the clinician to decipher unless the patient is studied in a controlled environment. See Figure 1.20.

Genetic Versus Environmental Control

The ECM (genetically programed to respond universally and generally) controls the immediate local nongenetic nonspecific homeostatic response to the information of the entering environmental trigger(s), while the cell controls and supplies the genetic pattern of the dynamic homeostatic response. The human genome project activity and other research endeavors have provided a much better understanding of the polymorphic nature of the human genome, as well as, many acquired factors for the varied response.[113,114] For example, identical external stimuli even though they may be generally similar can produce markedly different specific homeostatic responses in different individuals based upon each individual's genetic uniqueness. **However, different types of homeostatic responses can also occur due to the patient's nongenetic part of biochemical individuality, total body pollutant load, and the state of nutrition at the time of exposure**. These factors will alter homeostatic responses immensely at times, allowing for mild to extremely abnormal responses in the patients with chemical sensitivity and/or chronic degenerative disease. These nongenetic factors account for the initial responses due to entry of noxious foreign information. Although the body's overwhelming responses are due to environmental triggers, genetics, and especially epigenetics play a significant role. For example almost every significant genetic defect whether major or minimal

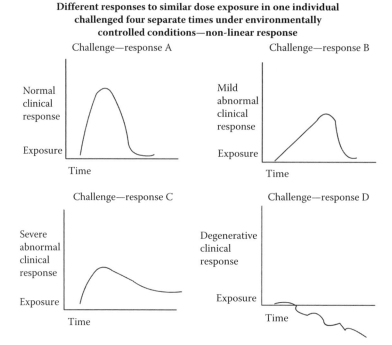

FIGURE 1.20 Same dose exposure at the same time of day on four separate days for four different challenges. Alterations of the response are due to the total body pollutant load and state of nutrition of the individual as well as the spontaneous molecular diversity. On that particular day all of these conditions result in nonlinearity of response. Biochemical individuality will play a significant role in these people. (From EHC-Dallas, 2005.)

appears to make the individual prone to chemical sensitivity. People with sickle cell anemia and trait, Mediterranean anemia or trait, of Gilbert's disease have all been prone to chemical sensitivity. However, there are no known one or two genetic defects that always cause chemical sensitivity.

Specifically the human DNA has approximately 25,000 genes identified. Genes themselves need instructions on what to do and where and when to do it. A human liver cell contains the same DNA as a brain cell, yet somehow it knows to code only these proteins needed for functioning of the liver. Those instructions are found not in the letter of the DNA itself but on it in an array of chemical markers and switches. These are collectively known as the epigenome and they are along the length of the double helix (Figure 1.21).

The epigenetic switches and markers in turn help switch on and off the expression of particular genes. The epigenes are complex hardware that induce the DNA to manufacture a variety of protein, cell types. The epigenome is just as critical to the development of humans as is the genome. The epigenome is sensitive to cues in its environment. Jirtle and Waterland.[115] showed that by changing the diet of agouti mice (genetically fat and yellow) to one rich in methyl donors (onions, garlic, beets), they could change the offspring to slim and brown mice. By epigenetics they had methylated the critical agouti gene dramatically changing the mice's offspring. They had added a chemical to their genes (agouti gene) and that of the offspring. The chemical switch dimmed the gene's deleterious effects from fat and yellow to a mouse normal, brown and slim. Researchers are finding that an extra bit of vitamin, a brief exposure to a toxin, or even an added dose of mothering can tweak the epigenome and thereby alter the gene in ways that affect an individual's brain for life.

Recently, it has been shown that other epigenetic signals from the environment can be passed down from one to several generations without changing a single gene sequence. It is well established that environmental effects like radiation, which can alter the genetic sequence of sex cells DNA can leave a mark on subsequent generations.

The Physiologic Basis of Homeostasis

Epigenome and its environmental susceptibility

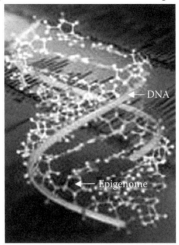

FIGURE 1.21 DNA with Epigenome.

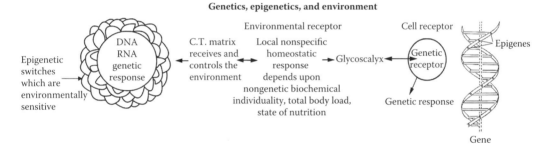

FIGURE 1.22 Relationship of genetics and environmental responses (From EHC-Dallas, 2002.)

It is known that the environment in the mother's womb can alter the development of a fetus, but now, also diet, behavior, and surroundings can work their way into the germ line and echo far into the future. Apparently inhalation, ingestion, or absorption of environmental toxins are possible causes for altering genes. **"Gene as fate" has become conventional wisdom, but it is now outdated by epigenetics.** We appear to have a measure of control over our genetics. Epigenetics introduces the concept of free will to our ideas of genetics. The epigenome can change its response to the environment throughout an individual's lifetime. This principle is paramount in the case of the patient with chemical sensitivity and/or chronic degenerative disease because this means that if a gene or epigene is able to be altered an individual may well be able to correct it by other nutritional and/or environmental manipulation. Unlike genetic mutations epigenetics are potentially reversible. Szyf[116] has shown that even in cancer epigentics can change the switches and stop its progression. Others have published data on animal subjects suggesting an epigenetic component to inflammatory diseases like rheumatoid arthritis, neurodegenerative disease, and diabetes. Diets may be tailored on the genetic makeup of the individual. Fang[117] has shown that the polyphenol epigallocatechin-3-gallate (EGCG) in green tea can alter the epigentics to change cancer. Products in soy and other compounds have also been shown to fight cancer. Jirtle[118] says that there are multiple epignomes in an individual (Figure 1.22).

Basement Membrane

Basement membranes are an important part of the matrix because they also are involved in harmonic communication as well as being barriers and storage areas. See Figure 1.23. They are a dynamic interface between the epithelium of the skin and other end organs or the mucous membranes and

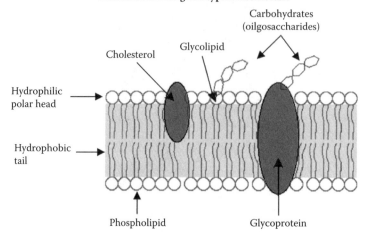

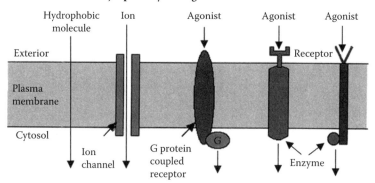

FIGURE 1.23 (See color insert following page 364.) A cell needs to communicate with its environment so that it can make appropriate responses. The external signal may enter a cell via four major pathways. 1. Hydrophobic molecules, 2. 1- Ion channels, 3. G-protein-coupled receptors, 4. Enzymes. (From Schematic Drawing of a Typical Membrane, http://www.web-books.com/MoBio/Free/Ch1B; Major Pathways for Signals to Enter a Cell, http://www.web-books.com/MoBio/Free/images/Ch6A1.gif. Free images.)

the matrix. They also interface with the endothelium of the capillary walls and the matrix of the end organ, creating a dynamic sieve. The basement membranes contain lipids and proteins. The lipids consist mainly of phospholipids and cholesterol. In the plasma membrane and some organelle membranes, proteins and phospholipids are attached to carbohydrates, which form glycoproteins and glycolipids.

The basement membrane is a calcium reservoir rich in vitamin C, which snares free radicals and is another part of the molecular sieve.[119] It consists of networking protein laminin, which is 20% of the carbohydrate content with 4–6% of this being sialic acid. The networking protein laminin is the basement membrane. It binds the epithelium to the collagen type 4 and 5 and heparin and heparin sulfates of the basement membrane. **The basement membrane is extremely important for the normal growth of the epithelium**. The basement membrane also covers all different types of cells of the end organ, including that of smooth muscle cells, striated muscle fibers, cardiac muscle cells, Schwann cells, terminal naked axons, etc. The basement membrane's underlying vascular endothelium and end-organ parenchymal cells has a special important role as a molecular sieve. **Alteration of the basement membrane results in alteration of function of the organ and if the pollutants are not contained leads to organ damage, which is seen in the end stage** of chemical sensitivity and/or chronic degenerative disease. **Even tumors may occur as an aberrant function of the basement membrane**. For example, multiple genes, including an abnormal p53, detected in colon cancer,[119] and more recently, alterations in the matrix enzyme metallo proteinase (MMP)[119] have been associated with polyp development. The interrelationships between these alterations polyp development and the progression to carcinoma are not well understood. Changes in these interrelationships are being **studied. MMPs generally function to degrade PGs and matrix glycoproteins**. This process of remodeling of the ECM is an integral part of normal tissue growth and differentiation. However, unregulated degradation of the ECM in the process of carcinogenesis may lead to an advantage for the cancer cell. Loss of basement membrane integrity may correlate with an increase probability of distant metastasis and poor prognosis. Therefore, over exposure in the ECM of the MMPs may be one part of the multistep processes by which the neoplastic cell can proliferate and metastasize. To date three MMPs have been most associated with colorectal adenomas and carcinoma: MM-2, MM-7, MM-9.[119]

MMP-7 over expression is an early event in the carcinogenic cascade as normal gut mucosa progresses to adenomatous cells and tumors. As changes, adenomas acquire the ability to become invasive adenocarcinoma, the MMP-7 remains elevated, while both MMP-2 and MMP-9 are over expressed. Epithelial tumors are composed of tumor cells and surrounding stroma. The site of MMP over expression/activation in the tumor seems to be different, depending on the individual MMP. Biologically, the location of MMP over expression may be an indication of a process intrinsic to the cancer cell or due to normal surrounding stromal cells being induced to express or secrete MMPs by some local or systemic noxious stimuli. The pathological changes seen in cancer cells may have a direct effect on disordered signaling pathways and are intrinsic to the cancer cell. This early change may allow abnormal cells to overcome normal cell–cell growth inhibition. Similarly, this phenomenon may be the first step in proteolysis of the basement membrane leading to all tumor invasion.

An additional mechanism by which MMPs are prepared to increase the local invasiveness is to facilitate angiogenesis. Endostatin significantly reduced the invasive properties of tumor and normal endothelial cells primarily, by decreased activation and expression of MMP-2. Taken together, MMP-2 and potentially other MMPs appear to have a role in angiogenesis and thus, may be blocked by angiogenesis inhibition. (See Vascular chapter in Volume 2 - Mechanisms of Chemical Sensitivity and Chronic Degenerative Disease, section on angiogenesis.)

Regulation of and by Isoosmia, Isotonia, and Isoionia

As a major part of the ebb and flow of dynamic spontaneous homeostatic maintenance physiology, the CTM is the regulator of isoosmia, isotonia, and isoionia. The dynamics of homeostasis of these factors keeps dynamic spontaneous change within physiologic ranges, allowing energy efficient bodily function. See Figure 1.24.

Isoosmia. One of the three giant legs of the stool of dynamic homeostasis is isoosmosis. **Isoosmosis**. This is the net difference of water between a region of high solute concentration and one that has a lower solute concentration. The higher the solute concentration is, the higher the water concentration becomes and conversely. Isoosmosis occurs when the two are in dynamic equilibrium. Osmotic pressure is the amount of pressure that must be applied to prevent net diffusion of water through a membrane, which is directly proportional to the concentration of osmotically active particles. Isoosmia is 300 mOsm/L ± 2–3%, sodium makes up 94% of the solute while sugar and urea each make up 3% individually.[120] Disturbance of the vascular membrane and matrix by chemical overload results in the mild peripheral and periorbital edema seen in the chemically sensitive. **This phenomenon is partially due to membrane changes resulting in altered osmosis**. The mild edema occurs after a pollutant stimulus and reverses when the stimulus is withdrawn.

Isotonia. This aspect of homeostasis often shows a simple schema of the different fluid states of isoosmia being hypertonic, hypotonic, and isotonic. At times, large osmotic pressures can develop

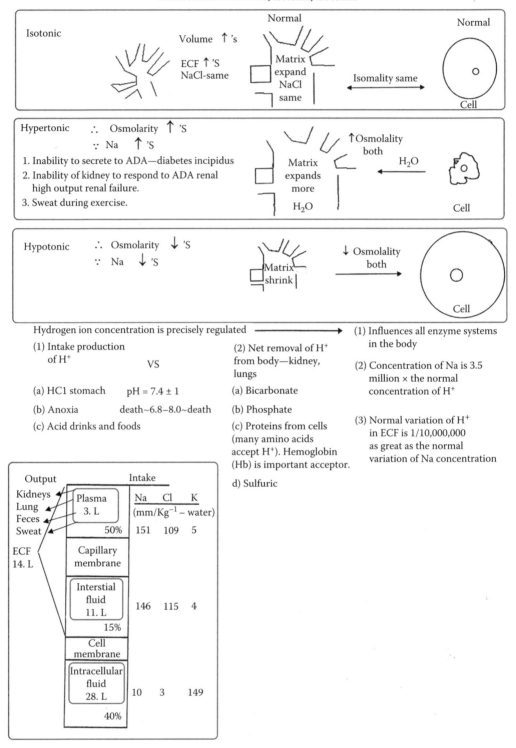

FIGURE 1.24 Homeodynamics of Isotonia, Isoionia, Isoosmia. (Modified from Guyton, A.C. and Hall, J.E., *Textbook of Medical Physiology*, 9th ed., Philadelphia: WB Saunders Co., 265, 1996.)

across cell membranes. **For each millimole concentration gradient of an impermeable solute, one has 19.3 mmHg**.[121] Excess solute or excess fluid will result in some states causing dehydration or excess water phenomenon respectively, whereas isotonia is equilibrium.

Isoionia. The relative amounts of extracellular fluid distributed between plasma and interstitial spaces are determined mainly by the balance of hydrostatic and colloid osmotic pressure across the capillary membrane. The distribution of fluid between intracellular and extracellular compartments is determined mainly, by the osmotic effect of the small solute (e.g., sodium, chloride, etc.) and the ECF's protein content. The reason for fluid distribution is that cell membranes are highly permeable to water but rather impermeable to small ions such as potassium, sodium, and chloride. Therefore, water moves across cell membranes rapidly so that the intracellular fluid remains isotonic with the extracellular fluid. The homeostatic regulation of extracellular fluid osmolarity and sodium concentration are closely linked because sodium is the most abundant ion in the extracellular compartment. Plasma sodium concentration is normally regulated within close limits of 140 + 5 mEq liter with an average concentration of about 142 mEq/liter. Osmolarity averages about 300 mOsm/liter and seldom changes more than 2–3%.[122]

By far the most abundant substance to diffuse through a cell membrane is water. **Enough water diffuses in each direction through a red cell membrane per second to equal 100 times the volume of the cell itself**.[123] Yet, normally, the amount that diffuses in two directions is precisely balanced so that not even the slightest net change of water content occurs. The volume of the cell remains constant when in dynamic homeostatic balance. This balance of water allows for excellent, rapid, and efficient communication. If under certain conditions a concentration difference of water occurs, then there is a net movement by osmosis of water into or out of the cell with hypertonia or hypotonia resulting.[124] This sieve effect also means that nutrients and water soluble toxics can be caught in the filter causing pollutant injury and loss of dynamic function. Usually edema will be the clinical sign.

IF in the CTM has potential spaces to handle excess or deficient fluids. The hydrostatic pressures of their potential spaces are almost the same. However, there is some variability of the pressures, i.e., the plasma is 7–8 mmHg, joints are 3–5 mmHg, and IF spaces may also have different pressures (i.e., intraperitoneal space 8 mmHg, joint synovial space 4–6 mmHg, epidural space 4–6 mmHg.)[125]

The plasma has osomalarity of 301 mOsm/liter. The main components are albumin 80%, globulin 20%, and fibrinogen 0.3%.[126] A large subset of chemically sensitive patients with autonomic nervous system dysfunction have blood pressure readings between 90/70 and 70/50.These patients have been found to have a 10–15% low circulatory plasma volume, which can be corrected by administering normal saline and albumin.

Electrostatic Tone

The basic electrostatic tone established by the isoionia, isoosmia, and isotonia of the matrix[127] and redox reactions reacts to every change in the ECM with deviations of electrical potential. The received information that is encoded in the PG/GAG-water complex of the matrix can inform the cell membrane as a potential deviation of the glycocalyx. If this potential deviation is strong [128] enough (information selection) leads to a cell reaction via depolarization of the cell membrane (e.g. muscle or nerve cells). Alternatively, other cell types via activation of both secondary messengers (cyclic adenosine monophosphate, inositol triphosphate) and many others, transmit coded information in the basic substance of the CTM to cytoplasmic enzymes. This information lands in the cell nucleus and can finally come into contact with the epigenetic and genetic material of the cell nucleus at the appropriate site. This information is followed by transcription of the appropriate DNA part (gene) in the various types of RNA. After the information is transferred into the cytoplasm the various RNA types in the cytoplasmic reticulum start to translate the information into products individual to the cell.[129]

This harmonic orchestration of the dynamics of homeostasis distributed through the communication system is disturbed in chemical sensitivity (periodic homeostatic disturbance) and chronic degenerative disease (aperiodic homeostatic disturbance). These areas of disturbance are evidenced by the severe reaction locally, regionally, distally, and centrally. These interference fields will react to the next minute environmental exposures very easily. We have seen these kinds of reactions in individual cases, e.g, odor sensitivity to perfumes, phenols, formaldehyde, etc., and especially in the chemically sensitive patients (periodic homeostatic disturbance). We have also seen minute glucose alteration in the diabetic (chronic aperiodic homeostatic disturbance) due to previously damaged tissue (interference fields). The normal dynamics of the homeostatic communication system act locally, regionally, distally, and centrally throughout the body by disseminating information and reciprocally receiving feedback information effortlessly from the distal parts.[130] **These two homeostatic feedback areas of the ground regulation system (GRS) (CTM and brain [hypothalamus]) must be finely tuned and work synchronistically in order to have normal energy efficient homeostasis**. Therefore, manipulation of this system positively by reducing the total pollutant load and constantly inducing proper nutrition is paramount in restoring and maintaining wellness to the patients who are developing chemical sensitivity (periodic homeostatic disturbances) and/or chronic degenerative diseases (aperiodic homeostatic disturbances). The chemically sensitive patient (periodic homeostatic disturbances) and those with chronic degenerative disease (aperiodic homeostatic disturbances) appear to have lost their function for the normal dynamics of homeostatic regulation at several points (locally, regionally, or centrally) in the normalization process thus, making it difficult to rapidly respond normally and energy efficiently to even a very minute stimulus. The clinician sees this phenomenon even with exposure to ambient doses of perfume, car exhaust, or disinfectants, or with intake of excess oxygen, etc., when observing a patient with chemical sensitivity. If this downward spiral of loss of homeostatic regulation occurs, information transmission is altered; metabolism (vascular spasm and hypoxia) and the tissue will be changed. **The chemically sensitive and chronic degenerative diseased patient will eventually develop tissue changes [NSMR and colloid changes (gel to sol and sol to gel)] then lapse into fixed-named autonomous disease eventually, developing end-organ failure**. Since the connective tissue is the largest system in the body engulfing and underlying all cells, organs, and tissues, its collection and dissemination of information is ubiquitous. The size of the connective tissue system is paramount in regulating homeostasis for if the downward spiral occurs, information collection and dissemination will be impaired.[131] Since the GRS is an open system responding to every environmental stimuli it is unstable by necessity in order to handle the multiple stimuli that enter the body.

The changes with the information infusion are difficult to see anatomically but they can be measured by electromagnetic analysis of the skin and matrix by specific cutaneous thermography. New techniques are being developed to measure this change (see Cardiovascular chapter in Volume 2 - Mechanisms of Chemical Sensitivity and Chronic Degenerative Disease for more information). To understand the leading role of the function of the ground regulation system in the pathogenesis of chronic degenerative diseases (aperiodic homeostatic disturbances) and chemical sensitivity (periodic homeostatic disturbance) clearly, the GRS is the source of bioinformation not only at the cellular level but also at the humoral system and the nervous system. The significant fact is that **the function of the GRS itself can be altered by every functional disturbance of any tissue from either external or internal stimuli**. This knowledge is essential particularly with regard to the multiple dynamic informational and metabolic feedback homeostatic loops without which chronicity and degeneration are not easily understood.

The time factor (chronicity) plays a significant role in the function of the GRS. Without informational and metabolic feedback mechanisms the effects of long lasting minimal stresses (foci disturbance or interference fields)[132] cannot be explained or understood. **Every normal GRS response that has a short-lasting stimulus (e.g., piperine), results in partial depolarization of the PG/GAGs with a rapid repolarization**.[133] However, some stimuli are long lasting due to the kinetics and half-lives of the pollutants involved. Pollutant half-lives may last from only a day (e.g., hexane), to half-lives that last months (e.g., α-acetoamino fluorine, approxiamtely 67 days), and even half-lives lasting years (e.g., DDT, dioxin). If the tissue is nutritionally depleted or not returned to normal homeostasis due to kinetics and or half-life of the offender(s) or is inflamed after noxious stimuli enters the body, an additional secondary short stimulation emanating from the resulting localized inflammatory foci (disturbance or interference fields) can cause a lasting depolarization. **This lasting depolarization eventually leads to structural tissue changes adversely influencing the function of the GRS**. At the end of this tissue degeneration process, the transformation of the colloid is in the direction of the gel or sol depending upon the original state. These transformations result in metamorphosis of the ground substance depending on which was the original colloid. Any areas will become fibrotic or develop dense matrix. However, in some cases like mechanical stress in joints the change is from a sol to a gel. In this particular situation, this transformation means that there is an alteration toward the direction of biological inactivity as this occurs in every other tissue when its surface charge is lost.

Storage Capacity of the Matrix

The matrix may have storage capabilities that can both aid and disturb the dynamics of homeostasis. PGs are able to store carbohydrates (sugars), proteins (nitrogens), fats (fatty acid), and water in domain. Excess protein can be stored in the form of collagen and PGs in the ground substance while excess calories are stored as triglycerides in matrix compartments as well as fat cells of obese people. Excess CHO is stored in muscle and liver in the form of glycogen. Excess CHO leads to the formation of PG in the subcutaneous and interstitial tissue and is stored as PG thus, the entire ground substance has a capacity to be a protein and CHO store, which in excess will disturb the dynamics of homeostasis. Sugar can be bound by nonenzymatic glycosylation to proteins, collagen, elastin, PG/GAGs, albumin, myelin, cell membrane, etc. The excess glycosylation of the basement membrane and of the rest of the CTM alters the resulting nutrient diffusion and waste removal. **As the matrix canuliculi become clogged and these excesses of protein appear, inflammation of the matrix and vessels occurs. This "slag" can be removed only by protein fasting.** Fasting is used to great advantage in most patients with chemical sensitivity and/or chronic degenerative disease. (See Nutrition chapter in Volume 4 - Mechanisms of Cardiovascular Disease and Chemical Sensitivity.) Fasting seems to restore basic physiology and eliminate symptoms.

Influence of Matrix Vesicles on Ground Regulation

A little-observed but still a physiologically and pathologically important homeostatic regulation principle of the ECM is the shedding of vesicular elements of connective tissue, Nerve and the defense cells in the ECM.[73,134] These vesicles disintegrate under the release of a large number of biologically active substances that apart from the vesicle contents (including proteolytic and hydrolytic enzymes and cytokines) originate from the breakdown of the vesicle membrane (including prostoglandins and leukotrienes.[134] Pischinger and Heine have shown that there is a regulation of tumor by the ECM by the tumor matrix vesicles. See Figure 1.25.

Using tissue cultures and transgenic mice, Bissell et al. have shown that the CTM is the regulator of mammary carcinoma.[135–137] They found when using a normal and cancerous strain of cells in vitro that they could turn off the cancerous cells by restoring the three-dimensional CTM. Even though a cancer gene was present they could render the cancer dormant by changing the phenotype where the cancerous genotype would remain but be ineffective. Bissell et al. found that if they destroyed the CTM with metalloproteinase stromolysin-1, cancer cells would develop.[135–137] These studies show that the rule is genes can be influenced by the cells microenvironment (ECM) and that an aberration in the microenvironment can play a role in the initiation of carcinoma where the cancer cell looses its ability to sense its microenvironment properly.

Laminin (which is part of the basement membrane) interlocks with receptors at the cell surface. Integrins transport chemical and physical signals across the cell membrane. The presence of laminin to bind integrin receptors made the receptor 50–100 times more active then when laminin was

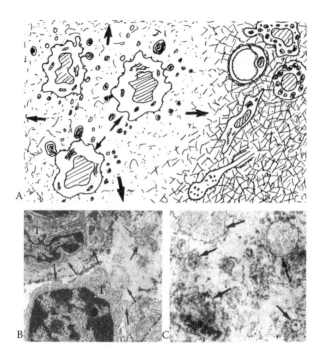

FIGURE 1.25 Regulation of tumor extracellular matrix by tumor matrix vesicles. (A) In the left half of the picture there are 3 tumor cells with matrix vesicles being formed. The extracellular matrix has broken down into smaller fragments. The arrows show the spread of this principle into the surroundings as well as the mutual influence of the tumor cells on one another. The right side of the picture shows a normal extracellular matrix with typical cell components; (B) and (C) show Scirrhous breast carcinoma, (B) Formation of tumor matrix vesicles (arrows) on the surface of tumor cells. X 5,000, (C) Degenerating tumor matrix vesicles in the tumor extracellular matrix (arrows). x 7,500. (Reprinted from Pischinger and Heine, *Matrix and Matrix Regulation,* Haug International, Brussels, 1991. Combined Figures, p. 29, 30. With permission.)

absent.[138,139] Certainly, the loss of the ECM basement membrane cripples cell differentiation and morphology.

From these studies, breast cancer like most other chronic degenerative diseases seems to be a failure in communication between the cell and its microenvironment, particularly the ECM.

Autonomic Control

The autonomic nerves are connected to the CNS centrally through the spinal cord and the hypothalamus and end in the ECM sieve distally. (See Chapter 2 for details.) There are also local reflex arcs. This proteoglycan and glucoaminoglycan meshwork of the CTM accompanies and often surrounds the arterioles and capillaries **thus resulting in a local anatomical communication between nerves and vessels**. The autonomic nerve endings are always approximately < 0.2 micron from a cell and thus, any changes in the nerve will influence the local cell particularly the lymphocytes, the cells in the lymph nodes, and blood vessels. If a sensory autonomic nerve fiber is irritated, a reflex is set off that modifies the local and often the central blood supply, but especially the microcirculation.[140] Chronically, with a more severe but still moderate stimulus there is sympathetic adrenergic induced vascular spasm (seen in ischemic muscular pains after exercise, blue hands and feet), while stronger stimuli or activation of other autonomic pathways (sympathetic-cholinergic) cause vasodilatation (red face, hands, etc.). We have seen some patients who had different segmental responses of blue hands and red faces. One segment may have a sympathetic adrenergic response of blue hands and feet, while others have a sympathetic cholinergic response manifesting as red face.

When the tolerance threshold of the circulatory sympathetic system is exceeded, pain is produced. Pain can excite motor and other autonomic nerve fibers to such an extent that they respond with the contraction of the adjacent musculature and vascular spasm. **The reduction in the blood supply then continues to lower the irritation threshold of the sensitized nerves and receptors even more, until pain, muscle contraction, and inadequate blood supply occur concomitantly**. This condition is followed by metabolic disturbances in the tissues, which is then followed by increased pain combined with reduced defensive capability. All responses become exaggerated in the organs and nerves past this injury. These nerves become sensitized. The sensitization allows for a lowered threshold of triggering and, thus, nerve hypersensitivity.

The defense mechanism, especially the T-cells and gamma globulin is reduced in quantity and function in the patient with chemical sensitivity and/or chronic degenerative disease. These changes interact in a vicious cycle, which prevents or at least delays the autogenous healing process. Immune cells come and go through all these areas and to the ANS, which is also connected to the lymph nodes thus, affecting lymphatic and immune function. These anatomic and physiological facts demonstrate the presence of a most complex communication system both locally and throughout the body for maintaining the dynamics of homeostasis. Of course, these systems (autonomic and cellular) interconnect to and from the endocrine glands. These glands, in return, send hormones to all pervasive information-gathering matrix receptors affecting their functional capacity. **Thus, a connection exists between the local ECM, the immune, the central nervous, and the endocrine amplification systems, locally, regionally, and centrally** (Figure 1.26). Due to this anatomy there is a system not only for a basically near automatic local homeostatic regulation but also for regional and central regulation.

Once the information alarm of noxious stimuli entry is sounded, the GRS goes into action as a part of every defense response and, if necessary, when the local homeostatic control is exceeded (when total body load is too high or chronic) **the GRS initiates trophic neurogenic changes**. Eventually these changes are followed by more trophic changes and/or an inflammatory process with a litany of subsystem homeostatic responses, which occur at the matrix, cellular, and subcellular levels. Thus, this vast system is responsible for regulating all basic vital functions and, therefore, it is a holistic response system. The GRS responds to all stimulation totally but not uniformly, since stimuli may be at varying degrees of intensity and chronicity. It functions as a reciprocal (feedback) bio information

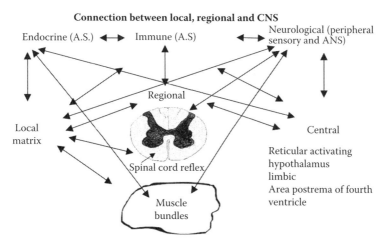

FIGURE 1.26 Connection between local, regional, CNS and amplification system with integration of the dynamic homeostatic mechanism and its amplification systems. (From EHC-Dallas, 2002.)

system from the internal and external environment as the system not only broadcasts and receives information throughout the body, which has ramifications locally but it also orchestrates regional and central homeostasis. Sometimes the dynamics of homeostasis are understood but often they are so complex that the clinician only partially knows the processes that keep the body in dynamic equilibrium and the repair mechanisms that heal the body's dysfunction. We, as clinicians, often do empirical procedures that are known to influence homeostasis but until recently we really did not know how or why the process worked. Any environmental overload including bacterial,[141] viral,[142] parasitic,[143] physical,[144] or chemical[145] will trigger the GRS. Pischinger,[3] Hauss,[146] Heine,[4] Rea,[147] Randolph,[148] and von Ardenne[149] have shown that chronic chemical overload not only influences homeostasis but also, if in excess, causes dyshomeostasis.[150] We now have a fair understanding of how the near automatic dynamic stabilizing process works to keep the body functioning in spite of massive and constant environmental assaults.

It is now obvious that the local homeostatic response is controlled or influenced by epigenetics, genetic, environmental incitants, the sieve phenomenon, and the vascular, peripheral sensory nerves, autonomic and immune response. These processes can be altered in the patient with chemical sensitivity and/or chronic degenerative disease causing a vast array of symptoms.

Vascular Function—Homeostasis

The dynamics of homeostasis of the blood vessels and their surrounding intracapillary matrix is dependent upon (1) intracapillary pressure, (2) colloid osmotic pressure, (3) interstitial osmotic pressure, and (4) interstitial hydrostatic pressure (Table 1.2). The blood vessels especially the vascular walls are a part of the communication system as described on the previous pages.

Diffusion

By far, the most important means by which substances are transferred between plasma and the IF is by diffusion. Again, this process allows for widespread communication throughout the body because of its intense properties, as well as, the vascular molecular sieve effect. A tremendous number of water molecules and dissolved particles diffuse back and forth through the capillary wall, providing a continual dynamic mixing between the IF and the plasma and other intravascular contents. Again, this transfer and mixing allows for detailed communication. Diffusion results from thermal motion of water molecules and dissolved substances in the fluid. The different particles may move first in one direction and then another, moving randomly in every direction. **Fluid can also diffuse through the capillary cell membranes without having to go through the pores**. Water-soluble substances (sodium, chloride, glucose) diffuse only through pores. The capillary wall is extremely dynamic manifesting the sieve communication effect. Despite the fact that not more than 1/1,000 of the surface area of the capillaries is represented by the intracellular clefts between the endothelial cells, the velocity of thermal molecular motion in the clefts is so great that even this small surface area is sufficient to allow tremendous diffusion of water soluble substances through these cleft pores (see Figures 1.26 and 1.27). Again this fact allows for widespread rapid communication capabilities. **The rate that water molecules diffuse through the capillary membrane is about 80 times as great as, the rate at which plasma flows linearly along the capillary**. Thus,

TABLE 1.2
Vascular Homeostasis

1. Intracapillary pressure	3. Interstitial osmotic pressure
2. Colloid osmotic pressure	4. Interstitial hydrostatic pressure

Source: From EHC-Dallas, 2002.

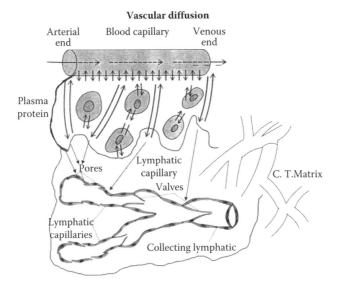

FIGURE 1.27 Water diffusion 80 times the transit time from arterial to venous capillary, causing massive leak and fluid interchange. Ten percent goes to the lymphatic system. (Modified from Guyton, A.C. and Hall, J.E., *Textbook of Medical Physiology*, 9th ed., Philadelphia: WB Saunders Co., 185, 1996.)

rapidity of communication is assured considering this rate and areas of the blood vessel. The water of the plasma is exchanged with the water of the IF 80 times before the plasma can travel the entire distance of the capillary. See Figure 1.27.

This great volume of fluid exchange, which occurs in such a short time span in the end capillaries, allows for significant variations of water and water-soluble contents without causing disruption in the dynamics of homeostasis. However, because of the variations rapid communication is present. Water-soluble substances will also pass through these pores and channels and can clog them mechanically if the contents are too large. If the toxics contained in the fluid are smaller than the pores, the toxics can be diffused by traveling through the vessel wall to the CTM. Thus, the toxics can be sent through the vessel wall to the matrix, emphasizing another communication apparatus where the sieve effect can be triggered.

Vascular Sieve

The permeability of the capillary pores for different substances varies according to their molecular diameters, thus giving another molecular sieve effect that is similar to that of the matrix. See Figure 1.28.

The capillaries in different tissues have extreme differences in their permeabilities, which will change communication capabilities.[151] **For instance, the membrane of the liver capillary sinusoids is so permeable that even plasma proteins pass freely through the walls almost as easily as water and other substances.** This phenomenon is good for rapid communication however, this process also allows more toxics contained in the protein to go to the liver cells. More toxics present in these cells can aid in driving detoxification. On the other hand too many toxics can cause damage to the liver cells. In another instance the permeability of the renal glomerular membrane for water and electrolytes is about 500 times the permeability of muscle capillaries. Pollutant damage to different areas causes a different set of symptomatology, as well as different alterations in physiology, damaging the dynamics of local and distal homeostatic mechanisms. This damage results in altered information transferred through the body, altering metabolic function that leads to resultant tissue changes.

Lipid soluble substances like carbon dioxide, alcohols, benzene, toluene, xylene, chlorinated solvents, hexane, and pentane have similar permeability as do the water-soluble substances like glucose,

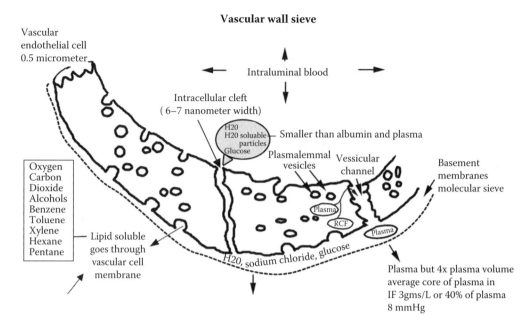

FIGURE 1.28 Vascular wall sieve. (From EHC-Dallas, 2002. Modified from Guyton, A.C. and Hall, J.E., *Textbook of Medical Physiology,* 9th ed., Philadelphia: WB Saunders Co., 184, 1996.)

TABLE 1.3
Simultaneous Biopsies of Fat and Blood in 30 Chemically Sensitive Patients

	Fat	Blood
Hexane	5–20 ppb	0–5 ppb
Tetrachloroethylene	5–20 +	0–5 ppb
Clorinated pesticides	5–20 +	0.1–.3 ppb

Source: From EHC-Dallas, 1995.

---

which can go through the vascular cell membrane, whereas sodium must go through intracellular pores. Again, these are routes of communication that can be easily altered.

However, vascular and lymphatic channels for protein permeability are about the same in the glomeruli and muscle, while the liver has higher protein permeability. This fact allows the random distribution of toxics throughout the body but does not assure uniform deposition even though communication is intact. This nonuniform occurrence has been observed and confirmed by fat biopsies at the Environmental Health Center–Dallas.[152] From the observations of the EHC-Dallas toxic and allergic depositories also occur in nutrient depleted areas that are already damaged like old scars. It has been seen in fat and other tissue biopsies that there is an abundance of toxics in fat cells or in areas of concentration of lipids like in the nervous and cardiovascular systems (Table 1.3). This type of normal dynamics of homeostasis will cause or exacerbate periodic homeostatic disturbances (chemical sensitivity) and (chronic degenerative disease) in individuals.

In addition, where higher vascular permeability for protein occurs, toxics can damage the proteins and attach to them creating haptens, which trigger susceptibility to allergic responses. Both mechanisms have been seen in the periodic homeostatic disturbances (chemically sensitive) and the aperiodic homeostatic disturbances (chronic degenerative diseased) patients.[153]

As shown previously, the rate of diffusion of a substance through any membrane is proportional to the concentration difference between two sides of the membrane (concentration gradient). This difference allows communication across membranes. As shown in the CTM section, the greater the difference between the concentrations of any given substance on the two sides of the capillary membrane, the faster the exchange is for the net movement of the substance in one direction through the membrane, restoring the dynamics of homeostasis thus, allowing rapid communication. The concentration of oxygen in the blood is normally greater than in the IF. Therefore, large quantities of oxygen normally move from blood toward tissues. In addition, the concentration of carbon dioxide is greater in the tissues than in the blood, which causes carbon dioxide to move into the blood and be carried away from the tissues. The same principle occurs with the distribution of lipophilic toxics that are seen in the chemically sensitive. Therefore, every time there is a toxic exposure to a chemically sensitive or chronic degenerative diseased individual the distribution occurs throughout the body to end-organ tissues, often causing varied responses and symptoms but always adding to the total body pollutant load if these substances are not detoxified. If the tissues are too overloaded communication will be altered. This distribution will not be uniform as evidenced by our fat biopsies performed in the chemically sensitive patients and chronic degenerative diseased patients seen at the EHC-Dallas. **Conversely, during times of stress, epinephrine will mobilize fats to be converted to sugars for quick energy. However, the lipophilic toxic chemicals attached to cell membranes and fat molecules will also be mobilized causing an internal redistribution including consequent metabolic responses with a repetition of old or a generation of new symptoms as the communication is altered**. Water-soluble substances tend to be converted from the liposoluble lipophiles. This conversion to the less toxic water solubles occurs in the end organ because of the processes of enzyme (i.e., glutathione peroxidase reaction for sulfur conjugation of hydrocarbons) and nonenzyme detoxification (i.e., glutathione binds to mercury without the involvement of enzymes). These conversions allow for precise detoxification and rapid changes in communications.

The rates of diffusion through the capillary membrane of most nutritionally important substances are so great that only a slight concentration difference is enough to cause more than adequate transport between the plasma and IF. For instance, **the concentration of oxygen in the IF immediately outside the capillary is probably no more than 1% less than the concentrations in the blood and yet this 1% difference causes enough oxygen to move from the blood into the interstitial spaces to supply all of the oxygen required for metabolism**. Again this change is enough for rapid communication to occur.

Clearly, a gradient occurs for vitamins, minerals, and amino acids in normal individuals again emphasizing rapid communication. **Chemically sensitive and/or chronic degenerative diseased patients appear to replace these nutrients rapidly in the serum; however, total cellular replacement, especially in the end organ, appears to take a much longer time probably due** to being held up in the cell membrane as described previously. Often we have seen chemically sensitive and chronically diseased patients at the EHC-Dallas who even though they have normal plasma levels of nutrients, when simultaneous measurements of intracellular levels are performed usually show a nutrient deficiency or excess. This gradient appears to be different for each individual vitamin, mineral, and amino acid in the chemically sensitive and chronic degenerative diseased patients since our comparative studies of plasma and intracellular nutrients are very different showing disparity between the plasma and intracellular tissue (Table 1.4).

Usually, in these dyshomeostatic patients, there is intracellular deficiency or excess of nutrients when the serum is normal. This discrepancy may have to do with membrane distribution quotients and solvent damage to the intracellular membranes found in the chemically sensitive and chronic degenerative diseased patients. In fact, other than immediate intravenous intervention (which does not work rapidly in all cases), complete intracellular replacement takes a period of months or years to correct in the severely damaged chemically sensitive or chronically diseased patient. This slow repletion may be caused by factors such as absorption, reduced state of the hypermetabolism of renal

TABLE 1.4
Mineral Levels in Whole Blood (Plasma, Red Blood Cells) in ppb ($n = 40$) (reference range in brackets)

Mineral	Plasma	Whole Blood	RBC
K	166	1670 (1290–2040)	3250 (2540–3720)
Na	3240 (3100–3600)	1920 (1750–2250)	—
Fe	1.53 (0.65–1.80)	—	926 (7–1050)
P	124	358 (300–440)	565 (480–750)
Ca	91	51 (38–60)	18 (8–31)
Mg	19	32 (28–44)	44 (36–64)
Cu	1.26	0.97 (0.61–1.28)	0.69 (0.52–0.89)
Mn	0.001	0.009 (0.005–0.017)	0.017 (0.007–0.03)
Co	—	0.0005	0.0004
Mo	0.0017	0.0015 (0.0007–0.0060)	0.0010 (0.0005–0.002)
Se	—	—	0.36 (0.19–0.38)
Li	—	0.0027 (0.0003–0.02)	—
Cr	—	—	0.24 (0.012–0.07)
Toxic Metals			
Al	—	0.17	0.07
Cd	—	0.0005	0.0010 (< 0.005)
Pb	—	0.158	0.029 (< 0.09)
Ni	—	0.0026	—
Be	—	—	0.0034
B	—	—	0.052 (0.005–0.11)
Sb	—	0.0003	—
As	—	—	0.0025 (< 0.01)

Source: From EHC-Dallas, 2002.

tubules, and stopping nutrient leaks. All of these areas need further exploration when considering the physiology of the patients with chemical sensitivity and/or chronic degenerative disease.

Microcirculation

The microcirculation has a special place not only in oxygen and nutrient distribution, but also in communication. This is because the micro vessels go almost everyplace in the body and can open and close to many unique substances and environmental stimuli. **Another physiological phenomenon that one needs to fully comprehend about the microcirculation is that blood does not flow continuously through capillaries. Instead it flows intermittently turning on and off every few seconds or minutes. It is pulse like. However, communication appears to flow constantly unimpaired or augmented by this phenomenon.**

This intermittent flow phenomenon is so often altered and misunderstood in clinical medicine and surgery. Although intermittent flow is physiologic and enhances oxygen and nutrient extraction it makes the dynamics of homeostatic regulation (of pulsed random control system theory) much more difficult and energy depleting once it is disturbed. We have observed this disturbance in a chemically sensitive or chronic degenerative diseased individual once dyshomeostasis occurs. This energy depletion may be due to the shunting process, poor mitochondrial diffusion, or some yet to be understood phenomenon. In this situation, sphincter nutrients like magnesium and calcium concentration may be altered by noxious stimuli, or if oxygen depletion occurs, results in vascular bed dysfunction. Any of these alterations prevents muscle relaxation and thus sphincter opening, which can result in improper oxygenation of the tissue around the end vessel.[152] **According to von Ardenne, this hypoxia will cause tissue changes such as capillary endothelial swelling resulting in a temporary vessel occlusion, which causes end vessel shunting of oxygenated blood away from the target tissue, and results in more diffuse local hypoxia accompanied by metabolic and tissue changes** (i.e., the conjugation detoxification process becomes inefficient, antipollutant enzyme production is insufficient, and lysozome production increases).[154] This finding accounts for the elevated PvO_2 (35–70 mmHg) found in the antecubital veins of the chemically sensitive and chronic degenerative diseased individuals. The resultant hypoxia in the end capillary area would then decrease the detoxification functions allowing for not only lactic acid buildup often seen in chronic fatigue (CFIDS) patients, but also a buildup of other toxics (e.g., pesticides, car exhaust) caused by oxygen dependent detoxification processes, i.e., cytochrome P-450 or plain oxidation often seen in the chemically sensitive patient. Hauss[154] has shown that toxics can increase the vascular endothelial swelling. If prompted the NSMR results in a vicious down hill cycle of hypoxia creating more endothelial swelling and even vessel wall changes such as scarring or plaque build up. The positive homeostatic control characteristic comes into play with a vicious downward cycle for a prolonged period after noxious stimuli entry until end-organ failure occurs or the mechanical function is interrupted. The concentration of oxygen in the tissues mainly affects the opening and closing of the precapillary arterioles and the normal function of precapillary sphincters. See Figure 1.29.

Normally, when the rate of oxygen utilization is great, the intermittent periods of blood flow occur more often with the duration of each period lasting longer, thereby allowing the blood to carry increased quantities of and extraction of oxygen as well as other nutrients to the tissues. When this phenomenon occurs, the dynamics of homeostatic function work efficiently. Also, communication is efficient. It is obvious that disturbance of this type of homeostatic mechanism in the chemically sensitive or chronic degenerative diseased patient results in nutrient deficiencies and anaerobic metabolic dysfunction, as well as communication difficulties.[154] Due to these metabolic changes, tissue changes eventually follow. Mitochondrial function is slowed or stopped and energy depletion occurs. Hauss[154] demonstrated how tissue changes occur when hypoxia or pollutant entry is present and then the NSMR is activated.

Hauss[154] **has further shown that toxic exposures through inhalation and ingestion of substances such as lead, mercury, and organic hydrocarbons, do disturb the microcirculation. Reaction to these substances also results in local tissue hypoxia, which triggers the NSMR. This response also results in stimulating fibroblasts to produce PG/GAGs in an attempt to reassert homeostatic control.**

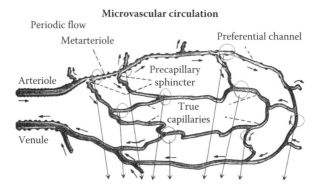

FIGURE 1.29 Image shows potential areas of endothelial swelling, resulting in shunting of blood away from endorgan tissues leaving them hypoxic, with the potential for dehydration and a marked change in electrostatic tone. (Modified from Guyton, A.C. and Hall, J.E., *Textbook of Medical Physiology*, 9th ed., Philadelphia: WB Saunders Co., 184, 1996.)

This activation causes fibroblasts to increase the output of PG/GAGs. This excess of PG/GAGs results in clogging of the matrix canuliculi, which according to von Ardenne, can then trigger vasculitis.[155] At the EHC-Dallas, thousands of cases of environmentally triggered vasculitis have been seen to fit Hauss's and von Ardenne's criteria and unless the noxious stimuli are removed this pathological progression results.

This NSMR appears to be instigated early in arteriosclerosis or connective tissue and vascular inflammation as well as fibrosis.[154,155] This NSMR reaction in the patient with chemical sensitivity and/or chronic degenerative disease may also be involved in the emergence of alteration of the dynamics of homeostasis resulting in abnormal lung function parameters following exposure to pollutant, other noxious stimuli, and after physical exertion. (See Respiratory chapter in Volume 2 - Mechanisms of Chemical Sensitivity and Chronic Degenerative Disease.) In the NSMR, many cells become reactionary, including endothelial cells of the vascular intima, fibroblasts, smooth muscle cells of the vascular media, etc. This mesenchymal reaction supersedes that of all other cell reactions in the organism. If noxious stimuli is not removed it becomes a dyshomeostatic reaction. For example, **the stimulation of the mesenchyme in the aortic wall is followed by incorporation of lipids into the macrophages, causing more dyshomeostasis and eventually results in arteriosclerosis.** This lipid incorporation into the mesenchyme of the vessel walls and the quantity and extent of lipid deposits depend on the strength and duration of the pollutant stimuli, thus the total body pollutant load.[152] The structural changes in the vessel wall usually regress to a point after the sclerogenous noxae has been removed, thus the dynamics of homeostasis can be restored.[152] However, if the sclerogenous irritation or oxygen deficiency lasts for a long time (reduced clearance capacity due to energy deficiency), then the nondecomposable metabolic waste products (collagen, fiber bundle, calcification, etc.) accumulate in the way shown on electromicrographs of all with resultant pathology. The structural damage to the vessel wall caused by the breakdown products no longer recedes (irreversible stage). (See wing biopsies, Figure 2.17.) At this stage as communication is slowed the increase in diffusion time (oxygen, nutrients, etc.) in both the peripheral and the central vessel wall begins to alter the metabolism. For example, the diffusion through the length of a capillary may decrease from 80 to 30 times, thus altering metabolism and communication. **Oxygen therapy plus strict fasting followed by dietary alteration under environmentally controlled conditions will often reverse this early arteriosclerotic process up to the certain point.** This point is where structural vessel wall changes (e.g., aneurysm, arterial stenosis) occur that are irreversible.

A similar structural change in response reaction in the fine capillaries causes the vessel wall to change in diabetes. In diabetes, there is a basement membrane thickening due to protein deposits, resulting in development of micro angiopathies. This process of the NSMR narrows the vessel wall lumen directly hampering local tissue oxygen supply, resulting in a vicious downward dyshomeostatic cycle by the positive feedback characteristic. A similar process can also occur in the membranes of the myelin sheath leading to the neuropathies in diabetics. According to von Ardenne,[155] these pathogenic basement membranes thickening can be broken down by multistep oxygen therapy. However, again it appears that if structural changes occur, resulting in fibrotic scars, the nerves and basement membranes are irreversibly damaged. This type of therapy will not work.

The chemically sensitive and chronically diseased patient appears to have severe problems with the oxygenation at the tissue level.[156] When studying the microcirculation in the chemically sensitive, there appears to be a higher demand for oxygen, probably due to the active need for detoxification at the end tissue level, which alters the capillary sphincter mechanism function over that seen in the normal individual. Capillary sphincter response appears altered with blood shunting in different areas of the end-organ tissue. This shunting results in more dyshomeostasis and hypoxia prompting a vicious downward cycle by the positive feedback homeostatic characteristic. It is not uncommon to see anticubital vein PO_2 as high as 40–60 mmHg in the chemically sensitive and the chronically diseased patients as compared to normal controls of 22 ± 5 mmHg.[156,157] This inability to extract sufficient oxygen results in inefficient metabolic breakdown of toxics by the oxygen (at the end capillary level), which is driven by cytochrome detoxification systems, resulting in inadequate

The Physiologic Basis of Homeostasis

detoxification by oxidation, acetylation, acylation (peptide conjugation) sulfonation, methylation, and glucuronide processes.[158] In addition, decreased tissue oxygen will lead to more tissue deterioration causing a strain on the homeostatic repair mechanism eventually, triggering the NSMR.

This deficient oxygenation phenomenon also explains the weaknesses often seen in the chemically sensitive and chronically diseased patient due to mitochondrial dysfunction and excess energy demand necessary for dyshomeostatic reactions. (See section on lipids.)

Vascular Tone

Besides normal permeability, there is normal homeostatic tone of the vascular tree, which keeps fluids on an even plane. Most factors that change blood vessel tone can be related to both large and small vessels even in the microcirculation. It should be remembered that **1% or 5000 m² of the body mass is (subject is body mass not blood vessels) blood vessels**; therefore, both local and distal factors can control and affect vascular tone. Distally, these factors controlling vascular tone can be autonomic stimuli, or humoral stimuli including antiadherence platelet factors. Local factors that control dilatation are excess blood volumes, mild hypoxia, heat, and oxygen. Release of mediators such as nitric oxide, adenosines, and prostacycline dilate the vascular tree. Of course, one type of sympathetic response puts out neuroendocrine substances like epinephrine and another like acetylcholine, which react on the vasoreceptor causing constriction in the former and dilatation in the latter. Mild modified sympathetic stimulation occurs because there is no parasympathetic innervation of the blood vessels. Normal vasoconstriction can be caused by a blood volume deficit (seen in a subset of patients with chemical sensitivity), by cold, by pain, by over reactive emotions, by norepinephrine releases, or by direct sympathetic stimuli through the autonomic nervous system (Figure 1.30).

Communication is an integral part of smooth vascular function regulating tone precisely. Vasomotor tone is a complex interaction involving connective tissue content with its matrix response around the vessels and vascular endothelium. Though complex the communication system is rapid and efficient. The vasomotor tone of vascular smooth muscle is due to neuromyogenic mechanisms,

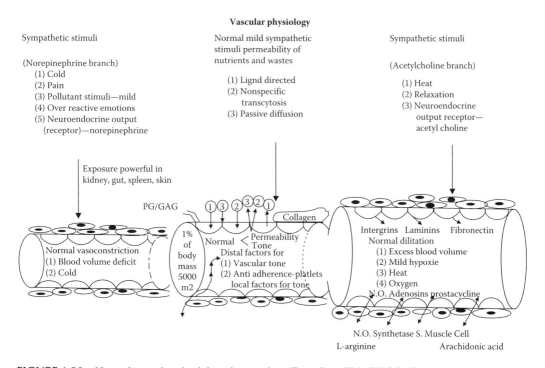

FIGURE 1.30 Normal vascular physiology is complex. (From Rea, W.J., EHC-Dallas.)

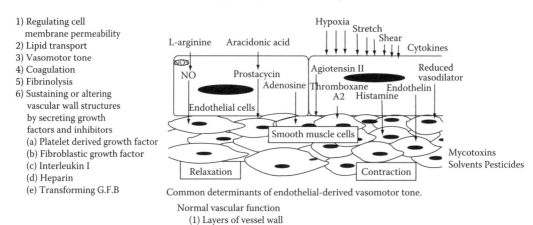

FIGURE 1.31 Surgery Journal. (Modified from Selke, F.W., Boyle, Jr. E.M. and Verrier, E.D., *Ann. Thorac. Surg.*, 62, 1222–1228, 1996.)

metabolic action, autoregulatory forces, extravascular compressive forces and possibly others we don't understand yet. (See Figures 1.30 and 1.31, and Table 1.5.)

Normal vascular tone means some sympathetic nervous system stimulation that allows just enough vasoconstriction to allow for strong vasoconstriction or dilatation as needed in order to obtain and maintain physiologic adjustment responses. The normal vascular responses are dependent upon the layers of the vessel walls especially the endothelial and CTM, the passing blood elements, and neuronal elements. The vessel wall is strongly influenced by heat, cold, blood volume, the state of the autonomic nervous system, pH, and the status of O_2 in the area (especially hypoxia). The vascular responses are extremely essential in the body's communication system because of adaptation to alterations in these factors.

Metabolically there are many relaxant factors including nitric oxide, carbon dioxide, prostacycline (Eicosinoid), adenosine, lack of free radicals, lactic acid, citrate, acetate, histamine CCR fam, serotonin (certain levels), lack of glucose, sodium, potassium, magnesium increase, citrate, bradykinin, hydrogen ion decrease with pH increase, vitamin deficiency, acetate, and prostaglandins.

Several other factors are known to increase vascular tone. These include endothelin (the most potent vasoactive peptide), leukotrienes especially thrombaxane A_2, angiotenson II, free radicals, histamine, serotonin, cycloxygenase dependent constricting factor, norepinephrine, epinephrine, ADH or vasopressin (formed in the pituitary and thalamus), calcium increase, prostaglandins and a decrease in pH.

Frequently, the vascular tone is so fragile in the chemically sensitive and chronic degenerative diseased patients that stable responses are difficult to obtain. **Virtually every patient with chemical sensitivity and most patients with chronic disease have abnormal vascular tone and responses**. Due to these abnormalities eventually, tissue changes will occur. This primary response is usually vasospasm with entry of minute noxious stimuli. Vasospasm could be local, regional, or distal. However, there is a small subset of patients with chemical sensitivity who flush and burn clearly, from abnormal dilatation. This subset of patients is intolerant of the heat, whereas the larger vasospastic group constantly fights vasospasm and is sensitive to the cold.

Lymphatics

The lymphatic system represents an accessory route by which fluid can flow from the interstitial spaces into the blood playing a significant role in the communication system both in the regulatory

TABLE 1.5
Vascular Factors

Relaxant Factors		Contraction Factors
Nitric oxide	Serotonin-hormetic factors	Endothelin
Carbon dioxide	Sodium	Leukotrienes esp. Thrombaxane
Prostacycline (Eicosinoid)	Potassium	A_2
Adenosine	Magnesium increase	Angiotenson II
Lack of free radical	Bradykinin	Free radicals
Lactic acid	pH increase	Histamine—hormetic factors
Citrate	Vitamin deficiency	Serotonin—hormetic factors
Acetate	Lack of glucose	Cycloxygenase dependent constricting factor
Histamine-hormetic factors	Prostaglandins	Norepinephrine epinephrine
		ADH—vasopressin
		Calcium increase
		Prostaglandins
		Decrease in pH

Source: Extracted from EHC-Dallas.

dynamics of normal homeostasis and in the dyshomeostasis in the chemically sensitive and chronically diseased individual. Lymphatics and the autonomic nervous system have a special relationship, which allows for precise communication between the immune, endocrine, and nervous system. Lymphatics can carry proteins and large particulate matter away from the tissue spaces neither of which can be removed by absorption directly into the capillary. **This removal of proteins from the interstitial spaces is an essential function without which we would die in about 24 hours.**

In the lymphatics in the pathologic state there also is an increase in concentration of the large molecules such as polyaromatic hydrocarbons such as seen emanating from charcoal grilled food, coal dust, silica, asbestos inhalation, diesel fumes, etc. These toxic substances have been found in lymph nodes of exposed sensitive and nonsensitive chronically diseased individuals certainly causing alteration in communication with a disruption of the harmonic lymph function, of the dynamic function of subunit homeostasis, and eventually, of centrally directed homeostasis. Most of the chemically sensitive and chronic diseased patients have chronic mild peripheral and periorbital edema, which certainly has to do with the alteration of dynamic harmonic interstitial connective tissue and lymphatic flow homeostasis and dyshomeostasis. Certainly, excess levels of hydrocarbons may alter the matrix homeostatic response, causing osmotic changes and thus, edema through this mechanism. These toxic substances found in lymph channels and nodes can damage T-lymphocyte and possibly B-lymphocyte replication when they present as a foreign antigen. Thus, the distal effect of immune function and the dynamics of homeostasis certainly are disrupted in the chemically sensitive patient as is evidenced by the usual lowering of T_8 (CD_8) suppressor cells.[159] Table 1.6 REF:T-System Cells—CS III p. 1340)

The status of the lymphatic system in each part of the body is extremely important for maintaining communication in the dynamics of holistic homeostasis because subunit homeostatic damage can influence the generalized integrated central homeostatic mechanism. **Every tissue in the body has lymph drainage except the superficial portions of the skin, the CNS, deeper portions of peripheral nerves, endomysium of muscles, and bones.** Even these tissues have minute interstitial channels, prelymphatics, through which the IF can flow thus, maintaining communication with the rest of the body. All IF goes to lymph vessels except in the brain where it goes into the cerebrospinal fluid. One may find higher concentrations of toxics in the nervous system not only because of the lipid content of the brain and lipophilic of many xenobiotics but also due to **the size of the**

TABLE 1.6
Immunological Data of Chemically Sensitive Patients[a] with Vascular Dysfunction and Normals (Mean Values and Differences)

	Patients (70 Persons)	Normals (60 Persons)	Difference Patients to Normal	Significance (p)
WBC (#/mm^3)	7010.00 ± 290.00	7560.00 ± 220.00	No	> 0.05
L (#/mm^3)	2420.00 ± 90.00	2770.00 ± 91.00	Smaller	< 0.005
L (%)	36.20 ± 1.10	37.30 ± 1.10	No	> 0.2
T_{11} (#/mm^3)	1780.00 ± 75.00	2080.00 ± 74.00	Smaller	< 0.005
T_{11} (%)	72.40 ± 1.00	75.20 ± 0.80	Smaller	< 0.05
T_4 (#/mm^3)	1090.00 ± 48.00	1160.00 ± 43.00	No	> 0.1
T_4 (%)	43.70 ± 1.00	42.20 ± 0.70	No	> 0.1
T_8 (#/mm^3)	560.00 ± 30.00	740.00 ± 38.00	Significantly smaller	< 0.0001
T_8 (%)	22.90 ± 0.70	25.80 ± 38.00	Smaller	< 0.005
T_4/T_8	2.20 ± 0.15	1.70 ± 0.06	Significantly larger	< 0.0001
B (#/mm^3)	220.00 ± 19.00	270.00 ± 23.00	No	> 0.05
B (%)	9.40 ± 0.6	9.40 ± 0.6	No	—

Source: From Rea, W.J., Johnson, A.R., Youdim, S., Fenyves, E.J., and Samadi, N., T and B lymphocyte parameters measured in chemically sensitive patients and controls, *Clin. Ecol.*, 4, 11–44, 1986. With permission.

[a] Suppressor T cells are depressed significantly in chemically sensitive patients with vascular dysfunction.

prelymphatic channels, which will mechanically block certain large toxic molecules inhibiting their egress from the brain. One suspects that some of these channels may be mechanically narrowed or occluded in the chronic diseased patient like the diabetic. Here, communication is impaired. EHC-Dallas studies in periodic and aperiodic vascular homeostatic disturbances by specific thermography showed hyper-responses or little to no responses in the lymphatics of the head and neck. (See Figure 1.32A through C.)

Approximately one-tenth of the fluid coming from the arterial capillaries enters the lymphatics (2–3 liters/day). Therefore, damage to lymph channels can cause severe damage to the dynamics of orderly homeostasis and its communication system. Lymph fluid consists not only of water but also electrolytes and is the only microchannel that can take up and return high molecular weight substances like protein to the central circulation. **Capillaries cannot reuptake proteins. They can only return those proteins that do not leave the capillary.** Lymphatics can absorb high molecular weight substances because of the anatomically different flap of the lymphatic endothelia, which mechanically allows their entry. (See Figure 1.32A and B.) This fact gives the communication system a unique view on the metabolism of these substances thus, furthering the presence of information distribution.

Protein in the IF and lymphatics is 2 gm/dl except in the liver where it is 6 gm/dl and intestine at 4 gms/dl. Since normally two-thirds of all lymph is retrieved from the liver and gut, the thoracic duct (mixture from whole body) has a protein concentration of 3–5 gm/dl. Lymph from the gut is also responsible for absorption of fat. There is up to 1–2% fat in the thoracic duct after a fatty meal. Due to this anatomical makeup of the lymph channel bacteria, viruses, parasites, cancer cells, and organic hydrocarbons can travel through the lymphatics[158] telegraphing information throughout the body.

There are 100 milliliters of lymph flow through the thoracic duct in resting humans per hour and perhaps another 20 milliliters flow into the circulation per hour through other lymphatic

The Physiologic Basis of Homeostasis

channels, making the total estimated lymph flow 120 ml/hour or between 2–3 liters per day. This phenomenon allows the body to have a slower communication system than through the vascular tree but perhaps the information is of higher or different significance.

The lymph channel in the CTM initially has a low flow with a minus 6 mmHg pressure to a high flow with a pressure increase (from minus 6 mmHg to 0 mmHg [atmospheric]).[160] **This**

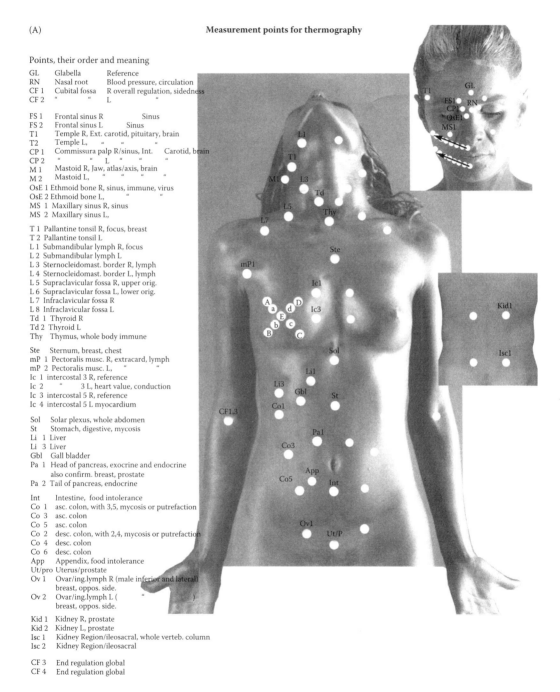

FIGURE 1.32A Measurement points and index for regulation thermography scanning. (From Thermography USA. 9057 Soquel Dr. AB, Aptos, CA 95003 USA. Tel:800.685.6689. Fax: 831.685.1128. www.thermography.net. With permission.)

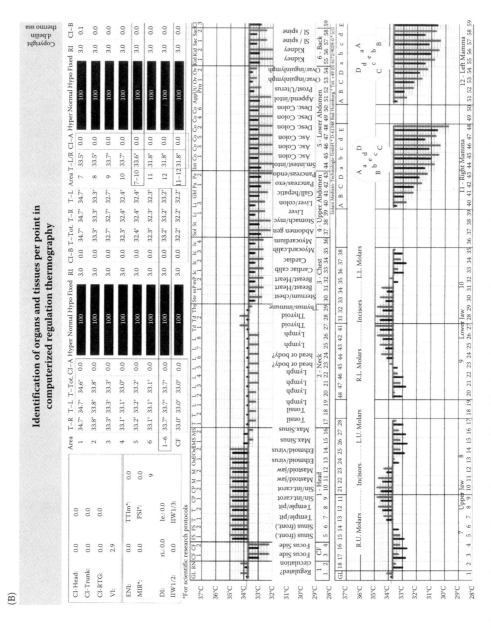

FIGURE 1.32B Thermography chart. Data print out and identification of organs and tissues per point in computerized regulation thermography. (From Beilin, D., Identification of Organs and Tissues per point in Computerized Regulation Thermography, Thermo USA, 2000.)

The Physiologic Basis of Homeostasis

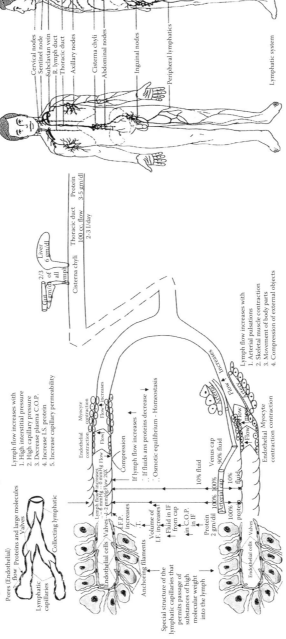

FIGURE 1.32C Homeostasis of lymph flow. (From EHC-Dallas, 2005.)

pressure increase gives a twentyfold increase in lymph flow. There is an increased lymph flow with high interstitial pressure and/or decrease plasma colloid osmotic pressure with increased interstitial protein or an increase in permeability of the capillaries in order to maintain fluid and protein homeostasis. The body has sensors for all of these phenomena and thus, can telegraph its acquired information to any part.

If interstitial pressure gets to 1–2 mm/Hg above atmospheric pressure, lymph flow due to tissue compression will not increase. If this lack of flow lasts too long, dyshomeostasis will occur in the local, then regional, and eventually into the central homeostatic mechanism.

Lymphocytic channels have a pumplike action due to the smooth muscles in the wall in order to maintain orderly fluid homeostasis. When the lymphatic channels become full and distended their pumplike action propels lymph flow forward. All major lymphatics have valves, which prevent back flow. Lymph can also be pumped by external pressure of the surrounding muscle contraction, movement of parts of the body, arterial pulsations, and compressions from external objects outside the body. Lymphatic capillary endothelium by contraction may also be capable of pumping lymph forward until it reaches the muscular lined lymphatics.

At the EHC-Dallas and Buffalo, many chemically sensitive and chronically fatigued patients have difficulty exercising. Each time they exercise they feel a great deal of pain for the next few days and they have many more reactions with an increase in environmental vulnerability caused by the accumulation of metabolic waste products like lactic acid. Therefore, a deep lymphatic massage as a part of their treatment could help to make them feel better. It appears that mobilization of toxics from the lymphatics by these two therapeutic techniques, exercise and lymphatic massage, plays a major role in recovery from their illness. However, if performed in the severely chronically ill chemically sensitive patient, this lymphatic massage technique will make them worse, which is probably due to extra release of toxics and the definite release of lactic acid.

Since the lymphatic system is also an overflow mechanism, it plays a central role in controlling the concentration, the volume, and the pressure of the IF, thus, being one of the significant regulators of harmonic, low energy requiring, efficient homeostasis. Thus, it plays a significant part in communicating this information to the rest of the body.

Considering these facts, the homeostatic mechanism of the matrix and IF is clear. First, the small amounts of proteins leave from the arterial capillary into the IF. Venous capillaries reabsorb only minute amounts; therefore, proteins tend to accumulate in the IF of the matrix and this, in turn, increases the colloid osmotic pressure of the IF. The increasing colloid osmotic pressure results in favor of fluid filtration into the interstitium. Proteins pull fluid osmotically into the interstitium, which increases both the IF volume and the IF pressure. The increasing IF pressure greatly increases the rate of lymphatic flow. **This flow in turn carries away excess volume, excess debris and excess protein accumulated in the spaces**. Once the IF protein concentration reaches a certain level it causes a comparable increase in IF volume and correspondingly IF pressure. At this point returning of protein and fluid by way of the lymphatics becomes great enough to balance exactly the rate of leakage from the blood capillaries. Therefore, the quantitative values of all these factors though dynamic reach a steady state of homeostasis. They will remain balanced at these levels even though they are fluctuating until something changes the rate of protein leakage and fluid from the blood capillaries.[160] This phenomenon is a masterful example of the automatic communication in bodily function.

Intravascular Content

The intravascular components are another aspect of the communication system of the dynamics of homeostasis. These consist of the endothelial walls and its components, the red blood cells, white blood cells, platelets, and plasma. These substances must be in the constant state of homeostasis but, at times, serious complications occur such as inflammation and/or clotting. **This homeostatic communication process is often deregulated in the chemically sensitive or chronic degenerative diseased patient as evidenced by the clinical picture often seen with inflammation characterized** by the spontaneous bruising, petechiae, purpura, Raynaud's phenomenon, peripheral, and periorbital edema, etc. The intravascular content is made up of intra and extracellular components that make the intravascular area a microcosm of intra and extracellular communication dynamics of the holistic homeostatic mechanism.

According to most authors, the phenomenon of oxidation with free radical generation in nature is the core mechanism of molecular cellular injury in all disease.[161–168] Oxidation is the process by which organisms disintegrate. If left unchallenged, the body will continue to oxidize, until cell death and cellular disintegration occurs. Since at all times the body will tend to oxidize, the homeostatic mechanisms come into play to counteract this oxidation thus communicating this degradation to the adjustment mechanism. This oxidation process appears to be spontaneous requiring little to no energy. Therefore, the homeostatic receptors are always alert for this sequestering of toxics in certain tissue areas. Ali,[169] as well as physicians at EHC-Dallas, have shown that vitamin C can prevent and reverse oxidatively induced erythrocyte membrane deformation in patients with chronic fatigue once this change is communicated to the adjustment mechanism. **Vitamin C will also disassociate plasma and cellular aggregates caused in vitro by norepinephrine, collagen, and ADP.**[170] Ali also found coagulopathy in microscopic findings in ischemic heart disease.[171–188] He observed evidence of cardiac myocytolysis in cardiomyopathy[171,172] and myocardial fibrosis associated with iron,[173–175] calcium,[178] and oxalate[177,178] deposits in hemodialysis patients.[179,180] These finding appear to be similar to Selye's original studies on calciphylaxis.[7] Similar morphologic observations were seen in vascular intimal proliferative changes and other vascular alterations in hemodialysis patients.[179,180] He also documented anatomic and enzymatic evidence of ischemic myocardial injury unaccompanied by occlusive coronary artery disease.[182,183] Ali has found microclots in the serum of sick people with loosely congealed blood elements and fibrin.[189] He feels these are energetic molecular homeostatic deregulations of the redox phenomenon.[189] These findings agree and augment the findings of von Ardenne,[149] Hauss,[154] and Sampson.[190]

Ali[189] performed high-resolution phase-contrast microscopic studies of unstained peripheral blood smears in diverse clinicopathologic states such as advanced coronary artery disease, congestive heart failure, arrhythmias, and poorly controlled diabetes and in smokers. These findings complement those of Heine, Achard, Mauriac, Sampson, Storti, and Schroder.[192] All of them did blood smears of normals versus hypoxic hypotensives showing alterations of the peripheral blood. Heine,[191] Schroder,[192] and Ali[189] similarily showed numerous pathological changes with hypoxia. The dynamics of many homeostatic mechanisms come into play to keep intravascular components in homeostatic balance. It has been shown that excess body stores of iron,[193,194] copper,[195,196] and mercury[197,198] are the risk factors of ischemic heart disease, while protective effects have been seen with antioxidants such as vitamin C,[199,200] vitamin E,[201,202] beta carotene,[203] and coenzyme Q10[204,205] against oxidative coagulopathy.

The dynamics of homeostatic balance between the pro oxidant environmental triggers that force the intravascular contents toward inflammation and clotting and the antioxidant nutrients that protect from inflammation and clotting is an extremely fine line. This need for balance keeps the physician and the chemically sensitive and chronic degenerative diseased patients working hard to achieve a broad range of normal physiological homeostasis. Precise communication is necessary and performed through the GRS.

So that permanent clotting does not occur (Figure 1.33). There is a clotting-unclotting equilibrium (CUE) in humans. This equilibrium occurs constantly. When communicated properly, the equilibrium maintains homeostasis of the intravascular tree.

This general homeostatic mechanism has been observed to be working efficiently in healthy humans[189] while it is altered in people with artificial or severely diseased organs. The presence of free radicals particularly the superoxide ion (O_2) plus the hydroxyl radical (OH^-) are normally produced during daily living from exposure to oxygen and by pollutant intake. The body and particularly the intravascular space must be able to prevent or dissolve the early microclots through antioxidants, redox absorbers, and if the assault is strong enough, fibrinolysins and heparin in order to maintain

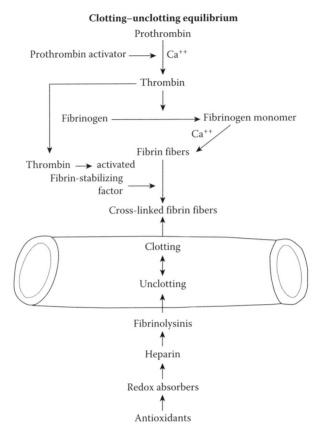

FIGURE 1.33 A balance must be constantly achieved to maintain and obtain health with no clotting. (Based on EHC-Dallas, 2005.)

TABLE 1.7
Superoxide Equation

		Fe		
O_2^1	$H_2O_2 + O$	Cu $\rightarrow$	$HO + HO + O$	$\downarrow$
Superoxide	$H_2O_2 + Fe^{2t}$	$\rightarrow$	$Fe^{2t} + HO + HO$	$\downarrow$
	$H_2O_2 + aH^-$	$\rightarrow$	$Q + H + OH + HO$	$\downarrow$
				Tissue Destruction

healthy intravascular homeostasis. These and other substances (i.e., calcium levels) make up the dynamics of intravascular homeostatic balance.

Accentuations of normal physiologic processes by noxious stimuli generated oxidants require stronger antioxidant measures in order to maintain the dynamics of intravascular homeostasis. The superoxide ion is a relatively weak oxidant and owes most of its destructive potential to its ability to generate hydrogen peroxide by reacting with molecular oxygen (Table 1.7).

Hydrogen peroxide, in turn, generates highly toxic OH⁻ radicals in the presence of transition metals such as iron and copper, which can trigger the clotting mechanism, resulting in intravascular clotting.

Ali has shown that congealed plasma and microclots have been reversed by antioxidant therapy thus, restoring intravascular homeostasis before a fatal injury occurs.[189] **This antioxidant therapy would be a physiologic extension and enhancement of the normal spontaneous homeostatic**

clotting/unclotting mechanism. Oxidation requires neither an expenditure of energy nor outside cues but is clearly accelerated by noxious environmental incitants. Oxidation is the process of how all of nature disintegrates or decays. However, the opposite on the redox equation is reduction **and this process of reversing oxidation requires the expenditure of energy**.[189] If utilized constantly reduction creates a costly energy drain for the body in order to maintain homeostasis.

Reduction can often be interfered with by the introduction of toxics as seen in the chemically sensitive and chronic degenerative diseased individuals thus, altering homeostasis and orderly efficient detoxification function. This interference by toxics may signify the onset of a pathological process not an extension of an exaggerated dynamics of the physiological homeostatic response. The glutathione replenishing pathway in the body is an example of a pathological process, which occurs with normal or low-level xenobiotic exposure, where oxidized glutathione must be converted into reduced glutathione to replenish the functional reduced glutathione pool. If a heavy metal such as mercury, lead, or cadmium is the oxidizing agent for glutathione, the reduction replenishing process becomes paralyzed, resulting in dyshomeostasis. The reduction cannot occur and results in body damage to one of its sulfur detoxification mechanisms. Since glutathione has a multitude of metabolic functions, a cascading effect can occur, causing the communication of severe homeostatic dysfunction throughout the body. **Thus, a balance between pro-oxidant pollutants or other noxious incitants and antioxidant nutrients must occur in order to prevent severe dyshomeostasis with intravascular inflammation and eventually clotting. This balance must be communicated throughout the body to prevent clotting**.

It has been shown that a state of circulating blood encompassing structural abnormalities involving erythrocytes, granulocytes, and zones of congealed plasma occurs in the early stages of oxidative coagulopathy after pollutant or microorganism assault, resulting in disturbed dynamics of homeostasis. According to Heine,[4] this normal process of leukolysis, RBC, platelet lysis, and congealed plasma is the normal way that the body maintains intravascular homeostasis (Figure 1.34).

By producing the protein buffers, disintegration of each of these components results from the need for protein buffers to keep the pH neutralized, which is around 7.4. However, acceleration of this process of lysis occurs with pollutant stimuli. Fibrin clots and thread formation with platelet entrapment occur in the intermediate stages. Finally, microclot and microplaque formation occur in the late stages of oxidant injury. Oxidative coagulopathy begins with activation of plasma enzymes and leads to oxidative permutations of plasma lipids, proteins, and sugars and is not merely confined to recognized coagulation pathways. Nonetheless, the trend is toward the altered dynamics of homeostasis. We have seen positive changes in cholesterol and plasma lipids after a period of avoidance of pollutants and heat depuration therapy with mobilization of toxics and restoration of homeostasis. **It appears that this process of oxidative coagulopathy represents one of the core pathogenic mechanisms of homeostatic deregulation seen in environmentally triggered and chronic diseases that leads to injury of the ECM, cell membranes, and intracellular organelle such as the mitochondria**.[189] Ali has shown that in early alteration of the intravascular tree there is a filamentous coagulum (fibrin needles) and lumpy coagulum with the early stages of plasma changes due to oxidative stress. Zones of congealed plasma in close vicinity of the white blood cells were found, clearly caused by environmentally noxious incitants, which induced informational changes where a metabolic conversion of tissue changes from sols to gels and gels to sol occurred. Such zones of plasma solidification also were commonly observed surrounding microbes, as well as primordial life forms. Congealed plasma also surrounded microcrystals encountered in oxidative coagulopathy (presumably composed of oxalates, arates, cholesterol and other substances precipitated by acidosis associated with oxidative coagulopathy), while congealed areas were small and discrete in mild cases. Large and confluent areas of plasma consolidation were seen in nearly every microscopic field in advanced cases of diseases such as arteriosclerosis. **At this stage the orderly dynamics of homeostasis is gradually turned into life-threatening pathology**. One can see that these formations may well telegraph the impending clot or plaque seen in some cases of arteriosclerosis. Certainly oxidative stress outside of the realm of normal homeostasis will lead to pathogenic information channeled to the altered dynamics of homeostasis triggering abnormal metabolic, physiologic, and finally tissue

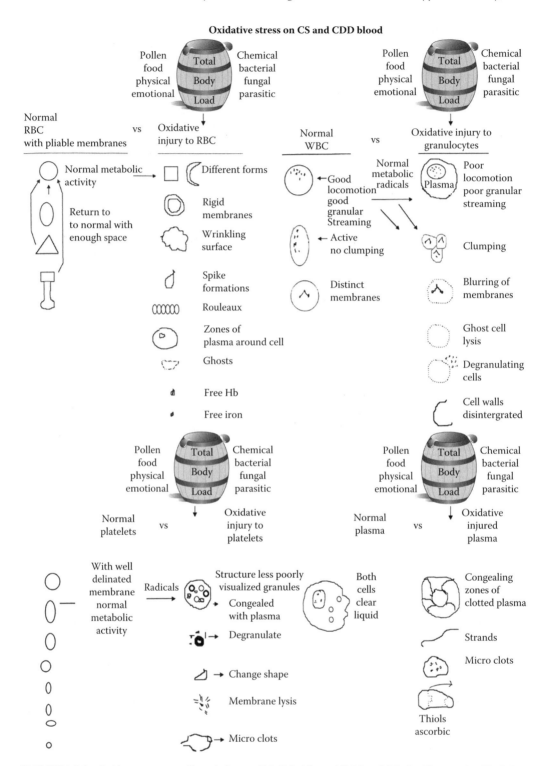

FIGURE 1.34 Oxidant stress on cells and plasma. (Modified from Ali, M. and Ali, O., *J Integrative Medicine*, 1, 1–112, 1997.)

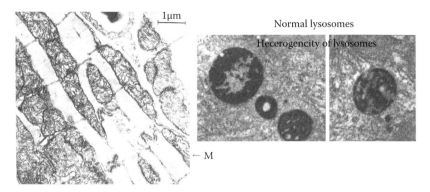

FIGURE 1.35 Developing of lysosomes is rare in normal cardiac tissue. It is common in the damaged heart. (Modified from von Ardenne, M., *Oxygen Multistep Therapy: Physiological and Technical Foundations*, trans. By P. Kirby and W. Krüger, George Thieme Verlag Stuttgart, New York, 113, 1990.)

changes. The lysozomes discovered by De Duve[206] are membrane limited organics that fulfill various functions including autolysis of the cells, damage to neighboring cells and foreign cells, digestion of foreign substances, and digestion of the cells own waste substances. The primary lysozymes are formed from the endoplasmic reticulum and the Golgi apparatus and contain (with great stability against their aggressive nature) a mixture of more than 35 enzymes, mostly **hydrolases,**[207] **whose activity optimally is achieved at a pH of 5.6**. When dyshomeostasis occurs the cell drops its pH to 5.6, the lysozomal membrane breaks open and the enzymes reach the ECF; adjacent cells may then be damaged or die.[208,209] Physiologically digestion of foreign substances taken up by the cell also occurs in secondary lysozomes, which are formed by a fusion of the foreign substance and the secondary lysozyme entrapped in the intracellular vacuoles (Figure 1.35).[206]

Digestion of the cells own waste products also occur in the secondary lysozomes. The lysozome content of the cell and the spectrum of enzymes contained in the lysozomes vary according to cell type.[211] **Most of the body's cells including heart muscle cells in a healthy dynamic homeostatic stable heart have few to no lysozomes, while ischemic hearts on the other hand contain many lysozomes**.[211–213] Excess liberation of the lysozymes will propagate dyshomeostasis.

Ionescu,[214] using various redox assays including redox electrodes connected to a high-resistance potentiometer (BJL-ultra weak chemoluminescience analyzer) with a high-sensitivity sector, measured the redox potential in 48 severe eczema patients, 23 psoriasis patients, and 15 chemically sensitive and tumor patients. He compared these with a control group of normal individuals.

The serum-E_n –value of the eczema, psoriasis, and chemically sensitive patients showed a significant tendency toward metabolic acidosis, with results of -75 ± 15 m 1/1, -65 ± 10m V versus $-92 \pm$ mV in healthy controls. Nontreated cancer patients with solid tumors and metastasis generally displayed a severe oxidosis in serum ($E_n > -60$ mV) but normal levels after irradiation, cytostatics, and antioxidative treatments. Antioxidative nutrients (grape, grapefruit, beets, vitamins, minerals, amino acids, etc.) tend to move the patient toward redosis.

Cells

Although cell homeostasis differs for each cell type, there is a general homeodynamic response as they communicate with each other in all cells. This response is communicated with a harmonious interaction that maintains the dynamics of homeostasis because of the complexity of the major cells, which include erythrocyte, leukocyte, fibroblast, lymphocyte, mast cells, and the monocyte macrophage cells, with their respective environment. Each interaction will be discussed separately, or in concert with one of the other type of cells.

The communication of the **leukocyte, the fibroblast, and monocyte macrophages will be discussed first because they are part of the initial cellular local nonspecific homeostatic reaction and defense processes, which respond to noxious stimuli entering the body**. These cells are eventually followed to the injury site by the lymphocytes in the specific immune response lymphocyte phase.

Leukocyte Homeostasis

Normally, the body lives in symbiosis with many bacteria and toxicants, as all mucous membranes are constantly exposed to large numbers of these substances. Homeostasis is established by the communication of normal function with a constant dynamic steady state of interaction between the noxious stimuli and response factors, including mucous barriers, white blood cells, and macrophages. The mouth contains spirochetal, pneumococcal, and streptococcal bacteria and to a lesser extent so does the respiratory tract. The topical toxicant load in the respiratory tract is just the opposite of the mouth because it appears that more toxics are presented and available for absorption in the upper respiratory tree and the lung than the mouth. The gastrointestinal tract, especially the colon, is loaded with bacteria, and the eyes, urethra, and vagina have bacteria in them. This is true of the toxicant presentation also. There must be a constant interaction between the cells and the toxicant with balance being maintained by the dynamics of the local homeostatic mechanism.

Any decrease in the number of white cells, as is seen in many chemically sensitive patients, immediately allows invasion of the tissues with bacteria or toxics that are already present in the body. This invasion will either throw off the communication of the homeostatic balance or place a strain on the nutrient supply in order to maintain equilibrium. Therefore, there must be a homeostatic balance that keeps the body functioning well and communicating its state of health throughout. **Within two days, if the decrease of white blood cells continues or if there is stress on the bone marrow output or if leukolysis speeds up, ulcers appear in the mouth and colon or the person develops some form of respiratory infection**[215] **or local inflammation near the area of entry**.

Several subsets of chemically sensitive patients respond to noxious stimuli differently. After noxious environmental stimuli entry, one group develops recurrent mouth ulcers, especially *Herpes simplex,* but these can also be caused by other types of toxics, food incitants, or food additives. Another group responds with colon lesions; a third group develops recurrent respiratory infections; another subset develops recurrent cystitis; and finally, a large group with abnormal white blood cell counts has no infections. In this group it appears as if other mechanisms are functioning extremely well in order to ward off microbial-induced illness and maintain homeostatic balance. Clearly, in these patients, different subsets of homeostatic regionalization and compartmentalization of the injury has occurred, showing yet another way the integrated subsets of homeostatic mechanisms work to prevent overwhelming disease.

When one discusses the dynamics of cell homeostasis, one has to realize that cells differ according to their anatomy and function and that their life times may be altered due to their special function. Due to this variation communication varies. **For example, leukocytes multiply rapidly and die easily, while brain cells multiply slowly, if at all, and are very resistant to death**. There is a whole spectrum of cell lifetimes in between.

A characteristic of leukocyte function for homeostatic regulation is their capacity for physiologic lysis.[216] This lysis occurs in all white cells but more in the neutrophils because of their quantity. Each deviation of leukocyte homeostasis (even standard deviation) is due to WBC production and removal. According to Heine,[217] when leukocytes have a physiologic reduction, usually leukolysis has occurred. It has been reported that there is normally a 17% reduction of leukocytes between the arterial and venous side of the body. This means that billions of cells are being destroyed or sequestered rapidly every 24 hours in normal dynamic physiologic homeostatic balance (Figure 1.36).

In addition to the **protein phosphate buffers, leukolysis liberates lipids, polysaccharides, nucleic and amino acids, tissue hormones, oxido reductive compounds, protein like complexes, and a large spectrum of hydrolytic enzymes**. These substances activate the macrophages and mast

The Physiologic Basis of Homeostasis

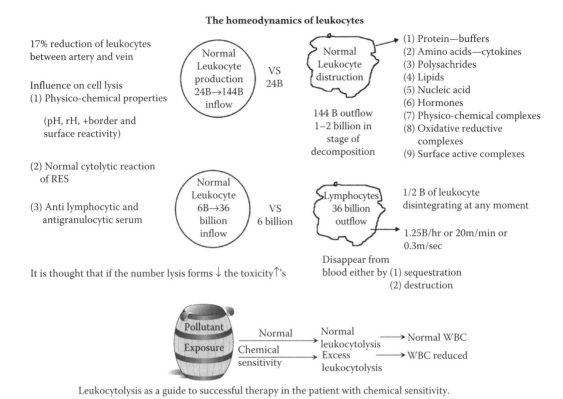

FIGURE 1.36 Balance of leukocytes. There is a 17% reduction between the cells on the arterial side and those on the venous side of the circulation. (From EHC-Dallas.)

cells among other things, while the rest are recycled to act as the nutrient supply for the fibroblast. The physiologic leukolysis is an important regulator in the formation and breakdown of the ECM. It is reasonable to include forms of leukocytes in the differentiation of the blood protein to establish the extent of leukolysis. In a healthy individual there are 5.7% lysis forms in the blood from a finger stick. With 5000 being the low number of peripheral leukocytes there are about 300 lysis forms per mm^3. This means that if there is an even distribution throughout the body with 5 liters of blood there are always 1–2 billion leukocytes in the process of decomposition. With lymphocytes being 25% of the total 5,000 leukocytes, and 1250 lymphocytes per mm^3 in a total blood volume of 5 liters there are 1,250,000,000 × 5 = 6,250,000,000 = 6.109, namely 6 billion lymphocytes. This becomes an inflow of 6 times the amount or 36 billion lymphocytes. Thus, 36 billion lymphocytes have to disappear from the blood to remain in constant dynamic homeostasis. This means thirty billion lymphocytes disappear every 24 hours, 1.25 billion per hour, 20 million per minute, or 0.3 million per second if one uses the 25% lymphocytes of total leukocytes as a basis. The total numbers of leukocytes that disappear can be estimated at four times the lymphocyte amounts, namely 1.2 million per second. If one remembers that the smear only shows prestages, namely elements that are already affected, the relationship of 1–2 billion cells in the stages of decomposition with 1.2 million cells actually disappearing per second as a probability. This lysis of leukocytes is an example of how the dynamics of homeostasis occurs and may continue dynamically throughout life. The disintegration appears to be cell necrosis and apoptosis due to oxidation, which helps maintain pH homeostasis by its protein-buffering capacity. **This phenomenon of leukolysis is usually accelerated in a subset of chemically sensitive patients who have white blood counts between 2000 and 4500 mm^3**. These low white counts emphasize how chronic exposure to noxious stimuli upset the dynamics of homeostasis in the chemically sensitive individual.

Leukocyte homeostasis is a very complex function, since lack of response may yield overwhelming infection or just inability of the system to allow proper bodily function while maintaining normal homeostasis. The special defense system for combating infectious agents or toxics is the leukocyte system, which defends either by phagocytosis and/or by forming antibodies with sensitized lymphocytes to destroy the invader. On the other hand, if a high level of noxious stimuli is encountered one will see the inflammatory reaction triggered in full force since this reaction appears to be the secondary defense against noxious invasion for restoration of the dynamics of homeostasis. If the chronic repetitious stimuli are strong and chronic enough to stimulate this inflammatory response, then continuation of dyshomeostasis occurs, resulting in chronic inflammation and changes in metabolism with eventual tissue changes if allowed to go unchecked.

Poly **Polymorphonuclear leukocytes or neutrophils are central in the innate-immune response. They represent the first line of defense against bacterial and fungal infection**. They kill marauding pathogens by both the production of reactive oxygen intermediary and proteolytic enzymes. In contrast to the beneficial responses, neutrophils have been shown to play an important role in the development of a number of inflammatory diseases, ranging from rheumatoid arthritis[218] to the acute respiratory distress syndrome.[219] These inflammatory diseases are mediated through inappropriate persistence of the phagocytes at the site of the inflammation. Ultimately, removal of these cells or inhibition of their activity is an essential step in the resolution of inflammation.

As terminally differentiated cells neutrophils look like prepackaged killers waiting for phagocytosis to occur. The cytoplasm is filled with granules containing antimicrobial and digestive compounds and is ready for translocation and emptying into the phagosome. The NADPH oxidase, the primary source of oxidants, is already synthesized and awaits a signal for assembly and activation. This preparedness allows a response within seconds. The readiness of this redox reaction makes the prevention of firing a very delicate matter, in that, if it does happen the CTM must have the capability of neutralizing the oxidant-generated free radicals, otherwise inflammation will occur. This capacity of the matrix to handle oxidant-derived free radicals depends upon previous injury and the nutritional integrity of the matrix. Of course, if the injury is too massive or chronic then chronic inflammation will occur. Bystander neutrophils (those not engaged in phagocytosis but standing ready) may aid in the termination of the inflammation by altering apoptosis.

The pathways and processes of oxidant production are well defined in the neutrophil. On stimulation, the **NADPH** oxidase is rapidly assembled in the plasma membrane leading to the production of O_2^- (singlet oxygen) on its external surface.[220] Successful ingestion of the targeted microorganism results in O_2^- release into the phagosomal space. They rapidly dismutate to H_2O_2, which can then be converted to HOCl by the granule enzyme, myeloperoxidase. Other secondary oxidants of the oxidative burst are chloroamines formed upon the reaction of HOCl or the amine groups. These oxidant products make a significant contribution to the anti microbiocidal arsenal of the neutrophil. Work from Dahlgren and colleagues suggest that NADPH oxidase activation occurs in granule membranes, increasing the likelihood of significant extra phagosomal oxidation.[221] If the recognition site becomes altered by excess pollutant exposure and there are no microorganisms to attack, then the damaging of the free radicals may be very devastating to the leukocytes, often killing them. **Many chemically sensitive patients have WBC counts in the 2000–3000 range until the pollutant load is lifted**.

Several processes in the neutrophil may be under redox control. These include the modulation of the transcription factor and therefore, cytokine production and release. Redox reactions may prime cells for subsequent activation and to affect cell-to-cell interaction.[222–224] Apoptosis is under redox control.[225] These mechanisms are defective in chronic granulomantaous disease (chronic degenerative disease) resulting in defects in antimicrobial killing.[226,227]

To mount an inflammatory response and fight a bacterial infection the apoptotic rates of neutrophils are significantly delayed.[228] **The intracellular redox potential of the cell has been shown to regulate this apoptotic cascade. The persistence of a neutrophil-mediated inflammatory response is due, in part, to a delay in their spontaneous rates of apoptosis**. Intracellular glutathione depletion appears to trigger neutrophil apoptosis and the termination of inflammation. Glutathione plays an important role in maintaining intracellular redox equilibrium.[229-231]

Ali et al.,[189] using a 15,000 times magnification high-resolution phase contrast darkfield microscope and usually working with live smears have shown that normal granulocytes show amoeboid movements with their locomotion provided by streaming of their granules into little protrusions of cytoplasm, which grow in size and become legs of the cells. Such healthy cells continuously change their configuration as they explore their microenvironment. Not uncommonly active phagocystosis and the engulfing of debris occur rapidly within seconds. These healthy cells use active and vigorous streaming usually followed by active engulfing and finally, followed by destruction of the engulfed toxics or bacteria. They maintain homeostasis efficiently with little energy requirement. The granulocytes if damaged by noxious overload (e.g., hexachlorobenzene) have a poor ability to engulf and eliminate a foreign substance.

In neutrophil homeostasis, the goal is to balance the numbers of neutrophils triggered by noxious incitants to the prevention of a release of a large amount and number of oxidants and cytokines, which would make less demand on antioxidants to counteract them. If the stimulus is too little, no triggering occurs, and an individual will have only mild pharmacological effects from the oxidants and cytokines and he/she may continue to maintain the dynamics of physiological homeostasis. This maintenance of balance depends upon the nutritional pool for maintaining cell vigor and the total load of pollutant derived oxidative or other types of stress.

Degenerating PMNs and lymphocytes also act as protein buffers, which help balance the pH, allowing for a normalization of acid–base balance. This degeneration apparently occurs by apoptosis since no inflammation is generated. Therefore, maintenance of the dynamics of physiological homeostasis occurs since there are many reactions that produce normal acids. **The constant need for buffering of the resulting acidosis from the oxidant and pollutant-stimulated assault in tissues appears to be one of the reasons the white blood count of the chemically sensitive and some chronically diseased individual tends to be low**. The homeostatic mechanism for leukolysis is out of balance due to constant oxidant stress and utilization in order to combat the unrelenting thrust toward oxidant-induced acidosis.

The earliest changes involving granulocytes seen by Ali[189] were clumping and loss of locomotion with cells lying limp in pools of plasma, with absence of granular streaming and amoeboid cytoplasmic streaming. In later stages of disintegration, granulocyte cytoplasm showed vacuolization zones of increased density and disintegrating membranes, most likely due to lysozyme release. Congealed plasma surrounds ruptured cells. **In early fixed-named diseases clear evidence of homeostatic dysfunction in the format of phagocytic dysfunction was observed on the majority of smears**. Actively mobile phagocytic leukocytes failed to actively engulf and digest primordial life forms (yeastlike organisms). These phagocytes appeared to have lost their previously vigorous function. Leukocytes in such situations were observed to approach clusters of primordial organisms, shrink, break, and move away.[189] This behavior is clearly foreign to normal functioning cells and the action was definitely a dyshomeostatic response. Clearly, this loss of function depends on the stage of degeneration when it occurs in pathological states. This degeneration may only be an accelerated oxidation of a normal physiological process of cell death or again, may be due to oxidant induced pollutant injury directly to the leukocytes. Probably both mechanisms come into play since it has been shown by studies that many toxicants like hexachlorobenzene will damage or prevent phagocytosis in granulocytes.[189]

Our studies at the Environmental Health Center–Dallas and Buffalo have shown the total load of toxics also retard or stop phagocytosis. We have seen some patients, especially a subset of chemically sensitive and chronically diseased patients with impaired phagocytosis, recurrent sinus, bronchial, or bladder infections, who require constant use of antibiotics. These patients often develop recurrent candida infections due to the overuse of antibiotics and immune suppression (Table 1.8).

PMNs can go only so far in the restoration of homeostasis. The formation and release of peripheral macrophages from the bone marrow takes 3–5 days to obtain a full complement with injured tissue.

TABLE 1.8
Phagocytic Index (PI) in a Subset (100) of Chemically Sensitive Patients Who Have Recurrent Infections

Bacteria[a]	PI (%)				
	50–59 (No. of Patients)	60–69 (No. of Patients)	70–79 (No. of Patients)	80–89 (No. of Patients)	90–95 (No. of Patients)
Staphlyloccus aureus	1	11	44	72	3
Streptococus epidermidis	0	8	38	61	5
Pseudomonas aeruginosa	0	4	41	59	2
Candida albicans	0	8	38	55	2

Improvement of Phagocytic Index in Two Chemically Sensitive Patients Whose Recurrent Infections Stopped with Treatment

Bacteria	Pretreatment		Posttreatment[b]	
	Patient 1	Patient 2	Patient 1	Patient 2
Staphlyloccus aureus	84.5	70.2	95	81.5
Streptococus epidermidis	84.5	69	95	81.5
Pseudomonas aeruginosa	84.5	69	95	81.5
Candida albicans	84.5	69	95	81.5

Source: From EHC-Dallas, 1988.

[a] Most had more than one microbe tested.

[b] Individual treatment: (1) avoidance of pollutants in air, food, and water; (2) injection therapy for biological inhalants and food; and (3) nutrient supplementation.

The local tissue released macrophages and the polymorphonuclear leukocytes (PMNs) will stay to restore homeostasis until the bone marrow macrophages arrive. These macrophages will continue the homeostatic process initiated by PMNs since the macrophages are able to engulf bigger and more toxic incitants along with PMN fragments something the granulocytes cannot do.

Eosinophils The eosinophils normally constitute about 2% of all the blood leukocytes. **Eosinophils are weak phagocytes and they exhibit chemotaxis.** In comparison to the neutrophils, it is doubtful that the eosinophils are of significant importance in the protection against the usual types of infection. However they **are part of the innate immune system response**.

On the other hand, eosinophils are often produced in large numbers in people with parasitic infections where they migrate into tissues diseased by parasites. Most of the parasites are too large to be phagocytized by the eosinophils or by any other phagocytic cells. Nevertheless, the eosinophils attach themselves by way of special surface molecules to the parasites, and they release substances that kill many of them. For instance, one of the most widespread infections in the world is schistosomiasis, a parasitic infection found in as many as one-third of the population of some developing countries; the parasite invades, literally, any part of the body. Eosinophils attach themselves to the juvenile forms of the parasite and kill many of them. They do so in several ways: (1) by releasing hydrolytic enzymes from their granules, which are modified lysozomes; (2) probably, by also releasing highly reactive forms of oxygen that are especially lethal; and (3) by releasing from the granules a highly larvacidal polypeptide (major basic protein). In the United States another parasitic disease that causes eosinophilia is trichinosis, which results from invasion of the muscles by the Trichinella parasite ("pork worm") after a person eats uncooked pork.

Eosinophils also have a special propensity to collect in tissues in which allergic reactions of the IgE type have occurred, such as in the peribronchial tissues of the lungs in people with asthma and in the skin after allergic skin reactions. This is caused partly by the fact that many mast cells and basophils participate in allergic reactions; these mast cells and basophils release an eosinophil chemotactic factor that causes eosinophils to migrate toward the inflamed allergic tissue. The eosinophils are believed to detoxify some of the inflammation-inducing substances released by the mast cells and basophils and probably, also, they help to phagocytize and destroy allergen-antibody complexes thus, preventing spread of the local inflammatory process. Frequently, eosinophils are lost from the blood in the patient with chemical sensitivity and/or chronic degenerative disease. Most likely this shift occurs toward the inflammatory process. **A similar group of chemically sensitive and/or chronic degenerative diseased patients show an increase in their eosinophil blood count.**

Basophils and Mast Cells   Mast cells are widely distributed in the body, and particularly, they are associated with the ECM capillaries. **They are selectively found in large numbers adjacent to blood or lymph node vessels but are most prominent immediately beneath the epithelial surfaces of the skin (especially the superficial zone and mucosal surfaces of the genitourinary, gastrointestinal and respiratory tracts.** Estimated circulation ranges from 4000 to 5000 per mm^3 in the lung; 7000–12,000 per mm^3 in the spleen, and 20,000 per mm^3 in the gastrointestinal tract.[233,234]

In peripheral tissue, mast cells are located in close vicinity to intraepithelial T-lymphocytes. There are 3 types of mast cells: (1) those cells containing only tryptase (predominant in the intestinal mucosa and alveoli); (2) a second type containing (s) tryptase, chymase, and carboxypeptidase (predominate in skin and intestinal submucosa); and (3) the third type contains chymase, which is found in the intestinal submucosa and nasal mucosa. All these enzymes are serine proteases, which when activated can destroy the CTM and if in excess destroy normal dynamic homeostasis. Interleukins 4,5,6, are found in mast cells. See Figure 1.37.[235]

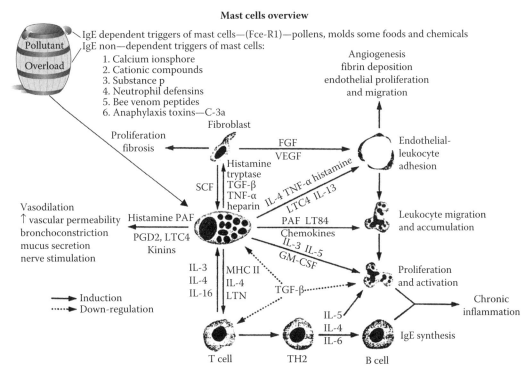

FIGURE 1.37 Mast cells can be triggered both by IgE and non-IgE dependent substances, causing both up and down regulation of physiology. (Modified from Pischinger, A., *Matrix and Matrix Regulation: Basis for a Holistic Theory in Medicine,* Ed. Hartmut Heine, Editions Haug International, Brussels, 47, 1991.)

Growing evidence suggests physiological roles for mast cells as protection against inflammatory damage in the intestinal mucosa because they release cytokines such as ILI and others, which have anti-inflammatory properties. **The mast cell can be regulated by nitric oxide, which has a protective effect on the intestines**. Again, there is a dynamic homeostatic balance present.

The **basophils** in the circulating blood are similar to the large mast cells located immediately outside many of the capillaries in the body. They **play a major role in tissue remodeling after injury**. Mast cells are one of the maintainers of the dynamics of homeostasis along with the fibroblast and monocyte macrophage systems. The origin of mast cells is from the progenitor cell of the basophil and CD34 agranular mononuclear cells under the influence of IL-3 and stem cell factors and IL-6. **Mast cells originating from the fibroblast stem cell are developed, activated, and supported by stem cell factor. The stem cell factor permits mast cell survival by suppressing apoptosis**. Mast cells go to the first phase of differentiation in the tissues. Both mast cells and basophils liberate heparin (glucoseaminoglycan) into the blood. Heparin is water-soluble, a nonprotein-based sulfate, and a polymer sugar stored in the enclosed membranous vesicles. This glucoseaminoglycan can prevent blood coagulation, as well as speed the removal of fat particles from the blood after a fatty meal. Heparin has many other functions including dampening or dissolution of allergic reactions, dissolution of platelet adherence, maintenance of cell groups and their adhering to cell membranes or their structural parts. When released from mast cells or given subcutaneously, sublingually, or transdermally, fibroblasts, and local macrophages engulf heparin. **Heparin thus, aids in the formation of the ECM (PG/GAG and collagen) in the fibroblast**. It also activates 50 enzymes thus, playing a role in the translation and transcription of DNA and RNA. **Heparin also promotes the breakdown of the anaphylactic complement component C_3 thereby, breaking the cycle of anaphylaxis**. It also helps the regulation of collagen synthesis and collagen fiber polymerization. Heparin inhibits the effect of interferon and modulates growth factor such as in plasminogen activity, angiogenesis factor, fibroblastic growth factor (fibrosis), platelet growth factor, and glycoprotein.[235] The mast cells and basophils also release histamine as well as, smaller quantities of bradykinin and serotonin. Indeed, it is mainly the mast cells in inflamed tissues that release these substances during inflammation. **Mast cells release proteases that cause gut anaphylaxis**.

The mast cells and basophils play an exceedingly important role in some types of allergic reactions because the type of antibody attached to their surface. For example, the IgE type has a special propensity to become attached to mast cells and basophils. Then, when the specific antigen subsequently reacts with the antibody the resulting attachment of the antigen to the antibody causes the mast cell or basophil to rupture and release exceedingly large quantities of histamine, bradykinin, serotonin, heparin, slow reacting substance of anaphylaxis, and a number of lysosomal enzymes. These in turn cause local vascular and tissue reactions that cause many, if not most, of the allergic manifestations. In addition to IgE and specific antigen, a variety of biological substances, including its products such as complements, and certain cytokines (i.e., I-L, 3, IL6), chemical agents (i.e., DDT), and physical stimuli (i.e., cold, low pH), can elicit the release of basophil and mast cell mediators. **The responsiveness of basophils and mast cells to individual stimuli varies. For example, gut mast cells are much more sensitive to neuropeptides or morphine than are the pulmonary mast cells**.[236] Moreover, these stimuli can induce a pattern of mediator release that differs from the case associated with Fce-RI dependent mast cell activation. Certain cytokines can directly activate mast cells or basophils and/or motivate the mediator release from the cell in response to IgE and antigen or other stimuli. However, the effect of individual cytokines are often vastly different in mast cells and basophils.[237-240] **Catecholamines produce significant stimulus for the degradation of mast cells**: thus, granules are released under conditions of increased stress either emotional or physical (environmental), or anything that can activate the sympathetic part of the autonomic nervous system. Low pH will also trigger degranulation. Mast cells are one of the prime targets in chemical sensitivity and in some chronic degenerative diseases. They appear to be altered in these situations. Once a mast cell releases its agents, the autonomic nervous system stimulation continues to occur and there is a viscous cycle of easier triggering of the mast cell release, which increases the nerve

The Physiologic Basis of Homeostasis

injury phenomenon of the autonomic nervous system. This phenomenon appears to occur in most of the patients with chemical sensitivity. As stated previously, 85% of the EHC-Dallas patients with chemical sensitivity measured by heart rate variability have a sympathetic response and are constantly releasing nor-epinephrine and potentially triggering the mast cells. This chronic triggering of the mast cell in chemical sensitivity may account for the constant response to entering reactions of pollutants like car exhaust, pesticides, alcohols, phenols, newsprint and perfumes.

Mast cells have been implicated in the production of human peritoneal adhesions in that they play a direct role in stimulating or modulating the myofibroblasts of the peritoneal cavity. The contraction of the proliferated granulation tissue in wound repair is a physiologic reaction that limits the exposed surface area and therefore, facilitates the wound repair process. Xu et al. found that a mast cell sonicate significantly enhanced the contraction of a three-dimensional collagen lattice in which peritoneal adhesion fibroblasts were embedded. In fibrotic disorders in which mast cell presence has been described, hyperplasia usually appears before a dense fibrotic tissue is established. Thereafter, there is a decrease in sustainable mast cells and the number remains constant.[241,242] This action would point to a clear involvement of mast cells not only at the onset of the process but also during later stages. In this study Xu et al. have shown that mast cells are a constant feature of peritoneal adhesions and that they influence myofibroblast proliferation, collagen production, and their contraction, indicating a role of mast cells both at the onset of fibrosis and at later stages. These observations also are in synchronicity with the fact that **mast cells are long living cells that have the property to be repeatedly activated and to resynthesize their mediators**[243] (Figure 1.38.)

The immune system is also involved in neutralizing foreign noxious stimuli. These functions include: IgE, i.e., neutralizing toluene disisocynate, cytotoxic responses for the metals like mercury, lead, cadmium, and the gamma globulin (from plasma cells) combined with complements forming immune complexes, and the T-cell system with the helper and suppressor cells and their natural killer cells, which neutralize and kill bacteria. These functions are elaborated in Chapter 3: Immune System.

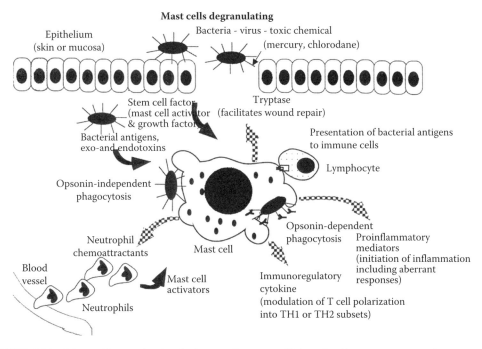

FIGURE 1.38 Mast cells can develop leaky membranes or if the stimuli are strong enough rupture the membranes causing degranulation. (Modified from Minireviewer Infection in Immunity, Soran, Abraham, Ravi, Melaviya, Sept., 3501–3508, 1997.)

Lymphocyte

The dominant morphologic alteration of lymphocytes in order to maintain homeostasis involves enlargement and lymphoblastic transformation. According to Ali,[189] in mild to moderate degrees of degeneration the lymphocyte nuclei lose their normal intense basophilic appearance and exhibit pale blue staining. Cytoplasmic vacuolation was an uncommon feature. In most advanced stages of oxidation, up to 90% of lymphocytes in most smears showed abnormal cytologic characteristics with a majority of the cells in lymphoblastic transformation.

In order to maintain homeostasis from early development until adulthood, lymphocyte populations increase in size. **In adults, the total number is constant due to balance between production, multiplication, sequestration, and leukolysis; thus dynamic homeostasis occurs**. The sizes of T- and B-cell populations are maintained independently. B-cells are produced by the bone marrow and are not under the influence of the thymus. However, the T-cells are under the influence of the thymus. The T-cells have numerous types including small and large T-cells, killer cells, CD_4 helper cells, and CD_8 suppressor cells. **The control of total lymphocyte number is probably independent of cell specificity but depends on an imbalance between demand, production, lympholysis, and apoptosis**.

The homeostatic control of cell numbers implies a kinetic steady state where cell production equals cell loss. The immune system must adapt to changes in the environment through selection of appropriate colonal replications. Selection of new specific types in all lymphoid compartments depends on the renewal rate of cells in that compartment. **Renewal rates depend upon cell production and death within as well as, cell input and output to and from other cellular compartments**. All of these parameters are reflected in the time that a cell survives within a certain compartment comprising lymphocyte life spans, dynamic homeostasis, selection, and competition.

Lymphocytes maintain dynamic homeostasis by entering the circulatory system continually along with the drainage of lymphocytes from lymph nodes and other lymphatic tissue. After a few hours, they pass back into the tissue by diapedesis and then reenter the lymph once again only to return to either lymph tissue or the blood repeatedly. The lymphocytes were originally thought to have a life span from days to months to years in contrast to the 3,000,000 platelets that are replaced every 10 days.[244] This long life span appears to be true. However, there are many difficulties in measuring these life spans.

Definitions of lymphocyte life spans vary according to different conceptual and experimental systems. There is agreement on the high rate of lymphocyte production and turnover in the bone marrow and thymus[245-247] and on the fact that peripheral mature lymphocytes contain both short-lived and long-lived populations with most lymphocyte reproduction and expansion occurring in the periphery.

The mechanism of cell renewal differs in T- and B-cells. B-cells are renewed from continuous production from the bone marrow[250] **while T-cells are mainly renewed by cell generation in the periphery**,[249,250] 30–40% of peripheral immunocompetent B- and T-cells are renewed every three days and most B-cells are renewed every 10 days.[251-261] A minority of B-cells lasts for three weeks.

Cell transfer experiments have shown that lymphocytic survival wanes with environmental influences. Thus, lymphocytic life span is not an intrinsic property of a cell. Short-lived B-cells appear to have a short life span as they fail to receive signals to induce them to differentiate and therefore they do not become long-lived cells. Equally long-lived B-cells appear to have received that signal.[261]

T- and B-cells are maintained separately. The total number of T-cells is independent of CD_4 and CD_8 lymphocytes.

Fibroblast and Macrophage Homeostasis (Reticuloendothelial)

As shown previously, physiologically, fibroblasts respond to any stimuli by the triggering of the nonspecific mesenchymal response with the resultant production of PG/GAGs. Thus, communication is propagated through this system. Most environmental substances are **known to trigger the fibroblast to produce PG/GAGs**, which further the body's ability to communicate. These triggereing substances **include not only noxious stimuli but physical phenomena like movement, exercise, pressure applied to the skin, fascia, or muscle, as well as needle puncture and trauma. Chemical** exposure, bacteria, viruses, external EMF, and free radicals also trigger the fibroblast to produce PG/GAGs.[78] The PG/GAGs form new CTM, replace worn out matrix, or if in excess clog the ECM canuliculi, which results in inflammation to the point where the microcapillaries can become inflamed and severely damaged. **The macrophages have to be in a homeodynamic balance with the fibroblasts in order to continue to take out excess PG/GAGs and other debris (including free radical damaged tissue and autocatalytic debris).** If the imbalance in favor of the fibroblast occurs, vasculitis follows, spreading even to the larger vessels, and causing clinical syndromes like thromboangiitis obliterans.[262] If the macrophages do not keep removing the fibroblast stimulated PG/GAG, inflammation occurs. This homeodynamic balance is maintained electrically at 280 millivolts with stimulation of both cells.[263] Macrophages also absorb electrons and heat from redox reactions, other oxidative reactions, and free radical emanation as does the CTM, which swings back and forth depending on thermal balance. The changing states of the CTM result in autocatalysis through the release of serine proteases, hydralases, peptidases, sulfatases, lysozymes from the macrophages, as well as glucose excess. A dynamic homeostatic balance between the fibroblasts and the macrophages does not occur if the excess noxious stimuli are present or if nutrient deficiency occurs. If excess noxious stimuli are present or if nutrient deficiency occurs, there will be excess production of PG/GAGs and free radicals resulting in inflammation. Once the threshold of inflammation is exceeded, a whole new sequence of events develops. (See defense and inflammation discussed further on.)

The fibroblast also secretes structural and other proteins. These proteins include collagen, laminin, fibronectin, immunomodulin, lysine, and proline rich elastin.[264] Again, there is communication so that homeodynamic equilibrium occurs with the fibroblast replacing and the macrophages removing these substances as fast as they are produced to maintain the dynamics of homeostasis. It is easy to see why the dynamics of the fibroblast-macrophage communication resulting in homeostasis is so important in preventing inflammation. This homeodynamic balance can easily be thrown off by the overloading with noxious incitants or nutritional deficiency, by the unstable environmental receptors of the GRS, or by the artificially weakened function of the macrophages induced by toxic overload.

A large portion of the monocytes, once they enter the tissues, turn into macrophages, become attached to the tissues, and remain attached for months or even years until a communication is signaled to perform specific protective homeostatic functions. These fixed-tissue macrophages have the same capabilities as the mobile macrophages to phagocytize large quantities of bacteria, viruses, necrotic tissue, or other foreign particles in the tissue. When appropriately stimulated by a noxious stimulus signal, they can break away from their attachments and once again, become mobile macrophages that respond to chemotaxis and all the other stimuli related to the inflammatory process. Thus, in order to maintain dynamic homeostasis the body has a widespread "monocyte-macrophage system" in virtually all areas of the body where communication signals of noxious stimuli entry occur.[265]

Reticuloendothelial System
The combination of monocytes, mobile macrophages, fixed-tissue macrophages, and a few specialized endothelial cells in the bone marrow, liver, spleen, and lymph nodes collectively make up the reticuloendothelial system and some of the innate immune system. These cells communicate with other macrophages and the rest of the immune system. However, almost all these cells originate from monocytic stem cells; therefore, the reticuloendothelial system is almost synonymous with the monocyte-macrophage system. This generalized phagocytic system should be remembered as a communication and defense system located in all tissues of the body. The phagocytic system occurs in those local tissue areas where large quantities of particles, including bacteria and viruses, toxins, toxics, and other unwanted substances must be destroyed. These areas include all of the orifices, e.g., nasal, respiratory, gastrointestinal, urethral, and vaginal areas, as well as the liver, spleen, and lymph nodes.

Normally the skin is impregnable to infectious agents but becomes vulnerable when the skin is broken; classic examples include cuts or eczema. When infection begins to invade the subcutaneous tissues, it communicates with a local receptor and if too many free radicals are generated local inflammation ensues. **The tissue macrophages (histiocytes) can divide *in situ* and form still** more macrophages, which can perform the usual functions of attacking and destroying the infectious agents as described earlier.

Essentially no particulate matter, such as bacteria, that enters the tissues can usually be absorbed directly through the capillary membranes into the blood. Instead, if the particles are not destroyed locally in the tissues because of their size they enter the lymph system and flow through the lymphatic vessels to the lymph nodes located in many areas or points along the course of the lymphatics. The foreign particles are trapped there in a meshwork of sinuses lined by tissue macrophages. The ebb and flow of lymphatics is very evident in the CS patients. With any minute exposure one observes swelling of the lymph nodes in the neck, groin, or axilla depending on where the exposure enters. The nodes in the neck are often swollen due to one or all of several exposures: the ingestion of foods to which they are sensitive, food and water contaminants, or inhalation of bacteria, virus, particulates, or toxics. **It appears that in many CS patients the upper lymph system is chronically overloaded as evidenced by specific cutaneous thermograph and physical palpation.** In addition, inhalation or ingestion of small amounts of substances that trigger swelling may last for several days. See Figure 1.39.[266]

Large numbers of macrophages line the sinuses and if any particles (toxic, bacteria, virus) enter the sinuses a local communication alarm is sounded that noxious stimuli have entered and the macrophages phagocytize them. General dissemination of noxious stimuli throughout the body is prevented. Once phagocytosis occurs an alarm is sounded that the mission is accomplished locally. At times, this system cannot keep up with the exposure and will become clogged (toxics, asbestos, silica, coal dust, bacteria, diesel, cancer cells) resulting in an alteration of the dynamics of regional, and even at times central homeostasis.

Another route by which invading organisms frequently enter the body is through the lungs. Large numbers of tissue macrophages are present as integral components of the alveolar walls. When the signal is sounded that a noxious stimulus has entered the macrophages can phagocytize particles that become entrapped in the alveoli. **If the particles are digestible, the macrophages can also digest them and release the digestive products into the lymph. If the particle is not digestible, the macrophages often form a "giant cell" capsule around the particle until such time, if ever, that it can be slowly dissolved.** Such capsules are frequently formed around tubercle bacilli, silica dust particles, carbon particles, and implanted artificial tissue or organs. In another example, sarcoid granuloma formation, which in our experience is triggered by mold exposure, can form a capsule.

Still another favorite route by which bacteria and other toxics invade the body is through the gastrointestinal tract. Large numbers of bacteria and toxics constantly pass through the gastrointestinal mucosa

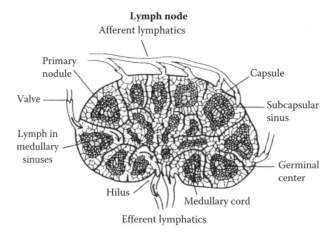

FIGURE 1.39 This Figure demonstrates the general organization of the lymph node, showing lymph entering through the lymph node capsule by way of afferent lymphatics, then flowing through the node to the medullary sinuses, and finally passing out of the hilus into the efferent lymphatics. (From Guyton, A.C. and Hall, J.E., *Textbook of Medical Physiology*, 10th ed., W.B. Saunders Company, Philadelphia, 396, 33–3, 2000. With permission.)

The Physiologic Basis of Homeostasis 79

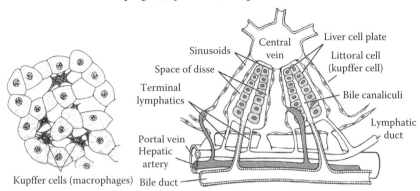

FIGURE 1.40 Macrophages (Kupffer cells) line the liver sinusoids and blood vessels in order to contain debris and toxics to maintain homeostasis. (Modified from Guyton, A.C. and Hall, J.E., *Textbook of Medical Physiology,* 10th ed., W.B. Saunders Company, Philadelphia, 396, 33–4, 2000.)

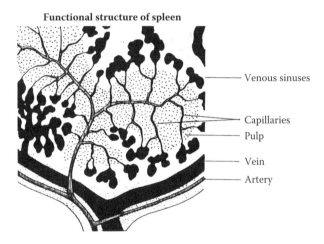

FIGURE 1.41 Spleen. (From Guyton, A.C. and Hall, J.E., *Textbook of Medical Physiology,* 10th ed., W.B. Saunders and Company, Philadelphia, 396, 33–5, 2000. With permission.)

into the portal blood. Before this blood enters the general circulation it must pass through the sinuses of the liver; these sinuses are lined with tissue macrophages (Kupffer cells) shown in Figure 1.40[267].

These cells form such an effective particulate filtration system that almost none of the bacteria from the gastrointestinal tract succeed in passing from the portal blood into the general systemic circulation. **Indeed, motion pictures of phagocytosis by Kupffer cells have demonstrated phagocytosis of a single bacterium in less than one-hundredth of a second.**[267] This speed and vigor of macrophage phagocytosis can be altered by pollutant overload, resulting in dyshomeostasis with the dissemination of altered information going rapidly throughout the body.

If an invading organism or other noxious stimulus does succeed in entering the general circulation, other lines of defense remain by the tissue macrophage system, especially by the macrophages of the spleen and bone marrow. In both these tissues, macrophages become entrapped by the reticular meshwork of the two organs, and when foreign particles or toxics come in contact with the meshwork they are phagocytized.

The spleen is similar to the lymph nodes except that blood instead of lymph flows through the tissue spaces of the spleen. See Figure 1.41[268] shows the spleen's general structure, demonstrating a small peripheral segment.

Note that a small artery penetrates from the splenic capsule into the splenic pulp and terminates in small capillaries. The capillaries are highly porous, allowing whole blood to pass out of the capillaries into the cords of the red. The blood then gradually squeezes through the trabecular meshwork of the cords and eventually returns to the circulation through the endothelial walls of the venous sinuses. Again, this anatomical make up demonstrates the filtering effect of the body. The trabeculae of the red pulp are lined with vast numbers of macrophages, and the venous sinuses are also lined with macrophages. This peculiar passage of blood through the cords of the red pulp provides an exceptional means for phagocytosis of unwanted debris in the blood, including especially old and abnormal erythrocytes, leukocytes, and platelets. **The proper balance of phagocytosis of sick and damaged cells along with the normal physiologic injury, which causes phagocytotic damage, results in dynamic homeostasis where the individual can function in a healthy manner.** Splenectomized individuals lose one large line of defense and often become susceptible to recurrent infections. Patients with long-standing sickle cell anemia develop splenic fibrosis thus, developing an auto splenectomy and have similar types of recurrent infections. One can see from the anatomical study of these organs that a series of filters is present in the body to remove noxious stimuli and substances. **The filters include the cell membranes, the CTM, the glomeruli and kidney, the liver, the spleen, lymph nodes, and the vascular wall, interstitial fluid, and lymphatics.** Any or all filters are capable of filtering out toxics, particulates, and other debris, which can chronically disturb the dynamics of homeostasis. Reduction of the toxic total pollutant load can help keep these dynamics of homeostasis functioning well.

At times macrophages can act as a suppressor cell specifically inhibiting the toxicity of the tumor.[269] At other times, the suppression if toxicity is damaged often by chemical exposure, resulting in a loss of suppression.[270]

Erythrocyte Homeostasis

The most common abnormalities from the oxidative injury in chemically sensitive and chronic degenerative diseased patients studied by Ali[189] consisted of lack of erythrocyte plasticity, irregularities of its outlines, and clumping. Von Ardenne[271] has shown a similar response. Many erythrocytes showed surface wrinkling, teardrop deformity, sharp angulations, and spike formations. In later stages, zones of plasma congealing were seen around many erythrocytes. In more advanced cases, an increasing number of erythrocytes showed shrinkage and filamentous outgrowths extending from their membranes. Such filamentous outgrowth covers the entire surface of cells to produce a Medusa-like appearance. Other cells appeared as ghost outlines. Zones of congealed plasma surrounded many cells.[189] Simpson[270] found similar findings when he studied unstained blood under the electron microscope. With total deprivation of oxygen for a short interval, the cell remains fully active. This interval can be prolonged by cooling as has been shown on studies of aortic cross clamping during open heart surgery in an ischemic heart.[271] However, if the hypoxia lasts too long irreversible damage occurs. The organ will develop irreversible damage and fail to function with structural disintegration occurring. The reversible phase of hypoxia has much influence on red blood cells. The oxygen deficiency causes capillary endothelial swelling and local reduction of blood flow to the tissues. It is difficult if swelling is accompanied by a reduction in pH. This pH drop is caused by a transition to anaerobic metabolism in tissue surrounding capillaries. Due to lactic acid there is a rearrangement of the architecture of the red and white cell membranes. Therefore, a change in information results in a change in metabolism, which is followed by a change in anatomy. **With a drop of pH from 7.4 to 6.5, there is an alteration in the endothelial cell membranes affecting the microcirculation synergistically. There is also a large reduction of flexibility of erythrocytes with adhesion to the capillary endothelium and rouleaux formation, resulting in jamming of the micro vessel.** The formation of microthrombi follows or the local microcirculation flow decreases, resulting in a small area of hypoxia. In addition, there is an increased stickiness of erythrocytes with nonenzymatic adhesion of fibrogen to the endothelium. Erythrocyte membrane rigidity occurs further blocking the microcirculation since their sizes range from 2 to 8 micron, whereas the microcirculation vessels diameter are 3–4 microns. This blocking causes shunting and hypoxia. The vicious downward cycle and the positive feedback characteristic of homeostasis occur with eventual total thrombosis. Reversibility of this downward process in the patient with chemical sensitivity and/or chronic degenerative disease is accomplished by multistep oxygenation, and pollutant load reduction thus, restoring the dynamics of homeostasis.[271] (See Nutrition chapter in Volume 4 - Mechanisms of Cardiovascular Disease and Chemical Sensitivity for details.)

Platelet Homeostasis

According to all investigators, the earliest changes involving platelet dyshomeostasis were platelet clumping and loss of membrane detail. Enlarged platelets are seen frequently. Increased intensity of coagulopathy, degranulation, and lysis of platelets are common. Zones of congealed plasma, the beginning of soft clots, are pronounced and often extended to erythrocytes and leukocytes in the vicinity. Fibrin deposits occurred both as fibrin needles and amorphous masses surrounding lysed platelets and their capsules. Platelets are influenced by environmental stimuli. One example of this environmental exposure phenomenon was seen at the EHC-Dallas.

> **Case study.** A 45-year-old Latin American female, a native and resident of Honduras, developed thrombocytopenia with purpura and severe chemical sensitivity and was diagnosed with idiopathic thrombocythopenic purpura. When placed under environmentally controlled conditions and deadapted for four days with her total body pollutant load reduced, her platelet counts (that) ranged from 1000/mm^3 to 10,000/mm^3 rose to 150,000/mm^3. Each time the patient received an organochlorine pesticide exposure, her platelet count would drop to 10,000 or less. Once the patient achieved a chronic constant reduction in both her pesticide load and her total body pollutant load, the platelet homeostasis was restored with her counts ranging from 150,000/mm^3 to 175,000/mm^3. Since environmental treatment for the last four years was instituted, she remains well without taking medications. This graphic example of exaggerated environmentally triggered platelet dyshomeostasis allows us to understand the oxidant-induced degeneration of platelets and how total body pollutant load (drug induced or toxicant induced) can influence such changes, taking the platelet counts away from equilibrium. Table 1.9 shows a small number of patients with toxic induced platelet dyshomeostasis and reticular endothelial dysfunction.

This phenomenon should not appear strange to the clinician who knows that other environmental incitants such as some medication can cause the same type of platelet suppression.

TABLE 1.9
Environmentally Triggered Thrombocytopenia after four Days Deadaptation in the ECU with Total Load Decreased

Age/Sex/Race	Initial Platelets (mm^3)	After 1 week (mm^3)	Long-Term (1–10 years) (mm^3)	Triggering Agents Biological Inhalants	Food	Toxic Chemicals[a]
60 years/w/f	30,000	200,000	200–400,000	−	−	+
38 years/w/f	15,000	400,000	200–400,000	−	−	+
45 years/w/f	40,000	150,000	150–300,000	−	+	+
40 years/w/f	60,000	170,000	150–300,000	+	+	+
50 years/w/f	50,000	140,000	140–200,000	−	+	+
47 years/w/f	30,000	100,000	100–150,000	−	−	+
37 years/w/f	5,000	125,000	150–175,000	+	+	+
47 years/w/f	1,000	150,000	180–200,000	+	+	+
40 years/w/m	25,000	130,000	500–225,000	−	−	+
45 years/w/f	10,000	150,000	100–300,000	+	+	+
60 years/w/f	30,000	100,000	150–400,000	+	+	+

Source: From EHC-Dallas, 2001.

[a] Inhaled double-blind: phenol, < 0.0024 ppm; formaldehyde, < 0.2 ppm; ethanol, < 0.50 ppm; chlorine, < 0.33 ppm; pesticide 2,4DNP, < 0.0034 ppm.

Plasma Proteins

About three-quarters of the body's solids are proteins. These include structural proteins, enzymes, nuclear protein, proteins that transport oxygen, proteins of the muscle that cause muscle contraction, and many other types of proteins that perform functions both intracellularly and extra cellularly throughout the body. Any of these can be damaged, the volume can be altered, or the properties can be changed in the individual with chemical sensitivity or chronic degenerative disease.

The normal concentration of **amino acid** in the blood is between 35–65 mgs/dl. This is an average of about 2 mg/dl for each of the 20 amino acids, although some are present in far greater amounts than others. Since the amino acids are relatively strong acids, they exist in the blood supply principally in the ionized state, resulting from the removal of one hydrogen atom from the NH_2 radical. They actually **account for 2–3 Meq/L of the negative ions in the blood**. The precise distribution of the different amino acids in the blood depends to some extent on the entering proteins but the concentration of at least some individual amino acids are regulated by selective synthesis in different cells. **The dietary alterations in the chemically sensitive individual are prone to change this amino acid ratio level**. Protein products absorbed in the gastrointestinal tract are almost all in the form of amino acids. Blood levels rise only a few mgms/per deciliter. This slight rise occurs because absorptions occurs over a 2–3 hours period and after entering the blood the excess amino acids are absorbed within another 5–10 minutes by cells throughout the body. The turnover rate of the amino acids is so rapid that many grams of protein can be carried from one part of the body to another in the form of amino acids each hour. Amino acids are too large to be passively absorbed through the cell membrane therefore; they are actively transported both in and out. The transport is by a carrier mechanism but this is poorly understood.

If amino acids are in excess in the glomeruli of the kidney they will go to the renal tubules and spill into the urine. Almost immediately after entry to the tissue cell amino acids combine with another peptide linkage under the direction of the cells RNA and ribosomal system to form cellular proteins. Few free amino acids remain in the cell. **Lysomal enzymes digest the protein back into the cell when the amino acids are needed. Exceptions to lysomal enzyme digestion would be contractible proteins, chromosomal proteins, and structural proteins like collagen**. The liver, intestine, and kidney can store large amounts of protein.

Many steps of amino acid absorption, transport, storage, and the maintenance of appropriate blood levels are involved in maintaining appropriate homeostasis. Many of these can be altered in the chemically sensitive or chronic degenerative diseased individual. (See Nutrition chapter in Volume 4 - Mechanisms of Cardiovascular Disease and Chemical Sensitivity.) The most crucial step in transport appears to be the maintenance of appropriate blood levels. When the amino acids drop down below a critical level they are mobilized from the cells. **Hormones like growth hormone and insulin increase the formation of tissue proteases whereas adrenocortical glucocortical hormone increases the concentration of plasma amino acids.**

There is a constant interchange between cells and amino acids throughout the body. Each cell has an upper limit of how much protein it can store and above this limit the amino acids are used for energy. The major type of protein present in the plasma amino acids are albumin, globulin, and fibrogen. Albumin provides the colloid osmotic pressure in the plasma, which in turn prevents protein loss from the capillaries. There is a subgroup of patients with chemical sensitivity and/or chronic degenerative disease who are also borderline hypotensive. Their blood pressure runs from 90/60 to 70/50. They get dizzy upon arising from a sitting position, occasionally pass out, cannot tolerate the sauna, and have difficulty in obtaining and maintaining safe endpoints for food and mold injection therapy. Their plasma volumes are 10–15% constricted. These patients respond to intravenous saline and albumin (20–50 gms. I.V.)

Albumin also can neutralize toxics that enter the blood by binding them and sending them to the tissue detoxification mechanisms. Globulins perform a number of enzymatic functions and equally as important they are principally responsible for natural and acquired immunity that a person has to fight

The Physiologic Basis of Homeostasis

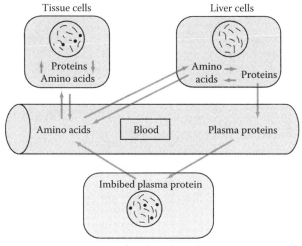

FIGURE 1.42 Reversible equilibrium among tissue proteins. (From Guyton, A.C. and Hall, J.E., *Textbook of Medical Physiology,* 10th ed., W.B. Saunders Company, Philadelphia, 794, 69–2, 2000. With permission.)

invading organisms. About one in 10 patients with chemical sensitivity is gamma globulin deficient and responds to injections of g-globulin. Fibrinogen polymerizes into long chain fibrin threads during blood coagulation thereby, forming blood clots that help response levels in the circulatory system.

It has been estimated that normally 400 gms of body proteins are synthesized and degraded each day as part of the continual state of flux of amino acids. See Figure 1.42[272] Even during starvation or severe debilitating diseases the ratio of total body tissue proteins to total plasma proteins in the body remains relatively constant at 33 to 1.

When cells are full with protein the excess amino acids are broken down, deaminated, and stored as fat or glycogen. These are then ready for quick energy. When an individual eats proteins a certain proportion of that person's body protein continues to be degraded into amino acids and then deaminated and oxidized. This degradation involves 20–30 gms of protein each day, which is an obligatory loss of protein. **Therefore, to prevent a net loss of protein from the body one must ingest a minimum of 20–30 gms of protein each day, and to be on the safe side, ingestion of a minimum of 60–75 gms is recommended**. Except for the obligatory degradation of protein each day the body uses almost entirely carbohydrates or fats for the energy demands as long as they are available. However, after several weeks of starvation when the quantity of stored CHO and fats begins to run out, the amino acids of the blood begin to be rapidly deaminated and oxidized for energy demands. From this point of starvation 125 gms of protein daily resulting in cellular function deteriorates perceptibly. CHO and fat are protein spares because they are utilized first over a preference of protein for immediate energy.

If it is clear that dynamic homeostasis is prevalent in protein absorption and metabolism, then pollutant entry into the blood stream can attach to the proteins, especially albumin, and create haptens, which will disturb intravascular metabolism. Patients with chemical sensitivity occasionally develop a catabolic state, loosing weight until they are down to 70–80 pounds. This is due to becoming totally food intolerant. These patients have to have intravenous hyperalimentation to survive. (See Nutrition chapter in Volume 4 - Mechanisms of Cardiovascular Disease and Chemical Sensitivity.)

Plasma Lipids

A number of chemical compounds in food and in the body are classified as lipids. These include neutral fat (triglycerides), phospholipids, cholesterol, and a few other compounds of lesser importance.

Chemically, the basic lipid compound of the triglycerides and of the phospholipids is fatty acid, which is simply a long chain hydrocarbon organic acids.

The triglycerides are used in the body mainly to provide energy for the different metabolic processes. They share this function almost equally with carbohydrates. Some lipids such as cholesterol, phospholipids, and small amounts of triglycerides are used throughout the body to form membranes of all cells of the body and to perform other cell functions. Therefore, **homeostatic balance must be maintained for energy and membrane integrity**. Both of these functions may be disturbed in the hypersensitive and chronic degenerative diseased patient.

Almost all fats in the diet are absorbed from the intestinal lumen into the intestinal lymph. During digestion most triglycerides are split into monoglyceride and fatty acids. Then while passing through the intestinal epithelial cells, the monoglycerides and fatty acids are resynthesized into new molecules of triglycerides that enter the lymph as minute, dispersed droplets (chylomicrons) having diameters between 0.08 and 0.6 micron. A small amount of protein apoprotein B is absorbed to the outer surfaces of the chylomicrons. This leaves the remainder of the protein molecules projecting into surrounding waters thereby, increasing suspension stability of the chylomicrons in the fluid of the lymph and preventing their adherence to the lymphatic vessel walls thus, allowing homeostasis to occur. Also, most of the cholesterol and phospholipids absorbed from the GI tract enter the chylomicrons. Although the chylomicrons are composed principally of triglycerides they also contain 9% phospholipids, 3% cholesterol, and 1% apoprotein B. From the gastrointestinal tract the chylomicrons are then transported upward through the thoracic duct and emptied into the circulatory venous blood supply at the juncture of the jugular and subclavian vein.

About an hour after eating a meal the fats of the chylomicron are removed by hydrolysis and the triglycerides are removed by the liver. Both are stored in the tissues and liver cells. Lipids are transported mainly as free fatty acids and some as glycerol created by the hydrolysis of triglycerides. On leaving the fat cells fatty acids ionize strongly to the plasma and the ionic portion combines immediately with albumin molecules of the plasma proteins. The fatty acids bound in this manner are free fatty acids or esterfied fatty acids to distinguish them from other fatty acids that exist in the plasma in the forms of esters of glycerol, cholesterol, or other substances. The plasma concentration of free fatty acids is about 15 mgm/dL, which is a total of only 0.45 gms of fatty acids in the total circulatory system. Even this small amount accounts for almost all the transport of fatty acids from one part of the body to the other. The rate of turnover of free fatty acids is very rapid as one-half of the plasma fatty acids are replaced by new fatty acids every 2–3 minutes.

Almost all normal energy requirements of the body can be provided by oxidation of free fatty acids without using any CHO or protein. Alteration of this mechanism in the patient with chemical sensitivity and/or chronic degenerative disease could account for much of the weakness and fatigue they experience. All conditions that increase the rate of utilization of fat or cellular energy also increases free fatty acid concentration in the blood. In fact, the concentration sometime increases fivefold to eightfold. A concentration of such a great increase occurs in the body especially in starvation and diabetes. In both cases the individual derives little or no energy from CHO. Under normal conditions only about three molecules of fatty acids combine with each molecule of albumin but **as many as 30 molecules of fatty acids can combine with a single molecule of albumin when the need for fatty acid transport is extreme**. This phenomenon shows how variable the rate of lipid transport can be under condition of different physiologic needs. In the post absorptive state after the chylomicrons are removed from the blood more than 95% of the lipids in the plasma are lipoproteins. The lipoproteins contain cholesterol, triglycerides, phospholipids, and protein C, which make up a total concentration of 700 mgm per 100 ml of plasma. Lipoprotiens are 3 types [2-] low [1-] intermediate and high density. The primary function of lipoproteins is to transport their lipid components in the blood.

Large quantities of fat are stored in the adipose tissue (modified fiboblasts) and the liver. The major function of the adipose tissue is storage of triglycerides until they are needed to provide energy elsewhere in the body. Large quantities of lipases are present in the adipose tissue and when the fat is needed back in the blood for energy they are activated to mobilize fat to the blood.

The Physiologic Basis of Homeostasis

There is a 2–3 week turnover of fat in adipose tissue, which emphasizes the dynamic state of storing fat.

The principle function of the liver in lipid metabolism is to degrade fatty acids into small compounds used for energy. The liver also functions to synthesize triglycerides from both CHO and some proteins as well as to synthesize other lipids from fatty acids especially cholesterol and phospholipids. **Degradation and oxidation of fatty acids occur only in the mitochondria**. Carnitine is the carrier substance to the mitochondria. Once transported to the mitochondria the fatty acids combine with coenzyme A to form acyl-CoA for energy formation. For every stearic fatty acid molecule that is split to form nine acyl-CoA molecule 30 extra hydrogen atoms are removed. In addition, for each of the nine molecules of acetyl-CoA that is subsequently degraded by the citric acid agate, 8 more hydrogen atoms are removed making an additional 72 hydrogens, which when added to the 32 hydrogen atoms makes 104 hydrogen atoms eventually released by degradation of each stearic acid molecule. Of this hydrogen group 34 are removed by fluoroproteins and 70 are removed by nicotinamide adenine dinucleotide NAD as NADH and H^+. These two groups of hydrogen atoms are oxidized in the mitochondria but they enter the oxidation system at different points so that **one molecule of ATP is synthesized for each of the 34 fluoroprotein hydrogens and 1.5 molecules of ATP are synthesized for each of the 70 NADH and H+ hydrogen atoms**. Combined this makes 34 plus 105, or a total of 139, molecules of ATP that are formed by the oxidation of hydrogen derived from each molecule of stearic acid. Another nine molecules of ATP are formed in the citric acid cycle itself, which is separate from the ATP released by the oxidation of hydrogen where the ratio is one for each nine acetyl-CoA molecule metabolized. Thus, **a total of 148 molecules of ATP are formed during the complete oxidation of one molecule of stearic acid**. When fatty acid chains split into acetyl-CoA two molecules of acetyl-CoA condense to form one molecule of acetoacetic acid, which is transported in the blood to other cells throughout the body where it is used for energy. The concentrations of acetoacetic acid, B-hydoxybutyric, and acetone create ketosis if too high. This condition occurs in starvation, diabetes, and sometimes in a person's diet that is composed almost entirely of fat.

Clearly, if toxic lipophilic xenobiotics are attached to fatty acids they can stop or partially inhibit the production of energy causing weakness and/or fatigue as we have seen in most chemically sensitive and chronic degenerative diseased patients. In addition, vitamin B deficiency as seen in the majority of chemically sensitive patients would have a similar effect. Often the administration of 1–2 cc of B-complex injection will give the chemically sensitivie patient a boost.

Neurological Aspects of Homeostasis

When nerves are involved in the patient with chemical sensitivity and/or chronic degenerative disease the law of nerve injury[273] comes into effect, making the distal nerve and end organ supersensitive to neurotransmitters. After injury in patients with chemical sensitivity and/or chronic degenerative disease, the nerves and organs distal to the injured nerve instead of calming down as they are supposed to remain and even became more supersensitive especially if they continue to be bombarded with chronic noxious stimuli overload.

This is discussed in the section on adjustments and in Chapter 2. See regional and central homeostasis.

POLLUTANT ENTRY AND THE BODY'S HOMEOSTATIC RESPONSE TO AND FATE OF THE NOXIOUS STIMULI

Total Environmental Load

The total environmental pollutant load is the key to understanding the first step of the dynamics of homeostasis in individuals with chemical sensitivity and chronic degenerative disease. See Figure 1.43. Here information input of noxious stimuli that will potentially enter the body is accumulated. The body may eventually have to respond if entry occurs.

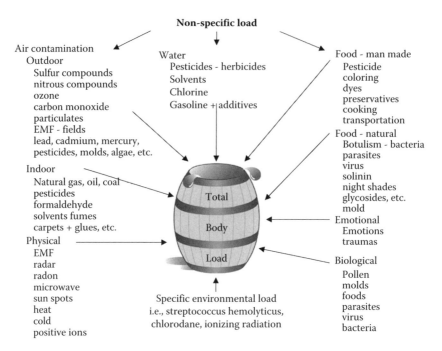

FIGURE 1.43 Total environmental load. (From EHC-Dallas, 2002.)

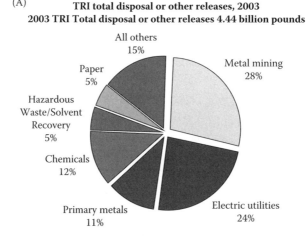

FIGURE 1.44A (See color insert following page 364.) The public has been exposed to toxic chemicals. (From Toxic Release Inventory. 2003. Total Disposal or Other Releases, EPA 260-R-05-001 May 2005.)

The total environmental load of contaminants in air, food, and water is rapidly expanding due to our increase in population, which results in more waste being deposited in our environment. **According to the U.S. EPA tracking program there was 4.44 billion pounds of contaminants released into the U.S. environment for the year of 2003** (last data reported). Of course these releases have occurred year after year and will continue in the foreseeable future. See Figure 1.44A through D.

Therefore, the body is constantly assaulted by not only the thousands of biological substances such as pollens, molds (thousands), foods (hundreds), bacteria (thousands), viruses (thousands), and

(B)

TRI On-Site and Off-site Disposal or Other Releases by State, 2003

	On-site disposal or other releases															
	On-Site Disposal to Class I Underground Injection Wells, RCRA Subtitle C Landfills, and Other Landfills				Other On-Site Disposal or Other Releases											
State	Class I wells Pounds	RCRA Subtitle C Landfills Pounds	Other on-site Landfills Pounds	Subtotal Pounds	Fugitive air Emissions Pounds	Point Source Air Emissions Pounds	Surface water Discharges Pounds	Class II-V Wells Pounds	Land Treatment Pounds	RCRA Subtitle C Surface Impoundments Pounds	Other surface Impoundments Pounds	Other Land Disposal Pounds	Subtotal Pounds	Total On-Site Disposal or Other Releases Pounds	Total Off-Site Disposal or Other Releases Pounds	Total On-Site and Off-Site Disposal or Other Releases Pounds
Alabama	0	7,739,581	10,578,722	18,318,313	10,050,670	48,028,693	7,805,767	8,305	147,954	0	14,642,205	464,575	81,148,169	99,466,482	18,982,157	118,448,639
Alaska	0	0	310	310	395,715	1,689,402	541,992	21,374,380	0	0	270,833,014	244,609,756	539,444,259	539,444,569	199,327	539,643,896
American Samoa	0	0	0	0	8,460	6	0	0	0	0	0	0	8,466	8,466	0	8,466
Arizona	0	1	5,854,841	5,854,842	1,063,942	3,605,620	6,965	0	167,809	0	6,854,241	30,027,815	41,726,391	47,581,233	632,299	48,213,532
Arkansas	3,303,538	295,473	2,439,167	6,038,178	6,279,944	14,018,639	5,419,186	93,481	501,024	1	2,039,402	43,324	28,395,002	34,433,180	6,168,633	40,601,813
California	0	25,491,811	608,973	26,100,784	3,946,761	14,238,606	4,617,780	30,086	675,350	5,134	130,552	444,069	24,088,337	50,189,121	7,682,581	57,871,702
Colorado	0	97	6,086,523	6,086,620	753,351	2,204,320	2,955,073	0	341	0	2,726,225	3,317,078	11,956,389	18,043,008	4,474,503	22,517,511
Connecticut	0	0	393	393	855,300	2,154,851	722,325	0	0	0	78	778	3,733,332	3,733,725	1,650,391	5,384,116
Delaware	0	0	1,252,178	1,252,178	290,367	7,003,978	918,650	0	9	0	8,178	48	8,221,230	9,473,408	4,139,175	13,612,583
District of Columbia	0	0	0	0	3,338	1	8,062	0	0	0	2,082	0	13,482	13,482	306	13,788
Florida	20,875,927	84	10,441,600	31,317,611	5,640,681	71,659,217	2,507,602	154,144	2,067,467	405,334	9,159,351	123,809	91,717,605	123,035,216	3,417,197	126,452,413
Georgia	0	90	522,944	523,034	9,084,590	90,203,030	9,573,187	0	431,299	35,472	13,660,727	748,427	123,736,732	124,259,766	2,391,094	126,650,860
Guam	0	10	10	22	1,926	162,669	71,627	0	0	0	7	3	236,233	236,255	2,048	238,303
Hawaii	0	0	0	0	180,874	1,900,899	364,067	2,670	19,681	0	0	229,586	2,697,777	2,697,777	419,791	3,117,568
Idaho	0	30,517,289	42,662	30,559,951	995,037	3,038,312	4,642,166	0	669,554	2,239,333	6,571,414	12,027,196	30,183,012	60,742,962	584,846	61,327,809
Illinois	794	23,219,451	4,464,352	27,684,597	8,628,864	47,405,033	7,221,378	0	477,550	29,315	8,987,454	447,772	73,197,365	100,881,962	31,519,211	132,401,173
Indiana	1,075,521	8,094,913	17,522,055	26,692,489	8,722,806	64,952,587	23,296,297	0	1,850,171	1,500	8,571,928	777,844	108,173,133	134,865,622	99,900,349	234,765,971
Iowa	0	0	1,889,630	1,889,630	3,791,907	18,311,363	3,274,619	0	750	0	2,611,789	39,658	28,030,086	29,919,715	-7,482,328	37,402,043
Kansas	470,746	33,000	3,535,497	4,039,243	2,641,059	10,557,093	4,021,144	500	419,554	0	3,795,286	127,585	21,562,221	25,601,464	3,262,448	28,863,912
Kentucky	0	0	15,143,069	15,143,069	6,024,718	51,653,886	2,986,815	2,971	333,583	334,515	5,381,917	921,961	67,640,366	82,783,436	7,798,069	90,581,505
Louisiana	35,904,030	7,050,318	7,348,228	50,302,576	12,230,196	42,697,721	11,303,522	0	46,587	289,599	4,062,499	148,267	70,778,392	121,080,968	5,764,045	126,845,013
Maine	0	0	838,723	838,723	686,531	3,611,216	3,334,311	0	0	0	0	9,557	7,641,615	8,480,338	824,733	9,305,071
Maryland	0	0	1,827,936	1,827,936	596,281	35,266,724	2,704,113	50,375	159,930	0	34,954	119,632	38,932,009	40,759,945	4,739,560	45,499,505
Massachusetts	0	0	41,739	41,739	741,565	5,297,813	68,806	0	39	0	80,638	670,421	6,859,282	6,901,021	2,112,307	9,013,328
Michigan	1,907,236	1,046,569	3,757,600	6,711,405	3,915,701	46,250,106	1,218,195	0	123,547	0	5,389,756	1,062,639	57,959,944	64,671,349	36,933,939	101,605,288
Minnesota	0	251	5,526,048	5,526,299	3,110,524	10,522,357	1,246,533	0	71,993	0	5,284,791	63,705	20,299,903	25,826,201	5,622,420	31,448,621
Mississippi	12,654,379	65	1,985,150	14,639,593	5,810,502	25,988,162	7,751,711	0	393,645	15,268	6,767,118	221,233	46,947,640	61,587,233	1,489,728	63,076,960
Missouri	0	199,911	2,187,364	2,387,276	2,811,105	24,870,136	2,620,282	0	12,617	544	40,355,980	21,326,736	91,997,402	94,384,677	8,114,819	102,499,496
Montana	0	81	22,971,757	22,971,839	597,835	3,486,141	49,172	209,716	933	0	10,445,012	6,475,019	21,263,828	44,235,667	970,916	45,206,583
Nebraska	0	0	7,740,974	7,740,974	886,910	5,985,779	18,177,388	0	678,512	1,000	755	277,302	26,007,647	33,748,621	17,722,548	51,471,169
Nevada	0	29,771,405	2,606,310	32,377,715	588,282	1,253,025	88,601	3	43	0	188,495,103	185,492,867	375,917,924	408,295,639	758,761	409,054,400
New Hampshire	0	0	0	0	157,081	5,217,148	86,194	0	0	0	22,805	165	5,483,393	5,483,393	460,749	5,944,142
New Jersey	0	104,607	33,841	138,448	1,840,951	10,484,189	4,148,642	5	6,583	0	6,551	188,018	16,674,938	16,813,386	6,310,629	23,124,015
New Mexico	18,605	0	5,296,008	5,314,613	300,082	647,660	62,237	0	326,660	0	1,270,605	9,831,265	12,438,509	17,753,122	146,129	17,899,251
New York	0	2,919,297	1,137,787	4,057,084	2,855,774	24,254,524	7,937,611	0	23,565	0	257	398,503	35,470,235	39,527,319	4,499,897	44,027,216
North Carolina	0	2,023	3,265,168	3,267,191	6,936,819	93,358,632	8,592,474	0	245,329	0	6,096,625	609,420	115,839,298	119,106,489	10,039,033	129,145,522
North Dakota	0	0	2,263,167	2,263,167	251,193	4,621,269	248,831	0	17,123	0	7,020,805	2,975	12,162,196	14,425,362	9,213,436	23,638,798
Northern Marianas	0	0	0	0	324	5,701	0	0	0	0	2	0	6,027	6,027	0	6,027
Ohio	29,289,527	7,261,217	16,604,064	53,154,807	9,765,195	123,297,398	6,716,124	0	499,432	307,465	12,485,002	178,440	153,249,057	206,403,864	45,193,295	251,597,159
Oklahoma	1,307,526	1,757,603	1,814,396	4,879,525	2,467,942	13,460,673	3,451,751	0	13,387	1	547,749	524,632	20,466,134	25,345,659	4,622,081	29,967,740
Oregon	0	13,208,609	11,729,305	24,937,914	2,538,300	10,784,137	2,452,054	0	187,648	12,337	503	2,350	15,977,329	40,915,243	1,213,679	42,128,922
Pennsylvania	0	11,433	5,711,204	5,722,637	5,871,513	85,113,791	9,684,378	0	12,327	1,790,272	309,090	5,508,710	108,290,080	114,012,717	52,905,725	166,918,443
Puerto Rico	0	13,533	250	13,783	942,381	7,081,569	24,365	0	0	0	0	3,500	8,051,815	8,065,598	734,652	8,800,250
Rhode Island	0	250	0	250	236,638	390,168	6,947	0	0	0	0	0	633,753	634,003	258,275	892,279
South Carolina	0	62,318	2,387,601	2,449,919	6,720,209	45,438,879	3,473,022	0	37,514	85	2,944,266	327,757	58,941,732	61,391,651	22,316,357	83,708,008
South Dakota	0	0	5,533,657	5,533,657	656,431	807,101	3,199,143	101	665	0	23,556	5,350	4,692,347	10,226,004	94,109	10,320,113
Tennessee	0	54,792	18,926,296	18,981,088	11,030,628	77,689,173	2,410,104	0	26,760	0	22,765,611	2,457,322	116,379,598	135,360,685	7,178,908	142,539,594
Texas	85,633,991	3,619,830	24,058,727	113,312,547	35,163,397	56,502,383	21,670,263	42,000	3,090,383	11	3,684,426	1,953,398	122,106,261	235,418,808	26,485,939	261,904,747
Utah	0	7,672,552	12,978,810	20,651,362	1,857,825	7,268,894	56,978	0	292,689	0	130,057,726	78,732,740	218,266,853	238,918,214	3,077,073	241,995,287
Vermont	0	0	4	4	19,824	45,025	136,856	0	0	0	7	5	201,716	201,721	144,981	346,702
Virgin Islands	0	0	0	0	124,893	831,616	355,871	0	597	0	10,124	0	1,323,101	1,323,101	7,916	1,331,017
Virginia	0	344	3,317,037	3,317,381	5,233,866	45,714,310	8,198,926	0	203,048	0	1,887,312	244,389	61,481,850	64,799,231	9,431,206	74,230,438
Washington	0	633,020	191,652	824,673	1,560,909	11,772,038	1,397,187	0	29,989	0	5,035,903	517,707	20,313,733	21,138,405	1,741,529	22,879,934
West Virginia	10	7,844	7,700,777	7,708,631	3,840,246	75,083,656	4,194,526	87	597,291	21,088	4,937,887	342,718	89,017,499	96,726,130	5,443,463	102,169,593
Wisconsin	0	1,677	655,225	656,902	2,873,281	21,768,989	4,595,302	0	812,747	49,651	21,242	302,494	30,423,707	31,080,609	19,717,618	50,798,227
Wyoming	7,960,400	2,909	7,064,110	15,027,419	513,877	1,638,924	10,990	0	1,564	4,341	1,019,873	12,290	3,201,858	18,229,277	1,034,080	19,263,357
Total	200,402,228	170,794,270	267,883,840	639,080,339	205,095,324	1,381,295,231	222,628,110	21,968,824	15,675,243	5,542,266	817,040,382	612,362,811	3,281,608,191	3,920,688,530	518,031,287	4,438,719,817

Note: This information does not indicate whether (or to what degree) the public has been exposed to toxic chemicals. Therefore, no conclusions on the potential risks can be made based solely on this information (including any ranking information). For more detailed information on this subject refer to *The toxics release inventory* (*TRI*) and factors to consider when using TRI data document at www.epa.gov/tri/tridata.

Note: On-site disposal or other releases include underground injection to class I wells (Section 5.4.1), RCRA subtitle C landfills (5.5.1A), other landfills (5.5.1B), fugitive or non-point air emissions (5.1), stack or point source air emissions (5.2), surface water discharges (5.3), class II-V wells (5.4.2), land treatment (5.5.2), RCRA subtitle C surface impoundments (5.5.3A), other surface impoundments (5.5.3B) and other land disposal (5.5.4). Off-site disposal or other releases include from section 6.2 underground injection to Class I wells (M81), RCRA subtitle C landfills (M65), other landfills (M64, M72), storage only (M10), solidification/stabilization — metals and metal category compounds only (M41 or M40), wastewater treatment (excluding POTWs) — metals and matal category compounds only (M62 or M61), RCRA subtitle C surface impoundments (M66), other surface impoundments (M67, M63), land treatment (M73), other land disposal (M79), underground injection to Class II-V wells (M82, M71), other off-site management (M90), transfers to waste broker — disposal (M94, M91), and unknown (M99) and, from section 6.1, transfers to POTWs (metals and metal category compounds only).

Does not include off-site disposal or other releases transferred to other TRI facilities that reported the amounts as on-site disposal or other releases.

FIGURE 1.44B Toxic Release Inventory. 2003. On-site and Off-site Disposal or Other Releases, by State. EPA 260—R-05-001 May 2005.

(C) **TRI On-Site and Off-Site Disposal or Other Releases, by State, 2003: Mercury and Mercury Compounds**

	On-site disposal or other releases															
	On-Site Disposal to Class I Underground Injection Wells, RCRA Subtitle C Landfills, and Other Landfills				Other on-site disposal or other releases											
State	Class I Wells Pounds	RCRA Subtitle C Landfills Pounds	Other On-site Landfills Pounds	Subtotal Pounds	Fugitive Air Emissions Pounds	Point Source Air Emissions Pounds	Surface Water Discharges Pounds	Class II-V Wells Pounds	Land Treatment Pounds	RCRA Subtitle C Surface Impoundments Pounds	Other Surface Impoundments Pounds	Other Land Disposal Pounds	Subtotal Pounds	Total On-site Disposal or Other Releases Pounds	Total Off-site Disposal or Other Releases Pounds	Total On-site and Off-site Disposal or Other Releases Pounds
Alabama	0.00	95,000.00	376.61	95,376.61	1,075.58	5.355.71	58.68	5.00	1.70	0.00	1,254.70	194.10	7,945.47	103,322.08	4,359.37	107,681.45
Alaska	0.00	0.00	5.90	5.90	11.30	199.30	0.40	6,246.00	0.00	0.00	63,668.19	36,981.37	107,106.56	107,112.46	4.42	107,116.88
Arizona	0.00	0.00	392.00	392.00	43.46	1,823.78	0.00	0.00	0.00	0.00	3,347.43	2,782.44	7,997.11	8,389.11	1,000.08	9,389.19
Arkansas	0.00	0.00	68.27	68.27	0.41	1,334.76	4.79	0.00	2.20	1.00	10.60	0.27	1,354.02	1,422.29	87.06	1,509.35
California	0.00	130,252.40	585.80	130,838.20	29.96	4,676.63	2.71	0.16	0.00	0.80	0.00	29.30	4,739.57	135,577.77	24,691.35	160,269.12
Colorado	0.00	0.00	654.13	654.13	36.70	756.71	1.08	0.00	0.00	0.00	9.10	1,781.80	2,585.39	3,239.52	468.60	3,708.12
Connecticut	0.00	0.00	0.00	0.00	6.62	110.87	0.32	0.00	0.00	0.00	0.00	0.00	117.81	117.81	5,148.52	5,266.33
Delaware	0.00	0.00	374.90	374.90	747.00	338.80	16.31	0.00	0.00	0.00	32.00	0.00	1,134.11	1,509.01	2,174.33	3,683.34
District of Columbia	0.00	0.00	0.00	0.00	0.00	0.50	0.00	0.00	0.00	0.00	0.00	0.00	0.50	0.50	0.00	0.50
Florida	0.00	0.00	2,958.70	2,958.70	2.11	3,269.94	15.14	0.00	0.10	0.00	1,035.34	161.04	4,483.67	7,442.37	281.48	7,723.85
Georgia	0.00	0.00	2.59	2.59	563.33	3,389.54	36.23	0.03	0.74	0.00	1,150.18	0.01	5,140.07	5,142.66	42.81	5,185.47
Hawaii	0.00	0.00	0.00	0.00	0.00	367.18	3.75	2.60	0.00	0.00	0.00	0.00	373.53	373.53	38.12	411.65
Idaho	0.00	322.00	614.00	936.00	0.01	620.42	0.00	0.00	0.00	0.00	323.00	429.40	1,372.83	2,308.83	0.00	2,308.83
Illinois	0.00	23.00	1,786.27	1,809.27	16.53	7,005.99	18.92	0.00	56.95	0.00	514.60	110.00	7,723.00	9,532.27	84.990.87	94,523.14
Indiana	0.54	813.00	1,208.66	2,022.20	30.01	9,384.45	194.40	0.00	159.22	0.00	831.16	112.05	10,711.27	12,733.47	2,142.01	14,875.48
Iowa	0.00	0.00	56.60	56.60	0.16	3,216.14	6.15	0.00	0.00	0.00	82.00	4.60	3,309.05	3,365.65	806.60	4,172.25
Kansas	0.00	0.00	92.70	92.70	1.04	2,817.00	1.54	0.00	0.00	0.00	216.10	0.00	3,034.68	3,127.38	195.95	3,323.33
Kentucky	0.00	0.00	1,108.46	1,108.46	39.90	3,597.85	71.75	4.04	8.60	9.80	1,150.40	581.00	5,463.34	6,571.80	291.23	6,863.03
Louisiana	0.00	11,000.00	75.23	11,075.23	1,908.08	2,613.46	63.53	0.00	0.10	0.00	2,121.52	0.20	6,706.89	17.782.12	9,339.83	27,121.95
Maine	0.00	0.00	31.30	31.30	0.00	90.11	1.53	0.00	0.00	0.00	0.00	0.00	91.64	122.94	12.23	135.17
Maryland	0.00	0.00	18.50	18.50	1.24	1,863.60	0.22	0.11	511.00	0.00	64.00	0.10	2,440.27	2,458.77	491.84	2,950.61
Massachusetts	0.00	0.00	0.00	0.00	0.00	209.60	0.37	0.00	0.00	0.00	4.30	0.00	214.27	214.27	3,009.73	3,224.00
Michigan	0.00	0.00	424.84	424.84	3.38	2,876.70	25.38	0.00	0.00	0.00	471.90	88.60	3,465.96	3,890.80	460.84	4,351.64
Minnesota	0.00	0.51	786.20	786.71	40.03	1,846.34	0.32	0.00	0.00	0.00	137.70	1.40	2,025.79	2,812.50	457.46	3,269.96
Mississippi	1,155.30	4.60	224.60	1,384.50	0.15	872.37	128.15	0.00	0.00	0.00	5,471.70	0.00	6,472.36	7,856.86	68.45	7,925.31
Missouri	0.00	2.49	165.30	167.79	1.35	3,953.87	5.70	0.00	0.00	0.00	315.63	122.34	4,398.89	4,566.68	2,819.75	7,386.43
Montana	0.00	0.00	704.80	704.80	1.16	1,061.83	0.51	5.10	4.30	0.00	9.70	449.00	1,531.59	2,236.39	46.20	2,282.59
Nebraska	0.00	0.00	1,003.86	1,003.86	7.17	560.28	0.01	0.00	0.00	0.00	0.00	5.90	573.36	1,577.22	4,588.88	6,166.10
Nevada	0.00	4,089.00	456.10	4,545.10	162.65	4,645.08	9.10	1.40	0.00	0.00	1,370,398.60	5,234,558.40	6,609,775.23	6,614,320.33	0.06	6,614,320.39
New Hampshire	0.00	0.00	0.00	0.00	0.00	136.90	0.00	0.00	0.00	0.00	5.00	0.00	141.91	141.91	41.28	183.19
New Jersey	0.00	6.00	0.00	6.00	1.12	1,314.59	9.60	0.00	0.00	0.00	0.00	39.00	1,364.31	1,370.31	1,159.61	2,529.92
New Mexico	0.95	0.00	918.60	919.55	2.74	1,352.19	1.30	0.00	0.00	0.00	93.29	1,079.50	2,529.02	3,448.57	51.29	3,499.86
New York	0.00	801.90	796.00	1,597.90	0.14	1,785.90	144.73	0.00	0.00	0.00	0.00	0.70	1,931.47	3,529.37	931.36	4,460.73
North Carolina	0.00	0.00	777.30	777.30	248.01	3,875.67	23.96	0.00	0.00	0.00	768.90	10.00	4,926.54	5,703.84	546.55	6,250.39
North Dakota	0.00	0.00	103.40	103.40	0.10	2,546.13	0.30	0.00	0.00	0.00	132.10	0.00	2,678.63	2,782.03	417.42	3,199.45
Ohio	55.00	0.00	1,870.06	1,925.06	1,218.94	9,180.27	24.05	0.00	50.10	32.00	1,753.77	2.15	12,261.28	14,186.34	7,816.22	22,002.55
Oklahoma	0.00	8,805.00	72.30	8,877.30	7.54	1,651.41	4.79	0.00	5.00	0.00	44.91	3.90	1,717.52	10,594.82	616.52	11,211.34
Oregon	0.00	1,340.20	25.45	1,365.65	10.78	793.06	0.56	0.00	0.32	0.00	0.00	0.00	804.72	2,170.37	143.57	2,313.94
Pennsylvania	0.00	0.10	4,900.41	4,900.51	30.55	10,345.20	72.09	0.00	10.70	22.20	173.00	1,471.70	12,125.43	17,025.95	9,577.03	26.602.98
Puerto Rico	0.00	0.00	0.00	0.00	0.10	455.72	9.28	0.00	0.00	0.00	0.00	0.00	465.11	465.11	36.97	502.07
Rhode Island	0.00	0.00	0.00	0.00	0.00	0.00	0.00	0.00	0.00	0.00	0.00	0.00	0.00	0.00	2.50	2.50
South Carolina	0.00	0.00	261.28	261.28	109.84	1,943.19	1,563.49	0.00	0.00	0.00	842.30	0.29	4,459.10	4,720.38	177.34	4,897.72
South Dakota	0.00	0.00	597.47	597.47	0.13	230.75	0.00	0.00	0.00	0.00	0.10	0.00	230.98	828.45	1.00	829.45
Tennessee	0.00	183.53	1,324.11	1,507.64	1,048.04	2,550.12	78.21	0.00	0.00	0.00	552.90	58.91	4,288.18	5,795.82	11,140.40	16,936.21
Texas	111.00	1,611,00	3,924.17	5,646.17	48.90	13,448.60	55.08	0.00	1.00	0.00	2,709.61	154.00	16,417.19	22,063.36	5,574.58	27,637.94
Utah	0.00	3,948.60	1,230.62	5,179.22	27.02	1,015.49	27.90	0.00	12.30	0.00	1,717.30	12,117.01	14,917.02	20,096.24	3,712.59	23,808.83
Vermont	0.00	0.00	0.00	0.00	0.00	0.00	0.00	0.00	0.00	0.00	0.00	0.00	0.00	0.00	0.00	0.00
Virgin Islands	0.00	0.00	0.00	0.00	0.00	202.00	0.00	0.00	2.00	0.00	3.00	0.00	207.00	207.00	70.00	277.00
Virginia	0.00	0.00	847.90	847.90	0.71	2,298.26	29.17	0.00	0.00	0.00	351.60	25.83	2,705.57	3,553.47	718.27	4,271.74
Washington	0.00	0.00	0.30	0.30	0.10	320.18	8.83	0.00	65.54	0.00	215.00	32.00	641.66	641.95	1,128.10	1,770.05
West Virginia	0.00	9.00	3,211.33	3,220.33	1,048.37	4,295.15	24.20	0.55	295.30	0.00	1,575.83	15.39	7,254.79	10,475.12	1,219.42	11,694.54
Wisconsin	0.00	0.00	72.26	72.26	1,055.77	2,705.01	4.56	0.00	1.00	0.00	0.03	25.81	3,792.20	3,864.46	2,962.35	6,826.80
Wyoming	0.00	0.00	438.44	438.44	0.85	1,914.31	0.00	0.00	0.10	0.00	140.25	79.77	2,135.29	2,573.73	68.11	2,641.84
Total	1,322.79	258,212.33	35,547.72	295,082.85	9,589.05	133,218.93	2,748.10	6,264.99	1,188.27	65.80	1,463,698.74	5,293,509.28	6,910,283.16	7,205,366.00	196,130.52	7,401,496.53

Note: This information does not indicate whether (or to what degree) the public has been exposed to toxic chemicals. Therefore, no conclusions on the potential risks can be made based solely on this information. (including any ranking information). For more detailed information on this subject refer to *The toxics release inventory* (*TRI*) and factors to consider when using TRI data document at www.epa.gov/tri/tridata.

Note: On-site disposal or other releases include underground injection to class I wells (Section 5.4.1), RCRA subtitle C landfills (5.5.1A), other landfills (5.5.1B), fugitive or non-point air emissions (5.1), stack or point source air emissions (5.2), surface water discharges (5.3), Class II-V wells (5.4.2), land treatment (5.5.2), RCRA subtitle C surface impoundments (5.5.3A), other surface impoundments (5.5.3B) and other land disposal (5.5.4). Off-site disposal or other releases include from section 6.2 underground injection to Class I wells (M81), RCRA subtitle C landfills (M65), other landfills (M64, M72), storage only (M10), solidification/stabilization—metals and metal category compounds only (M41 or M40), wastewater treatment (excluding POTWs)—metals and matal category compounds only (M62 or M61), RCRA subtitle C surface impoundments (M66), other surface impoundments (M67, M63), land treatment (M73), other land disposal (M79), underground injection to Class II-V wells (M82, M71), other off-site management (M90), transfers to waste broker—disposal (M94, M91), and unknown (M99) and, from section 6.1, transfers to POTWs (metals and metal category compounds only).

Does not include off-site disposal or other releases transferred to other TRI facilities that reported the amounts as on-site disposal or other releases.

FIGURE 1.44C Toxic Release Inventory. 2003. On-site and Off-site Disposal or Other Releases, by State: Mercury and Mercury Compounds. EPA 260—R-05-001 May 2005.

Tri On-Site and Off-Site Disposal or Other Releases, by State, 2003: Lead and Lead Compounds

	On-site disposal or other releases															
	On-Site Disposal to Class I Underground Injection Wells, RCRA Subtitle C Landfills, and Other Landfills				Other on-site disposal or other releases											
State	Class I Wells Pounds	RCRA Subtitle C Landfills Pounds	Other On-Site Landfills Pounds	Subtotal Pounds	Fugitive Air Emissions Pounds	Point Source Air Emissions Pounds	Surface Water Discharges Pounds	Class II-V Wells Pounds	Land Treatment Pounds	RCRA Subtitle C Surface Impoundments Pounds	Other Surface Impoundments Pounds	Other Land Disposal Pounds	Subtotal Pounds	Total On-Site Disposal or Other Releases Pounds	Total Off-Site Disposal or Other Releases Pounds	Total On-Site and Off-Site Disposal or Other Releases Pounds
Alabama	0.00	661,000.00	2,282,570.38	2,943,570.38	19,245.84	29,468.43	6,984.22	2,900.00	8,263.52	0.00	286,368.80	28,121.48	381,352.29	3,324,922.67	580,714.96	3,905,637.63
Alaska	0.00	0.00	304.00	304.00	8,470.38	3,331.35	126.60	7,245,519.00	0.00	0.00	90,781,017.00	81,682,033.70	179,720,498.03	179,720,802.03	3,751.48	179,724,553.51
Arizona	0.00	1.00	91,536.10	91,537.10	16,062.76	14,296.84	11.01	0.00	0.00	0.00	2,045,244.37	2,588,945.16	4,664,560.14	4,756,097.24	88,240.54	4,844,337.78
Arkansas	0.00	36,006.80	44,501.83	80,508.63	1,888.14	8,393.03	1,757.74	0.00	8,899.05	0.00	18,900.00	5,721.02	45,558.98	126,067.61	285,135.35	411,202.96
California	0.00	6,222,046.00	134,546.25	6,356,592.72	4,472.87	13,013.40	1,091.09	4.50	484.58	31.70	27.70	156,058.91	175,184.75	6,531,777.47	1,411,119.97	7,942,897.44
Colorado	0.00	96.80	100,002.70	100,099.50	1,376.98	4,426.46	44.55	0.00	0.00	0.00	2,131,970.07	3,201,407.40	5,339,225.46	5,439,324.96	135,609.22	5,574,934.18
Connecticut	0.00	0.00	393.00	393.00	968.29	1,812.06	235.05	0.00	0.00	0.00	3.97	5.20	3,024.27	3,417.27	206,365.62	209,782.90
Delaware	0.00	0.00	18,808.00	18,808.00	260.61	3,460.25	279.40	0.00	0.00	0.00	2,831.00	48.00	6,879.26	25,687.26	63,827.05	89,514.31
District of Columbia	0.00	0.00	0.00	0.00	0.01	0.00	0.00	0.00	0.00	0.00	0.00	0.00	0.01	0.01	290.01	290.01
Florida	0.00	8.89	257,909.70	257,918.59	5,596.45	19,473.28	1,549.09	0.00	720.65	2,853.44	176,203.43	42,253.49	248,649.82	506,568.41	137,538.45	644,106.87
Gerorgia	0.00	1.00	16,353.27	16,354.27	6,941.05	17.935.44	1,891.67	0.26	630.60	0.00	327,148.05	335,652.23	690,199.29	706,553.56	86,871.69	793,425.26
Guam	0.00	0.00	0.00	0.00	2.60	178.92	1.00	0.00	0.00	0.00	6.40	3.40	192.32	192.32	0.00	192.32
Hawaii	0.00	0.00	0.00	0.00	770.20	4,325.84	21.24	3.50	0.00	0.00	0.00	95,737.81	100,858.59	100,858.59	6,604.50	107,463.09
Idaho	0.00	4,410,017.00	29,098.61	4,439,115.61	211.49	1,430.70	780.39	0.00	18.04	34,037.00	971,022.82	901,158.00	1,908,658.44	6,347,774.05	6,313.38	6,354,087.43
Illinois	0.00	1,259,641.60	70,197.59	1,329,839.19	14,657.04	30,536.75	7,881.00	0.00	4,342.33	0.00	235,346.99	6,015.70	298,779.80	1,628,619.00	1,191,914.44	2,820,533.43
Indiana	1,600.00	767,572.73	544,498.53	1,313,671.26	19,171.14	138,943.99	5,478.02	0.00	28,958.30	0.00	468,418.41	79,901.45	740,871.31	2,054,542.57	5,005,206.94	7,059,749.51
Iowa	0.00	0.00	18,077.70	18,077.70	2,816.31	21,960.35	1,700.07	0.00	0.00	0.00	30,661.68	3,847.07	60,985.49	79,063.19	447,104.89	526,168.07
Kansas	26.00	0.00	66,132.30	66,158.30	1,040.91	7,218.44	68.45	0.00	0.00	0.00	22,943.41	31,789.00	63,060.21	129,218.51	84,546.36	213,764.87
Kentucky	0.00	0.00	273,646.99	273,646.99	10,942.52	25,728.66	2,763.48	2,540.30	2,690.50	1,885.00	344,110.20	391,661.00	782,321.66	1,055,968.65	425,283.87	1,481,252.52
Louisiana	0.00	510,000.00	170,530.44	680,530.44	2,700.94	13,508.44	38,389.54	0.00	1,228.27	984.00	363,654.30	28,317.90	448,783.39	1,129,313.83	174,146.88	1,303,460.71
Maine	0.00	0.00	10,493.85	10,493.85	64.51	1,340.80	1,048.85	0.00	0.00	0.00	0.00	1,242.33	3,696.48	14,190.33	10,134.64	24,324.97
Maryland	0.00	0.00	82,338.70	82,338.70	630.40	4,463.50	192.38	9.00	2,285.80	0.00	1,200.00	13.482.22	22,263.30	104,602.00	93,247.96	197,849.96
Massachusetts	0.00	0.00	1,680.83	1,680.83	1,158.18	3,335.48	26.62	0.00	39.00	0.00	2,714.00	1,527.33	8,798.60	10,479.43	250,684.57	261,164.00
Michigan	1.00	213,934.80	132,028.04	345,963.84	6,982.75	41,764.04	4,034.05	0.00	268.30	0.00	87,572.90	16,088.00	156,710.04	502,673.88	502,473.82	1,005,147.71
Minnesota	0.00	0.00	164,325.30	164,325.30	1,470.86	11,206.91	987.91	0.00	982.00	0.00	38,951.80	347.23	53,946.71	218,272.01	498,856.76	171,128.76
Mississippi	253,323.50	60.00	44,667.53	298,051.03	508.39	10,121.67	2,050.48	0.00	1,650.00	0.00	15,425.66	11,758.60	41,514.79	339,565.82	212,054.01	551,619.83
Missouri	0.00	60,221.00	63,348.50	123,569.50	150,897.87	80,129.94	4,282.54	0.00	160.00	125.00	13,304,189.62	7,273,616.41	20,813,401.37	20,936,970.87	4,528,837.08	25,465,807.95
Montana	0.00	81.46	12,040,219.20	12,040,300.66	3,378.92	1,751.60	395.40	3,132.58	190.00	0.00	2,545,879.21	1,466,771.26	4,021,498.97	16,061,799.63	12,263.77	16,074,063.40
Nebraska	0.00	0.01	66,256.00	66,256.01	2,361.19	8,114.06	112.52	0.00	0.00	0.00	0.25	4,546.08	15,134.10	81,390.11	165,201.78	246,591.89
Nevada	0.00	7,937,695.30	483,858.40	8,421,553.70	2,198.71	2,597.03	66.00	1.40	0.00	0.00	4,230,134.60	29,125,782.68	33,360,780.42	41,782,334.12	109,917.82	41,892,251.94
New Hampshire	0.00	0.00	0.00	0.00	57.42	616.84	104.10	0.00	0.00	0.00	500.00	165.19	1,443.55	1,443.55	87,346.07	88,789.62
New Jersey	0.00	2,148.00	681.00	2,829.00	568.53	6,498.53	451.94	0.00	232.79	0.00	0.00	46,691.20	54,442.98	57,271.98	701,196.08	758,468.07
New Mexico	0.00	0.00	128,669.00	128,669.00	428.76	878.20	123.21	0.00	0.00	0.00	54,147.31	665,691.07	721,268.55	849,937.55	15,253.87	865,191.43
New York	0.00	539,031.47	49,061.26	588,092.73	1,635.42	17,669.55	13,243.87	0.00	22.80	0.00	0.00	60,570.80	93,142.44	681,235.17	299,072.67	980,307.85
North Carolina	0.00	0.00	106,546.84	106,546.84	4,375.50	15,515.58	2,754.02	0.00	882.73	0.00	204,651.70	268,338.66	496,518.19	603,065.03	374,436.09	977,501.12
North Dakota	0.00	0.00	31,304.80	31,304.80	18.31	7,988.68	521.19	0.00	0.00	0.00	41,867.60	471.90	50,867.78	82,172.58	61,820.93	143,993.51
Northern Marianas	0.00	0.00	0.00	0.00	1.10	0.50	0.00	0.00	0.00	0.00	1.65	0.00	3.25	3.25	0.00	3.25
Ohio	0.00	2,091,817.50	704,239.12	2,796,056.62	21,176.66	64,198.23	6,938.12	0.00	7,459.00	7,427.00	295,056.15	3,325.00	405,580.16	3,201,636.78	2,882,931.78	6,084,568.56
Oklahoma	357.00	142,049.00	33,687.43	176,093.43	1,618.85	17,037.55	454.48	0.00	677.50	0.00	2,897.82	40,150.70	62,836.89	238,930.32	122,494.16	361,424.49
Oregon	0.00	1,319,725.60	36,565.35	1,356,290.95	14,116.21	3,516.20	2,981.39	0.00	952.27	0.00	493.00	90.00	22,149.07	1,378,440.02	52,552.35	1,430,992.37
Pennsylvania	0.00	10,662.90	313,964.80	324,627.95	23,328.09	66,941.36	5,327.95	0.00	6.00	1,285,000.00	11,040.50	27,355.20	1,418,999,09	1,743,626.79	3,376,344.08	5,119,970.87
Puerto Rico	0.00	0.00	0.00	0.00	31.00	3,270.26	955.90	0.00	0.00	0.00	0.00	2,822.62	7,079.78	7,079.78	10,973.80	18,053.58
Rhode Island	0.00	0.00	0.00	0.00	36.84	40.38	14.92	0.00	0.00	0.00	0.00	0.00	92.14	92.14	16,060.81	16,152.95
South Carolina	0.00	0.00	48,105.10	48,105.10	3,534.99	13,484.17	2,578.66	0.00	27.30	0.00	81,105.40	154,072.31	254,802.83	302,907.93	158,054.34	460,962.27
South Dakota	0.00	0.00	3,299,176.38	3,299,176.38	2,629.87	134.39	673.12	0.00	0.14	0.00	163.00	5,100.00	8,700.52	3,307,876.90	1,982.77	3,309,859.67
Tennessee	0.00	45,302.00	399,984.59	445,286.59	2,961.91	17,426.21	4,608.85	0.00	2,509.70	0.00	6,057,203.00	20,771.81	6,105,481.48	6,550,768.06	310,761.62	6,861,529.68
Texas	65,062.00	304,402.85	374,852.09	744,316.94	8,712.14	57,906.98	4,388.78	0.00	3,084.90	2.00	764,083.17	195,509.18	1,033,687.16	1,778,044.10	1,110,197.94	2,888,202,03
Utah	0.00	2,966,829.34	335,502.78	3,302,332.12	2,939.52	10,644.97	280.02	0.00	1,000.00	0.00	5,918,865.74	78,006,804.34	83,940,534.59	87,242,866.71	303,335.74	87,546,202.45
Vermont	0.00	0.00	4.46	4.46	12.27	133.55	9.74	0.00	0.00	0.00	5.70	0.00	161.26	165.72	9,816.76	9,982.48
Virgin Islands	0.00	0.00	0.00	0.00	0.00	646.00	0.00	0.00	300.00	0.00	61.00	0.00	1,007.00	1,007.00	187.00	1,194.00
Virginia	0.00	0.00	154,146.60	154,146.60	5,891.49	21,744.93	1,811.88	0.00	3,036.62	0.00	62,880.50	121,290.87	216,656.29	370,802.89	450,167.12	820,970.01
Washington	0.00	633,001.37	7,358.84	640,360.21	720.60	4,119.99	2,572.32	0.00	25.03	0.00	153,603.00	111,329.94	272,370.88	912,731.09	90,676.41	1,003,407.50
West Virginia	0.00	178.00	416,841.43	417,019.43	523.10	6,415.65	2,885.00	86.88	11,824.00	0.00	240,302.48	14,908.02	276,945.13	693,964.56	298,583.69	992,548.25
Wisconsin	0.00	927.10	32,758.40	33,685.50	4,572.00	14,573.62	1,219.35	0.00	0.00	0.00	4.79	36,428.71	56,798.47	90,483.96	540,141.91	630,625.87
Wyoming	0.00	0.00	101,470.36	101,470.36	2,139.38	5,414.25	17.01	0.00	19.60	1,412.00	34,266.50	6,025.30	49,294.04	150,764.40	19,462.56	170,226.96
Total	320,369.50	30,134,460.01	23,783,244.37	54,238,073.88	389,276.27	881,084,19	139,166.18	7,254,197.42	93,871.32	1,333,757.14	132,355,146.43	207,281,752.88	349,728,251.81	403,966,325.69	28,018,108.38	431,984,434.07

Note: This information does not indicate whether (or to what degree) the public has been exposed to toxic chemicals. Therefore, no conclusions on the potential risks can be made based solely on this information (including any ranking information). For more detailed information on this subject refer to *The toxics release inventory* (*TRI*) and factors to consider when using TRI data document at www.epa.gov/tri/tridata.

Note: On-site disposal or other releases include underground injection to class I wells (Section 5.4.1), RCRA subtitle C landfills (5.5.1A), other landfills (5.5.1B), fugitive or non-point air emissions (5.1), stack or point source air emissions (5.2), surface water discharges (5.3), Class II-V wells (5.4.2), land treatment (5.5.2), RCRA subtitle C surface impoundments (5.5.3A), other surface impoundments (5.5.3B) and other land disposal (5.5.4). Off-site disposal or other releases include from section 6.2 underground injection to Class I wells (M81), RCRA subtitle C landfills (M65), other landfills (M64, M72), storage only (M10), solidification/stabilization—metals and metal category compounds only (M41 or M40), wastewater treatment (excluding POTWs)—metals and metal category compounds only (M62 or M61), RCRA subtitle C surface impoundments (M66), other surface impoundments (M67, M63), land treatment (M73), other land disposal (M79), underground injection to Class II-V wells (M82, M71), other off-site management (M90), transfers to waste broker—disposal (M94, M91), and unknown (M99) and, from section 6.1, transfers to POTWs (metals and metal category compounds only).

Does not include off-site disposal or other releases transferred to other TRI facilities that reported the amounts as on-site disposal or other releases.

FIGURE 1.44D Toxic Release Inventory. 2003. On-site and Off-site Disposal or Other Releases, by State: Lead and Lead Compounds. EPA 260—R-05-001 May 2005.

parasites but also by the more than 80,000 manmade organic and inorganic chemicals as well as physical phenomena of heat, cold, geomagnetic storms, etc. Also, manmade electromagnetic radiation emanating from cell phones, power lines, television, computers, etc. are present in ever-increasing numbers.

Environmental science and monitoring devices are now just becoming practical for certain facets of analysis of environmental pollutions. See Figures 1.45 A through G. The EPA has developed a series of monitoring stations across the U.S., which gives the clinician the ability on a daily basis to evaluate the patients and the environments for nitric, nitrous, nitrogen dioxide, carbon monoxide, sulfur dioxide, ozone, methane gas, and particulate matter (< 2.5μ). See Figure 1.45A.

Unfortunately data for the other heavy metals like Hg and Cd, etc., are available only over monthly or yearly periods. The same is true for volatile organic pollutants. There is a website for organics, www.scorecard.org, which gives estimates for organic air emitters and the companies with the highest emissions. See Figure 1.45B.

There are also maps and data showing where herbicides and pesticides are used. See Figure 1.45C.

An example of the specific and nonspecific total environmental load is the partial analysis of outdoor air as part of the total environmental load as shown in Figure 1.45A. One can see an example of total environmental load when the kinetics might be overwhelmingly dominant for one pollutant (e.g., ozone). However, the combination of analysis of Dallas, Texas or any city air for pollutants may take on an entirely different kinetic picture due to the combined toxicities of the six pollutants. In addition to these six pollutants, there are thousands of other pollutants that may make the kinetics impossible to find.

Part of the total outdoor environmental load also encompasses organic pollutants and particulate matter. An example of the ambient air of a city like Los Angeles versus a clean area like Maverick County, Texas, is shown in Figure 1.45B. **One can see that those who live in a clean area have a massive reduction in environmental and thus, total body pollutant load of organic chemicals their bodies have to contend with and to process than do those individuals who live in polluted areas**.

The total environmental pollutant load is further enhanced by the massive use of pesticide and herbicides as found in the state of Florida, Oregon, and the farming states of the Midwest. Figure 1.45C. Molds, bacteria, viruses, terpenes, and methane gas make up the majority of the earth's natural pollutants contributing to the total environmental load.

At best we now have the partial analysis of what consists of outdoor air pollution and how it is a part of the total environmental load. See Figure 1.45D.

Indoor air pollution also makes up part of the total environmental load. There are many toxic substances in indoor air including pesticides, natural gas, formaldehyde, etc. See Figure 1.45E.

Here, more complicated kinetics will occur because of the varied nature of indoor pollutants combined with outdoor pollutants. Remember that 50% of outdoor pollutants make up indoor pollutants. An example of this would be the entry of ozone from outdoors, which is then combined with indoor terpenes from pine wood (d-limonene and pine terpene) to form more toxic substances indoors. However, natural gas, oil, pesticides, formaldehyde, and solvents make up the majority of home pollution.

Water pollutants make up a significant part of the total environmental load since people bathe and drink polluted water. See Figures 1.45F1 and 1.45F2.

Food pollutants are a significant part of the total environmental load. See Figure 45G.

The entry of water pollutants (through drinking and skin absorption) and food pollutants occurs as well, which makes the combined kinetics of entering pollutants even more difficult to discern, especially without the use of environmental control units and principles. A clinical example of this kinetic variation would be when a person goes outside on a day when carbon monoxide levels are high; he develops shortness of breath and a cough as a result of the pollutant exposure. He goes inside, which does relieve his symptoms, but then he develops headache and fatigue due to the combined kinetics of the entering outdoor pollutants of carbon monoxide and ozone plus the terpenes from the indoor construction (d-limonene and pine terpenes). He subsequently, becomes ravenously hungry due to the combined pollutant exposure, so he drinks polluted water and eats polluted (nonorganic) food.

The Physiologic Basis of Homeostasis 91

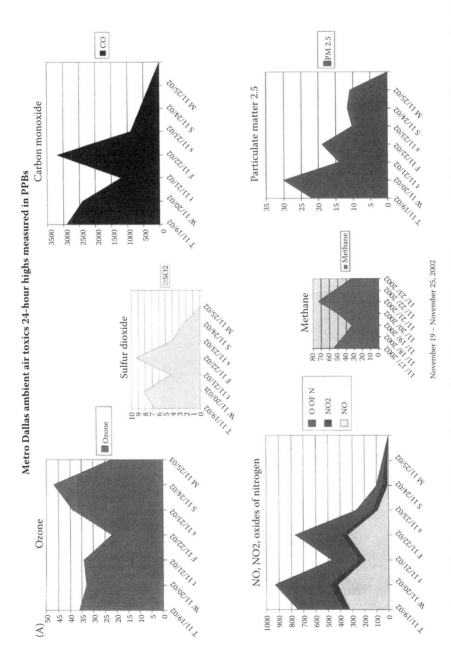

FIGURE 1.45A (See color insert following page 364.) Metro Dallas ambient air toxics 24-hour highs measured in PPBs. (From EHC-Dallas and local EPA.)

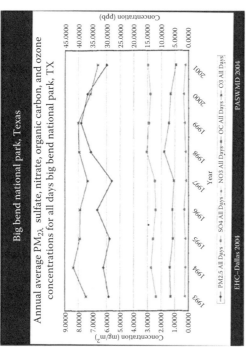

FIGURE 1.45B Three areas in the U.S. showing environmental pollutants. Los Angeles is one of the most polluted areas in the country. Sedona, Arizona, is free of hydrocarbons but has a large amount of metals present in the air. Big Bend, Texas, is one of the least-polluted areas in the United States. (From www.scorecard.org. EHC-Dallas, 2004.)

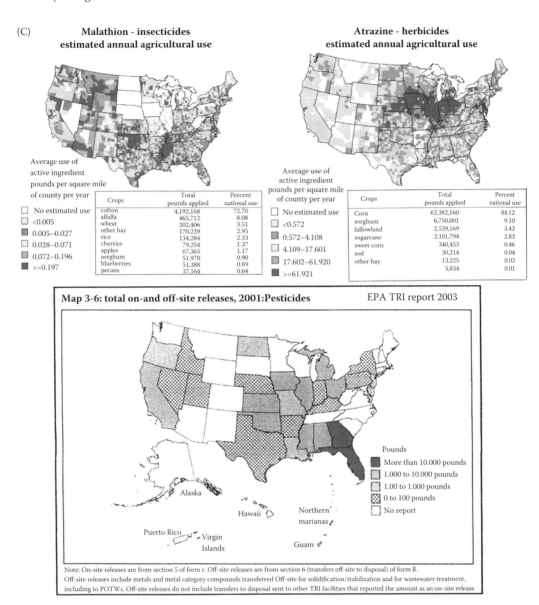

FIGURE 1.45C (See color insert following page 364.) Pesticide and herbicide U.S. maps. (From EPA TRE Report, 2003.)

Due to the combined kinetics of the four areas of exposure (indoor and outdoor air, food, and water contaminants) he then develops abdominal pain in addition to his fatigue and headache.

The weight of all of this environmentally generated pollutant load once it has entered the body, will place a stress on the dynamics of the body's homeostatic mechanism, at times modifying function, and at other times just increasing the total nonspecific total body pollutant load. This excess load will prime the body for eventual dysfunction of homeostasis. As the clinician evaluates and decreases the total body load of the patient he/she has a greater advantage to obtain a status of good health.

Not only is the combined cumulative dose of all these pollutants critical but also the hormetic properties of noxious stimuli are extremely important. These effects on homeostasis depend upon the multiplicity and chronicity of exposure of the substances involved, the kinetics of the substance(s) (half-life, toxicity, virulence), the timing of substance(s) measurement in the body, and where the

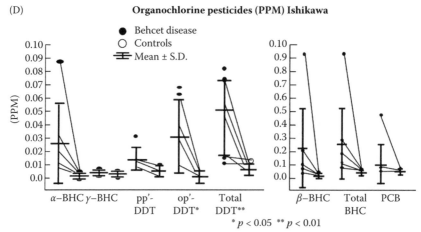

BHC: Benzene hexa chloride   PCB: Polychlorinated biphenyls
All measurements were carried out in 1976–77.
All organochlorine pesticides except Rindane (Chlordane) were banned in 1970.
Therefore, control value at this time is about 1/10 of the value at 1976.
All cases were fresh patients with Behcet disease, no previous treatment.

FIGURE 1.45D Sural nerves in Behcet's Disease—histopathological, electronmicroscopical, and X-ray microanalytical studies. (From Ishikawa, S. et al., Environmental pollutants in blood of Behcet's disease-organochlorine pesticide and PCB, Study on Etiology, Treatment and Prevention of Behcet's disease, Behcet's Disease Research Committee of Japan, Ministry of Welfare, Japan, 156–60, 1973.)

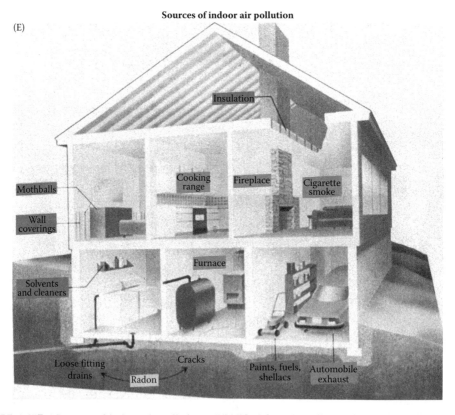

FIGURE 1.45E Sources of indoor air pollutions. (Modified from Rea, W.J., *Chemical Sensitivity, Vol. II. Sources of Total Body Load,* Lewis Publishers, Boca Raton, FL, 1994.)

The Physiologic Basis of Homeostasis

(F1)

Potential areas of water contamination

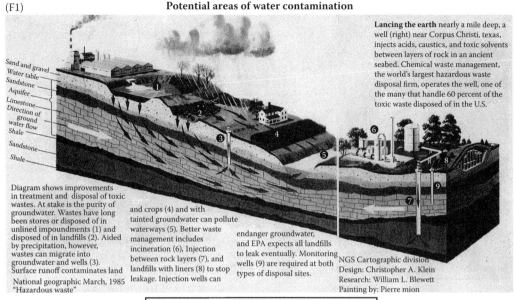

Chlorination of water

Water	PPM $ChCl_3$	PPM $ChCl_2Br$	PPM $ChClBr$	PPM $ChBr_3$
Spring	26	3	1	10
Well	5	0	0	7
City	29,000	100	12,600	5,700

Plastic components that leach into liquids
from the container vs. glass containers

This phthalate extraction study was performed to show the presence of certain phthalate compounds in plastic containers, tygon tubing, and our own phthalate kit glass bottle.

Five different samples were prepared:

1. Tall plastic bottle 100 ml WL-1. Used for drugs of abuse urine collection (1).
2. Short plastic bottle 60 ml graduated, sterile royaline - used for drugs of abuse urine collection (2).
3. Kit plastic bag: speci-grad SPA - R/J - 94. Used to ship our drug and toxic kits. Pat# 4,932,791.
4. Glass bottle (phthalate kit) 30 ml amber glass/green lid. Used for phthalate, organophosphate and herbicides urine collection.
5. Tygon plastic tubing: formulation: R-3603, I.D. = 1/8, 16 GB2 O.D.= 1/4. 14-169- IE lot # LC007U. A 1 1/2 inch piece cut and placed in urine in a glass culture tube and shaken for about 3 hrs.

Contamination by containers

Blank urine was added to samples 1–4 and shaken gently for a period of a period of approximately 18 hr. The samples were then prepared as written in the toxics' SOP. A2 L aliquot was analyzed by GS/MSD #4 in SIM mode. The results are shown below.

Phthalate compound	1. Tall plastic bottle	2. Short plastic bottle	3. Kit plastic bottle	4. Amber glass bottle	5. Tygon plastic tubing
Dimethylphalate	18.8	—	25.6	—	30.9
Diethylphthalate	270.8	—	45.1	—	10.6
Dibutylphthalate	—	74.8	53.5	—	29.3
Butylbenzylphthalate	—	—	—	—	—
Di-2-ethylhexylphthalate	—	—	—	—	162.7
Di - Octylphthalate	—	—	—	—	—
Analyst: MYS		All values in mg/ml (ppb)			
Source: Laseter, J.L. 1995. Personal communication					

FIGURE 1.45F1 (See color insert following page 364.) Pollutants of water. (*National Geographic*, "Hazardous Waste," March 1985; From EHC-Dallas, 2002; Laseter, J.L., Plastic versus Glass Containers. Personal communication.)

(F2)

Volatile organic chemical blood test

VOST compound	Mountain valley in glass	Mountain valley in plastic	Spring house in glass	Spring house in plastic
Benzene	0.01	0.03		0.04
Toluene	0.03	0.09		0.12
Ethylbenzene				
Trimethylbenzene		0.03		
Xylene				
Styrene				
Dichloromethane	0.33	0.05		0.92
Chloroform				0.19
Carbon tetrachloride	0.01	0.01	0.01	0.01
Bromoform			0.02	0.02
Bromodichloromethane		0.01		0.15
Dibromochloromethane				0.11
Trichloroethane				
Tetrachloroethane				
Trichloroethylene		0.17		0.14
Tetrachloroethylene		0.11		0.07
Chlorobenzene				
Dichlorobenzene				
Benzaldehyde				
Methylmercaptan				

Note: Results are reported in μg/L = ppb, and are based on GC/MS data.
Source: Laseter, J. L. 1985. Personal communication.

FIGURE 1.45F2 Volatile organic screening test (VOST) for selected EHC-Dallas spring water samples: Glass versus plastic containers. (Adapted from Laseter, J.L., AccuChem Lab, Personal communication, 1985.)

substance(s) are deposited (end-organ response). An example shown by Hatthakit,[274] is the administration of piperine intravenously. The mean arterial blood pressure is decreased by 10–30 mm Hg that is followed by an increase above normal blood pressure of 50–60 mm Hg with a gradient return to normal in a few minutes. No significant excretion of piperine is possible in this short time. However, sequestration by binding protein can and does occur on this time scale. The drug Fluoxetine does not cause these effects. Therefore, unlike the interaction of the body's neutralizing substances with Fluoxetine, overcompensation in response to piperine in the organism occurs in spite of the presence of the chemical agent (although not directly at the site of action). Most likely piperine triggers relaxation of the smooth muscle with resulting vasodilation to which the homeostatic overcompensation occurs by either a release of renin in the kidney, or sympathetic vasoconstriction, or both.[275] Clearly, the two mechanisms are coupled in homeostasis and as such related but the mechanisms are different with neither one related to the kinetics of the causative agent.

The contribution of Calebrese and Baldwin[276,277] was to undertake the enormous task of evaluating the literature for evidence of hormetic effects using a priori defined criteria. Without this ambiguous documentation of the widespread presence of hormetic properties, it would have been difficult to claim the general ability of hormetics as being rooted in the two fundamental biological principles of homeostasis and optimization of function.

An interaction of any exogenous agent, including nutrients, water, food, drugs, and other chemicals as well as heat, cold, and physical injury, will trigger a homeostatic response that aims at destroying dynamic equilibrium. Dynamic homeostasis is based on numerous feedback loops, it is wholly based on the simple principles of stimulation or direct versus indirect inhibition. Indirect stimulation is due to abolishing an inhibitory signal. Indirect inhibition is caused by termination of a stimulatory signal. Signals and their time courses are controlled by the kinetics of their signaling agents or by the dynamics of their effects, whichever represents their rate-determining process. No

The Physiologic Basis of Homeostasis

(G)

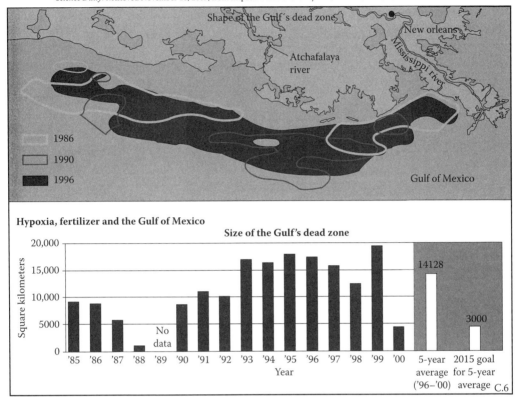

FIGURE 1.45G (See color insert following page 364.) Toxins in foods. (From Pangborn, J.B. and Smith, B., On Man and His Environment in Health and Disease, Presented at the 13th Annu. Int. Symp., Dallas, Feb. 23, 1995.)

matter how extensively the kinetics of the chemical is studied, its effect cannot be understood if the dynamic half-life is longer than its kinetic half-life and vice versa. For example, the half-life of the DNA adduct of 2-aceto-aminofluorene (2AAF) is about 67 days whereas its kinetic half-life is in the order of hours.278 Therefore, infrequent exposure every 14 days will result in accumulation of DNA damage to a steady state even though internal exposure to 2AAF will never be in a steady state (large dynamic AUC but small kinetic AUC-area under the plasma concentration time curve).

In the case of total environmental load, studying the process of injury allows us to understand some of the dynamics of the effect. However, recovery from injury is almost always slower than the process of injury itself and thus, one must understand the details of homeostasis. Hence, kinetics of the combined specific and total nonspecific environmental load will be what determines the dynamics of the effect as in the case of 2AAF. For example, studying the multitude of effects of dioxins on the homeostasis will not help us to understand the overall dynamic effects of dioxin exposure because each effect is driven by the long kinetic half-life of the compounds.

Both of these aspects, the quantity or weight of all substances (total body load combined) and the quality, i.e., virulence or toxicity of the specific incitant (specific body load as well as the total body pollutant load) must be evaluated in each patient who has distinct homeostatic disturbances.

(H)

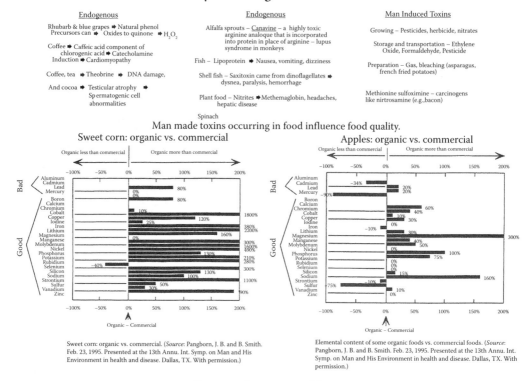

Sweet corn: organic vs. commercial. (*Source*: Pangborn, J. B. and B. Smith. Feb. 23, 1995. Presented at the 13th Annu. Int. Symp. on Man and His Environment in health and disease. Dallas, TX. With permission.)

Elemental content of some organic foods vs. commercial foods. (*Source*: Pangborn, J. B. and B. Smith. Feb. 23, 1995. Presented at the 13th Annu. Int. Symp. on Man and His Environment in health and disease. Dallas, TX. With permission.)

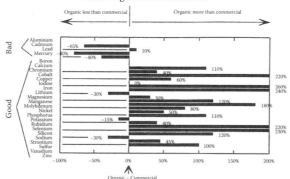

Pears: organic vs. commercial. (*Source*: Pangborn, J. B. and B. Smith. Feb. 23, 1995. Presented at the 13th Annu. Int. Symp. on Man and His Enviroment in health and disease. Dallas, TX. With permission.)

FIGURE 1.45H (See color insert following page 364.) Toxins in foods.

TOTAL BODY LOAD

The total body pollutant load is the sum of all pollutants in air, food, and water including other noxious stimuli that the body incorporates from the intake of air, food, and water.[279] These include bacteria, virus, fungus, food, chemicals, and physical phenomenon, e.g., heat, cold, EMF.[280] It is clear that if the nonspecific total body load is exceeded generally (by years of working in a refinery, living in the city, working in a toxic office building, etc.), or even specifically (as bacteria in a wound, an individual sprayed with pesticides, a victim of a chemical explosion), the homeostatic mechanism will be activated. If the total body load is in excess but not large enough to cause immediate death, energy and/or a nutrient demand and eventually, a drain of both will occur, which results in altered physiology and function. This drain is due to the demand for excess metabolic adjustment and excess function in order to neutralize the triggered effects. If the exposures are chronic this drain will cause either delayed repair of the damaged tissue or dyshomeostasis with secondary dyshomeostatic tissue changes and disinformation, which is then sent throughout the system. Thus, the easier triggering with each insult results in a downward spiral toward fixed-named autonomous disease, e.g., asthma, gastritis, thrombophlebitis, cardiac arrhythmias, until end-organ failure occurs.

When evaluating the potential effect of the nonspecific total environmental load upon homeostatic function, it is, to a point, irrelevant whether the noxious stimuli are chemical, mechanical, thermal, bacterial, viral, parasitic, etc. **When studying chronic body dysfunction, the quantitative aspect of the stimulus (the weight of the total body pollutant load) appears to be far more important in triggering reactions than is the quality**. However, both are important. An example of the effects of the nonspecific environmental load occurs when an individual may not be able to eat red meat, especially beef, when he is in the city. His nonspecific total body load is high from a concentrated load of pollutants. He may develop flulike symptoms, vertigo and joint pains from eating beef. Whereas, when he goes to the seashore where his total body pollutant load is reduced, due to better air with fewer pollutants, his symptoms and signs disappear and then, he has no reactions from eating beef. In this case, the total environmental load is very significant in maintaining health.

This same principle is seen in the patient who has recurrent infections. When the total environmental and body load is decreased, the infections disappear. When the load is increased, an infection is likely to start.

The quality or type of incitant and its kinetics and hormetic effects, however, is also important in that certain specific substances are more toxic or virulent than others thus creating an increase in a specific load effect (most of the field of medicine is dedicated to this fact) as well as the nonspecific total load effect. For example, the organochlorine pesticide, heptachor epoxide, is much more toxic than ethanol; although, both can trigger the dyshomeostatic response given the appropriate load and circumstances. Another example is the Staphylococcus aureus, which may trigger a more severe tissue reaction and disease process than the Staphylococcus epidermidis but again, under the right circumstances an excess load of the latter (though very unusual) can cause disease (e.g, hypersensitivity causing chronic recurrent sinusitis).

Local Receptors

Local Homeostatic Responses and Information Reception

The early dynamic response of homeostasis is initially carried out in an efficient manner locally through the body's receptors especially through the CTM and peripheral sensory and autonomic nerves, which perceive and telegraph the entry information through a vast intermeshed communication system of the body. The receptors exist initially in the skin and mucous membranes and in the CTM, which triggers multiple subsystems geared for a local response (i.e., CTM changes, cellular defense mechanisms, etc.) and enhance all attempts of the body's physiology to contain and continue to localize the noxious stimuli. If the stimuli are chronic and therefore, are usually multifactorial, the entry impulses may branch out of its local area of containment. When a noxious stimulus is not contained locally, the effects are followed by attempts at regional or segmental containment often limited by fascial boundaries and spinal reflexes. In some cases, the body's response is due to the excess stimuli where the defense physiology is mandated for survival to activate integrative adjustments and responses controlled by the central homeostatic mechanism. See Figure 1.46. However, local responses are paramount because this is the area where the noxious incitants enter. At this entry point the body makes massive attempts to neutralize the noxious stimuli for maintenance of energy efficient function.

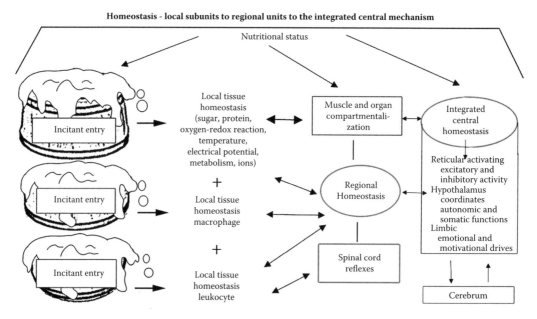

Receptors going to specific areas of the brain

FIGURE 1.46 (See color insert following page 364.) Receptors to the brain. Most receptors have connective tissue matrix as part of their anatomy. (From EHC-Dallas.)

FIGURE 1.47 The initiation of the homeostatic mechanism. (From EHC-Dallas.)

Other receptors besides those of the CTM and sensory and autonomic nerves are the epithelial receptors and those of the mucous membranes, the eyes, noses, and the ears. These receptors transmit their acquired information directly through dedicated neurofilaments to the specific designated areas in the nervous system, particularly the brain where the rest of their information is integrated centrally. These dedicated epithelial receptors include the somatosensory, mechanico- (stretch, pressure), nocio- (pain), chemo-, thermo-, and electromagnetic receptors, which connect directly to specified areas of the brain that have a specific function.

The response of these receptors is, then, integrated into the central homeostatic mechanism in the brain. See Figure 1.47.

Local responses (i.e., from an insect bite or needle puncture causing local pain and edema and/or pollen overload causing edema and rhinorrhea) **are dictated by the magnitude of the injury (the kinetics of the entering substance stimulating, inhibitory, half-life), by the integrity of the tissue involved where entry of the noxious stimuli occurs, and by the state of nutrition in the tissue.**

The functional unit (receptor) that receives and collects environmental information (either positive or negative) transmits information of incitant entry and orchestrates the regulation of the dynamics of homeostasis, including the defense systems both locally and throughout the body, according to Pishinger[3] and Heine,[4] is the GRS. This reception process usually occurs in the CTM, the skin and mucous membranes.

Receptors are part of the GRS and should be considered first when considering noxious incitant entry. Though many people in the medical field have different opinions most agree that the receptors are proteins or protein sugars (PGs, GAGs) interacting with extracellular physiological signals and converting them into extracellular and intracellular responses. **Receptors may be any functional macromolecular component of an organism to which a noxious incitant binds**. This binding would occur to toxic and nontoxic chemicals, bacteria, virus, parasites, and drugs in the ECM or cells. The environmental receptors occur particularly in the ECM and in the cells it contains (leukocytes (PMN) lymphocytes, mast cell, fibrocyte, and macrophage). **The most important concept is that a receptor receives a signal and transduces it to an effector mechanism**.

The properties of the skin, mucosa and the CTM that make it a good receptor are that it is an open system, which responds to every stimuli either external or internal. **The CTM receptor makeup is comprised of a gel solution surrounded by PG/GAG or protein sugars**. The gel of the matrix is environmentally reactive because it can expand, contract, and control changes not only of the weather but also with any noxious stimuli. Therefore, a spring effect occurs once the CTM receives the noxious stimuli. In addition, since the matrix is made up of predominantly sugar, it is structurally diverse. This diversity allows for molecular instability, which is in spontaneous flux. This flux causes frequent electrical changes in the matrix, allowing it to be labile. Therefore, reception of noxious stimuli is easily detected, and because of the spontaneous changes there is nonlinearity of responses. Receptors can change local temperatures due to their absorption of free radicals, which generate the heat.

Receptors can be: (a) enzymes, usually spanning the cell membrane one time in response to binding a ligand and usually increases the phosphorylation of intracellular proteins (e.g., the tyrosine residues, occasionally serine, or threonine residues) or the vital regulatory components are phosphorylated. Also, (b) receptors can be activated ligand transmembrane ion channels; (c) transducer the G protein, which often can couple promiscuously with several G proteins who in turn are not particularly faithful to a single receptor. The end results are affected by the G protein stimulating one of a variety of pathways. Several examples would be activation of adenyl cyclases, phospholipases or even ion channels (permitting certain ions from entering or leaving cells), which would give varied clinical responses upon subsequent similar challenge. In other words, the separate challenges would give nonlinear sequences. This promiscuity of response to an environmental stimulus would account for various perception and adjustment responses. G protein coupled receptors are important for the activity of most peptides such as hormones and/or eicosanoids biogenic amine (e.g., catecholamines). See Table 1.10.[281]

Also, (d) receptors located within the cell (transcription factor) that once were bound to a ligand, **either increase or decrease DNA transcription either by binding DNA or by modulating the effects of histamine**.

Once we start looking at receptors in detail, and especially how they respond when exposed to different ligands that bind to them, things start to happen. Exposing receptors to different molecules, singly or in combination, can give rise to a vast array of effector responses allowing the body to adjust and neutralize the noxious stimuli. Invariably, exposing the receptor to the endogenous ligand(s) normally found in the body gives rise to an agonist response—the normal "'positive'" effect of the endogenous ligand is duplicated.[281] In fact, the body can often identify substances not normally found in it, which have either a more specific effect on a particular receptor than does the normal ligand, a greater affinity for a particular receptor, or even a greater biological effect (greater intrinsic activity).

It is usually possible to identify specific antagonists that antagonize the effect of agonists. Such antagonism is commonly reversible by increasing the concentration of agonist (competitive

TABLE 1.10
Types of Receptor

	Receptor Enzymes		Ligand-Gated Ion Channels	G-Protein – Coupled Receptors	Transcription Factors
	Tyrosine phosphorylases	Other (Thr, Ser, gyuanyl cyclase)			
Number of transmembrane domains	1		4	7	-n/a-
Examples	Insulin, EGF, PDGF, some lymphokines	TGF beta (Ser Thr), atrial natriuretic peptide	Nicotinic cholinergic receptor, $GABA_A$, Gln, Asp & Bly receptors	Many receptors: most peptide hormones, eicosanoids, biogenic amines e.g., catecholamines	Receptors for steroid hormones, thyroid hormone, vitamin D, retinoids

Source: http://www.anaesthetist.com, Receptors—A brief note, 17 November 1999. With permission.

antagonism) but sometimes an antagonist is found that blocks the receptor by binding at a separate site and this is not overcome by increasing the concentration of agonist. The latter is noncompetitive antagonism.[281]

A partial agonist is somewhere between both of the receptors.[281] In the absence of an agonist, the partial agonist may do a very good job of stimulating the receptor. However, if one of these partial agonists is added to a preparation where the receptor is already exposed to a decent concentration of a full agonist, then the partial agonist will sit on the receptor and interfere, causing less agonist activity. As with an antagonist, this effect will be reversed by increasing the concentration of the agonist. The effects of the full and partial agonist are subadditive.

Even more peculiar is the inverse agonist.[281] It is difficult but not impossible to explain how this works. If one takes a preparation and adds an agonist, one sees a particular effect. Add an inverse agonist instead, and one sees the opposite of the effect. For example, benzodiazepines like diazepam cause sedation and drowsiness through their agonist effects at the $GABA_A$ receptor, while inverse agonists at this receptor cause hyperexcitability and even seizures. Apparently, the chemical by its hormesis effect has dictated the response in the receptor. **Many environmentally sensitive patients complain of the opposite effect of a medication when they take it**. Often when they are hoping to be sedated by a drug they become agitated.

Receptor characteristics are the following: (1) the ligand-binding site must be saturable, and half of this maximal saturation should be at a concentration similar to the physiological concentration of the ligand; (2) specific binding of the physiological ligand is desirable, although structurally similar ligands may also bind and the 'wrong' stereoisomer must not bind with a similar affinity, i.e., this binding of similar ligands may cause severe or subtle body dysfunction because there are so many chemicals in the environment that have similar structures; (3) the organism must contain the binding site for the ligand.[281]

Receptor expression is usually dynamic. **Chronic stimulation of receptors often results in decreased numbers of receptors, while understimulation causes an increase in the number of receptors**. There are many practical examples of this process, for example long-term stimulation of beta-1 and beta-2 receptors causes their numbers to decrease (with profound effects on asthmatics and people in heart failure), while **denervation of some structures may cause up-regulation and modification of receptors to a point where even mild stimuli cause substantial responses** (for example in achalasia, and also in spinal cord injury). Such changes are not necessarily **just** alterations in numbers of receptors-types, but the activities of receptors may also change. Over- or understimulation of one receptor can also have profound effects on other receptors (so-called heterologous desensitization). Receptors are found not only on most cells but also (appear to be overwhelmingly) in the CTM. One can see how the total body load and the specific body load would not only change reception perception but also lead to nonlinear responses. **This phenomenon of chronic noxious stimuli giving receptor depression or increase would also explain why time in the environmentally controlled unit (ECU) with the total environmental load reduced would increase the number and activity of receptors rendering the patient temporarily hypersensitive until equilibrium occurs.** Homeostasis of the receptors and their function under stimulation will increase the number and function of receptors, resulting in temporary hypersensitivity. Both of these situations are seen in the partially treated patients with chemical sensitivity and/or chronic degenerative disease who temporarily become supersensitive to multiple peripheral environmental incitants.

The important fact about the receptors of the ECM peripheral sensory and autonomic nerves is that the matrix permeates the entire extracellular space of the body, which surrounds every cell. These receptors of the extracellular matrix react universally, but not uniformly to each external entering or internal perturbation by noxious stimuli. Thus, this tissue has a special place in receiving incoming external information and regulating local homeostasis.

Matrix Receptor

The entire field of physiologic activity for the dynamics of homeostasis starts within the biological terrain of the ECM and the extracellular fluid (ECF), which according to Pischinger and Heine,[3] are the centers of the ERS of the body. The three-dimensional architecture of the CTM as opposed to linear architecture can influence and even alter genetic expression.[4,136] Even the cutaneous receptors will go through or be influenced by the CTM. Thus, the reception, collection, integration, and distribution of information for the regulation of the dynamics of homeostasis are found in the GRS, whether it is local, regional, or central. **The CTM components, which include PG/GAGs, collagen, elastin, fibronectin, laminin, integrin, etc., link up with the glycocalyx (the membranous cell sugar film) at the interphase of the cell surface. The glycocalyx allows communication between the cell and the CTM.** See Figure 1.48.

Thus, the external environmental receptors and the sugars of the ECM are embedded in the lipids and proteins of the cell membrane becoming glycolipids and glycoproteins and forming an additional individual surface film. This film contains all of the cell membrane receptors, cell adhesion molecules (CAMs, integrins, adherence, selectin), blood group substances, and histocompatability antigens.

The glycocalyx components are interactively related to the depolarization of muscles and nerves and the secondary messengers at the inner side of the membrane of the cells of other organs. Secondary messengers (cAMP, cGMP, inositol phosphate [all containing monosaccaharides]) and the cytoskeleton are essential components of the intracellular information line in which data impulses containing information may be quickly directed from the CTM to the cell and vice versa.[78] Then dynamic homeostatic adjustment responses occur locally, regionally, and centrally. Each facet will be discussed separately. See Figure 1.49.

In addition to skin and mucosal receptors, including the ears and eyes, the local primary communication network of the ECM centers around the fibroblasts, macrophages, and mast cells as well as the peripheral sensory and autonomic nerves. **The primary receptor and communication network is also part of the loose areolar tissue that is part of the local connective tissue, which is produced by the fibroblasts and made up of collagen fibrils, PGs and loose GAGs along with the water ordered by these substances.** Anatomically, this meshwork (fibroblast and PG/GAGs) makes up one-half of the connective tissue of the body with the other half consisting of predominantly collagen. Fibroblast, macrophages, and PG/GAGs, respond to electromagnetic signals emanating from entry due to pressure, chemicals, bacteria, viruses. etc. (Figure 1.50).

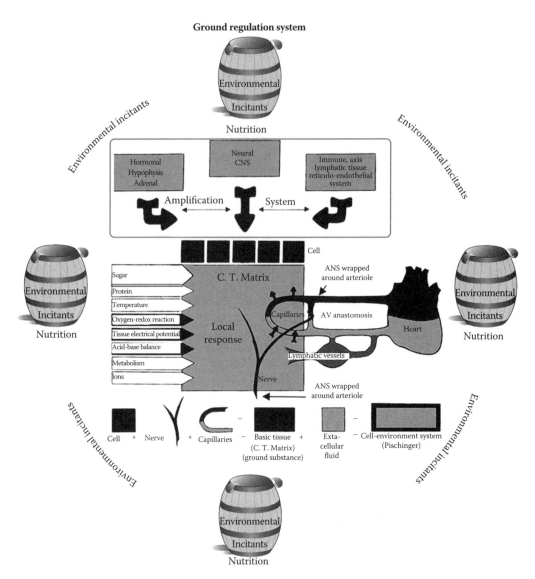

FIGURE 1.48 Ground regulation system—Cell-connective tissue—Environmental system. Note the autonomic nerve wrapped around the end blood vessel. This anatomical fact is crucial for understanding the neurological involvement in the dynamics of homeostasis as well as neural therapy, acupuncture and specific intradermal provocation neutralization therapy. (Modified from Dosch, P., *Manual of Neural Therapy: Accoding to Huneke (Regulating Therapy with Local Anesthetics)*, 11th (revised) Edition, First English Edition, Karl F. Haug Publishers, Heidelberg, 112.)

PG/GAGs have an electronegative charge produced by the sulfates, amino acids, and sialic acid (neuraminic acid), which when stimulated send impulses forming a dynamic harmonic resonance communication network locally as well as throughout the body as shown on the section in communication.

Mast cells also makeup the cells of the matrix and are integrated with the macrophage and fibroblast functions. Other cells, besides the fibroblasts, in the local area are fixed-tissue macrophages, mast cells, and free-running macrophages. All of these cells communicate by growth factors and cytokines with the fibroblast, other PG/GAGs, collagen, and specific immune cells like the lymphocyte. They have been discussed in detail in this chapter. **The macrophages and fibroblasts must**

The Physiologic Basis of Homeostasis

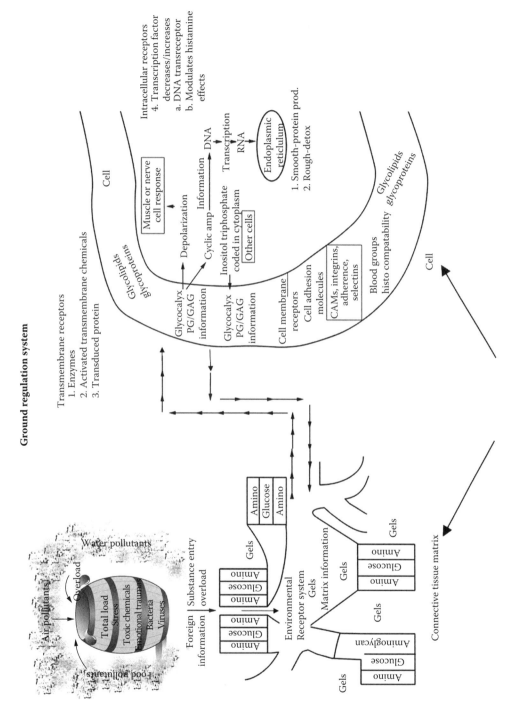

FIGURE 1.49 Matrix-cell-reception feeding into the general communication system. (From EHC-Dallas, 2002.)

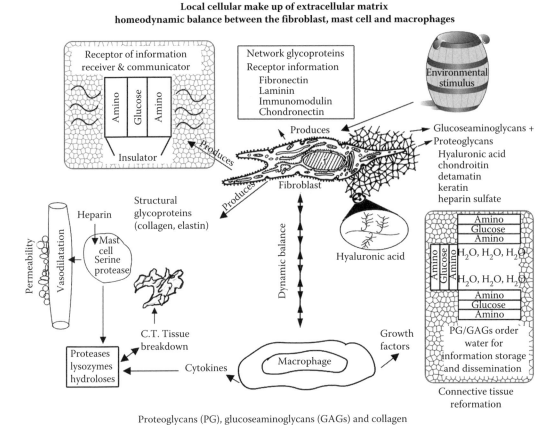

FIGURE 1.50 Makeup of extracellular matrix, homeodynamic balance between the fibroblast and the macrophages, proteoglycans (PG), glucoseaminoglycans (GAGs), and collagen. (From EHC-Dallas, 2002.)

be in a dynamic balance with each other in order to obtain and maintain the orderly dynamics of homeostasis, thus, maintaining health.

The PG/GAGs order the water to allow memory[282] and electromagnetic coherence[282] for proper depolarization[282] and repolarization.[282] This depolarization and repolarization process receives information from the arrival of incoming environmental stimuli. After receptor perception PG/GAGs, spread incoming information locally between the matrix, the macrophages, mast cells, and other fibroblasts throughout the body. These receptors receive reciprocal (feedback)[283] (on how to respond and adjust to the entering stimulus) information for the homeostatic adjustment process. This information comes from the local surrounding CTM and the local cellular reactions. If the injury intervention does not occur then amplification of responses through the regional, autonomic nervous system reflexes, as well as from the integrated central homeostatic mechanism in the brain, and unaffected parts of the body follows. When the PG/GAGs are triggered, they send coherent impulses by rapid depolarization that conveys information locally and if necessary throughout the body.[284] See Figure 1.51.

Coherent impulses triggered by noxious stimuli entry are driven locally for the response phase. If the total local pollutant load exceeds the capacity of the local homeostatic adjustment and defense mechanism to handle it the coherent impulses travel regionally and eventually to the central homeostatic mechanism in the brain.

Collagen, forming the other 50% of the connective tissue **is also produced by the fibroblast but at a slower rate than the PG/GAGs**. Collagen is a structural protein as well as part of the communication system. After leaving the fibroblast, the collagen molecule forms fibrils and fibril

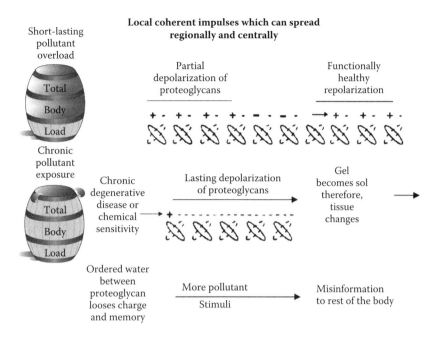

FIGURE 1.51 Decrease in pollutant overload by avoidance, fasting, and rotary diet. Proteoglycans retain their charge stability and prevent tissue and information changes. (Modified from Rea, W.J., *Chemical Sensitivity, Volume IV. Tools of Diagnosis and Methods of Treatment,* 2044, CRC Press, Lewis Publishers, 1997. With permission.)

bundles, which become part of the local CTM with the PG/GAGs, thus acting as a support structure of the local matrix and also becoming repair tissue after severe tissue injury gaps occur. Elasticity is produced by interpenetration of the collagen fibril with the PG/GAGs, opposing matrix deformation while maintaining matrix volume, as well as, elastin insertion. These compounds along with the water or IF make up the bulk of the local ground substance. Once the information of noxious stimuli entry occurs and is perceived by the receptor, collagen becomes involved in the communication process to eventually trigger the tissue repair mechanism if sufficient tissue damage has occurred. This entering information relates to modeling and remodeling tissue, forming scars, cell attachment and differentiation, chemotactive agents, antigens, and as integral components in central pathological processes.

Adjustment Responses

Local Cell and Matrix Reactions for the Adjustment Response

The adjustment mechanism is a process that is driven by the communication system where the body compensates for the entry of noxious stimuli by a series of local matrix and cellular micro changes. **Usually early adjustment occurs daily without inflammation and scarring. This process can be very broad or very narrow yet it is virtually automatic**. A transient tachycardia or bradycardia would be a clinical example of an adjustment reaction. Adjustments can be the result of early defense parameters (e.g., excess thirst, heat conservation, etc.) or just local minor physiologic aberrations (e.g., transient increase in respiration, sweating, etc.)

Most adjustments are within the physiologic range. Adjustments may occur in physiology, i.e., including changing pH, increase in Ca^{++} and decrease Mg^{++}, and increase in metabolism, multiple passes of toxics through the liver and lungs, increase in rate of detoxification (e.g., conjugation mechanisms are speeded up), autonomic nerves system changes, immune enhancements, change in receptors, more macrophages produced and released, etc.

Local responses and micro adjustments are controlled by pH changes, electrolyte flux (especially sodium, potassium, calcium, magnesium, lithium), electrical changes, redox reactions, and fibrocyte-macrophage and leukocytes influx with mediator and growth factor release as well as, the kinetics of the incitant. If the incitant is chronic and exceeds the containment by the local response, then regionalization occurs (i.e., triggering of appendicitis or calf muscle spasms).

The dynamics of local homeostasis are also influenced strongly by the state of nutrition. The amount of the nutrient pool present influences the prompt automatic response for neutralizing the sudden influx of entering noxious stimuli and adjusting the body's physiology to continue smooth function is necessary. If the nutrition is faulty the automatic response may be inadequate and the body will malfunction. Thus, the effects of noxious stimuli will spread throughout the system creating conditions for chronic illness.

To understand the dynamics of normal local homeostasis in order to prevent disease one must be able to trace the path of noxious environmental stimuli from their sources in the environment (in air, food, and water) to **their point of entry into the body. This is followed by the information perception of a foreign presence by the ERS in the surface areas and by the CTM. The sequence of events that occurs, following noxious incitant entry, results in physiologic micro adjustments (i.e., pH decrease by cleavage of sulfate, acetates, etc.), macrophage release from surrounding tissue by the increase in Ca^+ and decrease in Mg, antipollutant enzyme activation (superoxide dismutase, catalase glutathione, peroxidase induction) increase in the microcirculation, and enhancement of oxygen extraction, etc. These physiological adjustment changes neutralize the incitant and correct its entry changes, allowing the normal dynamics of homeostasis to continue**.

If the noxious stimuli are in excess either by sheer volume or by kinetic overload chronic physiologic dysfunction results first locally and then, generally, trophic (neurogenic responses) and eventually inflammation occurs. Vascular spasm with resultant hypoxia occurs after the initial vasodilatation and around the wound periphery with a dissipation of the energy. Often tissue changes occur, resulting in the dysfunction and dissolution of the CTM by proteases, lysozymes, sulfatases, hydrolases released by the macrophages.

This change of physiologic transmission of information to an actual pathological metabolic or near pathological response is paramount in understanding how information can influence tissue response. After the noxious stimulus has been neutralized, the repair mechanism is then activated. A change of the body's fluids from sols to gels and gels to sols and then the NSMR occurs (i.e., for scar formation and healing). An example of this healing process is basic pathophysiology, which occurs in acute coronary thrombosis ([see Cardiovascular chapter in Volume 2 - Mechanisms of Chemical Sensitivity and Chronic Degenerative Disease]), resulting in fibroblast (with a proliferation of PGs, GAGs, and collagen) and macrophage activation (release of cytokines and growth factors).

The environmental insult triggers a change in information, which stimulates proliferating fibroblasts to produce more glucoseaminoglycans, proteoglycans, and collagen. These are for repair of a wound, but if no wound is present, the entering noxious stimuli can result in pathological responses. Even at times changes occur that transform fibroblasts into myocytes (which are found in some individuals with blood vessel inflammatory and atherosclerotic damage).

If the macrophages cannot keep up with the chronically stimulated fibroblast output to engulf and destroy the excess PG/GAGs, the NSMR continues. This reaction induces troph (neurogenic) edema followed by inflammation from the production of proinflammatory cytokines by the macrophages and leukocytes. These cytokines stimulate more tissue changes, eventually resulting in automatic triggering of fixed-named, irreversible disease e.g., arteriosclerosis, kidney failure, Parkinson's disease, heart, lung, and liver failure, angioedema, spondylosis) until end-organ failure occurs.

Once one understands the dynamics of each step in the progression toward disease (incitant entry, local reception, adjustment, defense, and repair process, regional, neurovascular, and the integrated CNS homeostatic regulation), the clinician can clinically evaluate the patient's stage of homeostasis

The Physiologic Basis of Homeostasis

or dyshomeostasis. Strategies can then be developed to stop the downward slide to end-organ failure or death and, often, can reverse the downward slide of the body back to good health.

Biological Rhythms and Periodic Signals of Normal Homeostasis

The dynamic normal physiologic changes of homeostasis fluctuate and may be rhythmic, e.g., spontaneous cellular oxidation resulting in membrane changes occurs with rhythm. The rhythm of this open system is provided by intrinsic biochemical processes in the cells, as shown in, e.g., the rhythmic synthesis of ATP by mitochondria.[78] The total body immune system modulation fluctuates, which makes people more vulnerable at certain times of the day or month or year. Outside the cell matrix molecules have thermic fluctuations. **Some fluctuations in part are due to heat generated from metabolic reactions and free radical absorption and other fluctuations partially are due to external environmental changes**. Normally, molecular rhythms finally merge into large overriding rhythms, like rhythms that occur in seconds (e.g., cardiac beat), minutes (e.g., intestinal peristalsis), hours (e.g., sleep-wake rhythm), days (e.g., healing, regeneration) and years (e.g., growth). In addition, the body is at the mercy of weather cycles (365 days, seasonal, and 24-hour day/night cycles), which must be integrated with the intrinsic biorhythm, which again, is a multifunctional nonlinear intrinsic system response. An example of how intrinsic biological rhythms integrate with external rhythms is found in the study of clams and how they open their shells. If the clams are transported from the east coast of the U.S. to the west coast the shell opening time biorhythm is changed by four hours. This change, influenced by external rhythms also, appears to occur in humans.[285]

Schuman radiation of 7.8 Hertz arising from oscillation from the ion flux sent between the North and South poles is another example of an environmental rhythm with which the body must synchronize. The individual with periodic homeostatic dysfunction (chemical sensitivity) and/or those people with aperiodic homeostatic dysfunction (chronic degenerative disease) have damaged receptors and/or their responses are altered. Those with periodic homeostatic disturbances appear to be labile and thus, overreact to the external environmental stimuli (i.e., a weather front); whereas, either patient with homeostatic dysfunction or chronic disease has alteration in the normal dynamic homeostatic response. The near automatic dynamic homeostatic responses occur but have different avenues of reception, adjustments, and neutralization capabilities thus, altering the individual's biorhythms.

Health and stability of an organism depend mostly on a synchronization of all biorhythms, which make up the dynamics of homeostasis.[286] It takes an extremely healthy and integrated organism to be able to handle not only the normal but also abnormal stimuli that the body is bombarded with by living in the twenty-first century. **The recovery of individually defined biorhythms is, therefore, the objective of all biological and medical therapies**. The entry of a noxious substance into the body is not simply eliminated directly by mechanical means alone but rather indirectly eliminated by the mobilization of the organism's integrated homeostatic self-healing powers, by the marshalling of many parts of the response metabolism, and by the changes in physiology that will neutralize the stimuli. Since there is a societal increase in chronic disease, a manipulation of biorhythms is being developed for therapy and prevention, which stresses the importance of a behavior adapted to the biological phases.[287] According to Heine, normal periodic homeostatic feedback processes in biological systems serve, above all, to organize vital processes in space and time and help to render data transmission more reliable. For example, rhythms make a space and time organization of biological processes possible. They allow self-synchronization, thereby guaranteeing a temporal coordination of various vital processes and, in particular, an isolation of incompatible ones. For example, the blood pressure of an individual is 110/80 with a normal range of 100/70–120/85, and when the blood pressure changes to 200/140 an obvious adjustment of homeostasis is demonstrated. Thus, an incompatible parameter is isolated immediately, allowing the clinician to see subtle and in this case, gross dyshomeostatic changes. Rhythms, moreover, permit an exact forecasting of repetitive events like the vital signs of blood pressure, temperature, pulse, and respiration, etc. Abnormalities of any steady dynamic state parameters over a period of time support the observation that pathology is occurring (BP, pulse, temperature, etc.), again allowing the clinician the opportunity for early therapeutic intervention.

Periodic signals support the efficiency of intracellular communication between all organelles and also between cells by a wall of cell adhesion molecules permeable to ions ("gap junctions"). **Gap junctions act as electrochemical synapses and are dependent upon vitamin A.**[291] Since the body's natural frequencies never are exactly fixed small frequency alterations may cause frequency-modulated data transmission, which may or may not trigger the proper messages for triggering the homeostatic mechanism.

Triggering often depends upon the lability of the tissue. Therefore, if there is previous damaged tissues, such as interference fields (disturbance foci) with increased lability, the threshold for triggering may be lower thus, allowing depolarization to occur as stated previously, and when pain occurs the tolerance threshold of the autonomic nerve fibers is rapidly exceeded. A clinical response occurs with motor nerve involvement, resulting in something like muscle spasm. The reduction in blood supply to the particular area that the autonomic nerve fibers supply continues to lower the triggering threshold of the sensitized nerves and the receptors until metabolic changes occur. Thus, this vicious downward cycle may be triggered by small natural frequency alterations. As opposed to linear systems however, frequency and amplitude cannot vary independently in nonlinear systems[289] and must be coherent. **When coherence occurs there is an almost instantaneously energy efficient signaling throughout the body, resulting in energy efficient dynamic harmonic homeostasis**. The CTM, for example, may be physically compared to pendulums interconnected by nonlinear springs. When suitable energy is fed into this nonlinear system, e.g., an impulse on one side will be distributed through the system and may cause resonance in various parts of the system. Therefore, temporarily isolated responses will then influence other, neighboring partial responses and so on[290] until a general response occurs or the isolated response dissipates. Thus, the energy efficient physiologic subunit and central integrated normal homeostatic response occurs and the body functions properly.

Regional Homeostatic Adjustment

The progression of pollutant entry and injury occurs when the pollutant or its response cannot be contained by the local injury adjustment or defense mechanism, i.e., leukocytes, macrophages, etc. Then the regional mechanism becomes involved and physiologic adjustments have to be made. See Figure 1.52.

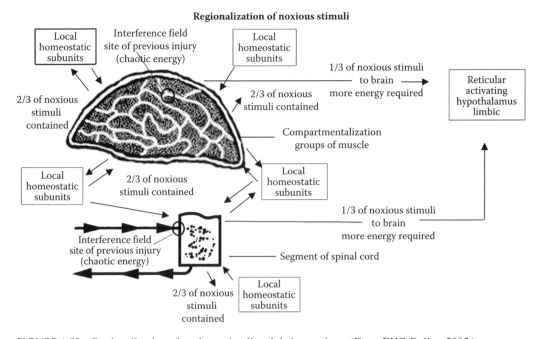

FIGURE 1.52 Regionalization of noxious stimuli and their reactions. (From EHC-Dallas, 2002.)

The regional homeostatic mechanism is activated by integration of information from the functioning of local homeostatic subunits suggesting that local containment is impossible. The regional response does not involve the central homeostatic mechanism for area adjustment responses. An example of regionalization would be when there is an injury in an isolated muscle but the muscle responses occur in the whole integrated muscle bundle, i.e., the rhomboid muscle segment is injured and a series of the surrounding muscles in the rhomboid area of the shoulder respond. Another example would be a reflex response of an area like that of an internal organ such as the appendix. When the organ is damaged there is a visceral, dermal, muscular, sensory, autonomic, spinal reflex, resulting in an increase in pain, local temperature, muscle spasm, and peritoneal irritation. This regional homeostatic adjustment mechanism comes into play when the local response is exceeded chronically and can adjust for responses the local reaction cannot handle. These regional responses seem very important in the containment of dyshomeostatic spread of information and thus, adjustment response in the hierarchy of homeostatic functions. These types of regional responses also conserve energy when they do not involve the central homeostatic control. These adjustment responses can be viscera-viscera, viscera-muscular, or viscera-cutaneous. (See Neuromuscular chapter in Volume 2 - Mechanisms of Chemical Sensitivity and Chronic Degenerative Disease.) All these adjustment responses are extremely important in gastrointestinal regulation and when in physiologic limits to conserve energy while functioning efficiently.

Huneke, Dosch,[291,292] and Kellner[296] showed that with an excess incitant entry the segmental or regional areas develop interference fields or centers of irritation of tissue, usually not limited to one unit. These interference fields will modify normal adjustment responses often resulting in dyshomeostasis. These fields occur due to excessive pathogenic stimuli or local nutrition depletion. According to Kellner,[293] these fields are characterized by small areas, infiltrated by small lymphocytes. Often, according to Huneke,[294] 30% of the influence of those fields reaches beyond the local segment and therefore, is capable of setting off dysfunction in any part of the body. **Two thirds of the noxious stimuli received by the receptors and other adjustment responses are contained regionally**. Containment of an injury in an area, usually, helps ward off systemic disease. Any point in the body can become an interference field where damaged tissue may exist thus, creating a hypersensitive area in the body. When this damaged tissue phenomenon occurs, the regional area containing an excess total load cannot be eliminated thus, leaving this regional area in the vulnerable state of unphysiological, permanent excitation causing a local interference field. Normal adjustment responses are then altered causing more regional and distal dyshomeostasis. Huneke[294] found that interference fields stand outside the segmental order, sending out impulses in the nerves in responses, which can become pathogenic. This observation follows the law of denervation in that when a nerve is damaged, the end organ and distal nerve become hypersensitive. Any focus or interference field is a permanent source of irritation because it burdens the homeostatic regulatory system and continuously forces the body to increase the additional stress load, which drains energy from efficient function and, thus, alters normal adjustment processes.

Normally, the signal transmission code relies on directed, constantly changing impulse frequencies, according to Huneke.[295] **The disturbances and impulses produced by an interference field are always without aim or purpose, uniquely bound by chaos, yet, specifically localized to a particular segment in the body**. Due to this chaotic phenomenon the interference field transmits false information through the communication system, which, usually, misleads the regional and at times, the central homeostatic regulatory mechanism. This altered function in turn interrupts the purposeful collaboration between different homeostatic control circuits. This interruption leads to inhibition of the body from functioning as a whole, thereby, causing pain, weakness, fatigue, and in many cases supersensitivity.

Segmental areas become important because no organ lies in only a single segment. **The damaged area of the interference field usually involves overlapping segments causing disturbances in communication**. Thus, reactions may cause deviations from normal responses to many segments, due to the anatomy and physiology of the muscles, sympathetic chain, and the spinal cord.

Generally, **segmental reactions occur on the same side of the body where the initiating response and organ is located**. An example would be the patient who has deposited a pollutant high in the sympathetic nerve trunk on the left side. He will have pain and tenderness all along the sympathetic chain in the neck, chest, and abdomen all on the left side. This function demonstrates that regionalization has occurred.

A pain reflex produces a response primarily in the segment to which the system responds by efferent somatic and autonomic nerves. In the somatic field, muscle spasm is produced, which may extend to becoming a block in the regional segment. For example, one may have spasm of a small muscle segment in the leg, and the whole region suddenly produces a spasm like a "Charlie horse." **Autonomic reaction may take a variety of forms depending upon the patient's reactive state. There may be more or less vasomotor effects, changes in skin temperature, and circulatory disturbances, such as microcirculation blocks with shunting of blood away from designated areas in the underlying tissues**, thus resulting in more tissue damage and possibly inflammation. Interference disturbances in internal organs can also occur and, finally, along with these disturbances, interference fields within the neurohumoral regulatory mechanism may be present.

Due to the vast communication properties of the body an interference field or a dyshomeostatic focus produces a change not only in the cell environment but also in the reactive capacity of individual organs and the organism as a whole. Segmental regional involvement often occurs and is a deterrent to global body dysfunction. The principle of isolating the noxious stimuli to prevent global dyshomeostatic responses applies here as it does with local incitant containment.

After chronic dyshomeostatic metabolic alteration from chronic noxious stimuli exposure, tissue and other communication properties changes will eventually occur. The spreading of dysfunction of homeostasis in the ground communication regulation system does not take place suddenly, but gradually, with informational and metabolic feedback changes from the neural and humoral system. **This compartmentalized local and regional homeostatic dysfunction occurs because of the difficulty in transmitting electrical bioinformation across septa, fascia, and muscular membranes**. The homeostatic dysfunction in a compartmentalized area is like a closed circuit that can be broken by transmitting information via the loose areolar tissue surrounding lymphatics, arteries, and veins,[296] which then allows exit of the information out of the compartment. **Compartmentalization of homeostatic dysfunction gives a great advantage to the physician because he has a chance of keeping disinformation from spreading systemically thus, preventing total body chronic disease or chemical sensitivity**. In other words, generalized chronic fatigue, fibromyalgia, and vascular dysfunction can be prevented and reversed at this stage. There are many examples of compartmentalization such a regional muscle spasm and shortening, localized headache, blephorospasm, hoarseness, etc. **The role of the clinician is always to keep the noxious stimuli and their responses localized**. Reduction of the total body load by fasting and good environmental control will help localize that response. What is interesting is although fasting will decrease the total body load, receptor sensitivity will increase temporarily for a few days rendering the patient more sensitive to lower-intensity environmental pollutants and foods.

Central Homeostasic Adjustment

A reciprocal relationship exists between the reactive responses of the body's homeostatic feedback subsystems and the total body's central integrated homeostatic adjustment mechanism, including, the reticular formation, the hypothalamus, the limbic system, the pineal gland and area postrema of the fourth ventricle. (See Figure 1.53.) Informational feedback loops for homeostatic adjustment make the dynamics of homeostasis work and also their complexity allows a large area for potential dysfunction, resulting from excess of both external and internal stimuli, nutritional depletion, alteration in receptor numbers and sensitivity, and thus, energy depletion ensues. The adjustment phase of the holistic homeostatic mechanism works well as long as the components function normally and nutrition is complete. If the automatic dynamic homeostatic functions are out of the bounds of normalcy, they can aggravate and eventually cause autonomous

The Physiologic Basis of Homeostasis

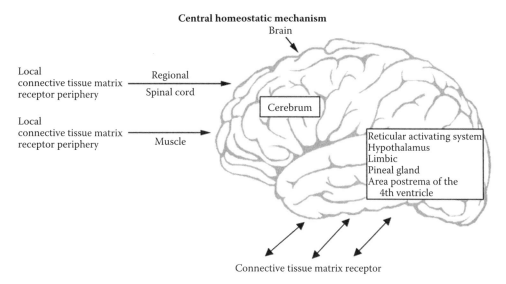

FIGURE 1.53 The central homeostatic mechanism consists of the integrated function of the reticular activating mechanism, the hypothalamus, and the limbic system. (From EHC-Dallas, 2002.)

self-stimulation of the physiologic process. This abnormality will initiate and propagate disease because of the disinformation that is consistently sent throughout the GRS, which in turn triggers more physiologic dysfunction. Each part of this integrated central homeostatic mechanism will now be discussed.

1. The **reticular formation**, located in the brainstem, **controls autonomic and somatic integration, and awareness**. Many adjustment responses occur when this mechanism is functioning normally. There are bulbar excitatory centers such as in the substantia nigra, locus ceruleus, nuclei of the raphe, and the gigantocellular neurons that receive excitatory signals from the periphery. **The level of activity of the brain stem excitation and therefore, the level of activity of the entire brain is determined, largely, by the type of sensory signals that enter the brain from the peripheral communication system**. As the receptor system becomes more active and thus more sensitive, the sensory areas of the brain's homeostatic mechanism become more sensitive to theses signals. Pain signals incite the excitatory area and therefore excite the brain to attention. Some chemically sensitive patients have reported that to maintain a high level of concentration, like when writing papers or creating music, etc., they have to be a little bit uncomfortable. This state of being uncomfortable is not to the point of pain, but rather, seems to create enough excitatory impulses in order for them to become super alert and more perceptive. Normal adjustment responses have occurred.

Normal physiologic adjustment is exceeded when others need stimulants like coffee, tea, soft drinks, etc., to keep them alert. Not only do the excitatory signals pass to the cerebrum from the bulbar reticular excitatory area of the brain, but feedback signals, also, return from the cerebrum back to bulbar region for adjustment responses. Therefore, anytime the cerebral cortex becomes activated by either thought processes or brain motor processes, signals are sent back from the cortex to the brain stem excitatory areas, which in turn send still more excitatory signals to the cortex. This normal adjustment feedback signal helps to maintain the level of excitation of the central cortex or even enhance it. This circuit is a positive feedback characteristic for normal adjustment that allows activity in the cerebrum to support still more activity thus leading to an active mind. See Figure 1.54.

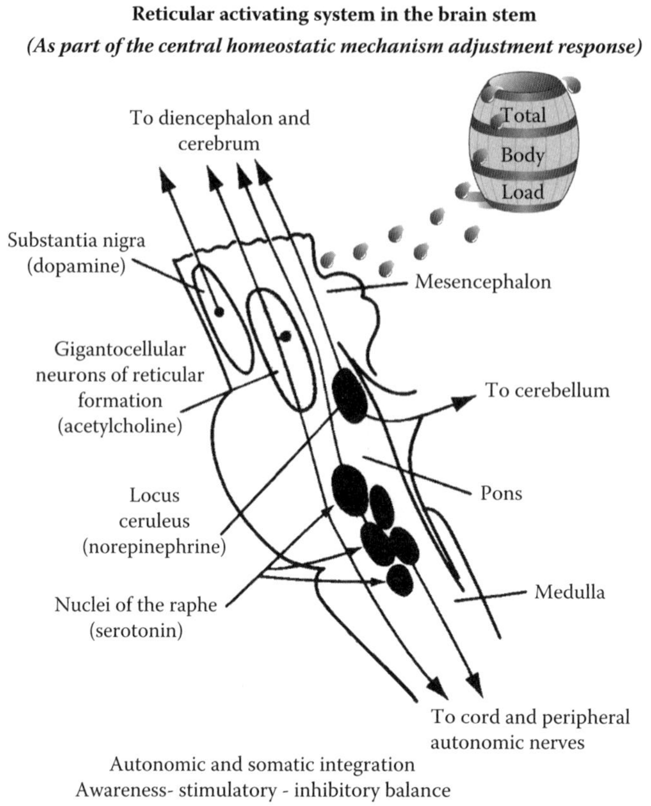

Reticular activating system in the brain stem
(As part of the central homeostatic mechanism adjustment response)

FIGURE 1.54 The bulbar excitatory area can be triggered by various motor, sensory, and pain inputs to activate the central homeostatic mechanism. (Modified from Guyton, A.C. and Hall, J.E., *Textbook of Medical Physiology*, 10th ed., Philadelphia: WB Saunders Co., 680, 2000.)

The reticular inhibitory area is located medially and ventrally in the medulla and can decrease activity in the upper brain stem, superior portion of the brain stem, and pons for normal adjustment. One of the mechanisms is to excite serotonergic neurons of the raphe to secrete serotonin.

Aside from direct neuronal control of brain activity by the transmission of nerve signals, is neurohormonal control. This neurohormonal control can give several adjustment reactions. This neurohormonal response includes the secretion of norepinephrine from the locus ceruleus; norepinephrine stimulates most areas of the brain except for a few inhibiting receptors. The substantia nigra region sends fibers to the putamen and the caudate nucleus where they secrete dopamine. The gigantical cellular neurons of the reticular excitatory area send fibers upward into the brain and downward into the spine and to the area where their terminals secrete acetylcholine.

2. The **hypothalamus controls autonomic and somatic integration of the dynamics of the homeostatic responses for physiologic adjustment**. It is the core of the central regulation of the dynamics of homeostasis. The hypothalamus has a two-way communication pathway with all of the limbic system. It also sends output signals (1) downward to the brain stem to the reticular areas of the mesencephalon, pons, medulla, mamillary bodies, and to the peripheral nerves; (2) upward toward areas of the diencephalon and cerebrum; and (3) into the hypothalamic infundibulum to partially control most of the secretory function of the anterior and posterior pituitary glands. The hypothalamus for normal adjustment regulates cardiovascular responses (i.e., increased blood pressure), pupillary dilatation, body

temperature (shivering), gastrointestinal stimulation, hunger, satiety, thirst, neuroregulatory control, reflexes, oxytocin, including uterine contractibility, male ejaculation, water conservation, bladder constriction, decreased heart rate and blood pressure, vasopression release, panting, sweating, thyrotropin inhibition (Figure 1.55) and many other functions. It is the master control of the central automatic response, as well as regulation of body water.

3. The **limbic system controls emotional behavior and** motivational drives. Triggering tries to give easy adjustment and is elaborated on in Chapter 2 and Neuromuscular chapter in Volume 2 - Mechanisms of Chemical Sensitivity and Chronic Degenerative Disease. This system, if involved, can account for a spillover of emotional reactions like rage, hunger, thirst, etc., where abnormal stimuli and the dynamics of those reactions vary widely.

The limbic system is particularly involved in reward centers, punishment center, rage, learning, behavioral awareness, consolidation of long-term memories, **conversion from short to long-term memory and olfactory integration of smells**. Along with many other functions, it is also the motivational center. Chronic pollutant damage often leads to a severe inertia, which plagues the patient with chronic fatigue or chemical sensitivity once their motivational drive is gone. This inertia often prevents them from getting well until the pollutant is eliminated and nutrition is restored. Also, most patients with cerebral chemical sensitivity have short term memory loss. This condition may be due to damage in the limbic system as well as the temporatl lobes, which correlate with short-term memory loss. Triple camera SPECT brain scans show alterations of the temporal lobes. See Figure 1.56.

These five components of the brain (reticular activating system, hypothalamus, limbic system, pineal gland, and area postrema of the fourth ventricle) make up the integrated automatic central homeostatic control system, which triggers the adjustment response in order to control the orderly

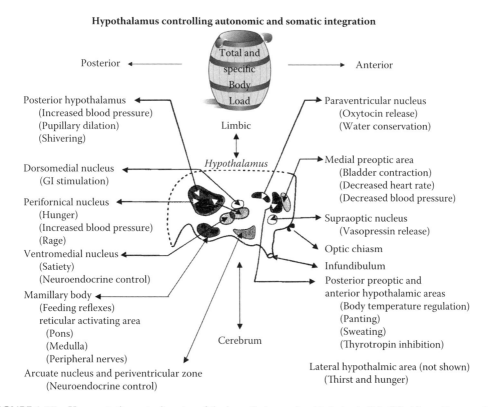

FIGURE 1.55 Homeostatic control center of the hypothalamus (sagittal view). (Modified from Guyton, A.C. and Hall, J.E., *Textbook of Medical Physiology*, 9th ed., Philadelphia: WB Saunders Co., 754, 1996.)

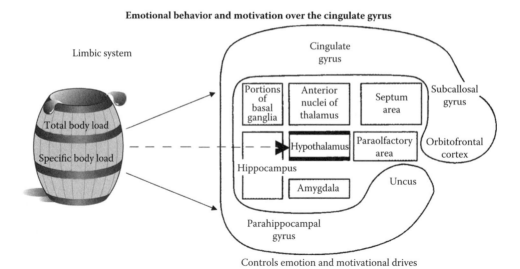

FIGURE 1.56 Limbic system showing the key position of the hypothalamus, which aids in integrated homeostatic function. (Modified from Guyton, A.C. and Hall, J.E., *Textbook of Medical Physiology*, 9th ed., Philadelphia: WB Saunders Co., 753, 1996.)

dynamics of homeostasis. These components emit adjustment responses to regulate the dynamics of homeostasis centrally as much as the CTM does peripherally. For example, a noxious substance enters the body through the respiratory tract triggering the local homeostatic response of the CTM. Part of the noxious substance goes up the nose to the olfactory nerves and triggers the nerve to send information regarding the substance through impulses to the hypothalamus (and the rest of the central homeostatic system). Here, the homeostatic subsystem of automatic central control is triggered for adjustment to the noxious stimuli. The rest of the noxious substance goes down into the bronchial tubes irritating them and then this part of the inhaled substance is absorbed across the alveoli into the blood. Here, the dynamic fluctuating adjustment responses in three more homeostatic subsystems are triggered—one is the bronchial, the second is the alveolar, and the third is the vascular. The subsystem impulses are then integrated into the central homeostatic mechanism, which results in a further physiological adjustment response.

There are literally thousands of other routes of fluctuating homeostatic adjustment processes that are triggered, i.e., blood pressure regulation, pulse changes, alteration in the respiratory rate, electrical movement, temperature, etc. Each subsystem is in charge of its own local dynamics of homeostasis and, when possible, each one isolates the noxious stimuli locally then no further information reaction occurs. If isolation and neutralization of the incitants is impossible locally and overwhelmingly uncontained, messages are then sent further into the body for more adjustment responses. A coordination of the homeostatic subsystem responses through the GRS to the central homeostatic mechanism follows. These reactions are harmonic, energy-efficient, integrated, and generalized, with a reciprocal automatic adjustment response occurring, which results in a dynamic total body homeostatic adjustment. Integrated information is sent via feedback loops through the whole physiologic system of the body, causing alterations and adjustments in order to neutralize the noxious stimuli or utilize the positive stimuli.

Cerebral Function and Its Relationship to Automatic, Dynamic Homeostasis

The cerebral hemispheres surround and connect with the limbic system, which surrounds and connects with the hypothalamus. **Therefore, the cerebrum's function can influence but not totally change the integrated dynamic homeostatic adjustment response**. As stated previously, these local, regional, and central homeostatic mechanisms have near automatic feed back signals for homeostatic adjustment. In addition, connections from the reticular activating system, the hypothalamus, and the limbic system, also occur between the somatic responses of the body and the mental function of the cerebrum. They both convey information, which triggers homeostatic responses to and from each other. This interconnection between the somatic nerve function of the peripheral tissue and mental activity of the cerebrum means, for example, that the basic homeostatic regulation in the cartilage and bones is interconnected to the central homeostatic regulation that, in addition, involves the cerebrum. Therefore, homeostasis could be influenced by mental conflicts or positive thinking of the cerebrum, by local central brain physiologic alterations, such as traumatic injury, or hypoxia, as well as local peripheral cartilage somatic injury and vice versa. For example, information, and the dynamic homeostatic response can be directed either way, where the patient may complain of joint pain and mental fogginess, or the other way, where he may experience no joint pain and have mental clarity. These adjustment functions could be regulated through the reticular activating system, which has to do with alertness, and then modified through the cerebral cortex. During clinical evaluation, these dynamic homeostatic feedback interconnections must also be considered when evaluating the status of the dynamic function of the homeostatic mechanism and the condition of the total body pollutant load (bacteria, viruses, chemicals, physical and psychological stimuli). Therefore, the automatic function of the homeostatic mechanism can, at times, be influenced by cerebral function. Just as with altered nutrition or excess total body pollutant load, cerebral function with a concentrated attempt at altering homeostasis, can influence, but usually cannot override the dynamics of normal homeostasis. **Automatic homeostatic function cannot, however, be controlled entirely by the cerebral functions**. Examples of the attempts of individuals to will away reactions to favorite foods, colognes, etc., were often seen while observing patients under controlled conditions in the environmental control unit of the hospital or clinic. As hard as these individuals tried, they could not abort a reaction, although at times, they could partially modify the response. More than 95% of the time the reaction occurred just as it did on previous challenges. Yogis have been known to slow heart rate and metabolism with years of training, but they still can not totally control automatic homeostatic functions.

4. The area postrema of the fourth ventricle has to do with food and is discussed in GI chapter in Volume 2 - Mechanisms of Chemical Sensitivity and Chronic Degenerative Disease. This area is poorly understood but has only a four cell barrier like the hypothalamus and pineal gland. It clearly has functions that affect the homeostatic adjustment mechanism.
5. The pineal gland is discussed extensively in Chapter 4, Endocrine System, and therefore, will not be discussed here. However, since **melatonin has eight times the restorative power to the immune system than any other substance in the body it is highly significant in the homeostatic mechanism**. A large subset of patients with chemical sensitivity has insomnia, which could alter immune functions.

Principles and Facts about Adjustment Responses
Environmental Principals of Homeostasis and Dyshomeostasis
Understanding some of the basic scientific facts about the dynamics of homeostasis allows us to see how chemical sensitivity (periodic homeostatic disturbance) and chronic degenerative disease (aperiodic homeostatic disturbance) occurs. Also, it shows us how dyshomeostasis can allow the propagation from acute into chronic degenerative disease processes, which eventually if they are untreated may become fixed-named and often, irreversible, endstage, end-organ disease (e.g., arthritis, asthma, gangrene, chronic degenerative disc disease, arteriosclerosis, etc.)

There are **nine principles** that the physician must understand in order to comprehend the altered physiology and clinical response that occurs when the dynamics of the homeostatic mechanism are being disturbed. These include the total environmental pollutant load in relation to the total body load (Randolph, Rea), the adaptation phenomena of Selye,[7] Adolph,[297] Randolph,[5] the phenomena of bipolarity of response, Randolph,[5] the biochemical individuality of response (Williams),[38] the phenomena of spreading, Randolph,[5] Pischinger,[3] Rea,[298] the switch phenomena of Savage,[299] the hormetic vs. linear-dose effect (Calebrese), and the law of nerve damage (Cannon). Each of these will be discussed separately, realizing that they may occur simultaneously and are clearly orchestrated by the homeostatic mechanisms through the GRS.

1. Total Environmental Load—previsously discussed.
2. The Linear dose and Hormetic effect—previously discussed.
3. The Law of Nerve Damage Causing Hypersensitivity—previously discussed and elaborated on in Chapter 2.
4. Total Body Pollutant Load (Figure 1.57)—previously discussed.
5. Adaptation

The fifth principle that one must understand when studying the dynamics of homeostasis is the principle of adaptation.[300,301] **Adaptation is an acute survival mechanism the body uses in order to combat a new exposure. If the noxious stimulus is constant and chronic an adjustment occurs to a new set point, including increased levels of metabolism, immune response, and enzyme detoxification function occurs**. (See Figure 1.58). When exposure continues over time and metabolism, immune response, and enzyme detoxification breakdown, maladaptation occurs. Maladaptation is the intermediate precursor state that occurs before fixed-named disease or death develops.

If the exposure is too long, becoming chronic, the body moves from the alarm stage where cause and effect can be proven clinically, to the stage of masking or adaptation where the adjusted dynamic physiologic process can still neutralize the exposure but this next adaptation is due to an increase in metabolism that is needed for a response. Clinically, cause and effect is not evident. Symptoms occur, usually of fatigue and pain, but the cause is unknown. In this case, since the body is at a higher set point of metabolism in order to neutralize the excess total body load, the symptoms and signs do not specifically relate to cause and effect as they do in the alarm phase of adaptation. In the **masked stage or adaptation phase the causes are masked, the body gets "used" to the exposure, and due to the increased energy demand used in the higher metabolism, nutrient drain results**. If the nutrient drain is not replaced by supplementation and chronic noxious stimuli

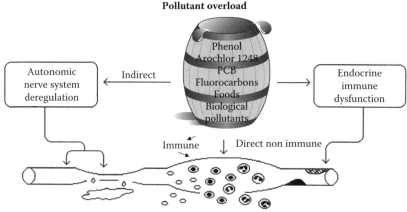

FIGURE 1.57 Barrel—Total body load—The total of all incitants entering the body through air, food, and water triggering the damage to the blood vessels. (Modified from Rea, W.J., *Chemical Sensitivity, Volume I: Principles and Mechanisms,* Lewis Publishers, Boca Raton, FL, 1992. With permission.)

The Physiologic Basis of Homeostasis 119

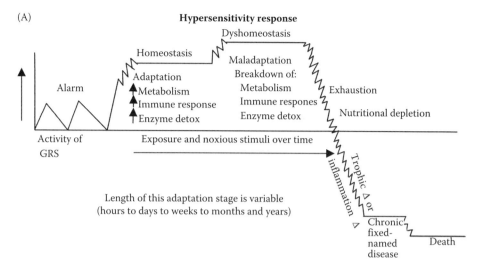

FIGURE 1.58A Adaptation response. (EHC-Dallas, 2005.)

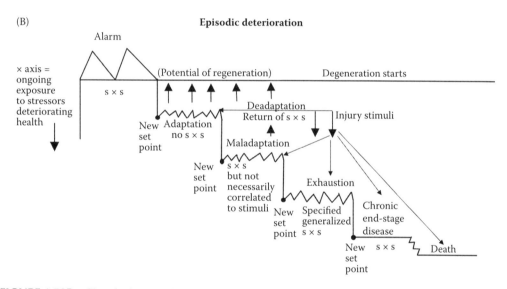

FIGURE 1.58B Chronic degenerative response. (From EHC-Dallas, 2005.)

exposure continues, there is an increase in the total body load of pollutants with more chemical sensitivity and/or chronic malfunction, because the detoxification systems are functioning poorly. Then the stage of exhaustion occurs with nutrient, immune, metabolic depletion followed by tissue breakdown and eventually, fixed-named autonomous disease (e.g., lupus, arteriosclerosis, rheumatoid arthritis, vasculitis) occurs. We and others[298–302,5,7] have observed that for optimum health one should keep the body in the alarm stage with full nutrient pools so that awareness of the entry of specific environmental triggers can be seen or felt and avoided or eliminated. Those triggers that cannot be avoided will be neutralized rapidly by the body's defense mechanisms that are full of the nutrient fuels necessary for a robust response. **It is clear from our long-term studies of over thirty years that individuals who live their life in the adapted state initially, seem to be at a greater advantage over those who have to live in the hypersensitive state because nothing appears to bother them.** However, over a long period of time, since well-adapted individuals avoid very few noxious incitants, they have a greater tendency to manifest aperiodic end-stage disease like cancer,

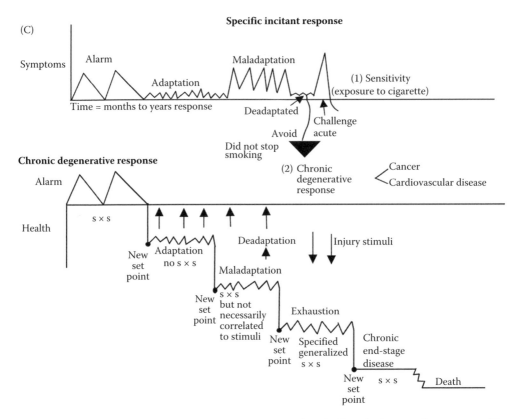

FIGURE 1.58C (1) Specific incitant response—cigarette smoke. (2) Chronic degenerative response. (From EHC-Dallas, 2005.)

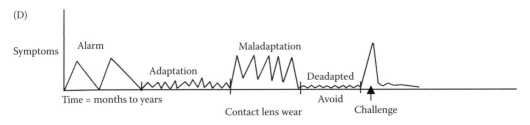

FIGURE 1.58D Specific incitant response—contact lens. (From EHC-Dallas, 2005.)

or complications of arteriosclerosis, like coronary insufficiency, renal failure, or stroke. The individual who spends his life in the nonadapted alarm stage and avoids many various incitants, continues to have energy with less chronic disease, lives a full life, and dies of old age.

6. Bipolarity

The sixth principle one must understand when studying homeostasis is the principle of bipolarity. **Bipolarity occurs when an incitant enters the body and produces a stimulatory phase followed by a withdrawal or overcompensation phase** (Figure 1.59).

When the incitant leaves or is neutralized, there is an increased response of the metabolism, the immune system, and the enzyme detoxification systems, in order to neutralize the entering noxious incitants along with this total body response. This increase in the metabolism results in an overcompensation response with a slow turn off of the neutralizing mechanisms thus, hangover-like symptoms (i.e., headache, fatigue, foggy thinking, etc.) may occur. Clinically there is a stimulation of the

The Physiologic Basis of Homeostasis

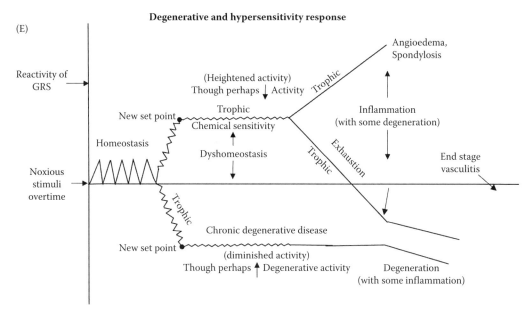

FIGURE 1.58E Overall hypersensitivity and chronic degenerative response. (From EHC-Dallas, 2005.)

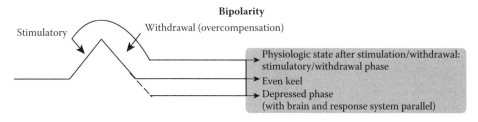

FIGURE 1.59 Bipolarity. (Modified from Rea, W.J., *Chemical Sensitivity, Volume I: Principles and Mechanisms,* Lewis Publishers, Boca Raton, FL, 1992. With permission.)

brain with the patient feeling too good (high). If the patient misinterprets this feeling as normal, he will continue to take in more contaminants, thus increasing the total body pollutant load further and disrupting the dynamics of homeostasis. If the chronic incitant(s) are removed acutely, the patient has a withdrawal or overcompensation phase in which he may develop symptoms similar to a hangover. This clinical response to incitants after acute removal appears to be due to a slow, down regulation or a turning off of the response metabolism and the detoxification systems. If the patient's metabolism is past the alarm stage and is in the stage of masking or exhaustion, a downward depression of the metabolism response phase occurs and due to a positive feedback characteristic may continue to change the metabolism. This type of change depletes the immune and enzyme detoxification systems and then can cause tissue changes until end-organ failure occurs. Some environmental incitants (e.g., CO, CCL4,) bypass the stimulatory phase causing immediate inhibition of function, thus, resulting in an immediate depressive phase.

Hormetic properties may come into play causing a low-dose stimulatory effect and a high-dose inhibitory effect on the dynamics of homeostasis. Although depending on the kinetics of the noxious substance, inhibition may initially occur at low doses and stimulation at high doses.

7. Biochemical Individuality of Response

The seventh principle one must understand that affects the dynamics of homeostasis is the biochemical individuality of response (Figure 1.60).

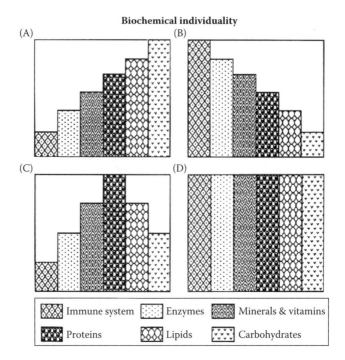

FIGURE 1.60 Biochemical individuality—uniqueness depends upon genetics, state of nutrition and total fetal body load, total current body load, as well as the state of nutrition at the time of the exposure. (A) decreased immune and normal carbohydrates (B) normal immune and decreased carbohydrates (C) decreased immune and normal minerals and vitamins (D) all normal (From Rea, W.J., *Chemical Sensitivity, Volume I: Principles and Mechanisms,* Lewis Publishers, Boca Raton, FL, 1992. With permission.)

In this stage, each individual responds to an incitant in his/her own way. This response is dependent upon the genetics their polymorphism, epigenetics, their current state of nutrition, the amount of toxics and the state of nutrition taken up by the fetus during the period of gestation, and the total body load, including the state of nutrition at the time of exposure. The individual can raise or lower the total body load depending on the amount of toxics introduced and the state of nutrition in the form of carbohydrates, fats, proteins, enzymes, and coenzymes. An example of the individual response would be when four people are exposed to the same dose of a toxic inhalant (i.e., natural gas leak). One person develops nausea and vomiting, another develops cardiac arrthymia, another experiences premenstrual syndrome, and another develops asthma.

8. Spreading
The eighth principle used in understanding the dynamics of homeostasis from environmental insults is the principle of spreading (Figure 1.61).

In the early stage of metabolic change, an individual who is exposed to a toxic load or substance in which the dynamics of homeostasis are altered returns to normal with the removal of the substance (alarm reaction). However, if the pollutant load continues to be greater, or more frequent, the previously damaged tissue becomes more labile, resulting in easier triggering of the metabolic adjustment process by related (methyl alcohol, ethyl alcohol, etc.) and then nonrelated substances. These substances (i.e., phenol, hydrolizine, mold, and beef) may continue to provoke the dynamic responses of the homeostatic system until the whole adjustment process collapses. Thus, one finds spreading of noxious stimuli to brother, sister, and closely related compounds to eventually non-related ones, which may trigger the same or various other systems to induce, local or regional dynamics of homeostasis to occur. In addition, less and less toxic substances continue to trigger the system until substances commonly thought to be nontoxic such as food will trigger a response,

The Physiologic Basis of Homeostasis

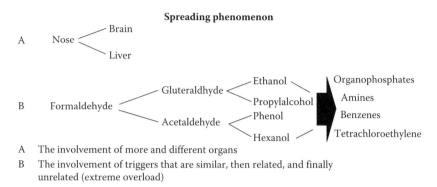

A The involvement of more and different organs
B The involvement of triggers that are similar, then related, and finally unrelated (extreme overload)

FIGURE 1.61 Spreading. (Modified from Rea, W.J., *Chemical Sensitivity, Volume I: Principles and Mechanisms,* Lewis Publishers, Boca Raton, FL, 1992. With permission.)

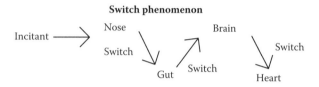

FIGURE 1.62 This is an example of the switch phenomenon, which often occurs after acute noxious stimuli challenge or over a lifetime of noxious stimuli. (From Rea, W.J., *Chemical Sensitivity, Volume I: Principles and Mechanisms,* Lewis Publishers, Boca Raton, FL, 1992. With permission.)

thus, causing a severe disruption of homeostasis. When this severe disruption and alteration occurs, the dyshomeostasis spreads to other organs causing more and varied symptoms unique to other organs.

9. **Switch**
   The last principle in understanding the dynamics of homeostasis in the presence of an increased total body load is the switch phenomena. As local and regional homeostatic disruption continues to occur, there will be temporary and, at times, permanent restoration of homeostasis in one affected end organ, while local and/or regional disruption occurs in other areas of the body. This newly sensitized area is usually due to secondary homeostatic disturbance foci (interference fields) formed by local altered metabolism and prior damaged tissue from earlier environmental insults. Therefore, an individual's responses to a total load of incitants may switch from organ to organ (changes symptoms and signs to the appropriate organ response; Figure 1.62). For example, after an initial exposure to xylene, an individual will have a runny nose. This excess flow stage later may develop into sinusitis and even later develop into asthma. Then symptoms may shift on into bloating and diarrhea. We have seen this response hundreds of times after challenge tests under environmentally controlled conditions in our hospital wing. This response can occur, after a short time or even months or years, resulting in a switch to another organ. Often organs or tissues become overloaded, damaged, or sensitized thus altering the dynamics of the homeostatic response. Therefore, a clinician must take a life long history to establish where the patient is in relation to homeostasis.

Facts about Normal Function of the Dynamics of Homeostasis

There are a few facts one should keep in mind when studying the dynamics of homeostasis related to periodic (hypersensitivity) and aperiodic (chronic degenerative) dysfunction. **First, the entire body contains 100 trillion cells with the red cells being the most abundant calculated to be around**

35 million. Total dynamic homeostasis of these cells, which occurs through the GRS, demands that communication is an orchestrated harmonic energy efficient functional response whether it is local, regional, or central. **Second**, although the many cells of the body differ markedly from each other, **all of them have basic characteristics that are alike**,[303] which helps in an efficient harmonic dynamic homeostatic response and aids in the understanding of function and positive manipulation in order to maintain the normal dynamics of homeostasis. For instance, in all cells oxygen combines with the breakdown products of carbohydrates, fat, or protein to release energy required for cell function, which allows for the harmonic integration of the dynamic cell action. Furthermore, the general mechanisms for changing nutrients into energy are basically the same in all cells, and all the cells also deliver end products of their chemical reactions into the surrounding fluids of the CTM, allowing for energy efficient transmission of information, and/or integrated homeostasis. This phenomenon again allows for positive generalized manipulation of the system so that one can intervene at any area of the body to help obtain and maintain homeostasis.

Third, almost **all cells have the ability to reproduce and when cells of a particular type are destroyed due to one cause or another; the remaining cells of this type often generate new cells** until the supply is replenished through the integrated homeostatic response mechanism. The nervous system may be an exception. However, new data suggests that even some nerve stem cells may divide particularly in the hippocampus.[307] The reproduction of cells requires a series of signals, which are regulated by the normal homeostatic control mechanism.

Fourth, about 56% of the adult human body is fluid; two-thirds of which is intracellular (IC) and one-third is extracellular (ECF). This fact is important since the ECF is part of the ERS, which covers vast areas in the body. This ECF is in motion and its distribution allows not only for great information reception, but also information distribution. **Fifth**, the ECF is in constant motion throughout the body being rapidly transported in the circulating blood and tissue fluid by diffusion through the capillary walls, which allows for rapid distribution of communicators and defenders. **Sixth**, the ECF contains ions and nutrients needed for the maintenance of cellular life, which is regulated by the GRS. A sufficient supply of these nutrients is required for efficient function and as a ready response pool to supply the defenders with neutralizing agents (such as plasma, macrophages, etc.), which counteract noxious stimuli entry. **Seventh, all cells live in essentially the same environment, which can be changed by local, regional, or central injury**. Local containment of injury is most desirable (e.g., ice pack on a burn or traumatic bruise, cleaning of a wound) because it minimizes energy demand for adjustment responses. Since all cells live within a similar environment, communication and information dissemination can be rapid and universal, although perhaps not uniform. **Eighth, cells are capable of living, growing and performing their special functions so long as the proper concentration of oxygen, glucose, different ions, amino acids, fatty substances, and other constituents** are available in the internal environment, whose function must be integrated through local, regional, and the central homeostatic mechanism. **Ninth**, the **ECF has large amounts of sodium, chloride, and bicarbonate while intracellular fluid has large amounts of potassium, magnesium, and phosphate ions**. Again, maintenance of the proper ratio of these ions depends on the nutrient supply to maintain the integrity of the membranes and their ion pumps, through harmonic energy efficient integration of the dynamics of homeostasis. Dysfunction occurs, i.e., with excess solvent exposure, which disturbs the lipid membranes, resulting in damage to ionic flux and produces edema. **Tenth, the connective tissue makes up 40% of all the body tissue surrounding every cell** and is a dominant force in the body's largest communication system (the GRS). Eighty-seven percent of the dermis and gastrointestinal tract is connective tissue, emphasizing the availability of this tissue to receive environmental information.[305] **Eleventh**, the body is spontaneously disintegrating due to oxidative stress, which is counteracted through homeostatic adjustment mechanisms, using antioxidant nutrients, oxidant-stimulated neutralizing breakdown tissue (protein buffer), and direct chemical neutralization (i.e., reduced glutathione, acid base, neutralization reactions, etc.). When combating these oxidative stresses, the body does extremely well if left alone when maintaining a reduced total body noxious stimuli load and by keeping itself in

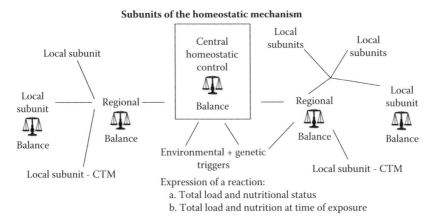

FIGURE 1.63 Myriads of subunits of the dynamic homeostatic mechanism. (From EHC-Dallas, 2002.)

the deadapted alarm phase using proper exercise and taking in clean, less polluted air, food, and water. Nutrient fuels through an organic food diet and nutrient supplementation, must be constantly supplied to replace the deficient ones required by vigorous homeostasis. If the pollutant or noxious stimuli overload continues to increase, stress will occur, thus placing a burden on the efficiently functioning homeostatic mechanisms. If further pollutant strain is placed on the homeostatic mechanisms, informational changes, metabolic changes, then tissue changes occur (sol to gel and gel to sol; NSMR), followed by fixed-named diseases, which eventually become autonomous, and finally result in end-organ failure. **Twelfth, the architecture of the three-dimensional CTM can influence and even alter the genetic response of cells**. At times, it can change the cells into malignant ones and even reverse the malignancy.[136] When one studies the dynamics of homeostasis and thus, the dynamics of chemical sensitivity and chronic degenerative disease, one realizes that the body has thousands of dynamic homeostatic control subsystems ranging from local to regional to distal (central) areas of the body, which are subunits of the orchestrated central homeostatic mechanism in the brain (Figure 1.63).

According to Guyton,[303] the most intricate of these are the genetic control systems that operate in all cells to control intracellular as well as extracellular function. There is a division of opinion on this idea since Pischinger,[3] Heine,[4] Bissal,[136] and others feel that the extracellular matrix architecture and its GRS influence and control the triggering of the genetic expression, thus emphasizing the environmental influence upon the cells and their genetics. It should be emphasized that epigenetic switches on the genes can influence and change environmentally induced changes of gene responses.[304] However, it appears there is a delicate balance between both the environment and genetics influencing all bodily functions. Clearly, genetics often control the expression of responses (although the hormetic characteristics of the noxious incitant can also control the expression of responses), but the substances in the environment are what trigger genetic action. It is quite clear that few genetic responses occur spontaneously without environmental trigger(s). Also, the human body has to deal with its surroundings at all times and thus, must take into account the variability of the total environmental pollutant load and other noxious substance load. Proper and orderly homeostatic responses depend upon the total body pollutant load and the fullness or depletion of the body's nutrient pools. The CTM, acting as the ERS for the cell is an integral part of the energetically open system of the organism. The normal response and, also, the pathological response of the genetic material in the cell depend on the function of the CTM as the major environmental receptor system, and on the total body pollutant load. Therefore, it is important that the clinician understand the facts relating to environmental exposures and its ERS, the connective tissue matrix, in order to enhance the body's homeostatic response for optimum function through the GRS. Regardless of these divergent opinions of environment versus genetics as the mechanism of body function, all agree that thousands of homeostatic subunit control mechanisms must be integrated harmoniously in order to obtain optimum efficient dynamic homeostasis (when noxious stimuli enter the body) and thus, optimum mental and physical energy as well as creativity.

The master control of the dynamics of homeostasis is intimately related to an open (by necessity) system because of the need for easy triggering, and the somewhat unstable ERS of the CTM. These two mechanisms together cause hypersensitivity in individuals. The ERS of the connective tissue matrix gives local and distal reception and feedback responses and triggers the genetic expression of the cell in order to obtain and maintain efficient homeostasis. **There are four characteristics of the homeostatic control mechanisms that need to be understood when studying the body's response to noxious incitants and thus, chronic hypersensitivity (chemical sensitivity and chronic degenerative disease).**

The Characteristics of the Dynamic Homeostatic Communication Mechanism

These characteristics are the (1) negative feedback control response; (2) the gain phase of the control response; (3) the positive feedback control response; and (4) the feed forward control (See Table 1.11). Each will be discussed separately.

The first characteristic of the homeostatic control mechanism is then a negative feedback process, which is negative to the initiating stimuli, i.e., increased ECF carbon dioxide causes increased respiratory rate through which negative feedback causes a decrease in peripheral carbon dioxide. In other words, the response to a high concentration of CO_2 is one that causes a decrease in the concentration of CO_2, which is negative to the initiating stimulus. Conversely, if the carbon dioxide falls too low this fall causes a feedback increase in the concentration. This response is also negative to the initiating stimuli.[307] Another example is that high blood pressure sends a message to the vasomotor center in the brain to decrease peripheral resistance, thus blood pressure is decreased; or vice versa, a lowered blood pressure causes a series of reactions that promote an elevated blood pressure. Therefore, in general, if some factor becomes excessive or deficient, a control system initiates a negative feedback response that consists of a series of changes that direct the response toward a certain mean value, thus maintaining the dynamics of homeostasis.[307] This process does not always work efficiently in the state of periodic homeostatic dysfunction (chemical sensitivity) or aperiodic homeostatic dysfunction (chronic degenerative disease), thus producing bodily dysfunction and energy drain. **Often the optimal dynamics of homeostasis are difficult to obtain and maintain because of excess energy needed to return to the mean value when there is an excess total body xenobiotic load or nutritional deficiency, resulting in periodic or aperiodic homeostatic disturbances**. This homeostatic deterioration is due to depletion of the nutrient pools due to excess energy requirements of the feedback adjustment systems. Eventually homeostatic deterioration occurs to the point of irreversibility.

The second characteristic of the homeostatic control mechanism is the gain of the control system. The degree of effectiveness with which a homeostatic control system maintains constant conditions is determined by the gain of the negative feedback controls. The gain-phase response describes the characteristics and stability of a control system (e.g., negative feedback occurred with 180° phase difference between stimulation and response). Positive feedback is for a phase difference

TABLE 1.11
Characteristics—The Dynamic Homeostatic Communication Mechanism

1. Negative feedback	3. Positive feedback
2. Gain of the control system	4. Feed forward control and adaptive control

Source: From EHC-Dallas, 2002.

of 0° or 360°, and if the gain is also greater than a unit, instability results. Once the feedback loop is closed, of its nature, all cause-effect information is lost. **Clinically, chronic noxious stimuli overload results in gain phase and masking**. This gain is not 100% but will be much less, like one-half to one-third[308] of the original body function. Thus, less than 100% gain characteristic is similar to the collateral blood supply in arterial occlusion, in that the collateral blood supply is likely to keep alive that organ that the blocked vessel supplies. However, there would not be enough blood supply for optimum function.

Therefore, **homeostatic responses due to chronic noxious incitant stimuli at times may be somewhat less than ideal, resulting in an accumulative deficit**. If the chronic stimuli are not removed, the gain of the negative feedback control (homeostasis) responses results in informational, metabolic, and eventually, tissue changes.[307] Another example of < 100% gain characteristic is as follows: If oxygen demand stress is placed on the cumulative deficit of < 100% gain characteristic of homeostasis, homeostatic regulatory disintegration with the onset of fixed-named disease can eventually result. Two examples of the < 100% gains of the negative control feedback mechanisms follow. In the first example, assume that a large volume of blood is transfused into a person whose baroreceptor homeostatic control system is not functioning (i.e. heart transplant patient or a phenol damaged auto sympathectomy or block, chemically sensitive patient with neuropathy, or a diabetic patient with neuropathy). The arterial pressure rises from normal levels of 100 mm Hg to 175 mmHg. Then, assume the same volume of blood is injected into the same person when the baroreceptor mechanism is functioning where the pressure rises to only 125 mmHg. Thus, the feedback mechanism caused a correction of −50 mmHg, which is from 175 mmHg to 125 mmHg. There remains an increase in pressure of + 25 mmHg, called the error, which means the homeostatic control system is not 100% effective in preventing change. The closed loop gain of the system is then the total effect divided by the error. Thus, without the baroreceptor system, the total rise is 75 mmHg, with baroreceptor sites affecting the error, the rise is + 25 mmHg. Therefore, the gain of the person's baroreceptor system for control of arterial blood pressure is a minus 25 (−50/25) mmHg. The blood pressure increase is such that only one third as much would occur if this control system were not present.

The gains of some other physiological homeostatic control systems are much greater than that of the baroreceptor system. For instance, as the second example, the gain of the system controlling body temperature is about −33°F net gain. Therefore, with the best gain plus chronic stimuli, the temperature will not return to normal, but will be much closer to normal than the blood pressure in the aforementioned example. Therefore, one can see that the homeodynamics of temperature control mechanism is much more effective than the baroreceptor control mechanism. Thus, if in certain patients like the chemically sensitive who are usually cold, with temperatures ranging from 89 to 97°F, the homeostatic control mechanism is constantly being thrown off. The dynamics of homeostasis will be difficult to obtain and maintain even under the best of circumstances in these chemically sensitive individuals because so much energy is spent to equalize or stabilize the dynamics of the homeostatic adjustment response. This characteristic may well explain one reason why the chemically sensitive are so fragile at maintaining equilibrium and are frequently disturbed by small doses of a pollutant. Their homeostatic gain is less than ideal and therefore, these patients are hypothermic. If the chronic stimulus continues and the percent gain continues to be less than the normal homeostatic response, informational and then metabolic changes will occur, followed by tissue changes, and eventually end-organ failure.

The positive feedback mechanism is the third characteristic of homeostatic control. The body uses positive feedback infrequently because positive feedback can be tied to vicious downward cycles. Two good examples of positive feedback follow. The first is the initiation of the clotting mechanism, which continues until bleeding is stopped, however, if the clotting cascade continues, past this point, intravascular coagulation will occur and damage the organism, eventually costing its life. The second is the generation of neural impulses that become constant and if the stimulus is not withdrawn, pathology will occur to the point of atrophy or severe malfunction in the form of constant pain or spasm. Often one sees a vicious downward cycle of the positive feedback mechanism in the chemically sensitive individual who is attracted, almost addictively, to acquire more harmful exposures.

These exposures continue to overstimulate and tear down the homeostatic mechanisms, exacerbating the chemical sensitivity until the patient is nonfunctional. **Positive feedback is better known as a vicious downward cycle**. However, a mild degree of positive feedback can be compensated by the negative feedback control characteristics and therefore, a vicious cycle fails to develop. For instance, if the person has bled one liter instead of two liters, the normal negative feedback mechanisms for controlling cardiac output and arterial pressure would overbalance the positive feedback and a person would recover. If, however, the person bled two (2) liters, there would be a vicious downward cycle with decreased cardiac output and blood pressure until the individual died.[309]

Positive feedback is beneficial in vessel laceration with clotting and in childbirth with uterine contractions to stop bleeding and, yet, allow for the birth of the baby. Another important use of positive feedback is for the generation of nerve signals. When a membrane of a nerve fiber is stimulated, this causes a slight leakage of sodium ions through the sodium channels in the nerve membrane in the fiber's interior.[307] The sodium ions entering the fiber then change the membrane potential, which in turn causes more opening of channels, more change of potential, still more opening of channels, etc. Thus, from a slight leakage of sodium in the beginning, there is an explosion of sodium leakage that creates the nerve action potential. This action potential, in turn, exits the nerve fiber further along its length with the process continuing until the nerve signal goes all the way to the end of the nerve. In each case, where positive feedback is useful, the positive feedback itself is part of an overall negative feedback process. The positive feedback of the clotting process is a negative feedback process for maintaining normal blood volume. The positive feedback process that causes nerve signals allows the nerves to participate in literally thousands of negative feedback nervous control systems.

The fourth characteristic of homeostasis is the principle of feed forward control and adaptive control. When there is a rapid movement, the body uses feed forward control (a rapid reflex) [in control theory this would be velocity acceleration control] since the impulse does not have enough time to get to the brain. The brain receives the impulse after the fact and determines if the appropriate response was carried out. If not, it adapts and changes for the next such movement. In the chemically sensitive or chronically diseased patient, there is an overstimulated feed forward control mechanism leading to chronic maladaptation, which will lead to nutritional deficiency from the excess work caused by the adaptation, and thus, more and increased organ dysfunction occurs.

These characteristics hold true for the thousands of homeostatic subunit control systems operating throughout the body, with the genetic expression and epigenetic switches guiding the general response; but the incitant triggering of the environmental receptor and the homeostatic response systems of the matrix and GRS are of paramount importance for activating and influencing genetic expression, which will use any of the aforementioned characteristics to function. Regional operation of end-organ, dynamic homeostatic subsystems must be integrated with central homeostatic function. The respiratory and neurological responses that regulate CO_2 in the matrix are one example of regional integration. Another example is the regulation of glucose concentration in the ECF by the liver and pancreas; still another example is the helpful regulation of sodium, potassium, phosphate, and hydrogen ions, along with others, in the ECF as performed by the kidneys. All of these subunit homeostatic functions will exhibit their own characteristics in order to have orderly dynamic central homeostasis. In all this functioning, there is a grand hierarchy of control systems.

Homeostatic dynamics involves numerous subunit homeostatic phenomena that must be constantly monitored and kept in balance because of their ever-changing characteristics.

Proteoglycans and glucoseaminoglycans (PG/GAGs) of the CTM (connective tissue) are the generators and guarantors of dynamic isoosmia, isotonia, isoionia. In addition, the dynamics of homeostasis is dependent upon many (thousands) other subunits at various levels. Many examples of dynamic homeostatic layers and processes are evident, such as the antipollutant enzymes including superoxide dismutase, glutathione peroxidase, and catalase, which initially combat the noxious stimuli once they enter the body, aiding the matrix to neutralize pollutants from the environment.

The matrix also incorporates and neutralizes free radicals from redox reactions in order to maintain the range of the dynamics of homeostasis. The GRS also balances the ability of the matrix to repair the damage from noxious stimuli, which includes the ability of the matrix to contain heat as well as the heat generated from the production of free radicals. Other examples include the ability of the matrix to contain energy versus the ability to generate energy from ATP breakdown; the ability to balance hormone releasing factor against hormone inhibitory factors; the ability to neutralize epinephrine versus the ability of the body to produce it; the formation of sol balanced with the formation of gel; hydration versus dehydration; the stimulation of nerve impulses versus inhibition of nerve impulses; enthalpy (molecular organization) balances with entropy (molecular random distribution); autocatalysis balances with order and reorganization; sufficient quantity of anticoagulant (fibrinolysis, heparin) balances with the ability to clot (fibrinogen/thrombin). In addition, energy production balances with energy loss, endocrine balance versus imbalance; ability to repair tissue balances with tissue destruction, acid balances with base, and others. The list goes on and on, and there is not enough space in this book to cover them all.

The quality and quantity of antipollutant enzymes are altered in chemical sensitivity,[310,311] **resulting in less ability to detoxify and consequently, throwing the dynamics of homeostasis off balance**. The constituents of ECF have narrow normal ranges and maximum limits, which, if exceeded, may cause death, sometimes rapidly, or sometimes slowly. Some examples of these constituents are excess temperature of 10–12°F., which results in death, excess potassium above six, which results in rapid death, and low pH below 7.0, also resulting in death, etc.

Homeostatic Worsening and Secondary Foci   If the gain characteristics are near normal reactivity, the regulatory quality of homeostasis will initially have only local effects because of the lability of electromagnetic properties and coherence of the GRS, as well as the ability of the tissue released histocytes (macrophages) containing cofactors. Prolonged stimuli will cause less gain ± and will affect the entire organism, triggering and eventually damaging the dynamics of homeostasis. See Figure 1.64.

Once the chronic stimuli is entrenched with the more sensitive damaged fascial injury, every damaged homeostatic regulatory process predisposes the existing conditions toward further damage, whether it is in local, humoral, or neural. Thus, the dynamics of the homeostatic mechanism are then made more vulnerable to triggering by new environmental incitants.

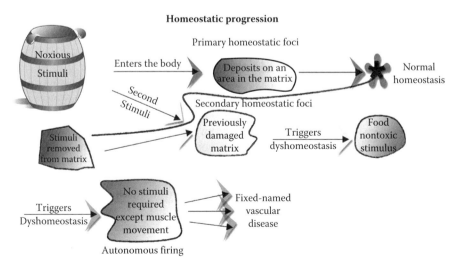

FIGURE 1.64   **(See color insert following page 364.)** Primary and secondary foci. Homeostatic triggering homeostasis and then dyshomeostasis. (From EHC-Dallas, 2002.)

Again, this homeostatic worsening is due not only to the characteristic of less than 100% gain but also to the hormetic properties of the noxious incitant. If the peripheral or central nerves become involved, then the law of nerve injury[273] would further contribute to this homeostatic worsening. Due to this down regulation of each individual dynamics of the homeostatic process, the interaction of a variety of homeostatic regulatory subsystems and eventually the integrated global homeostatic network would be affected. An example of this phenomenon would be an individual who smells the odor of a toxic substance on a food. He ingests the food. The toxic substance migrates to the olfactory nerve triggering the homeostatic mechanism in the nose and hypothalamus. The ingested part goes to the small intestine and liver, which then triggers the local homeostatic processes of the GI tract and liver. Since all four responses are less than 100% due to the gain characteristic (therefore, *metabolic alteration*) the integrated total body homeostatic response will be altered severely.

Homeostatic Control Mechanism An example of how the communication system controls homeostasis follows. The regulation of oxygen and carbon dioxide concentration in the ECF is peripheral and local (both in the case of CO_2 and oxygen), generally, being two of the major control mechanisms necessary for dynamic homeostasis. **Often tissue hypoxia occurs in the chemically sensitive and chronic degenerative patients**[312] **because of the failure of extraction of oxygen** due to shunting of blood away from critical parts of the capillary bed. Therefore, it is necessary to understand the dynamics of this normal physiologic homeostatic mechanism to enhance therapy. Since oxygen is one of the major substances required for chemical reactions in the cells, the body has a special homeostatic control mechanism to maintain an almost exact and constant oxygen concentration in the ECF. This mechanism depends primarily on the chemical content of hemogloblin. Hemogloblin (Hb) combines with oxygen as blood passes through the lungs. When oxygen is already abundant in the tissues, as blood passes through the end tissue capillaries, Hb, because of its own affinity for oxygen does not release it. If the oxygen concentrations are too low, sufficient oxygen is released to restore adequate tissue oxygen concentrations. Thus, the dynamics of homeostatic regulation of oxygen concentration in the tissue is located principally in the hemogloblin itself. However, if there is a large area of hypoxia occurring chronically, temporary vascular occlusion occurs in parts of the capillary bed causing low tissue O_2 and high capillary and venous O_2. Due to this normal physiological phenomenon, one can rapidly correct a tissue oxygen defect by elimination of the vascular pathological endothelial swelling, thus giving more blood (Hb) flow to the hypoxic tissue (by systemic oxygen administration). The endothelial swelling caused the shunting of blood away from the local tissue in damaged patients with periodic (chemical sensitivity) and aperiodic (chronic degenerative disease) homeostatic disturbances.

Carbon dioxide concentration in the ECF is regulated in a different way. Carbon dioxide is a major end product of the oxidative reaction in cells. If all the carbon dioxide formed in the cells should continue to accumulate in tissue fluids, the mass action of carbon dioxide itself would soon halt all energy giving reactions of the cells. However, the high carbon dioxide concentration in the blood excites the respiratory center in the brain causing an individual to breathe rapidly and deeply. This response increases the expiration of carbon dioxide and its removal from the blood and ECM. This process continues until the concentration of carbon dioxide returns to normal. Chemical exposure in the chemically sensitive can directly trigger the respiratory center causing hyperventilation and/or hypoventilations and thus a disturbance in the body's metabolism and homeostatic adjustment mechanism. **Often, chemically sensitive patients enter the Environmental Health Center–Dallas, with either a constant tachypnea or underbreathing, indicating they have homeostatic disturbances**. They are weak and often have high CO_2 and low peripheral venous tissue extraction of oxygen as evidenced by a high antecubital vein PvO_2 (30–65mm Hg). Clearly, they have deregulated dynamics of CO_2 and O_2 homeostasis.

Another basic homeostatic subunit control mechanism that is needed for survival of the organism through homeostasis is the regulation of arterial blood pressure. Several homeostatic subsystems contribute to the regulation of arterial pressure, being both local and distal in their responses. One of these, the baroreceptor system, is a simple and excellent example of a dynamic function of a homeostatic subunit control mechanism. In the wall of the bifurcation of the carotid arteries, as well as the arch of the aorta, are many baroreceptors, which are stimulated by stretch of the arterial wall. When the arterial pressure becomes elevated, the baroreceptors send barrages of nerve impulses to the medulla of the brain. Here the impulses inhibit the vasomotor center, which in turn decreases the number of impulses transmitted through the sympathetic nervous system to the heart and blood vessels. The decrease of these impulses causes diminished pumping activity by the heart and increased ease of blood flow through the peripheral vessels both of which lower the blood pressure toward normal. Conversely, a fall in arterial pressure releases the stretch fibers allowing the vasomotor center in the brain to become more active than usual thereby, causing the arterial pressure to move back to normal. This mechanism is lost during heart transplantation due to the nerve severance necessary for the transplant thus, altering to a point, this type of reflex homeostatic control.[313] **This mechanism also appears to be impaired in a subset of the chemically sensitive and/or chronically diseased individuals who develop drop attacks, indicating severe periodic homeostatic disturbances**. It is well known that excess phenol exposure can cause a temporary block in the sympathetic pathway or even result in autosympathectomy, which probably explains one reason why this subset of aperiodic homeostatic disturbed (chemically sensitive) patients has the problem of drop attacks.[314] (See Chapter 2 and Cardiovascular chapter in Volume 2 - Mechanisms of Chemical Sensitivity and Chronic Degenerative Disease for more information.)

Phenol can also cause chronic atrial fibrillation if applied in higher doses to the atrial epicardium in dogs[314] because it damages the sympathetic neurotransmission and the nerve impulse flow in the right atrium, which causes a heterogeneous variability of sensitivity to electric impulses in the atrial myocardium with shortened endocardial refractory periods.[315]

A combination of sympathetic and parasympathetic impulses must stimulate the epicardium and myocardium for the atrial fibrillation to occur. This same principle applies with environmentally triggered atrial fibrillation in humans as we have seen many phenol and solvent exposed (and isolated in their blood) homeostatic disturbances in our patients with atrial fibrillation and paroxymal atrial tachycardia. Since the nerve is damaged it becomes more and more sensitive distally and is easily triggered by nontoxic substances like food, pollens, some molds, and to a subset group of chemicals.

The best strategy for restoring the dynamics of homeostasis in any case of environmental assault (e.g., mercury, lead, pesticide, bacteria) upon the body is to withdraw the stimuli by decreasing the total body load and, if possible, eliminate the specific triggering agents. However, when possible, one must strengthen the nutritional status to decrease the body's total pollutant load by more efficient metabolism. The GRS is part of each of these nutrient's movements. This is because all nutrients (protein, CHC, lipids, glucose, oxygen) and waste products (carbon dioxide, urea) have to move across the molecular sieve[315] of the matrix from the basement membrane of the vascular wall, through the matrix, across the basement membrane of the end organ, pass through the cell, and then return back. In any case, in our experience, **excellent nutrition helps stabilize but, usually, does not cure periodic (chemical sensitivity) or aperiodic (chronic degenerative disease) homeostatic disturbances unless there is a decrease in the total body or specific tissue, pollutant, or other noxious stimuli load**. Only reduction of total body load over a period of time will stop aperiodic and periodic homeostatic disturbances.

Due to the aforementioned facts, nonchaotic energy, e.g., food, can send messages throughout the body via the reciprocal (feedback) information system of the GRS. Thus, the total body knows of local nutrient demand, excesses, and deficiencies, and can act accordingly to these signals as it does with a foreign intruder. In a subgroup of chemically sensitive patients, often with chemical exposure, lipotrophic factors are triggered causing a laying down of fat or conversely, the body can develop malnutrition due to failure of absorption by altered permeability of the gut and renal leaks of nutrients. In this subset of patients, a lipopenic factor appears to come into play. **Again,** misinformation acquired from excess ingestion of water contaminants and poor quality food, chemically contaminated foods, extraneous chemicals and other inhaled substances, as a result of chronic environmental pollutant exposure[316] can cause a slowing and altering of the dynamics of homeostasis. The body becomes energy inefficient usually, resulting in weakness and fatigue or dyshomeostasis with a pathologic process beginning to form. Excess wastes like urea, ammonium, and carbon dioxide can also signal the body to alter the dynamics of the homeostatic response.[317] Sometimes this response can occur to the body's detriment as well as to its benefit.

Defense System

Redox System Latency

The defense reaction usually involves release of free radicals to kill the microorganism or respond to chemical entry. It is a redox burst where the peroxide radical is liberated from the phagocyte and the OCI reaction occurs with a respiratory burst. This local liberation of a free radical can destroy whatever is around it including normal tissue. In Heine's opinion, the structural combination of water and sugar biopolymers represent the oldest information and defense system of oxygen breathing cells. These sugar polymers are suitable for helping the latent inflammatory readiness of the connective tissue to obtain a level of homeostasis, **so the body is not harmed while the redox system is taking up and giving off a normal amount of electrons**.[318] Due to this redox system phenomenon of the CTM, every situation that alters the electrical tone of the ECM can be encoded and reciprocally spread and processed locally, and, at times, throughout the body. At the same time, excess extracellular electron and protons in the form of oxygen and hydroxyl radicals, triggered by noxious incitants entry, which appear in every enzyme-guided transformation reaction (i.e., $H_2O + O_2$, Fe, Cu, $HO + HO \pm O_2$) in the body can be neutralized by the water and sugar polymers. **The resulting heat from these reactions is necessary for the further stimulation of biological processes.** The homeostatic balance between excess pollutant generation and this matrix neutralization process must always be maintained for the body to function normally. If it is not, inflammation will occur.

Clearly, obtaining energy from oxygen is necessary for life. In this process oxygen generates free radicals and thus, potential inflammation promoting free radicals has to be neutralized before the free radicals can damage local tissue. The energy that is released from the antioxidative enzyme processes and nutrients (superoxide dismutase, glutathione peroxidase, ascorbic acid, vitamins A and E), that neutralize these free radicals can be taken up by the water and sugar polymers of the ECM, which not only leads to a cooling of the heat generated from the local redox reactions but also simultaneously creates the availability of energy for the maintenance of homeostasis. The efficient energy liberation is paramount to optimum daily function. The damaging effects of the free radicals are thus buffered by the CTM.

High-energy electrons react in a similar way to the redox reaction, which to a significant extent, originates from the oxidative breakdown.[318] The electron and proton displacements that appear in enzymatic oxygen metabolism, mainly lead to the formation of multiple radicals. **With this normal process in place, therefore, energy is taken into the physiological redox potential of the organism via the ground substance.** If the enzymatic stages responsible for the electron and proton transfer are disturbed, which can begin locally, e.g., through an inadequate blood supply or direct destruction, **the result is a production of radicals with a resultant local tissue vulnerability**.

If the injury is chronic, due to excess nonspecific total or specific body load, the resulting nonphysiological alteration of the redox potential of the ground substance elaborating excess of free radicals leads to the dangerous state where chronic inflammatory disease may occur resulting in chemical sensitivity, arteriosclerosis, diabetes, etc., and often, generation of tumors. Emphasis is placed on the fact that **neutralization of free radicals after the microorganism has been killed or the entry chemical has been dissipated is essential for termination of the defense reaction. For example** an individual is bitten by a mosquito. His skin wheals but the reaction is rapidly neutralized in about 15 minutes. The wheal disappears and the reaction is terminated.

The Early Defense Mechanism

As shown previously, since the GRS is the communication system of the body it is also an integral part of the defense mechanism because of its need for adjustment and survival processes. Here, a message of injury can occur locally, regionally, distally, or centrally, and includes controlling and attracting local and distal physiological change elements such as immigration or transmigration of leukocytes, basophils, macrophages, fibroblasts, and anything else that sets off the body's defense mechanism. When the dynamics of homeostasis are strained and exceeded by chronic noxious stimuli that constantly trigger the defense mechanism, tropic changes chronic inflammation, periodic (chemical sensitivity) and eventually aperiodic (chronic degenerative diseases) homeostatic diseases can result. If the dynamics of homeostasis can halt the defense mechanism from being constantly triggered, disease will not result. **Eventually, in chemical sensitivity, the defense system may become triggered more easily until nontoxic substances, such as foods, can perpetuate or propagate the inflammation**.[319,320] The details of the generation and propagation of inflammation will be discussed later.

Emphasis should be placed on the diminution of the hypersensitivity of the defense system triggering by reduction of the total body's noxious incitant load. **The body's resistance has interconnected levels with inherited resistance and barrier systems (e.g., skin and mucous membranes) as the two original levels of defense**. This defense also includes the immune system with its unspecific and specific components (i.e., surface IGA) in the most advanced level; the mucosa-associated lymphatic system (MALT) acts as the mediator, as do other area immune systems and innate functions. These all can be enhanced by good nutrition and reduction of the total body pollutant load.

Resistance is thought to be genetically fixed and therefore, always is in evidence and passively effective (humans are, for example, naturally resistant to the cattle plague virus [innate immunity]). However, in treating thousands of chemically sensitive and chronic degenerative diseased patients over the last thirty years, we at the EHC-Dallas and Buffalo have found that the dynamics of the homeostatic resistant responses could be significantly altered by environmental and nutritional manipulation. For example, we have had patients with recurrent infections all year around. Once their total body pollutant load was reduced and their nutrition was restored their response systems were kept in the alarm stage, their infections disappeared, and for the following 25 years they had no infections. Their phagocytic index initially ranged from 50 to 70%. This range returned to 95–100% after environmental manipulation returns the immune system to vigorous resistance to any response that was environmentally induced (Figure 1.65A and B).

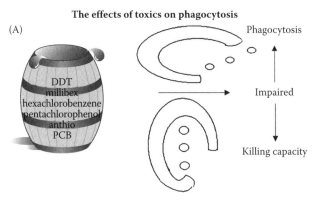

FIGURE 1.65A Effects of toxics on leukocytes and macrophages. Impaired phagocytosis and killing capacity. (From EHC-Dallas, 2002.)

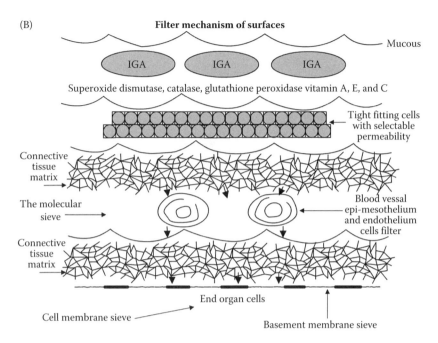

FIGURE 1.65B The filtering mechanism of the body. (From EHC-Dallas, 2005.)

The genetically dictated response system appears to be altered by the environmental changes, which could then be optimized by reduction of total body load, enhancement of nutrition, and the removal of specific triggering agents. It has been known that species resistance occurs in humans when exposed to other animals. However, it has now been shown that a portion of patients who have porcine heart valves develop antibodies to the porcine virus and when vulnerable may become ill with the disease[321] again, suggesting that environmental overload can influence innate genetic immunity.

The integrity of surface barriers is extremely important in the maintenance of the dynamics of homeostasis in that they can keep the total environmental load (virus, bacteria, toxins, and trauma) from penetrating the body and causing a strain on the system. The **maintenance of resistance of the surface barriers depends on the ability to keep the epithelial cells and oily layer of skin and mucous producing cells of the mucosa in vigorous condition** as well as, to keep the integrity of the mucosa of the gastrointestinal, urinary, genital, and respiratory tracts robust. In addition, areas of the skin and mucosa that utilize vitamins A, C, and E, surface IGA, and the antipollutant enzymes superoxide dismutase, glutathione peroxidase, and catalase, must be constantly supplied with sufficient nutrients in order to maintain robust resistance. The amount of exposure from the external environment will depend on the total environmental pollutant load (biologics, physical, and chemical components) and the specific quantity and quality of incitants. A reduction of this total and specific environmental load is paramount in maintaining health by protecting excess strain on the dynamic homeostatic mechanism. (See section "Total Body Load.")

When a noxious substance penetrates the surface barriers the process of homeostasis is activated and the adjustment mechanisms are triggered. The homeostatic adjustment process occurs at all levels of the body from skin and mucous membranes to the local CTM to every cell, cell membrane, and subcelluar or biochemical neutralizing mechanism. If the volume of the noxious substance is too great or too toxic, inflammation will occur (see section on inflammation). If the amount of the noxious substance or the level of toxicity is below a given threshold, **the homeostatic adjustment processes will function well and no inflammation occurs. This is the optimum maintenance level needed to prevent disease and continue robust health.**

The Physiologic Basis of Homeostasis

TABLE 1.12
Stages of the Normal Cellular Homeostatic Control Mechanism

1. Histiocytic (fixed tissue macrophages)
2. Microphage—PMN-activated
3. Monocyte-Macrophage
4. Lymphocytes

Source: From EHC-Dallas, 2002.

The barrier system of the skin and mucous membranes functions as an immediate reflex mechanism. It keeps microorganisms, as well as, toxics away from the primarily sterile and less toxic internal body regions and keeps them static at potential entry sites. Figure 1.65 B. The ECM is a secondary barrier system since as a molecular sieve; it is selective for the crossing of molecules and ions.[78] It keeps out those substances that are too large, such as bacteria, viruses, and many toxic chemicals. The same is true for the blood vessel walls, the lymph system filters, and the liver and splenic filters as well as the parenchymal cell membrane filter. **The barrier resistance functions in the body as a series of mechanical filters, which can stop the noxious stimuli at any or multi levels**. These filters include the skin and mucosa, the connective tissue matrix at the entry site, the blood vessel walls, the CTM after the blood vessels at the end-organ site, the lymphocytes and macrophages, the basement membrane of the end organ, and the end-organ cell membrane. See Table 1.12.

The process of local homeostasis after injury (bacteria, virus, toxic chemicals, and trauma) occurs in the following manner. **The tissue response can be viewed in four phases. These include the histiocytic, microphage, monocyte, and lymphocytic phases**. Though they occur as a continuum, each will be discussed separately.

Tissue Response to Local Entry (Initially Subinflammatory)

Tissue Bound Histiocytic (Macrophage) Defense Phase

The initial change after noxious stimuli entry is followed by a nonspecific and specific cellular defense reaction. The nonspecific response initially occurs. If this response is not satisfactory the specific immune response occurs on a continuum.

Initially in the histiocytic phase, which is nonspecific, the tissue bound histocytes (macrophages) are released by a pH decrease due to the liberation of sulfates, carbonyls, acetates and other acids cleaved by the noxious incitant from the CTM. An attempt to wall off the initiating substance whether it is bacteria, virus, parasite, foreign particles, or chemicals is made by the released tissue bound histocytes (macrophages). Also, in this phase, there is an increased permeability of the capillary wall due to late vasodilation, which allows edema to occur, thus diluting the toxics or biological incitants.[322] Vasoconstriction then occurs, creating vascular spasm in order to initiate homeostasis if the patient has a traumatic wound; or a vasoconstriction also occurs if the patient has entering noxious stimulus that creates free excess radicals and a decrease in pH due to hypoxia and acid products.

The calcium increases, the magnesium decreases, and a release of ATPases occur to support the conditions necessary to supply the initial energy for the dynamic homeostatic processes. This burst of energy has profound influence on the course of the illness. Often as in an ant bite the entry is overcome and in a short time after the toxic is neutralized, the reaction terminates. If the dynamic homeostatic energy is weak disease spreads or local inflammation is initiated. If it is strong no disease occurs and the reaction terminates.

If the injury is contained, no further reaction occurs. However, if the injury persists a little longer or is more intense, the weak signals are sent passively throughout the body.

An example of weak energy resulting in a poor response is emphasized in the following case:

Case study. A patient with chronic fatigue tries to exercise. She will not be able to do much work immediately because she already has chronic acidosis and excess total body pollutant load. The acquired increased pollutant load from the resultant exercise will make an increased demand for more oxygen in the already marginally O_2 supplied tissue, making the existing acidosis worse and causing greater incapacitation than existed before exercise was initiated. The homeostatic mechanism (already overstressed) will not be able to respond correctly and more energy will be required and the patient becomes nonfunctional. Here the < 100% gain characteristic occurs with metabolism being already marginal because the patient has chronically poor function. This chronically poor function will further decrease her overall function because metabolism cannot perform up to standard. The patient is then exposed to new incitants, like a bacterial entry through a cut. The patient's leukocytes and macrophages cannot initiate a vigorous response and the infection not only gets a foothold after entry, but also spreads rapidly, causing severe sepsis with soft tissue abscessing. As the abscess is finally brought under control, chronic osteomyelitis occurs. This condition smolders on for months, with the patient unable to eliminate the infection. Eventually the nutrient pool is depleted, and when the patient receives an inadvertent exposure to pesticide in her foods more resistance is lost. Fulminant infection occurs and the patient expires.

Microphage Phase

After the tissue bound histiocytic phase the microphage phase follows. In the microphage phase the PMNs and more local macrophages are activated. These cells migrate to the injured area and try to engulf the injuring substance. The vessels continue to leak, increasing local edema. Local edema not only dilutes the noxious substance, but allows the ingress of immunoglobulins released from the lysis of the plasma cells, which may be sensitized from a previous exposure to combat the injury. The reactions up to this stage are classified as nonspecific. In order to buffer the acidosis as well as to combat the noxious stimuli leukolysis increases over four times normal.[323]

Macrophage Phase

If the microphage phase is not followed by an adequate, rapid response to neutralize the noxious substance, a generalized systemic alarm signal occurs. This results in the specific macrophage phase with the release of the monocytic factor from the bone marrow and various lymph nodes and organs. This change to the monocytic-macrophage phase means that there is a change from the humoral shock phase to the antihumoral antishock conditions, which is a change from the prodromal phase to the active immunological phase. Humoral shock is resolved by the physiological increase and introduction of the monocytic factor (triple conjugated fatty acids), which attracts monocytes. Monocytes are attracted from the peripheral blood and bone marrow to migrate from distal areas to the local injury area, activating them to change into macrophages. This process takes from three to five days. The humoral-autonomic shock phase with monocytosis activates the entire defense system throughout the body. The lymphocytes are reduced in the blood and an increase occurs in the rate of leukocytolysis,[324] apparently to counteract the continued low pH. There is a reduction of venous blood oxyhemoglobin as a sign of peripheral blood oxygen utilization, due to the stress of the injury. More alpha and beta gammaglobulins are released from lysed plasma cells, increasing their concentration in the area of injury. These reactions are obvious in a healthy defense function but gradually disappear in the course of homeostatic regulating system disturbances like chemical sensitivity and/or chronic degenerative disease, resulting from overload of chronic noxious stimuli until they cannot be seen or reversed as the macrophage (monocytic) phase comes to an end. **Many toxics, such as hexachlorobenzene, are known to paralyze macrophage function**.[325] In a clean wound or mild toxic injury any inflammation generated is extremely mild, if there is any, and it dissipates in 4 to 5 days. If the entry area is contaminated moderate to severe inflammation occurs. (See section on the physiology of wound healing and section on inflammation.)

Lymphocytic Phase

Lymphocytic factor is released with the lymphocytes moving in, creating the lymphocytic phase. Finally, the injury may turn chronic, with lymphocytes infiltrating the area, initiating the *lymphocytic phase*. The specific immune cell response occurs. Lymphocytic factor will stimulate lymphocytes and increase cell-mediated immunity. This fact has been seen in 1500 patients receiving autogenous lymphocytic factor ALF that is isolated and developed at the Environmental Health

Center–Dallas. (See Treatment chapter in Volume 4 - Mechanisms of Cardiovascular Disease and Chemical Sensitivity.)

Immune processes are perception processes, which on a nonneuronal basis, differentiate between self and nonself, helping to maintain the dynamics of the homeostatic process. **Lymphocytes (B and T lymphocytes with subpopulations of helper, suppressor and killer cells) are capable of perception, memory, and recognition, which will allow for less energy drain and efficient homeostasis when encountering a foreign substance. According to Heine, because of their specificity,**[326] **lymphocytes may therefore, be considered as a "mobile brain" of the body, and macrophages and granulocytes as a "mobile intestine." The combination of "mobile brain and mobile intestine" will allow for clearer perception and neutralization of noxious substances in order to help maintain the dynamics of homeostasis**. (See Chapter 3: Immune System.) Thus, the lymphocytes can take part in the body's information collection system for homeostasis as environmental amplification receivers and responders, while the granulocytes, being part of the adjustment response, along with the macrophages, act as destroyers and purifiers of debris from a noxious stimulated area.

Of course, nutrient drain and even mental stress can change resistance. For example, Ader[327] has demonstrated that stress-inducing mental states (such as those nerves putting out epinephrine as measured by the heart rate variability test), which occur through the cerebral attachment of the central homeostatic control, also reduce resistance of the gut.[78] In such instances of homeostatic feedback function via the hormones of the hypothalamus-hypophysis-adrenal cortex-axis, leukocytes, peripheral nerves, and the whole basic homeostatic regulation are less resistant.[329] This same stress response occurs due to environmental noxious stimuli when autonomic neurologic responses are measured by pupillography and heart rate variability.

A characteristic of leukocyte function, of particular relevance for basic homeostatic regulation, is their capacity for physiological leukocytolysis.[3] In this case, neutrophilic granulocytes in dominating quantities are in the foreground. Each deviation of the dynamics of homeostasis (even standard deviations) is being opposed by self-destruction of leukocytes, under release of numerous biologically active substances (cytokines, prostaglandins, leukotrienes, proteases, hydrolases, DNA, RNA, etc.). This leukolysis occurs naturally, by normal body stress for buffering purposes, and unnaturally, by excess oxidant stress and inflammation. **Leukolysis (necrotic cell death), with cell and nuclear mitrochondrial and cell membrane disintegration, must not be confused with genetically conditioned "cell suicide," or apoptosis, whereby—in contrast to physiological leukocytolysis—a typical nuclear pyknosis without opening of the cell membrane is evident. The apoptotic cell just disappears without affecting the other cells**.[78] However, leukolysis may contain both types of mechanisms. (See Chapter 3: Immune System.) Thus, for good health with good homeostasis, a balance must occur between cell production, release, and physiologic cell destruction.

All homeostatic therapies, including antihomotoxic medicine, are suited to stimulate physiological leukocytolysis and to curb lysis in pathological surplus.[329,78]

Defense reactions bear a special relation to the intestinal mucosa. (See Gastrointestinal chapter in Volume 2 - Mechanisms of Chemical Sensitivity and Chronic Degenerative Disease.) This intestinal mucosa making up 75% of the body's defense response has the best-developed matrix in the body with corresponding high percentages of connective tissue cells (fibroblasts, lymphocytes, macrophages, granulocytes, plasma cells and—in particular mast cells) and embedded autonomic nerve fibers ("lymphoreticular tissue"). **The intestinal mucosa, similar to bone marrow, demonstrates that defense cells need the association with terminal autonomic nerve fibers in order to function properly.** See Figure 1.66. Thus, anatomical construction emphasizes the importance of the GRS as the communication system for homeostasis.

Immune Response

The immune system associated with intestinal mucosa constitutes the largest immune organ in the body. (See Chapter 3.) The interaction that is between matrix, leukocytes, and autonomic nerve fibers, controlled via neuroendocrine cytokines, correlates to the rhythmic cell molting of

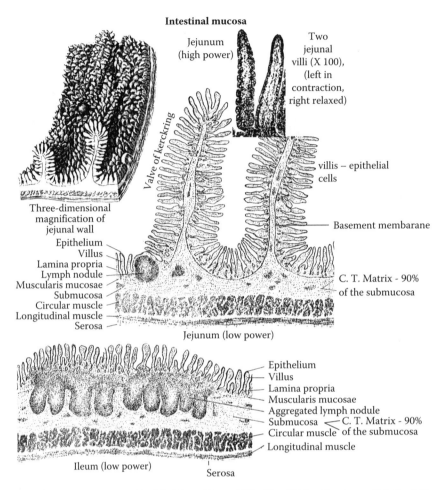

FIGURE 1.66 Intestinal mucosa. (Modified from Netter, F.H., *The CIBA collections of Medical Illustrations*, Vol. 3, New York: CIBA, 49, 1962. With permission.)

the intestinal epithelium. These functional relations, in turn, depend upon a healthy intestinal flora (ca. 10^{14} bacteria, fungi, and viruses plus 10^{14} the number of body cells). The intestinal flora may be damaged by disturbances in the basic dynamics of homeostatic regulation, even with adequate nutrition (e.g., by permanent mental strain, "stress," or chemical, bacterial, fungal [candida] or parasite overload). **The intestinal function is therefore, the potentially most often altered homeostatic disturbance factor of the organism, accounting for heartburn, gas, bloating, constipation, or diarrhea. Without robust dynamic intestinal health, one does not have robust dynamic distal homeostasis in any organ.** There are numerous examples of how intestinal health influences the rest of the body function. One example includes those patients with the HLA-B-27 genetic defect, which leads to ankylosing spondylitis. These are the patients whose symptoms, including the arthritis, are controlled almost exclusively by diet.[330] A second example is the patient who has irritable bowel syndrome along with chronic fatigue, which could be controlled by a technique of drinking nonpolluted spring water. Another example is the patient who has cardiac arrhythmias that are controlled by specific food avoidance. A further example is the asthmatic who can be controlled by dietary manipulation using specific food avoidance with a reduction of chemicals in foods.

The lymphatic system associated with the intestinal mucosa (GALT) is the largest part of the mucosa-associated lymphatic tissue (MALT) throughout the body and is, therefore, primary in maintaining the dynamics of homeostasis. It provides for an interphase between intestinal epithelium,

The Physiologic Basis of Homeostasis 139

autonomic nerve fibers, leukocytes (especially lymphocytes) and basic homeostatic regulation. These factors are the reasons to have balanced intestinal flora, digestion, absorption, and elimination, which must be maintained throughout life in order to maintain health.

Antigen absorption from the intestinal lumen into the gut associated lymphatic tissue (GALT) occurs via macrophage-like, specialized intestinal epithelial cells (M-cells or microvilli cells). **These cells phagocytize and process only unphysiological bacteria, viruses, and toxics.** Due to this process, antigenic determinants are transported back to the cell surface where they bind to MHC complexes (tissue compatibility antigen) of the glycocalyx. T- and B-lymphocytes, wandering interepithelially, absorb these determinants and thus, gather information on antigen conditions in the intestine. The lymphocytes return back to the lymph nodes of the mucosa, where they inform other lymphocytes accordingly. From then on, the B-lymphocytes start producing mucosa-protecting IgA and antigen-binding IgG. See Gastrointestinal chapter in Volume 2 - Mechanisms of Chemical Sensitivity and Chronic Degenerative Disease, for more information. The T-lymphocytes, via bloodstream and lymphatics, wander into the whole MALT to inform the defense cells therein.[331] Simultaneously, B- and T-lymphocytes can release a variety of hormones and neuropeptides. In addition, these hormones and neuropeptides are generated by the central hypophysis and the local gut chromaffin cells, which are irregularly interspersed with the mucosal cells (unicellular endocrine glands) throughout the gut. Receptors are generated for (e.g., ACTH, endorphins, somatostatin, substance P, VIP, IGF1 insulin-like growth factor, and perhaps many more). Consequently, GALT has nervous, as well as hormonal feedback, to the endocrine system completing a localized, regionalized, and distally centralized GRS. Thus the dynamics of homeostasis are maintained. As stated previously, the central homeostatic mechanism has attachments with the cerebral hemispheres, which allows mental as well as other environmental influences to impact the intestine and vice versa.[78,334] Generally speaking, exogenous and endogenous homotoxicoses (environmental and internal pollutions) may therefore have psychosomatic as well as somatopsychic origins. Since there is a two-way exchange; in most cases, it is impossible to define the onset stimulus due to the large number of possible causes, except when studying the individual under environmentally controlled conditions. In environmentally controlled conditions the total noxious stimulus load is decreased and the patient is in the deadaptive state thus, simplifying specific load cause and effect challenges (oral, inhaled, injected). However, it is always possible, generally, to evaluate the state of the total body pollutant load in any situation if the clinician understands the state of contamination of the surrounding environment. For more information, see Chapter 3.

Nonimmune Detoxification in Relation to Homeostasis and the Defense Mechanism

The basic nonimmune mechanisms for neutralizing noxious stimuli after they enter the body will now be presented. Here, if the total body load is in excess and chronic, one usually pays an energy and nutritional price for restoring dynamic homeostatic function.

Penetration of the noxious incitants occurs by invasion or intrusion through the mucosa into the CTM, which activates the GRS. An initiation, of the homeostatic response with its adjustment process is then triggered. Here local, regional, and central homeostatic responses are initiated and the early defense mechanism is activated as shown in the previous pages. **The CTM is also capable of neutralizing the free radicals often generated by the noxious stimuli and, as long as the pollutant load is not too great, inflammation will not occur. Once excess triggering is present, inflammation occurs.**

Nonimmune detoxification processes are those processes that clear the pollutant, bacteria, virus, chemical toxics, etc., from the body through the nonenzyme and enzymatic biochemical transformation systems, which are part of the homeostatic adjustment mechanism. This can be local with changes in pH, neutralization by chemical transformation, free radical neutralization, etc. Once penetration of the noxious incitant occurs, if the pollutant load is too great and the cellular defense has not neutralized, the incitant sequestration is the next line of defense. Sequestration can occur in the matrix or in any tissue or organ although muscle, liver, and brain seem predominant sequestration areas. **Sequestration is an acute survival mechanism in which the noxious incitants are** surrounded in order to prevent the dispersion of all of the noxious substances from being overwhelmingly exposed to the rest of the body all at once. Often an acute and large toxic substance enters the body triggering the sequestration process, which will respond to prevent acute death or severe injury. The price the body pays for this sequestration is prolonged internal exposure, often the generation of chronic illness, and depletion levels of the nutrients. For example, an individual is exposed to a near lethal dose of mercury. In order to survive initially the mercury is sequestered in several isolated areas of the body. As time goes on and in times of stress, the mercury is released causing chronic inflammation and then chronic disease. Nutrients will be lost and depleted in the sequestration process, and the body becomes more susceptible to new exposures. For example, the liver synthesizes and stores metallothionein for zinc metabolism; during cadmium exposure, metallothionein binds and sequesters large amounts of cadmium in the liver.[332]

The sequestration of cadmium tends to reduce its toxicity, but zinc metabolism becomes damaged in the process. Thus, zinc dependent detoxification systems function poorly and lead to decreasing detoxification of other xenobiotics (i.e., alcohol). This is one of the mechanisms by which the production of alcohol intolerance occurs in patients with chemical sensitivity. Furthermore, immune function may be altered by the zinc deficiency, which, in addition not only decreases other xenobiotic detoxification, but also alters immune function. With this decrease in detoxification, total body load increases, resulting in an increase in chemical sensitivity and/or chronic degenerative disease.

If the pollutant irreparably damages the cell, mitosis will occur to replace the dead cell. Cell damage is not always fatal because subcellular repair via biosynthetic routes is often accelerated by trauma, e.g., carbon tetrachloride causes a loss of ability to conduct and use substrate oxidation due to mitochondrial damage.[333] This damage is coupled with the synthesis of ATP and by increasing ATPase, cell metabolism intensifies, allowing healing to occur more rapidly.

Thus, the cellular repair process can restore damaged mitochondria and cell activity without necrosis and by increasing ATPase. Cells may also mutate to malignant and nonmalignant tumors, or become hypofunctional, i.e., organochlorine pesticides can induce breast cancer and impair phagocytosis.[325] These cell changes are apparently due to toxic properties of organochlorine pesticides, which can alter DNA, as well as its hormone mimicking properties, which will damage the estrogen receptors in the breasts. Different noxious stimuli produce different responses leading to susceptibility of a cell. For example, Staphylococcus causes neutrophil proliferation and thus, leukocytoclysis. Intracellular pathogens and extracellular organisms cause lymphocyte proliferation; parasitic invasion causes eosinophilia;[337] and most intestinal organisms cause a mixed response.[335] **Susceptibility of the cell to the toxin depends on the specialization of the cell, i.e., myocardial cells are susceptible to hypoxia, and rapidly dividing crypt cells in the intestines, are susceptible to DNA synthesis inhibitors like nitrogen mustard**.[336] **Unequal distribution of a toxic will alter susceptibility**; for example, toxins from the stomach are digested in the liver, which has the peptide enzyme to destroy toxins. In addition, the stomach has hydrochloric acid, which will destroy some substances. If the same toxic were in the lung, it might not be destroyed as quickly because little of this peptide exists in the lung and the resultant effect would be damage to the lung tissue. Another process that makes a cell more susceptible involves how each cell reacts to the noxious substance and then what each cell does with the noxious substance. For example, the cells of the liver have sufficient microsomal enzymes to destroy the incitant quickly, which is more than other cells in the body.[337] If the noxious stimuli penetrate the cell membrane, any of the subcellular elements can be damaged including the nuclei replication and fidelity of cell lines, which when disturbed give aberrant cells, e.g., tumors. Disturbance of DNA and RNA, altering of the activity of synthetases and mitochondrial dysfunction occur, which lowers ATP levels and produces chronic fatigue. **A pollutant driven switch from aerobic to anaerobic metabolism reduces energy generation 36 times; thus, accounting for chronic fatigue in some patients**. Lysozomes are strong nuclear phosphates, which can liberate toxic, enzymes. Peptides are disrupted by pyrogenic steroids, pachytene antibiotics, and other toxics like gypoxin. **Toxics may also alter xenobiotic catabolism in the rough endoplasmic reticulum and the smooth endoplasmic reticulum for protein generation**. Plasma membranes have micro heterogenicity with antigen recognition sites, which are disturbed by noxious stimuli, resulting in failure to recognize antigens of different foods, molds, pollens, bacteria, and viruses. When damaged, the patient becomes pan hypersensitive—as seen in some chemically sensitive patients.

Pollutant damage to the molecular target frequently occurs.[338] Damage to proteins can disrupt enzyme function, affecting nucleic acids, plus structural, carrier, storage, and regulatory proteins.[339] Transport enzymes can be damaged causing edema and intracellular imbalance, hormonal alteration, altered messenger activities, and imbalance of membrane inhibitors.[340]

Enzyme dysfunction can be widespread or local, i.e., the lactic dehydrogenase or cytochrome (p-450) damage will affect or eliminate certain types of detoxification, like those in conjugation with oxidative biochemical detoxification processes.[341]

Organophosphate or carbamate pesticide exposure will damage the inhibition or the alterations of the cholinesterase enzyme.[342] Nucleic acid in DNA is disrupted by irradiation and aromatic polycyclic hydrocarbons; both can cause tumor formation.[343] Damage to structural proteins may occur, causing weight loss.[344] The damage to carrier and storage proteins may occur, i.e., proteins can be affected by CO, NO_2, alkalines, alkamines, CN, and H_2S.[345] Informational or regulatory proteins may be injured by noxious stimuli causing alteration in the brain function and neurological responses, which can then rapidly result in mood swings, depression etc.[346]

Coenzymes are present in limited numbers and can be damaged by noxious incitants. For example, **free radical attack on NADPH leads to the rapid dysfunction, causing damage to the Krebs cycle function and energy generators of the ADP to the ATP.**

Integrity of the lipid membranes can easily be damaged by lipophilic organic hydrocarbons, solvents, etc. This damage can lead to auto transfusion of the xenobiotics that are attached to the membranes. This auto transfusion can cause a shifting of minerals into the cell in areas that should not have them, i.e., Ca,++ aluminum, manganese, lead, etc., can cause intracellular wall damage. The cell surfactants of the lungs may be destroyed, and blood cholesterol may be increased by this xenobiotic assault. Manganese damages protein.[347] Carbohydrates as an important component of the GRS and ECM of this system may be disturbed by xenobiotics causing inability to communicate.

Pollutant damage to the metabolism of foreign compounds can be caused by an increase in total body load or by an increase in specific pollutants. There is much noncatalytic reactive talk between nucleophils (an electron donor) and electrophils (electron acceptor) in which one is of exogenous origin and the other is of matrix or intracellular origin.[348] For example, sodium bicarbonate neutralizes hydrochloric acid in the stomach. Another example is the reduced glutathione's direct neutralization of 1,2 ethanes, mercury, cadmium, lead, or many other compounds (Table 1.13).

Other direct detoxification reactions include the following: cysteine directly neutralizing nitrooxides, sodium nitrate, and nitrobenzene compounds; NADPH neutralizing nitro benzenes, and benzopyrenes; albumin directly neutralizing benzopyrene and phenyl acetate; amino derivatives of Schiff' bases neutralizing biogenic amines; carbonyl will directly neutralize hydrazines and hydrazides to hydrazones; neutral aqueous solutions (e.g., plasma) neutralizes cocaine and some benzopyrenes; acid media (acid on the stomach) neutralizes many xenobiotics as does sodium bicarbonate; while salicylic acid and acetaminophen neutralize each other.[349] **The body has a great capacity to be protected by nonenzymatic chemical reactions.**[350]

Catalytic conversion of foreign compounds is made possible by partitioning the substructures (liphophilic chemicals) between phases (ECM, nucleus, mitochondria, rough endoplasmic reticulum, golgi apparatus, lyzosomes, and the hydrophilic cytoplasm.) **Present knowledge suggests that the major share of biotransformation is by enzyme systems, but one cannot discount nonenzyme neutralization of chemicals.**

The metabolism of foreign compounds usually occurs in the microsomal fractions of the rough endoplasmic reticulum (Figure 1.67) but some metabolism occurs in the mitochondria and other areas of the cell.

The cytochrome[350] system consisting of B_2, B_3, B_6, Fe, Zn, Mg, Se, Cu, and O_2 is needed for the detoxification, as it is the most common system involved. Other nutrients involved are vitamin C,

TABLE 1.13
Pollutant Overload Can Disturb the Following Type of Noncatalytic Reactions between Nucleophils and Electrophils in the Chemically Sensitive and Chronic Degenerative Diseased Individual

Electrophil		Nucleophil		Less Toxic End Product
1,2 disubstituted ethanes	+	Glutathione	→	Ethylene
Aryl nitroso compounds	+	Glutathione	→	Aryl amines, Aryl hydroxyl amines
Nitro oxide	+	Cysteine	→	S-Nitrocysteine
Sodium nitrite				
Nitro benzene	+	NADPH	→	S-Nitrocysteine
Benza pyrene	+	NADPH	→	Phenylhydroxyl-amine
Phenyl esters	+	Albumin	→	Bound + out
Phenyl acetate				
Biogenic aldehydes	+	Amino derivative of Schiff's bases	→	Alcohols → CO_2 + H_2
Biogenetic Ketones				
Hydrazines	+	Carbonyl group	→	Hydrazones
Hydrazides				
Cocaine	+	Neutral aqueous solutions (e.g., plasma)	→	Less toxic substances
Some benzopyrenes				
Xenobiotics	+	Acid media (e.g., stomach)	→	Less toxic substances
Xenobiotics		$NaHCO_3$	→	Less toxic substances
Reaction with each other				
Salicylic acid	+	Acetaminophen		Less toxic substances

Source: Rea, W.J., *Chemical Sensitivity, Vol I. Principles and Mechanisms*, CRC Press, Boca Raton, FL, 69(1), 1992. With permission.

folic acid, phosphotidyl choline, and bioflavanoids. In this process, oxidation (loss of electrons), which can be either enzymatic or nonenzymatic, is one of the prime detoxification processes. Occasionally this process also requires water and NADPH and, at times, even enzymes (oxidases). Reduction (gain electrons) requires lack of O_2 and NADPH, and reductase degradations use water, carbonyl-transferases, and carbonyl-esterases.

Conjugation reactions are often primary and require O_2. The five general categories include (1) **acetylation** (also needs acetyl coenzyme A, N-acetyl transferase, B_5, and a pH of 5.6-7.2); (2) **acylation** (peptide conjugation by carboxylase with the amino acids, glycine, glutamine, and taurine. These require acetyl-ligases and acetyl transferases, and ATP in the mitochondria. Excretion is by benzoic acid, hippuric acid and glycocides; (3) **sulfur conjugation**, which uses PAPS, glutathione transferases, inorganic sulfates, and cysteine or rhodanase; (4) **methylation,** which uses S-Adenosyl-L-Methionine dependant methyl transferase, B_{12}, folic acid, magnesium, and oxygen with a pH balance 7.3-8.2; (5) **gluconation** or **glucuronidation** in the endoplasmic reticulum, which in order to neutralize many amines, phosphates, histamine, serotonin, etc., uses glucuronic acid and uridine diphosphate (glucuronyl transferase), oxygen, and a pH of 7.8. Glucuronidation detoxifies the bulk of pollutants, as well as phenols and glucuronidation and has the highest capacity to detoxify phenol, but has a low affinity for the substance. It is possible for this system to induce bladder cancer if the circumstances are just right for arylamines and polycyclic aromatic hydrocarbons can occur through the pathway.

The Physiologic Basis of Homeostasis

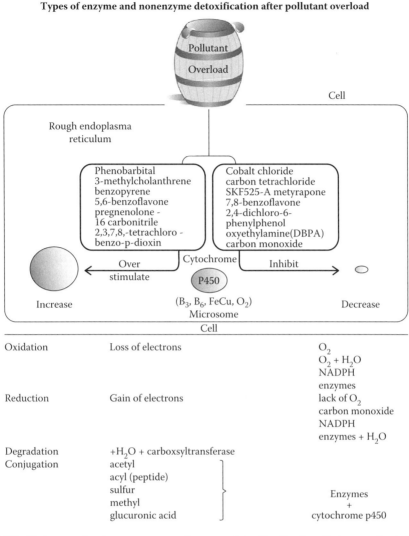

FIGURE 1.67 Redox reaction to nonenzyme and enzyme detoxification in pollutant overload. (From EHC-Dallas, 1995.)

Occasionally detoxification pathways produce toxic effects or bring out more toxics. For example, cyanide, or chlorinated or fluorinated hydrogen sulfide compounds will create a pseudo halogen trigger effect on the thyroid, causing thyroiditis. Another example of toxic effects is in the oxidation of organohalides, carbon disulfides, and the insecticides of carbonyl phosphothiomate that produces formaldehyde, carbon monoxide, and phosgene a nerve gas. These compounds can decrease enzyme activity and affect the function of the cytochrome p-450 system.

All of the defense mechanism is geared to be sub inflammatory. This situation can be accomplished if it is nutrient replete and the pollutant load is not too high. **Once these two conditions are violated inflammation occurs and a whole new situation of altered physiology and illness begins.**

Repair Mechanism—Healing

Wound healing is constantly and simultaneously occurring after release of proteolytic enzymes and growth factors by macrophages do their job of destroying damaged tissue. **Repair depends upon** the fibroblasts and the vigor and intactness of the NSMR and the body's state of nutrition. Many vulnerable areas in the body have nutrient blocks in their healing process. These wounded areas do not need tissue replacement but they do need nutrients in order to resume normal function. Other wounds are of a greater magnitude and actually need tissue replacement. Those wounds of greater magnitude trigger the NSMR from the fibroblasts to replace the gaps in normal tissue causing scarring, thus healing the tissue. **If too great or intense, of wound from trauma or a noxious incitant entry scarring always occurs but if the incitant is mild, scarring does not occur.** Each will be discussed.

Nonspecific Mesenchyme Reaction

The CTM is not only an epithelial mucosal support system and a dynamic mechanical filter system but because of aforementioned characteristics discussed in this section, it is, also, part of one of the prime communication systems and repair system of the body. This communication is a regulating factor in the dynamics of homeostasis, because the CTM, at the proper time and signal, releases wound-healing substances. **The healing wound contains the end capillary, the end autonomic nerve, the lymphatic, the cellular defense by macrophages, which release proleolytic enzymes to dissolve tissue, the repair mechanism by the fibroblasts (collagen, PG/GAGs), and mast cells.** The NSM reaction is extremely active sifting information for wound healing and then adjusting responses to accommodate and neutralize both external and internal noxious stimuli, which allows wound healing to occur.

This healing response can occur from an extremely minor injury to major wounds. It is constantly occurring with normal physiology due to the normal breakdown and the need for healing of the CTM. The majority of injuries that occur in the body are minor but they can become significant if they become chronic, which can result in major scarring and even end-organ failure. If local and regional homeostatic control is lost, major healing without scarring will not occur.

> **Case study**. A 43-year-old white female presented at the EHC-Dallas with severely swollen and ulcerated legs. She was barely able to walk and experienced excruciating pain in both legs. She showed 4+ pitting edema in each leg. Ulcers measuring 3 to 16 centimeters were present in both legs and there were multiple areas of spontaneous bruises and vascular pusticular eruptions. See Figure 1.68. Laboratory results showed decreased suppressor T cells, CD_3, CMI (2), altered autonomic function, and PvO_2—55 mm/Hg. The patient was placed on a rotary diet of organic food and told to avoid the foods to which she was sensitive after intradermal testing showed positive results. Further treatment included drinking glass bottle spring water and cleaning up her home environment. She was given antigen injections for food, molds, and chemicals to which she was sensitive. Intravenous and oral nutrition was administered and oxygen therapy at 8L for 2 hours per day was ordered. Over a period of three months the ulcers and edema completely cleared and she did well preventing recurrences.

The brain can influence the periphery but also the periphery can influence the brain. It has often been observed when an individual with pain or sequestered pollutants that result in muscle and fascial spasm is treated by some form of physical therapy at times, physiologically, metabolically, immunologically, and as information flows they immediately have more astute brain function, mental clarity, and energy. The clearing responses occur through the GRS when the dynamics of homeostasis are obtained and maintained. However, when the noxious stimuli are prolonged or increased, the NSMR augments for healing, causing a series of anatomical changes that will result in a change in physiology. **This NSM reaction can be aberrant due to altered homeostatic physiology, resulting in disease processes like arteriosclerosis, kidney, and lung failure**.

Taking into consideration the previously discussed anatomical and physiological facts, any noxious stimulus can trigger this system, which then sends information dissemination locally, regionally, and centrally, and with a series of metabolic changes, including vascular spasm. A domino effect results in an anatomical nonspecific mesenchysmal response, which alters the body morphologically thus, forming fibrous scars.[127] This response involves the fibroblast (as described

The Physiologic Basis of Homeostasis

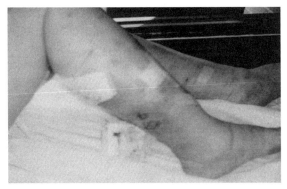

4+ Edema and ulcerated legs before treatment

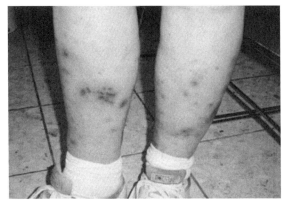

Ulcerated legs totally healed, edema gone
after 3 months of treatment

FIGURE 1.68 (See color insert following page 364.) Ulcerated legs healed after massive avoidance of pollutants and food. Other treatment included injection therapy with molds, chemicals, and foods to which she was sensitive; intravenous and oral nutrient therapy; sauna/physical therapy; oxygen therapy, and use of autogenous lymphocytic factor. (From EHC-Dallas, 2004.)

under cellular defense), which release more PG/GAGs, collagen, and the tissue macrophage, which releases growth factors including proteolytic enzymes. The circulating macrophages also release growth factors as do the monocyte and the granulocyte, which release proteases, and lysozymes. The mast cells, which release histamine and heparin, and the plasma cells, which release IGG, all are involved in the healing of tissue. **The noxious environmental stimulations to the aforementioned cells can result in a cascade of events resulting in connective tissue breakdown due to the release of proteases, i.e., matrix metalloproteinases (MMPs), lysozymes, hyrdolases from leukocytes, mast cells, and macrophages with subsequent repair by the NSMR, resulting from the macrophage, mast cells, and fibroblasts.**

With an acute large injury the connective tissue will be broken down in the area surrounding the injury. This area of injury will heal with the production of collagen and PG/GAGs from the fibroblasts causing a scar. **If the injury is mild no scar formation will occur. If the injury lasts a longer time, is deeper, and more intense, scar formation occurs.** Scarring continues to occur as repeated low levels of noxious stimuli enter the body. These areas of excess wounding tend to occur at the same place; interference fields are also found in the same organ until end-organ failure occurs.

An example of excessive wound healing would be a kidney that is constantly being damaged by a heavy metal like lead with wound healing by the NSMR occurring. Eventually one would have so much scar tissue that the kidney would be dead and end-organ failure would occur.

Physiology of Wound Healing

Although the wound-healing process differs slightly from tissue to tissue, the process is similar throughout the body. The result in almost all tissues is scar when the wound is severe enough. The goal of acute wound management is to facilitate the body's innate tendency to heal so that a strong but minimally apparent scar results. Generally, the normal wound-healing process cannot be accelerated.

The physiology of wound healing is described in phases and differs little from the initial cellular response just described. See Figure 1.69.[351] Although each of these phases will be discussed as a separate entity, the phases blend without distinct boundaries.

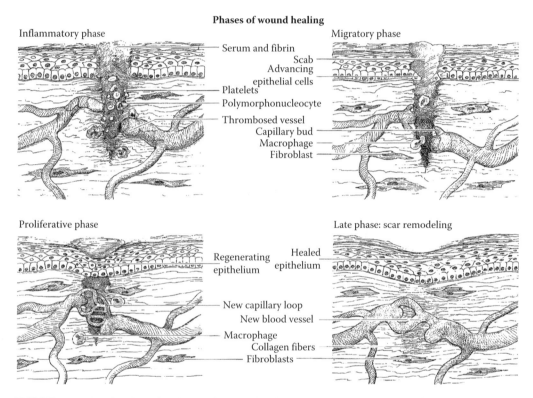

FIGURE 1.69 Depicted are the phases of wound healing. In the early phases (top, left), platelets adhere to collagen exposed by damage to blood vessels to forms a plug. The intrinsic and extrinsic coagulation cascades generate fibrin, which combines with platelets to form a clot in the injured area. Initial local vasoconstriction is followed by vasodilatation mediated by histamine, PGE_2, PGI_2, serotonin, and kinins. Neutrophils are the predominant inflammatory cells (a polymorphonucleocyte is shown here). In the migratory phase (top, right), fibrin and fibronectin are the primary components of the provisional wound matrix. Additional inflammatory cells, as well as fibroblasts and other mesenchymal cells, migrate into the wound area. Gradually, macrophages replace neutrophils as the predominant inflammatory cells. Angiogenic factors induce the development of new blood vessels as capillaries. Epithelial cells advance across the wound area from the basal layer of the epidermis. The fibrin-platelet clot may dehydrate to form a scab. In the proliferative phase (bottom, left), the advancing epithelial cells have covered the wound area. New capillaries form. The wound's strength grows as a result of steadily increasing production of collagen and glycosaminoglycans by fibroblasts. Collagen replaces fibrin. Myofibroblasts induce would contraction. In the late phase (bottom, right), scar remodeling occurs. The overall level of collagen in the wound plateaus; old collagen is broken down as new collagen is produced. The number of cross-links between collagen molecules increases, and the new collagen fibers are aligned so as to provide a gradual increase in wound tensile strength. New capillaries combine to form larger vessels. The epithelium is healed, although it never quite regains its normal architecture. (From Lawrence, W.T., Bevin, A.G., and Sheldon, G.F., ACS Surgery: Principles and Practice, 7 Acute Wound Care, WebMD Inc., Fig. 3, 2007. With permission.)

Hemostasis   Most wounds extend into the dermis, injuring blood vessels and resulting in bleeding. The process stimulates vasoconstriction in the injured vessels, mediated by epinephrine from the peripheral circulation and norepinephrine from the sympathetic nervous system. Prostaglandins, such as prostaglandin $F_{2\alpha}$ ($PGF_{2\alpha}$), and thromboxane A_2 are also involved. As the vessels contract, platelets adhere to the collagen exposed by damage to the blood vessel endothelium and form a plug. Platelet aggregation during the homeostatic process results in the release of cytokines and other proteins from the alpha granules of the cytoplasm of platelet. These cytokines include platelet-derived epidermal growth factor (PDGF), transforming growth factor-β (TGF-β), transforming growth factor-α (TGF-α), basic fibroblast growth factor (bFGF, fibroblast growth factor 2 [FGF2]), platelet derived epidermal growth factor (PD-EGF), and platelet-derived endothelial cell growth factor (PD-ECGF). Some of these cytokines have direct effects early in the healing process, and others are bound locally and play critical roles in later aspects of healing.

The extrinsic coagulation cascade is stimulated by a tissue factor released from the injured tissues and is essential for clot formation. The intrinsic cascade is triggered by exposure to factor XII and is not essential. Both coagulation cascades generate fibrin, which acts with platelets to form a clot in the injured area. **In a large wound, the superficial portion of this clot may dehydrate over time to produce a scab**. In addition to contributing to hemostasis, fibrin is the primary component of the provisional matrix that forms in the wound during early healing. Fibrin becomes coated with vitronectin from the serum and fibronectin derived from both serum and aggregating platelets. Fibronectins are a class of glycoproteins that facilitate the attachment of migrating fibroblasts as well as other cell types to the fibrin lattice.[352] **By influencing cellular attachment, fibronectin is a key modulator of the migration of various cell types in the wound**.[353,354] In addition, the fibrin-fibronectin lattice binds various cytokines released at the time of injury and serves as a reservoir for these factors in the later stages of healing.[355]

Inflammation   **Tissue damage at the site of injury stimulates the inflammatory response. This response is most prominent during the first 24 hours after a wound is sustained. In clean wounds, signs of inflammation dissipate relatively quickly, and few if any inflammatory cells are seen after 5–7 days**. Healing progresses as is shown in the section of the macrophage phase. In contaminated wounds, inflammation may persist for a prolonged period. **The signs of inflammation, originally described by Hunter in 1794,[356] include rubor (erythema), tumor (edema), calor (heat), and dolor (pain)**.

The signs of inflammation are generated primarily by changes in the 20–30μm diameter venules on the distal side of the capillary bed. In the first 5–10 minutes after wounding, the skin blanches as a result of vasoconstriction that contributes to hemostasis. The initial vasoconstriction is followed by vasodilatation, which generates the characteristic erythema. The **vasodilatation** is mediated by (1) vasodilator prostaglandins such as **PGE_2 and prostacyclin**, released by injured cells; (2) **histamine**, released by mast cells and possibly by platelets to a lesser degree; (3) **serotonin**, also released by mast cells; (4) **kinins**, the release of which is stimulated by the coagulation cascade; and possibly by other factors as well. As the blood vessels dilate, the endothelial cells lining the microvenules tend to contract and separate from one another, resulting in increased vascular permeability. Serum migrates into the extravascular space, giving rise to edema. Inflammatory cells initially adhere loosely to endothelial cells lining the capillaries and roll along the endothelial surface of the vessels. **The inflammatory cells eventually adhere to the vessel wall, in a process mediated by the β2 class on integrins, and subsequently transmigrate into the extravascular space**.[357] Chemoattractants stimulate the migration of inflammatory cells to the injured area. As monocytes migrate from the capillaries into the extravascular space, they transform into macrophages in a process mediated by serum factors and fibronectin.[358–360] After migration, the inflammatory cells must be activated before they can perform their biologic functions.

Neutrophils are the predominant inflammatory cell in the wound during the 2–3 days after wounding, but macrophages eventually become the predominant inflammatory cell in

the wound. Since monocytes are present in the serum in much lower numbers than neutrophils, it is not unexpected that they are rarely seen in the wound area initially. After appearing in the wound, both neutrophils and macrophages engulf damaged tissue, digesting them in lysosomes. After neutrophils phagocytize damaged material, they cease to function and often release lysosomal contents, which can contribute to tissue damage and a prolonged inflammatory response. Inflammatory cells and liquefied tissue are the constituents of pus, which may or may not be sterile, depending on whether bacteria are present. Unlike neutrophils, macrophages survive after phagocytosing bacteria or damaged material. **The shift to predominant inflammatory cell type within the wound from neutrophils to macrophages is at least in part due to macrophages' extended life span**. Macrophage-specific chemoattractants may also selectively attract macrophages into the wound.

In addition to phaocytosis, macrophages are capable of secreting MMPs that break down damaged tissue; they are also a primary source of cytokines that mediate other aspects of the healing process. **Experimental studies have demonstrated that neutrophils are not essential to normal healing**,[361] **whereas macrophages are necessary**.[362] These additional macrophage functions—especially their role as a cytokine source—most likely are what make them essential.

Migratory Phase   Many substances attract fibroblasts and other mesenchymal cells into the wound during the migratory phase, including many of the cytokines[352,363–365] (Table 1.14). It is not known which of them are most active biologically at different points after wounding. **The fibroblasts migrate along the scaffold of fibrin and fibronectin**, as mentioned. This migration involves the upregulation of integrin receptor sites on the cell membranes, which allows the cells to bind at different sites in the matrix and pull themselves through the scaffold. Migration through the provisional matrix is also facilitated by synthesis of MMPs, which help cleave a path for the cells. Additional cytokines stimulate the proliferation of mesenchymal cells important in the wound-healing process once these cells have been attracted into the would area[366,367] (see Table 1.14).

Angiogenesis   Angiogensis is also initiated in the migratory phase during the first 2 or 3 days after wounding. Before revascularization of the injured area, the wound microenvironment is hypoxic and is characterized by high lactic acid levels and a low pH. **Angiogenic factors stimulate the process of neovascularization. Some of the more potent angiogenic factors are derived from platelets and macrophages**.[368,369] (See Tables 1.14 and 1.15.) New vessels develop from existing vessels as capillaries. The capillaries grow from the edges of the wound toward areas of inadequate perfusion within the provisional wound matrix, where lactate levels are increased and tissue oxygen tension is low. The generation of new vessels involves both migration and proliferation of cells. Both cellular activities are modulated by the angiogenic cytokines. **A key aspect of endothelial cell migration is the upregulation of the α-β_3 integrin binding domain that facilitates the binding of the endothelial cells to the matrix**. Migrating endothelial cells produce plasminogen activator, which catalyzes the breakdown of fibrin, as well as MMps, which help create paths through the matrix for the developing blood vessels. When the budding capillaries meet other developing capillaries, they join and blood flow is initiated. As the wounded area becomes more vascularized, the capillaries consolidate to form larger blood vessels.

Epithelialization   **Epithelialization of skin involves the migration of cells from the basal layer of the epidermis across the denuded wound area**.[370] This migratory process begins approximately 24 hours after wounding. The migrating cells develop bands 40–80 Å wide that can be seen with electron microscopy and stained with antiactin antibodies. About 48 hours after wounding, the basal epidermal cells at the wound edge enlarge and begin to proliferate, producing more migratory cells. If the normal basement membrane is intact, the cells simply migrate over it; if it is not, they migrate over the provisional fibrin-fibronectin matrix.[371] As migration is initiated, desmosomes that link epithelial cells together and hemidesmosomes that link the cells to the basement membrane disappear.[372] Migrating

TABLE 1.14
Involvement of Cytokines in Wound-Healing Functions

Wound-Healing Function	Cytokines Involved
Neutrophil chemotaxis	PDGF
	IL-1
Macrophage chemotaxis	PDGF
	TGF-β
	IL-1
Fibroblast chemotaxis	EGF
	PDGF
	TGF-β
Fibroblast mitogenesis	EGF
	DG
	IGF
	TGF-β
	TGF-α
	IL-1
	TNF-α
Angiogensis, endothelial cell chemotaxis, mitogenesis	EGF
	Acidic and basic FGF (FGF1 and FGF2)
	TGF-β
	TGF-α
	VEGF
	PD-ECGF
Epithelialization	EGF
	Basic FGF (FGF2)
	PDGF
	TGF-β
	TGF-α
	KGF
	IGF
Collagen synthesis	EGF
	Basic FGF (FGF2)
	PDGF
	TGF-β
	IL-1
	TNF-α
Fibronectin synthesis	Basic FGF (FGF2)
	PDGF
	TGF-β
	EGF
Proteoglycan synthesis	Basic FGF (FGF2)
	EGF
Wound contraction	Basic FGF (FGF2)
	TGF-β

(Continued)

TABLE 1.14
Involvement of Cytokines in Wound-Healing Functions (Continued)

Wound-Healing Function	Cytokines Involved
Scar remodeling, collagenase stimulation	EGF
	PDGF
	TGF-β
	IL-1
	TNF-α

Source: From Lawrence, W.T., Bevin, A.G., and Sheldon, G.F., ACS Surgery: Principles and Practice, 7 Acute Wound Care, Table 4, WebMD Inc., 2002. With permission.

Note: EGF—epidermal growth factor FGF—fibroblast growth factor IGF—insulin like growth factor IL-1—interleukin-1 KGF—keratinocyte growth factor PD-GF—platelet-derived endothelial cell growth factor PDGF—platelet-derived growth factor TGF—transforming growth factor TNF—tumor necrosis factor VEGF—vascular endothelial growth factor.

TABLE 1.15
Cell Sources of Cytokines

Cell Type	Cytokines
Platelet	EGF
	PDGF
	TGF-β
	TGF-α
Macrophage	FGF
	PDGF
	TGF-β
	TGF-α
	IL-1
	TNF-α
	IGF-1
Lymphocyte	TGF-β
	IL-2
Endothelial cell	FGF
	PDGF
Epithelial cell	TGF-α
	PDGF
	TGF-β
Smooth muscle cell	PDGF

Source: From Lawrence, W.T., Bevin, A.G., and Sheldon, G.F., ACS Surgery: Principles and Practice, 7 Acute Wound Care, Table 5, WebMD Inc., 2002. With permission.

cells express integrins on their cell membranes that facilitate migration. As they migrate, they secrete additional proteins that become part of the new basement membrane, including tenascin,[373] vitronectin, and collagen types I and V. In addition, they generate MMPs to facilitate migration as noted.

When epithelial cells migrating from two areas meet, contact inhibition prevents further migration. The cells making up the epithelial monolayer then differentiate into basal cells and divide, eventually yielding a neoepidermis consisting of multiple cells layers. Epithelialization progresses both from wound edges and from epithelial appendages. Epithelial advancement is facilitated by adequate debridement and decreased bacterial counts, as well as by the flattening of rete pegs in the dermis adjacent to the wound area. **The epithelium never returns to its previous state**. The new epidermis at the edge of the wound remains somewhat hyperplastic and thickened, whereas the epidermis over the remainder of the wound is thinner and more fragile than normal. The rete pegs do not form in the healed area.

Proliferative Phase and Collagen Synthesis The proliferative phase of wound healing begins approximately five days after wounding. During this phase, the fibroblasts that have migrated into the wound begin to synthesize PGs and collagen, and the wound gains strength. Until this point, fibrin has provided most of the wound's strength. Although a small amount of collagen is synthesized during the first five days of the healing process,[374] the rate of collagen synthesis increases greatly after the fifth day. Wound collagen content continually increases for three weeks, at which point it begins to plateau.[375]

Although there are at least 18 types of collagen (discussed previously), the ones of primary importance in skin are type I, which makes up 80–90% of the collagen in skin, and type III, which makes up the remaining 10–20%. A higher percentage of type III collagen is seen in embryologic skin and in early wound healing. **A critical aspect of collagen synthesis is the hydroxylation of lysine and proline moieties within the collagen molecule**. This process requires specific enzymes as well as oxygen, vitamin C, α-ketoglutarate, and ferrous iron, which function as cofactors. Hydroxyproline, which is found almost exclusively in collagen, serves as a marker of the quantity of collagen in tissue. Hydroxylysine is required for covalent cross-link formation between collagen molecules, which contributes greatly to wound strength. **Deficiencies in oxygen or vitamin C or the suppression of enzymatic activity by corticosteroids may lead to underhydroxylated collagen incapable of generating strong cross-links. Underhydroxylated collagen is easily broken down**. After collagen molecules are synthesized by fibroblasts, they are released into the extracellular space. There, after enzymatic modification, they align themselves into fibrils and fibers that give the wound strength. Initially, the collagen molecules are held together by electrostatic cross-links as fibrils form. These cross-links are subsequently replaced by more stable covalent bonds. The covalent bonds form between lysine and lysine, between lysine and hydroxylysine, and between hydroxylysine and hydroxylysine;[376] the strongest cross-links form between hydroxylysine and hydroxylysine.

PGs, also synthesized during the proliferative phase of healing, consist of a protein core linked to one or more glycosaminoglycans. Dermatan sulfate, heparin, heparin sulfate, keratin sulfate, and hyaluronic acid are the more common PGs. The biologic effects of PGs are less well understood than those of collagen. They generally anchor specific proteins in certain locations and affect the biologic activity of target proteins. Heparin is an important cofactor of bFGF during angiogenesis. Other PGs most likely facilitate the alignment of collagen molecules into fibrils and fibers.

Wound Contraction Collagen has no contractile properties, and its synthesis is not required for wound contraction. During the **proliferative phase, myofibroblasts appear in the wound and probably contribute to its contraction**.[377] Myofibroblasts are unique cells that resemble normal fibroblasts and may be derived from them. They have convoluted nuclei, vigorous rough endoplasmic reticula, and microfilament bundles 60–80 Å in diameter. These microfilaments can be stained with antiactin and antimyosin antibodies. Many authorities believe that the myofibroblasts pull the wound together from the edges of the wound; however, others believe, on the basis of observation in collagen lattices, that it is the fibroblasts within the center of the wound that generate the force of wound contraction. To date, this issue has not been resolved. TGF-β is a potent stimulant of wound contraction in experimental models.[378] The wound edges are pulled together at a rate of 0.60–0.75 mm/day. The rate of contraction varies with tissue laxity. Contraction is greatest in anatomic sites where there is redundant tissue. Wound contraction generally continues most actively for 12–15 days or until wound edges meet.

Late Phase: Scar Remodeling **Approximately three weeks after wounding, scar remodeling becomes the predominant feature of the healing process**. Collagen synthesis is downregulated, and the wound becomes less cellular as apoptosis occurs. During this phase, there is continual turnover of collagen molecules as old collagen is broken down and new collagen is synthesized along line of stress.[379,380] Collagen breakdown is mediated by several MMPs, found in scar tissue as well as in normal connective tissues.[381] **At least 25 MMPs that affect different substrates have been identified**. The more common of these include MMP-1 (collagenase-1), MMP-2 (gelatinase A), and MMP-3 (stromelysin-1). The activity of these collagenolytic enzymes is modulated by several tissue inhibitors of metalloproteinases (TIMPs). During this phase, there is little net change in total wound collagen,[382] but the number of cross-links between collagen strands increases.

The realigned, highly cross-linked collagen is much stronger than the collagen produced during the earlier phases of healing. The result is a steady, gradual growth in wound tensile strength that continues for 6–12 months after wounding. **Scar tissue never reaches the tensile strength of unwounded tissue**, however. The rate of gain in tensile strength begins to plateau at six weeks after injury. The common clinical recommendation that patients avoid heavy lifting or straining for six weeks after laparotomy, hernia repair, or many orthopedic procedures is based on the time required for increased tensile strength.

Role of Cytokines in Wound Healing Wounding stimulates specific cellular activities in a consistent manner that is reproducible from wound to wound. Many, if not all of these cellular activities appear to be mediated by cytokines. The predictability with which cellular activities start and stop after wounding suggests that the cytokines mediating them are released in a closely regulated fashion; however, the details of this process have not yet been elucidated.

Numerous cytokines are known to be capable of mediating the major biologic activities involved in wound healing. Most of these activities can be mediated by more than one factor, and researchers have not yet been able to determine which factors are the most important stimulants of wound-healing functions in vivo. One possible explanation for the duplication in mediating functions is that factors with similar activities may act at different times in the course of the wound-healing process.

Cytokines are produced by platelets, macrophages, lymphocytes, endothelial cells, epithelial cells, and smooth muscle cells. Some cytokines, such as PDGF, are produced by several cell types,[383-385] whereas others, such as interleukin-2 (IL-2) are produced by only one cell type.[386,387] The cell of origin is a key variable that determines the time at which a factor will be present after wounding. Platelets, for example, release PDGF,[383] TGF-β,[388] and epidermal growth factor (EGF),[389] and it would be expected that these cytokines would be found in a wound soon after injury. Factors produced by several different cell types may be released by individual cell types at different times. For example, PDGF[383,384] and TGF-β,[390] which are produced by both platelets and macrophages, might be released by platelets soon after wounding and by macrophages at a later stage in the healing process.

The names of cytokines are frequently misleading. In many cases, they derive from the first known cell of origin or from the first function discovered (or hypothesized) for the factor. As a result, a polyfunctional factor may have a name implying that is has only one function, a factor produced by multiple cell types may have a name suggesting that it is produced by a single cell type, or a factor's name may lay claim to a capability that the factor does not have. For example, TGF-β does not have this capability; the name has not been altered.

Cytokines are also a promising tool in the biologic modification of the wound-healing process. Early experimental work was done with small quantities of factors extracted from biologic sources (e.g., platelets). Currently, recombinant technology can provide large quantities of highly purified material that can be used clinically. **It has been experimentally demonstrated that many of the cytokines are capable of accelerating wound healing in normal and healing-impaired models**. TGF-β has markedly increased wound-breaking strength in incisional wounds in rats soon after wounding.[385] BFGF has increased the strength of incisional wounds when injected on day 3 after wounding.[392] EGF has accelerated the closure of partial-thickness wounds in pigs when applied topically,[393] and it has accelerated collagen accumulation in a wound chamber model.[394] PDGF has accelerated healing in incisional wounds in rats when administered in a slow-release vehicle at the time of wounding.[395] Cytokines have also been observed to reverse healing deficits produced by diabetes,[396] steroids,[397] doxorubicin,[398] and radiation[399] in experimental models.

The positive results of these experimental studies encouraged the use of cytokines in clinical trials in humans. In an early human study, EGF accelerated the healing of skin graft donor sites.[400] In another study, it was applied topically to chronic nonhealing wound in a uncontrolled group of patients and was considered to contribute to improved healing in the majority.[401] Autogenous platelet extracts have been used on chronic nonhealing wounds as well, with good results.[402] In a better-controlled study, recombinant human PDGF-bb accelerated healing where applied topically to pressure sores in a randomized, double-blind, placebo-controlled fashion.[403] In another carefully controlled, randomized, prospective study, bFGF was also demonstrated to be efficacious as a topical wound-healing supplement for pressure sores.[404] PDGF-bb has been demonstrated to be efficacious and has been approved for use on diabetic ulcers.[405] It is being marketed as becaplermin (Regranex).

It is not known which factors will be most effective as healing adjuvants in either normal or impaired healing states. It would seem logical that addition of a combination of factors in a sequence mimicking that characteristic of normal healing would produce optimal effects when healing is unimpaired. When healing is impaired, it would seem logical to augment the quantity of whatever factors are lacking or present at reduced levels. However, much work remains to be done—first, to determine which factors are most critical in normal states, and, second, to determine which factors are lacking in impaired states so that the best use can be made of the recombinant factor now available.

A classic example of an aberrant mesenchymal reaction being chronically triggered is the fragile fibrous surface layer of an arteriosclerotic plaque, which when repeatedly triggered can easily result in an acute coronary plaque rupture and occlusion. In this case, a thin fibrous layer is deposited over the macrophage, myocytes, and cellular waste by the fibroblasts. One can see in Figure 1.70 the layers of extra fibroblast derived myocytes, and increase in ECM with deposition of cellular debris, lipids, calcium and the invasion of macrophages with the overlying fragile layer of fibrous tissue, which can easily rupture. This sequence of events demonstrates how the information

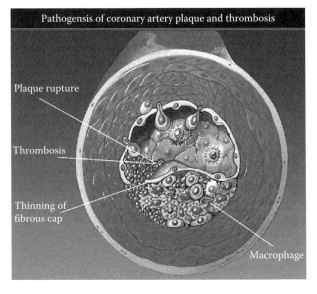

FIGURE 1.70 Aberrant nonspecific mesenchyme reaction in the coronary artery causing acute thrombosis with rupture. (Modified from Hauss.)

from entering foreign compounds can trigger the body's homeostatic mechanism to the point of tissue pathology if it is chronic. Conversely, if the chronic triggering is eliminated, by reducing the general total body pollutant load, and the specific sensory antigen, then, the plaque can be reversed. This reversal can occur only up to a point where it is no longer possible to reverse, as when a thick fibrous layer has been deposited. During the reversal process, new information of a less pollutant load is sent throughout the body. The result is a return of all the normal dynamics of the homeostatic process. Thus, by manipulation of this integrated system of adjustment responses, in an individual functioning in this homeodynamic state, a certain degree of dynamic homeostasis with a state of wellness can be achieved.

The information given and received through the GRS can give the signal to start a new repair, or remodeling (i.e., heart failure remodeling), with injury to the tissues. This signal is as frequently seen in the coronary plaque but with a firmer basis, including scarring (if the injury is severe). **Regular stable arteriosclerotic plaques have a strong macrophage and fibrous cover so they do not rupture**. This stronger fibrous cap is not seen in the previous acute coronary thrombosis effect. If this signal is neither given nor received, healing does not occur and a scar will not form. The alteration of incoming information from a foreign substance infusion can signal more break down of connective tissue and also receive growth factors to repair it. See Figure 1.71.

In healthy tissue, orderly information input from an entering foreign substance will restore the dynamics of homeostasis but in a chronically ill patient pathological tissue changes (interference fields) will occur and disseminate misinformation. These anatomical changes will occur due to the excess aberrant tissue one sees from a chronically triggered NSMR. **This dissemination of misinformation triggers secondary amplification system responses of the nonimmune enzyme detoxification systems, the immune, the neurological and/or endocrine systems in chemically sensitive or chronically degenerative diseased (i.e., spondylitis) individuals. This amplification will then allow costly (in the form of energy requirements), more complex, dyshomeostatic aperiodic and periodic responses**. In this state of amplification, homeostasis can occur but the adjustment response would be with more difficulty and with greater energy demands when the continued trigger of the NSMR occurs. Often, the continued misinformation generated by this reaction will prevent the dynamics of homeostasis from occurring at all, resulting in autonomously triggered pathology. **This response usually manifests as severe inflammation of "unknown cause."** The ability to obtain homeostasis often gets more difficult due to the anatomical build up of mesenchyme tissues, if the total body load stimuli (chemical, bacteria, viruses, trauma, etc.) are more severe, virulent, or more prolonged and/or chronic. The process

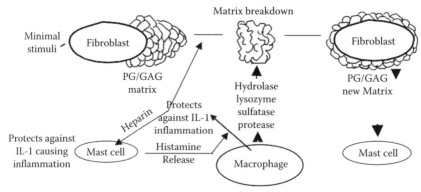

FIGURE 1.71 Normal dynamic function of the connective tissue matrix in dynamic homeostasis with fashioning breakdown and refashioning of tissue. When balanced because of mild stimuli or neutralization, no scar formation occurs. (From EHC-Dallas, 2003.)

of amplification including the aforementioned systems is clearly accentuated in late stages of the periodic (chemical sensitivity) and aperiodic (chronic degenerative diseases), homeostatic disturbances.

Another lesser response to the NSMR build up is the response seen through tissue build up in the fascia. Though the mesh network of the CTM is stretched out and invested in a membrane-like structure with little conductivity in the form of cylinders that surround the superficial fascia, **perforating neurovascular nerve bundles and their loose areolar connective tissue are easily influenced by environmental changes (due to the loose areolar tissue reactivating) such as static fields, airborne electricity charges (air ionization), and electromagnetic impulse fields**.[406] An observed fact previously mentioned is the response to weather changes that is found in many people. It has been shown that 30–40% of the population responds adversely to positive ion invasion before a storm by reacting with symptoms of depression, agitation, or lumbar sacral pain, joint aches, or swelling, suggesting that a hypersensitivity reaction has occurred in this type of connective tissue.[407,408] Because of previously altered electrical physiology (most likely due to changes in the NSMR) that results in sensor hypersensitivity, one can understand the perception of weather changes; positive ion increase with production of symptoms as seen just before a storm in many dyshomeostatic, environmentally sensitive patients. These hypersensitive patients who have for instance, weather responses are much worse than the general sensitive population. Often they can perceive emissions from high power lines, cell phone towers, computers, TV, etc., as well as the onset of weather fronts This observation shows the body's ability to respond to nonchemical, nonbiological noxious stimuli, again, and emphasizes the electrical nature of the human body.

Rate of Enzyme and Nutrient Deficiency

In addition to noxious stimuli overload (total body pollutant load), enzyme and nutrient deficiencies have been shown to propagate chemical sensitivity and chronic degenerative disease (i.e., acetylation defects, C-1-esterase deficiency, phenylketonuria, beriberi, scurvy), impairing wound healing. **Nutrients and enzymes are used as fuels for detoxification, physiologic adjustment processes, and wound healing**. The low levels of nutrients can result in relative total body pollutant overload since the neutralization mechanism for noxious stimuli is not as effective.[409,410,411] Consequently, when there is not enough fuel available to complete all responses efficiently, then wound healing is delayed. For example, a patient was found to have a malabsorption of nutrients after a chronic sublethal exposure to pesticides known to damage enzyme function (i.e.,cholinesterase). Later the patient developed acute and chronic atrial fibrillation as a seemingly, unrelated event. The blood analysis showed intracellular magnesium deficiency. In spite of many cardiac medications and external electrical defibrillation, the patient continued to deteriorate until the body's total magnesium deficiency was corrected by replacement. The patient then needed no medication to maintain normal sinus rhythm so long as the magnesium level was in the high range and the pesticide load was reduced, which kept the cholinesterase production adequate. Therefore, chronic disease persisted until the proper nutrient correction and incitant elimination was achieved. In this instance acquired nutrient deficiency propagated the degenerative disease process. Replacement of nutrients restored the healing process where the deficiency was totally corrected and the damage was reversed.

Many enzyme and nutrient deficiencies can occur with excess total body load of general noxious substances (i.e., solvents, aldehydes, phenols, pesticides), which specifically damage segments or suppress function by the weight of their total load and by specific (i.e., formaldehyde, chloropyrophos, chlorine, etc.) toxics, which damage specific biochemical processes.[412–415] Any of these combined with an enzyme or nutrient deficiency can result in a relatively excess total body load, causing dyshomeostasis. The other causes of nutrient deficiency that makes the individual susceptible to hypersensitivity and degenerative disease are varied. These include poor diet, high gastric pH (50% of people over 50 years old have hypo- or achlorhydria), selective malabsorption, overcompetition of chemicals for absorption versus nutrients (i.e., Isoniazide outcompete B_6 absorption and other selected chemicals outcompete other nutrients) imbalanced gut flora, hypermetabolism to neutralize chemicals, and the diuretic effect of toxics, which increase the excretion of many vitamins, minerals, and amino acids, resulting in diminished nutrient pools.

Nutrient replacement will stop early physiological blocks especially if the noxious stimuli are removed, allowing restoration of normal function.

HOMEOSTATIC DYSFUNCTION LEADING TO A DISORDERED HOMEOSTATIC RESPONSE

Periodic Response

As previously stated, homeostasis can be maintained only if the organism is working economically with good energy. See Figure 1.72. An intact control circuit(s) reacting normally and functioning at optimum control quality with negative feedback is able to cope with an additional noxious stimuli demand quickly and economically like the response shown. This normal response is periodic, which results in homeostasis.

When there is dysfunction of the control circuits, which may be due to any cause (i.e., nutritional deficiency, excess total body load; trauma, bacterial invasion, etc.), periodic and aperiodic deviations will occur in control quality. The hypersensitive response is a response exaggerated over the normal periodic physiologic response. These may be of several degrees of severity. In the case of regulatory lability (hypersensitive response, periodic homeostatic disturbance), any stimulus will produce a labile deviation of an excessive response.

Usually, a short period stimulus will produce a steeper gradient at a high volume with a return to baseline being excessive to the normal homeostatic response and requiring longer to recede. Another type of hypersensitive homeostatic dysfunction is the type where the homeostatic response overshoots the baseline response resulting in more symptoms before they right themselves.

Finally there is a type of response that continues to rise and each response never reaches the baseline until the whole homeostatic mechanisms deteriorates and the patient develops degenerative disease. Usually, trophic edema and subsequently degenerative disease develops before the degenerative inflammation occurs.

Early dyshomeostatic reactions are abundant daily in society where one sees anxiety attacks, attention deficit, dull or hyperactive behavior, learning disabilities, migraine, headaches, brain fog, lack of concentration, rhinorrhea, stuffed nose, fibromyalgia, and fatigue, bloating, gas, and heartburn, bronchial, bladder or intestinal spasm, spontaneous bruising, petechiae, etc. **Chronic**

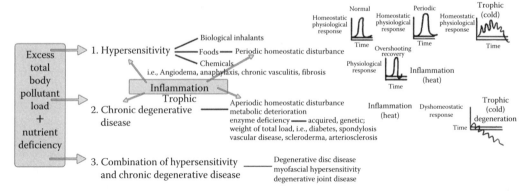

FIGURE 1.72 Types of physiologic changes from homeostasis to dyshomeostasis in degenerative disease. (From EHC-Dallas, 2002; EHC-Dallas, 2004.)

use of symptom suppressing medication is discouraged at this stage because it is the crucial time when the degenerative processes can be reversed. It is the responsibility of the clinician to find and eliminate the triggering agents in order to reverse degeneration. Early diagnosis, with elimination of the etiology and correction of the nutrient deficits, will often stop disease progression and revert the individual back to normal good health. Bergsmann et al.,[95] Perger,[416] Heine,[4] and Pischinger[3] have described two types of homeostatic dysfunction (disturbances) responses, which result in either a periodic deterioration (hypersensitive) of homeostatic function or an aperiodic one (degenerative).

These dyshomeostatic responses must be understood by the clinician for these early conditions are usually the only point in time where normal physiology can be restored easily to prevent disease from occurring and/or becoming fixed. When these disturbances occur chronically, informational changes, followed by metabolic alteration (end-organ hypoxia), results in tissue changes (sol to gel, gel to sol—NSMR). Eventually fixed-named, autonomous (generated by muscle movement) disease results (cancer, diabetes, angioedema, anaphylaxis, spondylosis, etc.) The dominance of either dyshomeostatic process (periodic or aperiodic) results in a chronically ill patient with whom the clinician will have to take sometimes minor and, at other times, extreme measures to restore the dynamics of normal energy efficient homeostasis. A periodic deterioration of the dynamics of homeostasis results in extreme lability of the homeostatic response mechanism, with hypersensitivity of the CTM and cells responding to more and more stimuli until the entire body becomes supersensitive.

In this case of periodic homeostatic disturbance, it appears that the orderly depolarization and repolarization of the PG/GAGs are changed with the loss of the harmonic resonance of the electrical impulses. Oscillation of the PG/GAGs becomes unstable, due to their constant stimulation, inhibition, and spring effect, driven by the excess chronic exposure to noxious stimuli. The PG/GAGs do not repolarize as early as in the normal response, requiring more energy to achieve dynamic homeostasis. The receptors multiply and become more supersensitive. Although the patient becomes more hypersensitive or physiologically drained, fatigue and pain will occur earlier, resulting in a fragility of response that can eventually incapacitate the individual.

The existence of chemical sensitivity (CS; periodic homeostatic deterioration), which was described by Randolph,[5] elaborated on by Rea,[417] and discussed by Pischinger et al.,[3] is present in most homeostatic disturbances, being dominant in a large subset of chronically ill patients. It is usually, in addition to being an isolated response, combined with chronic degenerative disease. **The periodicity occurs because each of the individual's reactions presents in an episodic nature and has a finite termination of the reaction**. In the normal dynamics of homeostasis for example, an incitant (i.e., pesticide) comes into the body and triggers a finite coherent energy efficient physiologic response. The periodic homeostatic deterioration results in an adjustment response similar to normal, however, there is much more difficulty in reaching the base line termination. The response is exaggerated and the downward limb often will overshoot before the normal termination is reached. As observed in controlled studies in the environmentally controlled unit (ECU), the leukocytes may be depressed transiently, as is the total serum complement or gamma globulin. Each parameter paralleling the clinical response will return to normal in 2–4 hours if the incitant is withdrawn, or neutralized, and if it is only mildly toxic. The clinical response parallels with a production of symptoms and signs, i.e., transient muscle ache, forehead pressure, runny nose all followed by sleepiness, fatigue or pain. **Thus, the term of periodicity in which an incitant creates a finite cause and effect reaction, corresponds to Selye's[7] alarm reaction and response**.

Nutrient depletion can occur with this type of response, if there is a persistent chronicity of noxious incitants entering the body and constantly triggering the homeostatic adjustment response subunits. In this case, the nutrient repletion mechanism cannot supply sufficient amounts of nutrients promptly in order to directly (i.e., acid base; nonenzymatic glutathione neutralization, reduction processes resulting from oxidation) or indirectly (i.e., enzyme detoxification or immune system macrophages) neutralize the incitants. For example, studies at the EHC-Dallas have shown 60%

of the 300 patients with chemical sensitivity studied had B_6 deficiency; and/or 30% vitamins C, B_1, B_2, B_3, B_5 deficiency; and 20% vitamin A and E deficiency.[418] A large portion of the patients with chemical sensitivity also had mineral, amino acid and lipid deficiency levels. This depletion then results in a vicious downward cycle of inadequate responses to other noxious stimuli. The periodic response can disturb the body after chronic incitant stimulation, by using more and more nutrients until the body becomes depleted. Then, with the nutrient depletion, inappropriate information dissemination and responses occur sequentially and often randomly. The information patterns change, which signal nonphysiological adjustment responses, then the chronically altered metabolism changes, and finally the tissues change. Chronic fixed-named end-organ autonomous disease may follow. An example of this periodic (chemical sensitivity) degeneration is as follows:

> **Case study.** An individual without glove or mask protection uses toluene, which he absorbs and inhales, working with it frequently to degrease his equipment (*informational input*). This bioinformation intake (intermittent reception), if chronic, changes the local, metabolic, and tissue responses into hypersensitivity reactions (accelerated adjustment responses). First, he develops rhinitis followed by sinusitis (*metabolic changes*). Then the spreading phenomenon occurs as he becomes increasingly sensitive (*metabolic and electrical changes*) to different ambient odors like newsprint, gasoline, car exhausts, and pesticide. Each exposure would give a finite reaction, but occurs with less and less intense stimuli (increase in periodic hypersensitivity and increase in number of receptors and receptor sensitivity). This patient had recurrent sinusitis for 5 years, which was confined until he developed a tachycardia (another *metabolic change*) and severe sinus pain, requiring antibiotics and eventually surgery for sinus obstruction (tissue changes) due to inflammation and fibrosis (tissue changes). Thus, the individual develops the informational change, followed by metabolic changes, and then tissue changes from the toluene exposure. Since the disinformation occurred to and from the tissues, the dyshomeostatic adjustment response occurred, altering tissues until fixed-named disease occurred. This hypersensitive state may exist for many years before fixed-named, autonomous disease is dominant.

In this case, when the chronic stimuli continued, similar entrainment into the connective tissue would be expected for the electrical frequencies of toluene, namely, 570 Hz, 86 KHz, 550 K Hz. Also, these have been found entwined by-cells in 400 ppm toluene for which the TLV-TWA is a 100 ppm.[419] In support of this finding, studies using local connective tissue imprints emanating from a frequency generator, have been made artificially on surgical specimens of normal CTM, and transferred by cell culture to the next generation,[420] which suggests that noxious incitants can imprint the connective tissue leaving a vulnerability in the tissue for deposition of the EMF imprint of each subsequent incitant. This damaged tissue area (interference field) would then be more vulnerable to the next incitant deposition or stimulation of much less intensity, often causing an abnormal dyshomeostatic response. These damaged secondary homeostatic foci can then become areas for deposition of each new round of noxious stimuli, thus, making the propagation of chronic illness easier.

Aperiodic Response

In the case of a slow sluggish (aperiodic) deviation, the response to a stimulus is delayed and slow but always with a downward response.

An aperiodic response is thought to be an eventful deterioration of homeostasis with a loss of periodicity. Clinically, this homeostatic response is slow or paralyzed unable to right itself and, thus, organ failure eventually occurs. In some cases the response overshoots the baseline. The initial homeostatic value is reached slowly or not at all.

In both types of control systems, time and energy may be wasted in responding to environmental stimuli and thus, the principle of economy of energy for the dynamics of homeostasis is lost and/or diminished. The patient becomes fatigued and develops weakness and pain.

The aperiodic homeostatic disturbance never quite returns to normal.[421] In itself, it is not a hypersensitive state like the periodic disturbance. This dyshomeostatic disturbance response does not have a clear finite incitant reaction, as evident in the periodic response. However, it does have an undulation of response from entering external noxious stimuli with a downhill course, which puts increased demands on the energy supply. It is slow and delayed. **This aperiodic response is due to genetic or acquired enzyme, coenzyme, immune, and nutrient, etc. deficiency or to local hypoxia and is usually triggered by the relatively overwhelming chronic environmental noxious stimuli load**. This incessant excess total body load relative to the body's response mechanisms to counteract or neutralize the noxious incitant, gives rise to chronic diseases like chronic degenerative intervertebral disc disease, diabetes mellitus, scleroderma, arteriosclerosis, benign and malignant tumors, etc.

Both homeostatic disturbances (periodic and aperiodic) can occur in the same individual, including spondylosis, diabetes mellitus, etc., but in different parts of the body. They can occur separately or they can occur concomitantly in the same organ or separate ones. This observation is demonstrated in a 41-year-old, white, female patient having CREST syndrome with gangrene of her left index finger from unrelenting vascular spasm.

Case study: When studied while in the deadapted state, within a less-polluted, environmentally controlled facility and fasting for 4 days, a reduced total body load occurred, allowing her vasospasm to disappear and her gangrene to clear. Then using individual challenge tests, she was found to have periodic reactions to the ingestion of beef, cow's milk, and pork; to the inhalation of formaldehyde, phenol, and chlorine; and to the inhalation or injection of molds, which also triggered the vascular response. Many other substances showed negative reactions upon challenge. She also developed other sclerotic tissue changes (aperiodic) that appeared to decrease with the overall reduction of the total body pollutant load and to heal with appropriate nutrient (vitamins, minerals, and amino acids) supplementation, suggesting damage to the detoxification system. This patient has done well over the last 15 years with occasional vascular spasm, which was always cleared by reducing the total body load, by fasting under environmentally controlled conditions. This case is an example of the combination of the periodic hypersensitivity and the aperiodic (nutrient deficiency) degeneration of the homeostatic response.

Thus, the periodic and aperiodic disruptions of homeostasis are the first constant changes that occur from normalcy. At this early stage, both processes appear to be reversible, and the clinician must take action to remove the noxious incitants and to correct the nutrition and physiology, so that fixed-named irreversible disease can be prevented.

SPECIFIC MECHANISMS OF INFLAMMATION AND FUNCTION OF NEUTROPHILS AND MACROPHAGES

Inflammation as discussed in the section on wound healing is a further extension of the normal removal process of bacteria, viruses, toxics, and free radicals discussed earlier in this chapter. Therefore, much of the discussion will appear redundant (see section on local response to injury; wound healing and inflammation) but the authors felt that this section should be emphasized due to the severity of response and the potential for severe tissue injury when chronic entry of noxious stimuli occurs causing inflammation. When tissue injury occurs, whether caused by bacteria, trauma, toxic chemicals, heat or any other phenomenon, multiple cytokines, proteases, lysozymes, etc., can cause dramatic secondary changes in the tissues that are released by the injured tissues. If the wound is mild, the dynamic homeostatic mechanisms come into play; neutralize the incitants in the tissue and release growth factors. If the wound is more severe and the defense cells cannot maintain physiologic homeodynamics, excess release of tissue growth factors occurs. These released factors include lymphokines, leukokines, growth factors, etc. The entire complex of injured tissue stimulates inflammation. (See Figures 1.73 and 1.74.)

Inflammation is characterized by (1) **vasodilatation of the local blood vessels** with consequent excess local blood flow and oxygen consumption, (2) **increased permeability of the capillaries**

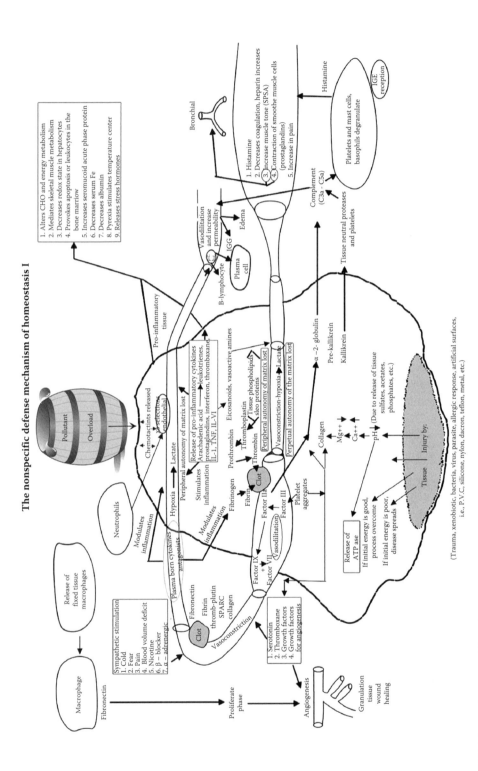

FIGURE 1.73 The nonspecific defense mechanism of homeostasis I—acute phase inflammation—disrupted homeostasis and inflammation. Entry pathway of pollutant to end-organ disturbance: (1) pollutant entry through mucous membranes or nose, bronchial tubes, lungs, gut, vagina, penis, bladder, or skin, (2) barriers include enzymes (S.O.D., glutathione peroxidase, catalase), antioxidant nutrients, or mechanical, (3) connective tissue matrix entry, (4) cell membrane barrier—vascular, nerve (peripheral and autonomic), (5) connective tissue matrix—surrounding end organ, (6) end-organ cells. (From EHC-Dallas.)

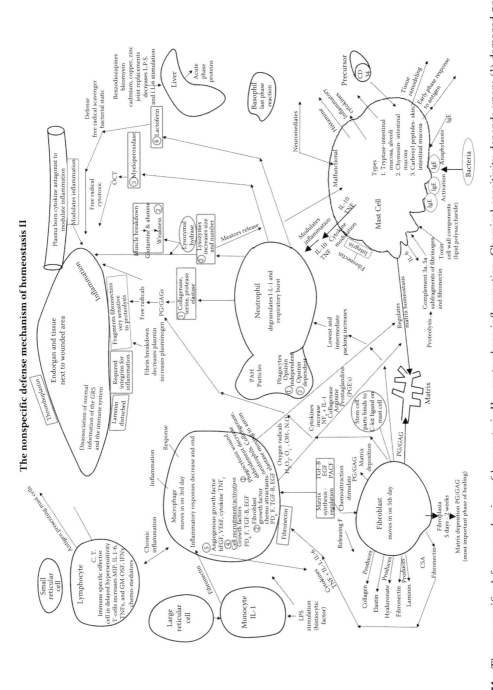

FIGURE 1.74 The nonspecific defense mechanism of homeostasis II—acute phase inflammation. Chemical sensitivity depends upon: (1) damaged pollutant entry site of any mechanical surface, (2) damaged antioxidant enzymes and nutrients, (3) damaged connective tissue matrix and ground regulation system, (4) damaged cell membranes of vascular and nerve systems, (5) damaged connective tissue matrix near end organ, (6) damaged end-organ cell membranes. (From EHC-Dallas.)

with leakage of large quantities of fluid into the interstitial spaces occurs, (3) often there is **clotting of the fluid** in the interstitial spaces because of excessive amounts of fibrogen and other proteins leaking from the capillaries, (4) **migration of large numbers of granulocytes and monocytes** into the tissue occurs, and (5) **swelling of the tissue cells** is present.

Some of the **many tissue products that cause the initial inflammatory reaction are histamine, bradykinin, serotonin, prostaglandins, several different reaction products of the complement system, reaction products of the blood-clotting system, and multiple hormonal substances (lymphokines) that are released by sensitized T cells**. Several of these growth factors (CSF) strongly activate the macrophage system, and within a few hours, the macrophages begin to devour the destroyed tissues; at times, the macrophages also further injure the still-living tissue cells.

One of the first results of inflammation is to "wall off" the area of injury from the remaining tissues. **The tissue spaces and the lymphatics in the inflamed area are so blocked by fibrogen clots that barely any fluid flows through the space. This "walling-off" process delays the spread of bacteria or toxic products.**

The intensity of the inflammatory process is usually proportional to the degree of local tissue injury. For instance, staphylococci invading the tissues liberate extremely lethal cellular toxins, which can be absorbed by the capillaries and lymphatics and spread throughout the body. As a result, inflammation develops rapidly, indeed, much more rapidly than the staphylococci themselves can multiply and spread. **Therefore, local staphylococcal infection is characteristically walled off rapidly and prevented from spreading through the body. On the other hand, streptococci do not cause such intense local tissue destruction**. Therefore, the walling-off process develops slowly while the streptococci reproduce and migrate. As a result, streptococci often have a far greater tendency to spread through the body and cause death than do staphylococci, even though staphylococci are far more destructive to the tissues. This same process occurs with local xenobiotic infiltration. Some types of toxics give a large local reaction, while others are much more virulent systemically, resulting in a rapid and at times slow onset of severe illness without much of a local reaction. However, if the stimuli are prolonged and chronic, one develops informational changes followed by metabolic and anatomic alteration. Some toxic chemicals cause large tissue reactions while others cause small reactions similar to tissue responses to bacteria. **This phenomenon is strongly emphasized in the different types of synthetic implants.**[422] For example, the vitek Teflon jaw implant gave severe local tissue damage while the body attempted to wall it off, whereas the stainless steel jaw implants and plates have less inflammation. Though they still cause a local and distal implant syndrome, they are much more benign when it comes to local tissue destruction.

The tissue macrophage is a first line of defense against infection as previously shown in the defense mechanism. **Inflammation occurs as an oxidative and prolonged response of the normal homeostatic response**. However, when inflammation is generated many more cytokines and growth factors are released, which will inflame the tissue and the blood vessels. Therefore, one will see the same elements of the normal homeostatic defense mechanism in this section, but in addition, one will further understand how the responses are prolonged and exaggerated by the intensity and chronicity of the noxious incitants. Within minutes after inflammation begins usually triggered by free radicals, the macrophages (tissue histiocytes) already present in the tissues are released due to a decrease in pH from cleavage of acetates, sulfates, etc., from the PG/GAGs. With exposure, macrophages, whether histiocytes in the subcutaneous tissues, alveolar macrophages in the lungs, microglia in the brain, or others, immediately begin their phagocytic actions. When activated by the products of infection and inflammation, the first effect is rapid enlargement of each of these cells. Next, many of the previously sessile macrophages break loose from their attachments and become mobile, forming the first line of defense against infection or toxics during the first hour or so (histocytic stage). The numbers of these early mobilizable macrophages often are not great.

The Physiologic Basis of Homeostasis

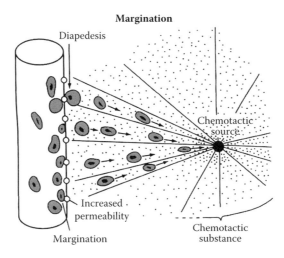

FIGURE 1.75 Margination. (Modified from Guyton, A.C. and Hall, J.E. *Textbook of Medical Physiology*, 10th ed., W.B. Saunders Company, Philadelphia, 396, 33–2, 2000. With permission.)

Within the first hour or so after inflammation begins, large numbers of neutrophils invade the inflamed area from the blood. This invasion is caused by products from the inflamed tissues that initiate the following reactions: (1) they alter the inside surface of the capillary endothelium, causing neutrophils to stick to the capillary walls in the inflamed area, as seen in Figure 1.75, margination; (2) they cause the endothelial cells of the capillaries and small venules to separate easily, allowing openings large enough for neutrophils to pass by diapedesis into the tissue spaces; (3) other products from the inflammation cause chemotaxis of the neutrophils toward the injured tissues.

Thus, within several hours after tissue damage begins, the area becomes well supplied with neutrophils, which make up the second line of defense (microphage stage). Since the blood neutrophils are already mature cells they are ready to begin immediately their scavenger functions for killing bacteria and removing foreign matter.

Also, within a few hours after the onset of acute, severe inflammation, the number of neutrophils in the blood sometimes increases fourfold to fivefold from a normal 5000 to 6000 to 15,000 to 25,000 neutrophils per microliter. The products of inflammation such as cytokines and growth factors that enter the blood stream cause neutrophilia. They are transported to the bone marrow where they act on the marrow capillaries and on the stored neutrophils to mobilize these immediately into the circulating blood. This makes more neutrophils available to the inflamed tissue area and slows leukocytolysis.

A second macrophage invasion of the inflamed tissue is a third line of defense (monocytic stage). More macrophages are necessary because they can phagocytize elements that the neutrophils cannot, including damaged neutrophils. Along with the invasion of neutrophils, monocytes from the blood enter the inflamed tissue and enlarge to become macrophages. However, the number of monocytes in the circulating blood is low. Also, the storage pool of monocytes in the bone marrow is much less than that of neutrophils. Therefore, the buildup of these macrophages in the inflamed tissue area is much slower than that of neutrophils, requiring several days to become effective. Furthermore, even after invading the inflamed tissue, monocytes are still immature cells, requiring eight hours or more to swell too such large sizes and develop tremendous quantities of lysozomes, only then acquiring the full capacity for phagocytosis. Yet, after several days to several weeks, the macrophages finally come to dominate the phagocytic cells of the inflamed area because of greatly increased bone marrow production of monocytes, as explained previously.

As already pointed out, macrophages can phagocytize far more bacteria and toxics (about five times as many) and far larger particles, which includes even neutrophils themselves, and large quantities of necrotic tissue, than can neutrophils. The macrophages also play an important role in initiating the development of antibodies.

The fourth line of defense is greatly increased production of both granulocytes and monocytes by the bone marrow. This production results from stimulation on the granulocytic and monocytic progenitor cells of the marrow. However, it takes 3–4 days before newly formed granulocytes and monocytes reach the stage of leaving the bone marrow. **If the stimulus from the inflamed tissue continues, the bone marrow can continue to produce these cells for months and even years, sometimes at rates of production 20–50 times normal**. This increased production results in a demand from the nutrient pools and if the stimuli become chronic causes vitamin, mineral and amino acid depletion.

Although more than two dozen factors have been implicated in the control of the **macrophage-neutrophil response to inflammation, five of these are believed to play dominant roles. They are (1) tumor necrosis factors (TNF^{a+b}); (2) interleukin-1 (IL-1); (3) granulocyte-Monocyte colony stimulating factor (GM-CSF); (4) granulocyte colony stimulating factor (G-CSM); (5) monocyte colony stimulating factor (M-CSF)**.

These factors are formed by activated macrophages and T-cells in the inflamed tissues and in smaller quantities by other inflamed tissue cells.

The cause of the increased production of granulocytes and monocytes by the bone marrow is mainly the three colony stimulating factors. One of which, GM-CSF, stimulates both granulocyte and monocyte production and the other two, G-CSF and M-CSF, stimulates granulocyte and monocyte production, respectively.

This combination of TNF, IL-1, and colony stimulating factors along with other important factors provides a powerful dynamic homeostatic feedback mechanism that begins with tissue inflammation and then proceeds to formation of defensive white blood cells and finally, removal of the cause of the inflammation with the restoration of homeostasis.

Tissue necrosis factors (TNFS) are pleotrophic cytokines produced in response to injury or infection. These molecules are considered to be primary effectors of the inflammatory and immune response. There are two distinct forms of TNF, TNFa, TNFb, TNFs play a necessary and beneficial role as mediators of resistance to virus, bacteria, fungi, and cancer cells. **It is important to note that an over production of these factors can lead to a variety of pathological conditions including wasting, systemic toxicity, and septic shock**. TNFx was expressed locally in 50% of the tumors studied being largely confined to macrophage like cells in the stroma. Two distinct TNFR receptors have been identified and cloned ($TNFR^{1+2}$). Both receptor sites are distinct. Immunologically soluble TNF receptor binding proteins have been found in human serum and urine that can neutralize the biological activity of TNF a and b. These soluble forms are TNF receptors, which have now been shown to arise as a result of the shedding of the extracellular domains of the receptors and concentrations of about 1.2 Ng/ml of healthy subjects. **Elevated levels of soluble TNF receptors have been reported in the serum or plasma with pathological conditions such as endotoxemia, mercury exposure, UIV infection, and malignancy**. These soluble receptors bind to TNF and inhibit its biological activity by their competing with cell surface receptors for TNF binding. This receptor binding protects other cells from the effects of TNF and localizing the inflammatory response. TNFR, and R_2 receptors were significantly higher in the cancer patients than the healthy controls.[421] The immune system reacts to an antigen from an inflammatory standpoint with a balancing act between the **release and activation of TNF and other pro inflammatory cytokines at one or more sites of the antigen presentation**. Also the production and biologic activation of their inhibitors occurs.

It now seems clear that soluble inhibitors of cellular immunity, specifically inhibitors of proinflammatory cytokines, are active in those clinical states in which antigenic tissues are tolerated by a seemingly intact immune system. **There is no longer any doubt that inflammation can be up regulated by removing these inhibitors and if the removal is sufficient cancer can be destroyed**. Similarly, pathogen and those antigens that characterize them immunologically can be a target of

The Physiologic Basis of Homeostasis

a more efficient up regulated immunological/inflammatory response so that a state of tolerance to foreign invaders can be reversed.

PROINFLAMMATORY CYTOKINES

There are two forms of the cytokine interleukin –1 (IL) IL1-X and IL-1B. These are cytokines produced primarily by mononuclear phagocytes and also by a number of other cells including skin keratocytes, epithelial cells, and some cells of the CNS.

These cytokines produce a wide variety of effects on numerous cell types. These also regulate the production of a great number of other proteins, including other cytokines, IL-2, TNF and colony stimulating factors. IL1X and IL1B are important mediators of inflammatory and immune response. **IL-1 appears early during the inflammatory reaction and plays a significant role in the productions of pathologic conditions including chronic inflammation, septic shock, and defects associated with hematopoesis.** The effects produced by IL-1 result from the binding of these cytokines to two distinct cell surfaces receptors, IL-1R types I and II. The type I receptor is found on B cells and polymorphonuclear leukocytes. The type 1 is capable of transducing the signal and can produce all biologic effects attributed to IL-1. Type II receptors are membrane bound and serve as the precursor of the soluble IL-1 binding factor, which is shed under appropriate circumstances to antagonize and modulate IL-1 activity.

Cells known to produce IL-1X receptors include monocytes, neutrophils, macrophages and fibroblasts. Cytokines known to up-regulate IL-1 production include IL-13, IL-6, IL-4, interferon gamma, granulocyte-macrophage colony stimulating factors (GMCSF), and transforming growth factor (TGFB). IL1X receptor is released in vitro during experimentally induced inflammation as a part of the natural cause of many diseases. It is thought to be part of a naturally occurring mechanism that limits and in most circumstances eliminates the extent of the potentially deleterious effects of IL-1 by blocking its activity with IL1x-receptors. Experimental study in rabbits pretreated with IL-1R has been shown to prevent death resulting from septic shock produced by lipopolysacharide (LPs) injection. IL-l rx prevents the development of immune complex mediated colitis and blocks cerebral spinal fluid inflammation induced by cerebral ventricular administration of IL-1.

INTERLEUKIN-6

IL-6 is a multifunctional cytokine produced by a variety of cell types including T-lymphocytes, monocytes/macrophages, fibroblasts, hepatic, vascular endothelial cells, cardiac myxomas, bladder cell carcinoma, myelomas, etc. The effects of IL-6 on different cell types are numerous and varied. These effects include the following: stimulation of B cell differentiation and antibody secretion; activation as a costimulant with phytohemaglutinin (PHA) to increase IL-2 production and IL-2 receptor expression, T-cells and enhancement of the differentiation of cytotoxic T-cells; growth factor for mature thymic and peripheral T-cells, myelomas, hybridomas; the induction of neuronal cell differentiation. These varied effects indicate that IL-6 plays a major role in the mediation of the inflammatory and immune responses provoked by infection, injury, and malignancy. Elevated levels of IL-6 are also associated with increased production of soluble IL-6R in many pathological conditions. Clearly, a delicate balance exists in order to maintain homeostasis.

When the neutrophils and macrophages engulf large numbers of bacteria or virulent toxics, necrotic tissue occurs, essentially, all the neutrophils and many, if not most, of the macrophages eventually die. If the toxics are not acutely virulent, most survive. After several days, a cavity is often excavated in the inflamed tissues that contain varying portions of necrotic tissue, dead neutrophils, dead macrophages, and tissue fluid, or pus. After the infection has been suppressed, the dead cells and necrotic tissue in the pus gradually autolyze over a period of days, and the end products are usually absorbed into the surrounding tissues until most of the evidence of tissue damage is gone. One sees pus, usually sterile, in much smaller areas at times almost pinpoint when toxic damage occurs and often only inflammation occurs. **In chemical sensitivity, one can see acnelike lesions with sterile pus after certain exposures especially chlorinated types and massive mold infestations**. Also, one sees punctuate pustular and some clear skin lesions on the face, chest, and trunk that have a small amount of tissue inflammation around them. **Chloracne is well known to occur after repeated chlorine exposures**. Eventually, if the stimuli become chronic, the lymphocytic phase occurs. The lymphocytes are in the matrix surrounding the blood vessels, usually, resulting in a lymphocytic vasculitis.

NEUROLOGICAL SYSTEM CONNECTION WITH THE CONNECTIVE TISSUE MATRIX AND INFLAMMATION

After folding out of the Schwann cell membrane and now only surrounded by a basal membrane, autonomic nerve fibers end blindly in the matrix. The distance to the next cells may reach from a few up to hundreds of nanometers. Typical synapses are not generated; however, the term *synapsis at a distance* is widely used. As stated previously, the changes in the blood vessels are partially dependent upon the innervation and the generation of nerve impulses in the form of leaks when the pollutant triggered nerve responds to the end vessel. Trophic (neurogenic) edema occurs locally and occasionally, generally. If the response is subinflammatory, which is often, it is not heat generated. However, the metabolic changes and the NSMR can by triggered by the mild vessel spasm and the resultant hypoxia. Often once the trigger of the cascade occurs, one will develop chronic degenerative disease or hypersensitivity with gross inflammation occurring.

The latent inflammability of the matrix depends on its intact innervation (for example, stroke patients do not develop rheumatoid arthritis on the paralyzed side). For the central inflammatory effect, the degranulation of mast cells[422] requires locally a sufficient release of adrenaline, noradrenaline, acetylcholine, neuropeptides, and cytokines out of the endings of autonomic nerve fibers. In this context, it is important that sensitive axons do not only show an afferent component as pain receptors but also efferent functions.[423] **The so-called axon reflex—transfer of an afference into the next branch of the same axon as efference has been known for a long time**.[78] Usually, this will cause trophic (neurogenic) edema without inflammation but if severe enough, triggers inflammation.

Viscero-sensitive axons, e.g., on mental or environmental noxious stimuli, are capable of releasing efferent pain neuropeptides like substance P and calcitonin-gene-related-peptide (CGRP), which have a degranulating effect on mast cells and a stimulative effect on cytokine release (e.g., IL-1, IL-6, TNF-) from macrophages (see Chapter 2). The resulting "neurogenic inflammation" may spread to further terminal sensitive axons, whereby the process may turn into positive feedback and spread further.[424] Tissue remodeling occurs by collagen and PG/GAG regeneration from fibroblasts (NSMR) where only electron-optically identifiable ultra fine scars may develop. These in turn cause the terminal axons to feel pain and so on. **Consequently, "neurogenic inflammation" seems to be very important for the generation of all types of soft tissue rheumatism (fibromyalgia)**[425,426] (See Chapter 2: Nervous System.)

Substance P may increase T lymphocyte proliferation and B lymphocyte differentiation in order to keep up the neurogenic inflammation. **In addition, the endings of sensitive nerves also function as immunoreceptors via axonal retrograde transport of antigen, antibodies, and toxics, thus, impulses reach the center are processed and finally modulate release caused by inflammation-relevant-released neuropeptides from peripheral terminal axons**.[425]

The GRS is also part of every defense action. If local dysfunction occurs with distal effects, then repeated environmental assaults can create and propagate either a trophic or inflammatory reaction. Some of this reaction may be neural and some intrinsic matrix. An example of this action is the insertion of the intra-arterial canula for cardiopulmonary bypass. This process immediately sends information that will ensure the release of both local and distal inflammatory mediators, which will cause almost immediate total trophic reaction and eventually body inflammation from the bypass.[426] If the bypass is short in time, the trophic edema or the inflammation is turned off immediately. If the bypass is prolonged, the patient will have a prolonged course (recovery phase) to turn off the inflammation before recovery occurs. If the inflammation is too severe, the patient may not recover. It appears that much hypersensitivity and chronic degenerative disease may be triggered by the nonheat generating nuerogenic response, but if the degree of noxious stimuli is too forceful, inflammation occurs.

SUMMARY

In summary, along with the skin and mucosa, the CTM is the ERS. The GRS consisting of the CTM, fibroblast, macrophages, mast cell, leukocyte, end capillary vessel, and autonomic nerves, is a global information system for regulating the dynamics of homeostasis in the body. There is an intermeshing network consisting of the thousands of local subunits feeding into regional centers where, if all those subunits do not handle the incoming noxious stimuli peripherally the subunits become integrated into the brain by the reticular activating system, the hypothalamus, the limbic system, the pineal gland, and the area postrema of the fourth ventricle. This system integrates those thousands of local and regional dynamic homeostatic subunits into a smoothly functioning energy efficient process, which allows the body to maintain itself against a hostile environment. Periodic homeostatic disturbances (chemical sensitivity) and aperiodic homeostatic disturbances (chronic degenerative diseases) are ways the body responds to excess chronic noxious stimuli. These periodic and aperiodic disturbances are the first signs of dyshomeostasis and result in informational changes that are followed by metabolic alterations (hypoxia), which will eventually lead to tissue anatomical changes (changes for sol to gel and gel to sol). These changes are followed by the nonspecific mesenchymal reaction with loss of autonomy of the CTM and with resultant spontaneous autonomous triggering of the dyshomeostatic reactions by muscle movements. The progression moves to fixed-named autonomous end-organ, end-stage disease. In attempts to prevent end-organ failure, knowledge of the process that occurs after noxious stimuli enter the body is necessary. One's knowledge of this process must be the guide to move through the onset of early end-stage disease and, eventually, see the manifestations to fixed-named autonomous diseases. It is this knowledge that offers us the greatest potential with all of the possibilities in treatment for preventing and reversing early homeostatic dysfunction.

REFERENCES

1. Calabrese EJ. 2001. Overcompensation stimulation: A mechanism for hormetic effects. In *Critical Reviews in Toxicology* 31(4&5). Ed. RO McClellan. Boca Raton, FL: CRC Press LLC. 425.
2. Calabrese EJ. 2001. Prostaglandins: Bipasic dose, Part II, Mechanistic foundations of biphasic dose responses in pharmacological and toxicological systems. In *Critical Reviews in Toxicology* 31(4&5). Ed. RO McClellan. Boca Raton, FL: CRC Press LLC. 471–624.
3. Pischinger A. 1975. *Matrix and Matrix Regulation: Basis for a Holistic Theory in Medicine.* Ed. H. Heine. Eng. trans. N MacLean. Brussels, Belgium: Editions Haug International.
4. Bissel MJ, Hall HG, and Parry G. 1982. How does the extracellular matrix direct gene expression? *J. Theor. Biol.* 99: 31–68.
5. Randolph TG, Ed. 1962. *Human Ecology and Susceptibility to the Chemical Environment.* Springfield, IL: Thomas.
6. Rea WJ. 1997. *Chemical Sensitivity, Vol. IV, Tools of Diagnosis and Methods of Treatment.* Boca Raton, FL: Lewis Publishers. 2185.
7. Selye H. 1946. The general adaptation syndrome and the disease of adaptation. *J. Allergy* 17:23.
8. Lee C. 1961. A new test for diagnosis and treatment of food allergies. *Buchanan County Med. Bull.* 25:9.
9. Rinkle HJ. 1949. Inhalant allergy, I, Whealing response of skin to serial dilution testing. *Ann. Allergy* 7:625–630.
10. Podell RN. 1983. Intracutaneous and sublingual provocation and neutralization. *Clin. Ecol.* 2:13–20.

11. Forman RA. 1981. Critique of evaluation studies of sublingual and intracutaneous provocative test for food allergy. *Med. Hypotheses* 7:1019–1027.
12. Scadding GK, and Brostoff J. 1986. Low-dose sublingual therapy in patients with allergic rhinitis due to house-dust mite. *Clin. Allergy* 16:483–491.
13. Pfeiffer GO. 1970. Sublingual procedures. *Trans. Soc. Ophthalmol. Otolaryngol. Allergy* 11:104.
14. Björkstén B and Dewdney JM. 1987. Oral immunotherapy in allergy—Is it effective? *Clin. Allergy* 17:91–94.
15. Leng, X Fu YX, Ye ST, and Duan SQ. Jan. 1990. A double-blind trial of oral immunotherapy for *Artemisia* pollen asthma with evaluation of bronchial response to the pollen allergen and serum-specific IgE antibody. *Ann. Allergy* 64:27–31.
16. Taudorf E, Laursen LC, Lanner A, Björkstén B, Dreborg S, Søborg M, and Weeke B. Aug. 1987. Oral immunotherapy in birch pollen hayfever. *J. Allergy Clinical Immunology* 80:153–161.
17. Van Nierkerk CH and DeWet JI. 1987. Efficacy of grass-maize pollen oral immunotherapy in patients with seasonal hay-fever: A double-blind study. *Clin. Allergy* 17:507–513.
18. Taudorf E, Laursen LC, Lanner A, Björkstén B, Dreborg S, Søborg M, and Weeke B. Aug. 1987. Oral immunotherapy in birch pollen hayfever. *J. Allergy Clin. Immunol.* 83(3):589–594.
19. Tari MG, Mancino M, and Monti G. 1990. Efficacy of sublingual immunotherapy in patients with rhinitis and asthma due to house dust mite: A double-blind study. *Allergo. Immunopathol.* 18(5):277–284.
20. Moller C, Dreborg S, Lanner A, and Björkstén B. 1986. Oral immunotherapy of children with rhinoconjunctivitis due to Birch Pollen allergy. *Allergy* 41:271–279.
21. Wortman F. 1977. Oral hyposensitization of children with pollinosis of house-dust asthma. *Allergo. Immunopathol.* 5(15):15–26.
22. Holt PG, Vines J, and Britten D. 1988. Sublingual allergen administration, I, Selective suppression of IgE production in rats by high allergen doses. *Clin. Allergy* 18:229–234.
23. O'Shea JA and Porter SF. 1981. Double-blind study of children with hyperkinetic syndrome treated by multi-allergen extract sublingually. *J. Learning Disabilities* 14:189–90.
24. Felziani V, Marfisi RM, and Parmiani S. 1993. Rush immunotherapy with sublingual administration of grass allergen extract. *Allergo. Immunopathol.* 21(5):173–178.
25. Sabbah A, Hassoun S, Le Sellin J, André C, and SicardH . 1994. A double-blind, placebo-controlled trial by the sublingual route of immunotherapy with a standardized grass pollen extract. *Allergy* 49(5):309–313.
26. Troise C, Voltolini S, Canessa A, Pecora S, and Negrini AC. 1995. Sublingual immunotherapy in *Parietaria* pollen-induced rhinitis: A double-blind study. *J. Investig. Allergol. and Clin. Immunol.* 5(1):25–30.
27. Hansel FK. 1957. Optimal dosage therapy in allergy and immunity. *Ann Otol. Rhino.* 66:729–742.
28. Lee CH, William RI, and Binkley EL. 1969. Provocative testing and treatment for foods. *Archives of Otolaryngology* 90:113.
29. Miller JB. 1977. A double-blind study of food extract injection therapy: A preliminary report. *Ann. Allergy* 38(3):185.
30. Miller JB. 1976. Food allergy: Technique of intradermal testing subcutaneous injection therapy. *Trans. Am. Soc. Ophthalmol. Otolaryngol. Allergy* 16:154–168.
31. Rapp DJ. 1978. Double-blind confirmation and treatment of milk sensitivity. *MJA* 1:571.
32. Rapp DJ. 1979. Food allergy treatment for hyperkinesis. *J. Learning Disabilities* 12:42–50.
33. Rapp DJ. 1978. Weeping eyes in wheat allergy. *Trans. Am. Soc. Ophthalmol. Otol. Allergy* 18:149–150.
34. Boris M, Schiff M, Meiforf S, and Inselman L. 1983. Bronchoprovocation blocked by neutralizing therapy (Abstract). *J. Allergy Clin. Immunol.* 71:92.
35. McGovern JJ. 1980. Correlation of clinical food allergy symptoms with serial pharmacologic and immunologic changes in the patient's plasma (Abstract). *Ann. Allergy* 44:57–58.
36. Darlington LG. 1983. Food allergy and rheumatoid disease. *Ann. Rheum. Diseases* 42:219–220.
37. Monro J. 1983. Food allergy in migraine. *Proc. Nutr. Soc.* 42:241.
38. Rea WJ, Podell RN, Williams ML, Fenyves EJ, Sprague DE, and Johnson AR. 1984. Elimination of oral food challenge reaction by injection of food extracts. *Arch. Otolaryngol.* 110:248–252.
39. King WP, Rubin WA, Fadal RG, Ward WA, Trevino RJ, Pierce WB, Steward JA, and Boyles JH. 1988. Provocation-neutralization: A two-part study, Part I, The intracutaneous provocative food test: A multicenter comparison study. *Otolaryngology: Head and Neck Surgery* 99(3):263–271.
40. King WP, Rubin WA, Fadal RG, Ward WA, Trevino RJ, Pierce WB, Steward JA, and Boyles JH. 1988. Provocation-neutralization: A two-part study, Part II. Subcutaneous neutralization therapy: A multicenter comparison study. *Otolaryngology: Head and Neck Surgery* 99(3):272–277.

41. Gerdes K. 1989. Provocative-neutralization testing: A look at the controversy. *Clinical Ecology* 6:21–29.
42. Pischinger A. 1983. Das System der Grundregulation, Grundlagen ür eine ganzheitsbiologische Theorie der Medizin. 4. Aufl. Karl f. Haug Verlag, Heidelbert.
43. Rea WJ. 1997. *Chemical Sensitivity, Vol. IV, Tools of Diagnosis and Methods of Treatment.* Boca Raton, FL: Lewis Publishers. 2481–2540.
44. Rozman KK, and Doull J. 2003. Scientific foundations of hormesis, Part 2, Maturation, strengths, limitations, and possible application in toxicology, parmacology, and epidemology. In *Critical Reviews in Toxicology* 33(3&4):451.
45. Calabrese EJ. 2005. Homesis presents a good model for toxicological risk assessment—And it's not homeopathy. *The Scientist.* 22, Fig. 1.
46. Rea WJ. 1992. *Chemical Sensitivity, Vol. I, Mechanisms of Chemical Sensitivity.* Boca Raton, FL: Lewis Publishers. 7–46.
47. Rea WJ. 1997. *Chemical Sensitivity, Vol. IV, Tools of Diagnosis and Methods of Treatment.* Boca Raton, FL: Lewis Publishers. 2187.
48. Calabrese EJ and Baldwin LA. 2001. Hormesis: A generalizable and unifying hypothesis. 253. In *Critical Reviews in Toxicology* 31(4&5). Ed. RO McClellan.
49. Rea WJ. 1992. *Chemical Sensitivity, Vol. I, Mechanisms of Chemical Sensitivity.* Boca Raton, FL: Lewis Publishers. 155.
50. Rea WJ. 1996. *Chemical Sensitivity, Vol. III, Clinical Manifestations of Pollutant Overload.* Boca Raton, FL: Lewis Publishers. 1597.
51. Rea WJ. 1996. *Chemical Sensitivity, Vol. III, Clinical Manifestations of Pollutant Overload.* Boca Raton, FL: Lewis Publishers. 1727.
52. Linsenmayer TF. *Cell Biology of the Extracellular Matrix.* Chapter 1. Ed. E Hay. New York and London: Plenum Press.
53. Pischinger A. 1991. *Matrix and Matrix Regulation: Basis for a Holistic Theory in Medicine.* Ed. H Heine. Brussels, Belgium: Editions Haug International. 127–128.
54. Guyton AC and Hall JE. 1996. *Textbook of Medical Physiology,* 9th ed. Philadelphia: WB Saunders Co. 265.
55. Woodburne RI and Burkel WF. 1994. *Essentials of Human Anatomy,* 9th ed. New York: Oxford Univeristy Press. 21, Fig. 15.
56. Pischinger A. 1991. *Matrix and Matrix Regulation: Basis for a Holistic Theory in Medicine.* Ed. H Heine. Brussels, Belgium: Editions Haug International. 44.
57. Pischinger A. 1991. *Matrix and Matrix Regulation: Basis for a Holistic Theory in Medicine.* Ed. H Heine. Brussels, Belgium: Editions Haug International. 49.
58. Heine H. 1988. Akupunkturtherapie—perorationen der oberflächlichen Körperfaszie durch kutane Gefäb-Nervenbündel. *Therapeutikon* 4(1988b):238–244.
59. Croley TE and Carlson M. 1991. Histology of the Acupuncture Point. *American Journal of Acupuncture.* 19(3):247–253.
60. Bergsmann O and Bergsmann R. 1988. Projecktionssymptome. Reflektorische Krankgeitszeichen als Grundlage für holistiche Diagnose und Therapie. Facultas Universitätsverlag, Wien.
61. Heine H and Anastasiadis P. 1992. *Normal Matrix and Pathological Conditions.* New York: Gustar Fischer Verlag. 70.
62. Pischinger A. 1975. *Matrix and Matrix Regulation: Basis for a Holistic Theory in Medicine.* Ed. H Heine. Eng. trans. N MacLean. Brussels, Belgium: Editions Haug International. 198.
63. Pischinger A. 1975. *Matrix and Matrix Regulation: Basis for a Holistic Theory in Medicine.* Ed. H Heine. Eng. trans. N MacLean. Brussels, Belgium: Editions Haug International. 73.
64. Yamada KM. Fibronectin and other structural proteins. In , 2nd ed. Ed. ED Hay. New York, London: Plenum Press. 21983.95–114
65. Pischinger A. 1975. *Matrix and Matrix Regulation: Basis for a Holistic Theory in Medicine.* Ed. H Heine. Eng. trans. N MacLean. Brussels, Belgium: Editions Haug International. 74.
66. Pischinger A. 1975. *Matrix and Matrix Regulation: Basis for a Holistic Theory in Medicine.* Ed. H Heine. Eng. trans. N MacLean. Brussels, Belgium: Editions Haug International. 185.
67. Athenstaedt H. 1974. Pyroelectric and piezoelectric properties of vertebrates. *Ann. New York. Acad. Sci.* 238:68–110.
68. Pischinger A. 1975. *Matrix and Matrix Regulation: Basis for a Holistic Theory in Medicine.* Ed. H Heine. Eng. trans. N MacLean. Brussels, Belgium: Editions Haug International. 25.
69. Schulz-Ruhtenberg C. 1997–98. Academy for Applied Regulation-Thermography. D-61348 Bad Homburg. Schöne Aussicht 8a.

70. Ishikawa S, Naito M, and Inabe K. 1970. A new videopupillography. *Opthalmologica* 160:248.
71. Heart Rhythm Instruments, Inc. 2002. Quantitative Assessment of the Autonomic Nervous System: Based on an Analysis of Heart Rate Variability. 173 Essex Avenue, Metuchen, NJ 08840.
72. Pischinger A. 1975. *Matrix and Matrix Regulation: Basis for a Holistic Theory in Medicine*. Ed. H Heine. Eng. trans. N MacLean. Brussels, Belgium: Editions Haug International. 45–47.
73. Heine H. 1987. Die Grundregulation aus neuer Sicht. Ärztezeischr.f.Naturheilverf. 28:909–914.
74. Heine H and Anastasiadis P. 1992. *Normal Matrix and Pathological Conditions*. New York: Gustar Fischer Verlag. 77.
75. Heine H and Anastasiadis P. 1992. *Normal Matrix and Pathological Conditions*. New York: Gustar Fischer Verlag. 153.
76. Heine H and Anastasiadis P. 1992. *Normal Matrix and Pathological Conditions*. New York: Gustar Fischer Verlag. 81.
77. Heine H and Anastasiadis P. 1992. *Normal Matrix and Pathological Conditions*. New York: Gustar Fischer Verlag. 78–79.
78. Heine H. 1997. Lehrbuch der biologischen Medizin. 2. Auflage Stuttgart: Hippokrates.
79. Nicolis G and Prigonine I. 1987. Die Erforschung de Komplexe. München und Zürich: Piper.
80. Pischinger A. 1975. *Matrix and Matrix Regulation: Basis for a Holistic Theory in Medicine*. Ed. H Heine. Eng. trans. N MacLean. Brussels, Belgium: Editions Haug International. 52–53.
81. Blumenfeld LA. 1981. *Problems of Biological Physics*. Berlin-Heidelberg-New York: Springer.
82. Pischinger A. 1975. *Matrix and Matrix Regulation: Basis for a Holistic Theory in Medicine*. Ed. H Heine. Eng. trans. N MacLean. Brussels, Belgium: Editions Haug International. 105.
83. Smith CW and Best S. 1989. *Electromagnetic Man: Health and Hazard in the Electrical Environment*. London: JM Dent.
84. Pischinger A. 1975. *Matrix and Matrix Regulation: Basis for a Holistic Theory in Medicine*. Ed. H Heine. Eng. trans. N MacLean. Brussels, Belgium: Editions Haug International. 147.
85. Heine H and Anastasiadis P. 1992. *Normal Matrix and Pathological Conditions*. New York: Gustar Fischer Verlag. 153.
86. Fröhlich K. 1984. Zitiert nach Popp. Vortag an der Jahrestagung der DAH, Bad Nauheim.
87. Popp FA. 1983. Neue Horizonte in der Medizin. Heidelberg: Karl F. Haug Verlag. From A Pischinger. 1991. *Matrix and Matrix Regulation: Basis for a Holistic Theory in Medicine*. Ed. H Heine. Brussels, Belgium: Editions Haug International. 105.
88. Klina. 1981. From A Pischinger. 1991. In *Matrix and Matrix Regulation: Basis for a Holistic Theory in Medicine*. Ed. H Heine. Brussels, Belgium: Editions Haug International. 105.
89. Klina. 1987. From A Pischinger. 1991. In *Matrix and Matrix Regulation: Basis for a Holistic Theory in Medicine*. Ed. H Heine. Brussels, Belgium: Editions Haug International. 105.
90. Bergsmann O and Bergsmann R. 1988. Projecktionssymptome. Reflektorische Krankheitszeichen als Grundlage für holistische Diagnose und Therapie. Facultas Universitätsverlag, Wein. From A Pischinger. 1991. In *Matrix and Matrix Regulation: Basis for a Holistic Theory in Medicine*. Ed. H Heine. Brussels, Belgium: Editions Haug International. 105.
91. Rea WJ, Pan Y, Fenyves EJ, et al. 1991. Electromagnetic field sensitivity. *J. Biolelectricity*. 19(1&2):241–56.
92. Balasz EA and Gibbs PA. 1970. The rheological properties and biological function of hyaluronic acid. In *Chemistry and Molecular Biology of the Intercellular Matrix*, Vol. 2. Ed. EA Balasz. New York, London: Academic Press. 1241–1254.
93. Buddecke E. 1971. Grundris der Biochemie. 2. Aufl. W. de Gruyter, Berlin.
94. Pischinger A. 1975. In *Matrix and Matrix Regulation: Basis for a Holistic Theory in Medicine*. Ed. H Heine. Eng. trans. N MacLean. Brussels, Belgium: Editions Haug International. 23.
95. Bergsmann O and Bergsmann R. 1988. Projecktionssymptome. Reflektroische Krankheitszeichen als Brundlage für holistische Diagnose und Therapie. Facultas Universitätsverlag, Wien.
96. Pischinger A. 1975. In *Matrix and Matrix Regulation: Basis for a Holistic Theory in Medicine*. Ed. H Heine. Eng. trans. N MacLean. Brussels, Belgium: Editions Haug International. 48.
97. Guyton AC and Hall JE. 2000. *Textbook of Medical Physiology*, 10th ed. Philadelphia: WB Saunders Co. 272–273.
98. Athenstaedt H. 1974. Pyroelectric and piezoelectric properties of vertebrates. *Ann. New York. Acad. Sci.* 238:68–110. From A Pischinger. 1991. In *Matrix and Matrix Regulation: Basis for a Holistic Theory in Medicine*. Ed. H Heine. Brussels, Belgium: Editions Haug International. 102.
99. Bergsmann. 1983. From A Pischinger. 1991. In *Matrix and Matrix Regulation: Basis for a Holistic Theory in Medicine*. Ed. H Heine. Brussels, Belgium: Editions Haug International. 107.

100. Pischinger A. 1991. In *Matrix and Matrix Regulation: Basis for a Holistic Theory in Medicine*. Ed. H Heine. Brussels, Belgium: Editions Haug International. 107.
101. Pischinger A. 1991. In *Matrix and Matrix Regulation: Basis for a Holistic Theory in Medicine*. Ed. H Heine. Brussels, Belgium: Editions Haug International. 14.
102. Heine H and Anastasiadis P. 1992. *Normal Matrix and Pathological Conditions*. New York: Gustar Fischer Verlag. 7.
103. Heine H and Anastasiadis P. 1992. *Normal Matrix and Pathological Conditions*. New York: Gustar Fischer Verlag. 6.
104. Alexander CM and Werbzena. Chapter 8. In *Cell Biology of the Extracellular Matrix*. Ed. E. Hay. New York–London: Plenium Press. 255–256, 284.
105. Pischinger A. 1991. In *Matrix and Matrix Regulation: Basis for a Holistic Theory in Medicine*. Ed. H Heine. Brussels, Belgium: Editions Haug International. 87–88.
106. Wing L. 2003. Sydney, Australia. Personal communication.
107. Pischinger A. 1991. In *Matrix and Matrix Regulation: Basis for a Holistic Theory in Medicine*. Ed. H Heine. Brussels, Belgium: Editions Haug International. 191.
108. Guyton AC and Hall JE. 2000. *Textbook of Medical Physiology,* 10th ed. Philadelphia: WB Saunders Co. 163.
109. Heine H. 1999. Homotoxicology: A Synthesis of Medical Schools of Thought on a Scientific Basis. *Institut für Antihomotoxische Medizin und Grundregulations forschung Bahnackrstrabe* 16 D-76532 Baden-Baden, Germany. 9.
110. Sharon N and Lis H. 1993. Kohenhydrate und Zellerkennung. Spektrum der Wissenschaft. Heft 3:66–74.
111. Heine H. 1999. Homotoxicology: A Synthesis of Medical Schools of Thought on a Scientific Basis. *Institut für Antihomotoxische Medizin und Grundregulations forschung Bahnackrstrabe* 16 D-76532 Baden-Baden, Germany. 8.
112. Heine H. 1999. Homotoxicology: A Synthesis of Medical Schools of Thought on a Scientific Basis. *Institut für Antihomotoxische Medizin und Grundregulations forschung Bahnackrstrabe* 16 D-76532 Baden-Baden, Germany. 10.
113. U.S. Department of Energy Office of Science, Office of Biological and Environmental Research, Human Genome Program. 2003.
114. Watters E. 2006. DNA is not destiny. The new science of epigenetics rewrites the rules of disease, heredity, and identity. *Discover* 27(11):33–37.
115. Waterland RA and Jirtle RL. 2003. Transposable elements: targets for early nutritional effects on epigenetic gene regulation. *Mol. Cell. Biol.* 23:5293–5300.
116. Szyf M. Department of Pharmacology and Therapeutics at McGill University in Montreal, Canada. From E Watters. 2006. DNA is not destiny. The new science of epigenetics rewrites the rules of disease, heredity, and identity. *Discover* 27(11):33–37.
117. Fang MZ, Wang Y, Ai N, et al. Nov. 15, 2003. Tea Polyphenol (−)-Epigallocatechin-3-Gallate Inhibits DNA Methyltransferase and Reactivates Methylation-Silenced Genes in Cancer Cell Lines. *Cancer Research* 63:7563–7570.
118. Jirtle RL and Skinner MK. 2007. Environmental epigenomics and disease susceptibility. *Nat. Rev. Genet.* 8:253–562.
119. Heslin MJ, et al. 2001. Role of Matrix Metallo Proteinases in Colorectal Carcinogenesis. Dept. of Surgical Oncology University of Alabama. *Ann. Surg.* 233(6):786–792.
120. Guyton AC and Hall JE. 1996. *Textbook of Medical Physiology,* 9th ed. Philadelphia: WB Saunders Co. 303.
121. Guyton AC and Hall JE. 1996. *Textbook of Medical Physiology,* 9th ed. Philadelphia:WB Saunders Co. 304.
122. Guyton AC and Hall JE. 1996. *Textbook of Medical Physiology,* 9th ed. Philadelphia:WB Saunders Co. 359.
123. Guyton AC and Hall JE. 1996. *Textbook of Medical Physiology,* 9th ed. Philadelphia:WB Saunders Co. 45.
124. Guyton AC and Hall JE. 1996. *Textbook of Medical Physiology,* 9th ed. Philadelphia:WB Saunders Co. 304–305.
125. Guyton AC and Hall JE. 1996. *Textbook of Medical Physiology,* 9th ed. Philadelphia:WB Saunders Co. 189.
126. Guyton AC and Hall JE. 1996. *Textbook of Medical Physiology,* 9th ed. Philadelphia:WB Saunders Co.. 191.
127. Hauss WH, Junge-Hülsing G, and Gerlach G. 1986. Die unspezifische Messsenchymreaktion. Thieme, Stuttgart.

128. Pischinger A. 1991. *Matrix and Matrix Regulation: Basis for a Holistic Theory in Medicine.* Ed. H Heine. Brussels, Belgium: Editions Haug International. 19.
129. Review by H Heine und G Schaeg. 1979. Informationassteurerung in der vegetativen Peripherie. Zschr. Hautkr. 54:590–597.
130. Rea WJ. 1996. *Chemical Sensitivity, Vol. III, Clinical Manifestations of Pollutant Overload.* Boca Raton, FL: Lewis Publishers. 1476.
131. Rea WJ. 1996. *Chemical Sensitivity, Vol. III, Clinical Manifestations of Pollutant Overload.* Boca Raton, FL: Lewis Publishers. 1738.
132. Pischinger A. 1991. *Matrix and Matrix Regulation: Basis for a Holistic Theory in Medicine.* Ed. H Heine. Brussels, Belgium: Editions Haug International. 89.
133. Pischinger A. 1991. *Matrix and Matrix Regulation: Basis for a Holistic Theory in Medicine.* Ed. H Heine. Brussels, Belgium: Editions Haug International. 198.
134. Heine H. 1988. Anatomische Struktur der Akupunkturpunkte. Dtsch. Zschr. Akup. 31:(1988a)26–30.
135. Srelrow A, Friedman Y, Ravanpay A, Daniel CW, and Bissell MJ.1998. Expression of Hexa-1 and Hexa-7 is regulated by extracellular matrix-dependent signals in mammary epithelial cells. *J. Cell Biochem.* 69(4):377–391. Erratum in *J. Cell Biochem.* 71(2):310–320,
136. Roskelley CD and Bissell MJ. The dominance of the microenvironment in breast and ovarian cancer. *Semin. Cancer Biol.* 202;12(2):97–104.
137. Bascom JL, Fata JE, Hirai Y, Sternlicht MD, and Bissell MJ. 2005. Epimorphin overexpression in the mouse mammary gland promotes alveolar hyperplasia and mammary adenocarcinoma. *Cancer Res.* 65(19):8617–8621.
138. Hynes RO. 1983. Fibronectin and its relation to cellular structure and behavior. In *Cell Biology of Extracellular Matrix,* 2nd ed. Ed. ED Hay. New York, London: Plenum Press. 295–334.
139. Heine H. 1999. Homotoxicology: A Synthesis of Medical Schools of Thought on a Scientific Basis. *Institut für Antihomotoxische Medizin und Grundregulations forschung Bahnackrstrabe* 16 D-76532 Baden-Baden, Germany. 42.
140. Guyton AC and Hall JE. 2000. *Textbook of Medical Physiology,* 10th ed. Philadelphia: WB Saunders Co. 140.
141. Maillard JY. 2005. Antimicrobial biocides in the healthcare environment: Efficacy, usage, policies, and perceived problems. *Ther. Clin. Risk. Manag.* 1(4):307–320.
142. Xatzipsalti M, Psarros F, Konstantinou G, et al. 2008. Modulation of the epithelial inflammatory response to rhinovirus in an atopic environment. *Clin. Exp. Allergy* 38(3):466–472.
143. Hoberg EP, Polley L, Jenkins EJ, et al. 2008. Integrated approaches and empirical models for investigation of parasitic disease in northern wildlife. *Emerg. Infect. Dis.* 14(1):10–17.
144. Rea WJ, Pan Y, Fenyves, et al. 1991. Electromagnetic filed sensitivity. *J. Bioelectricity* 10(1&2):241–256.
145. Rea WJ. 1994. *Chemical Sensitivity, Vol. II, Sources of Total Body Load.* Boca Raton, FL: Lewis Publishers.
146. Hauss WH. 1961. Über die Entstehung und Behandlung rheumatisher Erkrankungen. *Hippokrates* 32(17):678.
147. Rea WJ. 1997. *Chemical Sensitivity, Vol. IV, Tools of Diagnosis and Methods of Treatment.* Boca Raton, FL: Lewis Publishers. 2046.
148. Randolph TG. 1962. *Human Ecology and Susceptibility to the Chemical Environment.* Springfield, IL: Charles C. Thomas.
149. von Ardenne M. 1990. *Oxygen Multistep Therapy: Physiological and Technical Foundations.* Trans. P Kirby and W Krüger. New York: George Thieme Verlag Stuttgart.
150. Pischinger A. 1991. In *Matrix and Matrix Regulation: Basis for a Holistic Theory in Medicine.* Ed. H Heine. Brussels, Belgium: Editions Haug International. 83.
151. Guyton AC and Hall JE. 1996. *Textbook of Medical Physiology,* 9th ed. Philadelphia: WB Saunders Co. 158.
152. Guyton AC and Hall JE. 1996. *Textbook of Medical Physiology,* 9th ed. Philadelphia: WB Saunders Co. 159.
153. Rea WJ. 1996. *Chemical Sensitivity, Vol. III, Clinical Manifestations of Pollutant Overload.* Boca Raton, FL: Lewis Publishers.
154. Hauss WH. 1970. Rolle der Mesenchymzellen in der thogenese der Arteriosklerose. Doc. angiol. 2:11.
155. von Ardenne M. 1990. *Oxygen Multistep Therapy: Physiological and Technical Foundations.* Trans. P Kirby and W Krüger. New York: George Thieme Verlag Stuttgart. 110.
156. Rea WJ. 1997. *Chemical Sensitivity, Vol. IV, Tools of Diagnosis and Methods of Treatment.* Boca Raton, FL: Lewis Publishers. 2554.

157. von Ardenne M. 1990. *Oxygen Multistep Therapy: Physiological and Technical Foundations.* Trans. P Kirby and W Krüger. New York: George Thieme Verlag Stuttgart. 291.
158. Rea WJ, Johnson AR, Youdim S, Fenyves EJ, and Samadi N. 1986. T and B lymphocyte parameters measured in chemically sensitive patients and controls. *Clin. Ecol.* 4:11–44.
159. Rea WJ. 1992. *Chemical Sensitivity, Vol. I, Mechanisms of Chemical Sensitivity.* Boca Raton, FL: Lewis Publishers. 47–154.
160. Guyton AC and Hall JE. 2000. *Textbook of Medical Physiology,* 10th ed. Philadelphia: WB Saunders Co. 172–173.
161. Ali M. 1983. *Spontaneity of Oxidation in Nature and Aging.* Monograph, Teaneck, New Jersey.
162. Ali M. 1985. The agony and death of cell. *Syllabus of the Instruction Course of the American Academy of Environmental Medicine.* Denver, Colorado.
163. Ali M. 1987. *Intravenous Nutrient Protocols in Molecular Medicine.* Monograph. NJ: Institute of Preventive Medicine, Bloomfield.
164. Ali M. 1990. Molecular basis of cell membrane injury. In *Syllabus of the Instruction Course of the American Academy of Environmental Medicine.* Denver, Colorado.
165. Ali M. 1990. Spontaneity of oxidation and molecular basis of environmental illness. In *Syllabus of the 1991 Instruction Course of the American Academy of Environmental Medicine.* Denver, Colorado.
166. Ali M. 1992. Spontaneity of oxidation and chronic disease. In *Syllabus of the Instruction Course of the American Academy of Environmental Medicine.* Denver, Colorado.
167. Ali M. 1997. Oxidative Coagulopaty. In *Syllabus of the Capital University of Integrative Medicine.* Washington, DC.
168. Ali M. 1993. Hypothesis: Chronic fatigue is a state of accelerated oxidative molecular injury. *J. Adv. Med.* 6:83–96.
169. Ali M. 1990. Ascorbic acid reverses abnormal erythrocyte morphology in chronic fatigue syndrome. *Am. J. Clin. Pathol.* 94:515.
170. Ali M. 1991. Ascorbic acid prevents platelet aggregations by norepinephrine, collagen, ADP and ristocetin. *Am. J. Clin. Pathol.* 95:281.
171. Ali M, Prasad PV, Rigolosi R, et al. 1982. Cardiac myocytolysis: pathologic basis of cardiomyopathy in dialysis patients (Abstract). *Kidney International* 21:161.
172. Ali M and Fayemi AO. 1982. Cardiac myocytolysis in cardiomyopathy. In *The Pathology of Maintenance Hemodialysis.* Springfield, IL: Charles C. Thomas. 132–138.
173. Ali, M. 1980. Hemosiderosis in hemodialysis patients: an autopsy study of 50 cases. *JAMA* 244:343.
174. Ali M, AO Fayemi, R Rigolosi, et al. 1982. Failure of serum ferritin levels to predict bone marrow iron content after intravenous iron-dextran therapy. *Lancet* 652.
175. Ali M and Fayemi AO, and Braun E. 1982. Dissociation between hepatosplenic and marrow iron in liver cirrhosis. *Arch. Pathol. Lab. Med.* 106:200–204.
176. Ali M and Fayemi AO. 1982. Arterial and myocardial calcification. In *The Pathology of Maintenance Hemodialysis.* Springfield IL: Charles C. Thomas. 130–132.
177. Fayemi AO and Ali M, and Braun E. 1979. Oxalosis in hemodialysis patients: a pathologic study of 80 patients. *Arch. Pathol. Lab. Med.* 103:58–62.
178. Ali M and Fayemi AO. 1982. Cardiac oxalosis. In *The Pathology of Maintenance of Hemodialysis.* Springfield, IL: Charles C. Thomas. 146–147.
179. Ali M and Fayemi AO. Arterial intimal thickening. In *The Pathology of Maintenance Hemodialysis.* Springfield, IL: Charles C. Thomas. 175–185.
180. Fayemi AO and Ali M. 1979. Intrarenal vascular alterations in hemodialysis patients: a semiquantitative light microscopic study. *Hum. Pathol.* 10:685–692.
181. Ali M and Fayemi AO. 1979. The pathology of end-stage renal disease in hemodialysis patients. *Isr. J. Med. Sci.* 5:901–909.
182. Ali M, Braun E, Laraia S, et al. 1981. Immunochemical LD assay for myocardial infarction. *Am. J. Clin. Pathol.* 76:426–429.
183. Ali M, Laraia S, Angeli R, et al. 1982. Immunochemical CK-MB assay for myocardial infarction. *Am. J. Clin. Pathol.* 77:573–579.
184. Ali M. 1992. The basic equation of life. In *The Butterfly and Life Span Nutrition.* Denville, NJ: The Institute of Preventive Medicine Press. 225–236.
185. Ali M. 1993. *The Ghoraa and Limbic Exercise.* Denville, NJ: The Institute of Preventive Medicine Press.
186. Ali M. 1989. *The Cortical Monkey and Healing.* Bloomfield, NJ: The Institute of Preventive Medicine.
187. Ali M. 1996. *What Do Lions Know About Stress?* Denville, NJ: Life Span.
188. Ali M. 1997. *Healing, Miracles and the Bite of the Gray Dog.* Denville, NJ: Life Span.

189. Ali M and Ali O. 1997. AA oxidopathy: The core pathogenetic mechanism of ischemic heart disease, Part I. *J. Integrative Medicine* 1:1–112.
190. Sampson J. Personal communication.
191. Pischinger A. 1991. In *Matrix and Matrix Regulation: Basis for a Holistic Theory in Medicine.* Ed. H Heine. Brussels, Belgium: Editions Haug International. 31.
192. Schröder HJ. 1959. Gesetzmässigkeiten bei der Nekrobiose und Autolyse der weissen Blutzellan und ihre biologische Bedeutung. Habil. –Schr. Med. Fak. D. Univ. Hamburg (Hier weitere Literatur).
193. Sullivan JL. 1981. Iron and the sex difference in heart disease risk. *Lancet* 1:1293–1294.
194. Salonen JT, Nyyssonen K, Korpela H, et al. 1992. High stored iron levels are associated with excess risk of myocardial infarction in eastern Finnish Men. *Circulation* 86:803–811.
195. Salonen J, Salonen R, Seppanen K, et al. 1991. Interactions of serum copper, selenium, and low density lipoprotein cholesterol in atherogenesis. *BMJ.* 302:756–760.
196. Salonen R and Salonen JT. 1990. Progression of carotid atherosclerosis and its determinants: a population based ultrasonography study. *Atherosclerosis.* 81:33–40.
197. Salonen JT, Seppanen K, Nyyssonen K, et al. 1995. Intake of mercury from fish, lipid peroxidation, and the risk of myocardial infarction and coronary, cardiovascular, and any death in eastern Finnish men. *Circulation* 91:645–655.
198. Mattila KJ, Nieminen MS, Valtonen VV, et al. 1989. Association between dental health and acute myocardial infarction. *BMJ* 298:779–781.
199. Enstrom JE, Kanim LE, and Klein MA. 1992. Vitamin C intake and mortality among a sample of the United States population. *Epidemiology* 3:194–202.
200. Nyyssonen K, Parvianen MT, Salonen R, et al. 1997. Vitamin C deficiency and risk of myocardial infarction: Prospective population study of men from eastern Finland. *BMJ.* 314:634–638.
201. Stephens NG, Parsons A, Schofield PM, et al. 1996. Randomised controlled trial of vitamin E in patients with coronary disease: Cambridge Heart Antioxidant Study (CHAOS). *Lancet* 347:781–386.
202. Stampher MJ, Hennekens CH, Manson JH, et al. 1993. Vitamin E consumption and the risk of coronary disease in women. *N. Eng. J. Med.* 328:1444–1449.
203. Shaish A, Daugherty A, O'Sullivan, et al. 1995. Beta-carotene inhibits atherogenesis in hypercholesterolemic rabbits. *J. Clin. Invest.* 96:2075–2082.
204. Folkers K, Littarru GP, Ho L, et al. 1970. Evidence for a deficiency of coenzyme Q10 in human heart disease. *Inter. J. Vitamin Nutr. Research* 4:380–390.
205. Folkers K, Watanabe T, and Kaji M. 1977. Crtique of coenzyme Q10 in biochemical and biomedical research and in ten years of clinical research in cardiovascular disease. *J. Mol. Med.* 2:431–460.
206. deDuve Cand Wattiaux R. 1996. Functions of lysosomes. *Ann. Rev. Physiol.* 28:435.
207. Pischinger A. 1991. In *Matrix and Matrix Regulation: Basis for a Holistic Theory in Medicine.* Ed. H Heine. Brussels, Belgium: Editions Haug International. 195.
208. von Ardenne M. 1970/71. *Theoaretische und experimentelle Grundlagen der Krebs-Mehrschritt-Therapie,* 2nd ed. (containing the O_2MT concept). Berlin: Volk und Gesundheit.
209. von Ardenne M and Krüger W. 1979. Local tissue hyperacidification and lysosomes. In *Lysosomes in Applied Biology and Therapeutics,* Vol. VI. JT Dingle, PJ Jacques, IM Shaw. North Holland, Amsterdam.
210. Porter KR and Bonneville MA. 1965. *Einführung in die Feinstruktur von Zellen und Geweben.* Berlin: Springer.
211. von Ardenne M. 1990. *Oxygen Multistep Therapy: Physiological and Technical Foundations* Trans. P Kirby and W Krüger. New York: George Thieme Verlag Stuttgart. 113.
212. von Ardenne M. 1990. *Oxygen Multistep Therapy: Physiological and Technical Foundations* Trans. P Kirby and W Krüger). New York: George Thieme Verlag Stuttgart. 114.
213. von Ardenne M. 1977. Lysosomenbildung als Reizantwort des Herzmuskels auf O_2-Mangel. Cardiol. bull. *Acta cardiol.* 14/15:165.
214. Ionescu G, Merk M, and Bradford R. 1999. Simple Chemiluminescence Assays for Free Radicals in Venous Blood and Serum Samples: Results in Atopic, Psoriasis, MCS and Cancer Patients. *Forschende Komplementämedizin* 6:294–300.
215. Guyton AC and Hall JE. 2000. *Textbook of Medical Physiology,* 10th ed. Philadelphia: WB Saunders Co. 399.
216. Pischinger, A. 1957. Das Schicksal der Leukozyten. Z. Mikr.-anat. Forsch.63, 169–192.
217. Heine H and Anastasiadis P. 1992. *Normal Matrix and Pathological Conditions.* New York:Gustav Fischer Verlag. 80–81.
218. Mark B and Winterbern CC. 2002. Redox Regulation of Neutrophil Function. *Antioxidants and Redox Signaling.* 4(1–3) Forum Editorial. Mary Ann Liebert Inc.

219. Robinson J, Watson F, Bucknall RC, and Edwards SW. 1992. Activation of neutrophils reactive oxidants production by synovial fluid from patients with inflammatory joint disease: soluble and insoluble immunoglobulin aggregates activate different pathways in primed and uinprimed cell. *Biochem. J.* 46:2401–2406.
220. Repine JE. 1992. Scientific perspectives on adult respiratory distress syndrome. *Lancet* 339:466–469.
221. Babior BM. 1999. NADPH oxidase: an update. *Blood* 93:1464–1476.
222. Karlsson A and Dahlgren C. 2002. Assembly and activation of the neutrophil HADPH oxidase in granule membranes. *Antioxid. Redox. Signal* 4:49–60.
223. Pricop L and Salmon JE. 2002. Redox regulation of Fcγ receptor-mediatedd phagocytosis: implications for host defense and tissue injury. *Antioxid. Redox. Signal* 4:85–95.
224. Swain DS, Rohn TT, and Quinn MT. 2002. Neutrophil priming in host defense: role of oxidants as priming agents. *Antioxid. Redox. Signal* 4:69–83.
225. Wany Q and Doeschuk CD. 2002. The signaling pathways induced by neutrophil-endothelial cell adhesion. *Antioxid. Redox. Signal* 4:39–47.
226. Kasahara Y, Iwai K, Yachie A, et al. 1997. Involvement of reactive oxygen intermediates in spontaneous and mediated apoptosis of neutrophils. *Blood.* CD 95 (Fas/APO-1) 8915:1748–1753.
227. Gallin JI and Buescher ES. 1983. Abnormal regulation of inflammatory skin responses in male patients with chronic granulomatous disease. *Inflammation* 7:227–232.
228. Morgentstern DE, Gifford MAC, Li L, et al. 1997. Absence of respiratory burst in X-linked chronic granulomatous disease mice leads to abnormalities in both host defence and inflammatory response to *Aspergillus fumigatus*. *J. Exp. Med.* 1854:207–218.
229. Ward C, Wong TH, Murry J, et al. 2000. Induction of human neutrophil apoptosis by nitric oxide donors: evidence for a caspase-dependent, cycle-GMP-independent mechanism. *Biochem. Pharmacol.* 59:305–314.
230. Fadeel B, Ahlin A, Henter J-I, Orrenius S, and Hampton MB. 1998. Involvement of caspases in neutrophil apoptosis: regulation by reactive oxygen species. *Blood* 92:4808–4818.
231. Hart SP, Ross JA, Haslett C, and Dransfield I. 2000. Molecular characterization of the surface of apoptotic neutrophils: Implication for functional down regulation by phagocytes. *Cell. Differ.* 5:493–503.
232. Kane DJ, Sarafian TA, Anton R, et al. 1993. Bcl-2-inhibition of neural death: decreased generation of reactive oxygen species. *Science* 262:1274–1277.
233. Wasserman SI. 1989. Mast Cell-mediated inflammation in asthma. *Ann. Allergy* 63:546–550.
234. Minireviewer Infection in Immunity. Sept. 1997. Soran, Abraham, Ravi, Melaviya. 3501–3508.
235. Pischinger A. 1991. In *Matrix and Matrix Regulation, Basis for a Holistic Theory in Medicine*. Ed. H Heine. Brussels, Belgium: Editions Haug International. 47.
236. Xu X, et al. 2002. Role of mast cells and myofibroblasts in human peritoneal adhesion formation. *Ann. Surg.* 266(5):593–601.
237. Scott-Coombes DM, Vipond MN, and Thompson JN. 1993. General surgeons' attifues to the treatment and prevention of abdominal adhesion. *Ann. R. Coll. Surg. Engl.* 75:1213–128.
238. Thompson J. 1998. Pathogenesis and prevention of adhesion formation. *Dig. Surg.* 15:153–157.
239. Powell DW, Mifflin RX, Valentich JD, et al. 1992. Myofibroblasts. I. paracrine cells important in health and disease. *Am. J. Physiol.* 277:C1-C19.
240. Gruber BL, Kew RR, Jelaska A, et al. 1997. Human mast cells activate fibroblasts: tryptase is a fibrogenic factor stimulating collagen messenger fribonucleic acid synthesis and fibroblast chemotaxis. *J. Immunol.* 158:2310–2317.
241. Claman HN and Jaffe BD. 1985. Chronic graft-vs-host disease as a model for scleroderma. II. Mast cell-depletion with deposition of immunoglobulins in the skin and fibrosis. *Cell. Immunol.* 94:73–84.
242. Persinger MA, Lepage P, Sinard JP, et al. 1983. Mast cell numbers in incisional wounds in rat skin as a function of distance, time and treatment. *Br. J. Dermatol.* 108:1799–187.
243. Levi-Schaffer F, Gare M, Shalit M. 1990. Unresponsiveness of rat peritoneal mast cells to immunologic reactivation. *J. Immunol.* 145:3418–3424.
244. Levin RH, Barrett PVD, Cine MJ, Berlin NI, Freireich EJ. 1964. Platelet Therapy and Red Cell Defect in Aplastic Anemia. *Arch. Intern. Med.* 114(2):278–283.
245. Ryser JE and Dutton RW. 1977. The humoral immune response of mouse bone marrow lymphocytes. *Vitro Immunol.* 32(5):811–817.
246. Fulop G, Gordon J, and Osmond DG. 1983. Regulation of lymphocyte production in the bone marrow. I. turnover of small lymphocytes in mice depleted of B-lymphocytes by treatment with anti-IgM antibodies. *The J. Immun.* 130(2):644–648.

247. Osmond DG. 1985. The ontogeny and organization of the lymphoid system. *J. Invest. Dermat.* 85:25–93.
248. De Boer RJ, Freitas AA, and Perelson AS. 2001. Resource competition determines selection of B-cell repertoires. *J. Theor. Biol.* 212:333–343.
249. Tanchot C and Rocha B. 1997. Peripheral selection ot T-cell repetoires: the role of continuous thymus output. *J. Exp. Med.* 186(7):1099–1106.
250. Wakim LM, Waithman J, Van Rooigen N, et al. 2008. Dendritic cell-induced memory T-cell activation in nonlymphoid tissue. *Science* 219(5860):198–202.
251. Laurence J. 1993. T-cell subsets in health, infections, disease, and idiopathic CD_4+ T-lymphocytopenia. *Ann. of Internal Med.* 119(1):55–62.
252. Bosio CM and Elkins KL. 2001. Susceptibility to secondary francisella tularensis live vaccine strain infection in B-cell-deficient mice is associated with neutrophilia but not with defects. II. Specific T-cell-mediated immunity. *Infect. Immun.* 69(1):194–203.
253. Calzascia T, Pellegrini M, Hall H, Sabbagh L. 2007. TNF-α is critical for antitumor but not antiviral T-cell immunity in mice. *J. Clin. Invest.* 117(12):3833–3845.
254. Kawaguchi S. 2005. B-cell reconstitution by transplantation of B220+ CD 117+ B-lymphoid progenitors into irradiated mice. *Immunology.* 114(4):461–467.
255. Sager H, Bertoni G, and Jungi TW. 1998. Differences between B-cell and macrophage transformation by the bovine parasite, theileria annulata: a clonal approach. *J. Immun.* 161:335–341.
256. Kosmas C, Stamatopoulos K, Stayroyianni N, Tsanaris N, and Papadaki T. 2002. Anti-CD_{20}-based therapy of B-cell lymphoma: state of the art. *Nature* 16(10):275–281.
257. Thomas M, Kremer C, Ravichandran K, Rajervsky K, and Bender T. 2005. C-myb is critical for B-cell development and maintenance of Follicular B-cells. *Immunity* 23(3):275–281.
258. Neron S, Suck G, Ma X-Z, et al. 2006. B-cells proliferation following CD_{40} stimulation results in the expression and activation of src protein tyrosine kinase. *Int. Immun.* 18(2):375–387.
259. Fulcher DA and Basten A. 1997. B-cell life span: A review. *Immun & Cell Biol.* 755(5):446–455.
260. Veldman JE and Deuning FJ. 1978 Histophysiology of cellular immunity reactions in B-cell deprived rabbits: an X-irradiation model for elineation of an "isdated T-cell system." *Virchows Archiv B Cell Path.* Zell-patholgie. 28(1):203–216.
261. Maclennan, ICM. 1998. B-cell receptor regulation of peripheral B-cells. *Curr. Opinion Immunol.* 10(2):220–225.
262. Pischinger A. 1991. In *Matrix and Matrix Regulation: Basis for a Holistic Theory in Medicine.* Ed. H Heine. Brussels, Belgium: Editions Haug International. 72.
263. Guyton AC and Hall JE. 2000. *Textbook of Medical Physiology,* 10th ed. Philadelphia: WB Saunders Co. 395.
264. Heine H and Anastasiadis P. 1992. *Normal Matrix and Pathological Conditions.* New York: Gustar Fischer Verlag. 77–91.
265. Guyton AC and Hall JE. 2000. *Textbook of Medical Physiology,* 10th ed. Philadelphia: WB Saunders Co. 394.
266. Guyton AC and Hall J E. 2000. *Textbook of Medical Physiology,* 10th ed. Philadelphia: WB Saunders Co. 396, Fig. 33-3.
267. Guyton AC and Hall JE. 2000. *Textbook of Medical Physiology,* 10th ed. Philadelphia: WB Saunders Co. 396, Fig. 33-4.
268. Guyton AC and Hall JE. 2000. *Textbook of Medical Physiology,* 10th ed. Philadelphia: WB Saunders Co. 396, Fig. 33-5.
269. Baldwin RW and Price MR. 1972. *Nature* 238:185–186. Blocking of lymphocytes mediated cytotoxicity by rat hepatic cells by tumor specific analysis antibody complex. *Proc of National Academy of Science* USA. 1981. 78.
270. Simpson LO. 1993. The effects of saline solutions on red cell shape: a scanning-electron-microscope-based study. *Br. J. Haematol.* 85(4):832–834.
271. von Ardenne M. 1990. *Oxygen Multistep Therapy: Physiological and Technical Foundations* Trans. P Kirby and W Krüger). New York: George Thieme Verlag Stuttgart. 71–74.
272. Guyton AC and HallJE. 2000. *Textbook of Medical Physiology,* 10th ed. Philadelphia: WB Saunders Co. 794, Fig. 69-2
273. Cannon WB and Rosenblueth A. 1949. *The Supersensitivity of Denervated Structures, A Law of Denervation.* New York: MacMillan.
274. Hatthakit U, Shida Pang KK, and Piyachaturawat P. 1994. Cardiovascular effects of piperine in anaesthetized dogs. *Asia Pacific J. Pharmacol.* 9:79–82.

275. Calabrese EJ. 2001. Overcompensation stimulation: a mechanism for hormetic effect. *Crit. Rev. Toxicol.* 31:425–470.
276. Calabrese EJ and Baldwin LA. 2001. Special Issue: Scientific Foundations of Hormesis. In *Critical Review in Toxicology* 31(4&5)351–695. Ed. RO McClellan. Boca Raton, FL: CRC Press LLC.
277. Calabrese EJ and Baldwin LA. 2001. Special Issue: Scientific Foundations of Hormesis. In *Critical Review in Toxicology*. 33(3&4):213–449. Ed. RO McClellan. Boca Raton, FL: CRC Press LLC.
278. Rozman KK and Doull J. 2001. Paracelsus, Haber and Arndt. *Toxicology* 160:191–196.
279. Rea WJ. 1992. *Chemical Sensitivity, Vol I, Principles and Mechanisms.* Boca Raton, FL: Lewis Publishers. 19.
280. Rea WJ and Brown OD. 1987. Cardiovascular disease in response to chemicals and foods. In *Food Allergy and Intolerance.* Ed. J Brostoff and SJ Challacombe. London: Bailliere Tindall. 737–783.
281. http://www.anaesthetist.com. 17 November 1999. Receptors—A brief note.
282. Pischinger, A. 1991. In *Matrix and Matrix Regulation: Basis for a Holistic Theory in Medicine.* Ed. H Heine. Brussels, Belgium: Editions Haug International. 52–53.
283. Pischinger, A. 1991. In *Matrix and Matrix Regulation: Basis for a Holistic Theory in Medicine.* Ed. H Heine. Brussels, Belgium: Editions Haug International. 96.
284. Rea WJ 1997. *Chemical Sensitivity, Vol. IV, Tools of Diagnosis and Methods of Treatment.* Boca Raton, FL: CRC Press, Lewis Publishers. 2044.
285. Hildebrandt G. 1985. Therapeutische Physiologie, Grundlagen der Kurortbehandlung in Balneologie. Medizinische Klimatolgie. Heidelberg: Springer Verlag.
286. Hildebrandt G, et al. 1993. Chronobiologische Aspektke des Schmerzes. In Stacher A (Hrsg.): Ganzheitsmedizin und Schmerz. Dritter Wiener Dialog. Wien: Facultats. 40–61.
287. Lemmer B. 1984. Chronopharmakologie. Tagesrhythmen und Arzneiwirkung. Stuttgart: Wissenschaftliche Verlagsgesellschaft.
288. Satok NT. 2005. Vitamin A: storing stellate cells in the human newborn vocal fold. *Ann. Otol. Rhinol. Larygol.* 114(7):517–524.
289. Pöppe, CHR. 1995. Reste von Ordnung im Unendlich-Dimensionalen. Spektrum der Wisenschatt. Heft 9:22–32.
290. Heine H. 1999. Homotoxicology: A Synthesis of Medical Schools of Thought on a Scientific Basis. Institut für antihomotoxishce Medizin und Grundregulations for schung. Bahnackrstrase 16. D-76532. Baden-Baden, Germany. 10–12.
291. Hunecke F. 1953. Krankheit und Heilung anders gesehen. Köln: Staufen-Verlag.
292. Dosch P. *Manual of Neural Therapy: Accoding to Huneke (Regulating Therapy with Local Anesthetics),* 11th (revised) ed.. First English ed. Heidelberg: Karl F. Haug. 105–107, 112.
293. Kellner G. 1963. Die Wirkung des Herdes auf die Labilität des humoralen Sytems. Österr. Zschr. Stomat. 60:312.
294. Dosch P. *Manual of Neural Therapy: Accoding to Huneke (Regulating Therapy with Local Anesthetics),* 11th (revised) ed.. First English ed. Heidelberg: Karl F. Haug. 112.
295. Dosch P. *Manual of Neural Therapy: Accoding to Huneke (Regulating Therapy with Local Anesthetics),* 11th (revised) ed.. First English ed. Heidelberg: Karl F. Haug. 101.
296. Nordenström, BEW. 1983. Biologically closed electric circuits. Nordic Medical.
297. Adolph EF. 1956. General and specific characteristics of physiological adaptations. *Am. J. Physiol.* 184:18.
298. Rea WJ. 1992. *Chemical Sensitivity, Vol. I, Principles and Mechanisms.* Boca Raton, FL: Lewis. 33–34.
299. Savage GM. 1884. *Insanity and Allied Neuroses: Practical and Clinical.* Philadelphia: Henry C. Leas' Son and Co.
300. Rea WJ. 1992. *Chemical Sensitivity, Vol. I, Principles and Mechanisms.* Boca Raton, FL: Lewis. 21.
301. Rea WJ. 1992. *Chemical Sensitivity, Vol. I, Principles and Mechanisms.* Boca Raton, FL: Lewis. 22.
302. Rea WJ. 1992. *Chemical Sensitivity, Vol. I, Principles and Mechanisms.* Boca Raton, FL: Lewis. 23.
303. Guyton AC and Hall JE. 2000. *Textbook of Medical Physiology,* 10th ed. Philadelphia: WB Saunders Co. 4–7.
304. Lin T, Islam O, Heese K. 2006. ABC transporters, neural stem cell and neurogenesis: a different perspective. *Cell Res.* 16(1)857–871.
305. Pischinger, A. 1991. In *Matrix and Matrix Regulation: Basis for a Holistic Theory in Medicine.* Ed. H Heine. Brussels, Belgium: Editions Haug International. 30.
306. Swiderek H, Logan A, and Al-Rubeai M. 2008. Cellular and transcriptomic analysis of NSO cell response during exposure to hypoxia. *J. Biotechnol.* 134(1–2):103–11.

307. Guyton AC and Hall JE. 2000. *Textbook of Medical Physiology,* 10th ed. Philadelphia: WB Saunders Co. 7.
308. Guyton AC and Hall JE. 2000. *Textbook of Medical Physiology,* 10th ed. Philadelphia: WB Saunders Co. 2.
309. Guyton AC and Hall JE. 2000. *Textbook of Medical Physiology,* 10th ed. Philadelphia: WB Saunders Co. 8.
310. Rea WJ. 1996. *Chemical Sensitivity, Vol. III, Clinical Manifestations of Pollutant Overload.* Boca Raton. FL.: Lewis. 1478.
311. Rea W J. 1997. *Chemical Sensitivity, Vol. IV, Tools of Diagnosis and Methods of Treatment.* Boca Raton, FL.: Lewis. 2239.
312. Rea WJ. 2000. Personal communication.
313. Braith RW, Mills RM Jr., Wilcox CS, et al. 1996. Fluid homeostasis after transplantation: the role of cardiac denervation. *J. Heart Lung Transplant.* 15(9):872–880.
314. Olgin JE, Sih HJ, Harnish S, et al.1998. Heterogeneous atrial denervation creates substrate for sustained atrial fibrillation. *Circulation.* 98(23):2608.
315. Pischinger, A. 1991. *Matrix and Matrix Regulation: Basis for a Holistic Theory in Medicine.* Ed. H Heine. Brussels, Belgium: Editions Haug International. 177.
316. Rea WJ. 2001. Personal communication.
317. Guyton AC and Hall JE. 2000. *Textbook of Medical Physiology,* 10th ed. Philadelphia: WB Saunders Co. 4.
318. Levine St A and Kidd MP. 1985. *Antioxidant Adaption: Its Role in Free Radical Pathology.* San Leandro, CA: Biocurrent Division.
319. Pischinger A. 1975. *Matrix and Matrix Regulation: Basis for a Holistic Theory in Medicine.* Ed. H Heine. Eng. trans. N MacLean. Brussels, Belgium: Editions Haug International. 24.
320. Rea WJ. 2002. Personal communication.
321. Leyh R, Wilhelmi M, Walles T, et al. 2003. A cellularized porcine heart valve scaffolds for heart valve tissue engineering and the risk of cross-species transmission of porcine endogenous retrovirus. *J. Thor. Cardio. Surg.* 126(4):1000–1004.
322. Pischinger A. 1975. *Matrix and Matrix Regulation: Basis for a Holistic Theory in Medicine.* Ed. H Heine. Eng. trans. N MacLean. Brussels, Belgium: Editions Haug International. 162.
323. Pischinger A. 1975. *Matrix and Matrix Regulation: Basis for a Holistic Theory in Medicine.* Ed. H Heine. Eng. trans. N. MacLean. Brussels, Belgium: Editions Haug International. 165.
324. Pischinger A. 1975. *Matrix and Matrix Regulation: Basis for a Holistic Theory in Medicine.* Ed. H Heine. Eng. trans. N MacLean. Brussels, Belgium: Editions Haug International. 39.
325. U.S. Department of Health and Human Services, Public Health Service. Agency for Toxic Substances and Disease Registry. September 2002. Toxicological Profile for Hexachlorobenzene. 1600 Clifton Rd. NE, E-29, Atlanta, GA 30333.
326. Heine H. 1999. Homotoxicology: A Synthesis of Medical Schools of Thought on a Scientific Basis. Institut für antihomotoxishce Medizin und Grundregulations for schung. Bahnackrstrase. D-76532 Baden-Baden, Germany. 45.
327. Ader R, Felten DL, Cohen N. 2006. *Psychoneuroimmunology,* 4th ed. 2 Volumes. China: Academic Press.
328. Heine H. 1999. Homotoxicology: A Synthesis of Medical Schools of Thought on a Scientific Basis. Institut für antihomotoxishce Medizin und Grundregulations for schung. Bahnackrstrase D-76532 Baden-Baden, Germany. 10–11.
329. Draczynski TH. 1997. Das Phänomen der physiologischen Leukozytolyse. In *Extracellular Matrix and Groundregulation System in Health and Disease.* Ed. H Heine and M Rimpler. Stuttgart: G. Fischer. 139–150.
330. Ebringer, A. 1983. The crosstolerance hypothesis. HLA-B27 and ankylosing spondylitis. *Br. J. Rhuematol.* 22(Suppl.2):53–66.
331. Regoli M, et al. 1995. Uptake of a gram-positive bacterium (Streptococcus pneumoniae R36a) by the Mcells of rabbit Peyer's patches. *Ann. Anat.* 177:119–124.
332. Rea WJ. 1992. *Chemical Sensitivity, Vol. I, Principles and Mechanisms.* Boca Raton, FL: Lewis. 58.
333. Rea WJ. 1992. *Chemical Sensitivity, Vol. I, Principles and Mechanisms.* Boca Raton, FL: Lewis. 102.
334. Thorn GW, Adams D, Braunwald E, Isselbachner KF, and Petersdorf RG. 1977. *Harrison's Principles of Internal Medicine,* 8th ed.. New York: McGraw-Hill. 112.
335. Thorn GW, Adams D, Braunwald E, Isselbachner KF, and Petersdorf RG. 1977. *Harrison's Principles of Internal Medicine,* 8th ed.. New York: McGraw-Hill. 1531, 1518, 1348.

336. Gray P, Lewis K, Masta A, Phillips D. 1994. Modulation of mustard toxicity by taurine. *Biochem. Pharmacol.* 47(3):581–583.
337. Guyton AC and Hall JE. 1996. *Textbook of Medical Physiology,* 9th ed. Philadelphia: WB Saunders Co. 883.
338. Rea WJ. 1992. *Chemical Sensitivity, Vol. I, Mechanisms of Chemical Sensitivity.* Boca Raton, FL: Lewis. 57–61.
339. Rea WJ. 1992. *Chemical Sensitivity, Vol. I, Mechanisms of Chemical Sensitivity.* Boca Raton, FL: Lewis. 65–66.
340. Rea WJ. 1992. *Chemical Sensitivity, Vol. I, Mechanisms of Chemical Sensitivity.* Boca Raton, FL: Lewis. 187.
341. Rea WJ. 1992. *Chemical Sensitivity, Vol. I, Mechanisms of Chemical Sensitivity.* Boca Raton, FL: Lewis. 56.
342. Rea WJ. 1992. *Chemical Sensitivity, Vol. I, Mechanisms of Chemical Sensitivity.* Boca Raton, FL: Lewis. 91.
343. Cook JW, Hewett CL, and Hieger I. 1933. The isolation of a cancer producing hydrocarbon from coal tar: Parts I, II, and IV. *J. Chem. Soc.* 395.
344. Rea WJ. 1992. *Chemical Sensitivity, Vol. I, Mechanisms of Chemical Sensitivity.* Boca Raton, FL: Lewis. 66.
345. Brabec MH and Bernstein IA. 1981 Cellular, subcellular, and molecular targets of foreign compounds. In *Toxicology: Principles and Practice,* Vol. I. Ed. AL Reeves. New York: John Wiley & Sons. 36.
346. Rea WJ. 1992. *Chemical Sensitivity, Vol. I, Mechanisms of Chemical Sensitivity.* Boca Raton, FL: Lewis. 67.
347. Nishida Y. 2003. Elucidation of endemic neurodegenerative diseases: A commentary. *Z. Naturforsch. CJ.* 58(9–10):752–758.
348. Rea WJ. 1992. *Chemical Sensitivity, Vol. I, Mechanisms of Chemical Sensitivity.* Boca Raton, FL: Lewis. 68.
349. Rea WJ. 1992. *Chemical Sensitivity, Vol. I, Mechanisms of Chemical Sensitivity.* Boca Raton, FL: Lewis. 70.
350. Rea WJ. 1992. *Chemical Sensitivity, Vol. I, Mechanisms of Chemical Sensitivity.* Boca Raton, FL: Lewis Publishers. 71–94.
351. ACS Surgery: Principles and Practice. 7 Acute Wound Care. Fig. 3. 2007 WebMD Inc.
352. Grinnell F, Billinghan RE, and Burgess L. 1981. Distribution of fibronectin during wound healing in vivo. *J. Invest. Dermatol.* 76:181.
353. Clark RAF, Folkvord J, Wertz RL. 1985. Fibronetcin as well as other extracellular matrix proteins mediate human keratinocyte adherence. *J. Invest. Dermatol.* 84:378.
354. Grinell F. 1984. Fibronectin and wound healing. *J. Cell. Biochem.* 25:107.
355. Wysocki AB and Grinell F. 1990. Fibronectin profiles in normal and chronic would fluid. *Lab Inves* 63:825.
356. Turk JL. 1994. Inflammation: John Hunter's "A treatise on the blood, inflammation and gun-shot wounds." *Int. J. Exp. Pathol.* 75(6):385–395.
357. Ley K. 1992. Leukocyte adhesion to vascular endothelium. *J. Reconstr. Microsurg.* 8:495.
358. Newman SL, Henson JE, and Henson PM. 1982. Phagocytosis of senescent neutrophils by human monocyte-dervived macrophages and rabbit inflammatory macrophages. *J. Exp. Med.* 156:430.
359. Proveddini DM, Deftos LJ, and Manolagas SC. 1986. 1,25-Dihydroxyvitamin D3 promotes in vitro morphologic and enzymatic changes in normal human monocytes consisten with their differentiation into macrophages. *Bone* 7:23.
360. Wright SD and Meyer BC. 1985. Fibronectin receptor of human macrophages recognizes sequence Arg-Cly-Asp-Ser. *J. Exp. Med.* 162:762.
361. Simpson DM and Ross R. 1971. Effects of heterologous anti-neutrophil serum in guinea pigs: hematologic and ultrastructural observations. *Am. J. Pathol.* 65:79.
362. Leibovich SJ and Ross R. 1975. The role of the macrophage in wound repair: A study with hydrocortisone and anti-macrophage serum. *Am. J. Pathol.* 78:71.
363. Seppa H, Grotendorst G, Seppa S, et al. 1982. Platelet-derived growth factor is chemotactic for fibroblasts. *J. Cell. Biol.* 92:584.
364. Gauss-Miller V, Kleinman H, Martin GR, et al. 1980. Role of attachment factors and attractants in fibroblast chemotaxis. *J. Lab. Clin. Med.* 96:1071.
365. Grotendorst GR, Chang T, Seppa HEJ, et al. 1982. Platelet-derived growth factor is a chemoattractant for vascular smooth muscle cell. *J. Cell. Physiol.* 113:261.
366. Stiles CF, Capone GT, Scher CD, et al. 1979. Dual control of cell growth by somatomedins and platelet-derived growth factor. *Proc. Natl. Acad. Sci. USA* 76:1279.

367. Leibovich SJ and Ross R. 1976. A macrophage-dependent factor that stimulates the proliferation of fibroblasts in vitro. *Am. J. Pathol.* 84:501.
368. Thakral KK, Goodson WH III, and Hunt TK. 1979. Stimulation of wound blood vessel growth by wound macrophages. *J. Surg. Res.* 26:430.
369. Knighton DR, Hunt TK, Thakral KK, et al. 1982. Role of platelets and fibrin in the healing sequence as in vivo study of angiogenesis and collagen synthesis. *Ann. Surg.* 196: 379.
370. Van Winkle W Jr. 1986. The epithelium in wound healing. *Surg. Gynecol. Obstet.* 127:1089.
371. Clark RAF, Lanigan JM, DellaPelle P, et al. 1982. Fibronectin and fibrin provide a provisional matrix for epidermal cell migration during wound reepithelialization. *J. Invest. Dermatol.* 70:264.
372. Gipson IK, Spurr-Michaud SJ, and Tisdale AS. 1968. Hemidesmosomes and anchoring fibril collagen appear synchronously during development and wound healing. *Dev. Biol.* 126:253.
373. Mackie EH, Halfer W, and Liverani D. 1988. Induction of tenascin in healing wounds. *J. Cell. Biol.* 107:2757.
374. Cohen IK, Moore CD, and Diegelman RF. 1979. Onset and localization of collagen synthesis during wound healing in open rat skin wounds. *Proc. Soc. Exp. Biol. Med.* 160:458.
375. Peacock EE Jr. 1984. *Wound Repair,* 3rd ed. Philadelphia: WB Saunders Co.
376. Veis A and Averey J. 1965. Modes of intermolecular crosslinking in mature and insoluble collagen. *J. Biol. Chem.* 240:3899.
377. Rudolph R, Guber S, Suzuki M, et al. 1877. The life cycle of the myofibroblast. *Surg. Gynecol. Obstet.* 145:389.
378. Montesano R and Orci L. 1988. Transforming growth factor beta stimulates collagen-matrix contraction by fibroblasts: Implications for wound healing. *Proc. Natl. Acad. Sci. USA* 85:4894.
379. Madden JW, Peacock EE Jr. 1971. Studies on the biolboy of collagen during wound healing: III. Dynamic metabolism of scar collagen and remodeling of dermal wounds. *Ann. Surg.* 174:511.
380. Forrester JC, Zederfeldt BH, Hayes TL, et al. 1970. Wolff's law in relation to the healing skin wound. *J Trauma* 10:770.
381. Riley WB Jr, Peacock EE Jr. 1967. Identification, distribution and significance of a collagenolytic enzyme in human tissue. *Proc. Soc. Bio. Med.* 214:207.
382. Witte LD, Kaplan KL, Nossel HL, et al. 1978. Studies of the release from human platelets of the growth factor for cultured human arterial smooth muscle cells. *Circ. Res.* 42:402.
383. Martinet Y, Bitterman PB, Mornex JF, et al. 1986. Activated human monocytes express the c-sis proton-cogene and release a mediator showing PDGF-like activity. *Nature* 319:158.
384. Shimokado K, Raines EW, Madtes DK, et al. 1985. A significant part of macrophage-derived growth factor consists of at least two forms of PDGF. *Cell* 43:277.
385. Walker LN, Bowen-Pope DF, Ross R, et al. 1986. Production of platelet-derived growth factor-like molecules by cultured arterial smooth muscle cells accompanies proliferation after arterial injury. *Proc. Natl. Acad. Sci. USA* 83:7311.
386. Barbul A, Knud-Hansen J, Wasserkrug HL, et al. 1986. Interleukin 2 enhances wound healing in rats. *J. Surg. Res.* 40:315.
387. DeCunzo LP, MacKenzie JW, Marafino BJ Jr., et al. 1990. The effect of interleukin-2 administration on wound healing in Adriamycin-treated rats. *J. Surg. Res.* 49:419.
388. Assoian RK, Komoriya A, Meyers CA, et al. 1983. Transforming growth factor-β in human platelets: identification of a major storage site, purification, and characterization. *J. Biol. Chem.* 258:7155.
389. Personen K, Viinikka L, Myllyla G, et al. 1989. Characterization of material with epidermal growth factor immunoreactivity in human serum and platelets. *J. Clin. Endocrinol. Metab.* 68:486.
390. Assoian RK, Fleurdelys BE, Stevenson HC, et al. 1987. Expression and secretion of type β transforming growth factor by activated human macrophages. *Proc. Natl. Acad. Sci. USA* 84:6020.
391. Mustoe TA, Pierce GF, Thomason A, et al. 1987. Accelerated healing of incisional wounds in rats induced by transforming growth factor-β. *Science* 237:1333.
392. McGee GS, Davidson JM, Buckley A, et al. 1988. Recombinant basic fibroblast growth factor accelerates wound healing. *J. Surg. Res.* 45.
393. Brown GL, Curtsinger L III, Brightwell JR, et al. 1986. Enhancement of epidermal regeneration by biosynthetic epidermal growth factor. *J. Exp. Med.* 163:1319.
394. Laato M, Niinikoski J, Lebel L, et al. 1986. Stimulation of wound healing by epidermal growth factor: a dose-dependent effect. *Ann. Surg.* 203: 379.
395. Pierce GF, Mustoe TA, Senior RM, et al. 1988. In vivo incisional wound healing augmented by platelet-derived growth factor and recombinant c-sis gene homodimeric proteins. *J. Exp. Med.* 167:974.

396. Tsuboi R and Rifkin DB. 1990. Recombinant basic fibroblast growth factor stimulates wound healing in healing impaired db/db mice. *J. Exp. Med.* 172:245.
397. Pierce GF, Mustoe TA, Lingelbach J, et al. 1989. Transforming growth factor-β reverses the glucocorticoid-induced wound-healing deficit in rats: possible regulation in macrophages by platelet-derived growth factor. *Proc. Natl. Acad. Sci. USA* 86:2229.
398. Curtsinger LJ, Peitsch JD, Brown GL, et al. 1989. Reversal of Adriamyscin-impaired wound healing by transforming growth factor-beta. *Surg. Gynecol. Obstet.* 168:517.
399. Mustoe TA, Purdy J, Gramates P, et al. 1989. Reversal of impaired wound healing in irradiated rats by platelet-derived growth factor-BB. *Am. J. Surg.* 158:345.
400. Brown GL, Nanney LB, Briffen J, et al. 1989. Enhancement of wound healing by topical treatment with epidermal growth factor. *N. Engl. J. Med.* 321:76.
401. Brown GL, Curtsinger L, Jurkiewicz MJ, et al. 1991. Stimulation of healing of chronic wounds by epidermal growth factor. *Plast. Reconstr. Surg.* 88:189.
402. Knighton DR, Ciresi K, Fiegel VD, et al. 1990. Stimulation of repair in chronic, nonhealing, cutaneous ulcers using platelet-derived would healing formula. *Surg. Gynecol. Obstet.* 170:56.
403. Robson MC, Phillips LC, Thomason A, et al. 1992. Recombinant human platelet-derived growth factor BB for the treatment of chronic pressure ulcers. *Ann. Plast. Surg.* 29:193.
404. Robson MC, Phillips LG, Lawrence WI, et al. 1992. The safety and effect of topically applied recombinant basic fibroblast growth factor on the healing of chronic pressure sores. *Ann. Surg.* 216:401.
405. Steed DL. 1995. Diabetic Ulcer Study Group: Clinical evaluation of recombinant human platelet-derived growth factor for the treatment of lower extremity diabetic ulcers. *J. Vasc. Surg.* 21:71.
406. Heine H. 1999. Homotoxicology: A Synthesis of Medical Schools of Thought on a Scientific Basis. Institut für antihomotoxishce Medizin und Grundregulations for schung. Bahnackrstrase 16. D-76532 Baden-Baden, Germany. 40.
407. Sulman FG. 1976. *Health, Weather and Climate.* Basel, Switzerland: S. Karger AG.
408. Rea, WJ. 1994. *Chemical Sensitivity, Vol II, Sources of Total Body Load.* Boca Raton, FL: Lewis Publishers. 1011.
409. Rea WJ. 1992. *Chemical Sensitivity, Vol. I, Mechanisms of Chemical Sensitivity.* Boca Raton, FL: Lewis Publishers.228–229.
410. Anderson KE and Kappas A. 1991. Dietary regulation of cytochrome P-450. In *Annual Review of Nutrition.* Palo Alto, CA: Annual Review Inc. 11:432.
411. Merliss RR. 1972. Phenol marasmus. *J. Occup. Med.* 14:55–56.
412. Rea WJ. 1992. *Chemical Sensitivity, Vol. I, Mechanisms of Chemical Sensitivity.* Boca Raton, FL: Lewis Publishers. 243, 257.
413. Vesell ES. 1973. Advances in pharmacogentics. *Prog. Med. Genet.* 9:291–367.
414. Alexanderson B, Evans DAP, and Sjöqvist. 1969. Steady state plasma levels of nortriptyline in twins: Influence of genetic factors and drug therapy. *Br. Med. J.* 4:764–768.
415. Vesell ES, Page JG, and Passananti GT. 1971. Genetic and environmental factors affecting ethanol metabolism in man. *Clin. Parmacol. Ther.* 12:192–201.
416. Perger F. 1956. Untersuchungen über den Wirkungsmechanismus der hochmolekularen Fettsäuren im Elpimed bei parenteraler Zuführung. *Med. Klin.* 51/31:1299.
417. Rea WJ. 1997. *Chemical Sensitivity, Vol. IV, Tools of Diagnosis and Methods of Treatment.* Boca Raton, FL: Lewis. 2017–2050.
418. Ross GH, Rea WJ, Johnson AR, Maynard BJ, and Carlisle L. 1989. Evidence for vitamin deficiencies n environmentally-sensitive patients. *Clin. Ecol.* 6(2):60–66.
419. Smith C. 2002. Personal communication.
420. Smith C, Rea WJ, and Griffith BB. 2002. Personal communication.
421. Lentz MR, and Gatanaga T, and Granger GA. 1990. Indenticator of TLRGT blocking factor in serum filtrates of human cancer patients. *Lympokine* 9(2):225–229.
422. Josefsson E, et al. 1991. Peripheral denervation suppresses the latephase of delayed-type hypersensitivity. *Int. Arch. Allergy Appl. Immunol.*95:58–63.
423. Weihe E, et al. 1991. Molecular anatomy of the neuro-immune connection. *Intern. J. Neuroscience.* 59:283–295.
424. Meggs WJ and Svec C. 2003. *The Inflammation Cure: How to Combat the Hidden Factor Behind Heart Disease, Arthritis, Asthma, Diabetes, & Other Diseases.* New York, NY: McGraw-Hill.
425. Heine H. 1985. Transkutane Heparintherapie. Selbstregulation antithrombotischer Zell-und gewebesaktivitäten. Medwelt 36:703–705.

426. Sprott H, et al. 1997. Collagen crosslinks in fibromyalgia. *Arthritis and Rheum.* 40:1450–1454.
427. Heine H. 1995. Grundregulation und rheumatischer Formenkreis. Ärztezeitschrift für Naturheilverfahren 36:415–426.
428. Pischinger A. 2007. *Matrix and Matrix Regulation: Basis for a Holistic Biological Medicine*, English translation by Ingeborg Eibl, North Atlantic Books.
429. Levine S and Kidd P. 1985. *Antioxidant Adaptation: Its Role in Free Radical Biochemistry*, San Francisco, CA: Biocurrents Press. EHC-Dallas.
430. Lawrence WT, Bevin AG, and Sheldon GF. 2002 . ACS Surgery: Principles and Practice. 7 Acute Wound Care. Table 4. WebMD Inc.
431. Pischinger A. 1991. In *Matrix and Matrix Regulation: Basis for a Holistic Theory in Medicine*, Ed. H. Heine, Eng. Trans. N. MacLean, Brussels, Belgium: Editions Haug International. 21.
432. Pischinger A.1991. In *Matrix and Matrix Regulation: Basis for a Holistic Theory in Medicine*, Ed. H. Heine, Eng. Trans. By Ingeborg Eibl, North Atlantic Books. 25.
433. Woodburne RT. 1957. *Essentials of Human Anatomy*, New York: Oxford University Press. 21.
434. Guyton AC and Hall JE. 1996. *Textbook of Medical Physiology*, 9th ed., Philadelphia: WB Saunders Co. 185.
435. Guyton AC and Hall JE. 2000. *Textbook of Medical Physiology*, 10th ed., Philadelphia: WB Saunders Co. 680.
436. Guyton AC and Hall JE. 1996. *Textbook of Medical Physiology*, 9th ed., Philadelphia: WB Saunders Co. 754.
437. Guyton AC and Hall JE. 1996. *Textbook of Medical Physiology*, 9th ed., Philadelphia: WB Saunders Co. 753.
438. Netter FH. 1962. Digestive system, Part II. Lower Digestive Tract. *The CIBA collections of Medical Illustrations*, Vol. 3, New York: CIBA, 49.

Section

Amplification Systems: Neurological, Immune, and Endocrine

2 Nervous System

INTRODUCTION

If noxious incitant entry is not contained by the local homeostatic mechanism of the skin, mucosa, local sensory and autonomic nerves and connective tissue (CT) matrix, the amplification systems are activated in an attempt to maintain dynamic equilibrium. The nervous system is integrated as one part of the dynamic homeostatic amplification system along with the endocrine and immune systems making up the other parts.[1] The CT matrix through the ground regulation system (GRS) relays information about noxious incitant entry along with the status of local homeostatic control to these other systems.[2] If the noxious stimulus input is too high or too chronic for adequate local control and containment, the amplification systems are brought into action to neutralize the incitant. These amplification responses occur almost instantaneously once local control is exceeded and many are reflex actions.[3] **If left unchecked, constant noxious stimulus can lead to hypersensitivity or chronic degenerative disease, not only through deregulation of the immune parameters and depletion of the nutrient pool, but also through a nerve-altering phenomena in the nervous system**.

The nervous system is composed of three major parts: the sensory input portion, the central nervous system (CNS) (integrative portion), and the motor output portion anyone of which can be altered causing chemical sensitivity and chronic degenerative disease. The autonomic nervous system (ANS) plays a highly significant role in the dynamics of homeostasis, as shown in Chapter 1. **It is significant as being a part of the spinal and central reflexes, as well as the integrative portion of the CNS's homeostatic mechanism. This central homeostatic mechanism consists of the reticular formation, the hypothalamus, and the limbic system, all of which regulate the near automatic dynamics of homeostasis**.[4] **The pineal gland and the area postrema of the fourth ventricle also appear to be involved in homeostasis. If this homeostatic mechanism is constantly triggered, nutritional depletion occurs, resulting in chemical sensitivity and chronic degenerative disease**. For monitoring purposes these areas have only four cell layers versus six in the other areas of the brain.

The ANS is part of the local, regional, and central GRS, which enables proper energy efficient homeostatic responses. This integration into the GRS occurs locally, regionally, and centrally in that the terminal branches of the ANS engulf organs, lobules of organs, and even most T-cells (especially immune clusters when they are in, for instances, Peyer's patches).[5] The ANS is both an immediate response system (local and regional reflexes) and a response amplification system once the local homeostatic control is exceeded. Thus, the ANS plays a central role in the dynamics of homeostasis. The ANS is linked anatomically and physiologically with the endocrine and immune systems making reciprocal feedback loops for the three amplifications systems, where each can strongly influence the other when chronic noxious stimuli are present. **The speed at which these nerve responses occur is often extremely rapid and occurs faster than any immune or endocrine response**.

For the body to respond properly to noxious incitant exposure, these neurological subsystems, the ANS, peripheral nervous system (PNS), and CNS, must be in dynamic homeostatic balance. However, in cases of periodic homeostatic disturbances (chemical sensitivity)[6] and aperiodic homeostatic disturbance (chronic degenerative disease),[7] the dynamic balance between these systems becomes impaired. The nervous system itself often becomes not only the secondary, but also the primary target organ for noxious incitant injury. Regardless of the severity of the initial injury, once local homeostatic control is lost, chronicity leads to amplification of the homeostatic response. Once the ANS is involved, a dyshomeostatic response follows, where sensory receptor neuronal circuits and sensory receptors detect the state of the surrounding area. The neurological response may be through local, regional, or central reflex arches.[8]

The ANS is involved with normally localized accentuated reactions. However, these reactions can occur regionally or globally, which, when necessary, aids and overrides the local homeostatic response. An individual with periodic homeostatic disturbances (chemically sensitive individuals) and/or those suffering from aperiodic homeostatic disturbances (chronic degenerative disease) usually do, but not always experience direct involvement of the nervous system. This system is a prime target for deposition of many toxic inorganic (e.g., lead, mercury, cadmium) and organic chemicals (e.g., car exhaust, solvents, pesticides). These chemicals can either deregulate or severely damage the orderly homeostatic response causing trophic (neuropathic) edema[9,10] and or inflammation.[11] If the gastrointestinal tract is involved, the autonomic nervous system is activated because it is intimately related to most of the gastrointestinal tract's absorptive function and especially its immune function.[12,13] These disturbances of the ANS from excess entry of noxious substances usually occur from nerve damage, which follows the law of deinnervation. This law says, "Once injury to a nerve occurs the whole cell or organ distal to the injury becomes supersensitive to the nerve stimulation or neurotransmitter release.[14,15] This response increases even as the amount of stimulation decreases."[16,17] As this regional or organ supersensitivity occurs, it propagates a downhill negative dyshomeostatic characteristic, which can lead to end-organ failure.

Once local homeostatic control is lost, the ANS normally becomes involved in any generalized homeostatic reaction since it is part of the GRS, which is the prime communicator with the rest of the nervous system and the body. In order to prevent not only early brain homeostatic dysfunction but also end-stage, irreversible peripheral or central nerve damage, the clinician must be cognizant of the potential effects of the total pollutant level on the brain and other parts of the CNS, PNS, and the ANS. The clinician must attempt to contain pollutants locally upon entry within local dynamic homeostatic control. In particular, intake of heavy metals[18] (e.g., lead, cadmium, mercury, aluminum, and others), pesticides,[19] and solvents,[20] all of which are able to easily disrupt the dynamics of homeostasis, should be reduced. Once local homeostatic control is lost, these toxic substances can dramatically affect the patient who has neurological involvement, resulting in dyshomeostasis.[21] For example, the dyshomeostatic reactions of xenobiotics block neutralizing amino acids (i.e., glutathione),[22] destroy enzymes (i.e., cholinesterase),[23] and damage membranes (i.e., xylene), which then damages other cell and organelle membranes and the cell membrane's sodium or calcium pump.[3]

This chapter emphasizes the environmental aspects of the amplification responses of both the autonomic and voluntary nervous systems. When these systems are overloaded by chemicals or other noxious stimuli, hypersensitivity and chronic degenerative disease states result. This chapter emphasizes the influence of these systems on anatomy and the dynamic physiology of homeostasis. This chapter especially emphasizes the ordered and disordered function that takes place once local incitant entry and the capacity of local homeostasis has been exceeded. In addition, we attempt to demonstrate how to evaluate early incitant entry and injury in conjunction with local homeostasis. This concept of dyshomeostasis is presented in order to establish a method for prevention of environmentally induced illness and chronic degenerative disease in the autonomic, central, and peripheral nervous systems.

Today's knowledge educates us to the fact that the functioning of the human body is a complex interrelationship between genetics, epigenetics, nutrition, and environmental inducers and propagators. **The most practical and cost-effective way of managing the dynamics of homeostasis is to learn ways that the toxic environment can be manipulated and reduced in the patients' favor** in order to maintain orderly energy efficient homeostasis and prevent irreversible end-stage disease. The final purpose of this chapter is to introduce the integration of the acupuncture points and their energy flow as they correlate and function with the ANS and GRS. The energy system appears to

Linear and Biphasic Effects of Noxious Stimuli Entry

We can have a typical linear dose/response relationship for noxious incitants but in many chemicals (possibly 50%) there is a biphasic or hormetic response. As stated in Chapter 1, the linear effect of chemicals means that there is a dose dependent increase in its effect; one has high-dose inhibition by the specific chemical that becomes a noxious stimulus entry into the body. However, the biphasic hormetic response with low-dose stimulation and high-dose inhibition (the hormetic effect—J curve) may also be just the opposite with low-dose inhibition and high-dose stimulation (U curve). This nonmonotonic dose response, although not considered hormetic by some, is still just as important.

When analyzing neurological responses the biphasic effect of the toxic chemical that has entered the body as well as the linear characteristics has to be considered along with the potential biphasic response of the neurotransmitters that are triggered. This observation creates a complex milieu of responses that may well cloud the clinical picture. **Linear as well as J- or U-type hormetic responses may result in some disruption of the homeostatic mechanism causing an overcompensation reaction or it may be that two different types of receptors may be triggered at the same or different times giving two different responses.** See Table 2.1.

Table 2.1 shows a partial list of receptors system displaying a biphasic dose-response relationship. Most of these substances are neuropeptides which may be triggered by noxious environmental stimuli. Both have been found to alter glutathione levels (GSH). Lead may induce a compensating response in highly sensitive CNS cells such as the astrocytes.[5-8] According to Legare,[5] low-dose exposure to lead may initiate damage to the astrocytes by affecting discrete subcellular targets such as mitochondria thereby altering cellular metabolism and function.

TABLE 2.1
A Partial Listing of Receptor Systems Displaying Biphasic Dose–Response Relationships

Receptor System Displaying Biphasic Dose–Response Relationships

Adenosine	Nitric oxide
Adrenergic	NMDA
Bradykinin	Opioid
CCK	Platelet-derived growth factor
Corticosterone	Prolactin
Dopamine	Prostaglandin
Endothelin	Renal binding
Epidermal growth factor	Somatostatin
5-HT	Spermine
Human chorionic gonadotrophin	Testosterone
Muscarinic	Transforming growth factor β
Neuropeptides[1]	Tumor necrosis factor α

Source: Reprinted from Calabrese, E.J. and Baldwin, L.A., *Critical Reviews in Toxicology*, 31(4&5), 353–424, 2001. With permission.
Note: [1]For example, substance P and vasopressin.
Abbreviations: CCK = cholecystotkinin; 5-HT = 5-hydroxytryptamine (serotonin); NMDA = N-methyl-D-aspartate

In 1976 and 1977, Matchett[9,10] published his dissertation on the effect of acute ethanol administration on spontaneous locomoter activity and how this was affected by different doses. This research principally was concerned with how the occurrence of biphasic dose response relationship may be mediated by catecholamines.[11]

Serotonin also has been shown to have biphasic responses.[12] Dopamine has also been shown to have biphasic responses.[13] Also, cholinergic cells are affected by ethycholine aziridinun ion causes biphasic responses. Toxic chemicals that can trigger biphasic response in the nervous system include lead and cadmium. Injection therapy is a very powerful tool in the treatment of individuals with chemical sensitivity and/or chronic degenerative disease. We use this hormetic principle and facts in treatment by the intradermal injection technique for their neutralization of symptoms of toxics and their subsequent dyshomeostatic reactions.

RECEPTORS

Input to the nervous system is provided by sensory receptors that detect such sensory stimuli as touch, sound, light, pain, cold, and warmth. Many receptors can have a stimulatory effect while others in the same area may produce an inhibitory effect. Therefore, receptors may respond to a linear dose stimulus while others may produce a biphasic response. There are five basic types of sensory receptors[15] and one generalized receptor: Usually these are in the sensory nerves and go to the dorsal root ganglia of the spinal cord. (1) mechanoreceptors which respond to compression, stretching, or pressure; (2) thermoreceptors, which detect temperature changes; (3) nocioreceptors (pain receptors), which detect physical or chemical damage occurring in tissues; (4) electromagnetic receptors, which detect light in the eye; however, it should be noted that electromagnetic frequency (EMF) radiation can be received by most T-cells, especially the skin, as we have seen in the electromagnetically sensitive patient who when blindfolded, and will still react to EMF (electromagnetic frequency) stimuli; (5) chemoreceptors, which detect taste in the mouth, smell in the nose, oxygen levels in the blood, osmolarity of the body fluids, and carbon dioxide concentration. (Chemoreceptors appear to be one of the most significant group of receptors). See Figure 2.1. (6) Nonspecific receptors, as well as other types of receptors (i.e., for drugs, anesthetics, toxics), seem to aid the body in awareness of most new stimuli in the environment. They also appear to be a part of every cell and the CT matrix. These types of receptors may be local and may or may not directly attach to the nervous system. **Of course, as shown in Chapter 1, the CT matrix is a prime receptor for the body and will not be discussed here**.

Sensory receptors detect different types of sensory stimuli by differential sensitivities. Each receptor is sensitive to only one type of stimulus (exception the nonspecific receptor), and is almost nonresponsive to any other stimuli, thus sending dedicated receptor lines for information input

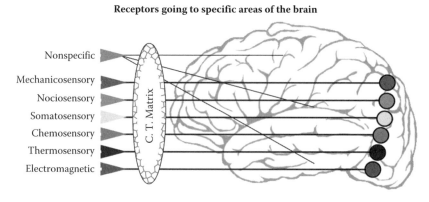

FIGURE 2.1 (EHC-Dallas, 2002.)

directly to the brain. Each has a nerve tract (labilize principle) via the peripheral sensory nerve to the anterior horn of the spinal cord and upto the brain receptor centers, which ends at a different point in the brain exactly where that respective center is located. Because of this anatomy and physiology, the CNS appears to be always aware of local incitant exposure and entry, but does not always have to react upon the signals. This lack of response is because many local homeostatic responses still occur, neutralizing the noxious stimuli. **All sensory receptors have one thing in common; they immediately affect the change in the electrical membrane potential**. The cause of the change in the membrane potential is basically a change in membrane permeability which allows ions to diffuse more or less through the membrane, thereby changing the transmembrane potential. The maximum amplitude of most sensory receptor potentials is about 100 millivolts. This is about the same maximum voltage recorded in action potentials and is the change in voltage when the membrane becomes maximally permeable to sodium ions. See Figure 2.2.

A special characteristic of all sensory receptors is that they adapt either partially or completely to any constant stimulus at any period of time.[17–24] This masking phenomenon allows the noxious stimuli to increase or persist, which then increases the total body pollutant load and suppresses the particular stressed nerve, altering homeostasis and ultimately causing nutrient depletion. Generally, fatigue and pain result. Each receptor adapts in its own way. In the eye, the rods and cones adapt by changing the concentration of their light-sensitive chemicals. The pacinian corpuscle in the skin is a viscoelastic structure so that when a light or strong force is suddenly applied to one side of the corpuscle, this force is instantly transmitted by the viscous component of the corpuscle directly to the same side of the central nerve fiber thus, eliciting the reception potential. Then, within a few hundredths of a second, the viscous fluid redistributes. The second effect of the pacinian corpuscle's adaptation is accommodation, which slows and vertically stops the sodium ion flow in the central nerve. If the cause is not withdrawn, the patient's system adjusts to the pressure.[25] The patient clearly pays the price for this adjustment with depleted nutrition, eventually chronic pain, and finally, degenerative disease.

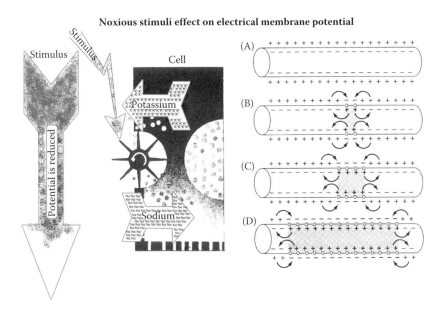

FIGURE 2.2 Change in membrane permeability to different ions yields conduction current. Every stimulus depolarizes the cell and produces stimulus patterns. The membrane resting potential (MRP) is lowered by permanent irritation and when oxygen metabolism is disturbed. (Modified from Dosch, P., *Manual of Neural Therapy*, 11th Ed., 51, 1984; Guyton, A.C. and Hall, J.E., *Medical Physiology*, 10th ed. Philadelphia: WB Saunders Co. 60, 2000.)

The rate at which receptors are stimulated changes when the stimulus strength changes. This receptor reaction predicts in one's mind the state of the body for the next few seconds or even the next few minutes. The receptors of the semicircular canals in the ear detect the rate at which the head begins to turn. Using this information when a person runs around a curve they can predict how much they will have to turn their bodies within two seconds to keep from losing balance.[26]

Some carotid and aortic baroreceptors may take up to two days to adapt completely. **This phenomenon is seen in the early chemically sensitive patient who is sensitive to weather fronts.** This patient will have symptoms for up to 2–4 days when a cloud cover moves in. After these few days, the patient appears to get "used to" (masked or adapted) this new set of environmental circumstances and can function better than he or she did in the early days of the new cloud cover. Again, such adaptability is a double-edged sword, because the patient improves some with the adaptation, yet the price of nutritional strain must be ultimately paid and in the end is costly to the system.

The CT receptors have been discussed in Chapter 1 and will not be discussed here.

AUTONOMIC NERVOUS SYSTEM

Anatomy and Physiology of the Peripheral Autonomic Nervous System

The peripheral ANS, as an efferent system, is made up of neurons that lie outside the CNS and which are concerned with visceral innervation. See Figure 2.3A. Both sympathetic and parasympathetic systems have preganglionic neurons in the brain and spinal cord. The afferent limbs of autonomic reflexes may lie in any afferent nerve. The preganglionic sympathetic fibers are myelinated, and leave the spinal roots as white rami communicants and synapse in the ganglia. Unmyelinated postganglionic fibers rejoin the anterior spinal roots, although some sympathetic fibers traverse the ganglia and synapse in more peripheral ganglia. See Figure 2.3B and C.

The neurotransmitter at all preganglionic terminals is acetylcholine which is not paralyzed by atropine.[27] **This nicotinic effect produces peripheral neuromuscular junction effect of skeletal muscle and a central effect within ganglia of the CNS. This nicotinic effect results in sympathetic overactivity with neuromuscular dysfunction including tachycardia, hypertension, dilated pupils, muscle fasciculation, muscle weakness, vasoconstriction, and hypoxia.**[27] **The action of acetylcholine at the distal end of the cholinergic postganglionic fibers is paralyzed by atropine.**[27] **The muscarinic effect produces peripheral and CNS responses and results in parasympathetic overactivity with bradycardia, miosis, sweating and salivary secretions, blurred vision, lacrimation, gastric and pulmonary secretions.**[27] Noradrenalin is the postganglionic neurotransmitter of the sympathetic nerves.

Following an incitant entry a dynamic local homeostatic response develops (see Chapter 1). Normal local autonomic response occurs with the changes in the cellular and CT matrix (e.g., increases pH, monocyte release from tissues, magnesium decreases calcium increases, etc.). If these responses are insufficient to neutralize a noxious incitant entry, then further changes occur. Changes will occur at the microphage stage (see Chapter 1) alerting the regional and central homeostatic mechanism through the autonomic nerves and CT matrix but only minimally.

Once the macrophage phase is entered due to more noxious or more intense stimuli, a generalized alarm sounds in the central homeostatic mechanism (reticular activating system, hypothalamus, and/ or limbic system, pineal gland and area postrema of the fourth ventricle). Chemically sensitive and some chronic degenerative diseased individuals (periodic homeostatic dysfunction) first experience a local reaction in response to pollutants or other noxious incitant exposure. This entry may then yield a hyper- or hypo (inhibiting)- sensitive response. These patients appear to experience their first systemic response through the ANS once the dynamics of local homeostatic control is exceeded.

Nervous System

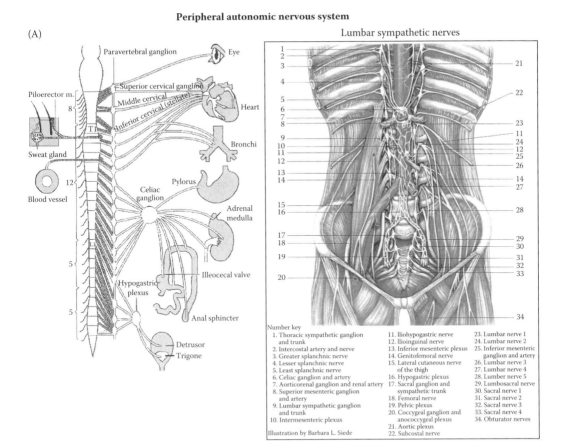

FIGURE 2.3A Peripheral autonomic nervous system. (Modified from Guyton, A. C., and Hall, J. E. *Medical Physiology,* 10th ed. Philadelphia: WB Saunders Co., 698, 2000; Siede, B.L., *Surgical Rounds Anatomical Chart,* 239, 2006.)

In the chemically sensitive and chronic degenerative diseased patient, signs and symptoms will manifest which will then cause awareness of the triggering of the ANS. Symptoms include adrenalin rushes of jitteryness, fatigue, insomnia, and food deregulation. **Frequently, symptoms of inhibition, stimulation, and/or dysfunction of the ANS are misinterpreted by the clinician as "functional" and nonobjective. In fact these responses can be understood as objective because these symptoms are the result of normal or exaggerated ANS triggered by environmental stimuli**. This understanding is not always easy because of the complexity of the ANS response due to the many circuits and receptors in the ANS, the variety and intensity of the noxious stimuli, and the state of the nutrient pool. All of these factors will modulate intensity and duration of response. The misinterpretation of the autonomic responses by the clinician is a gross mistake. For often, diagnosis of pollutant or other noxious incitant entry and injury resulting in dyshomeostasis at this initial stage is the only time that permanent nerve damage or other bodily injury can be prevented.

Triggering of the ANS results in an array of early warning signs and symptoms, which should not be ignored by the clinician or patient. This array of symptoms and signs includes cold hands and feet, livido reticularis, fatigue, anxiety, jitteriness, and insomnia. When patients' signs and symptoms are observed and studied under environmentally controlled conditions using objective measuring devices, like the pupillograph[28] or cardiac internal frequency analysis (heart rate variability),[16] for the first time, one can objectively analyze autonomic function and dysfunction before autonomic failure occurs. Although these machines do not measure all areas of the ANS,

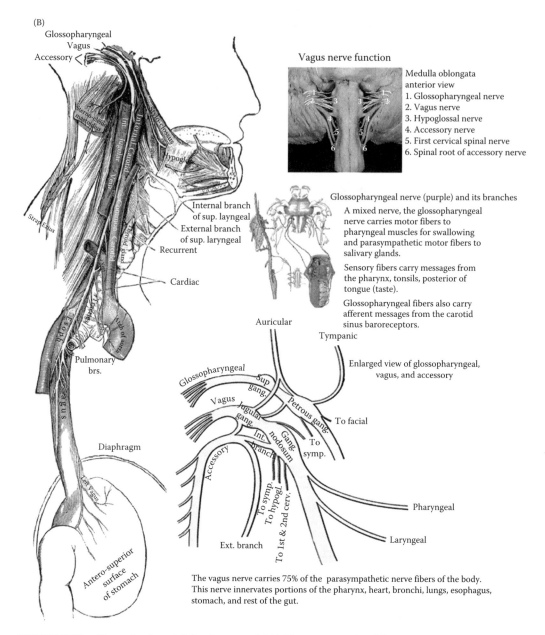

FIGURE 2.3B (See color insert following page 364.) Vagus nerve. (Modified from Gray, H., *Anatomy of the Human Body*, 1918; Medulla oblongato anterior view from *The Anatomy Project*, Pathenon Publishing Group, 1997.)

they go a long way in aiding the clinician in evaluating this system. We have performed these objective autonomic tests on over 3000 patients, demonstrating ANS involvement in most dyshomeostatic reactions, whether they are in chronic degenerative disease patients or patients with chemical sensitivity. See Tables 2.2 and 2.3.

Table 2.2 shows parasympathetic reactions occurred in 60% of the patients while 27% had sympathetic reactions. This is in contrast to the heart rate variability which records 82% sympathetic reactions. The head may have more parasympathetic innervations. However, once the clinician is familiar with the clinical responses of the ANS, patients' signs and symptoms, particularly when

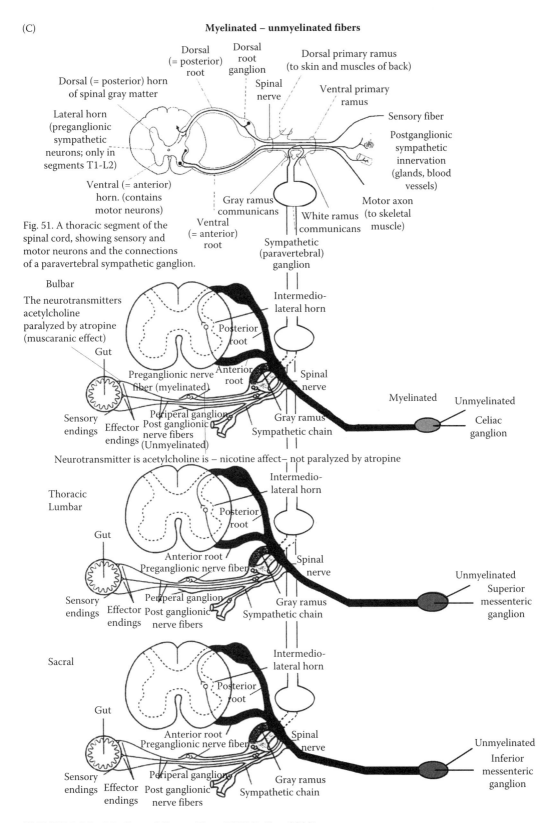

FIGURE 2.3C Myelinated fibers. (From EHC-Dallas, 2009.)

TABLE 2.2
Autonomic Nervous System Changes in 3000 Patients with Chemical Sensitivity and Chronic Degenerative Disease as Measured through the Eye by Pupillography (Iskikawa's Method)

Sympatholytic	24%	
Cholinergic + Sympatholytic	17%	60%
Cholinergic	19%	
Sympathomimetic + Cholinolytic	24%	
Cholinolytic	3%	27%
Nonspecific change	12%	12%
Normal	1%	

Source: Based on EHC-Dallas, 2002.

TABLE 2.3
Autonomic Nerve Measurement Using Heart Rate Variability by the Method of Riftine—81 Patients

	Pt	%
Sympathetic ↑ + Parasympathetic ↓	65	82
Sympathetic ↑	4	5
Sympathetic → Parasympathetic ↓	6	8
Sympathetic + Parasympathetic ↓	1	1
Parasympathetic ↑	1	1
Autonomic Balance	2	3
	81	**100**

Source: Based on EHC-Dallas, 2002.

observed under less-polluted environmentally controlled conditions, are often sufficient to identify objectively the pollutant-triggered homeostatic responses of the ANS. Whether these responses are normal, prolonged, or dysfunctional, as seen in the periodic homeostatic disturbances (chemically sensitive) or the aperiodic homeostatic disturbances (chronic degenerative diseased individual), they are very important in clinical evaluation and treatment of the patients. **Responses in total autonomic failure can be easily determined due to their gross adverse responses which are often irreversible and thus, untreatable.**

Understanding the anatomy and function of the ANS is necessary in understanding many of the symptoms and signs in chemically sensitive or chronic degenerative diseased patients, since many of their responses may be due to direct stimulation or inhibition of the ANS after local homeostatic control is lost. These responses will definitely affect the body's attempts at stabilizing the dynamics of homeostasis. The early symptoms and signs that are produced by the entry into the body of any noxious incitant, specifically pollutants, may be caused by either prolonged responses of the normal autonomic function, as seen in periodic homeostatic regulation, or they may be the result of pathological injury, which then, would depend upon the severity, chronicity, state of nutrition, and previous history of injury to that particular area of the body. **Prolonged normal homeostatic responses are more common and are, therefore, more easily reversed than pathological injury** (**i.e., surgical sympathectomy or medical sympathectomy like in sickle cell disease or phenol exposure**). In total autonomic failure, the condition may be end-organ disease with reversibility impossible to achieve. However, in a pathological state, if no significant scar tissue has formed or if the scar can be treated by direct injection therapy, it is possible to reverse some injury conditions. Once the pollutants and noxious stimuli have been removed in order to stop the trophic (neurogenic) changes or inflammation, normal tissue often reappears.[27] Often, these pollutants and stimuli can be removed from the tissue by avoidance programs, heat depuration, yoga, and exercise therapy, which allows for the return to normal or nearly normal compensated function.

Once local homeostatic control has been exceeded, noxious incitants (i.e., toxic chemicals, bacteria, virus, mold, and trauma) can stimulate or block large portions of the sympathetic or parasympathetic nervous system. **If the sympathetic system is predominantly involved in an acute pollutant-induced response, a mass discharge of the sympathetic nerves occurs resulting in a flight or fight mode (jittery anxiety attack) and, if chronic, followed by chronic fatigue.**[29] See Figure 2.4. **If the parasympathetic nerve (vagus) is stimulated and there is a lack of sympathetic stimulation, sudden drop attacks, bradycardia, gut hypermotility, gut acidity, and/or loss of erection may occur.**[30]

If long-term chronic noxious stimuli are present and constantly stimulating the sympathetic nervous system, one sees hypersensitivity, episodes of anxiety, chronic fatigue, often insomnia, and peripheral microcirculation vasoconstriction with regional hypoxia in chemically sensitive and/or chronic degenerative diseased patients.

In contrast to sympathetic stimulation and responses, the parasympathetic nervous system has a series of reflexes that are very specific and regional.[31] When this system is triggered, regional dysfunction usually follows. **The parasympathetic nervous system leaves the brainstem with the occulomotor nerve (III) leaving the upper medulla; trigeminal (V), the facial nerve (VII) leaving the pons, the glossopharyngeal (IX) and vagus nerves (X) leaving the lower medulla, and the pelvic nerve leaving the lumbar area occasionally L-2,3, but usually L4,5) and the sacral area (S-1,2,3,) of the spinal cord.**[32]

Approximately 75% of all parasympathetic fibers are in the vagus nerves passing to the entire thoracic (lung and bronchi constriction, heart bradycardia, etc.) and abdominal viscera (increased peristalsis). It has been recently discovered that some parasympathetic nerves innervate the blood vessels of the brain.[33] Parasympathetic nerves in cranial nerve III flow to the pupillary sphincters of the eye for constriction and ciliary muscles constriction.[34] Parasympathetic nerves in the trigeminal nerve (V) can be responsible for facial pain.[35] Parasympathetic fibers flow from cranial nerve VII to the lacrimal and nasal (causing excess secretion), and submandibular glands, and fibers from cranial nerve IX flow to the parotid gland causing excess salivation.[36] The sacral parasympathetic fibers congregate in the pelvic nerves, which leave the sacral plexus on each side of the spinal cord with L-4, 5 and S 1, 2, 3 levels and distribute peripheral nerves to the descending colon, rectum (increased tone, sudden urgency to defecate when sensitized), bladder (sudden urgency to urinate with increased tone when sensitized), and lower portion of the uterus and external genitalia (which causes erection).[37] The regionalization of the parasympathetic responses to noxious stimuli may explain why, at first exposure, many chemically sensitive and/or chronic degenerative diseased patients experience various and multiple responses, such as isolated cardiac, gastrointestinal, pupillary, and genitourinary dysfunctions. Studies at the EHC-Dallas show that 60% of the initial homeostatic disturbances are parasympathetic in nature or sympatholytic (24%) or a combination of both when studied by pupillography. The results are not the same when studied by heart rate variability, which demonstrates almost an 82% sympathetic response "heart rate variability". See Table 2.3.[16] This discrepancy may be explained not only by the anatomical differentiation in the regions measured but also by the sensitivity of the instruments used. The heart rate variability measures the vagus and splanchnic nerves predominately, whereas the pupillography measures many more parasympathetic responses especially of the head and neck as well as the sympathetic responses. These facts emphasize the complexity of the analysis of the ANS responses and the complexity of the ANS.

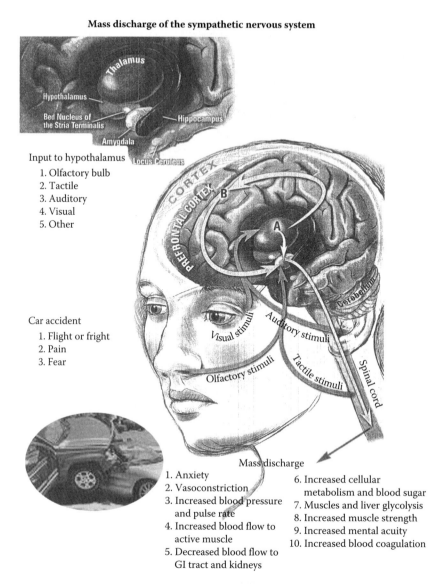

FIGURE 2.4 (Modified from Vieweg, W. V., Julius, D. A., Fernandez, A., et al., *AJM*, 119, Fig. 1. 387, 2006.)

In some instances, localized sympathetic responses do occur because of localized pollutant exposure. The responses may be a little more complex since the regionalization of the sympathetic system is not distributed to the same part of the body, as are the somatic spinal nerve fibers from the same segments. Instead, the sympathetic fibers from the cord segment T-1 generally pass up the sympathetic chain to the head, from T-2 into the neck, from T-3, T-4, T-5, and T-6 into the thorax, from T-7, T-8, T-9, T-10, and T-11 into the abdomen, T-12, L-1, and L-2 into the legs.[37] The distribution of sympathetic nerves is determined partly by the position in the embryo where the organ originates, thus making for a more complex anatomy. Most systemic blood vessels, especially those of the abdominal viscera and the skin of the limbs, are constricted by sympathetic stimulation due to the fact that there is no parasympathetic innervation on the blood vessels except for some brain vessels. Parasympathetic stimulation has almost no effect on the blood vessel, except possibly to dilate certain restricted areas such as the blocked area of the face and brain.

When the local homeostatic control is exceeded, **these sympathetic responses are** triggered, **resulting in vasoconstriction of skin blood vessels, which produces cold hands and feet, as seen in most chemically sensitive patients. However, one may see flushing in the face as a result of vasodilation, which probably results from cholinergic release from sympathetic nerves but perhaps could be triggered by the parasympathetic innervation of the blocked area of the face since some brain blood vessels are innervated by the parasympathetics and some are not**.

However, recent studies have shown that there is parasympathetic innervation for the carotid and cerebral blood vessels.[18,19,22] These studies have shown that noxious stimuli will change and especially dilate the cerebral microvascular bed.

Normal sympathetic tone occurs in the blood vessels, keeping them at about one-half of their potential maximum diameter. If it were not for the continued background sympathetic tone, the sympathetic system could cause only vasoconstriction, never vasodilatation.

As previously mentioned, in many instances, the sympathetic nervous system responds by mass discharge. This frequently occurs when the hypothalamus is activated by flight or fright, fear, or pain. The stress or alarm response occurs throughout the body, resulting in increased arterial pressure and increased blood flow to active muscles, with concurrent decreased blood flow to the GI tract and the kidneys. Increased cellular metabolism occurs throughout the body with increased blood glucose concentration, glycolysis in the liver and muscles, and increased muscle strength, mental activity, and blood coagulation. **This discharge, though usually finite, within a few minutes can occur in a wave after wave manner in some chemically sensitive and chronic degenerative diseased patients, thus developing anxiety followed by a fatigue phenomenon**. This recurrent state has been measured in numerous patients by the heart rate variability analysis (see Table 2.3), which shows a superadrenalin effect with the parasympathetic being decreased and the sympathetic being increased.

A mass discharge of sympathetic nerves is frequently seen in some chemically sensitive and/or chronic degenerative diseased patients following chronic noxious stimuli exposure and after local homeostatic control is exceeded. In fact, a large number of the chemically sensitive have a chronic superadrenalin effect when measured by heart rate variability. This supersensitivity may be due to partial nerve damage by pollutant injury. The sympathetic nervous system is increased and the parasympathetic system is decreased. Initially, this response will result in agitation, nervousness, anxiety, and excitability, as well as the aforementioned responses discussed in the previous paragraph. However, with chronic noxious stimuli exposure, the tissue fatigues severely and the patient develops generalized weakness and fatigue. This response appears to be a fatigue in the adrenalin effect, which initially is stimulating, but with excess stimulation, results in receptor and tissue response fatigue. The fatigue effect may also be due to faulty neuro feedback from the mitochondria of the brain and muscles. Regardless of the mechanism involved the individuals with chemical sensitivity and/or chronic degenerative disease usually have chronic fatigue.

A sympatholytic response (24%), which is similar to a cholinergic response as measured at EHC-Dallas by the iris corder (pupillography), and (heart rate variability), also often occurs. **These sympatholytic effects are often combined with cholinergic effects (17%), or isolated cholinergic (19%) effects occur. These different types of responses, which are often the result of pesticide or solvent exposure, generally are the opposite of a sympathetic response. These cholinergic responses cause decreased arterial pressure and decreased blood flow to active muscles. This decreased blood flow, in turn, inhibits large muscle exercise, as is often seen in the chemically sensitive and/or chronic degenerative diseased patient, especially those who experience chronic fatigue**. Concurrent with the decrease in arterial pressure and reduced blood flow to active muscles is an increase in blood flow to other organs. In the muscle, decreased rates of cellular metabolism, decreased glucose concentration, decreased muscle glycolysis, decreased muscle strength, decreased muscle activity, and decreased blood coagulation occur. **Basically, the**

sum of these effects is inactivation of the body (with the exception of the gut), reducing the ability of chemically sensitive and/or chronic degenerative diseased individuals to respond to both environmental and other stressors. An example of this type of sympatholytic response is a 55-year-old artist who developed both peripheral voluntary and autonomic nervous system dysfunction and failure after his local and central homeostatic mechanisms had been exceeded, which made walking difficult for him.

Case study. This patient's abnormality was accompanied by difficulty in breathing and talking, due to vocal cord nervous dysfunction. His arterial blood pressure would fluctuate between 140/80 and 50/20 mmHg, depending upon pollutant exposure. At times, he would pass out when his blood pressure was low and would finally recover with administration of a liter of intravenous saline. This patient was a Shop and Art teacher. On the job, he had been exposed to various types of solvents, paints, and lacquers, which altered his homeostasis and other metabolic functions, subsequently changed his tissues, and eventually fixed his disease to become autonomous. He gradually deteriorated until he became incapacitated.

Challenge testing at the EHC-Dallas revealed positive reactions to molds, pine terpene, tree terpene, grain smut, grass smut, *Candida albicans*, cigarette smoke, chlorine, ethanol, formaldehyde, women's cologne, men's cologne, orris root, news material, and phenol. Two foods, oatmeal and beef, caused hypotension. He had high magnesium in his blood, low cholesterol, and low creatinine phosphokinase. His triple-headed camera single photon emission computerized tomography (SPECT) brain scan showed locally reduced blood flow and brain function in a fairly large area involving the parietal lobe, temporal lobe asymmetry, and a pattern consistent with neurotoxicity, which revealed hot and cold areas in the cerebral hemispheres. Iris corder (pupillography) examination showed cholinergic and sympatholytic reaction indicating a predominance of parasympathetic stimulation. Heavy metals in his blood were below reference range. Toxic organic chemicals in this patient's blood showed high benzene, 2.2 ppb (C < 1.0 ppb); 2-methylpentane, 13.8 ppb (C < 1.0 ppb); 3-methylpentane, 32.4 ppb (c < 1.0 ppb); and n-hexane, 10.7 ppb (C < 1.0 ppb). Blood gases showed his PaO_2 was low, at 74 mmHg (C = 75 to 100 mmHg). Base excess was high at 2.8 mmol/L (C = −2.0 to 2.0 mmol/L). Plasma cortisol, norepinephrine, serotonin, and mineral corticoids were all within normal range. CAT and MRI scans of chest, abdomen, and head for tumors were negative.

After exposure, this patient would experience falling blood pressure, which would start slowly at around 130 to 120/80 and decrease to 100/60 to 80/50 to 60 to 40/0 mmHg. At these times, this patient would develop fuzzy thinking and then pass out. This falling blood pressure would become life-threatening in that several times he was found comatose without blood pressure. He was treated with flurocortisone, extra salt, constant excess saline (oral and intravenous), leg-wrapping with a gravity-suit, a rotary diet, and a less-polluted environment. This patient presented a classic case of a chemically induced dyshomeostatic ANS failure.

This patient had a case of neurally mediated hypotension (NMH) that was a result of his autonomic failure.

CENTRAL CONNECTION OF THE AUTONOMIC NERVOUS SYSTEM WITH THE HYPOTHALAMUS

Our studies have shown that the ANS response is secondary to earlier triggers of the homeostatic mechanism when it is overloaded by noxious stimuli. Whether the autonomic dysfunction manifests as a sympathetic or parasympathetic response, it is clear that not all of the dysfunction originates in the peripheral nerves although all peripheral information if strong enough is fed to the brain or spinal cord. **Many of the dysfunction actions may originate centrally in the damaged hypothalamus, but they can also occur in the reticular formation and limbic systems and also in the pineal gand and area postrema of the fourth ventricle**. As shown in Chapter 1, the hypothalamus is the highest autonomic center in the brain derived from the olfactory bulb and is the focal point in triggering the central incitant-driven homeostatic responses seen in the periodic homeostatic disturbances (chemically sensitive) and aperiodic homeostatic disturbances (chronic degenerative diseased patient), after local and regional homeostatic control is exceeded.

Nervous System

The hypothalamus is thought to be the seat for orchestrating the dynamics of homeostasis for the body being the central environmental monitoring and autonomic response system. It appears however, that **its function is the coordination center for many of the dynamic homeostatic subsystem responses triggered by not only odorous environmental pollutants received through the olfactory nerve and noxious incitant impulses received by other peripheral afferent autonomic nerves but also by other pollutants received through the blood from the rest of the body**. Chapter 5 of this book will show that the reticular formation and the limbic system are also involved. The hypothalamus also has functional connections with the forebrain and brainstem as well as the pituitary and pineal glands in addition to the olfactory and peripheral afferent nerves (Figure 2.5).[23]

When the hypothalamus is triggered by noxious incitants, it may activate any or all of the organs or other areas in the body, resulting in varied symptomatology. When all of these organs or areas are triggered, it is more difficult for the dynamic function of homeostasis to occur, but in the early periodic homeostatic disturbances (chemically sensitive) or aperiodic homeostatic disturbances (chronic degenerative diseased patient), it does happen. The hypothalamus is close to, or contains, the supraoptic nucleus, the ventral medial nucleus, the preoptic nucleus, the paraventricular, posterior, dorsal medial ventromedial nuclei, the mammillary body, and the interpendular nucleus. Pollutants can enter the olfactory nerve, or the nerve can be stimulated by impulses going to the superoptic nucleus and then stimulating the hypothalamus. **Anterior stimulation of these nuclei produces responses that correspond roughly to parasympathetic responses, while posterior stimulation causes sympathetic responses**.[39] As shown previously, stimulation that produces sympathetic responses results in increased arterial pressure, blood glucose concentration, glycolysis, muscle strength, mental activity, rate of coagulation, and blood flow to active muscles with decreased blood flow to organs that are not needed for rapid activity, i.e., the gut, kidneys, etc.

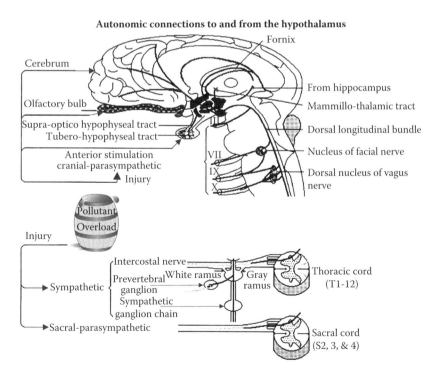

FIGURE 2.5 The frontal lobe and cornu ammonis send messages to one hypothalamic nucleus. Afferents from the olfactory bulb go to the supraoptic nucleus and middle fore brain bundle. Exits are from mammillary body and also down spinal nerves. (From EHC-Dallas, 1996.)

Damage to the olfactory nerve is common in the patient with chemical sensitivity rendering the hypothalamus sensitive to impulses coming from the nose and thus easier triggering may occur. This same sensitivity may occur from the afferent autonomic nerves of the gut. In addition, direct damage to the hypothalamus may occur from toxic exposure via the blood stream, again resulting in a hypersensitive response from all the organs connected with the hypothalamus either centrally or peripherally.

Both sympathetic and parasympathetic responses are seen in almost all pollutant-overloaded patients with chemical sensitivity and/or chronic degenerative disease. Frequently, a moderately severe chemically sensitive individual will experience bursts of energy characterized as a sympathetic response that may last for a few minutes to hours and then completely give out, resulting in exhaustion that may last for a day or two. This cycle is often repeated until the patient is incapacitated if the noxious incitants, especially pollutants, are not removed. **This overload of noxious stimuli can result in permanent lesions in the hypothalamus leaving the body more vulnerable to lower doses of environmental stimuli or nontoxic stimuli like food. Alternately, as our studies have shown, the majority of early cases of autonomic dysfunction is reversible given the proper nutritional replacement and decrease in total body pollutant overload. Therefore, only a temporary metabolic block has occurred**. Certainly, once the hypothalamus becomes sensitized by nerve injury, rapid firing with lower threshold triggering can occur with secreted intensity of stimuli creating a vicious cycle of triggering from previously nonnoxious stimuli (foods, etc.) which keeps the ANS revved up.

In some cases, messages from the frontal lobe and cornu ammonis go to one hypothalamic nucleus, within the regional feedback, thus completing the dynamics of a homeostatic feedback loop. Also, afferents from the olfactory bulb go to the supraoptic nuclei and medial forebrain bundle. Stimulation of these anatomical pathways explains why the pollutant odor-sensitive chemically sensitive and/or chronic-disease patients have multiple ANS complaints with ultimate triggering of vascular as well as other smooth muscle systems. The ANS triggering produces a variety of responses, depending on the nature of the imbalance caused (Table 2.4).[40]

For example, generally, if the sympathetic nervous system is triggered, heart rate, blood pressure, sweat (due to the cholinergic stimulus of the sympathetic nerve), and pupil size all increase; with bronchial contractility, sputum, saliva, intestinal mobility, urine output, and uterine contractions all decreasing. The opposite is generally true when the parasympathetic nervous system is triggered. **The moderate to severely environmentally injured chemically sensitive and/or chronically diseased individual with severe dyshomeostasis, usually cannot sweat, and/or has difficulty in sweating, which apparently augments retention of pollutants and thus, increases chemical sensitivity**. This type of response suggests sympathetic cholinergic exhaustion but, clinically, represents severe autonomic deregulation. The responses become more complicated when either wing of the ANS is inhibited. The result is severe alteration in the dynamics of homeostasis.

Characteristically, individuals with chemical sensitivity and chronic degenerative disease have a more labile ANS apparently involving ANS injury which leads easily to increased dyshomeostasis. The ANS of these individuals is more easily triggered (except for sweating), physiologically disrupted, and difficult to maintain balance in the dynamics of homeostasis than the ANS of normal individuals. The ANS responses are more prolonged before recovery keeping the patient in a state of extended dyshomeostasis. Therefore, throughout their lives, these patients frequently experience symptoms of autonomic triggering.

The posterior lobe of the pituitary gland receives direct tracts from the supraoptic nucleus (sympathetic) and the anterior lobe nucleus tuberis (parasympathetic).[41] **Therefore, triggering or inhibition of the dynamics of either system results in neuroendocrine dysfunction in most parts of the body with local, regional, or central dyshomeostasis occurring**. Once an area is damaged, homeostatic feedback loops can also keep a cycle of endocrine dysfunction flowing, as is seen in many patients with late phase chemical sensitivity and chronic degenerative disease.

TABLE 2.4
Autonomic Effects on Various Organs of the Body

Organ		Effect of Sympathetic Stimulation	Effect of Parasympathetic Stimulation
Eye	Pupil	Dilated	Constricted
	Ciliary muscle	Slight relaxation	Constricted
Glands	Nasal	Vasoconstriction and slight secretion	Stimulation of copious (except pancreas) secretion (containing many enzyme-secreting glands)
	Lacrimal		
	Parotid	Lack of secretion	
	Submandibular	Lack of secretion	↑ in secretion
	Gastric	Lack of secretion	↑ in acid
	Pancreatic	Increase of secretion Stimulates enzyme	↓ enzyme
	Axillary	None	Stimulation
Sweat glands		Copious sweating (cholinergic)	None
Apocrine glands		Thick, odoriferous secretion	None
Heart	Muscle	Increased rate	Slowed rate
		Increased force of contraction	Decreased force of contraction (especially of atrium)
	Coronaries	Dilated (β_2); constricted (α)	Dilated
Lungs	Bronchi	Dilated	Constricted
	Blood vessels	Mildly constricted	Dilated—sympathetic/cholinergic response
Gut	Lumen	Decreased peristalsis and tone	Increased peristalsis and tone
	Sphincter	Increased tone (most times)	Relaxed (most times)
Liver		Glucose released	Slight glycogen synthesis
Gallbladder/bile ducts		Relaxed	Contracted
Kidney		Decreased output and renin secretion	None
Bladder	Detrusor	Relaxed (slight)	Excited
	Trigone	Excited	Relaxed
Penis		Ejaculation	Erection
Systemic arterioles	Abdominal	Constricted	None
	Muscle	Constricted (adrenergic α) Dilated (adrenergic β_2) Dilated (cholinergic)	
Skin	Mucosa	Constricted—complex	None—complex
Blood	Coagulation	Increased	None
	Glucose	Increased	None
Basal metabolism		Increased up to 100%	None
Adrenal meduallary secretion		Increased	None
Mental activity		Increased	None
Piloerector muscles		Excited	None
Skeletal muscle		Increased glycogenolysis	None
		Increased strength	None

Source: Modified from Guyton, A.C., *Human Physiology and Mechanisms of Disease,* Philadelphia: WB Saunders Co., 443, 1987.

Certainly, secondary foci such as deinervation myalgic hyperalgesia or peripheral neuropathy occur more frequently from firing more often on an already overloaded system.

When the pineal gland is constantly triggered insomnia occurs and is discussed in the endocrine chapter, Chapter 4. Also, the area postrema of the fourth ventricle, which relates to food regulation, is deregulated and will be discussed in Chapter 7, Gastrointestinal.

The autonomic outflow from the hypothalamus via the mammillary body goes back toward the cerebral cortex, to the lower vegetative centers, and the hypophysis cerebri.[41] This autonomic outflow, emanating from the hypothalamus and coming from the mammillary body, influences the parasympathetic centers, including cranial nerve VII to the face and cerebral vessels, cranial nerve IX to the pharynx and the region of the carotid sinus, and cranial nerve X,[41,42] which is in the medullary cardiovascular center triggering their homeostatic mechanisms. Over 80% of the parasympathetic discharge occurs through these nerves VII, IX, and X. Also, pollutant or other noxious stimuli exposure may influence the sacral center.

Often, chemically sensitive or chronically diseased patients are seen who have apparent triggering and homeostatic deregulation of the respiratory (central dyspnea), vascular (constant vasospasm), and temperature (hypothermia) control centers. Here, the basic dynamics of the respiratory and vascular homeostasis are not functioning properly usually because of sensitization from previous injury. These patients will present with breathlessness having normal pulmonary function and severe peripheral vascular deregulation with vasospasm (as measured by the heart rate variability machine), or they will present with high temperatures, including an inability to tolerate heat, or more commonly, they will present with low temperatures with inability to tolerate cold (as measured by specific dermal homeostatic regulation thermography).

The sympathetic trunks extend from the base of the cranium to the coccyx. The parasympathetic nerves go down the cord in the gray matter, through the white ramus, and extend either out the peripheral nerve or into the sympathetic ganglion to the viscera. The sympathetic trunk goes from the brain through the ganglion to the sacrum (Figure 2.6).

Some autonomically innervated tissues are supplied with the parasympathetic and sympathetic nerves, while others are supplied by sympathetic nerves alone, e.g., the adrenal and peripheral blood vessels, hair follicles, and sweat glands.[42] This complex distribution of autonomic fibers allows the extreme varied responses seen in a group of pollutant or other noxious stimuli triggered, chemically sensitive or chronic degenerative diseased individuals. It should be emphasized that the chemically sensitive patient with neurotoxic patterns on triple headed SPECT brain scans and with an inability to sweat may well have a pollutant-driven sympathetic response, which eventually will result in dyshomeostasis. In addition, this patient could possibly have a cholinolytic pattern triggered extracranially, since the cerebral vessels show a decrease in flow (which is reversible) in certain areas of the brain. **Vascular sympathetic nerves have both sympathetic and cholinergic fibers that also account for these patterns.** While this anatomic makeup allows for harmonious homeostasis, it can certainly complicate the dynamics of homeostatic responses if the stimuli are consistently strong or chronic and intense.

Reflex Pathways

There are several ANS reflex pathways. These include the cardiovascular, the gastrointestinal, urinary bladder, sexual, sweating, blood glucose concentration, gallbladder, and all renal reflexes.[42] As shown in Chapter 1 and 5, the reflex mechanisms with their central connections in the spinal cord and brainstem are responsible for the major portion of the automatic neural homeostatic regulation of visceral functions (Figure 2.7).[43]

Therefore, the rapid changes that are often seen in the chemically sensitive and/or chronic degenerative diseased dyshomeostatic patient may result from exposure to noxious stimuli (bacteria, virus, protozoa, toxic chemicals), which are known to produce a sudden onset of signs and symptoms

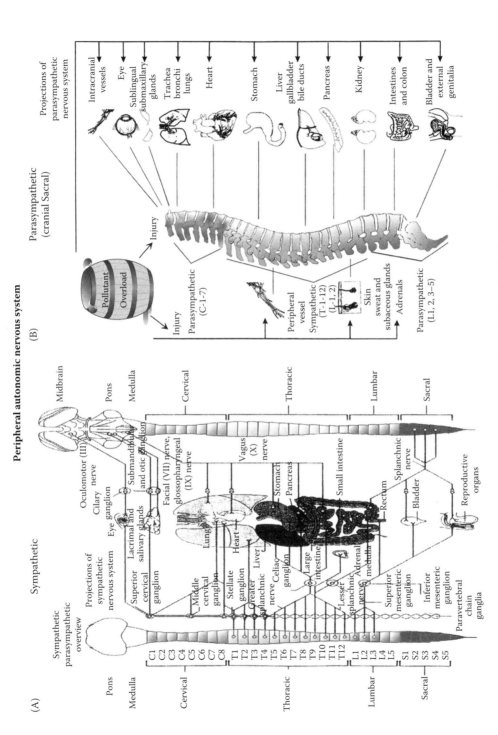

FIGURE 2.6 (See color insert following page 364.) (A) Sympathetic innervation of the spinal cord. (B) Cranial-sacral innervation of the parasympathetic nerves. (EHC-Dallas, 2002.)

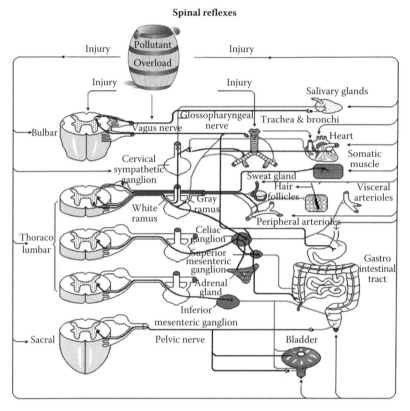

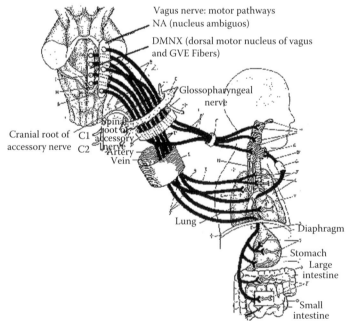

FIGURE 2.7 **(See color insert following page 364.)** Routes of reflex mechanisms that may be triggered after pollutant injury in the chemically sensitive and chronic degenerative diseased individual. Responses may be isolated to each single organ, i.e., gallbladder or appendix, or they may be generalized. (EHC-Dallas; Buffalo, 2002; Rea, W.J., *Chemical Sensitivity, Vol. III, Clinical Manifestatons of Pollutant Overload*. Boca Raton, FL: CRC Lewis Publishers. 1735, 1996. With permission.)

through simple viscero-cutaneous reflexes. **Regionalization is usually a function of the parasympathetic nervous system, while generalization is usually a sympathetic response. However, regionalization responses can occur through the spinal cord reflexes and, thus, through the sympathetic nerves if the noxious stimuli are weak**. In addition, because of this normal function due to the complexity of the ANS, exact definition of symptoms related to each limb of the ANS will be difficult to define at times. The dynamics of homeostatic function will work well within the normal local and spinal reflex mechanism if it is not overstressed by chronic entry of noxious stimuli.

In contrast to the local and regional reflexes, **certain higher order reflex correlation and feedback mechanisms are located in the brainstem in the reticular formation**.[44] These higher order reflexes round out and partially integrate the GRS response from the CT matrix for regulating the dynamics of homeostasis. This integration is further accomplished in the hypothalamus and the limbic system. These areas appear to be where the dynamics of the homeostatic subsystems are correlated with the integrated central homeostatic mechanism. **These areas also include neuron aggregates in the medulla oblongata, such as the respiratory and vasomotor centers, the center for carbohydrate metabolism, and the major portion of the hypothalamus**. All of these areas and their correlated mechanisms may be triggered or inhibited by chronic noxious incitant stimuli. These stimuli go through the GRS, causing functional homeostatic regulation to change to periodic and aperiodic homeostatic disturbances, which result in the dyshomeostatic response in the chemically sensitive or chronic degenerative diseased individual.

Impulses that arise in visceral organs are often conducted to these higher autonomic centers, mainly through pathways in the spinal cord and brainstem that are made up of short neurons and frequent sympathetic relays. Alternatively, they may come up the CT matrix, which is also a part of the GRS.

Dysfunction of the Autonomic Nervous System

All of these facts mentioned may partially explain the occurrence of symptoms and signs in some chemically sensitive and chronic degenerative individuals with early food and chemical triggering of the ANS via the gastrointestinal tract, genitourinary tract, nose, and lung. The less intense impulses may go up the extracellular matrix in the loose areolar tissue, or the more intense ones go through the sympathetic innervation of the blood vessels. Studies on a galvanometer show evidence for the impulses to travel up the extracellular matrix.[45] Perhaps more important is what happens when the CT matrix (i.e., areas around the blood vessels) is directly stimulated. Then patients have shown both less and/or more intense sensations which may result in mild or severe responses. Once triggering occurs in the blood vessel, the result is usually vasospasm, although triggering may also result in vasodilatation through the cholinergic pathway of the sympathetics. These responses can be the result of retrograde impulses from the extracellular matrix, parasympathetic or sympathetic impulses. Breathlessness, weakness, shakiness, and a ravenous desire to eat or drink may occur, **and, usually, these symptoms of weakness and shakiness are often clinically misinterpreted as hypoglycemia**. However, blood sugar measurements in hundreds of patients under environmentally controlled conditions at the EHC-Dallas have revealed normal sugars in all but two patients, suggesting neuroendocrine triggering. See Table 2.5.

Either a portion of, or the whole, neuroendocrine system may also be activated by pollutant or other noxious stimuli exposure, or, conversely, neuropeptides released from the end-organ neuroendocrine cells in response to pollutant exposure can activate the rest of the system. The following case report represents a classical ANS response to pollutant exposure resulting in dyshomeostasis.

> **Case study**. A 26-year-old white female entered the ECU with the chief complaints of shortness of breath, cold and blue hands and feet, episodes of bloating, an inability to sweat, migraine headaches, and spells of shaking. All of her symptoms were cleared after five days of fasting under environmentally controlled conditions. This, in itself, negated hypoglycemia since her blood sugar was 70 mg/

TABLE 2.5
Results of Neurotransmitters Hypersensitivity by Intradermal Skin Testing of 50 Chemically Sensitive Patients

		Dopamine	Epinephrine	Methacholine	Norepinephrine	Acetylcholine
No. (Patient)		44	23	41	42	17
Sex	F	32	18	30	30	13
	M	12	5	11	12	4
Age	Range	4–77	25–77	4–63	4–77	4–57
	Mean	4.20	37.9	39.7	40.6	42.3
Positive	No.	36	22	41	42	17
	%	82	46	100	100	100
Neutralizing dose-dilution (cc/#)	0.05/1	1 (3)				
	0.05/2	6 (17)			1 (2)	
	0.05/3	13 (36)	8 (36)	2 (5)	4 (10)	1 (6)
	010/3	2 (6)	1 (15)			
	0.15/3	2 (6)				
	0.05/4	5 (14)	10 (45)	14 (36)	19 (45)	5 (29)
	0.10/4	3 (8)	1 (5)	1 (2)	3 (7)	
	0.05/5	3 (8)	1 (5)	8 (20)	8 (19)	2 (12)
	0.10/5	1 (3)			2 (15)	1 (6)
	0.15/5		1 (5)	1 (2)	1 (2)	
	0.20/5					1 (6)
	0.05/6			7 (17)	2 (5)	4 (24)
	0.10/6			1 (2)	1 (2)	1 (6)
	0.15/6			2 (5)		
	0.05/7			2 (5)		
	0.10/7			2 (5)	1 (2)	
	0.20/7			1 (2)		
Symptoms triggered	NS	6 (17)	3 (14)	3 (17)	6 (14)	1 (6)
	MS	3 (18)	1 (5)	2 (5)	1 (2)	1 (6)
	GI	1 (3)	2 (9)	5 (12)	1 (2)	
	Skin	2 (6)				
	RS			2 (5)	2 (5)	

Source: Based on EHC-Dallas, 2001.
Notes: Percentage in parentheses
Abbrevations: NS = nervous system, MS = musculoskeletal system, RS = respiratory system, GI = gastrointestinal system
C. – 1/5 dilution

dL at the end of the fast. Subsequent testing revealed she was sensitive to six molds, 10 foods, and six chemicals. Oral challenge with individual foods revealed that cane sugar reproduced her bloating and shaking, while beets reproduced her shortness of breath. Inhaled, double-blind challenge of chlorine, < 0.33 ppm triggered discoloration and cooling of her hands and feet, reproduced her headache, made her unable to sweat, and caused her to bloat. Saline placebos elicited no response. Pupillography measurements of the ANS through the eyes revealed a sympathetic response. Intradermal tests for secretin and vasoactive intestinal peptide (VIP) also reproduced the aforementioned symptoms confirming neuroendocrine triggering.

Stimulation of the ANS results in release of neurotransmitters, adrenalin (epinepherine and norepinepherine) from the sympathetic system and acetylcholine from the parasympathetic system, which may amplify the local homeostatic response. Some sympathetic ganglia may liberate acetylcholine and some adrenalin, depending upon the intensity of the stimulus.

In general, stimulation of the sympathetic nerves results in excitation, while stimulation of the parasympathetic nerves results in inhibition of nerve impulses. **However, in the gastrointestinal tract, stimulation of the sympathetic nerves, which produces an amplification response, results in decreased gut motility, while stimulation of parasympathetic nerves increases gut motility and gut muscle tone**. For example, a chemically sensitive patient is exposed to a pollutant in his or her drinking water. **The local mucosal stimulus entry gives a mild response, but the patient immediately develops a sympathetic amplification response resulting in decreased motility of the gastrointestinal tract with severe bloating and distention, which may last for a few hours or days. See** Figure 2.8.

Peripherally he or she develops increased heart rate and blood pressure with severe vasospasm of the hands and feet. This type of homeostatic amplification response is frequently seen in a large portion of chemically sensitive or chronic degenerative diseased individuals after a pollutant challenge to the gut. Hours to days later, the patient will finally relieve the stimulated distention by passing and absorbing the gas with relief of the bloating and distention and return to the baseline dynamics of homeostatic function. These episodes can be incapacitating in the moderately severe chemically sensitive or chronic degenerative diseased patient because recovery time is so slow. The moderately severe patient with parasympathetic dysfunction after ingesting the offending food usually has diarrhea immediately, which clears the symptoms rapidly.

Although skeletal muscle is activated exclusively by the voluntary nervous system, smooth muscle can be stimulated to contract by multiple types of signals, including nervous signals, hormonal stimulation, muscle stretch, etc. The primary reason for the disturbance is that smooth muscle membrane contains many types of receptor proteins that can initiate the contraction process. The same process exists with toxic pollutants, foods, chemicals, etc. that could also trigger the contraction, while other receptor proteins, could inhibit smooth muscle contractions. Highly structured neuromuscular junctions are found in skeletal muscle but not in smooth muscle. Instead, the autonomic fibers that innervate smooth muscle branch diffusely on top of the CT matrix, which lies on top of a sheet of smooth muscle (Figure 2.9).

These fibers form diffuse junctions that secrete neurotransmitters, which are a nano-micrometer away from the muscle cells, which then diffuses to the cells. Even though there are many layers of muscle cells, the nerve fibers often innervate only the outer layers, thus the impulse travels from outer layer to inner layer by action potential conduction in the muscle mass or by subsequent

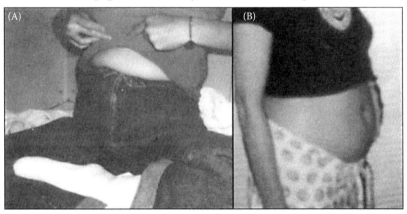

Sympathetic nervous system stimulation of the gut

FIGURE 2.8 (A) Inhaled challenge < 0.0034 ppm pesticide (2,4-DNP) after seven days in the deadapted state in the ECU with total load reduction. Bloating. (B) Increase of 20% in abdominal measurement from an intradermal injection of formaldehyde of 1/625 dilution. (EHC-Dallas, A-1980 and B-2002.)

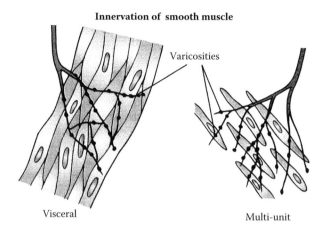

FIGURE 2.9 Note varicosities of the nerves, which secrete acetylcholine and epinephrine. No motor endplates occur in smooth muscle like they do in skeletal muscle. (Guyton, A.C. and Hall, J.E., *Medical Physiology*, 10th ed. Philadelphia: W.B Saunders Co., 90, Fig. 8–3, 2000. With permission.)

diffusion of the transmitter substance. **The axons that innervate smooth muscle fiber do not have motor endplates like in the skeletal muscle but have multiple varicosities, vacuoles, or granules that secrete neurotransmitters.**[46] In contrast to the vesicles of the skeletal muscles, the varicosities secrete both acetylcholine and epinephrine.[47] In a few instances, especially in the multi-unit type of smooth muscle, there are contact junctions which are only 20–30 nanometers from the muscle and function in the same way as skeletal muscle neuromuscular junctures. The latent period of conduction of the smooth muscle fibers is considerably shorter than the fibers stimulated by the diffuse junctions.[48]

Neurotransmitters, like acetylcholine and norepinephrine, which are secreted by the varicosities in these areas of smooth muscle can either stimulate or inhibit, depending upon the level of intensity and whether (acetylcholine, norepinephrine,) connects to excitatory or inhibitory protein receptors. If nerve damage (i.e., infection, solvents, pesticide, trauma, etc.) occurs then the law of denervation[49] applies, resulting in supersensitivity of the whole muscle that is innervated by that particular damaged nerve. Once the hypersensitivity to norepinephrine and acetylcholine occurs, exaggerated responses to the stimuli result. Generally, in the smooth muscle of the gut, acetylcholine excites nerves, while norepinephrine inhibits them. **The smooth muscle has far more voltage-gated calcium channels than does skeletal muscle, but fewer voltage-gated sodium channels. The calcium channels are responsible for the action potential, and, thus, contraction of the muscle.**

The preganglionic sympathetic nerve fibers contain acetylcholine, and, thus, noxious stimuli in this area results in an amplification of cholinergic stimulation, which contrasts to the postganglionic response to pollutants that is usually sympathetic and involves norepinephrine release. The postganglionic response does not occur in sweat glands, hair, or, in some instances, blood vessels, all of which may have a cholinergic response to pollutant stimulation through their sympathetic innervation. Therefore, if a chemically sensitive individual were exposed to an organophosphate pesticide and the dynamics of the local homeostatic response were overwhelmed, the result might be a confusing set of dyshomeostatic symptoms, as both the sympathetic and parasympathetic nerve impulses could be simultaneously triggered through pre- (parasympathetic) and post- (sympathetic) ganglionic fibers. In addition, sympathetic stimulation might inhibit the symptom-driven cholinergic responses of sweating, but cause vascular spasm with symptoms of cold hands and feet, which is seen in a large portion of our chemically sensitive patients. In addition, the varied responses seen in the pesticide-exposed chemically sensitive individual can be attributed to the triggering of these dual impulses. In addition to the release of acetylcholine or norepinephrine in response to pollutant or other noxious incitant exposure, ATP

Nervous System

and proper calcium channel and sodium channel function may be disturbed. Edema may also result with subsequent tissue hypoxia. Further, the mitochondria and ion flow of the ANS may be injured, as has been seen in some chemically sensitive individuals, which causes inappropriate release of these neurotransmitters and subsequent inappropriate amplified ANS response.

In general, most of the ectodermal structures such as the ducts of the liver, gallbladder, ureter, bladder, and bronchi are dilated due to sympathetic stimulation and/or are excited by parasympathetic stimulation, and become constricted. Sympathetic stimulation also causes the release of glucose from the liver, causing an increase in blood sugar, increase in glycogen from the liver and muscle, increase in muscle strength, increase in basal metabolic rate, and an increase in mental activity.

There used to be an anatomical dispute that the sympathetic fibers usually ended in the connective tissue in contrast to the parasympathetic fibers, which end in muscle fibers. However, it now appears that the fascia and loose areolar tissue are thin layers over the muscular insertions and the peripheral systemic nerves, and they enter the muscles through connective tissue (Figure 2.10).

Because of this anatomical fact, both fascial tightness and fascial spasms, which frequently result from pollutant exposure, are seen in most chemically sensitive and many chronic degenerative diseased individuals. These sequestered pollutant areas create primary and then secondary homeostatic focal injury (interference fields) that act as deposit areas for additional noxious and other nonnoxious stimuli. Eventually, this process triggers dyshomeostasis until finally, when the foci (interference fields) become supersensitive, spontaneous triggering occurs from simple muscle movement. In fact, following noxious incitant exposure, a common response in the chemically sensitive or chronically diseased individual is fascial spasm with fibromyalgia and fascial pain, especially in the back of the neck. In addition, generalized reactions throughout the body may occur. Frequently, osteopathic manipulation, massage, as well as laser, **or needle acupuncture, or integrated muscle stimulation (IMS), and** manual manipulation of acupressure points will release both the spasm, and thereby, release the sequestered pollutants with the autonomic responses. This release temporarily improves conditions in the chemically sensitive and chronic degenerative diseased individual. These procedures often help relieve the block in the homeostatic mechanism by allowing a flow of energy efficient responses that relieve the dysfunction.

Blood vessels, with their sympathetic innervation, contain both adrenergic and cholinergic fibers; however, most are sympathetic, while sweat and the piloerector apparatus contain only cholinergic fibers via the sympathetic nerves.[50] The sweat glands are stimulated primarily

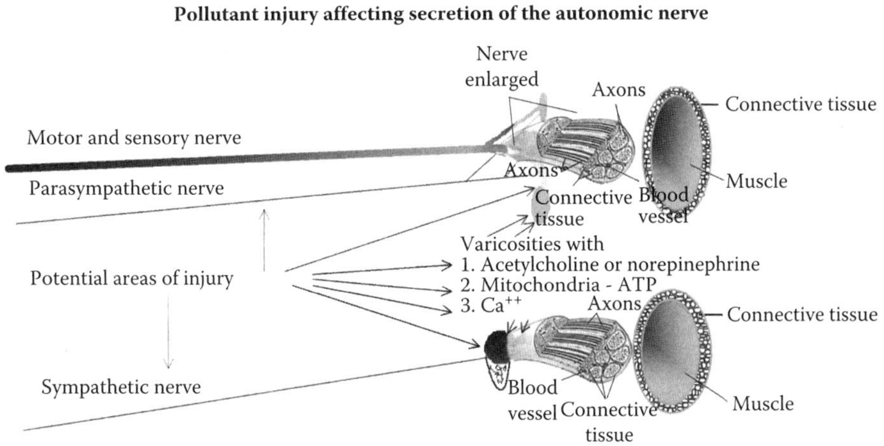

FIGURE 2.10 Autonomic nerve endings in the connective tissue matrix over muscles. (From EHC-Dallas.)

by centers in the anterior hypothalamus and preoptic area that are usually considered parasympathetic centers.[51] **Therefore, sweating could be called a parasympathetic function even though it is controlled by nerve fibers that anatomically are distributed by the sympathetic nervous system**. This anatomic fact might explain the phenomenon of vasospasm, resulting in cold hands and feet and a lack of sweating, observed in thousands of chemically sensitive and chronic degenerative diseased patients. This vasomotor response would then be an overstimulation of the sympathetic adrenergic response, but the lack of sweating would be a blocked sympathetic cholinergic response. In addition, the chemically sensitive and many chronic degenerative diseased individuals develop an inability to sweat and secrete sufficient quantities of sebum as well as an inability to respond rapidly to injury. Thus, dyshomeostatic responses are developed. Such a simultaneous response to pollutant exposure would be a sympathetic stimulated cholinergic response through the peripheral sympathetic nerve. **It should be noted that axillary gland output is parasympathetic nerve stimulation not the cholinergic part of the sympathetic seen in other sweat glands**. In some chronically ill patients, some of the dynamics of the dyshomeostatic responses (e.g., muscle fatigue, pain, brain fog) may, after a period of sensitization, be triggered by muscle massage or exercise. This occurs because the patients are at the stage of the degenerative process in which many more supersensitive, secondary homeostatic foci (interference fields) are triggered.

As previously stated, if the chronic stimuli continue a dyshomeostatic response occurs in the chemically sensitive and chronic degenerative diseased individual, and local homeostatic control is exceeded. In iris corder (pupillography) studies of pollutant effects on the ANS of our chemically sensitive and chronic degenerative diseased patients at the EHC-Dallas who have exceeded local homeostatic response mechanisms, we have demonstrated objective alteration of the ANS in excess of 90% of our patients. In addition, clinical peripheral responses, such as vasospasm in the hands and legs, distention or bloating of the gut, and bladder spasm, suggest that autonomic triggering approaches 100% in the chemically sensitive and chronically diseased individuals who have exceeded local homeostatic control. Recent pupillography studies at the EHC-Dallas have shown a 93% triggering of the ANS in contrast to the earlier 70% due to inclusion of an abnormal nonspecific pattern change which is seen in many chemically sensitive or chronic degenerative diseased individuals. Clinicians who do not thoroughly understand normal ANS anatomy and physiologic function are often confused by alterations that result from exposure to noxious stimuli. Misinterpretation of their homeostatic and dyshomeostatic responses often occurs. However, a better understanding of the anatomy and physiology of the autonomic nervous sytem leads to a precise explanation of most of the early complaints (e.g., bloating, gastrointestinal upset, rapid heart beat, arrhythmia, slow heart beat, urgency of urination, brain dysfunction) experienced by the chemically sensitive or chronic degenerative diseased patient before fixed disease occurs, but after local homeostatic control has been exceeded. In addition, with better understanding of this anatomy and physiology, the causes and effects of distal pollutant or noxious incitant injury can be measured and reproduced by pollutant challenge under controlled conditions, making possible avenues for prevention. (See Neuromuscular and G.I. chapter in Volume 2 - Mechanisms of Chemical Sensitivity and Chronic Degenerative Disease.)

Understanding of the anatomy and physiology of the ANS leads the clinician to the realization that any given chemical or combination of chemicals, which have exceeded local homeostatic control, may have a sympathetic, cholinergic, cholinolytic, sympatholytic, combined, or nonspecific response. This response depends on the following: the types of triggers, the location of triggering, the amount of the dose, the total load of other toxic chemicals and noxious stimuli, and the nutritional state of the body area. Some or all of these factors may be present in an individual at the time of exposure, and they can result in a homeostatic, or if the exposure is too prolonged, a dyshomeostatic response. **Organophosphate insecticides, which can trigger the local, regional, or general homeostatic mechanism, are an example of one group of those chemicals that usually generates a cholinergic or sympatholytic effect in the same chemically sensitive individual**.

Table 2.6 presents a more detailed breakdown of the functions of the ANS.

TABLE 2.6
Some Effects of Autonomic System Stimulation from Pollutant Injury Seen in the Chemically Sensitive or Chronic Degenerative Diseased Individual

Organ/Systems	Sympathetic		Parasympathetic	
	α-Receptor	β-Receptor	Cholinergic	Cholinolytic
Eye				
Radial muscle	Contraction	Relaxation for far vision		
Ciliary muscle	Dilation		Constriction	Dilatation
Heart		Increased heart rate	Decreased heart rate	Dilatation
Blood vessels	Constriction	Dilatation	Constriction	
Bronchial muscle		Relaxation	Constriction	
Stomach	Decrease in Motility	Decrease in motility	Increase in motility	Decrease in motility
Intestine	Dilatation		Increase in motility	
Sphincters	Contraction			
Urinary bladder	Contraction	Relaxation		
Detrusor muscle				
Trigone and sphincters			Decreased activity	
Sweat glands	Selective stimulation			
Uterus	Contraction	Relaxation		
Liver	Glycogenolysis		Spasm	
Muscle		Glycogenolysis		Relaxation
Insulin secretion	Inhibition	Stimulation		
Saliva secretion	Inhibition	Stimulation	Inhibition	Stimulation

Source: Based on EHC-Dallas. 1984.

Because of the presence of alpha and beta receptors and the presence of pollutant or other noxious incitant injury in the ANS, even more complex variations of autonomic nervous system responses occur, including sympathomimetic, sympatholytic, cholinergic, cholinolytic, and nonspecific responses.

Studies performed on breathing patterns in patients with ANS dysfunction may help us better understand the patterns of the dynamics of response in the chemically sensitive and/or chronic degenerative diseased patient.[3,52] Some of these patterns will be discussed in Neuromuscular chapter in Volume 2 - Mechanisms of Chemical Sensitivity and Chronic Degenerative Disease in more detail.[53] For example, McNicholas et al.[53] investigated the control of breathing in three patients with autonomic deregulation. These patients were free of respiratory and sleep-related symptoms. During wakefulness, their ability to reproduce a breath of a given tidal volume voluntarily did not differ from that of healthy control subjects. **However, defects in the metabolic control system were indicated in these patients by their lack of ventilatory response to acute hypoxia**. During sleep, the minute volume of ventilation in these patients with autonomic dysfunction was not significantly less than that of healthy subjects, but their pattern of breathing was highly irregular, resulting in coefficients of variation of respiratory variables that were two to three times greater than normal ($p < 0.05$). This irregularity persisted during slow-wave sleep, when the ventilatory pattern is, normally, highly regular. Similar problems of variability during sleep are seen in some chemically sensitive and chronic degenerative diseased patients when they are exposed to toxic chemicals or other noxious incitants. Once this irregularity in breathing pattern occurs, a great deal of energy is consumed, which then results in inefficient metabolism and an absence of wellness. After exposure, they often show symptoms of inefficient metabolism with symptoms of weakness, fatigue, pain, and inability to concentrate. These patients usually wake up fatigued. Throughout the treatment and study of many patients, we have observed that once the toxic load is reduced, this variability disappears and homeostasis is restored. Parkinson's patients used as nonhealthy controls did not have the previously cited irregularities. Therefore Parkinson's disease may represent an entirely different homeostatic dysfunction, as it appears to be a disease involving the substantia nigra. However, this disease is classified as an ANS dysfunction. **The combination of findings during wakefulness and sleep in those patients with autonomic dysfunction suggests a defect in the autonomic respiratory rhythm generator of the brainstem**. These findings also indicate that examination of the breathing pattern during sleep may be useful in the investigation of chemically sensitive and chronic degenerative diseased patients who are suspected of having autonomic dysfunction even in the absence of clinical respiratory abnormalities.

We, as well as Monro,[54] have seen a large subset of chemically sensitive and chronic degenerative diseased patients who have disturbed sleep patterns. Often, these dysautonomic patterns are eliminated in the ECU where the total pollutant load is decreased, and the dysautonomic patterns can then be reproduced by pollutant challenge. We have found monitoring sleep disturbance is a valuable clinical tool in following those who have been treated for chemical sensitivity and chronic degenerative disease. Often, as these patients' chemical sensitivity decreases and their homeostasis is restored, their sleep patterns improve.

In Monro et al's study,[54] 10 patients (3 children and 7 adults) with allergic rhinitis (AR) and sleep disturbance had a polysomnographic evaluation for one night. The sleep apnea syndrome (SAS) was diagnosed in each subject. A total of 625 apneic events were recorded. Of these, 57.5% were obstructive, 21.6% mixed, and 20.9% central apnea. The apnea occurred mainly in nonrapid eye movement (NREM) sleep (61%) and the remaining 39% occurred in rapid eye movement (REM) sleep. According to Monro, allergic rhinitis, complicated by sleep apnea disturbance, may be an important factor in the pathogenesis of idiopathic conditions such as essential hypertension.

These patients had not only allergic rhinitis, but also other symptoms, such as sinusitis, etc. Monro et al. feel that the "inability to breathe through the nose is an increasingly recognized cause of disordered breathing during sleep. Chronic mucosal swelling associated with an increased nasal airflow resistance could affect the function of the nasal receptors. This would cause a loss of neuronal input to respiration, which may lead to respiratory dysrhythmia with the disordered breathing events being "central and obstructive apnea."[54] They have had a similar experience to ours at the EHC-Dallas and Buffalo, where the triggering agents (usually food and chemicals) were defined, eliminated, or neutralized. When this was done, the patients saw improvement with not only their rhinitis but also with their sleep apnea.

Once the clinician recognizes that most of the autonomic dysfunction is environmentally triggered, emphasis should be placed on reducing the total nonspecific, as well as the specific, pollutant load (i.e., food, mold, and chemical incitants) in the chemically sensitive patient in order to prevent ANS failure. Early recognition and treatment of the environmental aspects of autonomic dysfunction leads to easy reversal of the homeostatic disturbance. As seen in the case report earlier in this chapter, once failure occurs, it appears to be impossible to reverse if fixed-tissue changes have also occurred. At times we have seen that those with temporary metabolic blocks can sometimes be reversed, as shown in injury of the patients with NMH.[55]

DENERVATION SUPERSENSITIVITY

Insight into the pathology of mild and severe dysfunction of the peripheral sensory nerve and the ANS can be observed in patients who have had partial or total sympathectomy or partial or total peripheral nerve damage. These patients have an extremely difficult, if not impossible, task of trying to balance their energy and information receptor system and maintain dynamic homeostasis. **In sympathectomy patients, the tone of the vascular and other smooth muscle is maintained by the adrenal output after the nerve is cut**. During the first week after sympathectomy, or if the parasympathetic is destroyed, the denervated organ becomes more sensitive to injected or endogenous norepinephrine or acetylcholine, respectively. This effect is demonstrated in Figure 2.11, which shows the blood flows in the forearm, before removal of the sympathetics, to be about 200ml/min. A test dose of epinephrine causes only a slight depression of flow.

Once the stellate ganglion is removed, normal sympathetic tone is lost. At first, the blood flow rises markedly because of the lost vascular tone, but over a period of days to weeks, returns to almost (100% gain) normal because of a progressive increase of the intrinsic vascular muscle tone, thus, compensating for the loss of sympathetic tone. Then another challenge of norepinephrine is administered. The blood flow decreases more than before, demonstrating that the blood vessels are about 2–4 times more responsive to norepinephrine than before the sympathectomy. **This denervation supersensitivity can occur in any organ, with the supersensitivity increasing as much as tenfold to one thousandfold.**[56] Supersensitivity to acetylcholine also occurs, where the total muscle becomes as sensitive as the motor endplate. We have seen a high percentage of patients who respond to this neurotoxic dose of acetylcholine, which confirms this supersensitivity.

The following case emphasizes the change that one might see with a permanent upper extremity partial sympathectomy with continued denervation supersensitivity.

Case Study. During an auto accident a 50-year-old, white female, received a traumatic injury to her right arm and neck. She was treated with physical therapy for over one year, but her arm and shoulder worsened until she could not move her arm. She had a right first rib resection and the right T_1, T_2, and T_3 sympathetic trunk was removed. The patient had immediate relief of her arm paralysis and became asymptomatic of pain. However, she immediately developed atrial fibrillation that persisted and was refractory to medication until she was treated with food elimination by one of the authors of this book (WJR).

Denervation supersensitivity is known to cause fibrillation of the pervading denervated muscle. In the case of the patient with the T_1, T_2, T_3 resection, it was found that she was sensitive to five foods, and, when these were eliminated from her diet, the atrial fibrillation disappeared. When they were placed back into her diet, the fibrillation returned. The patient has now remained free of atrial fibrillation without medication for over 24 months, as she continues to avoid the five foods to which she is sensitive. Clearly, the severance of the right T_1, T_2, and T_3 sympathetic trunk caused an imbalance in the ANS to the point that the autonomic response to the heart was not balanced. This denervation hypersensitivity increased to the point where it caused the heart to be triggered by normally, nonnoxious substances. Since there were other environmental triggers, namely the food to which the patient was sensitive, the ANS could not stabilize fast enough, giving a heterogeneous response to the right atrial myocardium and thus, causing atrial fibrillation. Heart rate variability in this patient revealed a decrease in

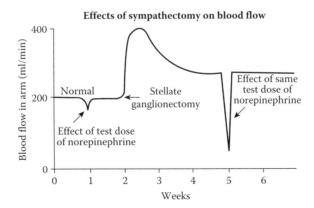

FIGURE 2.11 Graph traces effects of sympathectomy on blood flow and illustrates the law of denervation as the distal tissues temporarily, and at times permanently, become supersensitive to epinephrine, acetylcholine, and various other neurotransmitters. (Guyton, A.C. and Hall, J.E., *Medical Physiology,* 10th ed. Philadelphia: W.B Saunders Co., 705, Fig. 60.4, 2000. With permission.)

parasympathetic responses and an increase in sympathetic responses, thus giving a super adrenalin response, which primed the myocardium for multiple adverse responses such as arrhythmias.

This second case demonstrates how prolonged denervation supersensitivity can be devastating for the patient long term.

Case study. A 45-year-old white, female, talented artist had a left lumbar sympathectomy due to unrelenting vascular spasm in her left leg. This procedure saved the leg, but after the surgery, she developed the onset of food and chemical sensitivity. She reached the point of supersensitivity where she could not tolerate any oral nutritional supplementation, or injection therapy, or even absorption of the nutrient by placing it on her skin. Literally, she is homebound in her environmentally safe home. Also, attempts with homeopathy, acupuncture, and laser acupuncture failed because they all triggered more symptoms due to her supersensitivity.

Previous history showed that, as an infant, she had stayed at her birth weight for six months. At the age of 4, she developed allergies and sensitivity to sounds, food, fabrics, vitamins, and medicines. She had meningitis, at age 13, and also underwent an appendectomy with no major problems. At age 17, after an extreme reaction to penicillin given for infected tonsils, she had a tonsillectomy, which produced highly reactive responses to all medications including anesthetics. At age 22, following sensitivities to all of the above, she underwent varicose vein stripping. At age 23, she received traumatic multiple bone fractures (hip, pelvis, legs); at age 26 she underwent ear surgery following a hypersensitive reaction to sound vibrations and more sensitivities to drugs; at 28 she had a ski accident and broke her left leg, which resulted in life-threatening thrombophlebitis. She then developed arterial spasm of the top of left foot. Gangrene developed, and as a result of this, she underwent the sympathectomy of L5. Following this procedure, she had several breast surgeries, all of benign pathology. At age 43, she was diagnosed with Graves disease and treated with extremely small doses of propylthiouracil for three years. In 1992, shortly after the sympathectomy, she had an ear test that punctured her eardrum, resulting in a very serious bacterial and yeast infection. At this point, she became hypersensitive to the world around her, losing all normal body functions. Anaphylactic shock responses occurred as she became sensitive to many substances such as, chemicals, foods, and electronics, including TV, lights, iron, power lines, etc. She stated that just watching TV, her neck would collapse and her facial skin would swell and turn red as if she had a sunburn. After a car accident in 2000, her sensitivity to electromagnetics became worse from exposure to diesel fumes and other chemicals, which triggered seizures. Her sensitivity to electromagnetics increased with her allergic reactions to inhalants, and other pollutants.

Studies of this patient's ANS showed sympatholytic pattern, using pupillography, and a very high vasoconstricter tone of 0.92 (C-.63) with severe sympathetic stimulation as measured by heart rate variability. T-cell study results were T_{11} 1038 (C-1260-2650 mm^3), T_4 808 (C-652-1770 mm^3), T_8 245 (C-325-1050 mm^3) indicating immune suppression. She was not responsive to treatment with autogenous lymphatic factor, an immune modulator. The only treatment that appeared to help her was manual energy balancing and manipulation. Most attempts at therapy helped partially, but the treatment would not hold, presumably due to the effects from the left lumbar sympathectomy, which appeared to keep throwing her energy flow out of balance. This patient remains so sensitive 20 years after the sympathectomy that a few drops of norepinephrine on her skin will trigger myoclonic seizures starting in the left calf and progressing through her body. Conversely, a neutralizing dose of acetylcholine (1.625 dilution) dropped on her skin will temporarily calm her symptoms. This patient's supersensitivity often renders her so sensitive that her neutralizing endpoint doses continue to change.

This case represents an observed phenomenon that occurs with surgical or chemical sympathetic or parasympathetic nerve severance. This phenomenon involves an immediate dilatation of blood vessels with a sympathectomy, or a gastric atony when the parasympathetic vagus nerve is cut. During the first week or so after an autonomic nerve is severed, **the denervated organ becomes increasingly sensitive to norepinephrine or acetylcholine and this is, presumably, when food and chemical sensitivities develop.** In addition, the sensitivities appear to multiply, spreading to specific chemicals, molds, and even foods as triggering agents. In other words, the threshold of the environmental receptors for external environmental triggering has become lowered. This lower threshold allows smaller doses of toxic and then nontoxic incitants (i.e., foods, pollens, and some benign chemicals) to trigger symptoms. This phenomenon of denervation supersensitivity usually only lasts for a few weeks in the average patient, until adrenal output compensation occurs. Obviously, in some patients, this phenomenon is prolonged for weeks or even years, as shown in the aforementioned case. The cause of denervation supersensitivity is partially known as protein kinase phospholytion.

When the sensitivity of receptors in post sympathetic membrane receptors of the effector cell increases, sometimes many fold, the norepinephrine or acetylcholine is no longer released at the synapses. Therefore, sensitivity of the receptors is upregulated in order to perceive the exposure to the lesser amounts of the neurotransmitters. The increase in sensitivity is apparently an attempt to attach receptors to the remaining hormones. When a dose of the neurotransmitter is then injected into the circulating blood, the reaction is partly enhanced. At this point in the reaction, the entire muscle becomes as sensitive as the motor endplate. This increased sensitivity forces the muscle to go into spasm and/or causes early fatigue. Denervation supersensitivity originating from trauma or disease processes (i.e., diabetes, sickle cell, viruses, toxic chemical) can manifest in many ways. For example, Gunn[57] has shown that denervation supersensitivity may manifest in symptoms of chronic degenerative disc disease, chronic fatigue, and fibromyalgia, etc.

LAW OF DENERVATION

Normal nerve and muscle depend upon intact innervation to provide a regulatory or "trophic" (neurogenic) effect. Formerly, it was supposed that loss of the trophic (neurogenic) factor, through total denervation, led to "denervation supersensitivity." More recently, **it has been shown that any measure which blocks the flow of motor impulses, i.e., phenol, pesticide, solvents, etc., and deprives the effector organ of excitatory input for a period of time can cause "disuse supersensitivity"** in that organ, as well as in associated spinal reflexes. "Supersensitive" nerves and innervated structures react abnormally to stimuli, according to Cannon and Rosenblueth's law of denervation: The law states, "*When a unit is destroyed, in a series of efferent neurons, an increased irritability to chemical agents develops in the isolated structure or structures, the effect being maximal in the part directly denervated.*" This is the aforementioned sympathectomy or vagus nerve severance previously discussed.[58] In other words, when a nerve **is not functioning properly (neuropathy), it becomes supersensitive and will behave erratically**. This principle is fundamental and universal, yet it is not at all well-known or credited![58] Nerve damage is usually involved in most dyshomeostatic reactions involving the immune, neurologic, endocrine, and vascular systems.

Cannon and Rosenblueth[59] **recognized four types of responses indicating increased sensitivity. We have observed and measured these responses in most clinical dyshomeostatic reactions. The types of responses indicating increased sensitivity include the following**: (1) the amplitude of response is unchanged, but its time-course is prolonged (superduration of response), which one sees repeatedly in the early chemically sensitive patient; (2) the threshold of the stimulating agent is lower than normal (hyperexcitability), thus, creating food and chemical sensitivity; (3) lessened stimuli, which do produce responses of normal amplitude (increased susceptibility) without a lowered threshold, as seen in food and chemical sensitivity; and (4) the capacity of the tissue to respond is augmented (superreactivity). See Figure 2.12.

Cannon[59] and Gunn,[60] each independently, demonstrated that **supersensitivity can occur in many structures of the body including skeletal muscle, smooth muscle, spinal neurons, sympathetic ganglia, adrenal glands, sweat glands, and even brain cells. Because of this phenomenon, orderly homeostasis is disrupted. Furthermore, both Cannon and Gunn showed that denervated structures overreact to a wide variety of chemical and physical inputs including stretch and pressure**. We have seen this phenomenon repeatedly in the chemically sensitive and chronic degenerative diseased patient triggered by chemical agents (phenols, solvents, pesticides, etc.) and physical agents (i.e., cold, electromagnetic stimulations, weather front changes, stretch, pressure, etc.).

Chemical pollutants can trigger autoimmune disease like Devic's disease, also known as neuromyelitis optica (NMO), results in MS-like demyelinating lesions along the optic nerves and spine.

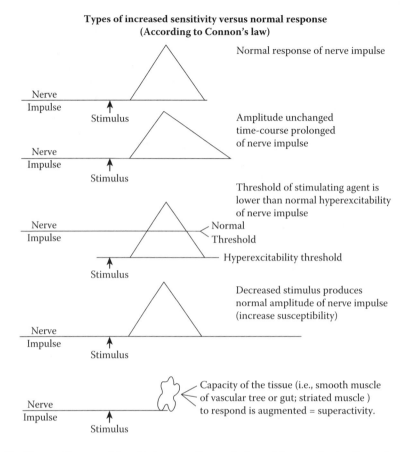

FIGURE 2.12 Types of increased Sensitivity according to the law of denervation leading to hypersensitivity to environmental responses (molds, terpenes, bacteria, viruses, neurotransmitters, foods, toxic and nontoxic chemicals). (EHC-Dallas.)

Affected individuals often experience rapid visual loss, paralysis, and loss of leg, bladder, and bowel sensation. Some lose their sight permanently. Devic's disease is diagnosed by the presence of a specific autoantibody referred to as NMO-IgG—in the blood. NMO-IgG sets off a chain of events that leads to a toxic buildup of a excitatory neurotransmitter glutamate. NMO-IgG binds to a protein that normally sops up excess glutamate from the synaptic cleft. When NMO-IgG is around, this sponge-like action is blocked, allowing glutamate to accumulate and downward spiral of cell death and demyelinating lesions results in chronic degenerative disease.

Of all the structures that develop **supersensitivity, the most common and significant is smooth muscle (vascular and gut) and striated muscle**. Apart from pain and tenderness that may occur within **striated muscle (possibly from the compression of supersensitive nocioceptors), neuropathy increases muscle tone and causes concurrent muscle shortening. Thus, when the gut is involved, cramps occur. Muscle shortening, in turn, can mechanically cause a large variety of pain syndromes such as neuropathy, lumbar disc pain, etc., by its** relentless pull on various structures.[58]

According to Gunn, muscle shortening is the key to myofascial pain of neuropathic origin. Stated differently, myofascial pain cannot exist in the absence of muscle shortening, for if there is no shortening, there is no pain. Gunn sometimes refers to myofascial pain as the "shortened muscle syndrome."[58] This phenomenon may explain why different therapies like chiropractic, massage etc. will temporarily help a patient with myopathic pain.

Muscle shortening is a fundamental feature of musculoskeletal pain syndromes. **According to Gunn, the term** *spasm* **is commonly used to describe muscle shortening in myofascial pain syndromes, but shortening is generally caused by classic contracture**. Spasm—that is, increased muscle tension with muscle shortening—comes from **motor nerve activity and is seen in electromyography (EMG) as continuous motor unit activity**. Spasm cannot be stopped by voluntary relaxation. However, **EMG examination** of a shortened muscle rarely reveals any motor unit activity.[60]

Classic *contracture,* on the other hand, is the evoked shortening of a muscle fiber in the absence of action potentials. In denervated, supersensitive skeletal muscle fibers, acetycholine slowly depolarizes muscle membrane, thus inducing electromechanical coupling with the consequent slow development of tension without action potentials. Since no action potentials are revealed by EMG, muscle shortening is most likely caused by contracture.[61]

It is therefore, best to avoid using the term "spasm" when describing striated muscle shortening. (It has recently been suggested that there is direct sympathetic innervation to **the intrafusal fibers of muscle spindles, and sympathetic stimulation can cause muscle tension in curarized animals that is blocked by alphadrenergic antagonists**.)[62] See Neuromuscular chapter in Volume 2 - Mechanisms of Chemical Sensitivity and Chronic Degenerative Disease for a detailed explanation. **Clinically, muscle shortening can be palpated as ropey bands within muscle.** The bands are seldom limited to a few individual muscles, but are present in groups of muscles according to the pattern of the neuropathy. In radiculopathy, bands are also present in paraspinal muscles.[61]

An important source of pain in musculoskeletal pain syndromes is from muscle shortening that mechanically stresses muscle attachments, causing conditions such as "bicepital tendonitis" or "lateral epicondylitis."[61]

Shortening of muscles that act across a joint increases joint pressure, upsets alignment, and can precipitate pain in the joint, i.e., arthralgia. Increased pressure upon spinal joints can cause the "facet-joint syndrome." **Muscle shortening can eventually bring about degenerative changes—i.e., osteoarthritis**.[61]

Shortening in paraspinal muscles acting across a disc space compresses the disc and can cause narrowing of the intervertebral foramina, either by indirectly irritating the nerve root (e.g., through pressure of a bulging disc), or by applying direct pressure on the root after it emerges.[61] One frequently sees this syndrome in the patient with chemical sensitivity and/or chronic degenerative disease.

A self-perpetuating cycle can arise: pressure on a nerve root causes neuropathy; neuropathy leads to pain and shortening in target muscles, including paraspinal muscles; shortening in paraspinal muscles further compresses the nerve root. (This self-perpetuating cycle is a vicious downward homeostatic characteristic circle of pain, with the pressure on blood vessels leading to ischemia and more pain.)[62]

Muscle bands are usually pain-free, but can become tender and painful, possibly by compressing intramuscular nocioceptors or microneuromas.[63] Gunn reported in patients with low back injury that the motor points of some muscles may be tender. Of 50 patients with low back "strain," 26 had tender motor points and 24 did not, while 49 of 50 patients with radicular signs and symptoms suggesting disc involvement had tender motor points. The one patient without such tender points had a hamstring contusion which limited straight leg raising. Of 50 controls with no back disability, only seven had mild tender points after strenuous activity; while 46 of another 50 controls with occasional back discomfort had mild motor-point tenderness. In all instances, the tender motor points were located in myotomes corresponding to the probable segmental levels of spinal injury and of nerve root involvement (when present). Patients with low back strain and no tender motor points were disabled for an average of 6.9 weeks, while those with the same diagnosis but tender motor points were disabled for an average of 19.7 weeks, or almost as long as the patients with signs of radicular involvement, who were disabled for an average of 25.7 weeks. Tender motor points may, therefore, be of diagnostic and prognostic value, serving as sensitive localizers of radicular involvement and differentiating a simple mechanical low back strain from one with neural involvement.

Focal areas of tenderness and pain are often referred to as "trigger points." See Neuromuscular chapter in Volume 2 - Mechanisms of Chemical Sensitivity and Chronic Degenerative Disease. When pain is primarily in muscles and is associated with multiple tender trigger points, the condition is referred to as myofascial pain syndrome.[64-68]

When muscle bands are fibrotic and painful, the condition is sometimes known as "fibrositis, fibromyalgia, fibromyositis" or "diffuse myofascial pain syndrome." The etiology of the syndrome is "unkown," but it has many clinical features of the radiculopathic group: pain and stiffness of long duration (more than three months), pain increased by physical or mental stress (the role of anxiety and emotional stress in causing muscle spasm and pain is well-known); multiple tender points; nerve compression and disc degeneration; soft tissue swelling; joint pain; and neuropathy.[69] In our experience, the pain, weakness, and fatigue can be influenced by supersensitivity to food, chemicals, pollens, dust, molds, and lack of oxygen, all of which would fit with our previously described nerve hypersensitivity.

Autonomic manifestations of neuropathy are vasomotor, sudomotor, and pilomotor changes, all of which have been previously discussed. Vasoconstriction induced neuropathic pain is generally differentiated from inflammatory pain by the fact **that neuropathic pain in affected parts of the body is perceptibly colder while the inflammatory pain causes affected areas to be warmer**. Retained catabolites from ischemia may exacerbate the neuropathic pain.[70] Increased sudomotor activity can occur and the pilomotor reflex is often hyperactive and visible in affected dermatomes ("goose bumps").[63,71]

These symptoms can be an interaction between pain and autonomic phenomena. A stimulus such as chilling, which excites the pilomotor response, can precipitate pain; vice versa, pressure upon a tender motor point can provoke the pilomotor and sudomotor reflexes.[72] Increased tone in lymphatic vessel smooth muscle, and increased permeability in blood vessels[73] can lead to local **subcutaneous tissue edema ("neurogenic" edema or "trophedema")**. This finding can be confirmed by the peau d'orange effect (orange peel skin) or by the "Matchstick" test: trophedema is nonpitting to digital pressure, but when a blunt instrument such as the end of a matchstick is used, the indentation produced is clear-cut and persists for minutes.[74] This simple test for neuropathy is more sensitive than EMG. Trophic changes such as dermatomal hair loss may also accompany neuropathy.[74]

Neuropathy and denervation affect the quality of collagen and other metabolic effectors of fibroblasts in soft and skeletal tissues due to activation of the nonspecific mesenchyme mechanism. (<100% gain mechanism of homeostasis). This activation is an important factor in chronic pain and degenerative conditions because replacement collagen has fewer cross-links and is markedly weaker than normal mature collagen.[75] Any form of stress—whether emotional or physical, whether extrinsic or intrinsic—causes muscle shortening. **The increased mechanical tension that muscle shortening generates hastens wear and tear because it pulls on degraded collagen** that provides the strength to ligaments, tendons, cartilage, and bone. Neuropathy expedites degeneration in weight-bearing and activity-stressed parts of the body, causing "spondylosis," "discogenic disease," and "osteoarthritis" among others. Such conditions are currently regarded as primary diseases, but Gunn thinks they are secondary to a radiculopathic process. **Radiculopathic pain conditions, therefore must be treated with some urgency in order to prevent a chronic degenerative disease**.[76]

In addition, to denervation supersensitivity, the vascular and neuroendocrine response of the ANS is extremely important in regulating the dynamics of homeostasis. Neuroendocrine responses occur in the adrenal medulla and other smaller clusters of neuroendocrine cells, which are found in crucial areas such as the GI tract, the carotid sinus area, and numerous vascular bifurcations, such as the organ of Zuker-Kandl at the aortic bifurcation and the carotid bifurcation for the baroreceptors. These areas are very specific for reciprocal information. For more information, see the neuroendocrine section that follows under the Endocrine System in Chapter 4. Abnormal vascular responses occur anywhere nerve damage manifests.

DENERVATION AND SPONDYLOSIS: NEUROPATHIC PAIN

Gunn[77] has observed that aperiodic dyshomeostasis resulting in degenerative disc disease is the result of denervation hypersensitivity. He noted that attrition of the dorsal nerve root was caused by a partial denervation neuropathy and included pressure, stretch, angulation, and friction. Etiologic agents can be due to trauma, toxics, infection, emotional stress, or tumors. **The result of these injuries is the partial denervation of the dorsal nerve root of the spinal cord which gives spasm of the end blood vessel next to the disc.** Since the disc depends mostly on diffusion for its nutrient supply, oxygen and other nutrient influx is altered. When these are decreased, the disc starts to degenerate, resulting in a painless prespondylosis state. The intervertebral muscles shorten, causing more vascular dysfunction and pain. As the radiculopathy persists, the discs degenerate. Spondylosis (the structural disintegration and morphologic alteration that occurs in an intervertebral disc) progresses until pain is severe. See Figure 2.13.

Ordinarily, spondylosis follows a gradual, relapsing, unremitting course that is silent unless and until symptoms are precipitated by an incident (traumatic, toxic, or metabolic) often so minor that it is missed by the patient. Spondylosis occurs due to vascular spasm followed by hypoxia and more spasm of the paraspinal muscle resulting in pressure on the disc. All gradations of spondylosis can exist, but early or incipient spondylotic changes, even when unsuspected, can cause radiculopathy.

An acute nerve injury will result in a short duration discharge while the same injury in an already damaged nerve (neuropathic) will result in a sustained discharge. Thus, for pain to become a persistent symptom, the affected fibers must be previously irritated or defective (secondary foci) as shown in the previous two case reports. Both of the previously described patients had previous damage to their vascular and nervous system before their sympathectomies. This previous damage factor appears to be the reason why some people develop severe pain after apparently minor injury and further explains, as Gunn[57] observed, why pain can continue beyond a reasonable period of time. Of course, if a patient breathes or ingests phenolic compounds (toluene, xylene, etc.) these may be deposited on the previously injured primary focus causing further irritation, inflammation, and even greater nerve damage.

Spondylosis increases with age; therefore, spondylitic pain is more common in middle aged individuals who have accumulated what Sola[66] has termed an injury pool or what Bergsman,[78] Berger,[52] and the Huneke brothers[79] would call interference fields with sensitive secondary foci. (See Chapter 1.) This injury pool or area is an accumulation of repeated major (trauma or toxic metabolic) or minor (food, pollen, mold sensitivities) injuries to a segment of nerve fibers leading to unresolved clinical residuals, which may or may not produce pain.

According to Gunn[80] and our observations, neuropathic pain is distinguished by the following: (1) pain in the absence of an ongoing tissue damaging process (overt); (2) there is a delay of onset

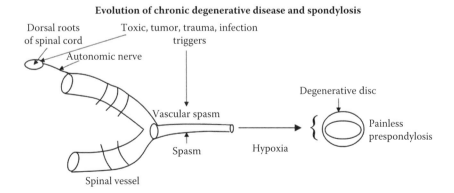

FIGURE 2.13 Evolution of chronic degenerative disease and spondylosis. (From EHC-Dallas.)

after a precipitory injury (incitant entry, adjustment, muscle constriction, loss of energy, vascular spasm, hypoxia, nonspecific mesenchyme reaction, tissue changes); (3) the presence of abnormal sensations such as burning or (searing) shearing pain (dysethesia) or deep aching pain which is more common than dysethesia pain in the muscoskeletal pain syndrome; (4) paroxysomal brief sharp or stabbing pain; (5) pain felt in the region of sensory deficit; (6) mild stimulus causing extreme pain (allodynias); (7) pronounced summation that also occurs with repetitive stimuli; and (8) loss of joint range or pain caused by mechanical effects of muscle shortening.[80] Any of the above features should raise the suspicion of neuropathic pain.[81,82]

Neuropathy is determined principally by clinical examination, as there can be nerve dysfunction without any detectible structural changes.[83] However, frequently a neurometer, pupillography, or heart rate variability machine may demonstrate this abnormality. **Most clinical neuropathies are of mixed pathology; both axonal degeneration and segmental demyelination can occur in varying degrees.**[80]

Routine laboratory and radiological tests are unhelpful in diagnosing radiculopathy,[82] but specials tests may be indicated, e.g., EMG to determine primary disease of muscle; radiology to exclude intraspinal tumors; and laboratory investigations to rule out abnormal immunologic response. EMG, before denervation, may only show increased insertion activity, and nerve conduction velocities can be normal, but F-wave latencies of the nerves may be prolonged.[83,84] Thermography may reveal altered nerve ending skin temperatures, but does not, by itself, indicate pain.[85]

Radiological findings of spinal degenerative changes, commonplace in the middle-aged, should not be dismissed as they can imply some degree of previous nerve damage and thus degenerative disease (aperiodic homeostatic disturbance).

Neuropathy is most often at root level (i.e., radiculopathy) when mixed sensory, motor, and autonomic disturbances occur. These are epiphenomena of radiculopathy, which will present in the dermatomal, myotomal, and sclerotomal target structures supplied by the segmental nerve. They are often symmetrical; even when symptoms are unilateral, latent signs of neuropathy may be mirrored contralaterally.[86,87]

Dysfunction need not include pain unless nociceptive pathways are involved: some neuropathies are pain free,[84] such as sudomotor hyperactivity in hyperhidrosis and muscle weakness in ventral root disease.

If and when pain is present, it is practically always accompanied by muscle shortening in peripheral and paraspinal muscles as well as in tender and painful focal areas in muscles ("trigger points")[86–88,64,66,68] with autonomic and trophic manifestations of neuropathy.[89,84,68]

Neurally Mediated Hypotension and Tachycardia

There appear to be many forms and degrees of neuromediated hypotension and neuromediated tachycardia, which are less severe, but incapacitating, in some chemically sensitive and chronic degenerative diseased patients. These will be discussed on the following pages.

The physiology of NMH and tachycardia has received a resurgence of interest.[90] **Three factors that predispose a person to the development of NMH are a low resting blood volume, excessive pooling of blood in the dependent vessels, and excessive loss of plasma volume during upright posture**,[91] all of which can decrease venous return to the heart. This response is similar to the patient in the case at the beginning of this chapter. When cardiac output is decreased in this setting, a reflex increase in sympathetic neural outflow and an increased secretion of epinephrine normally results in an increase in peripheral vascular resistance and in heart rate, thereby leading to maintenance of a normal blood pressure. However, with dysfunction of one limb of the ANS, this maintenance of blood pressure does not occur. A group of patients exists who develop dizziness, short-term memory loss, lack of concentration, postural hypotension, and, at times, syncope as demonstrated by the previous extreme case. In these patients with an inability to hold injection neutralization end points and/or to detox in the sauna we found most have abnormal responses to upright tilt testing as

Nervous System

well as abnormal heart rate variability. **There seems to be a failure in these patients to mobilize blood reflexly from the dependent splanchnic and limb vasculature, emphasizing ANS dysfunction**. Measurements on the heart rate variability machine show a failure of the microvascular bed to contract and, at times, a chronotropic response in various patients with neurally mediated hypotension at the EHC-Dallas and Buffalo is abnormal. See Figure 2.14.

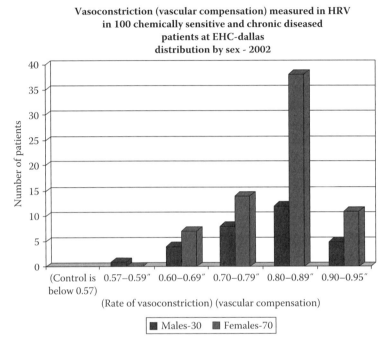

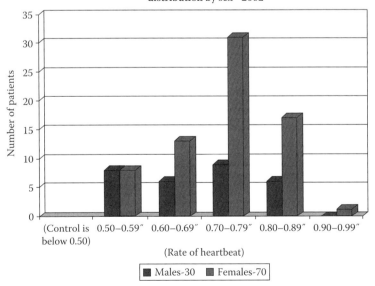

FIGURE 2.14 (See color insert following page 364.) Patients vascular compensation and chronotrophic responses are lost with severe autonomic stimulation as measured elevation from the supine to the upright position as measured by the heart rate variability machine by the method of Riftine.[16] (EHC-Dallas, 2002.)

The normal response to both upright tilt testing and exercise ought to involve prompt vasoconstriction, but in response to these stresses, several groups of patients have been identified with impaired vasoconstrictor responses in the forearm and splanchnic beds, and in microvascular flow to the skin.[92–94]

As shown in Figure 2.13, when measuring heart rate variability, there is a broad spectrum of microvascular flow with varied degrees of vasoconstriction, as a result of standing upright after lying flat. These studies suggest that the patient with NMH or tachycardia would have severe autonomic dysfunction. The etiology would be environmental triggers that cause nerve supersensitivity.

It is not known why the normal response to orthostatic stress fails to occur but in some partial denervation supersensitivity appears to be present. However our studies at the EHC-Dallas and Buffalo show ANS dysfunction, due to a myriad of triggering agents, including organophosphate pesticides, solvents, foods, and molds. These substances are known to affect the autonomic nerves and cause vascular deregulation, as measured by HRV and pupillography. The substances will either block or decrease output of norepinephrine, or cause a total sympathetic stimulus block. A case report of a 52-year-old white female exemplifies the problem when a known triggering agent is involved.

Case study: This woman presented with the chief complaint of sudden drop attacks. Over a 5-year period, she noted progressive recurrence of drop attacks, which finally interfered with her highly skilled job in the county court house. The attacks were occurring several times a day, at the time when she was hospitalized in the ECU. Her vital signs were stable until she had a drop attack. Then her blood pressure dropped to 40/0 until she regained consciousness. When her blood pressure returned to a normal value of 110/80, an autonomic nerve test via pupillography revealed an abnormal autonomic nerve pattern. Intradermal provocation skin test for silicon reproduced the drop attack. Blood toxics revealed hexane 0.5 ppb, plus xylene 2ppb. T_8 lymphocytes suppressor cells were low, at 160 mm^3 (C-300-400). She was secondarily sensitive to foods and molds, but none of them produced the drop attacks. The patient was taken to surgery, where the senior author removed her silicone breast implants. The patient had an uneventful postoperative course. She has been 5 years without drop attacks and works daily without medication. Her pupillography test returned to normal.

In the patient with NMH, the decreased venous return leads to an exaggerated sympathetic output, which in turn causes vigorous ventricular contractions. In the setting of an underfilled ventricular chamber, this increased inotropy triggers firing of mechanically sensitive C-fibers, which are normally activated under conditions of severe hypertension. The CNS response, appropriate for hypertension, involves withdrawal of sympathetic tone and relatively unopposed vagal tone, resulting in bradycardia. Vasodilation may occur due to loss in sympathetic tone. If the individual remains upright, the ultimate event can be either a profound drop in blood pressure and/or syncope, as shown in the previous case report.[95–97] Among the neuroendocrine changes that accompany the orthostatic intolerance are an increase in norepinephrine, vasopressin, β endorphin, and vasoactive intestinal polypeptide (reviewed by Benditt et al.).[95]

Several aspects of the hypothesis of mechanoreceptor-mediated bradycardia and hypotension have been questioned, and it must be acknowledged that this model may not adequately account for the development of NMH after cardiac transplantation where all nerves going to the heart are severed. However, the same changes may occur in these patients through the connective tissue matrix.[98]

Factors that can contribute to early activation of the reflex, according to Fortney et. al., include prolonged sitting or standing, mental arithmetic stress, emotional stress, vasovagal (including gory scenes and the sight of blood), a warm environment, and sodium depletion. We found these patients spend more time in bed because many had chronic fatigue syndrome. Studies earlier in this century[99] confirmed that both plasma volume and red blood cell mass are decreased with enforced, prolonged bed rest, which would be compatible with our present findings. The degree to which this sort of deconditioning occurs in chronic fatigue syndrome (CFS) has not been studied thoroughly, but any drop in plasma and red blood cell volume would further increase susceptibility to NMH: Intolerance of upright posture during an illness begets further orthostatic intolerance. Most people with chronic fatigue syndrome spend most of their time in a lying down position, which would aggravate the problems of NMH. Various medications (including phenothiazines, diuretics, vasodilating agents, tricyclic antidepressants—some of which have been used in the treatment of chronic fatigue syndrome symptoms and conditions associated with increased histamine release can also cause a decreased return of blood to the heart. Factors such as emotional stress, pain, exercise, and drug treatments (notable the β-agonist drugs, such as albuterol, used for asthma) can contribute to early activation of NMH through raising catecholamine concentrations. **We have seen many patients who are hypersensitive to pollens, dust, mold, foods, and chemicals trigger the NMH presumably through the release of histamine, but also, other possible vascular active mediators like VIPs**. We have been able to trigger the NMH response in some patients by using intradermal injections of VIPs. The other possibility of a mechanism of NMH is the hypersensitivity that results from denervation or partial denervation (Cannon's law ref.), which can occur from overexposure to phenolic compounds like xylene, toluene, solvents, etc. This phenomenon can result in a sympatholytic effect with resultant hypotension. Although cognitive stresses should elicit peripheral vasoconstriction, over half of subjects with neurally mediated syncope in a study by Manyari and colleagues,[100] had inappropriate lack of vasoconstriction when asked to perform mental arithmetic tasks. This fact would suggest a very hypersensitive state. Such inappropriate venous responses could provoke worse orthostatic tolerance in response to common, everyday cognitive stresses, and provide an attractive explanation for why some patients with NMH describe worse fatigue after reading or concentrating.[101] This mental gymnastic and fatigue phenomenon has been observed in the chemically sensitive for at least 30 years.

However, we want to emphasize that these types of abnormal neurovascular responses are similar to denervation hypersensitivity, which can be triggered by any noxious substance. Thus, **once the neurovascular damage has occurred, as shown throughout this book, the tissue would already be primed to act as a secondary focus of reactivity for cognitive or any other environmental stress**.

NEUROMUSCULAR HYPOTENSION AND CHRONIC FATIGUE

A low resting blood pressure in the absence of orthostatic symptoms does not represent disease.[102] However, large epidemiologic **studies have confirmed a consistent relationship between low blood pressure and fatigue**.[103,104] We have observed hundreds of patients with autonomic dysfunction who tended to have low blood pressure (80/50, 70/40, etc.) and markedly less physical and mental stamina, whether or not they had fatigue. In a study of 10,314 British civil servants, the lowest quartile of systolic blood pressure was associated with the highest prevalence of complaints of dizziness, tiredness, and somatic symptoms. The odds ratio for unexplained tiredness was 1.2 times higher for men and 1.33 times higher for women in the lowest quartile of systolic blood pressure compared with the highest. The mechanism of the association between low blood pressure and fatigue has not been well investigated in studies. Possibly, there is a partial denervation (whether mechanical or functional) hypersensitivity that affects the blood supply to various areas in the brain and peripheral blood vessels, which results in these complaints and we suggest that in these complaints the fatigue results in a central dyshomeostasis.

Chronic fatigue has long been described in those patients with orthostatic hypotension (or primary autonomic failure) in whom hypotension occurs rapidly within the first 5 minutes of standing or upright tilt testing.[105,106] In a prospective study conducted by Low and colleagues,[106] chronic fatigue was reported in 72% of patients with orthostatic hypotension (defined strictly as a ≥ 30 mm Hg drop in systolic or a 15 mm Hg drop in diastolic pressure in response to 5 minutes of upright tilt to 80°). Other frequent symptoms reported in this group included lightheadedness in 88%, cognitive difficulties (problems thinking and concentrating) in 47%, blurred vision in 47%, tremulousness in 38%, pallor in 41%, and anxiety in 29%. Such neurocognitive symptoms have been attributed to cerebral hypoperfusion, especially due to vascular spasm or endothelial cell wall swelling of the microcirculation, which causes shunting of oxygenated blood away from some areas of the end organ. In some epidemiologic studies, however, there is a relatively low prevalence of chronic fatigue in the setting of a substantial degree of hypotension. These facts emphasize that the relation between low blood pressure and fatigue is neither simple nor direct. Patients seen at the EHC-Dallas and Buffalo with similar clinical symptoms have been shown to have abnormal heart rate variability tests, triple-headed camera brain SPECT scans, and computerized balance tests. **They show hypersensitivity to food, mold, or chemicals upon challenges, again suggesting denervation hypersensitivity**.

In those with more delayed orthostatic hypotension[105] or with orthostatic tachycardia,[107,106] chronic fatigue has begun to receive more attention as a common presenting feature. In 1992, Streeten and Anderson[105] described 7 patients with delayed orthostatic hypotension, 6 of whom had chronic fatigue. Plasma volume and red blood cell mass were subnormal in the 3 patients in whom these assessments were made. After treatment with flurocortisone or octreotide, 3 of 4 patients who had substantial reduction or complete correction of the delayed orthostatic hypotension also experienced a substantial improvement in their chronic fatigue. Whether these subjects satisfied criteria for CFS was not recorded. These investigators speculated at the time, that the fatigue and exhaustion associated with other conditions, including the CFS, might result in some individuals from a failure to maintain blood pressure in the upright position.

In their series of 16 patients with postural othostatic tachycardia syndrome (POTS), 3 of whom also had hypotension during a brief tilt test, Schondorf and Low[107] identified 13 with chronic fatigue and 3 with postprandial bloating and delayed gastric emptying. All 16 patients experienced an acute onset of symptoms, and in 7 patients, the fatigue and lightheadedness appeared after an apparent viral infection. Symptomatic improvement in subjects with POTS has been noted with sodium loading, fludrocortisone, and other medications.[108–110] The fact that fatigue can be associated with NMH has been appreciated at least since 1932, when Sir Thomas Lewis described a young soldier in whom syncope and relative bradycardia developed during a venipuncture.[111] The soldier had a long history of exhaustion soon after starting to exercise, and the blood sample was being drawn to evaluate an episode of loss of consciousness that occurred while he had been on guard duty. After the venipuncture, he became hypotensive and unresponsive for several minutes, during which time his heart rate was 50–60 beats/min and his systolic blood pressure was 50 mm/Hg. He remained tremulous and tired for the next 36 hours, demonstrating that a long period of acute fatigue could follow a single episode of vasovagal or NMH.

Several comprehensive reviews of NMH are available.[95–97]

NMH has come to be regarded as the most common cause of recurrent syncope. It is seen with a somewhat greater frequency in women, with greater prevalence in those younger than 50, and evidence from small studies suggests a higher risk of NMH in those with a lower resting blood pressure.[110–112] The clinician must differentiate symptoms of NMH syncope from those of ventricular tachycardia, atrioventricular block, and neurocardiogenic syncope (i.e., associated with congestive heart failure). NMH can occur transiently. It is common after an acute infection,[111] and with some cases of trauma, but also can occur for years afterwards, as seen in some post polio patients, as well as other patients with anterior horn disease.[115] The routine physical examination and laboratory tests of NMH usually are normal, and hypotension is not detected in most instances unless the orthostatic stress is prolonged and individuals are prevented from using postural countermeasures. However, the heart rate variability tests and pupillography test for autonomic dysfunction are abnormal. There is an overlap between NMH and panic attacks probably due to excess lactic acid or adrenalin output, as well as magnesium deficiency.

According to Rowe and Calkins,[90] one of the patients who prompted their investigations into whether chronic fatigue could follow chronic early activation of this reflex pathway was a 16-year-old girl who had developed the insidious onset of fatigue 18 months earlier. She described becoming tired, shaky, lightheaded, and pale after walking more than 10 minutes, getting more tired if she stood quietly or sat upright, and having to lie down on her bed for 30 minutes after a shower. She said her head felt foggy and thick all the time, and her memory and reading comprehension had deteriorated. One notable physical finding was that her legs and arms developed a purple discoloration after a short period of quiet standing, a feature reported several decades ago in those with epidemic neuromyasthenia[116] and now commonly seen in environmentally triggered vasculitis and indicative of abnormal venous pooling. This mottling of the extremities is often seen with ANS dysfunction. A tilt table test performed to evaluate her recurrent presyncope was consistent with NMH. She had a normal blood pressure and heart rate at baseline but got lightheaded at 10 minutes, and at 20 minutes, had severe lightheadedness, pallor, and general discomfort. Her blood pressure decreased to 65/40, with an inappropriate but characteristic slowing of her heart rate.

Similar abnormalities were present in the next 6 adolescents who were evaluated for chronic fatigue, none of whom had experienced syncope.[117] Four of them met the 1988 criteria for chronic fatigue syndrome, and 4 out of the 7 total studied had a substantial improvement in symptoms when treated with increased intake of fluid, including sodium and medications directed against NMH. These improvements suggested that the symptoms of chronic fatigue syndrome could be approached, not from the dominant paradigm of viral infection and immune activation, but rather from the standpoint of autonomic dysfunction. The findings also prompted Rowe and Calkins[117] to examine in a larger group of patients whether NMH is an unrecognized cause of symptoms in chronic fatigue syndrome. It is clear from our studies at the EHC-Dallas and Buffalo that both immune and autonomic dysfunction are present in the patients with chronic fatigue and neurovascular dysfunction. We have demonstrated a low number of T-cells (especially T_8 suppressor cells) and low T-cell function in these patients.

To estimate the proportion of chronic fatigue syndrome patients with abnormal tilt tests and the proportion with a response to flurocortisone, Rowe and Calkins recruited 23 individuals who had been diagnosed with chronic fatigue syndrome. Features that exacerbated their fatigue included physical exertion, a hot shower, prolonged standing (such as waiting in line at the grocery store), and warm environments.[118] Participants also reported being more tired after a lightheaded episode. An abnormal drop in blood pressure in response to upright tilt testing was observed in 22 of 23 patients with chronic fatigue syndrome (96%), versus 4 of 14 healthy controls (29%). During the first 45 minutes of upright tilt to 70°, 16 patients (or 70%) with chronic fatigue syndrome developed hypotension, while all controls maintained a normal blood pressure. Perhaps more importantly, all 23 with CFS, but none of the controls, developed orthostatic symptoms during this first stage of tilt testing, suggesting that orthostatic intolerance may be a defining feature of the illness.

With open treatment of the NMH (using sodium loading and fludrocortisone), 9 of 19 patients (47%) reported substantial improvement in symptoms, which was defined carefully as a score of ≥ 7 on a 10-point wellness scale, along with similar degrees of improvement in activity and cognitive function. Seven of these patients also reported being at least somewhat better. To determine whether the subjective report of improvement in symptoms was associated with objective improvement in tolerance of upright tilt, 6 of the patients with an almost complete resolution of symptoms on therapy agreed to undergo repeat tilt testing. Five of the 6 had normal tilt test responses while on therapy, and the sixth had a marked improvement. Three others with mild improvements in symptoms continued to have abnormal tilt tests.[119]

One of the possible interpretations of these abnormal responses to tilt testing is that they represented a consequence of physical deconditioning, which would be assumed to be present in most, but not all, of those patients with chronic fatigue syndrome. Although this idea is a complicated question to study, the improvements in symptoms occurred within a few days in some patients, and in all patients, the improvement in orthostatic tolerance preceded return of their ability to exercise. Among those who have experienced a substantial improvement in symptoms, and in whom a resumption of more normal activity and exercise has been possible, attempts to stop medical therapy have been followed, within days, by a resumption of profound fatigue. **The results of drug**

treatment in this study are not very encouraging, since more than 50% did not respond. When dealing with this type of autonomic dysfunction, our studies have shown that one must define, eliminate, or neutralize as many specific triggering agents including molds, foods, and chemicals in order to stop the homeostatic dysfunction. **Furthermore, the long-term use of medication is fraught with many complications and should be used only as a last resort after failure of the environmental treatment.** See Table 2.7.

Bou-Holaigah et al.[120] state in their fibromyalgia study of 1996, that NMH is now recognized as the most common cause of recurrent syncope in those with structurally normal hearts[120-122] The disorder is more common among women[123,124] In those susceptible to NMH, symptoms of lightheadedness, warmth, diaphoresis and syncope can be precipitated by prolonged periods of upright posture, exercise, emotional stress, and warm environments[125-128] Recent studies have identified a strong association between syndromes of orthostatic intolerance, including NMH, and both chronic fatigue and chronic fatigue syndrome.[129,130,118-120] Chronic fatigue is almost universal among those with fibromyalgia,[131] and up to 70% of those with fibromyalgia satisfy formal criteria for the diagnosis of chronic fatigue syndrome.[132] This clinical overlap led Buchwald et al.[132] to hypothesize that those with fibromyalgia also have a high prevalence of NMH. In our experience they do not. **However we have found that fibromyalgia patients definitely have autonomic nervous dysfunction.** In this study by Bachwald, the investigators evaluated the clinical history and response to upright tilt testing of inpatients with fibromyalgia and in healthy controls and found that the fibromyalgia patients had abnormal tilt table responses.

TABLE 2.7
Blood Pressure and Heart Rate Responses to Upright Tilt Table Testing

	Fibromyalgia		Healthy Controls		p*
Baseline					0.62
Systolic BP	121	(19)	118	(15)	0.31
Diastolic BP	76	(11)	72	(9)	0.07
Heart rate	75	(13)	68	(10)	
At termination of tilt test					
Systolic BP	69	(35)	111	(33)	< 0.001
Diastolic BP	43	(17)	64	(17)	0.001
Heart rate	74	(33)	113	(30)	0.001

Results of Tilt Testing in Patients with Fibromyalgia and in Controls

	Abnormal** Stage of Tilt			Normal
Group	1	2	3	
Fibromyalgia	12	3	4	1
Controls	0	2	6	12

Source: From Bou-Holaigah, I., Calkins, H., Flynn, J.A., et al., *Clin Exp Rheum* 15, 239–240, 1997. With permission.

*All comparisons are unpaired T-tests. Values are expressed as mean (SD) Blood Pressure is in mm Hg.

**An abnormal response to upright tilt required syncope or pre-syncope in conjunction with a drop in systolic blood pressure of at least 25 mm/Hg, and no associated increase in the heart rate.

p < 0.001, fibromyalgia patients versus controls.

Bou-Holaigah et al.[119] also examined the prevalence of abnormal responses to upright tilt table testing in 20 patients with fibromyalgia and 20 healthy controls. Each subject completed a symptom questionnaire and underwent a three-stage upright tilt table test (stage 1: 45 minutes at 70° tilt; stage 2: 15 minutes at 70° tilt with isoproterenol 1–2 μg/min; stage 3: 10 minutes at 70° tilt with isoproterenol 3–4 μg/min). An abnormal response to upright tilt was defined by syncope or presyncope in association with a drop in systolic blood pressure of at least 25 mm/Hg and no associated increase in heart rate. During stage 1 of upright tilt, 12 of 20 fibromyalgia patients (60%), but no controls had an abnormal drop in blood pressure ($p < 0.001$). **Among those with fibromyalgia, all 18 who tolerated upright tilt for more than 10 minutes reported worsening or provocation of their typical widespread fibromyalgia pain during stage 1.** In contrast, controls were asymptomatic ($p < 0.001$). These results identify a strong association between fibromyalgia and NMH. Further studies will be needed to determine whether the autonomic response to upright stress plays a primary role in the pathophysiology of pain and other symptoms in fibromyalgia.[120] As previously shown in this chapter, Gunn[57] thinks it does.

Reflex Sympathetic Dystrophy (Causalgia)

Reflex sympathetic dystrophy (RSD) syndrome has been recognized since the Civil War. **It is an internal burning extremity pain (Causlgia) that occurs after an injury.** Some authors (Sautzmann) think that the common mechanism may be due to injury to the central and peripheral nervous tissue. Others[133–138] propose that sympathetic pain results from tonic activity in the myelinated mechanoreceptor efferents. Input causes tonic firing in the neurons that are part of the nocioreceptive pathway. Other investigators[134] propose a hypothesis that places the primary abnormality in the PNS. Usually, reflex sympathetic dystrophy occurs secondary to fractures, sprains, and trivial soft tissue injury.

Reflex sympathetic dystrophy occurs in 1–15% of the peripheral nerve injuries. Reflex sympathetic dystrophy has also been seen after head injury, stroke, polio, amyotrophic lateral sclerosis (ALS), myocardial infarction, polymyalgia rheumatism, operative procedures (e.g., carpal tunnel release), brachial plexus injury or pressure, cast/splint immobilization, and prolonged bed rest. Age and sex have no preference. Cardinal signs include pain, edema, stiffness, and discoloration. Pain occurs that is intense and burning, affecting the entire region. Pain will persist after stimulation (hyperpathia). Pain also occurs with light touch, for example, from a cotton swab or air blowing over the skin. Movement aggravates the pain, as does cold and low pressure weather fronts. Airplane ascent and descent make the pain worse. Edema (neurotrophic) is one of the earlier findings, while stiffness may occur anytime. Discoloration may vary from intense erythema to cyanotic, pale, purple, or gray coloring.

Lankford's[139] findings of secondary characteristics include: demineralization and osteoporosis among the most classic findings; pseudomotor changes varying from hyperhidrosis to dryness; marked temperature difference between affected and unaffected extremities; vasomotor instability most commonly manifested as decreased capillary refill; erythema may be a sign of increased capillary refill and should be compared with refill in an unaffected extremity; skin may develop a glossy, shiny appearance and in the late stages, trophic changes may involve a decrease in subcutaneous tissue. In reflex sympathetic dystrophy of the hand, nodules and thickening of the palmar fascia may develop.

Physically, three stages of reflex sympathetic dystrophy have been classified; however, the consensus panel recommended that staging be eliminated. **Stage 1** or early reflex sympathetic dystrophy is classified by pain that is more severe than would be expected from the injury, and it has a burning or aching quality. It may be increased by dependency of the limb, physical contact, or emotional upset. The affected area becomes edematous, may be hyperthermic or hypothermic, and shows increased nail and hair growth. Radiographs may show early bony changes. Duration is usually three months from onset of symptoms. Some patients remain in one stage or another for many months or even years. They may never progress or they may progress quickly to a later stage. It is important to remember that physical findings may be minimal, especially in those who remain in Stage 1 or progress slowly. **Stage II**, of established reflex sympathetic dystrophy, is defined by induration of edematous tissue. The skin becomes cool and hyperhidrotic with livedo reticularis or cyanosis present. Hair may be lost, and nails become ridged, cracked, and brittle. Hand dryness becomes prominent, and atrophy of skin and subcutaneous tissues becomes noticeable. Pain remains the dominant feature. It usually is constant and is increased by any stimulus to the affected area. Stiffness develops at this stage. Radiographs may show diffuse osteoporosis. The three-phase bone scan is usually positive with changes of osteoporosis. Duration is 3–12 months from onset. **Stage III**, or late reflex sympathetic dystrophy, is classified by the spreading of pain proximally. Although pain may diminish in intensity, it remains a prominent feature. Flare-ups may occur spontaneously. Irreversible tissue damage occurs. Skin is thin and shiny. Edema is absent. Contractures may occur. X-ray films indicate marked demineralization.[140–149] Reflex sympathetic dystrophy is considered by Gunn[63] and the authors of this book, to be a partial nerve injury, and thus it can be treated by the integrated muscle acupuncture therapy of Gunn, as well as by the avoidance of pollutants, replacement of nutrition, medically supervised sauna, massage, physical therapy, and oxygen therapy. Specific injection therapy for secondary food, mold, and chemical sensitivity will also be needed. Table 2.8 shows a small series of reflex sympathetic dystrophy patients at the EHC-Dallas who had reflex sympathetic dystrophy, mold, food, chemical triggering, and immune deregulation.

POLLUTANT INJURY TO THE EYE

Many chemically sensitive and chronic degenerative diseased patients have difficulty seeing. These patients experience vision problems such as blurring, tunnel vision, and light flashes. Floaters also bother many sensitive patients when they are in areas with high pollutant levels, such as underground parking garages, conference halls, public buildings, etc. Some patients have extreme light sensitivity. In addition, some chemically sensitive individuals suffer from direct pollutant damage that results in named eye diseases such as glaucoma, iritis, optic neuritis, cataracts, and others. Attention to the powerful effects of pollutant exposure on chemically sensitive individuals not only will aid treatment of the various fixed-named eye diseases that may result in select cases but also will often altogether eliminate their development.

Pollutant Injury to the Nervous System of the Eye

The eye has two significant types of innervation. The first is via the cranial nerves, and the second is through the autonomic nerves. Both systems may be significantly affected by pollutant exposure with subsequent injury resulting. However, often only autonomic block or injury occurs. Evidence obtained by pupillography shows the overwhelming number of responses in chronic degenerative diseased patients and chemical sensitivity patients are related to the parasympathetic nervous system, which is regional to this area through cranial nerves 3 and 7. These findings are in contrast to measurable autonomic dysfunction by the heart rate variability machine, which shows sympathetic dominance. This heart rate variability machine however, measures more vagal nerve and splanchnic nerve impulses rather than eye changes.

Chemically sensitive individuals with neurotoxic patterns displayed on the triple camera SPECT brain scan have a tendency not to sweat, in which the cholinergic part of the sympathetic response has been overridden by excess sympathetic stimulation of the eyes but the parasympathetic cholinergic response may still be manifested by pinpoint pupils and blurred vision.

Cranial Nerves

Several cranial nerves may be involved in pollutant injury to the eye. The most important, of course, is the optic nerve, since it is the nerve of vision. All these will be discussed in the following pages.

TABLE 2.8
Reflex Sympathetic Dystrophy Study in Six Patients (RSD), EHC–Dallas

No.	Name	Sex	Age	HRV Autonomic Nerves	Physical Fitness	Pupill-ography	Lab	Intradermal Test
1	JR	F	47		10.5 N-5.0		T_4 % increase, T_8 % decrease, T_8 decrease, $T_4 T_8$ increase, E. Chaffeensis IgG (−) IgM (−), Lyme disease IgG (−), IgM (−), EB virus IgG increase, EBNA IgG increase, estrone decrease, cholesterol increase LDL increase, PVO_2 49.1 mm/Hg, Lipase normal, antinuclear antibody (ANA) (−). Stool culture: *Klebsiella sp* (+), *Pseudomonas aeruginosa*	six chemicals 1. Formaldehyde 2. Phenol 3. Pentachlorinated- phenol 4. O.P. pesticide 5. Chlorine 6. Xylene
2	LG	F	17	Decrease in Parasympathetic system Increase in Sympathetic system	p.t. 11.5 N-5.0	Non specific change	*M.fermentans* (−), blood glucose increase. Sputum culture: Alpha streptococci (+), Diptherodides (+)	mold(+) mycotoxins, tree(+), grass(+), weed(+), terpene(+), T.O.E.(+), Candida (+). chemical(+), metal(+), food, peptides(+)
3	MR	F	18	Decrease in Parasympathetic system Increase in Sympathetic system	12.5	Non specific change	T_4 increase, lyme disease IgG (−), cholesterol increase, triglycerides increase, CO_2 decrease, Iron decrease, LD decrease, C-peptide increase IgG decrease, IGE increase, thyroid normal Nasal culture: *Staphylococcus epidermidis* (+), Diptherodides (+) Thermography: sinusitis, yeast fermentation, toxic liver, bladder, pancreas	mite(+), chemical(+), food(+)
4	JM	F	56	Clinical ANS dysfunction	—	—	T & B lymphocyte normal, PVO_2 25.7 mm/Hg, ANA (Hep-2) (−). Stool culture: *Caudia sp* (+), *Pseudomonas* (+)	chemical (−) food (−)

(Continued)

TABLE 2.8
Reflex Sympathetic Dystrophy Study in six Patients (RSD), EHC–Dallas (Continued)

No.	Name	Sex	Age	Autonomic Nerves (HRV)	Physical Fitness	Pupill-ography	Lab	Intradermal Test
5	PR	F	52	Clinical Autonomic Nervous System dysfunction	—	—	WBC decrease, lyme decrease, T_{11} decrease, T_4% increase, T_8 decrease, T_8% decrease, T_4T_8 increase, increase, ANA (+), venous duplex imaging (−), PVo_2 30, copper increase, mercury increase.	chemical(−)food (−) food (+), metal (+) candida (+)
6	VP	F	58	Decrease in Parasympathetic system Increase in Sympathetic system	10.6	Non specific change	T_{11} decrease, T_4 decrease, T_8 decrease, potassium decrease, glucose increase, manganese decrease, lithium decrease, Epstein Barr (EB) virus VCA increase, IBV nuclear antigen IgG increase, PVo_2 46.7 mm/Hg, ESR increase, α-ketoglutarate increase, fumarate increase, suberate increase, xuntharenate increase, 8-hydroxy-2deoxyguanosine increase, benzoate increase, 2-methyl bippurate increase. ANA (+). EKG: borderline abnormal, CMI 2/7 (+). **Thermography:** neurotoxic vasculitis, sinusitis, lymph rigid, thorax chaotic, enteropathy in liver, rigid over bladder/uterus, ovaries and right kidney—heavy metal likely. **Mastopathy bilateral**—rigid left breast at "C"	chemical (+) cotton (+)

Source: Based on EHC–Dallas.

Second Cranial Nerve

The optic nerve is, of course, the key nerve in pollutant injury of the eye with damage ranging from diminished vision to total blindness, as seen in some chemically sensitive patients. This damage can occur at any place from the optic disc along the nerve to the visual center in the brain. The optic nerve reacts to pollutants by degenerating during the active stage and with atrophy as the end result. The primary events are loss of a nerve and their myelin sheaths whereas the secondary events are reactive infiltration and gliosus. The histopathologic manifestations vary according to the acuteness or chronicity of the process.

Acute process of whatever cause shows replacement of myelin by masses of foamy and free-floating macrophages. These glitter cells are loaded with neutral fat and cholesterol. Such acute reactions occur with infarction, injuries, and poisons of the optic nerve.

Less intense reactions incite a similar but less intense reaction with replacement of the myelin by glia. As masses of macrophages disappear, locunae may be left in the optic nerve. The locunae may appear anywhere in the optic nerve and are nonspecific. Those in front of and behind the lamina cribosa have been most extensively documented classically associated with glaucomatous optic atrophy. They are collectively known as Schnaba's cavernous atrophy. The spaces are not empty. They are often filled with hyaluronic acid and are hyaluronidase sensitive, leading some authorities to interpret them as originating in the vitreous.

Chronic lesions of the optic nerve, such as those induced by progressive compression of the optic nerve or gradual disappearance of the retinal ganglion cells, show little or no lipid phagocytosis. Instead, there is simply collapse of the optic nerve framework and disorderly arrangements of glial cells. Electron microscopy shows variable degeneration of the myelinated nerve fibers.

Comparison of degenerative changes in the optic nerve and retina offers a unique opportunity to contrast white matter versus gray matter reactivity. Nowhere else in the nervous system are the two so neatly separate. In the case of infarcts, for instance, the brisk macrophage reaction in the optic nerve and the brain contrasts with that found in the retina, which is practically devoid of reactivity. **The contrast of myelin in the nerve and its absence in the retina is responsible for the difference**.

Papilledema

Papilledema connotes a swollen disc. This is classically associated with increased intracranial pressure. It leads to distention of the meningeal spaces about the nerve and, in the presence of a firm dura mater, a squeeze on the nerve itself. The back-up pressure results in leakage of serum into the nerve head and congestion of axoplasmic flow anterior to the lamina cibrosa. The swollen nerve head protrudes into the vitreous anteriorly and presses against the retina laterally, presenting a characteristic S-shaped configuration and, often, concentric folds, of the outer retinal layers. The angulated, marginal fibers are likely to become necrotic, thus accounting for the bizarre field defects that commonly accompany papilledema. Electron microscopy reveals and accumulation of mitochondria and axoplasmic particles in various stages of disintegration.

Drusen

Drusen of the optic nerve are hyaline, often calcareous, bodies of various sizes, situated in the prelaminar or nonmyelinated portion of the nerve head. Both eyes are usually affected. Except for an infrequent association with retinitis pigmentosa and a similarity to "mulberry" bodies with tuberous sclerosis, drusen are not known to be associated with other systemic or local conditions. Occasionally, they occur as a dominantly inherited trait. Their pathogenesis is unclear. They have no relationship to Drusen of Bruch's membrane, which unfortunately bear the same name.

Large Drusen of the nerve have a characteristic ophthalmoscopic appearance. They appear as translucent, globular bodies protruding from the disc and obscuring its margins. Small Drusen embedded in the nerve substance, on the other hand, may present only a subtle elevation of the disc simulating papilledema. Drusen are especially prevalent with small discs and characteristically obscure the physiologic cup. Unless sufficiently large or situated so as to encroach on the vessels, they cause no symptoms. Serious sequelae may result, however, from secondary venous obstruction with consequent papilloretinal hemorrhages, or they may cause optic atrophy through pressure on the adjacent nerve fibers. Fortunately neither occurrence is frequent, and most Drusen are discovered fortuitously in asymptomatic patients.

Histopathologically, Drusen of the nerve head are laminated, acellular bodies situated exclusively in the prelaminar or nonmyelinated portions of the nerve. Their basophilia varies with the degree of calcification. Just what tissue element gives rise to their formation is unclear, but once formed they continue to enlarge. The lamination suggests that growth occurs by slow accretion. Initially the nerve fibers are simply pushed aside rather than replaced. Rarely are they sufficiently large to occupy the entire nerve head or to cause blindness.

Cupping

Prolonged pressure within the eye (glaucoma) results in backward bowing of the lamina cribrosa. The vessels are pushed to the nasal side, and the nerve fibers appear to be impaled on the rigid scleral margin. Occasionally, retinal tissue is pulled into the cup or, alternatively, glial tissue fills the cup. Secondary venous occlusion is a frequent eventuality and presents a problem to the pathologist in deciding whether the glaucoma was the result or the cause of the vascular obstruction.

In contrast, to the sharply angulated and frequently deep glaucomatous cup, a shelving excavation occurs with optic atrophy due simply to loss of nerve fibers. In contrast to the glaucomatous cup, there is little or no displacement of vessels.

At times, cupping may occur in a normotensive eye. Called low-tension glaucoma, its pathogenesis is obscure. Unconfirmed suppositions postulate an inherently weak lamina cribrosa or insufficient vascular supply.

Vascular Optic Neuropathy

Prelaminar vascular optic neuropathy resulting from retinal artery occlusion causes an ischemic collapse of the capillaries in the nerve fiber layer of the nerve head. It is best visualized by fundus photography with contrast dye. The ophthalmoscopic counterpart is pallor of the disc. It is associated with either primary or secondary loss of the ganglion cells in the retina.

Paralaminar and retrolaminar vascular disease results from occlusion or impaired circulation of the posterior ciliary and adjacent choroidal vessels. The ocular manifestation is anterior ischemic optic neuropathy resulting, usually, from atherosclerosis or giant cell arteritis. Anatomically small discs and absence of physiologic cups may be predisposing factors for nonarteritic vascular occlusion.[152] The clinical counterpart of anterior ischemic optic neuropathy is swelling of the nerve head with complete or partial blindness.

In the case of arteritic neuropathy, the clinical diagnosis is usually confirmed by finding characteristic granulomatous changes in temporal artery biopsies, hence the name temporal arteritis. (See Vascular chapter in Volume 2 - Mechanisms of Chemical Sensitivity and Chronic Degenerative Disease.) The histopathologic abnormality consists of fragmentation of elastica, granulomatous infiltration of the vessel wall, and nodular necrosis of the muscularis. Fresh cases also show lipoidal myelinolysis of the optic nerve fibers, whereas late cases show simple collapse of the atrophic nerves and cavitation of the nerve (Schnabel's cavernous optic atrophy).[153]

Posterior ischemic optic neuropathy often masquerades, clinically under the erroneous diagnosis of "optic neuritis." Its histopathology has been rarely documented. Lacunar vacuolation may be one of the manifestations.

Hemorrhage into the meningeal sheaths surrounding the optic nerve may be included under the heading of vascular optic neuropathies. The predominant cause is subarachnoid hemorrhage from ruptures of intracranial aneurysms, wherein blood extends from the brain into the meninges around the optic nerve. It is accompanied by swelling and hemorrhage on the nerve head and surrounding retina. Yet it is by no means clear how the blood gets from the optic nerve sheaths into the eye. Rarely is blood found within the nerve itself.

Inflammation

Being in line with the posterior drainage path out of the eye, the optic nerve participates nonspecifically in any intraocular inflammation. On the other hand, several inflammatory diseases are especially likely to affect the optic nerves. These include herpes simplex,[154] Beçhet's disease,[155] sarcoid,[156,157] and Lyme disease. Indeed sarcoid of the optic nerve may be the initial manifestation of the systemic disease and then easily confused with a neoplasm.[158] Of the fungal infections, cyrptococosis is especially noteworthy because of its affinity for neural tissue, while inciting surprisingly little reactivity in the optic nerve.

Injuries

Injuries to the optic nerve may be of mechanical, chemical, thermal, or radioactive origin. Under the mechanical category are: (1) The orbital injuries inducing proptosis and avulsion of the optic nerve, and (2) skull injuries with fractures in the region of the optic foramen. These result from transaction of the nerve at the orbital apex. The avulsion of the optic nerve produces a chaotic ophthalmoscopic and pathologic picture of papillary exudates and hemorrhage. Fractures or contusions in the region of the optic foramen are relatively frequent cause of blindness and optic atrophy from injury.

Chemical injuries to the optic nerve are, in general, caused by the same agents that damage the white matter of the brain. These include lead, thallium, chloramphenicol, and ethylene glycol.[159] Most important because of their frequency are the optic neuropathies from the alcohols. Methyl alcohol notoriously causes acute blindness. Ethyl alcohol produces a more insidious visual loss (tobacco-alcohol amblyopia) but may be subacute in spree drinkers. In either case the nerves show, during the active stages, the lipid macrophages characteristic of myelinolysis and later simple optic atrophy. Of the thermal injuries, body burns may result in acute loss of vision after a variable latent period.

Radiation injuries to the optic nerve are real but have been infrequently documented pathologically. Most cases of radiation-induced optic atrophy result from chiasmal damage incidental to pituitary radiation. The visual loss comes on typically several months after the radiation, suggesting a vascular pathogenesis.

Third Cranial Nerve

The oculo-motor nerve innervates all eye muscles except those innervating the superior oblique muscle, the lateral external rectus muscle, the ciliary muscle, and the iris sphincter. Pollutant injury can occur to this nerve causing eye movement dysfunction.

Fourth Cranial Nerve

The trochlear nerve innervates the superior oblique muscle, which primarily rotates the eye outward and downward. Injury to this nerve will cause eye movement dysfunction.

Fifth Cranial Nerve

The trigeminal nerve (first and second branch) is responsible for sensation to the upper eyelid (first) and the lower lid (second) and for corneal sensation (second). Pollutant injury will derange this function.

Sphenopalatine Ganglion   The sphenopalatine ganglion (nasal ganglion, Meckel's ganglion, ptergopalatine ganglion) is the second division of the trigeminal nerve. It is a small **parasympathetic ganglion in the upper part of the sphenomaxillary fossa giving off four branches: the orbital, palatine, nasal, and pharyngeal nerves**. It is fluted reddish-gray ganglion measuring 6–8 mm. It is adjacent to the cranium outside the bowl near the sphenopalatine foramen. It is deep in the sphenopalatine fossa cranial and the mandibular nerve and close and later to (and below) the maxillary nerve (as it crosses the fossa).

The nerve fibers going into the sphenopalatine ganglion are a sensory nerve supply via the sphenopalatine nerve and the maxillary nerve (establishing a strong fine fiberlike connection between trigeminal fibers and parasympathetic fibers inside the ganglion). Parasympathteic connection is from the vagus nerve and three vagal ganglia in the brain stem. Parasympathetic nerves traveling piggy back on the N. petrosus major and fibers of the facial nerve originating at the ganglia geniculi (establishing a connection to the VII cranial nerve. The sympathetic connection is via the N. petrosus profundus establishing a connection with the superior cervical ganglion and the sympathetic nervous system.

Nerve fibers emerging from the sphenopalatine ganglion have connections to the parasympathetic fibers of the submandibular, lacrimal and parotid glands, the cilary ganglion, various functions of the eye, inner ear, olfactory nerve, lingual nerve (smell and taste), and the arterial blood supply to the brain.

The SPG helps with the secreto-motor function of the eustation tubes, larynx, pharynx, tonsils, mucous membranes of the nose and sinuses, soft and hard palate, lacriminal gland and sub mandibular and parotid glands. **A major role of the sphenopalatine ganglion is the regulation of brain blood vessels and the brain itself**. This regulation is due to the presence of vasoactive interstital peptides (VIP) and choline acetyltransferase for parasympathetic function and substance P (SP) for sensory nerve function. Both the internal carotid ganglion and intercarotid fibers can cause symptoms when activated such as edema, pain, dilatation, and low-grade encephalitis (e.g., cluster headaches, opthalmoplegic symptoms). **Treating the sphenopalatine ganglion can increase cerebral blood flow because of the postganglionic parasympathetic fibers innervating the vascular beds of the cerebral hemispheres**. Abnormal cranial parasympathetic out flow induces not only vasodilatation but also enhances protein extravasation and the release of proinflammatory cytokines. It also appears to mediate the intracranial hypersensitivity of chemically sensitive patients and is often responsible for sensitizing nocioceptive neurons in the spinal trigeminal nucleus.

Much varied symptomology has been attributed to dysfunction of the sphenopalatine ganglion. These include impaired cognition, involuntary and voluntary muscle spasm, head and neck facial pain, chronic fatigue, blindness and glaucoma, ophthalmic migraine, tinnitus, rhinitis, sinusitis, abdominal pain, diarrhea, shoulder, upper extremity, low back pain and sciatica, asthma, angina pectoris, hiccup and menstrual pain, hyper- and hypothyroidism, nasal congestion, sneezing, tearing, photophobia, hay fever, glossodynia and nausea, and inappropriate emotional states.[160] **The superficial location in the pharynx explains the extraordinary sensitivity of this ganglion to odor, chemical, and particles in the air.** Doty has described symptoms of an increased sensitivity/dysfunction of the parasympathetic sphenopalatine ganglion dysfunction caused by rose oil and MEE. Millquist has shown an increased cough with exposure to toxics.[161,162] The intricate connection to the vagus nerve and innervation of the saliva-producing gland predicts many of the digestive symptoms observed as the functional advance to dysfunction of the sphenopalatine ganglion.

We have observed seven patients who had involvement of the trigeminal nerve in addition to their peripheral neuropathy. These changes were recorded by a neurometer and pain in the area of the sphenopalatine ganglion and trigeminal nerve was a predominant part of the patient's symptomtology. All had ANS dysfunction when measured by heart rate variability and pupillography. It was determined that the trigeminal nerve and sphenopalatine ganglion were involved in a significant part of their neuropathic disease process.

Sixth Cranial Nerve

The abducens innervates the lateral external rectus muscle, which rotates the eyeball outward. Dysfunction resulting from pollutant damage to this area may occur.

Autonomic Nerves

Pollutant damage to the actions of the autonomic nerves of the eye appears to mirror much pollutant damage of the ANS throughout the body of the chemically sensitive and chronic degenerative diseased individual, except there appears to be more cholinergic responses through the eye. Once the ANS is triggered or inhibited by pollutants, changes in the eye responses will frequently follow, as evidenced by our studies with the iris corder, where 93% of systemic responses are reflected by changes in the ANS of the eye. These changes occur because the autonomic fibers both enter and leave the visual pathways near the chiasma (Figure 2.15) with the majority of fibers making connection in the supraoptic nuclei of the hypothalamus.

The supraoptic nucleus then makes elaborate connections and is in a position to relay visual and retinal impulses into the ANS. The nucleus then connects with the neuro hypophysis, the paraventricular nuclei, the mammillary bodies, and the nuclei. Efferent centrifugal fibers may also arise from the suproptic nucleus (*Chemical Sensitivity, Vol III* Chapter 26).[163] The retina hypothalamic pathways are involved with the photoneuric endocrine reflexes and circadian cycles and often malfunction in chemically sensitive individuals.

In general terms, the posterior part of the hypothalamus seems essential for pressure functions associated with sympathetic activity, where the anterior part of the hypothalamus seems to be devoted to parasympathetic activity. Pollution-induced stimulation of the posterior part of the hypothalamus induces vasoconstriction, heat production, increased metabolism, and pupillary dilatation.

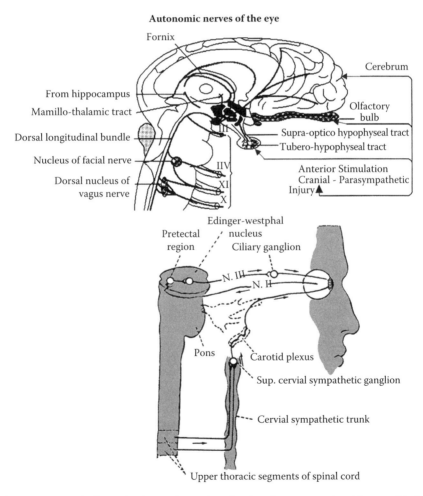

FIGURE 2.15 Autonomic fibers leaving and entering visual pathways near the chiasma with the majority of fibers making connection to the super optic nucleus of the hypothalamus. (Modified from Guyton, A.C., *Textbook of Medical Physiology*, 6th ed., WB Saunders, Philadelphia: 761, 1981.)

Of neuro-opthalmologic interest is the influence of the hypothalamus on the intrinsic muscles to the eye, the position of the lids, the secretion of tears, the vasomotor system, and the intraocular pressure. **Many chemically sensitive patients find a diminution or blurring of vision immediately upon entering a contaminated area such as a hotel or underground parking garage or polluted room, presumably due to triggering of these systems in the body as a result of pupil or accommodation spasm.**

Diminution of vision could be accounted for by an increase in intraocular pressure, which is independent of systemic blood pressure. In addition, a fall in the intraocular pressure associated with sympathetic activities, including local elevation of blood pressure, vasoconstriction, and pupillary dilatation, results from stimulation near the anterior column of the fornix. Changes such as these may be seen in chemically sensitive individuals upon pollutant exposure.

It has been suggested that a drop in intraocular pressure is related to decreased production of aqueous fluid, which is a consequence of decreased blood flow in the ciliary body that is probably due to local vasospasm. Stimulation of the cervical sympathetic nerve produces a similar effect and is more likely to be involved due to rapid decreases in pressure as seen in the treated chemically sensitive individual with glaucoma.

The autonomic elements in the brainstem are connected with the third, seventh, ninth, tenth, and eleventh cranial nerves, which are concerned with pupillary, lacrimal, salivary, cardiovascular, respiratory, alimentary, and other visceral activities. **These connections allow impulse communication with the eye and, thus, are a mirror for peripheral pollutant injury in many chemically sensitive and chronic degenerative diseased patients.**

The postganglionic pupillary fibers emerge from the first and second thoracic roots and the synapse in the supracervical ganglion. (See Figure 2.15 again.) Care has to be taken during cervical sympathectomy that the upper part of this ganglion is not injured, or a lid drop may occur. These fibers have a long pathway that is important diagnostically. For example, pancoast lung tumors impinging on the cervical plexus of the ANS can produce lid drop, and surgical removal of tumors can be hazardous, resulting in permanent lid drop. Often pollutant injury will cause a similar lid drop, which has been seen in some chemically sensitive patients treated at the EHC-Dallas. In the chemically sensitive individual, this drop suggests autonomic ganglion or nerve damage from the pollutant.

> **Case study.** A 53-year-old pediatrician gradually developed chemical sensitivity over a period of 10 years. Her main symptoms included bloating, constipation, chronic fatigue, nose bleeds and recurrent sinus, and bronchial infections. Physical exam showed a ptosis of the right eyelid and spontaneous bruises. It was noted that she would intermittently develop an eyelid lag on the right side. She was repeatedly worked up for superior sulcus lung tumor and sympathetic trunk damage. These work-ups were always shown to be negative.
>
> Laboratory tests showed ANS dysfunction. T_{11} and T_8 lymphocytes were low. Blood toxic profile showed hexane and 3-methylpentane. Inhaled challenge test showed organophosphate pesticide at ambient doses caused a repeated nosebleed. Toluene and xylene challenge caused the lid drop. Testing showed pollutant overload caused ANS dysfunction, which caused the intermittent lid lag.

In the eye, the peripheral ANS is divided into preganglionic and postganglionic parts separated from each other by synapses. The preganglionic parts are cholinergic. For example, their response could be initiated by acetylcholine, triggered by pollutant stimulation. The postganglionic pathways can be either cholinergic (parasympathetic) or adrenergic. For example, their response could be initiated by adrenalin (sympathetic)-triggered pollutant induction. The cholinergic response appears to be triggered more often than the sympathetic in the chemically sensitive and chronic degenerative diseased individual when measured by pupillography.

In the parasympathetic system of the eye, preganglionic fibers run with cranial nerves III, VII, IX, and X. The synapses take place in a peripherally situated ganglion for the ciliary muscle and sphincter of the pupil. The preganglionic fibers originate in the Edinger–Westphal nucleus in the

III nuclear region. They run with the third cranial nerve to the ciliary ganglion where they undergo synapsis. The postganglionic fibers run with the short ciliary nerves to the eyeball. (See Figure 2.15.) The parasympathetic fibers for the lacrimal gland originate in the nucleus of the IX cranial nerve and synapse in the sphenopalatine ganglion (Figure 2.16). The clinician will see two types of responses from the lacrimal gland. The first is repeated tearing, while the second is dry eyes. These responses will be due to stimulation or inhibition of the parasympathetic fibers. The synapses of the nerve fibers for the dilatation of the pupils take place in the superior cervical ganglion.

Certain chemicals and pollutants are known to stimulate the parasympathetic nervous system of the eye (Figures 2.17 and 2.18).

Many more chemicals can damage the parasympathetic nervous system. Usually, they are combined with other chemicals, and their toxicity depends upon the exact levels of the chemicals they are combined with. These contaminants include pilocarpine and mecholyl, eserine (or physostigmine), ubretid, and organophosphate insecticides. They cause constriction of the pupil, which is frequently seen in the chemically sensitive patient. In fact, over 40% of the chemically sensitive patients presenting at the EHC-Dallas have miosis. Known suppressors of the parasympathetic nervous system of the eye include atropine, tropicamide, and cyclogyl and botulinus toxin. Botulinus toxin will block motor nerves as well, since the effects of the toxin are not entirely parasympathetic. Exposure to botulinus toxin will allow the pupil to be dilated by suppression of the sphincter muscle. (See Figure 2.18 again.)

Certain substances are known to stimulate the sympathetic nervous system. These include epinephrine, phenylephrine, norepinephrine, tyramine, and cocaine. (See Figure 2.19.)

Other substances, including guanethidines, reserpine, d-methyl-dopa, and bethanidine, are known to suppress the sympathetic nervous system. (See Figure 2.19 again.) A large portion of chemically sensitive individuals (25%) have measureable sympatholytic effects when autonomic function is measured by pupillography.

Many toxic chemicals, including organochlorine insecticides as well as aliphatic and aromatic solvents, can cause an imbalance in the ANS. Depending upon dosage, timing, and combinations with other toxic chemicals, these substances may stimulate or suppress either part of the ANS. **It should be emphasized that the responses of the ANS may be physiological, pharmacological, or show toxic patterns**. The dynamics of homeostasis can then be maintained with the physician's awareness of these types of responses in the chemically sensitive and chronic degenerative diseased patient. This awareness allows for the early diagnosis and treatment of pollutant injury before it becomes permanent.

At the EHC-Dallas, we measured blood organochlorines and solvents in a series of chemically sensitive patients.[164] When studying them with the iris corder (pupillography) we found that the patients had predominantly cholinergic and sympatholytic changes in their ANS functions of the eye. Some patients had cholinolytic effects, and very few had sympathetic stimulation. (See Tables 2.9 and 2.10.)

NEUROGENIC VASCULAR RESPONSES TO POLLUTANT STIMULI

The nervous system response to noxious substances entering the body is immediate, with reflex-like rapidity once the hormetic effects are released the local homeostatic response is exceeded. The homeostatic response itself is immediate when it is local triggering the sensory peripheral nerve followed by the spinal sensory nerve, but it may be somewhat delayed, yet extremely rapid, when regional, distal, or a generalized homeostatic response occurs. This pollutant-triggered quick response is the immediate clinical response in the maladapted chemically sensitive and chronic degenerative diseased individual whose local homeostatic response is exceeded with overt and florid symptoms and signs. **This response is in contrast to the "adapted normal" individual whose homeostatic response is just as rapid, but imperceptible to cause and effect clinically**. The end blood vessel will be involved with local vasodilation

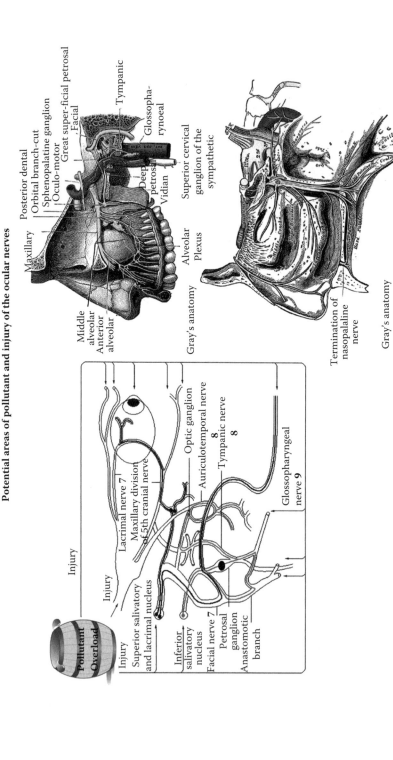

FIGURE 2.16 (See color insert following page 364.) Potential areas of pollutant injury of the ocular nerves and ocular autonomic nerves showing alveolar branches of superior maxillary nerve and spenopalatine ganglion of the trigeminal nerve. (Modified from Rea, W.J., *Chemical Sensitivity, Vol. III: Clinical Manifestations of Pollutant Overload*, CRC Press, Boca Raton, FL, 1889, 1996; Gray, H., *Anatomy of the Human Body*, Lea & Febiger, Philadelphia, 1918, New York, www.bartleby.com. 2000.)

Nervous System

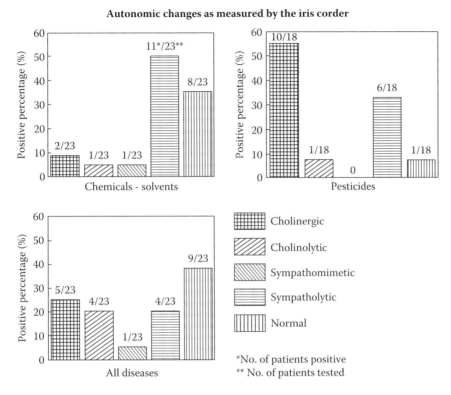

FIGURE 2.17 Autonomic changes as measured by the iris corder in patients with chlorinated pesticides and/or toxic volatile organic chemicals in their blood. (Shiragawa, S; EHC–Dallas, 1988; Rea, W.J., *Chemical Sensitivity, Vol. III,* 1890, Fig. 27.3, 1996.)

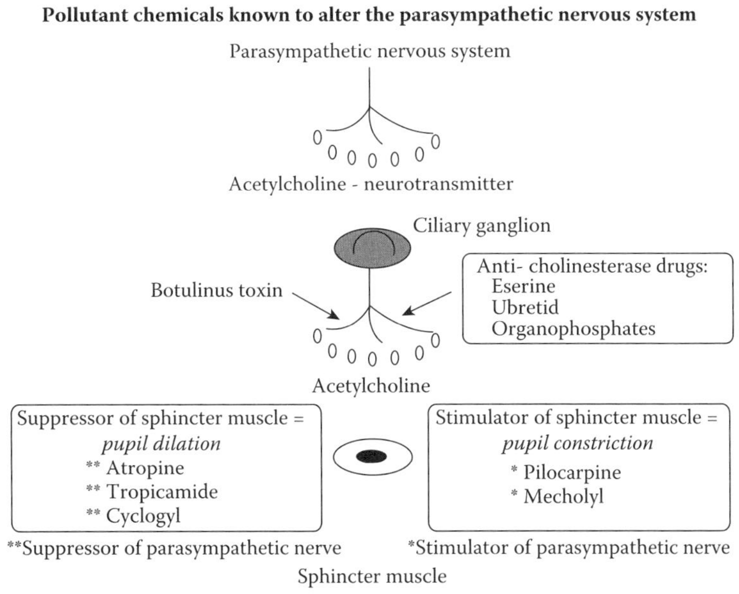

FIGURE 2.18 Pollutant chemicals that alter the parasympathetic nervous system. (Rea, W.J., *Chemical Sensitivity, Vol. III,* 1891, Fig. 27.4, 1996.)

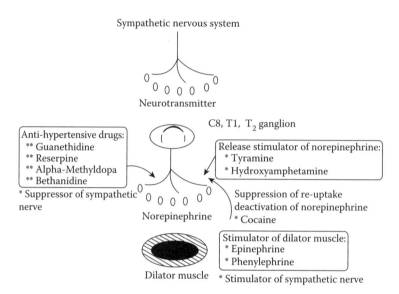

FIGURE 2.19 Pollutant chemicals known to alter the sympathetic nervous system. There are many more in various combinations and at certain levels giving a hormetic effect. (Rea, W.J., *Chemical Sensitivity, Vol. III*, 1891, Fig. 27.5, 1996.)

and a mild fluid leak occurring, thus causing neurogenic or trophic changes due to the peripheral autonomic nerves stimulation, as well as the end capillary circulation where edema occurs. This amplification response is apparently mediated through the PNS and the ANS (acting directly upon the end blood vessels), as the finding of autonomic dysfunctions evidenced by our studies using pupillography and heart rate variability tests.

When entering through the olfactory nerves, pollutants go directly to, or activate impulses or neurotransmitters, to the brain usually in the hypothalamus. In turn, the limbic system, including the amygdaloid and hypophysial pituitary axis, may be either activated (see Figure 2.2 again) or this system may be bypassed, with stimuli going directly to the anterior or posterior autonomic hypothalamic nuclei. **Triggering of the latter impulse pathways appears to be more common than triggering of the limbic system in the general or in noncerebrally involved chemically sensitive individuals. Triggering of the limbic system appears to be more frequent in the cerebral dysfunction of chemically sensitive and chronic degenerative diseased individuals.** Noxious stimuli occurring anywhere in the body often will affect the pupillary response, due to triggering via the ANS, as well as the microcirculation, which reflects dyshomeostasis of that area.

When the noxious stimulus occurs in the periphery, and once the local homeostatic response has been exceeded, there is a retrograde impulse to the dorsal nerve root ganglia (nodal points) through either the afferent fibers of the peripheral nerves (slow C or rapid delta A) or the gastrointestinal plexus.[165] The end microcirculation is then triggered by this reflex (Figure 2.20).

Any of 15 or more neurotransmitters can be released from the neuroendocrine cells of the gut causing varied homeostatic responses; however, the release of acetylcholine and or norepinephrine, which is more common, affects the microcirculation. (See Gastrointestinal chapter in Volume 2 - Mechanisms of Chemical Sensitivity and Chronic Degenerative Disease for more information.) In addition, there is often a local or regional response in contrast to a generalized one. Once this local and regional homeostatic control has been exceeded, and depending upon the severity or chronicity of a noxious incitant(s) and the state of the individual's nutrition, generalized responses will occur. For example, the sensory neurotransmitter SP (a neuropeptide) is released, causing immediate vasodilation and increased permeability of the microcirculation in the area of the nerve, which then leads to localized edema. This condition is often seen in the chemically sensitive or chronically

TABLE 2.9
Comparison of the Chemically Sensitive Groups and Healthy Volunteer Controls with Pupillography

Patient Types	A1*	CR*	T2*	VC*	T5*	VD*	Number	Mean Age
Chemical solvents	34.2 ± 7.6	0.47 ± 0.07	211.6 ± 30.1	43.4 ± 8.6	2438.7 ± 891.0[a]	10.6 ± 2.9	23	41.5
Chlorinated pesticide	28.6 ± 9.9[b]	0.45 ± 0.11	215.7 ± 53.3	37.8 ± 10.1[c]	2667.6 ± 1239.3[d]	8.4 ± 2.3[c]	20	45.4
Healthy volunteers control	36.9 ± 6.0	0.48 ± 0.07	205.4 ± 20.6	49.6 ± 8.5	1881.8 ± 555.5	12.1 ± 3.2	18	29.7

Source: From Shirakawa, S., Rea, W.J., Ishikawa, S., and Johnson, A.R., *Environ Med* 8(4), 121–127, 1992. With permission.

[a] $p < 0.03$ measured in the controlled environment (mean ± SD)
[b] $p < 0.006$ measured in the controlled environment (mean ± SD)
[c] $p < 0.001$ measured in the controlled environment (mean ± SD)
[d] $p < 0.02$ measured in the controlled environment (mean ± SD)

* Legend: **A1** = Initial pupil area (mm^2); **CR** = Constriction ratio A3/A1; **T2** = Time to half constriction (m/sec); **VC** = Maximum velocity of constriction (mm²/sec); **T5** = Time to recover to 63% of A3 after dilation from minimum state (m/sec); **VD** = Maximum velocity of dilation (mm^2/sec^2)

TABLE 2.10
Distribution of Autonomic Nerve Disturbance in Each Group as Measured by Pupillography

Autonomic Nerve Disturbance	Chemical Solvents	Chlorinated Pesticides	Total
Cholinergic	2 (13.3%)	7 (38.9%)	9 (27.3%)
Cholinolytic	1 (6.7%)	1 (5.6%)	2 (6.1%)
Sympathiomimetic	1 (6.7%)	0 (0%)	1 (3.0%)
Symphatholytic	11 (73.3%)	10 (55.5%)	21 (63.6%)
Totals	15	18	33

Source: From Shirakawa, S., Rea, W.J., Ishikawa, S., and Johnson, A.R., *Environ Med* 8(4), 121–127, 1992. With permission.

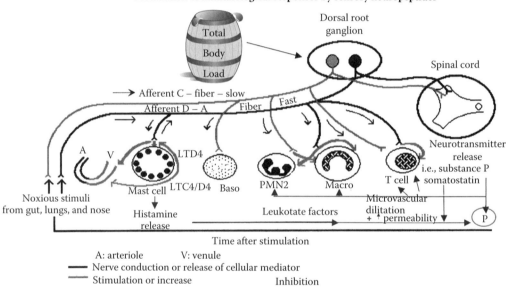

FIGURE 2.20 Modulation of immunological responses by sensory neuropeptides after pollutant stimuli, as seen in some chemically sensitive individuals. Retrograde impulses from the dorsal root ganglion. This is a complicated homeostatic feedback mechanism, which can involve neurotransmitter release. (Modified from Rea, W.J., *Chemical Sensitivity: Clinical manifestations of pollutant overload*, CRC Press, Boca Raton, FL, III, 1741, 1995.)

diseased patient. **The neurologically released SP then activates the non-IgE-mediated release of histamine via the mast cells, which then affects the microcirculation**. SP also stimulates leukotactic factors and leukotrienes, augmentating macrophage and neutrophil activity that prepare the body for a stronger local homeostatic response. Somatostatin (SOM) is released in other cells of the dorsal root, but it also can be released from the CNS and the pancreas. **SOM apparently acts as an antisubstance P in some of its function, while other functions are complementary**. Regardless of function, rapid tachyphylaxis frequently occurs. Recurrent reactions continue due to a limited amount of production of SOM, which suppresses nerve impulses triggered by noxious stimuli. SOM also prevents the recruitment of more basophils, while enhancing the activation of T-lymphocytes and, thus, down-regulating lymphokines. This action results in less damage to the vessel wall. **SOM inhibits release of histamine from the circulating basophil but not from the mast cells**.

Clearly, an increase in the total body pollutant load and constant barrage of specific noxious substances, which exceed local homeostatic control, can continue to trigger SP. With the resultant vascular and leukocytic components of inflammation (see Chapter 1) occurring, persistence of this inflammation results in the vasculitis usually seen in patients with chemical sensitivity and some patients with chronic degenerative disease. Here the local homeostatic and trophic responses have been exceeded so that inflammation occurs. This neurovascular induced inflammation may explain, in part, both the occurrence of mast cell hyperactivity and the ease with which these cells are triggered by varying and multiple stimuli of ever-decreasing doses.[166] This latter, "hair-trigger" response is frequently seen in the chemically sensitive patient whose local homeostatic control is exceeded and may partially account for the spreading phenomeon.[167] Once these events occur, release of leukotrienes, histamines, and possibly other chemical mediators will magnify the responses. These amplified responses, in turn, stimulate SP, creating a vicious downward dyshomeostatic cycle of response-ready triggering, which at times may be extremely difficult to terminate. This cyclic pattern of easy triggering is seen in patients with both far-advanced chemical sensitivity and chronic degenerative disease. Again, the vicious downward dyshomeostatic cycle emphasizes the characteristic of positive feedback of the homeostatic mechanism. In many instances, neuropeptide response has been hyperresponsive. For example, under environmentally controlled conditions in the ECU at the EHC-Dallas and using 15,000 individual, inhaled, double-blind challenges, we studied 2000 chemically sensitive patients who experienced reactions after pollutant challenge. We found that nerve and neuropeptide triggering in these cases was the primary metabolic response event, again emphasizing the denervation supersensitivity phenomenon.

Of course, when local homeostatic control is exceeded and once vascular dysfunction occurs, arterial spasm may follow with resultant local tissue hypoxia. Degeneration may further occur, with the release of free radicals and lysozymes, thus changing the metabolism. This degeneration then causes increased tissue damage and further inflammation. Other substances that affect vessel wall changes include serotonin and other vasoactive amines, epinephrine, and norepinephrine.

Once local homeostatic control has been exceeded after this cyclic pattern of easy triggering has been set in motion, the immune system may come into play, augmenting homeostatic responses further and leading to the propagation of the dyshomeostatic reaction. However, in some instances, immune triggering may be the primary event.

Pathogenesis of Neuroimmunological Mediators (Neuropeptide Triggering by Noxious Stimuli)

The multidirectional communication between the immunological, neurological, endocrinological, and local homeostatic control systems of the extracellular matrix provides opportunities for the coordination of specialized capacities for host defense of the dynamics of homeostasis. These multidirectional communications enhance amplification expression of the abnormalities underlying some neurological aspects of hypersensitivity reactions seen in the chemically sensitive and chronic degenerative diseased individual.[168–171] Lesions in one system lead to dyshomeostatic responses in the other systems, often exacerbating the dynamics of homeostatic responses and contributing to further pathology. **For example, injury to the hippocampus alters the numbers and functions of natural killer (NK) and T lymphocytes in the spleen, thymus, and blood, as the denervation supersensitivity increases**.[172–174] This injury or imbalance will cause short-term memory loss, as we have seen in hundreds of chemically sensitive patients at the EHC-Dallas. Often, lowering of the suppressor T-cells may be due to the postdenervation impulses emanating from the hippocampus and also peripheral nerves. These depressed cells may cause homeostatic dysfunction not only of the immune system, but also in various parts of the nervous system as seen in patients with chemical sensitivity and chronic degenerative disease. Depression may be directly related to low T-cells as we have seen in numerous chemically sensitive patients. **Peripheral neuroanatomic circuits may also be involved with distal feedback loops that mediate some effects of pollutants on** immunity. These circuits are recognized to be fibers of **sensory peripheral nerves, autonomic nerves and spinal nerves containing specific adrenergic, cholinergic, and peptidergic filaments. These nerve fibers end in zones of lymphocytes in the thymus, spleen, Peyer's patches, and bone marrow.**[175,176] Again this anatomical fact emphasizes both the amplified neuro influences upon homeostasis of the immune system and how pollutant or noxious incitant entry through these areas occurs. The entry of pollutants causes local, regional, and remote triggering of the dynamics of the homeostatic mechanism as well as their own hormetic effect, as is often seen in the individual who is chemically sensitive or has chronic degenerative disease. This anatomical fact again shows the broad scope of the GRS, and how rapidly communication and dissemination of information happens. See Figure 2.21.

The alteration of systemic communications (GRS) between elements of the dynamics of homeostasis in the immunological and neurological amplification systems is evident both during deficiency states and following the administration of an excess of a mediator from one of these systems, as occurs with pollutant or other noxious stimuli overload in the chemically sensitive or chronic degenerative diseased individual. The neural effects of noxious stimuli exposure on immune responses of homeostasis may depend on the native state of each system and the direction (inhibitory versus stimulatory) of the abnormality. **For example, antibody response to a foreign substance can be enhanced by epinephrine or by physical stimulation with enough intensity to increase the beta-adrenergic effects on T-cells.**[177] We have often seen depressed T-cells in individuals with chemical sensitivity or chronic degenerative disease attempting to exercise. As shown previously 82% of the patients with chemical sensitivity have a sympathetic response when the heart rate variability for autonomic function is measured. This sympathetic response should then stimulate an antibody response to counteract the entering foreign substance. The exercise-induced T-cell depression usually occurs after the cells have been repeatedly damaged by noxious stimuli making them secondary homeostatic foci (interference fields), which can trigger independently further and repeated adverse responses. In contrast, epinephrine induced or excercised induced stimulation and the elimination of sympathetic neural influences in athymic mice has been found to permit the development and generation of lymphocytes in thymic remnants.[178] A similar condition may occur in some individuals who have had excess epinephrine stimulation with chemical sensitivity and chronic degenerative disease. This fact would explain why these patients may have varied immune, neural, and endocrine homeostatic responses depending on the amount and quality of epinephrine they produce after the autonomic response to pollutant entry. Their symptom expression may be varied, which depends on their total body load, the state of nutrition, and the particular part of the nervous system that is triggered or inhibited at any given moment.

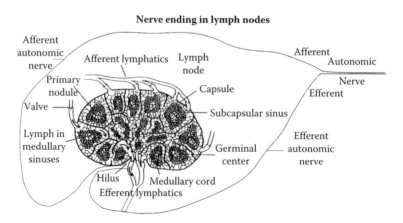

FIGURE 2.21 Picture of nerve ending in lymph nodes. (Modified from Guyton, A.C. and Hall, J.E., *Medical Physiology,* 10th ed. Philadelphia: WB Saunders Co., 93, 2000.)

Some laboratory observations help explain the extent of nerve involvement with the immune system that will influence homeostatic function in the individual with chemical sensitivity and chronic degenerative disease, after local homeostatic control has been exceeded. For example, experimentally induced primary **afferent denervation of lymph nodes of rats suppressed the response of antibody-producing cells to sheep erythrocytes repeated by challenges over 80% of the time**. Chronically, chemically exposed individuals with chronic degenerative disease or chemical sensitivity can have nerve impulse inhibition from excess phenols and toluene. Patients with surgical sympathectomy or anterior horn disease like polio will have alteration of the homeostatic balance as has been previously discussed. Benzene and xylene exposure may result in nerve impulse inhibition and subsequent immune dysfunction with production of antibodies to the cell. However, this suppression in animals was reversed by the administration of SP. We are not sure if the reversal occurs in humans, but, most likely, it does. In another instance, **the synaptic transmission in the superior cervical ganglion of immunized guinea pigs was potentiated specifically by homologous antigen through an HLA receptor-dependent effect of histamine released from mast cells in the ganglion**.[179] These experimental outcomes in animals suggest the ways immune deregulation may occur in individuals with chemical sensitivity and chronic degenerative disease. Clinical observation of thousands of patients with chemical sensitivity and chronic degenerative disease in controlled environments at the EHC-Dallas shows that an interaction between the nervous system and the immune system plays a significant part in their dynamic homeostatic dysfunction. For example, **phenol (often found in individuals with chemical sensitivity) can block the nerve responses to the lymph nodes and the T-lymphocytes, once local homeostatic control has been exceeded**. T-lymphocytes are already often suppressed in the chemically sensitive population. This blocking of nerve responses disrupts the dynamics of homeostasis. When this blockage of homeostasis occurs, these nodes and lymphocytes become depressed and function poorly. Particularly neck and, occasionally, groin nodes have been seen to rapidly swell after a minute dose of these noxious stimuli (phenol, benzene, xylene) by inhalation. The swollen nodes may last for weeks, even when the noxious substance is withdrawn. Conversely, as to T-cell inhibition by toxics many foreign substances exhibit a sympathetic (adrenalin) response. With repeated pollutant stimuli T-cells may be depleted from the lympatics and lymph nodes exhibiting a relative suppression of T-cells. This appears to create a burn-out phenomenon that is seen in many patients with chemical sensitivity and/or chronic degenerative disease. **As the total body pollutant load of the chemically sensitive or chronic degenerative diseased individual decreases, the nerve then responds again with unblocked impulses**. This phenomenon emphasizes the need for avoidance therapy in patients with chemical sensitivity and chronic degenerative disease in order to obtain optimum health. Peripheral sensory and some motor nerves seem to normalize about the same time that swelling in the nodes goes down, suggesting normalization of the node-nerve impulses. The T_8 suppressor cells increase, as do the dynamics of their homeostatic responses, which become more energy efficient and redevelop energy efficient resonating responses. This scenario is often seen at the EHC-Dallas and Buffalo when a patient with chemical sensitivity or chronic degenerative disease improves clinically.

Early immunological findings show that distinctive protein antigens are shared by immune and neural cells. (See Gastrointestinal chapter in Volume 2 - Mechanisms of Chemical Sensitivity and Chronic Degenerative Disease.) Thus, a communication network is set up as a part of the GRS and is essential for normal homeostatic function.

Neurons of the autonomic primary afferent and primary efferent nerves supply the skin, upper airways, lungs, and gastrointestinal tract and contain a range of functionally distinct mediators for both regulating the dynamics of homeostatic function and the body's defense. **The rapid vascular, smooth muscle, and secretory responses to neuropeptides released by noxious stimuli mimic mast cell-dependent, immediate hypersensitivity reactions and give clinical responses that simulate immune reactions**. We have seen many of these responses in some chemically sensitive individuals following a chemical exposure. Often they are neurological, mimicking the immune response with a resultant set of similar symptoms such as stuffed up or runny nose, asthma, or arrhythmias. Migraine headache in a patient exposed to formaldehyde is another example.

The neuropeptides (such as vasoactive intestinal peptide [VIP], peptide histidine isoleucine [PHI], and peptide histidine methionine [PHM], SP and other tachykinins, calcitonin gene-related peptide [CGRP], and SOM) that are found in human tissues have been detected in hypersensitivity reactions.

In some individuals with chemical sensitivity, symptoms resulting from deregulation, overproduction, or malfunction of these peptides have been reproduced by pollutant challenge, with disruption of homeostasis. The most prominent effects of VIP appear to be systemic, with pulmonary vasodilation and stimulation of epithelial and glandular secretion.[180] However, at the EHC-Dallas, we almost always see abdominal cramps reproduced on challenge with VIP. **VIP is also an adrenergic-independent bronchodilator of approximately fiftyfold greater potency than isoproterenol *in vitro*.**[181] The effects of the rest of these substances reflect cholinergic and adrenergic stimulations and may account for some of the varied vascular responses recorded in individuals with chemical sensitivity and chronic degenerative disease. Table 2.11 shows a series of VIPs studied in a group of chemically sensitive patients at the EHC-Dallas by intradermal provocation. The speed with which these substances are produced versus the speed with which they are metabolized must be balanced in order to maintain proper homeostasis.

The contributions of neuropeptides to immediate hypersensitivity reactions encompass direct effects not only on target tissue, but also on the activation and homeostatic regulation and deregulation of tissue cells and mast cells. This physiological fact demonstrates how homeostatic regulation occurs due to certain noxious stimuli. **Similar to lymph nodes in some tissues, 85% of subepithelial mast cells are in contact with or within 0.2 μm of peptidergic nerve fibers containing SP or CGRP** (Figure 2.22).[182] This anatomic fact explains the neuroimmune triggering seen in some individuals with chemical sensitivity and chronic degenerative disease and further emphasizes how they must have proper homeostatic control in order to prevent pathology. Since the autonomic nerves innervate both the lymph nodes and the mast cells, one may develop both normal and abnormal communications depending upon the pollutant load.

Neuropeptides influence development of tissue distribution, proliferation of synaptic responses, and cytotoxic activities of lympocytes in numerous experimental models,[170] again emphasizing the importance of clinically coordinating the responses of the neuroimmune axis.

TABLE 2.11
Intradermal Peptide Testing in Patients with Chemical Sensitivity

Formula	Concentrate	Sex	Age (years)	Diagnosis	No.	Positive (%)
$C_{147}H_{238}N_{44}O_{44}$ (vasoactive intestinal peptide)	250 μg/10mL	F:17 M:4	9 to 71[a]	GI	21	100
				NS	4	
				Respiratory	6	
				CV	8	
				Endocrine	5	
				MSL	7	
				GU	1	
$C_{189}H_{285}N_{55}O_{57}S$ (neuropeptide)	100 μg/10mL	F:15 M:4	9 to 71[b]	GI	19	74
				NS	6	
				Respiratory	3	
				CV	6	
				Endocrine	3	
				MS	4	

Source: Based on EHC-Dallas. 2004.

[a] Mean = 46 years.

[b] Mean = 42 years.

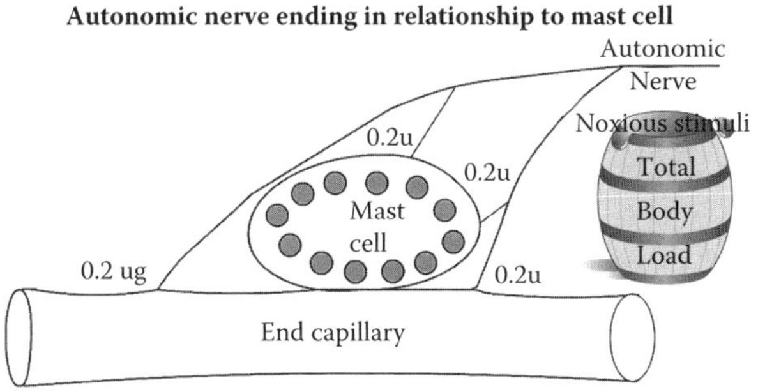

Nervous System

Autonomic nerve ending in relationship to mast cell

FIGURE 2.22 Nerve—mast cell. The autonomic nerve is never more than 0.2 micron from the capillary and the mast cell. (EHC-Dallas, 2002.)

A wide range of lymphocyte functions are inhibited or stimulated by neuroendocrine peptides such as alpha- and beta-endorphins, met- and leu-enkephalins, and adrenocorticotrophic hormone (ACTH). Liberation of these substances by toxic chemical stimulation at times accounts for the low suppressor T-cells seen in some individuals with chemical sensitivity after pollutant challenge. **If the endorphins are triggered, the sudden sleepiness seen in some patients with chemical sensitivity, also, occurs.** Proper dynamic homeostatic balance must be restored in this type of injury or a vicious downward cycle, " <100% positive gain" (as discussed in Chapter 1) characteristic of the homeostatic response occurs, eventually resulting in fixed-named, irreversible disease.

The effects of stress *in vivo* and of endogenous neuropeptides *in vitro* on NK and other cytotoxic lymphocyte activities have not been clearly defined, but reports of the adverse effects of some types of stress are beginning to appear.[183,184]

As shown previously, macrophages are normal cellular constituents of the homeostatic response in the nervous system just like other systems. After neural tissue injury by noxious stimuli (virus, bacteria, and chemical toxins) and in order to restore homeostasis, these cells increase in number and activity so that the damaged myelin is removed phagocytically before axonal regeneration.[185] At times this process is retarded. For example, many patients with chemical sensitivity and chronic degenerative disease who have **artificial implants made of substances such as silicon develop antimyelin antibodies, which then produce further homeostatic dysfunction and deterioration of the myelin before their disease process is arrested. It is only recently have macrophages and lymphocytes been recognized for their capacity to regulate neural growth, to inhibit nerve functions, and to induce a state of immune competence in some neural cells**. Murine and human T-lymphocyte functions also affect neuropeptides of the primary afferent, adrenergic, and cholinergic systems such as SP, SOM, VIP, and PHI/M.[170]

Increased concentrations of the macrophage product interleukin-1 (IL-1) was used after neural injury to stimulate the hypothalamic-pituitary axis,[186–188] to stimulate ACTH, leutinizing hormone (LH), thyroid stimulating hormone (TSH), and Somatotropins (GH), and to inhibit prolactin secretion. A case report of toluene overload and altered prolactin function demonstrates the concept of neuroendocrine triggering and imbalance in the individual with chemical sensitivity. (See Chapter 4: Endocrine System.)

The growth and maturation of oligodendrial cells and astrocyte cells are stimulated by proteins released from antigen and lectin-activated T-lymphocytes.[189,190] T-cells inhibited sympathetic innervation in the spleen of athymic mice as evidenced by the higher number of nerve fibers and higher content of norepinephrine after the introduction of thymocytes.[191] A similar circumstance of stimulation or inhibition of nerve and/or immune function may be seen in some individuals with chemical sensitivity and chronic degenerative disease who have solvents in their blood, depending on the type of pollutant involved and the nature of their autonomic homeostatic imbalance. Elimination of noxious stimuli usually restores T-lymphocytes and homeostatic balance, if the GRS is not entirely destroyed, which, because of its vastness (connective tissue matrix 40% of the body), is rarely the case.

Many other complex effects of immunological mediators are recorded in neural cells or regulated by neural factors. Included here are the inhibition by alpha melanocyte stimulating hormone on T-lymphocytes and neural cells.[179] Alpha melanocyte stimulating hormone inhibits the effects of the IL-1 cytokine on the stimulation of neural cell growth via prostaglandin E[189] and the activation of hypothalamic alpha adrenergic receptors by C5a (i.e., complement component).[193] The results from studies of animal models and some human diseases suggest that neuropeptides such as SP, SOM, CRP, and VIP, are involved in disease states of hypersensitivity and altered immunity, including the individual with chemical sensitivity and chronic degenerative disease. The most comprehensive findings come from animal models of arthritis and asthma. These studies showed elevated concentrations of neuropeptides in relation to both the development of disease and the prevention or reversal of elements of disease by neuropeptide antagonists, which are designed to restore homeostasis. **The results of these studies parallel our observations in individuals with chemical sensitivity and chronic degenerative disease that pollutant overload triggers responses from neuropeptide stimulation and elimination of the pollutants decreases the chemical sensitivity or chronic degenerative disease**. Here, homeostasis can be restored by either elimination of the stimuli or production of antagonists. **We have observed the elevation of IL-2 in a large set of patients with chemical sensitivity**. This elevation suggests noxious incitant triggering and certainly confirms immune derangement of the type just described.[194] See W.J. Rea, *Chemical Sensitivity, Volume IV*, Chapter 30, for more information. Indeed, homeostasis is disturbed in the patients with chemical sensitivity and chronic degenerative disease and, if unchecked, will lead to both a downward spiral of metabolism and tissue changes.

Laboratory evidence supports the presence of neuropeptide triggering. For example, repeated **application of a mild irritant stimulus to the hind paw of a rat resulted in persistent swelling and tenderness of the opposite hind paw.**[194]

This swelling was mediated through a neurogenic mechanism attenuated by ablation of either unmyelinated afferent nerves with capsaicin or sympathetic postganglionic efferent nerves by immunosympathectomy.[195] Clearly, the homeostatic feedback mechanism between the two paws, once chronically overloaded, caused a dyshomeostatic response, resulting in the edema. Once the homeostatic mechanism is activated and if the stimulus is removed, the pain and edema return to normal. A similar alteration may have occurred in chemically sensitive patients with arthritis and chronic degenerative diseased patients studied at the EHC-Dallas, since these patients developed immediate swelling upon pollutant challenge. Homeostasis was restored after the total body load was reduced and specific incitants were removed. The role of neural pathogenesis in the individuals with chemical sensitivity or chronic degenerative disease is supported by findings in the general population of several forms of immunologically induced arthritis caused by using a lower nociceptive threshold and a greater intraneuronal content of peptide neuromediators, such as SP.[196] **Meggs has shown upper airway inflammation, which appears to be neurologically triggered in some patients with chemical sensitivity**. If such a threshold were, in fact, present in individuals with chemical sensitivity and chronic degenerative disease, this lower threshold of triggering would then explain the secondary widespread triggering of the other systems. This fact correlates with Gunn's observations of denervation supersensitivity discussed earlier in this chapter. After a significant chronic pollutant injury where the lowered threshold occurs, especially in these patients with chemical sensitivity and chronic degenerative disease with arthritis and neurological dysfunction, a triggering of the other systems has been observed to occur. The triggering is most likely caused by the supersensitivity to multiple foods and chemicals. The lowered threshold is, further, an example of dyshomeostatic triggering, where the proteoglycan/glucoscaminoglycan (PG/GAG) of the matrix and other receptors become unstable and, thus, dyshomeostasis occurs. This dyshomeostasis then further triggers the nonspecific mesenchyme reaction with an increase in the proteoglycan and collagen output from the fibrocyte.

In the same models of immunological arthritis in rats, neonatal destruction of primary afferent neurons by thiomalate reduced and retarded the development of disease, whereas the infusion of SP into lower risk proximal joints increased the severity of joint inflammation and tissue damage.[195] This study certainly gives us a picture of the neural and neurotransmitter influence on the dynamics of homeostasis. In this type of response, the number of unmyelinated neurons in the saphenous nerves of rats was reduced by doses of gold sodium thiomalate, which alleviates immunologically induced arthritis, but the relative effects of thiomalate on afferent and sympathetic efferent fibers have not been determined.[195]

The importance of further defining the involvement of sympathetic neurons is emphasized by the observation that **immunologic arthritis was far worse in spontaneously hypertensive rats with a regulated high tonic sympathetic activity**[195] and by the finding that symptoms and joint function improved when patients with rheumatoid arthritis had sympathetic blockade with guanethidine.[53] At the EHC-Dallas, the joint abnormalities of patients with chemical sensitivity who had silicon jaw, breast, and other implants as well as arthralgias and arthritis appear to fit this neuroimmune triggering. These patients have toxic patterns on their Triple Camera SPECT brain scans, antimyelin antibodies, low suppressor T-lymphocytes, and cholinergic and sympatholytic patterns on their pupillography measurements while others exhibit the sympathetic response when measured by heart rate variablitiy. The implants clearly put a strain on the homeostatic mechanism (Figure 2.23).

Other cases of rheumatoid arthritis have been shown to be environmentally triggered by foods, molds, and toxic chemicals.[197–200] Some of these patients' symptoms may have been triggered neurologically as well as immunologically, again emphasizing the need for harmonic balance of these two systems, as well as the endocrine system, to maintain the dynamic homeostatic mechanism for health.

Similar experimental approaches have provided evidence for the roles of SP in rat and guinea pig models of asthma. Pretreatment of animals with specific antagonists of SP gives us an insight into the dynamics of homeostatic balance involving nerves and neurotransmitters. These antagonists significantly reduced the bronchospasm and respiratory mucosal edema evoked by SP that was released through vagal nerve stimulation, chemical irritation, or the application of capsaicin.[201] The sympathetic effect was allowed to function better. In addition, chronic treatment with capsaicin depleted the local stores of SP and reduced bronchoconstrictor responses to antigen in sensitized animals.[201] Clearly, chronic stimulation or inhibition of the nervous and endocrine systems by pollutants or other noxious stimuli, which leads to the alteration of the dynamics of homeostasis, causing dyshomeostasis, occurs in the chemically sensitive or chronic degenerative diseased individual. This chronic stimulation could cause constant irritation of these systems with ongoing release of some neurotransmitters and hormones, depletion of others and of the splanhic nerves; therefore, sympathetic effects with adrenal depletion or inhibitory effects of the vagus nerve could occur. Pollutant-triggered, continuous nervous system stimulation or inhibition leads to many of the symptoms observed in patients with chemical sensitivity or chronic degenerative disease. Also, dyshomeostatic signals are sent throughout the nervous system, allowing for easier triggering of more dyshomeostasis with a vicious downward cycle of positive feedback. In addition, there would be a spreading of these signals throughout the GRS, fostering further alteration of the dynamics of dyshomeostasis throughout the body.

In contrast to the lack of knowledge about the concentrations of neuropeptides in the lower airways, the time-dependent appearance of some neuropeptides in nasal lavage fluids has been established in relation to antigen challenge of allergic human subjects. SP,[203] SOM, CGRP, and histamine have been detected in nasal lavage fluids prior to antigen challenge. **In allergic patients, but not in normal controls, antigen challenge evoked threefold rises in histamine at 15 to 60 minutes**, 1.5 to 4.0–fold rises in CGRP at 15 minutes to 24 hours, and more than twofold rises in SOM at six hours, without altering the concentration of SP.[203] Thus, CGRP may make both early and late contributions to allergic nasal congestion, while SOM may regulate events in late hypersensitivity

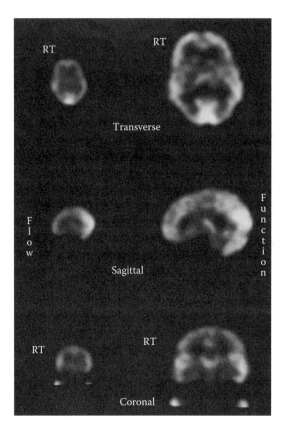

FIGURE 2.23 (See color insert following page 364.) Triple-headed SPECT camera brain scan of individual with breast implant. (From EHC-Dallas, 2005.)

reactions. The harmonic release and catabolism of these mediators is necessary for local, regional, and even general homeostasis. Although homeostasis can still occur with excess release of these aforementioned substances, more energy will be required to restore the dynamics of homeostasis. In order to achieve homeostasis, the body will have more nutrient utilization and subsequent drain, with less stamina from greater energy expenditure. Thus, more **fluctuations of response will occur, until homeostasis is obtained at a more costly energy drain**.

Specific lymphokines also can elicit the release of neuropeptides *in vivo*. Infusions of IL-2 in patients with metastatic carcinoma resulted in altered vascular permeability, causing edema and serosal effusions in association with an increase in the intraperitoneal concentration of SP to up to 500 mg.[194] A similar situation exists in some chemically sensitive patients at the EHC-Dallas who have edema triggered by pollutant exposure, resulting in dyshomeostasis. In these patients, IL-2s have been noted to be elevated. As the total body load is reduced, the IL-2 returns to normal when the dynamics of homeostasis are restored. In contrast, the subsequent intraperitoneal injection with lymphokine-activated killer cells in cancer patients elicited an increase in the concentration of CGRP in ascitic fluid.[204]

In other reports, the **high concentrations of neuropeptides found in extracts of lesional tissue or fluid** were the sole documentation of a possible pathogenetic homeostatic pathway. High levels of VIP, together with histamine and serotonin, were found in lung cells of rats exposed to another pollutant, asbestos, for one to six months according to protocols, leading to pulmonary fibrosis.[205]

SP but not VIP was high in the cutaneous fluid of patients with bullous pemphigoid and other vesicular dermatitides, whereas blisters induced on normal skin had no detectable SP.[206] **The elicitation of a classical wheal and flare response in human skin by SP, although the flare is** histamine-dependent, demonstrated the capacity of neuropeptides alone to mimic immediate hypersensitivity.[207] This response is certainly an example of another way to enter the GRS. At the EHC-Dallas, we have injected SP into the skin of some chemically sensitive patients, creating a wheal and flare. Injected SP also produced some systemic symptoms, which would not abate until somatostatin was injected directly into the SP-induced wheal, which then allowed generalized homeostasis to be obtained.

The possibility that immune and neuroendocrine systems communicate meaningfully in the reactions of host homeostatic defenses against pollutant injury in the individual with chemical sensitivity and in some chronic neuroimmunologic diseases in other populations is supported by findings of anatomic connections and common cellular and molecular pathways. These connections appear to be through the matrix, with the GRS being paramount. **Diverse neuropeptides stimulate and regulate critical elements of immediate hypersensitivity, inflammation, and immunity**, with specificity attributable to distinct subsets of leukocyte receptors, especially lymphocytes. These lymphocytes evoke the specific response of the body's defense system. Some lymphokines and monokines evoke primary neuroendocrine responses, stimulate neural cell proliferation, and activate astrocytes to a state of immune competence. In addition, leukocytes generate and secrete neuropeptides and peptides related to neuromediators, which must be metabolized properly in order to maintain homeostatic balance.

Clearly, deregulation of the dynamics of homeostasis of the nervous system can result from direct toxic chemical stimulation of the neuropeptides of the nervous system. This stimulatory action first triggers changes via the lymphocytes, or through the autonomic and peripheral voluntary nervous systems. This deregulation due to previously described routes, and also through the endocrine glands can be seen in the patient with chemical sensitivity. In other words, the GRS is present and functional throughout all these systems. These anatomical and physiological findings emphasize how the dynamics of homeostasis are obtained and maintained through the GRS. Once the homeostatic mechanism is triggered, the functions of this communication network explain many of the symptoms and signs that manifest as a result of the various system involvements in the patients with chemical sensitivity and/or chronic degenerative disease.

In the body's early response to pollutant stimuli, after the GRS is triggered, autonomic dysfunction and peripheral, neural-endocrine immune responses appear to be highly significant in the expression of the chemical sensitivity and chronic degenerative disease with the deregulation of homeostasis. The physician's ability to perceive and act in light of these various homeostatic and the dyshomeostatic responses may prevent fixed end-organ disease.

THE ACUPUNCTURE ENERGY FLOW SYSTEM (AES)

For centuries, ancient wisdom has accepted an energy system in and around the body that regulates health and disease. Unfortunately, until recently this system has not been well integrated into modern day physiology. Even at this point in our understanding, not all aspects of the acupuncture system are clear, and, in fact, modern medicine and physiology have virtually ignored this system. With the development of highly technical equipment, it has become evident that the acupuncture energy flow system (AES) can now be defined objectively, observed, measured, and manipulated positively, in order to balance the patient's health and prevent disease. We have decided to place some excerpts, techniques, and procedures in this book because patients are at times being helped by acupuncture and its modifications. As far as being integrated into modern holistic medicine, much of the scientific basis available is shown in Chapter 1. Acupuncture given its invasiveness and complexity is still difficult for the clinician as a practical part of his armormeteriam to prevent and treat early both types of homeostatic dysfunction. Hopefully, this will change as new knowledge evolves.

Acupuncture appears to be based on neural responses, especially through viscero-cutaneous reflexes of the dermal and visceral branches of the spinal nerves. This response is characterized by speed, usually of vascular constriction or dilatation, due to the end autonomic nerves being wrapped around the end capillary as shown in Chapter 1 by Croley. Many studies both in man and animals show distal capillary beds responding to topical treatment in another part of the body.[208]

Animal and human studies have shown that even cervical severance of the spinal cord showed no difference in the regional viscero-cutaneous, viscero-motor and viscero-visceral reflexes (discussed in Neuromuscular chapter in Volume 2 - Mechanisms of Chemical Sensitivity and Chronic Degenerative Disease and will not be discussed further here). These reflexes appear crucial in rapid acupuncture responses. Some body of opinion also suggests that intracellular microtubules and actinal fibers facilitate quantum wave communications both intracellulary and extracellulary.[208]

Acupuncture can influence cell membrane transports. Certainly, a major mechanism of action for acupuncture input on the body is the electrical semiconduction that permeates the electron rich surface of the fascia that invades every cell and organ of the body, thus allowing for establishment of homeostasis through the GRS. A strong electrical neuronal stimulation exists as well as a charged ionic fluid that circulates among the muscle fibers. **Neurohumoral effects are created, such as beta endorphin cascade, as well as release of other neurotransmitters**. These substances are VIP type peptides that act inside and outside the CNS. Also, myofascial and musculoskeletal regulation mechanisms are activated. These help relieve pain and fatigue by allowing normal dynamic homeostasis to occur.

According to Mann,[209] some of the acupuncture points and lines, particularly those on the back, have an effect on a specific organ and are in the appropriate dermatomes as seen in Figure 2.24 and 2.25. One of the difficult obstacles in understanding acupuncture that should be taken into account is that the Ancient Chinese language and its modern interpretation do not always mean the same as in modern discussion of (in Western Medicine) anatomy. They may say heart, liver, spleen, etc.[211] and mean something totally different than our concept of modern anatomy and physiology. This difficulty in communication appears to have been overcome by the experienced physician/acupuncturists of today. For example, Gunn[57] has modified the Chinese technique to fit with modern anatomy and has developed a rational protocol for needling damaged areas resulting from bone and muscle trauma, as well as from spondylosis and herniated discs. This IMS modifies the classic acupuncture to the point that the procedures become practical for every physician.

We will attempt to integrate the information and data from Gunn, Mann, and other authors into an anatomical and physiological basis for the optimum function of the acupuncture energy system to work.

The variation of the acupuncture points occurs on the arms, legs, and lower abdomen. There may be several reasons for this variation, but the two most important are, first, the discrepancy between the autonomic nerve trunk and the dermatomes and, second, the evolution of the embryological points (i.e., the kidney, bladder, and several organs though in different areas anatomically). All are derived from the embryological pouch, and thus have similar innervation and similar acupuncture numbers. See Figure 2.24.

Normally the dermatomes of the arms are from C_5–T_1, while the sympathetic dermatome obtained by stimulating anterior spinal root is T_2–T_9. This anatomical discrepancy helps to confuse one's understanding of functions when a point in the area is stimulated. When **the skin and deeper tissues are pierced by the acupuncture needles, both spinal nerves and the sympathetic nerves of the blood vessels are affected, causing a definite response**.

The series of acupuncture points on the back lateral to the association points are shown in Figure 2.25, from Mann.

These points have an effect on the same organ as its sister point at the same level, e.g., both B_{13} and B_{27} influence the lungs, as do the points on the governing vessel. The abdomen and front of the chest are traversed by the spleen, stomach, kidney, and the conception meridian, which have their effect mainly on the region of the body traversed. All four meridians where they cross the chest may be used to treat the lungs and heart, yet their abdominal course influences the abdominal viscera. **Therefore, on the back, a rough dermatomal pattern is preserved but ventrically it is more complex**.

Nervous System

Acupuncture dermatomes

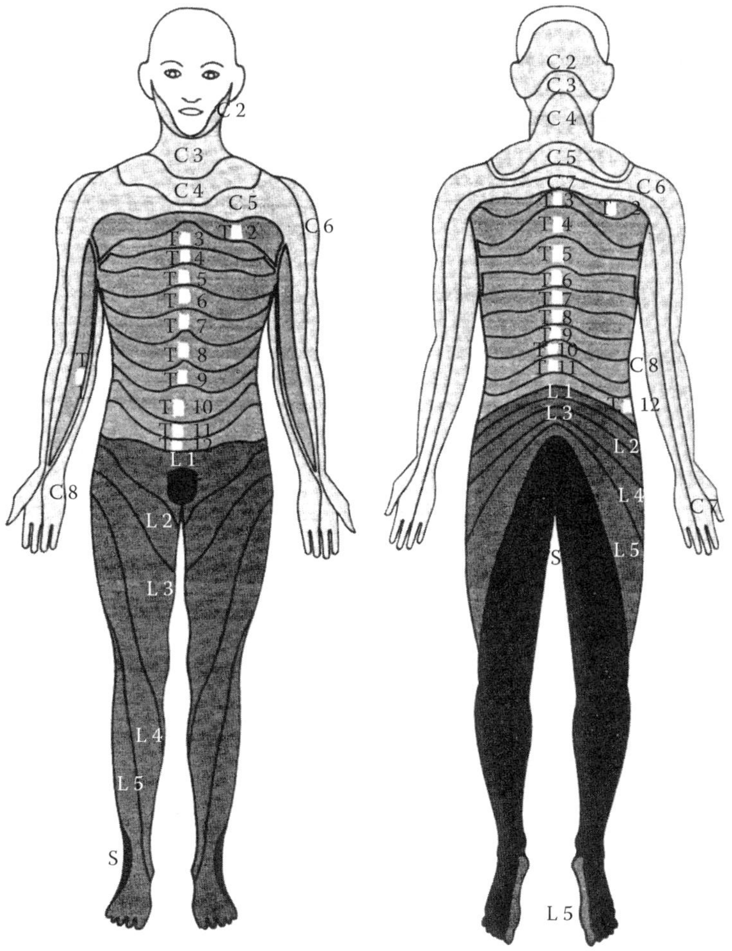

FIGURE 2.24 Head's zones (anterior and posterior aspect). The segments of the body are designated according to the segments of the spinal cord from which they are supplied. They are divided into the following groups in accordance with their nerve-exit points in the vertebral column: eight cervical or neck segments C1–C8, 12 chest-thoracic or dorsal segments T1–T12 (= D1 – D12), five lumbar segments L1–L5, five sacral segments S1–S5. (From Mann, F., Acupuncture: The Ancient Art of Chinese Healing and How It Works Scientifically, Random House, 66, 1972.)

The majority of the acupuncture points on the abdomen, thorax, and back are near the midventral and mid dorsal lines, which corresponds to the segmental reference of deep pain.[211] Figure 2.26.

The meridians of the lung, pericardium, and heart on the anterior surface of the arm correspond, at least approximately, to the appropriate dermatome. See Figure 2.27.

The meridians on the posterior surface of the arm—the large intestine, small intestine, and triple warmer—do not correspond to the appropriate dermatome. See Figure 2.28.

It should be noted that despite their name, stimulation of a large intestine acupuncture point influences the lung more than the large intestine and stimulation of a small intestine point influences the heart more than the small intestine, while stimulation of the triple warmer produces effects hard to catalog under a single organ.

The acupuncture points of the legs and head do not fit with what is known as dermatomes. These points on the legs are those of liver, gallbladder, kidney, bladder, spleen, and stomach. In all cases (except bladder), the dermatomes of these organs are on the trunk and not on the legs. It

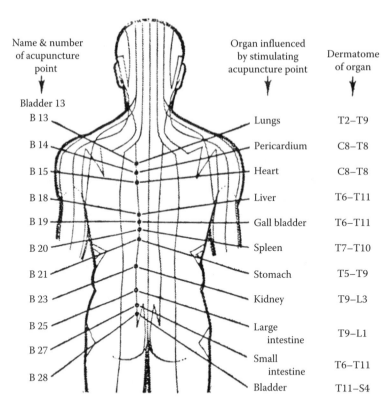

FIGURE 2.25 Hyperaesthetic zones in internal disease. Dermatomes derived from Hanse, K. and Schliak, H. Segmentale innervation, ihre bedeutung fur Klinik und Praxids 1962. Georg Thieme, Stuttgart; and other sources. (From ACUPUNCTURE by Felix Mann, copyright © 1962, 1971 by Felix Mann. Used by permission of Random House, Inc.)

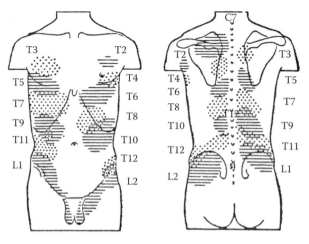

FIGURE 2.26 Segmental reference of deep pain when injecting hypertonic saline into the interspinous ligaments. Kellgren, J. H. On the distribution of pain arising from deep somatic structures with charts of segmental pain. (From ACUPUNCTURE by Felix Mann, copyright © 1962, 1971 by Felix Mann. Used by permission of Random House, Inc.)

Nervous System 255

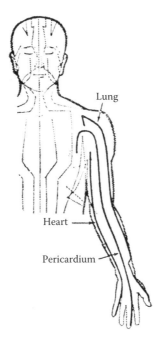

FIGURE 2.27 Meridians of lung, pericardium and heart on the anterior surface of the arm correspond approximately to the appropriate dermatome. (From ACUPUNCTURE by Felix Mann, copyright © 1962, 1971 by Felix Mann. Used by permission of Random House, Inc.)

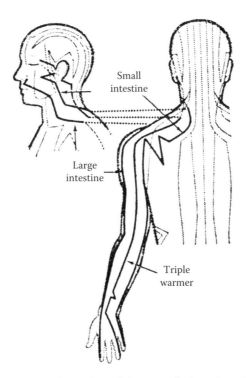

FIGURE 2.28 Meridians on the posterior surface of the arm—the large intestine, small intestine, and triple warmer—do not correspond to the appropriate dermatome. (From ACUPUNCTURE by Felix Mann, copyright © 1962, 1971 by Felix Mann. Used by permission of Random House, Inc.)

has been observed that stimulation of a leg acupuncture point effect on the appropriate organ, even though it may be 10 dermatomes away. Much evidence suggests that this different anatomical pattern on the posterior arm and its subsequent response may be via the spinal intersegmental long reflexes.[212,213] The response may as well be a fast extraspinal route in the sympathetic chain of the same side or even through the CT matrix.

Experimental evidence suggests intersegmental involvement. For example, Brown-Sequard conducted an experiment on two dogs that demonstrated the segmental and intersegmental pathways. One dog with cord severance at L_3 showed only bladder and rectal congestion when challenged with boiling water while the other dog with T_3 severance revealed all abdominal viscera congested when challenged.[213] **The distribution of the acupuncture points on the legs is such that each organ corresponds to several dermatomes, and each dermatome corresponds to several organs.** This problem can be partly resolved when it is realized that acupuncture points on the medial and anterior side of the thigh (liver, spleen, kidney meridians) do not have much effect on these organs but do affect the reproductive organs. Similar discrepancies occur with other points if the dermatological charts of Keegan and Garret are taken, which are obtained by charting the hyposensitivity from loss of a nerve root (Figure 2.29). The remaining acupuncture points fit more easily into a dermatomal pattern.

The kidney and bladder, which from an acupuncture point of view, function together and are dermatomes S_1 and S_2; the gall bladder and stomach are L_5; and the liver and spleen, which are hard to distinguish, are L_3 and L_4. Perhaps the long intersegmental spinal reflexes for the leg do not follow a dermatological pattern. Also, the sympathetic chain may be involved[213] showed this complexity in

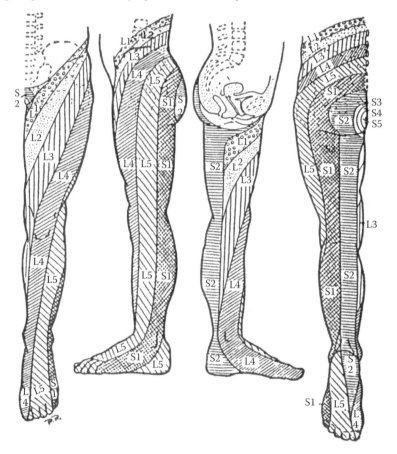

FIGURE 2.29 Acupuncture points on legs shows that each organ corresponds to several dermatomes and each dermatome corresponds to several organs. (From ACUPUNCTURE by Felix Mann, copyright © 1962, 1971 by Felix Mann. Used by permission of Random House, Inc.)

patients with severe pain encompassing several dermatomes. A pin prick to the trigger area might relieve pain on that dermatome or in several dermatomes or in just one other dermatome. After treatment, the impulse created by the pinprick can move out of a dermatome to relieve pain again, or it can move to another dermatome.[213]

The apportioning of the acupuncture points on the head to the various internal organs is hard to follow, both theoretically and in actual practice, though distinct effects actually occur. A series of experiments conducted by Koblank[214] showed that stimulation of a sharply defined area in the region of the superior concha of the nose caused cardiac arrhythmias in man, dog, and rabbit. **Severance of both vagus nerves resulted in ablation of the reflex for a few days, only to return with a weaker response**. Division of the maxillary nerve on one side resulted in ablation of the reflex when the nerve was stimulated on the same side, but the reflex persisted normally when the healthy side was stimulated. It was thought that there might be a connection with the vagal center in the brain, but another theory suggests that the connective tissue matrix was involved. **Koblank also showed that stimulation of the upper turbinates of the nose affected the heart, the middle turbinate affected the stomach, and the lower turbinate affected the reproductive organs**.[214]

At times, stimulating a few acupuncture points is effective in treating certain patients. Other times, stimulating any one of several meridians (encompassing a large number of acupuncture points) can be effective. Therefore, **a specific versus a general stimulus is used**. This specific response described under the intradermal injection neutralization phenomenon (see Treatment chapter in Volume 4 - Mechanisms of Cardiovascular Disease and Chemical Sensitivity) apparently takes place along the same line of nerve pathways. The generalized hypersensitivity state seems similar to the pain of a severe toothache when the entire same side of the face, arm, and upper chest are hypersensitive affecting the nerves in a large area. Hence, treatment would only require a needle placed anywhere in that area, which might include a large number of acupuncture points or any of several meridians. In other cases, a stimulus anywhere in a large area does not depend on hypersensitivity, but on the large number of neurons that have a final common pathway.[215]

Another principle involved in acupuncture is that a diseased organ has a lowered threshold of response and thus requires a small stimulus, while a large stimulus is needed for the healthy organ to respond. This fact correlates with the neuropathy hypersensitivity of Gunn,[57] the nerve injury supersensitivity of Cannon,[59] and the experiences observed at the EHC-Dallas and Buffalo. In the experience of Mann, not infrequently, a stimulus on any area as large as one to several dermatomes is sufficient for good results.

In all diseases, there exists one of three conditions: (1) tender areas that are spontaneously painful, (2) points tender under pressure, (3) points that show no pain under pressure but can be felt by the experienced physician.

Chinese literature describes about a thousand acupuncture points. Many more may exist. The meridians (nerve pathways) on one side of the body are duplicated by those on the other side. Plus two extra meridians exist, which run up the middle of the body. See Figure 2.30.

The number of acupuncture points along each of these meridians varies; the heart meridian, for example, has nine points on each side, the bladder meridian has 67, etc. (Figure 2.31).

The 12 organs and their associated 12 meridians encompass all parts of the body, with the exception of the head, sense organs, endocrine glands, and reproductive system. However, these can still be treated. Anything that happens along or near the course of a main meridian will influence the meridians of the organ that bears its name.

Acupuncture points and meridians also follow embryological development, i.e., when acupuncture points on the kidney meridian are stimulated, they affect not only the kidney, but also the embryologically related ovary, testicle, uterus, fallopian tube, and to some extent, the adrenal gland.

There are hundreds of embryological relationships in the body that make up an interdependence of different points to be utilized in acupuncture, which at times, makes treatment difficult.

Another example of this phenomenon is that one can treat a sore throat by stimulating lung point number 7. Positive results occur because both organs were part of the embryological alimentary tract.

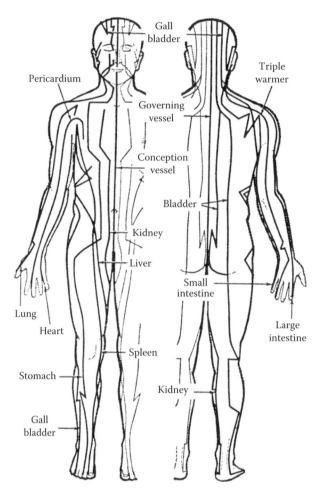

FIGURE 2.30 Twelve main meridians. (From ACUPUNCTURE by Felix Mann, copyright © 1962, 1971 by Felix Mann. Used by permission of Random House, Inc.)

The main meridian is in a line connecting various acupuncture points along the same pathway, but it has various subsidiary branches supplying areas of the body adjacent to it.

The dotted line in Figure 2.32 shows how the branch of the heart meridian transverses the lungs, great vessels, heart, diaphragm, and connects with the small intestine. Another part of this branch goes through the throat and eye. It is easy to see how the main meridian's sphere of influence is enlarged by its various branches. Other more specialized meridians are present, in addition to the 12 main meridians, which make practical understanding difficult. These include eight extra meridians, 12 muscle meridians, 12 divergent meridians and 15 connecting meridians.

Other energy flows occur that are difficult to relate to any specific area or meridian. For example, Chi is the vital life force energy whether it is in the growth of the plant, movement of an arm or thunder of a storm. To balance this force, the principle of opposites exists with negative and positive, or the yin and the yang. A relationship appears to be present between yin-yang and the meridians. The Chinese normally speak of the meridian in pairs, where they distinguish the members of each pair by reference to the arm or leg, thus, indicating the main location of the particular meridian, instead of the particular organ to which it is related.

In the actual practice of acupuncture, yin-yang occurs when a certain point is stimulated (e.g., the liver is tonified, or the liver or kidneys are upregulated) the heart (upregulated) will be tonified automatically, while the spleen is sedated.

Nervous System 259

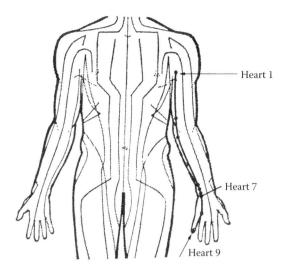

FIGURE 2.31 Heart meridian on left showing its acupuncture points. Heart 1 (under armpit) along the arm to heart 9 (at end of little finger). (From ACUPUNCTURE by Felix Mann, copyright © 1962, 1971 by Felix Mann. Used by permission of Random House, Inc.)

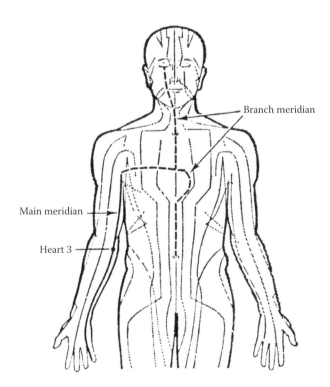

FIGURE 2.32 The dotted line shows that the branch of heart meridian traverses the branches to the lungs, goes to the great vessels, heart, diaphragm, and connects with the small intestine, while another part of this same branch goes to the throat and eye. (From ACUPUNCTURE by Felix Mann, copyright © 1962, 1971 by Felix Mann. Used by permission of Random House, Inc.)

Clearly, acupuncture in the twenty-first century has a provable scientific basis although there is still some difficulty in translating the ancient writing and concepts into modern physiology. IMS of Gunn is discussed in more detail in Treatment chapter in Volume 4 - Mechanisms of Cardiovascular Disease and Chemical Sensitivity. Gunn's protocol offers a more scientific modification of the Chinese acupuncture system.

NAET

Another variation of acupuncture techniques that shows promise for therapeutic intervention is the Nambudripod Allergy Elimination Technique (NAET). This technique is a blend of acupressure/acupuncture using kinesiology, chiropractic, nutritional, and allopathic disciplines to rebalance the body's bioenergy. These processes are supposed to eliminate the hypersensitivity part of chronic disease.

NAET is accomplished by having the patient hold the offending substance to which he or she is hypersensitive (and which has disrupted energy balance) while the technician stimulates the acupuncture points (with needle, finger, and/or pressure) until the underflow or overflow of energy is rerouted, rebalanced, and homeostasis returns. The endpoint (balance) is determined by restoration of muscle strength.

NAET has to be performed with each individual offending agent. Proper entry balance has to be constantly reinforced. There are promising case reports, but there is no series of numbers or data to substantiate this technique. However, it is included here because of its great potential as another modality that might help the hypersensitive stage of chronic disease.

VOLUNTARY CENTRAL NERVOUS SYSTEM

Noxious Injury to the Blood–Brain Barrier

The CNS is thought to be protected from noxious substances by the blood-brain barrier and the peripheral nerves are protected by the blood nerve barrier. The barrier prevents rendering of high levels and chronic exposure of some noxious stimuli to the brain and nerves, which allows the dynamics of orderly homeostasis to occur. Chronic noxious stimuli overload will trigger dyshomeostatic responses. Understanding of these barriers is based upon observations that some substances introduced into the body affect many soft tissue organs (e.g., liver, kidney, muscle), although they are excluded from the brain.[216] See Figure 2.33.

As a result of pollutant exposure, the blood-brain and peripheral nerve barriers may be altered in the chemically sensitive and/or chronic degenerative diseased individual, thus making the dynamics of efficient homeostasis more difficult and certainly requiring more energy to obtain it, if penetration of pollutants does occur. **Although some pollutants or other noxious stimuli may penetrate these barriers naturally without alteration of the barrier (e.g., anesthetics), the majority do not**. Prolonged anesthetics can disturb homeostasis, so with some types of toxics, this barrier becomes more permeable, resulting in entry of excess toxics into the brain and, thus, homeostatic dysfunction.

It has already been pointed out that the constituents of the cerebral spinal fluid are not exactly the same as those of the extracellular fluid (ECF) elsewhere in the body. Furthermore, many large molecular substances hardly pass at all from the blood into the cerebral spinal fluid or into the interstitial fluids of the brain, even though the same substances pass readily into the usual interstitial fluids of the body. Therefore, barriers, called the blood cerebral spinal fluid barriers and the blood brain barriers, exist between the blood and the cerebral spinal fluid and brain fluid, respectively. These barriers exist both at the choroid plexus and at the tissue capillary membranes in all the areas of the brain parenchyma **except in some areas of the hypothalamus, pineal gland, and area postrema of the fourth ventricle, where substances diffuse with ease into the tissue spaces**.

Nervous System

Pollutant injury to the blood-brain barrier

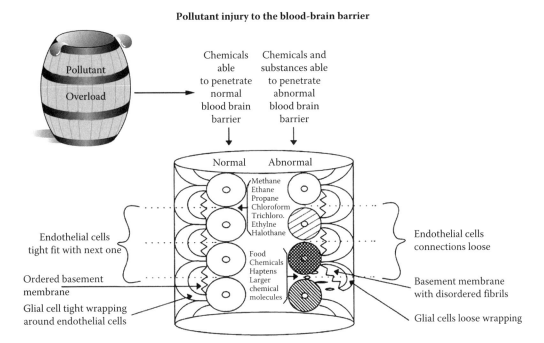

FIGURE 2.33 Pollutant injury to the blood-brain barrier. Some chemical compounds like anesthetics and small aliphatic hydrocarbons come across a normal blood brain barrier. (Modified from Rea, W.J., *Chemical Sensitivity, Volume III: Clinical Manifestation of Pollutant Overload,* Fig. 26.7, CRC Press, Boca Raton, FL, 1750, 1995.)

The ease of diffusion to these areas is important because they have sensory receptors that respond to different changes in body fluids, such as changes in osmolarity and glucose concentration. These responses provide the signals for feedback homeostatic regulation of each of these factors. This ease of diffusion also emphasizes how easy it is for noxious stimuli to trigger these areas of the brain's barriers. Again, the hypothalamus and pineal gland are very vulnerable to triggering by pollutant exposure in these organs, which results clinically in the varied autonomic endocrine triggering responses and sleep disturbances seen in early dyshomeostasis. The area postrema of the fourth ventricle is a target for circulating factor such as amylar, glucogon-like peptide, and the handling of food. Also, autonomic dysfunction can occur as a result of this deregulation

In general, the blood cerebrospinal fluid and the blood brain barriers are highly permeable to water, carbon dioxide, oxygen, and most lipid soluble substances such as alcohol and anesthetics. They are slightly permeable to the electrolytes such as sodium chloride and potassium and impermeable to plasma proteins and most nonlipid soluble large organic molecules. Therefore, the cerebrospinal fluid and blood brain barriers often make it impossible to achieve effective concentrations of therapeutic drugs, such as those containing protein antibodies and nonlipid soluble drugs in the cerebrospinal fluid or parenchyma of the brain. On the other hand, due to its permeability of lipid soluble substances, the brain is particularly susceptible to solvents like hexane, 2 and 3 methyl pentane, xylene, toluene, trichloroethylene, etc. found in the chemically sensitive patient with neurotoxicity. The result from the permeability of these toxics is altered brain function in a large subset of chemically sensitive patients, who present with short-term memory loss, difficulty in concentration, headaches, and dizziness. Frequently they cannot stand on their toes or walk a straight line with their eyes closed (positive stressed and tandem Romberg). They often act as if they are coming out of an anesthetic induced state, which is likely due to neuropeptide release, resulting from the triggering effect of sufficient residual quantities of chlorinated solvents in their brain. **The cause of the**

low permeability from the blood to cerebrospinal fluid and blood brain barrier is due to the manner in which the six-cell layers are joined by the tight junctions of the endothelial cells of the capillaries. The membranes of the adjacent endothelial cells are tightly fused with one another rather than having extensive slit-pores between them, as is the case in most other capillaries of the body. These tight junctions are dependent in part on vitamin A, which often is deficient in chronic disease. **The hypothalamus, pineal, and area postrema of the fourth ventricle only have four-cell layers and thus are more vulnerable to fluid changes.**

While the concept of the blood-brain barrier function is called into question by some physicians, there is general recognition that some substances are not normally excluded from the brain, e.g., anesthetics, analgesics, and tranquilizers. This concept becomes further suspect in light of clinical observations at the EHC-Dallas and at Buffalo. Here, patients in the deadapted state in the controlled environment revealed subtle brain reactions after the majority of challenge tests, whether the substances tested were foods, biological inhalants, or toxic chemicals, and independent of the final end-organ responses produced by these challenges. If this outcome does not disprove the existence of the blood-brain barrier, it does suggest that, at least **in the chemically sensitive and/or chronic degenerative diseased individual, the blood-brain barrier is less intact**. With leakage of fluid through the blood-brain barrier, toxics enter the brain cell, the GRS is triggered, and dyshomeostasis occurs. Blood brain barrier leaks certainly occur with some types of overload, as was evidenced by Haemings studies in which he showed the beef IgG put into animal stomachs ended up in their brains if they were overloaded.[217] If those same type leaks were to occur in the chemically sensitive patients (as they appear to), then, one could understand how some cerebral adverse reactions might occur from ingestion of certain foods. In the cerebral area where there is damaged tissue, it is preconditioned, allowing for a supersensitive triggering. If all of the noxious stimuli induced reactions could be triggered through the GRS, this blood brain barrier could react while intact and without being supersensitive. If the noxious stimulus is in excess chronically, dyshomeostasis will occur. In the milder cases of chronic exposure, the GRS is triggered by noxious stimuli, which may cause the homeostatic mechanism to be activated, but leaves the barrier intact. This triggering of the GRS would be another reason to explain the cerebral involvement of many chemically sensitive and chronic degenerative diseased patients.

Of course, with new technology like the triple camera SPECT brain scan, the uptake of technetium dye for the scan is glutathione dependent. The neurotoxic patients show areas of low uptake, thus emphasizing local glutathione deficiency and thus nerve cell dysfunction. Therefore, in all likelihood, the damage to glutathione is local in certain areas of the brain. However, this damage could be through excess triggering of the GRS due to chronic noxious stimuli, which depletes the glutathione generally, but unequally. These same patients may have excess uptake of dye in other areas, suggesting that pollutants have penetrated the normally impenetrable barrier, probably through a physiological or anatomical weakness.

Chemically sensitive individuals and especially dyshomeostatic patients appear to have a series of leaky barriers in the gut, lung, nasal membranes, blood vessels, subcellular mitochondria and other organelles, and the nervous system. These leaky barriers are caused by either hereditary mucosal defects or acquired injury from noxious stimuli, producing free radical oxidant stress, thereby accounting for blood membrane changes. Alterations in the blood-brain barrier are some of the reasons that the spreading phenomenon occurs in the chemically sensitive individual because it allows for more penetration of noxious stimuli, resulting in triggering of other parts of the brain. The dynamics of the homeostatic mechanisms are overburdened and nutrient fuels depleted, allowing the injury to occur. Since the PG/GAGs have become more fragile and hypersensitive, this situation allows for triggering by lesser and lesser noxious stimuli until a nontoxic substance like food can trigger the homeostatic mechanism. Increased permeability of the barrier allows for more and varied substances to cross and for more symptoms to occur. As the total cerebral toxic load increases, more alteration in the dynamics of homeostatic dysfunction will result. It appears, in general, that nonpolar, lipid-soluble compounds, i.e., xylene, toluenes, and pesticides, usually penetrate the blood-brain barrier more easily, while highly polar compounds like acetic acid or ammonia tend to be excluded. Many of the toxic chemicals found in the blood of patients with chemical sensitivity or chronic degenerative disease are lipophilic, including hexanes, pentanes, toluenes, xylenes, and chlorinated pesticides. Thus, because these chemicals are lipophilic, they are able to penetrate more rapidly, causing neurological and GRS dysfunction. In the immature brain, the "barrier" is generally not as effective, and toxic doses of some compounds, e.g., inorganic lead salts, mercury, and cadmium, may accumulate in the CNS of children. In contrast, the PNS of an adult develops marked effects from lead because the barrier prevents the chemicals from crossing to the CNS and, hence, effects are seen peripherally. A considerable amount of research and speculation regarding the anatomic features that relate to the functional barriers exists[216] because of efforts to define the blood-brain barrier clearly.

Three concepts are involved in the function of the blood-brain barrier.[216–219] First, endothelial cells of blood vessels form tight junctions, which may be dependent on vitamin A. The cells have no pinocytotic vessels in their cytoplasm, and they have no pores in luminal endothelial membranes as they do in the periphery. However, small molecules such as methane, ethane, and/or propane may penetrate to the neuron through the junctions and through the cytoplasm of the endothelial cells and glia.[216] These substances do bother chemically sensitive individuals, with even small exposures making them ill. Thus, the ease with which these substances penetrate the blood-brain barrier in the chemically sensitive individual does not necessitate a leaking barrier. They can enter anyway, triggering localized actions and/or even distal responses through the GRS.

Pinocytotic vesicles may appear in the endothelial cells of the CNS in various pathological conditions that increase the permeability of the blood-brain barrier and disrupt the dynamics of homeostasis. These vesicles may then transport some chemicals across the endothelial lining of small blood vessels in the CNS. When blood-brain barrier permeability is increased, some proteins (e.g., horseradish peroxidase, beef IgG, or chemicals combined with protein, such as Evans blue dye bound to albumin) may be observed in these intracellular vesicles.[219] Conditions that are known to increase vesicles are post irradiation brain edema,[220] ischemia,[221] and hypertension.[222] The latter two conditions are often observed in patients with chemical sensitivity or chronic degenerative disease. **If toxics hook onto proteins (especially albumin), they will also cause problems by generating haptens**. Thus, permeability increases in some chemically sensitive individuals and the spreading phenomenon occurs, allowing more toxic chemicals to enter the brain, which increases brain dysfunction and hypersensitivity while also continuing to trigger more dyshomeostasis. Also certain areas like the median eminence of the hypothalamus, the pineal gland, and the area postrema of the fourth ventricle have only four cell layers rather than six. This diminution of cells will decrease the mechanical barrier for toxics. **Second, the function of the blood-brain barrier may also be explained by the presence of glial cells wrapping around capillaries**. (See Figure 2.33 again.). Much of the brain capillary endothelium is invested with astrocytic processes, which compose a mechanical barrier to prevent noxious stimuli from reaching the neurons in many places. **However, in some brain areas, such as the median eminence of the hypothalamus, the pineal gland, and the area postrema of the fourth ventricle (which is in the upper medulla and important in monitoring the body's milieu, especially for sleep, wakefulness, and food abnormalities) along with the capillaries do not have glial wrappings, and, therefore, toxic substances are able to reach the neurons more readily**.[216] The majority of chemically sensitive patients show neurovascular instability, which may be due to ANS activation via the hypothalamus or due to direct injury to the hypothalamus. This instability may be a precursor of leaks in the barrier due to the continued vessel deregulation. Certainly dyshomeostais will occur, as animal[223] and human[219] studies suggest. The lack of glial wrappings in the medial hypothalamus leaves **the parasympathetic part of the hypothalamus slightly more prone to pollutant injury or other noxious incitant injury**. This finding correlates with our pupillography findings of predominant cholinergic and sympatholytic responses if a stimulatory injury occurs in a large group of chemically sensitive individuals, when measuring head and eye ANS function. It also would correlate with the sympathetic or super adrenalin effect in some patients where the parasympathetic is decreased and the sympathetic is increased, when these effects are measured by heart rate variability.

A **third** explanation of the blood-brain barrier involves the extracellular space, which may also act as a barrier. The extracellular basement membrane, which is part of the ECM extracellular membrane and thus part of the GRS, is located between the endothelial cells of the capillary, the glia, and neurons. **This membrane is an ordered fibrillar mucoprotein structure and may have unique properties in the brain, as it does in the kidney, where it serves as a molecular "sieve" in the transport of molecules needed for cell nutrition. It also functions to remove brain cell wastes and to regulate electro-osmotic flow of water, while simultaneously excluding other substances.**[216] If pollutant damage does occur, this matrix sieve may be altered, allowing other toxicants as well as an improper proportion of nutrients (as seen in some patients with chemical sensitivity) into the brain and prevent orderly removal of toxic wastes. Thus, an increase in dyshomeostatic responses occurs and ultimately, pathological responses. This alteration of the total barrier may partially explain the spreading phenomenon observed in some cases of chemical sensitivity and chronic degenerative diseased individuals.

In the PNS, the blood-neural barrier is variable, and peripheral neuropathy is seen in a large subset of patients with chemical sensitivity or chronic degenerative disease. For example, certainly the hexane and other solvents in these patients' blood will trigger peripheral neuropathy, as seen in over 1000 patients at the EHC-Dallas and Buffalo.

We present one series of 30 patients who had complaints of peripheral neuropathy, including numbness and tingling or loss or increase of sensation or motion. Sensory nerve conduction abnormality was present in all 30 patients, as measured by a neurometer. Associated symptoms included depression, with less intensity of headaches and less intensity of an increase in short-term memory loss, as compared with central neuropathy in 26% of the patients. These symptoms were secondary to the peripheral nerve symptoms and signs. Chronic fatigue, fibromyalgia, and bone pain were present in 57% of the patients. Respiratory symptoms included asthma and other types of bronchospasm in 40% of the patients. ENT complaints of tinnitus, hearing loss, and laryngeal edema were present in 40% of the patients. Cardiovascular symptoms included anaphylaxis, mitral valve disorder, and vasculitits in 27% of the patients; GI complaints of irritable bowel syndrome and malabsorption were seen in 17% of the patients.

Chemical sensitivity was present in all the patients; food sensitivity was present in 33%; biological inhalant sensitivity was present in 23%; and EMF sensitivity present in 3%. Immune deregulation in the form of abnormal T-cells and subsets was present in 37% of the patients (Table 2.12).

In addition, chronically diseased diabetic patients can develop peripheral neuropathy. In both the central and the PNSs, fenestrated epithelial cells have been found in areas that can be permeated by large molecules such as the aforementioned horseradish peroxidase. In contrast to the greater resistance of some areas of the CNS, the susceptibility of these barrier-free areas to some toxic substances, and even secondary food and biologic inhalant triggering, may be due to differences in ease of penetration of the toxicant through the spaces between the epithelial cells.[223–225] The absence of, or weakness in the blood-brain barrier, may account for the different symptomatology seen in some chemically sensitive individuals.[219,223–225]

Noxious Injury after Penetration of the Blood–Brain Barrier

After noxious substances penetrate the blood brain barrier and enter the brain tissue in patients with neurologically involved chemical sensitivity and/or the chronic degenerative disease their effects may still be difficult to determine due to the number of different cell types that potentially may be affected by the individual's total pollutant load, and also, due to the multiple specific incitants that trigger the GRS. However, triple-headed SPECT camera scans of the brain show disturbances of both flow and function, thus indicating a toxicity that results in disordered homeostasis with poor brain function in these disturbed areas (Figure 2.34).

TABLE 2.12
Test Results of 30 Patients with Peripheral Neurotoxicity Ages 29–74

	Percentage %		Percentage %
CMI—3 or below	50	**Intradermal Challenge**	23
		- Biological Inhalant Sensitivity	
T-cell function abnormal	87.5	Foods	33
T_4 lowered	43	Molds	95
T_4 increased	12.5	Orris root	78
T_8 lowered	25	Dust	73
T_8 increased	18.5		
B cells lowered	37		
		Cigarette smoke	70
Triple-Headed Brain SPECT Scan		Formaldehyde	70
(eight patients tested) had central toxicity	100	Algae	68
Pupillography measurement of	100	Trees, Weeds, Grass	68
autonomic nervous system (22 patients tested) showed abnormalities		Unleaded gas/diesel fuel	65
56 Neurometer tests–		Newsprint	61
40 tests were positive		Cologne	61
16 tests were negative		Terpenes	59
30 patients showed at least one abnormality	100	Petroleum derived ethanol	43
Chemical Sensitivity	100	Virus	41
Blood toxics, toluene, xylene, benzene components, chlorinated compounds, styrene, 2-methylentane, 3-methylpentane, n-Hexane, DDT, DDE, chlordane compounds, 2 hexachlorobenzene, B-BHC and Mirex.		Bacteria	36
		Chlorine	35
		Propane Gas	26

Source: Based on EHC-Dallas, 2008; Rea, W.J., Pan, Y., Griffiths, B., Fenyves, I.J., and Curtis, L., *J Nutr & Environ Med.*, 2008.

These SPECT brain scans are dependent upon glutathione peroxidase for uptake of the dye for the functional phase. Therefore, one can infer a deficiency of this enzyme in these areas of low or no function. **Different brain areas usually have different sensitivities to toxicants**. These variations reflect the unique biochemistry of the cells, as well as the differences in the degree of vascularization of brain areas[216] where the GRS may be triggered and the functional demands of pollutant exposure occur (Figure 2.35).

These differences would also show variations in the response of homeostasis. Homeostasis might be robust if sufficient nutrient pools were available for response and more difficult to obtain if the nutrient pools were insufficient. Due to anatomical variation, pollutants may cause greater damage to some areas than to others, accounting for the varied brain dysfunction seen in some patients with neurologically involved chemical sensitivity or chronic degenerative disease. **However, early signs of pollutant injury in this type of chemically sensitive or chronic degenerative diseased individual usually are those of short-term memory loss, minor episodes of confusion, inability to concentrate well, and mild clumsiness**. An inability to stand on the toes with the eyes closed or walk a straight line with the eyes closed (positive stressed and tandem Romberg) is often present. If one can eliminate or neutralize the noxious stimuli rapidly, thus preventing chronic stimulation, the dynamics of homeostasis can not only be obtained, but also maintained. Each anatomical variation will be discussed separately.

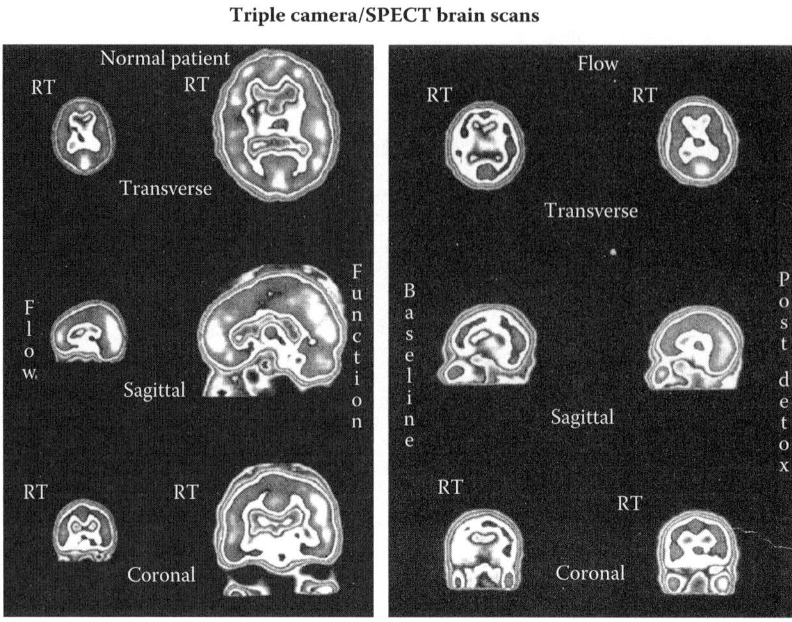

FIGURE 2.34 Triple-headed SPECT brain scans showing normal patient on the left two pictures, and a patient with neurotoxicity (pre- and posttreatment) on the right. Middle picture shows decrease in blood flow. Far right shows restoration of blood flow after detoxification (as seen by increase in shaded areas). (Courtesy of Simon, T.R., and Hickey, D., 1992; Rea, W.J., *Chemical Sensitivity, Vol. III: Clinical Manifestations of Pollutant Overload,* CRS Press, Boca Raton, FL, p. 1804, 1814, 1996.)

UNIQUE BIOCHEMISTRY

In the CNS, there are over 100 billion neurons, which are the nerve cells that make a nerve function. Electrical activity in the neurons is generally initiated in the dendritic process, and then propagated along axons to synapses where it is passed to other neurons. As shown earlier in this chapter, the incoming signals enter the neuron through synapses located mainly on the neuronal dendrites, but also on the cell body. For different types of neurons, there may be only a few hundred or as many as 200,000 synaptic connections from input fibers. Conversely, the output signal travels by way of a single axon leaving the neuron or many outputs. The special feature of most synapses is that the impulse passes in only one forward direction from axon to dendrites. See Figure 2.36.

Major structural features of neurons are thus related to their fundamental functioning, the reception, and transmission of information. Neurons are arranged into a multitude of differentially organized neural networks that determine the function of the nervous system. Therefore, different neurons may have unique biochemistry, which will respond differently to varying total body and specific body loads of noxious stimuli. Thus, responses will vary in individuals.

According to Zimmerberg and Chrernomordik[226] new knowledge in neurobiology is replete with examples of scientists using toxins that bind, neutralize, or cleave physiologically important cellular proteins. The inhibitory binding of saxitoxin and tetrodotoxin to the sodium channel, conotoxin and spider toxins to the calcium channel, α-bungartoxin to the acetycholine receptor, and clostridial toxins to SNAP (soluble NSF attachment protein) receptors (SNARE) proteins that facilitate high-fidelity membrane fusion occur. These toxins were not only instrumental in their respective isolation and investigation but enabled the dissection of basic cellular and physiological mechanisms. Rigoni et al.[227] report how the action of phospholipase A2, a neurotoxic component of snake venom that paralyzes the neuromuscular junction, reveals **a new regulatory mechanism for neurotransmitter release at the synapse. Lysophospholipids and free fatty acids, the hydrolytic products**

Nervous System

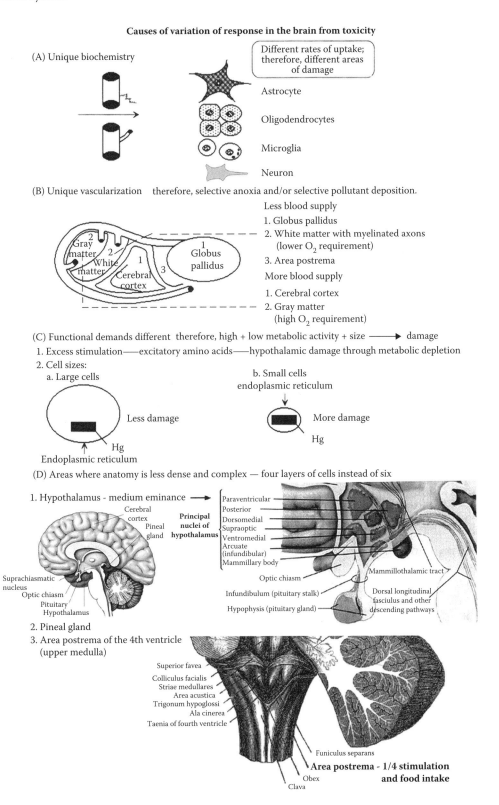

FIGURE 2.35 Variation in response due to unique biochemistry, amount of blood supply, functional demand, and size and layers of cells. (EHC-Dallas, 2005.)

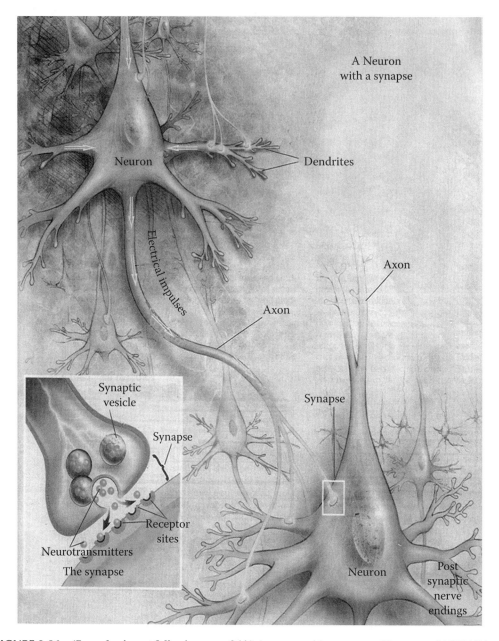

FIGURE 2.36 (See color insert following page 364.) A neuron with a synapse. (Courtesy of ADEAR.)

of this lipase, alter the energetics of the presynaptic membrane, thus affecting its disposition to bend and fuse with synaptic vesicles. This finding may not only explain the longstanding mystery of the molecular mechanism of action of presynaptic neurotoxins that have phospholipase A2 activity, but it also demonstrates the critical importance of membrane lipid composition for synaptic activity.

Phospholipase A2 hydrolyzes stable membrane lipids into lipids that cannot form bilayers. Rather, the lipid products form micelles (lysophospholipids) and inverted micelle-like structures (fatty acids) that reveal a positive and negative spontaneous monolayer curvature, respectively.[228,229] See Figure 2.37.

Nervous System

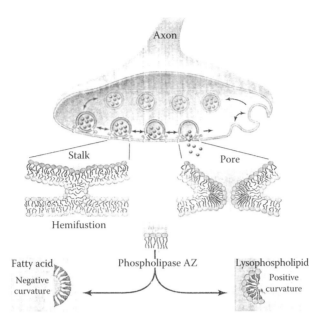

FIGURE 2.37 Lipids alter membrane bending at the synapse to control vesicle fusion and synaptic activity. Membrane fusion between a synaptic vesicle containing neurotransmitter molecules and the plasma membrane of a presynaptic neuron is hypothesized to proceed through the formation of a hemifusion intermediate followed by the formation and expansion of a fusion pore. Lipid mixing between the inner and outer leaflets of the vesicle membrane and plasma membrane (tan) at early fusion stages in likely restricted by the proteins surrounding the fusion site. The connection between contracting leaflets of two bilayers (the stalk) is favored by lipids that support negative leaflet curvature. In contrast, the curvature of the edge of the fusion pore, formed by distal bilayer leaflets brought together by hemifusion, is favored by positive curvature lipids. Phospholipase A2 cleaves lipids that form flat monolayers into fatty acids (negative curvature) and lysophospholipids (positive curvature). (Zimmerberg, J. and Chernomordik, L., Perspectives *Synaptic Membranes Bend to the Will of a Neurotoxin. Science* 310, 1626–1627, 2005. With permission.)

The current understanding of the molecular pathway of biological membrane fusion began with experiments in which these curvature-promoting lipids revealed curvature-promoting lipids revealed curvature-sensitive intermediates during calcium-dependent exocytosis, intracellular vesicle trafficking, and virus–host cell membrane fusion.[228,229] **In a "hemifusion intermediate"[230–235] the contracting leaflets of two apposing membrane bilayers merge. Negative-curvature lysophospholipids inhibit this process. In contrast, the opening of a fusion pore within the hemifusion structure (the pore connects the two aqueous environments delineated by the two apposing membranes) depends on the lipid composition of the distal monolayers of the membrane bilayers**. Opening of a fusion pore is inhibited by unsaturated fatty acids but promoted by lysophospholipids. Thus, biological membrane fusion is essentially lipidic in nature, but catalyzed by proteins. Therefore, solvents can disturb this function resulting in altered membrane activity.

Most importantly for neuronal function, if a fraction of the synaptic vesicles are even transiently hemifused with the presynaptic membrane, they are at the penultimate stage of exocytic fusion, awaiting fusion pore formation and poised to release transmitter very quickly.[236] Thus, hemifusion can explain the extremely fast kinetics of neurotransmitter release (< 100 μs after the intracellular calcium concentration rises) that characterize synaptic exocytosis. The prediction that synaptic vesicles at the active zone make transient hemifusion intermediates (see Figure 2.37) also can explain the spontaneous release of neurotransmitter from vesicles (miniature postsynaptic currents, or minis) that is a hallmark of synaptic activity. Hemifusion intermediates between a vesicle and a planar membrane also display spontaneously occurring fusion pores that link the aqueous interior of the vesicle with the aqueous space across the planar membrane.[237] At the synapse, it is not yet clear how lysophospholipids or fatty acids increase neurotransmitter release either by transient hemifusion intermediate between a synaptic vesicle and the presynaptic membrane at the active zone to directly increase the likelihood of fusion pore formations or indirectly by increasing the resting permeability of lipidic pores in the presynaptic membrane to calcium, thus enhancing calcium-triggered exocytosis. The continual presence of previously intravesicular proteins at the outer surface of the presynaptic membrane suggests that changes in membrane curvature block vesicle internalization and recycling. This may be due to phospholipase A2 activity and its hydrolytic products, which bend membrane bilayers outward[238] opposing invagination of the plasma membrane into a budding endocytic vesicle. Thus, **the energetics imparted by lipids to bilayers are critical to membrane dynamics. Therefore, interference of function can occur from exposure to solvents, heavy metals, and insecticides often seen in the blood of individuals with chemical sensitivity**. These substances can disturb alter or sever membranes prohibiting orderly synaptic function and thus orderly nerve conductivity.

General Principles of Neuronal Physiology

Some nerve fiber responses need to be transmitted to and from the CNS extremely rapidly, i.e., the momentary position of the legs for walking and running, while others can transmit at a slower pace. Therefore, the diameters of nerves range between 0.5 and 20 micrometers—with the larger diameters having the greatest velocities. The speed of these velocities is at times staggering. **The range of conducting velocities is between 0.5 and 120m/sec (in one second an impulse travels longer than a football field)**.

In general, nerve fibers are classified into larger myelinated A or small unmyelinated C fibers, with type A fibers being further divided into alpha, beta, gamma, and delta fibers. **The C fibers constitute more than one half of the sensory fibers in most peripheral nerves, as well as the postganglionic autonomic fibers**.

Signal intensity is one of the characteristics of each signal that is always conveyed. The **spatial summation** occurs by using increasing numbers of parallel fibers, while the **temporal summation** occurs by sending more action potentials along a single fiber. These are receptor fields that pick up strong stimuli. However, **when an individual develops neuropathy, the receptors and the receptor fields may become supersensitive due to partial denervation**, signaling the onset of chronic degenerative disease and/or chemical sensitivity.

The CNS is composed of hundreds to millions of neuronal pools, i.e., the cerebral cortex could be considered one large neuronal pool, as could the basal ganglia, specific nuclei of the thalamus, cerebellum, mesencephalon, pons, and medulla. In addition, gray and white matter would each separately be a part of a neuronal pool. **Each neuronal pool has its own special characteristic of organization that causes it to process signals in its own special way, thus allowing the total consortium of pools to achieve the multitude of functions of the nervous system resulting in central homeostasis**. Even though each pool may have a difference in function, all have general principles involved, which help to integrate their function. Understanding these principles will help the clinician to aid the patient in obtaining and maintaining the dynamics of homeostasis, and thus, prevent hypersensitivity and chronic degenerative disease. See Figure 2.38.

The **first principle** of neuronal function involves relaying of signals through the neuronal pool in which the input and output fibers meet. **Input dendrites arborize over a large area so they can synapse with the output dendrites or cell bodies of the neurons in the pool, making a field effect**. It must be emphasized that discharge of a single presynaptic signal almost never **causes an** action potential in a postsynaptic neuron. Instead, a large number of input signals must discharge on the same neuron, either simultaneously or in rapid succession, to cause excitation. Therefore, in

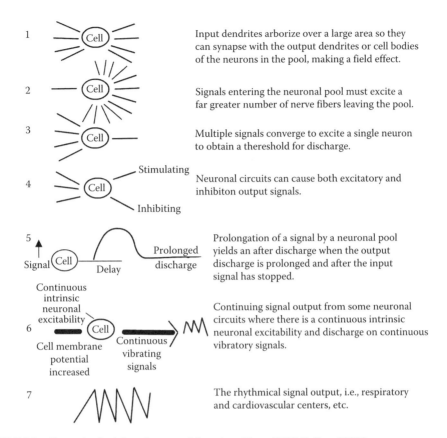

FIGURE 2.38 General principles of neuronal function. (From EHC-Dallas, 2002.)

problems of hypersensitivity, as in chemical sensitivity, one may have super threshold effects, facilitated effects, as well as subthreshold effects, the last of which require a small amount of external noxious stimuli to trigger a response. Sometimes incoming fibers inhibit neurons creating an inhibiting zone, but, usually, if the constant flow of noxious stimuli is not withdrawn, the supersensitivity of the receptor neuron or postsynaptic neuron is increased. The supersensitive neuron fires, with even less intense stimuli, if the constant flow of noxious stimuli is not withdrawn. This phenomenon is seen in many chemically sensitive people.

The **second principle** of neuronal function involves the divergence of signals passing through the neuronal pools. Here, **signals entering the neuronal pool must excite a far greater number of nerve fibers leaving the pool**. There is an amplifying response when the input spreads to an increasing number of neurons, i.e., corticospinal pathway is capable of exciting 10,000 muscle fibers. The second type of divergence is where an input signal is transmitted in two directions, e.g., to the cerebellum and thalamus. Again, if these fibers are rendered sensitive by for example, by trauma, solvents, or pesticides, they tend to fire more easily, and the patient starts on a downward slide to chronic degenerative disease.

The **third principle of neuronal function is convergence, where multiple input signals unite to excite a single neuron in order to obtain a threshold for discharge**. One can get a summation of information from multiple diverse sources such as peripheral nerves, protospinal fibers, corticospinal fibers from the cerebral cortex, and large pathways from the brain into the spinal cord. At times, the clinician may see imbalance of the nervous system in the chemically sensitive or chronic degenerative diseased patients due to this phenomenon; for example, multiple needle sticks in one area may give a single nerve output causing pain.

The **fourth principle of neuronal function is that neuronal circuits can cause both excitatory and inhibition output signals**. Sometimes, an incoming signal to a neuronal pool causes an output excitatory signal going in one direction, and at the same time, an inhibiting signal going elsewhere. This type of circuit is characteristically thought to control all antagonistic pairs of muscles. It is also a reciprocal inhibitor circuit for controlling the dynamics of homeostasis. This principle is important in preventing not only overexcitability in the muscles, but also in the brain. Early brain damage due to head trauma or toxic exposure can result in a hypersensitive patient who may exhibit diseases such as cerebral palsy, traumatic head injury, brain tumors, multiple sclerosis, Parkinson's disease, etc., and/or nonspecific responses. This hypersensitivity, which appears to be due to the head trauma, alternatively, may develop into chronic degenerative disease.

The fifth principle **involves prolongation of a signal by a neuronal pool yielding an after-discharge synapse either both when the output discharge is prolonged and/or after the input signal has stopped**. One sees this synaptic after-discharge in some extremely sensitive chemically sensitive patients, where the stimulus has been removed, but they continue to react for a period of time. Another type of after discharge occurs when a reverberatory or oscillatory discharge circuit occurs due to the positive feedback characteristic of homeostasis. Self-restimulation of circuits by anatomical feedback of the neurons when the facility inhibitory circuits become involved can occur.

The sixth principle of neuronal function involves **the continuous signal output from some neuron circuits where there is continuous intrinsic neuronal excitability and discharge on continuous vibratory signals**. The membrane potential of some neurons is normally high. It can be increased further by an overload of noxious incitants like viruses, bacteria, or toxic chemicals. At times we have seen chemically sensitive patients temporarily lose their sensitivity transiently during a viral infection only to have it return more severely afterward. Presumably, this loss of sensitivity is due to the "upregulation" of the nerve impulse thus "normalizing" the nerve physiology and/or causing a higher level of adaptation.

The final principle of neuronal function is **the rhythmical signal output (e.g., from respiratory centers in the medulla, pons, and cardiovascular centers, etc.)** Here, reverberatory circuits feed excitatory and inhibitory signals in a circular pathway from one neuronal pool to the next. Again, as emphasized in Chapter 1, the rhythm allows the clinician to rapidly diagnose abnormalities (i.e., on electrocardiogram (EKG), blood pressure, or temperature) and disturbed pathology.

Once these seven principles are understood, the state of the stability and instability of the circuits must be considered. In general, the amount of inhibition of impulses and fatigue of the circuits stop overstimulation. **However, the long-term sensitivities of synapses can be changed tremendously by downgrading the number of receptor proteins at the synaptic sites when there is overactivity and upgrading the receptors when there is underactivity**. This regulation is done by increasing or decreasing the receptor protein output from the smooth endoplasmic reticulum and golgi apparatus. If this process is disturbed, as in periodic (hypersensitive) or aperiodic (chronic degenerative diseases), the patient's conditions will be more fragile, easily causing alterations in the dynamics of homeostasis. This constant chronic over stimulation may cause up or down regulation of the receptors and their numbers, causing hypo- or hypersensitivity responses as seen in the chemically sensitive and chronic degenerative diseased individual.

Knowledge of receptor and neuronal physiology and environmental influences is extremely important in understanding the development of chemical sensitivity and chronic degenerative disease since it appears that the regulation of receptor and neuronal function proteins plays a critical role in the homeostatic process.

Unlike most other structures in the brain, the hippocampus is able to produce new neurons throughout adult life as it contains a population of neural stem cells. It is believed that **neurogenesis contributes to the primary task of the hippocampus, that of coordinating learning and memory**.[239] Radiation will destroy the stem cells and stop neurogenesis.

Two-way communication between neurons and nonneuronal cells, the glia, is essential for axonal conduction of synaptic transmission and information processing and, thus is required for neuronal functioning of the nervous system. Signals between neurons and glia cells include ion fluxes, neurotransmitters, cell adhesion molecules, and specialized signaling of the neuron. In contrast to the serial flow of information along chains of neurons; glia communicate with other glial cells through intracellular waves of calcium and via diffusion of chemical molecules. **By releasing neurotransmitters and other extracellular signaling molecules glia can effect neuronal excitability and synaptic transmission and perhaps coordinate activity areas in a network of neurons**.

Though neurons are very important in brain function, the **glia cells** are also extremely important. **They vastly outnumber the neurons by nine hundred billion to one hundred billion ratio**. The three types of glia cells, which make up 90% of the brain, are astrocytes (Schwann cells [peripheral], oligodendrites [central]), and microglia.

Microglia differ in their roles in the CNS function, resulting in variation of the dynamics of efficient homeostasis and in their sensitivity to toxic agents or other noxious stimuli. Their biochemistry is unique as well as their anatomy. Amorphous gray matter is packed with cells distinct from neurons. For example, *astrocytes,* the most numerous **glia cells** in the brain, are closely associated with neurons in gray matter and are **the nurse cells of neurons because they are thought to be essential in maintaining the stable microenvironment needed for neuronal function**. Astrocytes extend their tendrils in all directions toward synapses. They aid in nourishment and growth by bringing nutrients from the blood vessels. They also maintain a healthy balance of ions to the brain and ward off pathogens that evade the body's initial immune system.

The glia cells help the neurons with a transmission of signals. For example, astrocytes convert glucose to lactate to feed the neurons. They also protect the neurons by forming an oily, waterproof sheathing around the blood vessels, **which block water-soluble toxic agents from entering the brain and causing damage**. Because of their tendril type arms, astrocytes alter the number of neural synapses and their signaling strength. They may be architects of memory and learning, coaxing neurons to strengthen their connections along well-worn mental pathways. Some opinion suggests that astrocytes may contribute as much to communication as neurons do. In vitro evidence showed that signals across synapses of neurons were often sluggish or failed until astrocytes were placed in the culture. Then the number of signals increased seventyfold.[240]

The message strength at any one synapse does not increase when the astrocytes are added to the neurons in culture, but the number of synapses carrying the message does. The astrocytes **instruct the neurons to spread more synapses**. This influence of astrocytes essentially adds more eyes and ears to the neural communications system. **Astrocytes contain neurotrophic factor (brain-derived neurophic factor (BDNF)), which neurons need in order to grow**. More neurotrophic factor is secreted after injury so a route of repair is started. BDNF ups the firing rate and efficiency of neurons in the hippocampus, which is critical for learning and memory. Winter[241] showed the need for neurons and electricity in order for neurons to grow and integrate into implants. Nedergaard[242] showed that astrocytes talk back to neurons, thus emphasizing reciprocal feedback for obtaining and maintaining the dynamics of homeostasis.

Nedergaard[242] found that astrocytes respond to stimuli with a slow increase in the release of calcium. The calcium signal can move from one astrocyte to another through channels that directly connect to the interior of cells. Newman[243] actually demonstrated this calcium signal to go through the network of all astrocytes, seemingly spreading the message. **Astrocytes are part of a circuit that is responsible for neuroprocessing**. These glia cells participate in a second-to-second neural communication with the ongoing electrical transmission in neurons. **Astrocytes talk back to neurons by releasing glutamate (a signaling chemical at the synapses)**. There is a puzzling thought associated with the storing of memory. Why do messages go slowly along astrocytes and rapidly along neurons? It seems reasonable to suggest that the fast long distance signaling of neurotransmission might not build a sturdy mental pathway. In contrast, the slow system of astrocyte communication might maintain a localized route for longer memory storage. One can see that if either the neuron or the astrocyte were disturbed or damaged by an external noxious stimuli, the dynamics of homeostasis could be disturbed, actually causing dyshomeostasis. In addition, the unique biochemistry of the metabolism of each cell or cell group would cause a different response to pollutant entry.

Any pollutant damage to these cells creates dyshomeostatic reactions with instability of neuronal function. This type of damage is seen in a subset of individuals with neurologically involved chemical sensitivity as demonstrated by symptoms of memory loss, brain fog, lack of concentration, and altered judgment.[244] **Glial cells also communicate among themselves in a separate but parallel network to the neural network, influencing how well the brain performs and responds to noxious stimuli.**[245] See Figure 2.39.

Schwann cells are glia cells that surround synapses where a peripheral nerve meets muscle cells. These glia cells also surround nerves in the rest of the body.

When an axon fires, it creates voltage that is sensed in the neuron's membrane allowing for the influx of calcium. Initially after axon firing, there are no changes in the peripheral Schwann cells or central oligodendrocytes, but within 15 seconds calcium begins to influx into the cytoplasm. When excited, glial astrocytes release ATP. This substance grabs onto other astrocytes and opens up more calcium channels, which again release more ATP. The calcium and ATP influx travel to the nucleus causing various genes to switch on. By firing to communicate with other neurons, an axon could instruct the readout of genes in a glial cell and thus influence its behavior. ATP does not inhibit oligodendrocytes from forming myelin in brain tissue. However, adenosine alone stimulates the cells to mature and form myelin. Alteration of the ATP pathway may be one of the causes of weakness seen in the patient with chemical sensitivity.

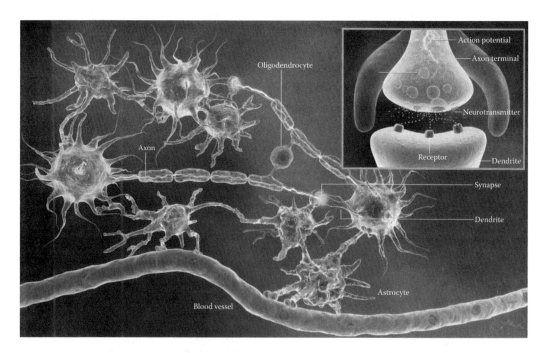

FIGURE 2.39 (See color insert following page 364.) Glia and neurons work together in the brain and spinal cord. A neuron sends a message down a long axon and across a synaptic gap to a dendrite on another neuron. Astrocyte glia bring nutrients to neurons, as well as, surround and regulate synapses. Oligodendrocyte glia produce myelin that insulates axons. When a neuron's electrical message [action potential] reaches the axon terminal (*insert*), the message induces vesicles to move to the membrane and open, releasing neurotransmitters [signaling molecules] that diffuse across a narrow synaptic cleft to the dendrite's receptors. Similar principles apply in the body's peripheral nervous system, where Schwann cells perform myelination duties. (Fields, D.R., *Scientific American*, 57, 2004. With permission.)

Nedergaard[242] has observed that synapses increased their electrical activity when adjacent astrocytes stimulated calcium waves. Such changes in synaptic strength are thought to be the fundamental means by which the nervous system changes in response to thought experience. Brain plasticity might play a role in the cellular basis of learning. **Thus, discrete astrocyte circuits in the brain coordinate activity with neuronal circuits helping to activate neurons and strengthen their signals at the synapses.** A chemical messenger, a protein, thrombopolastin, spurs synapse building. The proportion of glial cells to neurons increases as animals move up the evolutionary ladder.[245]

The *oligodendrocyte* in the CNS intrafascicular cells has a role similar to that of the peripheral Schwann cell and invests neuronal axons with spiral wrappings of myelin. Both the composition and form of the central and peripheral myelin differ, and the relation between the axon and the myelinating cell is also dissimilar: an oligodendrocyte may enclose several axons in separate myelin sheaths, whereas Schwann cells ensheath only one axon. See Figure 2.40.

Some oligodendrocytes are not associated with axons, and some of these are either an adult form of the oligodendrocyte precursor cell or a perineuronal (satellite) oligodendrocyte with processes ramifying about neuronal somata, with unknown functions. **These cells have a unique biochemistry and may respond differently to various noxious stimuli**.

The densities of oligodendrites vary from region to region, from heavily myelinated (i.e., pontocerebellar tract [estimated at 64,000/mm^3] to sparse and even unmyelinated,[248] fibers. **Again these groups would have a unique response to pollutant entry, thus the possibility is high of chemical sensitivity or chronic degenerative disease developing**. Oligodendrocytes express galactocerebroside, carbonic anhydrase II, myelin basic protein, phosphodiesterase, and transferrin.

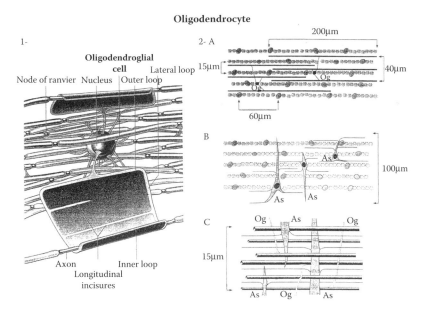

FIGURE 2.40 (1) Oligodendrocyte soma (top center) attached to numerous myelin sheaths that have been unfolded to various degrees to demonstrate the enormous surface area they occupy. Note also the displacement of oligodendrocyte cytoplasm to narrow ridges in the flattened sheet of myelin. (2) Arrangement of radial and longitudinal processes of oligodendrocytes (A) and astrocytes (B) to form a "woven" meshwork of processes (C). A. Shown against a background of the interfascicular glial nuclear rows, the types of radial (stem) and longitudinal (myelinating) processes of two oligodendrocytes (Og, black squares) are illustrated. B. Three astrocytes (As) illustrating typical radial and longitudinal processes. C. Higher magnification view of 'woven' meshwork of oligodendrocytic (Og) and astrocytic (As) processes. Typical dimensions: core-to-core distance between interfascicular glial rows, 15 μm; interastrocytic distance within a row, 60 μm; oligodendrocytic radial span, 40 μm; length of internode 200 μm; astrocytic radial span, 100 μm. (Modified from Suzuki, M. and Raisman, *G., Glia*, 6, 222–235, 1992. With permission.)

Antibodies can form to any of these molecules, giving unique responses. **We find many antimeylin antibodies in implant patients, autistic, and those patients who were exposed to neurotoxic solvents.**

Oligodendrites originate from the ventricular neuroectoderm and the subependymal layer in the fetus and continue to be generated from the subependymal plate postnatally. Some stem cells may migrate and seed into the white and gray matter to form a pool of adult progenitor cells, which may later differentiate to replenish lost oligodendrites and possibly remyelinate pathologically demyelinated regions. These altered areas will give altered responses to pollutant entry.

Microglia **are phagocytic cells with primary functions that resemble peripheral leukocytes**, but their response to toxicants has not been widely investigated.[223] Microglia are small dendritic cells found throughout the CNS, including the retina. They appear to derive from monocytes and/or their precursors, which invade the developing nervous system. Hematogenous cells pass through the walls of the parenchymal and the meningeal blood vessels and probably, also, invade neural tissue. Though amoeboid at first, they eventually lose their motility and transform into typical microglia (fixed tissue macrophages) bearing branched processes, which ramify in nonoverlapping territories within the brain. All microglia domains, defined by their dendritic fields, are equivalent in size and form a regular mosaic throughout the brain. The expression of micoglia-specific antigens changes with age, many becoming down regulated as microglia attain the mature dendritic form. These mature forms are stable, and turnover is rare. **Microglia form a stable network of antigen-presenting-cells throughout the brain and spinal cord**. Microglia are activated by traumatic and ischemic injury and in many diseases (i.e., Parkinson's, amyotropic lateral sclerosis (ALS), multiple sclerosis), and they become phagocytes. Certainly, if these cells are triggered by pollutant or other noxious stimuli or excess and/or chronic phagocytosis, nutrient deregulation and deficiency may occur. Thus, the dynamics of homeostatic regulation will be altered if the noxious stimulus is chronic.

The brain is an immuno-privileged tissue that exhibits dampened adaptive immunity in response to injury, infection, or tumor formation. However, the brain does exhibit a robust innate immune response thanks to the microglia, which defend against invading microorganisms and clean up by engulfing the debris of dying cells. In addition, the inflammatory mediators released by microglia during an innate immune response strongly influence neurons and their ability to process information.[246]

Microglia also appears to be pivotal in causing inflammation of the brain, just as macrophages are peripherally. Under normal conditions the hippocampus contains neural stem cells that give rise to adult stem cells. These cells have a glia-like morphology and vascular feet. These stem cells are thought to give birth to neuronal progenitor cells and astrocytes. Neuronal progenitor cells with a high proliferative capacity then give rise to granule cell neurons. During brain inflammation induced either by pollutants such as irradiation or by infection, microglia releases inflammatory cytokines. These cytokines, which include tumor necrosis factor-α (TNF-α), IL-1β, and IL-6, are secreted by activated microglia and invading macrophages. IL-6 interferes with the production of new neurons in adult brain tissue, perhaps by perturbing the hippocampal stem cells' microenvironment. **IL-6 may induce bipotent neural stem cells to generate more astrocytes than neuronal progenitor cells**. Alternately, IL-6 may cause a reduction in proliferation of neuronal progenitor cells or trigger them to undergo cell death. IL-6 and its downstream JAK-STAT signaling pathway have been implicated in the selective differentiation of cerebral cortical precursor cells into astrocytes.[247] Inhibition of neurogenesis by IL-6 may be due to increased production of astrocytes (or perhaps other glial cells) at the expense of neuronal precursor cells, since they share the same stem cell. Alternatively, inhibition of neurogenesis by IL-6 may be a consequence of the decrease in neuronal progenitor cell proliferation or an increase in the number of these cells undergoing apoptosis.[246] **Again, a delicate homeostatic balance occurs in the process of neurogenesis and prevention of inflammation.**

Microglia are usually helpful cells that may even supply the neurons with trophic factors such as BDNF. It seems that the brain's innate immune response to injury is a double-edged sword, acting simultaneously beneficially and detrimentally depending on the homeostatic mechanism and the chronicity and severity of injury. The fact that microglia secrete potent growth and differentiation factors such as BDNF and IL-6, with effects that reach far beyond the immune system, supports the idea that they may be central players in repairing brain tissue and maintaining its integrity. Moreover, **microglia may contribute to the rearrangement of neural connections and hence to the plasticity of the normal brain**.[246]

In individuals with chemical sensitivity or chronic degenerative disease, the variation and degree of vascularization account not only for varied GRS responses, but also for some of the variation in sensitivity of different brain areas to hypoxia and to exposure to toxic chemicals. For example, the globus pallidus has the same density of cell bodies as do other neural tissues, but it is more poorly vascularized than the cerebral cortex (hence the name pallidus, meaning pale). White matter is generally less vascularized than gray matter,[248] but the lower oxygen requirement of the myelinated axons, which make up much of the white matter, makes white matter generally less sensitive to pollutants than gray matter with its neuropil ratio.[216] Triple-headed camera SPECT brain scans on the subset of neurologically involved chemically sensitive patients at the EHC-Dallas usually show gray matter involvement, thus confirming this anatomical principle.

Functional demands on cells may cause different homeostatic and dyshomeostatic responses to toxicants (e.g., excitatory amino acids may damage hypothalamic neurons by causing excessive stimulation and metabolic exhaustion of the cells).[249] Cell exhaustion appears frequently in the patients with chemical sensitivity and partially accounts for the onset of various symptoms including weakness, difficulty in arising in the morning, and difficulty in responding to challenging stimuli.

Quantitative differences in essential cell components may make one cell type more sensitive than another type (e.g., small neurons such as granule cells in the cerebellum and visual cortex are preferentially killed or rendered dysfunctional when the whole brain is exposed to methylmercury). This reaction was seen in the Minamata, Japan, disaster where patients were damaged with methylmercury in the seafood they consumed.[250] The amount of cytoplasm and rough endoplasmic reticulum, which binds mercury or any xenobiotic, is less than in larger cells, and thus the small cells may be more likely overwhelmed by their effects.[251] Recycled reduced glutathione supplements are needed to neutralize the mercury. It may not be supplied rapidly enough, since mercury irreversibly binds glutathione to maintain homeostasis.

Studies of the microchemistry of the brain reinforce concepts of its structure as encompassing separate, diverse areas. **High concentrations of norepinephrine, serotonin, acetylcholine, and dopamine are found in various pathways of the central phylogenetic "old brain," which includes the hypothalamus, reticular formation, basal ganglia, and limbic systems**.[216] These pathways are the centers of homeostatic regulation. Pollutants, as well as psychogenic drugs, have now been shown to alter homeostatic microchemistry, thereby altering functions.[252,253] Many individuals with chemical sensitivity and chronic degenerative disease who are challenged with these substances experience exaggerated responses of function related to these old brain areas, once the dynamics of homeostasis are altered. However, administration of the appropriate neutralizing doses of the challenge substance such as serotonin, foods, molds, and certain chemicals, often curtails the reactions, suggesting that these individuals temporarily undergo an altered sensitivity resulting in altered homeostatic responses in these old brain areas. As shown previously, the central control of the dynamics of homeostasis is thought to be the hypothalamus. Since this area is easily injured, one usually finds homeostatic disruption of the hypothalamus in the individuals with chemical sensitivity and chronic degenerative disease.

Studies performed at the EHC-Dallas in controlled environments using intradermal provocation and neutralization techniques, as well as oral and inhaled challenge, on patients whose total load had been reduced in the ECU have revealed that in most people the symptoms are reproducible. Intradermal injection of histamine, serotonin, epinephrine, norepinephrine, dopamine, and methylcholine, or ingestion or inhalation of some toxic chemicals triggers varied dynamics of homeostatic responses. These challenge techniques under environmentally controlled conditions, along with future observations, will perhaps give us a clinical test for evaluating areas in the brain that are not properly functioning. However, experimental techniques for accurate assessment of substances that stimulate or suppress the transmitters are presently, too unsophisticated to allow generalizations to be made regarding the role of specific brain structures or areas and amine levels for specific function.

Usually, these responses from described challenges appear to be homeostatic. However, depending upon the severity of the noxious stimuli, its kinetics, and the chronicity of the incitant, one may see symptoms lasting for a long time or be on a continuum if their detoxification mechanisms are not functioning properly. The dynamics of homeostasis are then difficult to obtain and maintain.

Despite the limitations of present assessment techniques for use in the physician's office, some patients with chemical sensitivity and chronic degenerative disease can reliably and very precisely use their reactions to identify a pollutant to which they have been exposed. They may have, for example, one type of reaction after a histamine injection and another type after serotonin injection. Reactions may also vary when they eat certain foods or when they are exposed to certain inhaled toxic chemicals. Further, this select number of patients may be able to differentiate whether their reaction is histamine or serotonin induced. In our studies at both our clinics, it has become evident to us that injections of neurotransmitters, or injections that cause the release of endogenous neurotransmitters, can provoke a neurological response that gives a finite homeostatic reaction in many patients with chemical sensitivity and chronic degenerative disease.

Certain large cells in the CNS, such as **cortical and hippocampal pyramidal cells, cerebellar Purkinje cells, and motor cells in the ventral horn of the spinal cord**, have unusually large nuclei, and the DNA is predominantly present as euchromatin, the form of chromatin most closely associated with transcription.[254] These cells often have several nucleoli. **All these structural differences point to high metabolic activity in the cells and thus increased susceptibility to hypoxia and pollutant damage with the altered dynamics of homeostasis**.[216] The dyshomeostatic responses seen in the subset of patients with pollutant-triggered, neurologically involved chemical sensitivity are often those that relate to the functions of these cells of high metabolic activity.

Principles of Response after Toxic Exposure

In the individuals with chemical sensitivity and chronic degenerative disease, there are three principles that govern the responses of the nervous system to pollutants and thus, alter or disrupt the dynamics of homeostasis. These include the following: (1) **selective damage to one or more areas or components caused by selective exposure due to differences in ease of penetration through barriers**, (2) **selective anoxia or hypoxia via differences in blood flow and metabolic requirements of some elements**, or (3) **specific sensitivity resulting from qualitative or quantitative chemical differences in cell components**. Identification of the selective nature of the damage or homeostatic disruption from toxic agents is essential in determining the mechanism of action of toxicants. This selective nature has further value in analyzing the relation of brain structure to function[216] in various types of neurologically involved chemically sensitive and even chronic degenerative diseased patients. In addition, since these principles hold for pollutants, they also are valid for any noxious stimuli, which can disrupt or alter homeostasis. Once noxious stimuli enter the nervous system, one can see a rush to homeostasis. The goal is to keep chronicity of the load at a minimum to allow energy efficient homeostasis to occur. Prevention of injury occurs when the homeostatic mechanisms are allowed to function easily with adaptation being stopped.

Reversibility vs. Irreversibility of Cell Damage after Pollutant Exposure

The extent to which irreversible toxicity and sensitivity differ quantitatively (total body load) and in virulence (specific body load) from the therapeutic actions of drugs is appropriately discussed in

Mechanism of Acute Central Nervous System Injury

The ultimate event in environmentally triggered irreversible damage to the nervous system of the individual with chemical sensitivity and chronic degenerative disease can be described simply as neuronal death.[216] **It has always been thought that neurons cannot divide and be replaced**.[216] However, as previously shown in this chapter, new data suggests that there are stem cells in certain parts of the brain (i.e., hippocampus) that allow for regeneration of nerve cells. Until these stem cells were discovered, the only apparent exception for regeneration was the cilia of the olfactory nerve endings at the root of the nose, which are continually replaced.[258] These olfactory nerves are frequently damaged in the patients with chemical sensitivity. Though repair of this nervous tissue occurs easily, there frequently appears to be memory in the replacement tissues for certain chemicals, which allows the sensitivity to remain or even to progress. The memory suggests that the dynamics of homeostasis have returned to functional levels but that the gain characteristic (see Chapter 1) is in action, and there is a slight loss of equilibrium, which is compensated for by demanding more energy input in order to prevent dyshomeostasis from occurring. In other words, the duration of response is not quite as great, yet the energy required is increased, and homeostasis still occurs.

In the patient with chemical sensitivity and chronic degenerative disease in the early stages of illness, a condition appears to occur in which the cell membranes may be injured, though not severely enough to cause irreversible damage. This series of events is seen frequently posttrauma, as well as in the recently induced chemically sensitive individual. **Often, normal function may be restored for an individual even after considerable damage to the nervous system from exposure to toxic substances**. However, this process requires much more energy to obtain complete dynamic homeostasis. Redundancy of function in a population of neurons and plasticity of organization are the methods by which restoration of function is presumed to occur after the death of some neurons.[216] However, since it appears that nerve cells can regenerate from stem cells in the brain, the goal is to keep the redundancy and plasticity to a minimum. **These phenomena of redundancy and plasticity allow the process of adaptation to occur. Long-term adaptation should be guarded against, since it is a process that guarantees over time and without noxious incitant removal that total brain dysfunction and end-organ failure will occur**. The dynamics of homeostatic mechanisms become dysfunctional and, with a change in information input and nutrient drain, metabolism changes followed by tissue alterations. Usually, local hypoxia (sol to gel, gel to sol, nonspecific mesenchyme reaction) changes metabolism and tissues into autonomous irreversible fixed-named diseased tissue.

When acute end-organ disease (i.e., stroke) occurs, many tissue changes result. Necrotic cell death occurs in the core of the lesion where hypoxia is most severe, **and apoptosis occurs in the penumbra, where collateral blood flow reduces the degree of hypoxia**[259–262] (Figure 2.41). Apoptotic death is also a component of the lesion that appears after brain or spinal cord injury.[263–267] In chronic neurodegenerative diseases, it is the predominant form of cell death.[268–270]

As shown in Chapter 3, necrotic cell death, the lysozymes are released, destroying the cell membrane and its components from the outside. In apoptosis, the microcomponent membranes are destroyed, and the cell is destroyed from within.

In apoptosis, a biochemical cascade **activates proteases** that destroy molecules that are required for cell survival and others that mediate a program of cell suicide. During the process, the cytoplasm condenses, the mitochondria and ribosomes aggregate, the nucleus condenses, and chromatin aggregates. However, lysozymes are not released. After its death, the cell fragments into "apoptotic bodies," and chromosomal DNA is enzymatically cleaved to 180-bp internucleosomal fragments. Other features of apoptosis are a reduction in the membrane potential of the mitochondria, intracellular

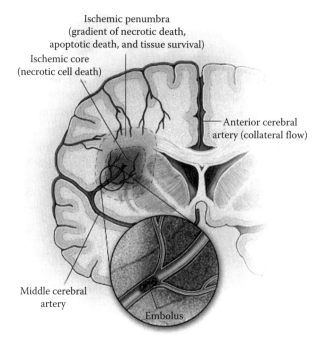

FIGURE 2.41 (See color insert following page 364.) An embolus in the bifurcation of the middle cerebral artery. The territory perfusion by this artery and areas with little or no collateral flow are subjected to extreme hypoxia and necrotic cell death. In the penumbra, where there is some degree of collateral blood flow, a gradient of tissue perfusion establishes a threshold among necrotic cell death, apoptotic cell death, and tissue survival. (Friedlander, R.M., *N Engl J Med.*, 348(14), 1366, 2003. With permission.)

acidification, generation of free radicals, and externalization of phosphatidylserine residues.[271–275] The rational development of target-based strategies for the treatment of diseases in which apoptosis is prominent requires an understanding of the molecular mechanisms of programmed cell death. As recently as 20 years ago, the mediators of this process were, for the most part, unknown. Beginning in 1993, a series of seminal studies of the nematode *Caenorhabditis elegans* identified several genes that control cell death.[276] In this worm, four genes are required for the orderly execution of the developmental apoptotic program. The ced-3, ced-4, and egl-1 genes harbor extra cells.[276–278] By contrast, ced-9-deficient worms have diffuse apoptotic cell death, indicating that this gene functions as an inhibitor of apoptosis. Metazoan homologues of ced-3 (caspases), ced-4 (Apaf-1), ced-9 (Bcl-2), and egl-1 (BH3-only proteins) have been identified.[277, 279–282]

The major executioners in the apoptotic program are **proteases known as caspases (cysteine-dependent, aspartate-specific proteases)**.[281,283] Caspases are cysteine proteases that are homologous to the nematode ced-3 gene product. The interleukin-1β-converting enzyme (also known as caspase 1), the founding member of the caspase family in vertebrates, was identified by its homology to ced-3.[277,279] Thus far, 14 members of the caspase family have been identified, 11 of which are present in humans.[277] Caspases directly and indirectly orchestrate the morphologic changes of the cell during apoptosis.

Caspases exist as latent precursors, which, when activated, initiate the death program by destroying key components of the cellular infrastructure and activating factors that mediate damage to the cells. Procaspases are composed of p10 and p20 subunits and an N-terminal recruitment domain. Active caspases are heterotetramers consisting of two p10 and two p20 subunits derived from two procaspase molecules (Figure 2.42). Caspases have been categorized into upstream initiators and downstream executioners. Upstream caspases are activated by the cell-death signal (e.g., tumor necrosis factor α, the Bcl-2 family), generating a truncated fragment with proapoptotic

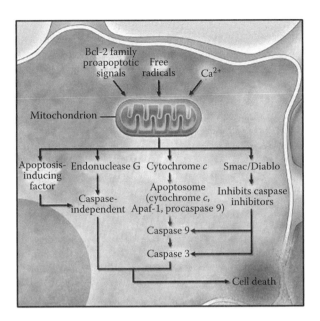

FIGURE 2.42 (See color insert following page 364.) Key mediators of the caspase pathway in the mitochondria. Three main signals cause the release of apoptogenic mitochondrial mediators: proapoptotic members of the Bcl-2 family, elevated levels of intracellular calcium, and reactive oxygen species. Four mitochondrial molecules mediating downstream cell-death pathways have been identified: cytochrome c, Smac/Diablo, apoptosis-inducing factor, and endonuclease G. Cytochrome c binds to Apaf-1, which, together with procaspase 9, forms the "apoptosome," which activates caspase 9. In turn, caspase 9 activates caspase3. Smac/Diablo binds to inhibitors of activated caspases and causes further caspase activation. Apoptosis-inducing factor and endonuclease G mediate caspase-independent cell-death pathways. (Friedlander, R.M., *N Engl J Med.*, 348(14), 1368, 2003. With permission.)

activity.[284] In addition to cytochrome c, other modulators of cell death within mitochondria are released during the apoptotic process.[285]

To control aberrant caspase activation, which can kill the cell additional molecules inhibit caspase-mediated pathways. Among these are proteins that are inhibitors of apoptosis. These inhibitors interact directly with modulators of cell death. For example, the X-linked inhibitor of apoptosis and the neuronal inhibitor of apoptosis are proteins in neurons that directly inhibit caspase 3 activity and protect neurons from ischemic injury.[286–288]

Caspases have a pivotal role in the progression of a variety of neurologic disorders. Despite the various causes of such disorders, the mechanism of cell death is similar in a broad spectrum of neurologic diseases.[289–291] However, the trigger of aberrant caspase activation in most of these diseases is not well understood. In acute neurologic diseases, both necrosis and caspase-mediated apoptotic cell death occur.[263,264,292–294] By contrast, in chronic neurodegenerative disease, caspase-mediated apoptotic pathways have the dominant role in mediating cell dysfunction and cell death.[295–298] A primary difference between acute and chronic neurologic diseases is the magnitude of the stimulus causing cell death. **The greater stimulus in acute diseases results in both necrotic and apoptotic cell death, whereas the milder insults in chronic diseases initiate apoptotic cell death.**

Ischemic stroke was the first neurologic disease in which the activation of caspase (caspase 1) was documented.[299] Moreover, inhibition of caspases reduces tissue damage and allows remarkable neurologic improvement.[299–301] Activation of caspases 1, 3, 8, 9, and 11 and release of cytochrome c have been demonstrated in cerebral ischemia,[302–305] and the Bcl-2 family has also been incriminated.[306,307] Mice that express a dominant-negative caspase 1 construct, or that are deficient in caspase 1 or caspase 11, have significant protection from ischemic injury.[299,303,308] Pharmacologic pretreatment of mice with a broad caspase inhibitor or with semiselective inhibitors of caspase 1 and caspase 3, or delayed treatment with a caspase inhibitor, protect the brain from ischemic injury.[301,292]

There is a pattern of combined necrotic and apoptotic cell death after ischemic or traumatic injury.[261,265–267,292,293] As already stated, in ischemia, necrotic cell death occurs in the core of the infarction, where hypoxia is most severe, and leads to abrupt cessation of energy supply and acute cellular collapse. Conversely, in the ischemic penumbra, the degree of energy deprivation is not as severe because collateral vessels supply the region with oxygenated blood. In this case, the cell must reach a critical threshold of injury to activate the caspase cascade. Before this threshold is reached, however, a compromise in neuronal energetics can cause cell dysfunction before cell death. See Figure 2.43.

Again, as shown throughout this book, **it is imperative to keep the patient in the alarm phase of Selye in order to prevent severe catastrophic or even chronic neurological diseases**. This will allow the patient to be vigilant to noxious stimuli entry and remove them as soon as possible. At the EHC-Dallas, we have observed over 15,000 patients with chemical sensitivity and chronic degenerative disease with neurological damage. In these patients, the neurons that were not killed usually returned to normal function once the total specific pollutant load was decreased and nutritional competence restored. Perhaps new brain cells were grown, but more likely, damaged cells were repaired. Of course, the patients were kept in the deadapted alarm stage with their total body pollutant load reduced to prevent further damage.

Conditioning is, in fact, dangerous due to masking of brain damage from apoptosis by the adaptation phenomena.[310] When some neurons die, other cells with the same function may be adequate to maintain normal activity, or, failing this, other neurons may assume the needed function. In situations where neither course is possible (for example, in extensive damage to a specialized population of neurons or brain nuclei), some loss of function inevitably results.[216]

However, **if mild damage occurs, adaptation restores normal function. The continued accommodation reaction of cells depends upon the degree to which metabolism is disturbed by the pollutants or other noxious incitants, as well as their distribution, the intake and maintenance of adequate detoxification and repair nutrients, and the ability to limit additional pollutant exposures**.

The average educational level of patients coming to the EHC-Dallas is 14 years and an IQ of 115.[311] Didriksen and Butler[312] have found that most of the dynamics of the homeostatic mechanisms come into play in these brain damaged highly intelligent patients. The brain function tests are designed for the average population with an IQ of 100. Therefore, many people with an IQ of 150–200 may show superior on these tests, rendering the tests useless to evaluate their brain function. However, it still is often possible to evaluate early brain damage in these people by finding what they could do previously. One can then explain the difference in function after challenge with noxious stimuli. Clearly, the dynamics of homeostasis are altered in these people with resulting brain fog, short-term memory loss, etc. Some degree of recovery of function usually occurs after nonfatal neurotoxic reactions, as the dynamics of the homeostatic mechanisms come into play. When cell death is not involved, the neurotoxic or sensitivity reaction lasts only until the toxic agent is removed or metabolized, or until cell constituents altered by the toxic exposure have regenerated. These latter cell changes caused by exposures account for the continued brain dysfunction seen in some neurologically involved patients with chemical sensitivity or chronic degenerative disease even though their total load has been reduced and many pollutants removed. **In these recently treated neurologically involved chemically sensitive or chronic degenerative diseased patients, proper distribution of replacement nutrients including oxygen may be initially difficult to achieve, until all areas of membrane transport and blood-brain barriers have healed**. Often in the brain, there appears to be a retraining mechanism involved in order to reestablish regular homeostasis. This recovery may also require both short- and long-term replacement nutrients. Reversible toxic and sensitivity reactions are often associated with pollutant exposure, but they also occur after therapeutic administration of drugs. Thus, neurotoxicity and sensitivity could be considered to include all undesired nervous

Nervous System

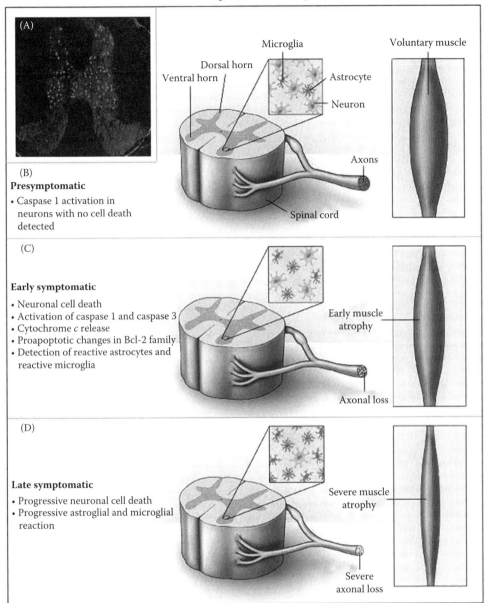

FIGURE 2.43 (See color insert following page 364.) Neurologic lesions in mice with ALS. Panel A shows activation of neuronal caspase 1 in an axial section of the spinal cord of a 90-day-old mouse with ALS (immunostained with a caspase 1 antibody). At this age, the mouse is at the beginning or the middle of the symptomatic stage. There is no caspase 1 activation in the dorsal horn or in the white matter. In the presymptomatic stage (Panel B), the earliest cell-death signal detected is the activation of neuronal caspase 1. At this stage, there are no overt signs of cell death or strong tissue reaction. In the early symptomatic stage (Panel C), there is widespread activation of caspase 1 and caspase3, release of cytochrome *c*, and proapoptotic changes in Bcl-2 family members. Ventral motor neurons and axons die, and reactive microgliosis and astrocytosis are present. As the disease advances, the findings described above become more overt (Panel D) and are accompanied by progressive muscle atrophy. (Friedlander, R.M., *N Engl J Med.*, 348(14), 1370, 2003. With permission.)

system effects of drugs, as well as other chemicals.[216] **At the EHC-Dallas studies of patients with chemical sensitivity in the ECU who were challenged in the deadapted state after reduction of their total body load reveal that the reactions and responses of these patients to pollutants are finite and clearly defined.** In most cases, these reactions usually last from five minutes to four hours, suggesting a pharmacological rather than a toxic effect (Figure 2.44).

However, this effect is clearly synchronous with the alarm stage reaction of Selye, suggesting that this homeostatic response is the physiological way the body deals with noxious homeostatic disturbing stimuli. Clearly, many medications, like antiseizure medications, penetrate the blood-brain barrier, thus helping the damaged area, but also they may cause dysfunction of other areas of the brain, acting similar to nonmedication chemicals. Gabapentin (neurotin), phenytoin (dilantin), valproic acid (depakote), carbamezapine (tegretol), and carboxamide are a few of the antiseizure medications that we have seen cause problems in patients with chemical sensitivity and chronic degenerative disease even though these medications aid in preventing seizures. Most of the antipsychotic drugs also may cause side effects, causing disturbances in the dynamics of homeostasis.

Generally, a distinction can be made between the pharmacological and toxicological effects that result from exposure to a drug or pollutant, but, at times, differentiation between the two is unclear in the individual of chemical sensitivity and chronic degenerative disease. **Characteristically, pharmacological effects are of short duration and completely reversible, allowing restoration of homeostasis in 3–4 hours (some cases up to 6–24 hours) while toxicological effects often include irreversible damage.** Of course, in some cases, there is a gray line between the two when the toxicological effect is severe. In this instance, dyshomeostasis would be permanent, causing energy drain and progressive downhill deregulation. However, it is clear that many reversible toxic changes occurring in neurons or glia cells during therapeutic use of drugs may be closely related to the mechanisms of therapeutic activity. Here a semblance of drug-induced homeostasis is restored only temporarily until the medication has worn off. Of course, this response is a masking response that often clinically covers up the homeostatic disturbing substances and allows progression of dyshomeostasis until fixed-disease occurs. For example, **the synaptic clefts between axons and dendrites of neurons are considered to be especially vulnerable to exogenous chemicals carried by the bloodstream, since the postsynaptic membrane is the site of receptors for chemical**

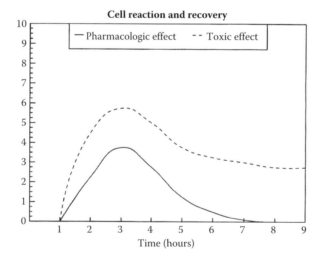

FIGURE 2.44 Graph shows effect pattern of the body's response to pollutants on the nervous system. Most pharmacological effects last for 2–4 hours. Toxic effects may never clear. (Modified from Rea, W.J., *Chemical Sensitivity, Volume III, Clinical Manifestations of Pollutant Overload*, CRC Press, Boca Raton, FL, 1757, 1996.)

transmitters in the nervous system. Many psychoactive drugs are thought to cause psychic changes by altering neuronal transmission, and many neurotransmitters are analogs of psychoactive drugs. Chemical overload creates homeostatic dysfunction that results in increased vulnerability of the total system with mechanisms and responses similar to those just described.

This vulnerability seen in some chemically sensitive patients after chemical exposure may involve the body's inability to block normal chemical transmitters going to postsynaptic receptors. **Either excess blocking or triggering of the transmitters may occur**. Also, the exogenous chemicals can actually act as false transmitters or affect concentrations of transmitters through their influence on synthesis, storage, release, reuptake, or enzymatic inactivation mechanisms. Vulnerability will also occur when homeostasis is disrupted enough that there are less nutrient fuels and response enzymes available in sufficient quantities to maintain the homeostatic mechanisms completely. Then the individual becomes more vulnerable because of the damaged dynamics of the homeostatic response mechanism. Due to the large number of possible areas of pollutant damage, it is difficult, at times, to pinpoint the exact area of neuronal injury for diagnosis and treatment in the subset of neurologically involved patients with chemical sensitivity or chronic degenerative disease. If pollutant or other noxious stimuli continue in individuals with chemical sensitivity or chronic degenerative disease, the pharmacological effects may then become toxic, probably due both to nutrient depletion of the end-organ cells and the structural damage caused directly by the chemical stimuli or the chemical alone.

The hormetic effect of the chemicals must also be considered when assessing toxicity. The dynamics of homeostasis are altered, and, with metabolic changes occurring, tissue changes will follow. This response may be true for most exogenous chemicals that trigger chemical sensitivity. These effects are proposed to occur in the areas of the CNS in which specific transmitter mechanisms are involved.[216] **These areas include the autonomic areas in the brain, such as the hypothalamus and amygdala, which normally have high concentrations of some biogenic amines, particularly serotonin, norepinephrine, dopamine, acetylcholine, and gamma amino butyric acid. Examples of compounds presumed to act therapeutically by altering neurotransmitters are monoamine oxidase inhibitors, cholinesterase inhibitors, reserpine, phenothiazines, and L-DOPA**.[216] The dynamics of homeostasis must be maintained in all areas of production. At times, with increased demands, these areas may be overstressed and thus depleted, causing adverse symptoms.

Organophosphate and some carbamate insecticides that are known cholinesterase inhibitors are also known to exacerbate and cause some chemical sensitivity and chronic degenerative disease. As contaminants contained in food and air these insecticides are able to cause pharmacologic, as well as neurotoxic effects in these synaptic areas due to constant exposure. These latter two insecticides can temporarily, and even permanently, alter the dynamics of homeostasis. This alteration is often seen in some patients with chemical sensitivity who had a large single exposure to organophosphates followed by daily low-dose exposures (in their foods and local environment).[313] In this particular situation, homeostasis is difficult to obtain and maintain. The energy demand is high, keeping the individual in a state of weakness and fatigue.

Other drugs, such as general anesthetics, appear to affect neurons generally, probably through a reversible effect on electrically excitable neuronal membranes.[216] General effects are also seen in halide-generated anesthetics, such as chloroform or halothane. Some of our patients exhibit this anesthetic-like effect with symptoms of sluggishness, sleepiness, weakness, and mental fuzziness. Once the incitant is withdrawn and after a period of detoxification, the dynamics of homeostasis are restored in the patients and they become well. Of our chemically sensitive patients studied in the mid- and late 1980s, 37% had detectable chloroform in their blood, which probably induced some anesthetic effect, accounting for those symptoms previously described. Also found in the blood of chemically sensitive patients are other toxic chlorinated and nonchlorinated hydrocarbons that are similar to, or used as, anesthetics. Since many of these hydrocarbons are solvents such as toluene, xylene, 1,1,1-trichloroethane, trichloroethylene, and tetrachloroethylene, one would expect and does find, neurotoxic and anesthetic effects in addition to the cardiac and liver changes. (See Table 2.13.)

These solvents are clearly disrupters of homeostasis that cause an energy drain and nutrient depletion in the patient. It is apparent from Butler's studies at the EHC-Dallas that as these chlorinated hydrocarbons leave the blood, mental function is usually restored.[314] At the EHC-Dallas, we have also observed that even if a patient has had 20 to 30 years of mental fuzziness from his or her chemical sensitivity or chronic degenerative disease, removal of the toxicants from his or her body will often result in normal or even supernormal mental function. The dynamics of the homeostatic mechanism are hard to destroy if the chronic exposure is not too severe and nutrition can be replaced. We have seen some intelligence quotients increase by as much as 30 points once proper treatment has been accomplished. (See Table 2.14.)

This improvement in mental function suggests a chronic pharmacological or superpharmacological effect, by one or more pollutants (usually multiple), that puts the brain in a state of somewhat suspended animation or suppression. As the pollutants are removed, the mental capabilities improve. For a period of time after the removal of pollutants, the patient usually is extremely supersensitive to exposure to single or multiple toxic chemicals at ambient doses. Brain function fluctuates with minute pollutant exposure until overall resistance to pollutants occurs, and the homeostatic mechanism is restored. Repair of the dynamics of homeostasis occurs as the individual replaces the nutrients for subcellular and cellular healing, and function is restored. During this vulnerable period, very astute brain function alternates with prior dull mental fogginess. This fluctuation has been observed in patients with all types of disturbances. It drains energy putting a high demand on the energy sources of the brain. **Therefore, the patient should be cautioned to avoid new exposures until the homeostatic mechanism has returned to normal for a period of time**.

NEUROLOGICAL EFFECTS OF TOXIC CHEMICALS

The National Institute for Occupational Safety and Health (NIOSH) lists neurotoxic disorders among the top 10 work-related diseases. This organization has estimated that 9.8 million people work with organic solvents, many of which are neurotoxic. Table 2.15 presents examples of these.

Approximately 20 million workers in the U.S. are exposed to toxic chemicals that can harm the nervous system. Further, another 20 to 40 million family members of these workers are probably

Table 2.13
Mechanism of Action—How Pollutant Overload Causes Neural Dysfunction

Blocking the access of normal chemical transmitters to postsynaptic receptors
Acting as false transmitters
Affecting concentrations of transmitters through effects on:
  Synthesis
  Storage
  Release
  Reuptake
  Enzymatic inactivation mechanisms
Damage to membrane and membrane potential
Damage to the repair mechanism

Source: Modified from Rea, W.J., *Chemical Sensitivity, Volume III, Clinical Manifestations of Pollutant Overload*, CRC Press, Boca Raton, FL, 1758, 1996.

TABLE 2.14
Effects of Environmental Treatment on Mental Function[a]

WAIS FSIQ Scores and Differences Before and After Environmental Treatment in the ECU			Bender scores and Differences Before and After Environmental Treatment in the ECU		
Before	After	Difference (+)	Before	After	Difference (−)
106	218	22	18	10	8
123	125	12	24	4	20
129	134	5	11	11	0
104	114	10	15	6	9
116	129	13	8	3	5
106	112	6	15	10	5
130	133	3	15	5	10
117	130	13	24	9	15
112	124	12	14	3	11
106	128	22	20	10	10
115	130	15	15	10	5
108	122	14	15	6	9
106	118	12	16	3	13
138	138	0	20	3	17
115	126	11	15	6	9
105	122	17	9	3	6
118	128	10	10	4	6
102	115	13	24	10	14
108	120	12	11	5	6
128	133	5	14	0	14
112	123	11	12	3	9
123	133	10	15	6	9
117	130	13	24	8	16
109	122	13	8	3	5
122	132	10			
mean = 115	mean = 126.36	mean = 11/36	mean = 15.5	mean = 5.9	mean = 9.6
$N = 25$		$p < 0.001$	$N = 24$		$p < 0.001$

Source: Based on Butler, J.R., Rea, W.J., Laseter, J.J., Deleon, I.R., Wright, S.G., and Milam, M.J., unpublished data, EHC-Dallas, 1986.

[a] Massive avoidance of pollutants in air, food, and water in the ECU using a rotary diet, intradermal injections, and nutritional supplementation.

exposed to contaminants that are carried home by the workers on their clothing, skin, hair, and body. However, the extent of exposure to toxic chemicals has only begun to be understood. We think these estimates are grossly conservative, since virtually every individual in American society is exposed frequently to some toxic chemicals via their air, food, and water. At these ambient, sublethal levels of chronic exposure, disturbances of homeostasis can occur. Once the dynamics of homeostasis are chronically disturbed, and if noxious stimuli are not removed, the prospects of chronic disease are extremely high.

Toxic chemicals are further present in paints, dyes, adhesives, degreasers, cleaning products, plastics, textiles, and inks. Millions of neurotoxic substances are released into the air each year,

TABLE 2.15
Clinical Manifestation of Neurotoxic Chemicals

Chemical	Skin Healing	Parasthesia	Tremors	Fatigue
Acrylamide		X	X	
Lead	X	X	X	X
Mercury		X	X	X
Insecticides		X		X
PBB	X	X		X
PCBSolvents	X	X	X	X
(xylene,toluene)	X	X	X	X

Source: Modified from Rea, W.J., *Chemical Sensitivity, Volume III, Clinical Manifestations of Pollutant Overload.* CRC Press, Boca Raton, FL, 1760–1761, 1996, EHC-Dallas,1992.

many of which are the same substances that we have discussed previously[316] and are found regularly in the blood of the chemically sensitive patient. See Table 2.16 and Figure 2.45A through C.

Also, not shown in Table 2.15 are the billions of pounds of toxic pesticides released into the air, food, and water each year. The environmental influences of symptom-named functional disturbances and subsequent fixed-named diseases will now be discussed.

Fixed-Named Disease

The final outcome of pollutant injury is cell death. Intake of pollutants if not reduced results in fixed-named diseases that often are irreversible. A discussion of the necrotic and aberrant apoptic cell death for chronic degenerative neurological disease is found in this section on pathogenesis and mechanics of injury to the CNS. However, if pollutant overload is reduced and local nutrition is restored early in the disease process, many of these named fixed-named disease processes can be markedly decreased, arrested, and, at times, reversed. Occasionally, these differences can be reversed with the dynamics of homeostasis being restored. Therefore, the clinician should strive to diagnose and treat environmentally induced illnesses as early in their course as possible whether they be due to periodic (hypersensitivity) or aperiodic homeostatic disturbances (chronic degenerative diseases).

Initially, functional molecular metabolic change manifests with a myriad of symptoms without definite single or multiorgan pathological changes. As chronic noxious stimuli continue, homeostatic disturbances become not only diffuse, but also fixed. Eventually pathological changes occur and fixed-named autonomous diseases such as multiple sclerosis, ALS or Parkinson's disease become apparent.

Fixed-named diseases known to have environmental influences include toxic neuropathy, Tourette's syndrome, Parkinson's, Alzheimer's, ALS, multiple sclerosis, Huntington's chorea, myesthenia gravis, brain cancer, and other nonmalignant brain dysfunctions. Generalized neuropathy is the only entity to be discussed here. Even though the rest are types of homeostatic dysfunction due to heredity and environmental influences, they are beyond the scope of this book. For more information, see W.J. Rea, *Chemical Sensitivity, Vol. III,* Chapter 26.[316]

Toxic Neuropathy

In 1856, Delpech[317] described the neurotoxicity of carbon disulfide (CS_2). He identified the symptoms that continue to be observed today.[318] In contrast to Delpech's observations, Charcot,[319] one of Freud's teachers, contended that these patients' complaints, symptoms, and signs should be more

TABLE 2.16
Chemicals Released into Dallas Air—2001

Substance	Tons
Acetone	248
Toluene	102
Xylene	65
Fluorocarbons	69
Ammonia	8
Trichloroethane	189
Sodium hydroxide	51
Hydrochloric acid	9

Substances in the Commercial Food Supply

Pesticides	Lead
Solvents	Cadmium
Herbicides	Mercury
Nitrogen fertilizers (Nitrates)	Nitrites

Substances Found in the Dallas Water Supply*

1,4-Dioxane	Diphenyl ether
2-Butanone,3-methyl	Ethene
Acetyl chloride, chloro	Ethyl acetate
Benzene, chloro	Hexane, 2,2,5-trimethyl
Benzyl butyl phthalate	Hexane, 2,2-dimethyl-Nitrogen dioxide
Bis (2-ehtylhexyl) phthalate	Propene 1-methoxy-2-methyl-Quinazoline,
Chloroform	8- methyl-Tetrachloroethylene
Diethyl phthalate	
Dimethyl ether	

Source: Based on EHC-Dallas, 2001.

* Samples are analyzed by GC/MS. The concentrations of chemicals found in the tap water are below ppm level and pesticides not analysed.

properly ascribed to hysteria (a totally nonprovable and still unproven hypothesis). These two views set the stage for the debate that ensued into the early twentieth century and may still be occurring in some uninformed groups today, usually through ignorance or possibly as a lame defense for polluters. Today, as neurological diagnostic techniques becomes increasingly precise and experimental evidence continues to be garnered strong evidence is emerging for the role of environmental triggers in the onset of fixed-named diseases. **Charcot's and Freud's ideas have become obsolete as the symptoms of neurotoxicity are clearly the result of a damaged target organ, the brain**. Mostly, the argument is no longer focused on whether or not toxic chemicals are able to cause illness. Instead, the debate now centers on identifying the levels of chemicals and individual circumstances necessary to cause harm by triggering dyshomeostasis along with inflammation, and eventually, if the stimulus cannot be removed, to cause end-stage damage. While we continue to reject notions of "safe" threshold levels of exposure, due to inadequate studies and the hormetic effect others continue to argue for arbitrary safe levels. The problem, of course, is that few studies take into consideration the hormetic effect of the chemical, the total body pollutant load, individual nutritional status, and biochemical individuality of the patient, focusing instead on one chemical rather than combinations.

Because there is a long, well-documented history of toxic neuropathy caused by exposure to CS_2, as well as documentation of other mental problems associated with exposure to various chemicals, today's mental health professionals would be wise to check for exposure to CS_2. They should also

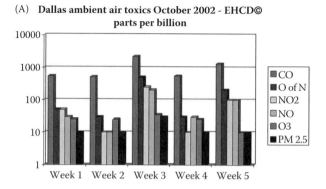

FIGURE 2.45A (See color insert following page 364.) Dallas Air, October 2002.

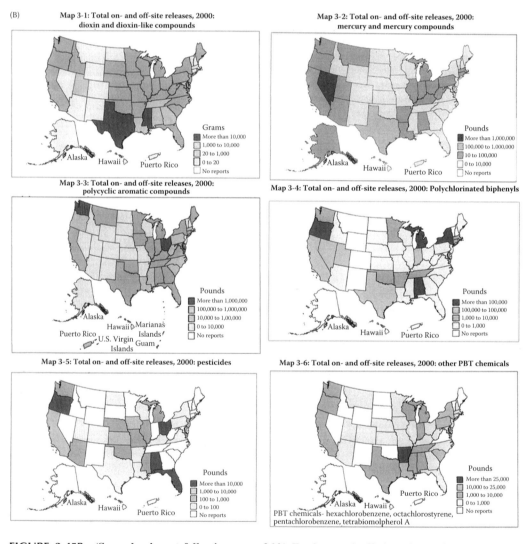

FIGURE 2.45B (See color insert following page 364.) Total on and off site releases 2000: Map 3-1-Dioxin and Dioxin-like compounds, Map 3-2-mercury and mercury compounds, Map 3-3-polycyclic aromatic compounds, Map 3-4-polychlorinated biphenyls, Map 3-5-pesticides, Map 3-6-other PBT chemicals. (Environmental Protection Agency)

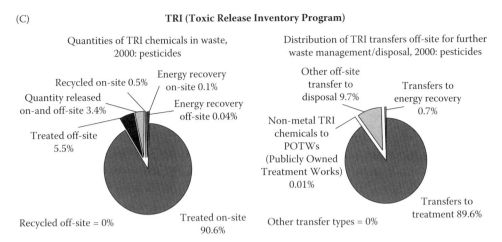

FIGURE 2.45C Quantities of TRI (Toxic Release Inventory) chemicals in waste, 2000: pesticides. Distribution of TRI transfers off-site for further waste management/disposal, 2000: pesticides. (From EPA, Toxics Release Inventory Data for PBT Chemicals, p. 73, Fig 3-14 (left) and p. 74, Fig 3-15 (right), 2000.)

evaluate for significant exposure to other toxic inorganic (Pb, Hg, Cd, Al) and organic (hydrocarbons, pesticides, etc.) chemicals, as well as organic solvents. When they encounter a patient who has symptoms such as hallucinations, headaches, lassitude, malaise, indifference, loss of memory, depression, nausea, chronic fatigue, psychomotor slowness, with decreased vigilance, productivity, and intellectual functioning, chronic fatigue, and paresthesias, toxic exposures will usually be found. Often these syndromes that are induced by exposure to organic solvents and inorganic heavy metals are accompanied by electrophysiological abnormalities. These abnormalities are demonstrated by triple-headed brain scans, posturography, pupillography, heart rate variability, and quantitative EEGs.

As with many aspects of neurological damage in patients with chemical sensitivity and chronic degenerative disease, generally, a clinician's lack of familiarity with the etiology of CS_2 toxic neuropathy might cause him to miss the presence of such damage, even if the other usual indicators, myocardial ischemia and hypertension, are present.[320,321] CS_2 is present around oil refineries and sour gas (sulphur containing) fields. It is used extensively in the rubber industry, as it provides elasticity, and it is essential in the manufacturing of viscous rayon and cellophane. Approximately 24,000 workers are estimated to be involved in such manufacturing.[318]

An example of the damaging effects of the H_2S is shown in Kilburn's study where 75% of 221 subjects were made ill by H_2S (a chemical brother of CS_2) exposure in Alberta, Canada, from 1969 through 1973.[322] Nearly all of them had headaches, altered behavior, confusion and vertigo. Agitation or somnolence were reported in 28%, nausea and vomiting in 22% and disequilibrium in 17%. It was clear from this study that symptomology emanated from a damaged brain.Other neurotoxic compounds, such as diethyldithiocarbamate, enhance the neurotoxicity of methylphenyletetrahydropyridine (MPTP) in mice. Savage et al.[323] evaluated the latent neurological effects of organophosphate pesticide poisoning. These authors found no significant differences between poisoned subjects and controls on audiometric tests, ophthalmic tests, electroencephalograms, or on the clinical serum and blood chemistry evaluation. Neuro-psychological tests of widely varying abilities, however, revealed apparent differences. Anger[324] found that, of the 588 compounds of known toxicological significance that are most frequently encountered in industry, 25% have documented neurotoxic effects, whereas less than 10% have been linked to cancers and fewer than 5% to cardiovascular toxicity. He further identified the varied neurotoxins emitted in the air and to which people can be exposed.[325]

Kilburn has reported a lot of leading chemicals associated with chemical encephalopathy in 266 patients, and he has documented the sources. These sources and types of chemicals are similar to the ones we have been reporting over the last 30 years, as one can see in Table 2.17. Also, there is clearly much objective data to show that brain damage has resulted from chronic toxic exposure to these substances.

Kilburn also found that 53.8% of the sway test performed with the patients' eyes closed was abnormal, and 39.9% of tests performed with the eyes open were abnormal. The sway test was by far the most objective finding in chemical encephalopathy. See Table 2.18.

These findings correlated with ours in which the stressed Romberg positive sway test, the pupillography, the computerized balance test and the triple camera SPECT scan correlated with chemical exposure 93% of the time. Kilburn also showed that 18 patients had seizures, and 15 had temporal lobe seizures, which correlated with our series findings.[326] Also, Kilburn showed the many complaints in the series of H_2S exposed workers were memory loss, fatigue, dizziness, and headache. See Table 2.19.[327]

Other findings from Kilburn's studies that correlated with the findings of EHC-Dallas studies include the following: Kilburn found similar neurological changes and complaints from an exposure of a refinery in Torrance, California.[328] He also found the same brain changes in children living around the refinery, as compared with controls in the Torrance, California. He found similar results of neurologic dysfunction in people who lived in the Brookhurst Casper,[329] Wyoming, area.[330] In addition, he found abnormalities in refinery workers in Houston.[331] A potassium creoylate (chlorine, creosote) aerosol in Alberton, Montana, also created neurotoxicity in 54 exposed subjects.[332] Again, objective brain dysfunctions by Kilburn were found. In Louisiana a leaky tank of 1000 liters of HCl

TABLE 2.17
103 Chemically Sensitive Patients with 366 Inhaled Double-Blind Challenges after four Days of Deadaptation with Total Load Reduced—Changes on Symbol Digit Modality (Cognitive Function, Short-term Memory Recall) Over two Standard Deviations

Chemical	Dose Level(PPM)	Total no. of Patients with Changes on Symbol Digit Modality Subtest[a]
Formaldehyde	< 0.20	24
Ethanol-petroleum derived	< 0.50	23
Phenol	< 0.002	20
Pesticide (2,4-DNP)	< 0.0034	18
Chlorine	< 0.33	15
Placebo	Saline —3 challenges	7[b]
1,1,1-Trichloroethane	ambient dose	3

Source: Based on EHC-Dallas. 1993.

[a] Seven patients reacted to challenge and placebo.

[b] Though challenged with three placebos, no patient reacted to more than one.

TABLE 2.18
Comparison Study—Kilburn–EHCD—Demonstrating Neurotoxic Effect on Balance

	Kilburn Sway Test 100 Patients	EHC-D Posturography-Martinez 100 Patients
eyes open	39.9 %	20%
eyes closed	53.8 %	80%

Source: Based on EHC-Dallas.

TABLE 2.19
Major Frequent Complaints (More Than One Time/Week by Patients) in H_2S Exposure—Kilburn

Symptom	Prevalence
Memory loss	11
Excessive fatigue	9
Dizziness	9
Headache	8
Decreased libido	8
Difficulty concentrating	5
Chest pain/tightness	5
Disorientation	5
Loss of strength	5
Nausea	4
Shortness of breath	4
Somnolence	2
Asthma	2
Cough	2
Sleep disturbed dreams	2
Depression sever	2
Blurred vision	2
Diarrhea	1
Syncope	1
Palpitations	1
Loss of appetite	1
Tinnitus	1
Body swelling and pain	1

Source: From Kilburn, K., *Chemical Brain Injury,* Van Nostrand Reinhold, New York, 96, 1998. With permission.

occurred next to a mobile home park. Forty-five exposed adults were compared with 41 unexposed adults with the exposed group demonstrating significant brain damage.[333]

Emissions from arsenic trioxide and arsenic acid plants occurred in an area of Bryan/College Station, Texas.[334] Kilburn studied 150 people in the area (Table 2.20).[335] The people developed peripheral neuritis. Many developed skin itching, fingernail changes, chest tightness, palpitations, and, again, brain dysfunction was a major finding.[336]

Kilburn also showed a series of patients who were exposed to chlorodane in routine spraying of an apartment complex. Two hundred and sixteen adult occupants were available for his study in 1994. Again, significant changes were found in objective brain function tests.[337] These findings are similar to results reported at the EHC-Dallas in pesticide exposed patients.

In another study in Labelville, Tennessee, Kilburn compared 98 people exposed to PCB contaminated soil and water with 58 unexposed people. The exposed group showed significant changes in brain function.[338] See Table 2.21 and Figure 2.46.

Trichloroethylene is known as the anesthetic that is rapidly absorbed into to the brain, liver, and heart. Chronic TCE exposure in 31 patients caused reduced nerve conduction.

Neighbors to a superfund site in Woburn, Mass developed blink reflex latency. Contamination of the Tucson water supply with TCE also caused brain damage.[338] Another study of Phoenix areas around microchip factories showed similar results, as did studies of Tinker Air Force base in

TABLE 2.20
Peripheral Neuritis in 75 Arsenic-Exposed Subjects and 18 Unexposed Compared by Analysis of Variance (ANOVA)

	Number	Now	Past	In Both	Ever	p
Feet tender or painful						
Exposed	42	13	14	6	44	
Unexposed	17	0	0	1	6	0.017*
"Pin and needles" in toes and feet						
Exposed	47	9	14	5	37	
Unexposed	17	0	0	1	6	0.038*
Numbness and tingling of legs and feet						
Exposed	43	16	11	5	43	
Unexposed	16	1	0	1	11	0.047*
Pain in hands						
Exposed	41	17	10	7	45	
Unexposed	17	0	1	0	6	0.008*
Tender calves						
Exposed	52	11	8	4	31	
Unexposed	17	0	0	1	6	0.121
Weakness of fingers						
Exposed	42	13	11	9	44	
Unexposed	18	0	0	0	0	0.002*
Periods when unable to stand						
Exposed	54	10	7	4	28	
Unexposed	18	0	0	0	0	0.023*
Skin peeling						
Exposed	52	7	13	3	31	
Unexposed	18	0	0	0	0	0.012*
Excessive sweating						
Exposed	49	12	8	6	35	
Unexposed	16	1	0	1	11	0.113

Source: From Kilburn, K., *Chemical Brain Injury,* Van Nostrand Reinhold, 158, New York, 1998. With permission.

* *p*-value statistically significant?Arsenic Study showing peripheral neuropathy in 75 exposed subjects and 18 unexposed compared by analysis of variance (ANOVA).

Oklahoma City and another study in Muscle Shoals, Alabama. Kilburn also showed brain damage in a group of people who had changes due to toluene rich vapors in Baton Rouge, Louisiana.[340]

In aluminum recycling plant in Muscle Shoals, Alabama, Kilburn studied 670 aluminum workers and compared them with less exposed 659 pipe fitters. He found brain and pleural abnormalities, as well as more sophisticated neurological changes in subgroups in the exposed workers.[340] Another of his studies involved a group of workers exposed to diesel exhaust, which found significant objective brain dysfunction.[341]

At the EHC-Dallas, we studied 100 patients (61 females, 39 males, 12–75 years old with an average age of 45) who had chronic chemical exposure in the workplace. Exposure to chemicals varied, but was mainly solvents and chlorinated pesticides in the ambient air. The patients used no protective gear because levels of the chemicals were thought to be safe. The main complaints were neurological symptoms in 49% of the patients (headaches, migraines, short-term memory

TABLE 2.21
Profile of Mood States Scores and Component Scores, Means, Standard Deviations, and Significance (p-values)

	98 Exposed		58 Unexposed		
	Mean	Sd	Mean	Sd	p
POMS Score	72.6	38.9	22.1	25.0	.0001*
Tension	18.5	7.3	8.9	4.6	.0001*
Depression	20.7	13.4	7.9	7.1	.0001*
Anger	15.9	10.4	7.7	6.5	.0001*
Vigor	11.6	6.2	17.0	6.2	.0001*
Fatigue	15.7	6.2	8.3	5.6	.0001*
Confusion	13.4	5.3	6.4	3.6	.0001*

Source: From Kilburn, K., *Chemical Brain Injury,* Van Nostrand Reinhold, New York, 184 (10.2), 1998. With permission.

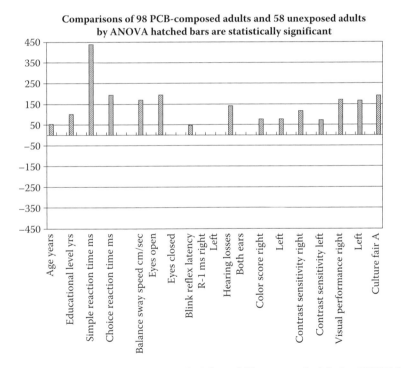

FIGURE 2.46 Comparisons of 98 PCB-composed adults and 58 unexposed adults by ANOVA hatched bars are statistically significant. (Kilburn, K., *Chemical Brain Injury,* 184 (Fig 10.2), Van Nostrand Reinhold, New York, 1998. With permission.)

loss, inability to concentrate, vertigo, light headedness) and musculoskeletal symptoms in 51% of the patients (fibromyalgia, fatigue, arthritis, arthralgia) who also had the neurological symptoms also (Table 2.22).

All patients were unable to stand on their toes with their eyes closed, thus losing their balance or unable to walk a straight line with their eyes closed (positive stressed or tandem Romberg). Associated symptoms included respiratory symptoms (shortness of breath, asthma, bronchitis)

TABLE 2.22
Main Complaints of 100 Patients with Central Neurotoxicity

	No.	%
Neurological Symptoms	100	100
(Headache, migraine, short term memory, inability to concentrate, vertigo, dizziness, depression, positive Romberg)		
Musculoskeletal Symptoms	51	51
(Fatigue, fibromyalgia, pain (generalized), arthritis, positive Romberg plus neurological sx		

Source: Based on Rea, W., Pan, Y., and Fenyves, I., *Abstract-Neurotoxicity: The Central Nervous System*; EHC-Dallas, 2002.

TABLE 2.23
Associated Diagnosis in Patients with Central Neuropathy

	No.	%
Respiratory Symptoms	35	35
(Asthma, shortness of breath, ARS, bronchitis)		
ENT Symptoms	35	35
(Dysphagia, throat swelling, hoarseness, laryngeal edema)		
G.I. Symptoms	18	18
(IBS, diarrhea, malabsorption)		
Cardiovascular	30	30
(Vasculitis, chest pain, hypertension, cardiomyopathy, angioedema, mitrovalve disease)		
Endocrine	9	9
(Hormone imbalance, PMS, hypothyroid problems, thyroiditis)		
Eye	2	2
(Cataracts, vision loss)		
Skin	1	1
(Rash)		
Candidiasis	4	4

Source: Based on Rea, W.J., Pan, Y., and Fenyves, I., *Abstract-Neurotoxicity: The Central Nervous System;* EHC-Dallas, 2002.

35%; ENT (hearing loss, tinnitus, hoarseness, laryngeal edema, dysphasia) 35%; GI (irritable bowel syndrome, malabsorption, diarrhea) 18%; cardiovascular (vasculitis, chest pain, hypertension, cardiomyopathy, angioedema, mitral valve disease) 30%; endocrine (ovarian imbalance, PMS, hypothyroid problems, thyroiditis) 9%; eye (cataract, vision loss) 2%; skin (rash) 1% and candidiasis 4% (Table 2.23).

All patients (100%) had some form of chemical sensitivity, which included six patients with implants (chin, breast, hip) and two with mercury toxicants. Food sensitivity occurred in 89% of patients, biological inhalant sensitivity in 76%, and EMF sensitivity in 3% of patients studied (Table 2.24).

Out of 81 patients tested for aliphatic hydrocarbons in their blood, 3-methylpentane in 89%, 2-methylpentane in 85%, n-hexane present in 79%, cyclopentane was 7%, 1% of patients showed n-pentane (Table 2.25).

Toxic solvent exposure was evident in 54 patients whose blood was measured: 1,1,1-tricholoroethane in 43%, toluene in 43%, benzene present in 24%, trimethylbenzene in 19%, dichloromethane

TABLE 2.24
Concomanant Etiological Diagnosis

	No.	%
Food Sensitivity	89	89
Biological Inhalant Sensitivity	76	76
Implant (chin, breast, hip)	6	6
Mercury Toxicity	2	2
EMF Sensitivity	3	3

Source: Based on Rea, W., Pan, Y., and Fenyves, I., *Abstract-Neurotoxicity: The Central Nervous System;* EHC-Dallas, 2002.

TABLE 2.25
Aliphatic Hydrocarbons in Blood of 81 Patients

	Above Detection < .1 ppb		Controls 100	
	Patients*	%	No.	%
3-methylpentane	72	89	10	10
2-methylpentane	69	85	8	8
n-Hexane	64	79	15	15
Cyclopentane	6	7	2	0
n-Pentane	1	1	9	9

Source: Based on Rea, W., Pan, Y., and Fenyves, I., *Abstract-Neurotoxicity: The Central Nervous System*; EHC-Dallas, 2002.

* Many patients had more than one solvent in the blood. (Average of 2.6 solvents in the blood.)

in 13%, xylene in 9%, chloroform in 7%, trichloroethylene in 7%, tetrochloroethylene in 7%, ethylbenzene in 2%, and dichlorobenzene in 2% of patients(Table 2.26).

Twenty-eight patients were measured for chlorinated pesticides. We found the following: 100% DDE, 45% trans-nonachlor, 31% tested positive for HCB, 25% oxychlordane, 18% with β-BHC, 14% Epoxide, 14% Dieldrin, 14% DDT, 7% DDD, 4% positive for Endrin, 4% positive α-BHC, 4% heptachlor, 4% with δ-BHC, and 4% Mirex (Table 2.27).

Skin testing for chemicals in 76 patients resulted in positive reactions in 96% for cologne, 68% for orrisroot, 64% for cigarette smoke, 55% for ethanol, 47% for formaldehyde, 42% for UNL-diesel fuel, 41% for phenol, 39% for chlorine, and 33% for new material, 26% for natural gas, 17% for propane gas, 11% for fireplace smoke, and 3% for office air (Table 2.28).

Intradermal skin testing of 57 patients showed the following positive reactions: mold 93%; algae, 70%; dust 74%; trees and grasses 74%; mite 74%; T.O.E. 51%; candida 49%; terpene 46%; dander 39%; cotton 28%; and smut 26%. Additionally, 42% of these 57 patients tested positive to weeds (74%), foods (98%), bacteria (51%), viruses (56%), and hormones (12%). See Table 2.29.

Sixteen patients underwent intradermal skin testing for metals. Positive reactions were as follows: nickel sulfate, 81%; zinc sulfate, 69%; gold, 38%; lead, 25%; aluminum, 19%; titanium, 19%; tin, 13%; copper, 6%; stainless steel, 6%; and silver, 2% (Table 2.30).

TABLE 2.26
Nonchlorinated and Chlorinated Solvents in the Blood of 54 Patients with Central Neuropathy

	Above Detection < .1 ppb		Controls 100	
	Patients*	%	No.	%
1,1,1-Trichloroethane	23	43	0	0
Toluene	23	43	0	0
Benzene	13	24	0	0
Trimethylbenzene	10	19	0	0
Dichloromethane	7	13	0	0
Xylene	5	9	1	1
Chloroform	4	7	0	0
Trichloroethylene	4	7	0	0
Tetrochloroethylene	4	7	0	0
Ethylbenzene	1	2	0	0
Dichlorobenzene	1	2	0	0

Source: Based on Rea, W., Pan, Y., and Fenyves, I., *Absract-Neurotoxicity: The Central Nervous System*; EHC-Dallas, 2002.

* Many patients had more than one solvent in the blood.

TABLE 2.27
Organochlorine Pesticides in the Blood of 28 Patients with Central Neuropathy

	Above Detection < .1 ppb		Controls 100	
	Patients	%	No.	%
DDE	28	100	10	10
Trans-nonachlor	13	46	0	0
HCB	9	31	0	0
Oxychlordane	7	25	0	0
β-BHC	5	18	2	2
Epoxide	4	14	0	0
Dieldrin	4	14	0	0
DDT	4	14	0	0
DDD	2	7	0	0
Endrin	1	4	0	0
α-BHC	1	4	0	0
Heptachlor	1	4	0	0
δ-BHC	1	4	0	0
Mirex	1	4	0	0

Source: Based on Rea, W., Pan, Y., and Fenyves, I., *Abstract-Neurotoxicity: The Central Nervous System*; EHC-Dallas, 2002.

TABLE 2.28
Intradermal Skin Testing for Chemicals of 76 Patients with Central Neuropathy

	Positive		Controls Position
	No.	%	No
Ladies/men's cologne	73	96	1
Orrisroot	52	68	0
Cigarette smoke	49	64	2
Ethanol	42	55	1
Formaldehyde	36	47	0
UNL—Diesel	32	42	0
Phenol	31	41	1
Chlorine	30	39	2
News material	25	33	1
Natural gas	20	26	0
Propane gas	13	17	0
Fireplace smoke	8	11	0
Silicone, Jet fuel, Office air	3	3	0

Source: Based on Rea, W., Pan, Y., and Fenyves, I., *Abstract-Neurotoxicity: The Central Nervous System*; EHC-Dallas, 2002.

Three hundred individual double-blind inhaled challenges on a smaller number of patients showed positive reactions for xylene 100%, ethanol 94%, chlorine 89%, phenol 88%, organophosphate pesticides 86%, formaldehyde 83%, and toluene 83%.

Total lymphocytes, T_{11}, T_4, and T_8 lymphocytes measured by flow cytometrically, showed a slight tendency for lower values. See Table 2.31.

Eighty-one patients were measured with triple-headed brain SPECT scan; all were positive for neurotoxicity. See Table 2.32.

Posturography was performed in 78 patients, with 79% having abnormal sensory organization and 44% showing abnormal motor organization. See Table 2.33.

Pupillography, using the iris corder for autonomic nerve system measurement, showed abnormalities in 84% of patients. Heart rate variability for measurement of ANS movement was performed on nine patients. All were abnormal. Specific cutaneous thermography was performed in 20 patients with 95% showing rigid (low) response and 5% showing hypersensitive response (Table 2.34).

Treatment consisted of a massive avoidance of pollutants in air, food, and water along with injection therapy for the hypersensitive state, oral and intravenous nutrition, sauna therapy, and autogenous lymphocytic factor as an immune modulator. A significant improvement rate (82%) occurred. This protocol promises to be a good possibility for treating central neuropathy. This study contributes to the understanding of the signs and symptoms, and, ultimately, the diagnosis of central neuropathy.

Descriptions of neurotoxicity can be found in Anger and Johnson.[342] According to Singer, the symptoms of chronic neurotoxicity caused by various substances include the following: personality changes (irritability, social withdrawal, amotivation [disturbance of executive function]); mental changes with short-term memory loss, inability to concentrate, mental slowness, sleep disturbance, chronic fatigue, headache, sexual dysfunction, numbness of hands and feet, lack of motor coordination, sensory disturbances, and psychosis, weakness, or depression. As one can see, this data is similar to Kilburn's and our studies mentioned in this book. Singer[343] further states that the majority of industrial workers who work with hazardous chemicals are probably exposed to neurotoxic agents (Tables 2.35 and 2.39).[343]

TABLE 2.29
Intradermal Skin Testing for Biological Inhalants in 57 Patients with Central Neuropathy

	Positive		Controls Position 30
	No.	%	No.
Food	56	98	22
Mold	53	93	1
Algae	40	70	3
Dust	42	74	1
Mite	42	74	2
Tree	42	74	0
Grass	42	74	1
Weeds	42	74	2
Viruses	32	56	0
T.O.E.	29	51	1
Bacteria	29	51	1
Candida	28	49	3
Terpene	26	46	2
Dander	22	39	1
Cotton	16	28	1
Smut	15	26	1
Hormones	7	12	2
Fabrics	2	9	0
Peptides	1	2	0

Source: Based on Rea, W., Pan, Y., and Fenyves, I., *Abstract-Neurotoxicity: The Central Nervous System*; EHC-Dallas, 2002.

TABLE 2.30
Intradermal Skin Testing for Metals in 16 Patients with Central Neuropathy

	Positive		Controls Position
	No.	%	No.
Nickel Sulfate	13	81	1
Zinc Sulfate	11	69	0
Gold	6	38	0
Lead	4	25	1
Aluminum	3	19	2
Titanium	3	19	1
Tin	2	13	0
Copper	1	6	0
Stainless Steel	1	6	0
Silver	3	2	0

Source: Based on Rea, W., Pan, Y., and Fenyves, I., *Abstract-Neurotoxicity: The Central Nervous System*; EHC-Dallas, 2002.

TABLE 2.31
300 Inhaled Double-Blind Challenges 30 pts. after 4 Days Deadaption with the Total Load Decreased

		Positive in %
Xylene	ambient	100
Petroleum-ethanol	.05 ppb	94
Phenol	.0050 ppb	88
Chlorine	.33 ppb	89
O-pesticide	.0034 ppb	86
Formaldehyde	.2 ppb	83
Toluene	ambient	83
1,1,1,Trichloroethane	ambient	80
Placebo-saline	—	10

Source: Based on EHC-Dallas, 2002.

TABLE 2.32
Triple Headed SPECT Brain Scan Measurements for Neurotoxicity in 81 Patients

	Positive in %	Controls # 25
SPECT Scan	100%	0

Source: Based on Rea, W., Pan, Y., and Fenyves, I., *Abstract-Neurotoxicity: The Central Nervous System*; EHC-Dallas, 2002.

Table 2.33 Posturography Results in 78 Patients with Central Neuropathy

	Percentage	Controls # 60
Abnormal sensory organization	79	0
Abnormal motor organization	44	0

Source: Based on Rea, W., Pan, Y., and Fenyves, I., *Abstract-Neurotoxicity: The Central Nervous System*; EHC-Dallas, 2002.

Bell has studied the role of panic disorder in toxicity.[345] **It has been shown that infusion of lactate will reproduce panic in sensitive patients**.[346] In addition, carbon monoxide[347] and cholecystokinin-4[348] can also produce panic. The patients with carbon dioxide hypersensitivity and panic disorder have a condition that simple air hunger or voluntary hyperventilation do not fully mimic.[347,349,350] People working and living in buildings with poor air quality with mildy elevated carbon dioxide levels will have shortness of breath, anxiety, and autonomic disturbance along with the panic attacks.

We have observed these patients in the environmental unit when CO_2 levels were 1000 ppm or higher and some had panic attacks. Patients with elevated cholecystokin (coiled-coil-based (CCB)) levels have also been observed to have panic attacks. Apart from excessive CCB release, (cholecystokinin (CCK)) peptide and/or genetic polymorphism may occur.

TABLE 2.34
Pupillograpaphy (ANS Measurements) in 82 Patients with Central Neuropathy

	No.	%	Controls Positive %
Normal	11	14	5
Abnormal			
Cholinergic	15	18	0
Sympatholytic	16	20	0
Cholinergic & sympatholytic	7	8	2
Cholinolytic	7	8	0
Sympathomimetic	3	4	0
Cholinolytic & sympathomimetic	7	8	0
Nonspecific change	16	20	3

Source: Based on Rea, W., Pan, Y., and Fenyves, I., *Abstract-Neurotoxicity: The Central Nervous System*; EHC-Dallas, 2002.

TABLE 2.35
Chemical Groupings in Which 10 or More Neurotoxic Chemicals are Found

Alcohols	Glycol derivatives
Alicyclic hydrocarbons	Halogenated aliphatic hydrocarbons containing Cl, Br, and I
Aliphatic and alicyclic amines	
Aliphatic carboxyl acids	Halogenated cyclic hydrocarbons
Aliphatic hydrocarbons	Ketones
Aliphatic nitro compounds, nitrates, and nitrites	Metals
Aromatic hydrocarbons	Nitrogen compounds
Cyanides and nitrites	Organic phosphates
Esters of aromatic monocarboxylic acids and monoalcohols	Organic phosphorous esters
	Organic sulfur compounds
Ethers	Phenols and phenolic compound

Source: Extracted from EHC-Dallas.

Recent studies in rats by Abou-Donia have shown that the pesticide N, N-Diethyl-meta-toluamide (DEET) plus stress can cause neuronal death.[351]

Abou-Donia[352] has stated that organophosphorus (OP) compounds are potent neurotoxic chemicals that are widely used in medicine, industry, and agriculture. The neurotoxicity of these chemicals has been documented in accidental human poisoning, epidemiological studies, and animal models. OP compounds have three distinct neurotoxic actions. **The primary action is the irreversible inhibition of acetylcholinesterase**, resulting in the accumulation of acetylcholine and subsequent overstimulation of the nicotinic and muscarinic acetylcholine receptors, resulting in cholinergic effects. Another action of some of these compounds, arising from single or repeated exposure, is a **delayed onset of ataxia, accompanied by a Wallerian-type degeneration of the axon and myelin in the most distal portion of the longest tracts in both the central and peripheral nervous systems**, and is known as OP ester-induced delayed neurotoxicity (OPIDN). In addition, since

Table 2.36
Metals and inorganic neurotoxic compounds

- Aluminum compounds
- *Arsenic and arsenic compounds
- Azide compounds
- Barium compounds
- Bismuth compounds
- Carbon monoxide
- Cyanide compounds
- Decaborane
- Diborane
- Ethylmercury
- Fluoride compounds
- Hydrogen sulphide
- *Lead and lead compounds
- Lithium compounds
- Manganese and manganese compouds
- Mercury and mercuric compounds
- *Methylmercury
- Nickel carbonyl
- Pentaborane
- Phosphine
- Phosphorus
- Selenium compounds
- Tellurium compounds
- Thallium compounds
- Tin compounds
- Substances that have been documented also to cause developmental neurotoxicity

Source: Modified from Grandjean, P. and Landrigan, P.J., Developmental Neurotoxicity of Industrial Chemicals, www.thelancent.com, 2006.

the introduction and extensive use of synthetic OP compounds in agriculture and industry half a century ago, many studies have reported long-term, persistent, chronic neurotoxicity symptoms in individuals as a result of acute exposure to high doses that cause acute cholinergic toxicity, or from long-term, low-level, subclinical doses of these chemicals. Abou-Donia attempts to define the neuronal disorder that results from OP ester-induced chronic neurotoxicity (OPICN), which leads to long-term neurological and neurobehavioral deficits. According to Abou-Donia, although the mechanisms of this neurodegenerative disorder have yet to be established, the sparse available data suggest that large toxic doses of OP compounds cause **acute necrotic neuronal cell death in the brain, whereas sublethal or subclinical doses produce apoptotic neuronal cell death and involve oxidative stress**.

OP compounds are chemicals that contain both carbon and phosphorus atoms.[352] They are derivatives of phosphoric (H_3PO_4), phosphorus or phosphonic (H_3PO_3), and phosphinic (H_3PO_2) acids. The biological action of OP compounds is related to their phosphorylating abilities. This is dependent on the electrophilicity (positive character) of the phosphorus atom, which is determined by its substituent groups. Steric factors of substituents also play a major role in determining the biological activity of these chemicals according to Abou-Donia. Lipid solubility is important because it enhances the ability of these compounds to cross biological membranes and the blood-brain barrier, leading to increased biological activity. **OP compounds are an**

TABLE 2.37
Neurotoxic Other Organic Substances

- Acetone cyanohydrin
- Acrylamide
- Acrylonitrile
- Allyl chloride
- Aniline
- 1,2-Benzenedicarbonitrile
- Benzonitrile
- Butylated triphenyl phosphate
- Caprolactam
- Cyclonite
- Dibutyl phthalate
- 3-(Dimethylamino)-propanenitrile
- Diethylene glycol diacrylate
- Dimethyl sulphate
- Dimethylhydrazine
- Dinitrobenzene
- Dinitrotoluene
- Ethylbis(2-chloroethyl)amine
- Ethylene
- Ethylene oxide
- Fluoroacetamide
- Fluoroacetic acid
- Hexachlorophene
- Hydrazine
- Hydroquinone
- Methyl chloride
- Methyl formate
- Methyl iodide
- Methyl methacrylate
- p-Nitroaniline
- Phenol
- p-Phenylenediamine
- Phenylhydrazine
- Polybrominated biphenyls
- Polybrominated diphenyl ethers
- *Polychlorinated biphenyls
- Propylene oxide
- TCDD
- Tributyl phosphate
- 2,2′2″-Trichlorotriethylamine
- Trimethyl phosphate
- Tri-o-tolyl phosphate
- Triphenyl phosphate

Source: Modified from Grandjean, P. and Landrigan, P.J., Developmental Neurotoxicity of Industrial Chemicals, www.thelancent.com, 2006.

economically important class of chemical compounds with numerous uses, such as in pesticides, industrial fluids, flame retardants, therapeutics, and nerve gas agents. Most modern synthetic OP compounds are tailor-made to inhibit acetylcholinesterase (AChE), an enzyme essential for life in humans and other animal species. Tetraethylpyrophosphate was the 1st organophosphate synthesized as an AChE inhibitor in 1854.[353] Later, dimethyl and diethyl phosphorofluoridates were synthesized.[353] During World War II, **OP compounds were developed primarily as agricultural insecticides, and later as chemical warfare agents**. The majority of OP insecticides are organophosphorothioates; nerve agents are organophosphonates or organophosphonothioates; industrial chemicals are typically organophosphates.[354] Organophosphates are also used as hydrolyic fluids in helicopters and jet engines. The fumes permeate the cockpit of these planes. See Table 2.40.

As stated previously by Abou-Donia, biologically, **OP compounds are neurotoxic to humans and other animals via three distinct actions: (1) cholinergic neurotoxicity, (2) OPIDN, and (3) OPICN**.

The primary action is the **irresversible inhibitions of acetyl cholinesterase resulting in accumulation of acetylcholine and subsequent overstimulation of the nicotinic and muscurinic acetyl choline receptors resulting in cholinergic effects**. These findings are shown earlier in this chapter by our iris corder measurements of the ANS.

Organophosphorus ester inhibition of AChE. OP esters inhibit AchE by phosphorylating the serine hydroxyl group at the catalytic triad site. The phosphoric or phosphonic acid ester formed with the enzyme is extremely stable and is hydrolyzed very slowly.[366] If the phosphorylated enzyme

TABLE 2.38
Neurotoxic Organic Solvents

- Acetone
- Benzene
- Benzyl alcohol
- Carbon disulphide
- Chloroform
- Chloroprene
- Cumene
- Cyclohexane
- Cylohexanol
- Cyclohexanone
- Dibromochloropropane
- Dichloroacetic acid
- 1,3-Dichloropropene
- Diethylene glycol
- N,N-Dimethylformamide
- 2-Ethoxyethyl acetate
- Ethyl acetate
- Ethylane dibromide
- Ethylene glycol
- n-Hexane
- Isobutyronitrile
- Isophorone
- Isopropyl alcohol
- Isopropylacetone
- Methanol
- Methyl butyl ketone
- Methyl cellosolve
- Methyl ethyl ketone
- Methycydopentane
- Methylene chloride
- Nitrobenzene
- 2-Nitropropane
- 1-Pentanol
- Propyl bromide
- Pyridine
- Styrene
- Tetrachloroethane
- Tetrachloroethylene
- *Toluene
- 1,1,1-Trichloroethane
- Trichloroethylene
- Vinyl chloride
- Xylene

Source: Modified from Grandjean, P. and Landrigan, P.J., Developmental Neurotoxicity of Industrial Chemicals, www.thelancent.com, 2006.

contains methyl or ethyl groups, the enzyme is regenerated in several hours by hydrolysis. On the other hand, virtually no hydrolysis occurs with an isopropyl group (e.g., sarin) and the return of AChE is dependent upon synthesis of a new enzyme. Phosphorylated AChE undergoes aging—a process that involves the loss of an alkyl group, resulting in a negatively charged monoalkyl enzyme.[352] OP compounds undergo detoxification by binding to other enzymes that contain the amino acid serine. **These enzymes include plasma butyrylcholinesterase (BChE)**[355,356] **and paraoxonase.**[357,358] Inhibition of AChE results in the accumulation of acetylcholine at both the muscarinic and nicotinic receptors in the CNS and the PNS. **Excess acetylcholine initially causes excitation, and then paralysis, of cholinergic transmission**, resulting in some or all of the cholinergic symptoms, depending on the dose size, frequency of exposure, duration of exposure, and route of exposure, as well as other factors such as combined exposure to other chemicals and individual sensitivity and susceptibility.

Human exposure. Human exposure—mostly via inhalation—to the OP nerve agent sarin was recently documented in two terrorist incidents in Japan. At midnight on June 27, 1994, sarin was released in Matsumoto City.[359] Of the 600 persons who were exposed, 58 were admitted to hospitals, where seven died. **Although miosis was the most common symptom, severely poisoned patients developed CNS symptoms and cardiomyopathy**. A few victims complained of arrhythmia and showed cardiac contraction. The second terrorist attack by sarin was in the Tokyo subway trains, at 8:05 a.m. on March 20, 1995, when a total of 5000 persons were hospitalized and 11 died.[360] Patients with high exposure to sarin in the Tokyo subway incident exhibited **marked muscle fasciculation, tachycardia, high blood pressure (nicotinic responses), sneezing, rhinorrhea, miosis, reduced consciousness, respiratory compromise, seizures, and flaccid paralysis**.[361] Patients with mild exposure complained of headaches, dizziness, nausea, chest discomfort, abdominal cramps, and

TABLE 2.39
Neurotoxic Pesticides

- Aldicarb
- Aldrin
- Bensulide
- Bromophos
- Carbaryl
- Carbofuran
- Carbophenothion
- α-Chloralose
- Chlordane
- Chlordecone
- Chlorfenvinphos
- Chlormephos
- Chlorpyrifos
- Chlorthion
- Coumaphos
- Cyhalothrin
- Cypermethrin
- 2,4-D
- DDT
- Deltamethrin
- Demeton
- Dialifor
- Diazinon
- Dichlofenthion
- Dichlorvos
- Dieldrin
- Dimefox
- Dimethoate
- Dinitrocresol
- Dinoseb
- Dioxathion
- Disulphoton
- Edifenphos
- Endosulphan
- Endothion
- Endrin
- EPN
- Ethiofencarb
- Ethion
- Ethoprop
- Fenitrothion
- Fensulphothion
- Fenthion
- Fenvalerate
- Fonofos
- Formothion
- Heptachlor
- Heptenophos
- Hexachlorobenzene
- Isobenzan
- Isolan
- Isoxathion
- Leptophos
- Lindane
- Merphos
- Metaldehyde
- Methamidophos
- Methidathion
- Methomyl
- Methyl bromide
- Methyl parathion
- Mevinphos
- Mexacarbate
- Mipafox
- Mirex
- Monocrotophos
- Naled
- Nicotine
- Oxydemeton-methyl
- Parathion
- Pentachlorophenol
- Phorate
- Phosphamidon
- Phospholan
- Propaphos
- Propoxur
- Pyriminil
- Sarin
- Schradan
- Soman
- Sulprofos
- 2,4,5-T
- Tebupirimfos
- Tefluthrin
- Terbufos
- Thiram
- Toxaphene
- Trichlorfon
- Trichloronat

Source: Modified from Grandjean, P. and Landrigan, P.J., Developmental Neurotoxicity of Industrial Chemicals, www.thelancet.com, 2006.

TABLE 2.40
Compound Cited in the Text and Their IUPAC Designations

Compound	IUPAC designation
Chlorpyrifos	*O,O*-diethyl *O*-3,5,6-trichloro-2-pyridyl phosphorothioate
Cyclosarin (GF)	*O*-cyclohexyl methylphosphonofluoridate
DFP	*O,O*-diisopropyl phosphorofluoridate
DEET	*N,N*-diethyl-m-toluamide
Diazinon	*O,O*-diethyl *O*-2-isopropyl-6-methylpyrimidin-4-yl hosphorothioate
Fenthion	*O,O*-dimethyl *O*-4 methylthio-m-tolyl phosphorothioate
Malathion	*S*-1,2-*bis*(ethoxycarbonyl)ethyl *O,O*-dimethyl phosphorodithioate
Methamidophos	*O,S*-dimethyl phosphoramidothioate
Permethrin	3-phenoxybenzl (1*RS*)-*cis-trans*-3-(2,2-dichlorovinyl)-2,2-dimethylcyclo-propanecarboxylate
Quinalphos	*O,O*-diethyl *O*-quinoxalin-2-yl phospho-rothioate
Tabun (GA)	*O*-ethyl *N,N*-dimethylphosphoamido-cyanidate
Sarin (GB)	*O*-isopropyl methylphosphonofluoridate
Soman (GD)	*O*-2,2-trimethypropyl methylphosphono-fluoridate
TOCP	Tri-ortho-cresylphosphate
VX	*O*-ethyl *S*-2-diisopropylaminoethyl methylphosphonothioate
VR	*O*-isobutyl *S*-2-diethylaminoethyl methylphosphonothioate

Source: From Abou-Donia, M.B. Organophosphorus Ester-Induced Chronic Neurotoxicity, Department of Pharmacology and Cancer Biology, Duke University Medical Center, Durham, North Carolina, 58(8), 485, 2003. With permission.

miosis. Interestingly, patients had pupillary constriction, even when their cholinesterase activity was normal. Furthermore, inhibition of red blood cell AChE activity was a more sensitive indicator of exposure than serum BChE activity.[362] The absence of bradycardia and excessive secretions—which are common in dermal or ingestion exposures—suggested that the major route of exposure to the sarin gas in these instances was via inhalation. The patients were treated with atropine eye drops for marked miosis, and with pralidoxime iodide (2-PAM).

Organophosphorus Ester-Induced Delayed Neurotoxicity (OPIDN)

Characteristics of OPIDN. According to Abou-Donia,[352] **OPIDN is a neurodegenerative disorder characterized by a delayed onset of prolonged ataxia and upper motor neuron spasticity as a result of a single or repeated exposure to OP esters**.[363–366] The neuropathological lesion is a central–peripheral distal axonopathy caused by a chemical transection of the axon (known as Wallerian-type degeneration), followed by myelin degeneration of distal portions of the long and large-diameter tracts of the CNS and PNS.[367] Incidents of OPIDN have been documented for over a century. The earliest recorded cases were attributed to the **use of tri-*o*-cresyl phosphate (TOCP) containing creosote oil for treatment of pulmonary tuberculosis in France in 1899**.[363,365,366] See Figure 2.47.

In 1930, TOCP was identified as the chemical responsible for an estimated 50,000 cases of OPIDN in the Southern and Midwestern regions of the United States[363,365,366] due to the use of OPIDN in bath tub gin (Jake Leg or ginger Jake).

More recently, Himuro et al.[368] reported that a 51-year-old man who was exposed to sarin during the Tokyo subway incident and survived its acute toxicity, then died 15 months later. Neuropathological

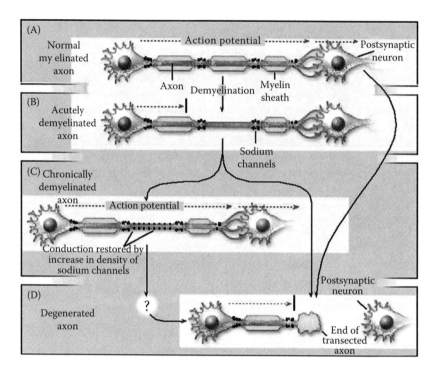

FIGURE 2.47 Demyelination and axonal degeneration in multiple sclerosis. In a normal myelinated axon (Panel A), the action potential (dashed arrow) travels, with high velocity and reliability, to the postsynaptic neuron. In actuely demyelinated axons (Panel B), conduction is blocked (black bar). In some chronically demyelinated axons that acquire a higher-than-normal density of sodium channels (Panel C), conduction is restored. Axonal degeneration (Panel D) by contrast, interrupts action-potential propagation in a permanent manner. (Waxman, S., *NEJM*, 338:5 (323), 1998. With permission.)

alterations and neurological deficits observed in this individual were consistent with the dying-back degeneration of the nervous system characteristic of OPIDN. This incident indicated that humans are more sensitive than experimental animals to sarin-induced OPIDN, inasmuch as it required 26–28 daily doses of LD 50 (25 μg/kg, i.m.) sarin to produce OPIDN in the hen.[369]

OPIDN has been divided into three classes: Type I is caused by the pentavalent phosphates and phosphonates, as well as their sulfur analogs; Type II is produced by the trivalent phosphites;[370] and Type III is induced by phosphines.[371,372]

All three OPIDN types are produced by OP compounds and characterized by central–peripheral distal axonopathy. Type II differs from Type I in terms of the susceptibility of rodents and the presence of neuropathological lesions in neuronal cell bodies.[371]

Type III OPIDN is not accompanied by inhibition of the neurotoxicity target esterase (NTE), thus casting further doubt on this enzyme as the target for OPIDN.[371,372]

Mechanisms of OPIDN. Early studies on the mechanisms of OPIDN centered on the inhibition of the esterases AChE[373] and BChE[374] by OP esters. Subsequent studies eliminated both enzymes as targets for OPIDN.[375] Johnson[376] proposed an NTE—an enzymatic activity preferentially inhibited by OP compounds capable of producing OPIDN as its target according to Abou-Donia. Despite numerous studies since the introduction of this concept 35 years ago, the NTE hypothesis has not advanced our understanding of the mechanism of OPIDN because: (a) evidence for the involvement of NTE in the development of OPIDN is only correlative; (b) it has not been shown how inhibition and aging of NTE leads to axonal degeneration; (c) NTE, which is present in neuronal and non-neuronal tissues and in sensitive and insensitive species, has no known biochemical or physiological function; (d) some OP pesticides that produce OPIDN in humans do not inhibit or age NTE;[377–380] and (e) phosphines that produce Type III OPIDN do not inhibit NTE.[371,372] However, the most convincing evidence against this hypothesis is the recent finding that NTE-knockout mice are sensitive to the development of OPIDN,[381–383] indicating that this enzyme is not involved in the mechanisms of OPIDN.

Protein kinases as targets for OPIDN. Because research on esterases did not increase our understanding of the mechanisms of OPIDN, Abou-Donia has been studying the involvement of protein kinase-mediated phosphorylation of cytoskeletal proteins in the development of OPIDN. These studies were prompted by the following observations: (a) Since OP compounds are effective phosphorylating agents, it is reasonable to expect that they would interfere with normal kinase-mediated phosphorylation of a serine or threonine group at the target protein. (b) The earliest ultrastructural alterations in OPIDN are seen mostly as aggregation and accumulation of cytoskeletal proteins, microtubules, and neurofilaments, followed by their dissolution and disappearance. (c) The structural and functional status of cytoskeletal proteins are affected significantly by protein kinase-mediated phosphorylation.

Anomalous hyperphosphorylation of cytoskeletal elements is associated with OPIDN, a neurodegenerative disorder characterized by distally located swellings in large axons of the CNS and PNS, with subsequent axonal degeneration. Central to his hypothesis is the observation that increased aberrant protein kinase-mediated phosphorylation of cytoskeletal proteins could result in the destabilization of microtubules and neurofilaments, leading to their aggregation and deregulation in the axon.[384]

Protein kinases are able to amplify and distribute signals because a single protein kinase can phosphorylate many different target proteins. Several protein kinases are turned on by second messengers. For example, calcium/calmodulin-dependent protein kinase II (CaM kinase II) is inactive until it is bound by the calcium-calmodulin complex that induces conformational changes and causes the enzyme to unfold an inhibitory domain from its active site.[385–387] Abou-Donia has demonstrated substantial increases in the autophosphorylation,[386,387] enzymatic activity,[388] protein levels, and mRNAs of CaM kinase II in hens treated with diisopropylphosphorofluoridate (DFP).[389] These aberrant alterations have resulted in increased phosphorylation of the following cytoskeletal proteins: tubulin, neurofilaments, microtubule associated proteins-2 (MAP-2), and tau proteins.[390–393] Increased activity of CaM kinase II can affect the stability of cytoskeletal proteins through post-translational modification. Phosphorylation of these proteins interrupts their interaction, polymerization, and stabilization, leading to their degeneration.[394–396] On the other hand, early studies identified transcription factors as critical phosphoproteins in signaling cascades. Immediate early genes control gene expression and therefore affect long-term cellular responses. Abou-donia et al. have demonstrated that the transcription of c-fos is elevated in OPIDN, perhaps through the activation of cAMP (adenosine monophosphate) response element binding (CREB), which is phosphorylated by CaM kinase II. Subsequent to c-fos activation,[394] we observed altered gene expression of CaM kinase II,[395] neurofilaments,[396] glial fibrillary acidic protein (GFAP), and vimentin.[397] Their results also showed an increase in medium (NF-M) and a decrease in low (NF-L) and high (NF-H) molecular weight neurofilaments in the spinal cords of hens treated with DFP.[396] This imbalance in the stoichiometry of neurofilament proteins interferes with their interaction with microtubules and promotes neurofilament dissociation from microtubules, leading to the aggregation of both cytoskeletal proteins.[398]

Immunohistochemical studies in nervous system tissues from TOCP- and DFP-treated hens demonstrated aberrant aggregation of phosphorylated neurofilament, tubulin, and CaM kinase II.[399]

A large number of published studies support the idea that long term, low-level (LTLL) exposure to OP esters may cause neurological and neurobehavioral effects. In order to differentiate these from other effects of OP such as the acute cholinergic episodes, intermediate syndrome and organophosphate induced delayed neuropathy (OPIDN), the term Chronic Organophosphate Induced Neuropsychiatric Disorder (COPIND) will be used primarily for the ease of reference. The question addressed in the particular review is whether LTLL exposure to OP may produce neurotoxicity. COPIND can be classified under two heading; those produced following one or more acute clinical cholinergic episodes, and those produced without such preceding attacks. With regard to the first group, there are a total of 11 studies,[400] all of which support the existence of a positive link between exposure to OP and neurotoxicity; six of these studies comprise descriptions of large numbers of cases without controls while five additional studies employ controls.[401] Appearance of neurotoxicity does not seem to be related to the number or the intensity of acute cholinergic attacks. With regard to the second group, three types of studies can be identified. First, there are five studies[401] using experimental animlas, all of which showed a positive link between OP and neurotoxicity. Second, a total of seven case studies without controls,[401] some involving large numbers of patients, concluded that there is a positive link between OP and neurotoxicity. Third, 19 studies investigated such a link using cases and control groups.[401] Of these, 15 studies (about 80%) showed a positive link and only four failed to identify any link between OP and neurotoxicity.

Only one study carried out blind without the identification of subjects or controls showed a positive link between OP and neurotoxicioty.[401] This blind study estimated the overall incidence of a form of neurotoxicity in people exposed to OP to be about 40 times higher than in the general population. The type of neurological involvement was unique and different from OP induced syndromes previously described. The profile of the neurological involvement was similar to that in COPIND whether or not preceded by acute cholinergic episodes, thus providing further evidence that theses two neuropathies probably share a similar mechanism. There is a characteristic pattern of involvement of 15 functional indicies of the autonomic nervous system examined in Goran's laboratory. There are, in addition, preferential anataomical sites of target organs affected, selective preservation of cholinergic function within the same neuropathy-positive site, and evidence of malfunction of cardiac chemoreceptors in patients exposed to OP. We have seen many patients who developed artrial fibrillation after exposure to OP pesticides. Each time they received a small dose of OP pesticide in the air or from food they would develop atrial fibrillation again. (See Cardiovascular chapter in Volume 2 - Mechanisms of Chemical Sensitivity and Chronic Degenerative Disease.)

The peripheral nerve involvement in OP exposure is predominantly sensory in nature affecting both small and large fibre populations. Neurobehaioral involvement of mainly cognitive dysfunction and other features are also described in orther studies. **The weight of current evidence is therefore very much in favor of the motion that chronic low-level exposure to OP produces neurotoxicity.** Criticisms leveled against this motion are unfounded and probably misconceived.

Organophosphorus Ester-Induced Chronic Neurotoxicity (OPICN)

Various epidemiological studies have demonstrated that individuals exposed to a single large toxic dose, or to small subclinical doses, of OP compounds have developed a chronic neurotoxicity that persists for years after exposure and is distinct from both cholinergic and OPIDN effects.[401]

This disorder has been variously referred to in the literature as: "chronic neuro-behavioral effects,"[360] "chronic organophosphate-induced neuropsychiatric disorder (COPIND),"[401] "psychiatric sequelae of chronic exposure,"[402] "central nervous system effects of chronic exposure,"[403] "psychological and neurological alterations,"[404] "long-term effects,"[405] "neuropsychological abnormalities,"[406] "central cholinergic involvement in behavioral hyperreactivity,"[407] "chorea and psychiatric changes,"[408] "chronic central nervous effects of acute organophosphate pesticide intoxication,"[409] "chronic neurological sequelae,"[410,411] "neuropsychological effects of long-term exposure,"[412] "neurobehavioral effects,"[413] and "delayed neurologic behavioral effects of subtoxic doses."[414]

Abou-Donia's review of the literature indicated that these studies describe a nervous system disorder induced by OP compounds that involves neuronal degeneration and subsequent neurological, neurobehavioral, and neuropsychological consequences. Abou-**Donia defines and describes this disorder, and refers to it as "organophosphorus ester-induced chronic neurotoxicity," or OPICN, he has made an excellent review, which follows.**

Characteristics of OPICN. OPICN is produced by exposure to large, acutely toxic—or small subclinical—doses of OP compounds. Clinical signs, which continue for a prolonged time (ranging from weeks to years after exposure), consist of neurological and neurobehavioral abnormalities.

Damage is present in both the PNS and CNS, with greater involvement of the latter. Within the brain, neuropathological lesions are seen in various regions, including the **cortex, hippocampal formation, and cerebellum. The lesions are characterized by neuronal cell death resulting from early necrosis or delayed apoptosis**. Neurological and neurobehavioral alterations are exacerbated by concurrent exposure to stress or to other chemicals that cause neuronal cell death or oxidative stress. Because CNS injury predominates, improvement is slow and complete recovery is unlikely.

OPICN following large toxic exposure to organophosphorus compounds. Several studies have reported that some individuals, who were exposed to large toxic doses of OP compounds, and who experienced severe acute poisoning and subsequent recovery, eventually developed the long-term and persistent symptoms of OPICN. Many of the adverse effects produced by OP compounds are not related to AChE inhibition.[415]

Individuals with a history of acute organophosphate exposure reported an increased incidence of depression, irritability, confusion, and social isolation.[416] Such exposures resulted in decreased verbal attention, visual memory, motoricity, and affectivity.[417] Rosenstock et al.[418] reported that even a single exposure to organophosphates requiring medical treatment was associated with a persistent deficit in neuropsychological functions. A study of long-term effects in individuals who experienced acute toxicity with OP insecticides indicated dose-dependent decreases in sustained visual attention and vibrotactile sensitivity.[410]

In another study, one-fourth of the patients who were hospitalized following exposure to methamidophos exhibited an abnormal vibrotactile threshold between 10 and 34 months after hospitalization.[418]

Callender et al.[411] have described a woman with chronic neurological sequelae following acute exposure to a combination of an organophosphorothioate insecticide, pyrethrin, piperonyl butoxide, and petroleum distillates. Initially, she developed symptoms of acute cholinergic toxicity. One month after exposure, she experienced severe frequent headaches, muscle cramps, and diarrhea. After three and one-half months, she developed numbness in her legs, tremors, memory problems, anxiety-depression, and insomnia. One year following exposure, she developed weakness, imbalance, and dizziness, and was confined to a wheelchair. Her symptoms were all characteristic of OPIDN. Twenty-eight months after exposure, she developed "delayed sequelae of gross neurologic symptoms," consisting of coarse tremors, intermittent hemiballistic movements of the right arm and leg, flaccid fasciculations of muscle groups, muscle cramps, and sensory disturbances.

Some victims of the Tokyo subway sarin incident, who developed acute cholinergic neurotoxicity, also developed long-term, chronic neurotoxicity characterized by CNS neurological deficits and neurobehavioral impairments.[361] Six to eight months after the Tokyo poisoning, some victims showed delayed effects on psychomotor performance, the visual nervous system, and the vestibule-cerebellar system.[419]

It is noteworthy that **females were more likely than males to exhibit delayed effects on the vestibular-cerebellar system**. Three years after the Matsumoto attack in Japan, some patients complained of fatigue, shoulder stiffness, weakness, and blurred vision. Others complained of insomnia, bad dreams, husky voice, slight fever, and palpitations. Colosio et al.[413] reviewed the literature on the neurobehavioral toxicity of pesticides, and reported that some individuals who were acutely poisoned with OP compounds developed long-term impairment of neurobehavioral performance. They also concluded that these effects were only "an aspecific expression of damage and not of direct neurotoxicity."

OPICN following subclinical exposures to organophosphorus compounds. Reports on OPICN in individuals following long-term, subclinical symptom producing exposures, without previous acute poisoning, have been inconsistent, mostly because of difficulty in the quantitative determination of exposure levels, but also because of problems with selection of controls. Several studies of workers exposed to low subclinical symptom producing doses of OP insecticides failed to show neurobehavioral alterations between pre- and postexposure measurements.[420–426] It has been suggested that the levels of exposure of subjects in these reports might have been below the threshold level needed to cause neurobehavioral deficits, and that studies of prolonged low-level exposures may eventually reveal neurobehavioral deficits.[413] Consistent with this opinion are the reports of impairment in neurobehavioral performance in individuals exposed to low levels of OP insecticides. **Professional pesticide applicators and farmers who had been exposed to OP pesticides showed elevated levels of anxiety, impaired vigilance and reduced concentration.**[427]

Kaplan et al.[428] reported persistent long-term cognitive dysfunction and defects in concentration, word finding, and short-term memory in individuals exposed to low subclinical levels of the organophosphorus insecticide chlorpyrifos. A significant increase in hand vibration threshold was reported in a group of pesticide applicators,[429] and significant cognitive and neuropsychological deficits have been found in sheep dippers who had been exposed to OP insecticides.[423,424] Male fruit farmers who were chronically exposed to OP insecticides showed significant slowing of their reaction time.[430] Female pesticide applicators exhibited longer reaction times, reduced motor steadiness, and increased tension, depression, and fatigue compared with controls.[425] Workers exposed to the OP insecticide quinalphos during its manufacture exhibited alterations in CNS function that were manifested as memory, learning, vigilance, and motor deficits, despite having normal AChE activity.[432] Rescue workers and some victims who did not develop any acute neurotoxicity symptoms nevertheless complained of a chronic decline in memory three years and nine months after the Tokyo attack.[432]

Pilkington et al.[406] reported a strong association between chronic low-level exposure to organophosphate concentrates in sheep dips and neurological symptoms in sheep dippers—suggesting that long-term health effects may occur in at least some sheep dippers exposed to these insecticides over their working lives.

Neurological and neurobehavioral alterations. Although the symptoms of OPICN are a consequence of damage to both the PNS and CNS, they are related primarily to CNS injury and resultant neurological and neurobehavioral abnormalities. Studies on the effects of exposure to OP compounds over the past half century have shown that chronic neurological and neurobehavioral symptoms include headache, drowsiness, dizziness, anxiety, apathy, mental confusion, restlessness, labile emotions, anorexia, insomnia, lethargy, fatigue, inability to concentrate, memory deficits, depression, irritability, confusion, generalized weakness, and tremors.[403,404,433,434]

Respiratory, circulatory, and skin dysfunction may be present as well in cases of chronic toxicity.[354] It should be noted that not every patient exhibits all of these symptoms. Gershon and Shaw[402] reported that most of the symptoms that develop after organophosphate exposure resolve within one year. Jamal[400] conducted an extensive review of the health effects of OP compounds and concluded that either acute or long-term, low-level exposure to these chemicals produces a number of chronic neurological and psychiatric abnormalities that he called "chronic organophosphate-induced neuropsychiatric disorder," or COPIND. Jamal recommended a multifaceted approach to the evaluation of the toxic effects of chronic, subclinical, repeated, low-level exposures to OP compounds; included were structural and quantitative analyses of symptoms and clinical neurological signs. In the present article, Abou-Donia's concept of OPICN encompasses structural, functional, physiological, neurological, and neurobehavioral abnormalities, including neuropsychiatric alterations or COPIND. OPICN may be caused by an acute exposure that results in cholinergic toxicity, or by exposure to subclinical doses that do not produce acute poisoning.

Neuropathological alterations. Petras[435] investigated the neuropathological alterations in rat brains 15–28 days following intramuscular injections of large, acutely toxic doses (79.4–114.8 μg/kg) of the nerve agent soman. He reported that the brain damage in all four animals that developed seizures was comparable to damage present in three of the four animals that exhibited only limb tremor. Neuropathological lesions were characterized by axonal degeneration, seen in the cerebral cortex, basal ganglia, thalamus, subthalamic region, hypothalamus, hippocampus, fornix, septum, preoptic area, superior colliculus, pretectal area, basilar pontine nuclei, medullary tegmentum, and corticospinal tracts. Although the mechanism of soman-induced brain injury was not known, Petras noted that the lesions did not resemble those present in experimental fetal hypoxia[436] or OPIDN.[367] These results are consistent with later findings obtained after acute soman exposure,[437,438] exposure to the nerve agent sarin,[439] and neuronal necrosis induced by the OP insecticide fenthion.[440] Petras[435] also indicated that soman-treated rats did not need to experience a seizure to develop lesions. Abdel-Rahman et al.[441] demonstrated neuropathological alterations in rat brain 24 hr after administration of an intramuscular LD 50 dose (100 μg/kg) of sarin. **Neuronal degeneration was present in the cerebral cortex, dentate gyrus, CA1 and CA3 subfields of the hippocampal formation, and the in Purkinje cells of the cerebellum**. Neuronal degeneration of hippocampal cells is consistent with OP compound-induced alterations in behavior, and cognitive deficits such as impaired learning and memory.[442–445] Furthermore, chronic exposure to OP compounds resulted in long-term cognitive deficits, even in the absence of clinical signs of acute cholinergic toxicity.[446,447]

Shih et al.[448] demonstrated that lethal doses ($2 \times$ LD 50) of all tested nerve agents (i.e., tabun, sarin, soman, cyclosarin, VR, and VX) induced seizures accompanied by neuropathological lesions in the brains of guinea pigs, similar to those lesions reported for other OP compounds in other species.[449–453] **Recent reports have indicated that anticonvulsants protected guinea pigs against soman- and sarin-induced seizures and the development of neuropathological lesions**.[454,455] Time-course studies also have reported that sarin-induced brain lesions exacerbated over time and extended into brain areas that were not initially affected.[438,456] This phenomenon emphasizes how the treatment with reduction of total body load and proper nutrition are necessary to halt the spread of brain lesions.

Kim et al.[457] found that that an intraperitoneal injection of 9 mg/kg ($1.8 \times$ LD 50) DFP in rats protected with pyridostigmine bromide and atropine nitrate caused tonic-clonic seizures, followed by prolonged mild clonic epilepsy accompanied by early necrotic and delayed apoptotic neuronal degeneration. Early necrotic brain injury in the hippocampus and piriform/entorhinal cortices was seen between 1 and 12 hours after dosing. On the other hand, typical apoptotic terminal deoxynucleotidyl transferase-mediated dUTP-X nick end labeling (TUNEL)-positive cell death began to appear at 12 hour in the thalamus. Daily dermal administration of $0.01 \times$ LD 50 of malathion for 28 days caused neuronal degeneration in the rat brain that was exacerbated by combined exposure to the insect repellent DEET and/or the insecticide permethrin.[458]

Correlation between neuropathological lesions and neurological and neurobehavioral alterations. Neuropathological changes—the hallmark of OPICN—could explain the neurological, neurobehavioral, and neuropsychological abnormalities reported in humans and animals exposed to OP compounds. A subcutaneous dose of 104 μg/kg soman induced status epilepticus in rats, followed by degeneration of neuronal cells in the piriform cortex and CA3 of the hippocampus.[453] Similar results have been reported in a variety of species.[448,459,460]

In another study, only those mice treated with a subcutaneous dose of 90 μg/kg of soman which developed long-lasting convulsive seizures exhibited the neuropathological alterations.[461] Twenty-four hours after dosing, numerous eosinophilic cells and deoxyribonucleic acid (DNA) fragmentation (TUNEL-positive) cells were observed in the lateral septum, the endopiriform and entorhinal cortices, the dorsal thalamus, the hippocampus, and the amygdala. **Animals that had only slight tremors and no convulsions did not show any lesions**.[461]

Guinea pigs given a subcutaneous dose of 200 μg/kg soman ($2 \times$ LD 50) developed seizures and exhibited neuropathological lesions in the amygdala; the substantia nigra; the thalamus; the piriform, entorhinal, and perirhinal cortices; and the hippocampus between 24 and 48 hours following injection.[453]

Male guinea pigs developed epileptiform seizures after receiving $2 \times$ LD 50 Subcutaneous doses of the following nerve agents: tabun (240 μg/kg), sarin (84 μg/kg), soman (56 μg/kg), cyclosarin (114 μg/kg), VX (16 μg/kg), or VR (22 μg/kg). The seizures were accompanied by necrotic death of neuronal cells, with the amygdala having the most severe injury, followed by the cortex and the caudate nucleus.[452] An intraperitoneal injection of 9 mg/kg ($1.8 \times$ LD 50) DFP caused severe early (15–90 min) tonic-clonic limbic seizures, followed by prolonged mild clonic epilepsy.[457] Necrotic cell death was seen one hour after DFP administration, primarily in the CA1 and CA3 subfields of the hippocampus and piriform/entorhinal cortices, and manifest as degeneration of neuronal cells and spongiform of neuropils. Whereas the severity of hippocampal injury remained the same for up to 12 hours, damage to the piriform/entorhinal cortices, thalamus, and amygdala continued to increase up to 12 hours. Furthermore, apoptotic death of neuronal cells (TUNEL-positive) was seen in the thalamus at 12 hours, and peaked at 24 hours. Rats that survived $1 \times$ LD 50 sarin (95 µg/kg) exhibited persistent lesions, mainly in the hippocampus, piriform cortex, and thalamus.[439]

Furthermore, brain injury was exacerbated over time; at three months after exposure, other areas that were not initially affected became damaged. A recent study has described the early neuropathological changes in the adult male rat brain 24 hours after exposure to a single intramuscular dose of 1.0, 0.5, 0.1, or $0.01 \times$ LD 50 (100 µg/kg) sarin.[441] Sarin at $1.0 \times$ LD 50 caused extensive severe tremors, seizures, and convulsions accompanied by damage involving mainly the cerebral cortex, the hippocampal formation (dentate gyrus, and CA1 and CA3 subfields) and the cerebellum. Damage was evidenced by (a) a significant inhibition of plasma BChE, brain region AChE, and M2 M-acetylcholine receptor ligand binding; (b) an increase in permeability of the blood-brain barrier; and (c) diffuse neuronal cell death coupled with decreased MAP-2 expression with in the dendrites of surviving neurons. The $0.5 \times$ LD 50 sarin dose did not cause motor convulsions, and only moderate Purkinje neuron loss. The 0.1 and $0.01 \times$ LD 50 doses of sarin caused no alterations at 24 hours after dosing. These results indicate that sarin-induced acute brain injury is dose-dependent.

In animals treated with $1 \times$ LD 50 sarin, both superficial layers (I–III) and deeper layers (IV–V) of the motor cortex and somatosensory cortex showed degeneration of neurons. In the deeper layers of the cortex, neuron degeneration was seen in layer V. Pyramidal neurons in layers III and V of the cortex are the source of the axons of the corticospinal tract, which is the largest descending fiber tract (or motor pathway) from the brain controlling movements of various contralateral muscle groups. Thus, sarin-induced death of layers III and V neurons of the motor cortex could lead to considerable motor and sensory abnormalities, ataxia, weakness, and loss of strength. Furthermore, disruption of the hippocampal circuitry because of the degeneration of neurons in different subfields can lead to learning and memory deficits. Lesions in the cerebellum could result in gait and coordination abnormalities. Because the severely affected areas (e.g., the limbic system, corticofugal system, and central motor system) are associated with mood, judgment, emotion, posture, locomotion, and skilled movements, **humans exhibiting acute toxicity symptoms following exposure to large doses of organophosphates may also develop psychiatric and motor deficits. Inasmuch as the damaged areas of the brain do not regenerate, these symptoms may progress and persist long-term**.[462–464]

These findings are in agreement with a recent study by Kilburn[465] that evaluated the neurobehavioral effects of chronic low-level exposure to the OP insecticide chlorpyrifos in 22 patients. Kilburn demonstrated, for the first time, an association between chlorpyrifos sprayed inside homes and offices and neurophysiological impairments in balance, visual fields, color discrimination, hearing reaction time, and grip strength. These patients also had psychological impairment of verbal recall and cognitive function, and two-thirds of them had been prescribed antidepressant drugs. In addition, the patients exhibited severe respiratory symptoms, accompanied by airway obstruction. Other chlorpyrifos-induced neurotoxicity incidents in humans have been reported.[466] These results are consistent with the report that daily dermal application of 0.1 mg/kg chlorpyrifos to adult rats resulted in sensorimotor deficits.[468]

Also, maternal exposure to a daily dermal dose of 0.1 mg/kg chlorpyrifos during gestational days 4–20 caused an increased expression of GFAP in the cerebellum and hippocampus of offspring on postnatal day 30.[468] A major component of astrocytic intermediate neurofilaments, GFAP is up-regulated in response to reactive gliosis resulting from insults such as trauma, neurodegenerative disease, and exposure to neurotoxicants.[470]

Mechanisms of OPICN. Recent studies have shown that **large toxic doses of OP compounds cause early convulsive seizures and subsequent encephalopathy, leading to the necrotic death of brain neuronal cells, whereas small doses produce delayed apoptotic death**. Pazdernik et al.[453] have proposed the following five phases that result in OP compound-induced cholinergic seizures: (1) initiation, (2) limbic status epilepticus, (3) motor convulsions, (4) early excitotoxic damage, and (5) delayed oxidative stress. The mechanisms of neuronal cell death in OPICN that appear to be mediated through necrosis or apoptosis, which may involve increased AChE gene expression, are discussed below.

Necrosis. According to Abou-Donia the large toxic doses of OP compounds that induce early seizures activate the glutamatergic system and involve the Ca2 + related excitotoxic process,[470,471] possibly mediated by the N-methyl-D aspartate (NMDA) subtype of glutamate receptors.[472,473] De Groot[454] hypothesized that accumulated acetylcholine, resulting from acute inhibition of AChE by OP compounds, leads to activation of glutamatergic neurons and the release of the excitatory L-glutamate amino acid neurotransmitter. This in turn produces increased depolarization and subsequent activation of the NMDA subtype of glutamate receptors—and the opening of NMDA ion channels—resulting in massive Ca2 + fluxes into the postsynaptic cell and causing neuronal degeneration. **Thus, glutamate-induced neuronal degeneration during seizures may occur as a result of lowering of the threshold for glutamate excitation at NMDA receptor sites.**

Activation of nitric oxide synthase, following stimulation of NMDA receptor sites, increases the level of nitric oxide, which functions as a signaling or cytotoxic molecule responsible for neuronal cell death.[474] As a retrograde messenger, nitric oxide induces the release of several neurotransmitters, including excitatory amino acid L-glutamate,[475] which alters neurotransmitter balance and affects neuronal excitability. The production of nitric oxide is enhanced in AChE-inhibitor–induced seizures.[476,477] Kim et al.[477] demonstrated the involvement of nitric oxide in organophosphate-induced seizures and the effectiveness of nitric oxide synthesis inhibitors in preventing such seizures.

Apoptosis. Neuronal degeneration caused by apoptosis or programmed cell death may have physiologic or pathologic consequences. Elimination of precancerous, old, or excess cells is carried out by apoptosis without injury to surrounding cells as seen in the necrotic process.[478] Small doses of OP compounds cause delayed neuronal cell death that involves free radical generation (i.e., reactive oxygen species [ROS]). Organophosphates that cause mitochondrial damage/dysfunction also cause depletion of ATP and increased generation of ROS, which results in oxidative stress.[478,480] ROS can cause fatal depletion of mitochondrial energy (ATP), induction of proteolytic enzymes, and DNA fragmentation, leading to apoptotic death.[479,481,482] These results are consistent with the DNA damage detected in the lymphocytes in peripheral blood in 8 individuals, following residential exposure to the OP insecticides chlorpyrifos and diazinon.[483]

The brain is highly susceptible to oxidative stress-induced injury for several reasons: (a) its oxygen requirements are high; (b) it has a high rate of glucose consumption; (c) it contains large amounts of peroxidizable fatty acids; and (d) it has relatively low antioxidant capacity.[481,482] A single sublethal dose of $0.5 \times LD\,50$ sarin, which did not induce seizures, nevertheless caused delayed apoptotic death of rat brain neurons in the cerebral cortex, hippocampus, and Purkinje cells of the cerebellum 24 hours after dosing.[466,484] Furthermore, rats treated with a single $0.1 \times LD\,50$ dose of sarin, and which did not exhibit brain histopathological alterations 1, 7, or 30 days after dosing, nevertheless showed apoptotic death of brain neurons in the same areas mentioned above, one year after dosing.[466,485] These results are consistent with the sensorimotor deficits exhibited by sarin-treated animals three months after exposure; the animals showed continued deterioration when tested six months after dosing.

Increased AChE gene expression. Recent studies have suggested that AChE may play a role in the pathogenesis of OPICN similar to that reported for Alzheimer's disease.[486,487] Abou-Donia has demonstrated that sarin induced the AChE gene in the same regions of the brain that underwent neuronal degeneration.[488] AChE has been shown to be neurotoxic in vivo and in vitro; it accelerates assembly of amyloid peptide in Alzheimer's fibrils, leading to cell death via apoptosis.[489] Some studies have demonstrated increased AChE expression in apoptotic neuroblastoma SK-N-SH cells after long-term culturing.[489] Brain AChE has been shown to be toxic to neuronal (Neuro 2a) and glial-like (B12) cells.[487] There are also reports that transgenic mice overexpressing human AChE in brain neurons undergo progressive cognitive deterioration.[490] These results suggest that sarin may provoke an endogenous cell suicide pathway cascade in susceptible neurons (e.g., in the caspase-3 pathway), resulting in the release of AChE into adjacent brain tissues. The aggregation of AChE initiates more apoptotic neuronal death. According to Abou-Donia, amplification of this cascade thus may result in the progressive neuronal loss that is the hallmark of sarin-induced chronic neurotoxicity. It is noteworthy that a common symptom of both OPICN and Alzheimer's disease is memory deficit, suggesting that the aging process may be accelerated following exposure to OP compounds in OPICN. We at the EHC-Dallas and Buffalo see many patients with chemical sensitivity have symptoms of short term memory loss that appear to recede with a decrease in total body load.

Other factors. The occurrence and severity of OPICN is influenced by factors such as environmental exposure to other chemicals, stress, or individual genetic differences. For example, cholinotoxicants such as organophosphates or carbamates—which do not have a positive charge and are capable of crossing the blood-brain barrier—act at the same receptors and thus exacerbate OPICN. Individuals with low levels of the plasma enzymes BchE[355,356,491] or paraoxonase[357,358] that act as the first line of defense against neurotoxicity (by removing organophosphates from circulation through scavenging or hydrolysis) are vulnerable to the development of persistent OPICN. According to Abou-Donia, all of these factors may be involved in development of chemical sensitivity.[492] Thus, prior chemical exposure, stress, or genetic factors might make individuals predisposed or susceptible to CNS injury upon subsequent exposure to other chemicals. This phenomenon is seen commonly in our clinics and is know as the second hit phenomenon.

Combined exposure to other chemicals that cause oxidative stress can intensify OPICN, which results from exposure to OP compounds.[458] Furthermore, according to Abou-Donia stress that also causes oxidative stress decreases the threshold level required to produce neuronal damage and results in increased OPICN following combined exposure to stress and organophosphates. Thus, OPICN may explain the reports that Persian Gulf War veterans showed a higher than normal propensity toward persistent neurological complaints such as memory and attention deficits, irritability, chronic fatigue, muscle and joint pain, and poor performance on cognition tests.[493-497] We commonly see this syndrome in the early onset of patients with chemical sensitivity.

A large number of these personnel were exposed to low levels of sarin during the demolition of Iraqi munitions at Khamisiya,[498,499] as well as to other chemicals and to stress.[500,501] Also, OPICN may explain the recent report that Persian Gulf War veterans are at an almost twofold greater risk of developing ALS than other veterans,[502] which is consistent with Haley's[503] suggestion that the increase in ALS is "a war related environmental trigger." Furthermore, OPICN induced by low-level inhalation of organophosphates present in jet engine lubricating oils and the hydraulic fluids of aircraft[504] could explain the long-term neurologic deficits consistently reported by crew members and passengers, although organophosphate levels may have been too low to produce OPIDN.[505] This report by Abou-Donia clearly documents the roles of organophosphates is disease.

Mercury (II) or Hg^{2+}, is neurotoxic and when exposed to normal brain tissue homogenates, is capable of causing many of the same biochemical aberrancies found in Alzheimer's diseased (AD) brain. Also, rats exposed to mercury vapor show some of these same aberrancies in their brain tissue. Specifically, **the rapid inactivation of the brain thiol-sensitive enzymes tubulin, creatine kinase and glutamine synthetase occurs on the addition of low micromolar levels** of Hg^{2+} or exposure to mercury vapor, and these same enzymes are significantly inhibited in AD brain. Further, extended Hg^{2+} exposure to neurons in culture has been shown to produce three of the widely accepted pathological diagnostic hallmarks of AD. These are elevated amyloid protein, hyperphosphorylation of Tau, and formation of neurofibrillary tangles. **The hypothesis is that mercury and other blood-brain permeable toxicants that have enhanced specificity for thiol-sensitive metals such as lead and cadmium that act synergistically to enhance to toxicity of mercury and organic-mercury compounds, like thimerosal that is found in vaccines** and other medicines. This hypothesis is also able to explain the genetic susceptibility to AD that is expressed through the APO-E gene family. Specifically, a reduction of APO-E genes types carrying cysteines decreases the ability to remove mercury and other thiol-reactive toxicants from the cerebrospinal fluid. This increases brain exposure to thiol-reactive toxicants and the risk of AD.[506]

SUMMARY

As dyshomeostasis develops in the nervous system, causes should be found and removed before the metabolic-induced tissue changes take place and cause autonomous, irreversible fixed-named diseases to occur. Even then the removal of toxics may not be enough due to the degenerative process set in motion due to toxic chemicals and their hormetic effect. Some of the degeneration can be stopped by using massive avoidance programs, subcutaneous injections for hypersensitivity, intravenous and oral nutrient supplementation, heat depuration, physical therapy and exercise, environmentally controlled conditions, and oxygen therapy (see Treatment chapter in Volume 4 - Mechanisms of Cardiovascular Disease and Chemical Sensitivity for more details)..

REFERENCES

1. Rea WJ. 1996. *Chemical Sensitivity, Vol. III, Clinical Manifestatons of Pollutant Overload*. Boca Raton, FL: Lewis Publishers. 1727.
2. Pischinger A. 1975. *Matrix and Matrix Regulation: Basis for a Holistic Theory in Medicine*. Ed. H Heine. Eng. trans. N MacLean. Brussels, Belgium: Editions Haug International.
3. Plum F. 1970. Neurological integration of behavioral and metabolic control of breathing. In *Breathing: Hering-Brener Centenary Symposium*. Ed. R. Porter. London: Churchill. 159–181.
4. McClellan RO. 2001. Special issue: scientific foundation of hormesis. *Criti. Rev. in Toxicol.* 31(4&5):420.
5. Legare ME, Barhoumi R, Burghardt RC, and Tiffany-Castglioni E. 1993. Low-level lead exposure in cultured astrolglia: identification of cellular targets with vital fluorescent probes. *Neurotoxicology* 14:267–272.
6. Raps SP, Lair JBC, Hert L, and Cooper AJ. 1989. Glutathione is present in high concentration in cultured astrocytes but not in cultured neuron. *Brain Res*. 493:398–401.
7. Holtzmon D, DeVries C, Nguyen H, Olson J, and Bensch K. 1984. Maturation of resistance to lead encephalopathy: cellular and subcellular mechanisms. *Neurotoxicity* 5:97–124.
8. Tiffany-Castiglioni E, Sierra EM, Wu J-N, Rowles TK. 1989. Lead toxicity in neuroglia. *Neurotoxicology* 10:417–444.
9. Matchett JA. 1976. The Effects of Acute Ethanol Administration on Central Catecholamines and Behavior in the Mouse, PhD Dissertation, the University of Kansas.
10. Matchett JA, and Erickson CK. 1977. Alteration of ethanol-induced changes in locomotor activity by adrenergic blockers in mice. *Psychopharmacology* 52:201–206.
11. DiSilvestro RA and Carlson GP. 1992. Inflammation, an inducer of metallothionein, inhibits carbon-tetrachloride-induced hepatotoxicity in rats. *Toxicol. Lett.* 60(2):175–181.
12. Baumgarten HG, Bjorklund A, Lachenmayer L, Nobin A, and Stenevi U. 1971. Long-lasting selective depletion of brain serotonin by 5,6-dihydroxytryptamine. *Acta. Physiol. Scand. Suppl.* 373:1–15.
13. Calabrese, EJ. 2001. Dopamine: Biphasic Dose Respones. *Criti. Rev. in Toxicol.* 31(4&5): 563–583.
14. Guyton AC and Hall JE (eds.). 2000. *Medical Physiology*, 10th ed. Philadelphia: WB Saunders Co. 60.
15. Gunn CC. 1996. *The Gunn Approach to the Treatment of Chronic Pain: Intramuscular Stimulation for Myofascial Pain of Radiculopathic Origin*. New York: Churchill Livingstone.
16. Riftine A. Quantitative Assessment of the Autonomic Nervous System based on an analysis of Heart Rate Variability. Heart Rhythm Instruments, Inc. 2337 Lemoine Ave, Suite 201, Fort Lee, NJ 07024.
17. Guyton AC and Hall JE (eds.). 2000. *Medical Physiology*, 10th ed. Philadelphia: WB Saunders Co. 531–532.
18. Suzuki N and Hardebo JE. 1991. Anatomical basis for a parasympathetic and sensory innervation of the intracranial segment of the internal carotid artery in man. *J. Neurolog. Sci.* 104(1):19–31.
19. Suzuki N, Hardebo JE, Kahrstrom J, and Owman CJ. 1990. Selective electrical stimulation of post-ganglionic cerebrovascular parasympathetic nerve fibers originating from the sphenopalatine ganglion enhances cortical blood flow in the rat. *Cereb. Blood Flow Metab.* 10(3):383–391.
20. Yanritzky D, et al. 2003. Wolff Award: Possible parasympathetic contributions to peripheral and central sensitization during migraine. *Headache* 43(7):704–714.

21. Hardebo JE and Suzuki N. 1991 The pathway of parasympathetic nerve fibers to cerebral vessels from the otic ganglion in the rat. *J. Auton. Nervous Syst.* 36(1):39–46.
22. Suzuki N and Hardebo JE. 1993. The cerebrovascular parasympathetic innervation. *Cerebrovasc. Brain Metab. Rev.* Spring 5(1):33–46.
23. Rea WJ. 1996. *Chemical Sensitivity, Vol. III, Clinical Manifestatons of Pollutant Overload.* Boca Raton, FL: Lewis Publishers. 1730.
24. Rea WJ. 1996. *Chemical Sensitivity, Vol. III, Clinical Manifestatons of Pollutant Overload.* Boca Raton, FL: CRC Lewis Publishers. 1732–1733.
25. Guyton AC and Hall JE (eds.). 2000. *Medical Physiology,* 10th ed. Philadelphia: WB Saunders Co. 531.
26. Guyton AC and Hall JE (eds.). 2000. *Medical Physiology,* 10th ed. Philadelphia: WB Saunders Co. 532.
27. Guyton AC and Hall JE (eds.). 2000. *Medical Physiology,* 10th ed. Philadelphia: WB Saunders Co. 708.
28. Ishikawa S, Naito M, and Inabe K. 1970. A new videopupillography. *Ophthalmologica.* 160:248.
29. Guyton AC and Hall JE (eds.). 2000. *Medical Physiology,* 10th ed. Philadelphia: WB Saunders Co. 82.
30. Guyton AC and Hall JE (eds.). 2000. *Medical Physiology,* 10th ed. Philadelphia: WB Saunders Co. 134.
31. Vieweg WV, Julius DA, Fernandez A, et al. 2006. Posttraumatic stress disorder: clincial features, patholphysiology, and treatment. *AJM* 119(May):387, Fig. 1.
32. Gray H. 1918. *Anatomy of the Human Body.* Philadelphia: Lea & Febiger, 12th Re-Edition by Warren H. Lewis. New York: Bartleby.com. 2000.
33. Suzuki N, Hardebo JE, Kahrstrom J, and Owman CJ, 1990. May. *Cereb Blood Flow Metab.* 10(3):383–391.
34. Gray H. 1918. *Anatomy of the Human Body.* Philadelphia: Lea & Febiger, 12th Re-Edition by Warren H. Lewis. New York: Bartleby.com, 2000. 539.
35. Gray H. 1918. *Anatomy of the Human Body.* Philadelphia: Lea & Febiger, 12th Re-Edition by Warren H. Lewis. New York: Bartleby.com, 2000. 517.
36. Gray H. 1918. *Anatomy of the Human Body.* Philadelphia: Lea & Febiger, 12th Re-Edition by Warren H. Lewis. New York: Bartleby.com, 2000. 222.
37. Gray H. 1918. *Anatomy of the Human Body.* Philadelphia: Lea & Febiger, 12th Re-Edition by Warren H. Lewis. New York: Bartleby.com, 2000. 316.
38. Gray H. 1918. *Anatomy of the Human Body.* Philadelphia: Lea & Febiger, 12th Re-Edition by Warren H. Lewis. New York: Bartleby.com, 2000. 502, 509.
39. Rea WJ. 1996. *Chemical Sensitivity, Vol. III,: Clinical Manifestatons of Pollutant Overload.* Boca Raton, FL: CRC Lewis Publishers. 1729.
40. Rea WJ. 1996. *Chemical Sensitivity, Vol. III, Clinical Manifestatons of Pollutant Overload.* Boca Raton, FL: CRC Lewis Publishers. 1732–1733, Tab. 26.1.
41. Rea WJ. 1996. *Chemical Sensitivity, Vol. III, Clinical Manifestatons of Pollutant Overload.* Boca Raton, FL: CRC Lewis Publishers. 1731.
42. Gray H. *Anatomy of the Human Body.* Philadelphia: Lea & Febiger, 1918. 12th Re-Edition by Warren H. Lewis. New York: Bartleby.com 2000. 316.
43. Rea WJ. 1996. *Chemical Sensitivity, Vol. III, Clinical Manifestatons of Pollutant Overload.* Boca Raton, FL: CRC Lewis Publishers. 1735, Fig. 26.3.
44. Rea WJ. 1996. *Chemical Sensitivity, Vol. III, Clinical Manifestatons of Pollutant Overload.* Boca Raton, FL: CRC Lewis Publishers. 1735.
45. Minutes of the Pasadena Meeting. Sept 18, 1923. *Phys. Rev.* 22(5):522–528.
46. Guyton AC and Hall JE (eds.). 2000. *Medical Physiology,* 10th ed. Philadelphia: WB Saunders Co. 91.
47. Guyton AC and Hall JE (eds.). 2000. *Medical Physiology,* 10th ed. Philadelphia: WB Saunders Co. 520–680, 708, 703–704, 8381.
48. Guyton AC and Hall JE (eds.). 2000. *Medical Physiology,* 10th ed. Philadelphia: WB Saunders Co. 92.
49. Guyton AC and Hall JE (eds.). 2000. *Medical Physiology,* 10th ed. Philadelphia: WB Saunders Co. 704–705.
50. Guyton AC and Hall JE (eds.). 2000. *Medical Physiology,* 10th ed. Philadelphia: WB Saunders Co. 699–700.
51. Guyton AC and Hall JE (eds.). 2000. *Medical Physiology,* 10th ed. Philadelphia: WB Saunders Co. 702–703.
52. Berger AJ, Mitchell RA, and Severinghaus JW. 1977. Regulation of respiration. *N. Engl. J. Med.* 297:138–142, 194–201.
53. McNicholas WT, Rutherford R, Grossman R, Moldofsky H, Zamel H, and Phillipson EA. 1983. Abnormal respiratory pattern generation during sleep in patients with autonomic dyfunction. *Am. Rev. Respir. Dis.* 128:429–433.

54. Choy R, Loveday M, Vesselinova-Jenkins CK, Monro J. 1986. Sleep apnoea syndrome in patients with allergic rhinitis. *Clini. Ecol.* 4(1):15–16.
55. Rea WJ. 1996. *Chemical Sensitivity, Vol. III, Clinical Manifestatons of Pollutant Overload.* Boca Raton, FL: CRC Lewis Publishers. 1727–1740.
56. Guyton AC and Hall JE (eds.). 2000. *Medical Physiology,* 10th ed. Philadelphia: WB Saunders Co. 705.
57. Gunn CC. 2002. *The Gunn Approach to the Treatment of Chronic Pain.* New York: Churchill Livingstone.
58. Gunn CC. 2002. *The Gunn Approach to the Treatment of Chronic Pain.* New York: Churchill Livingstone.6.
59. Cannon WB, Rosenblueth A. 1949. *The Supersensisivity of Denervated Structures.* New York: MacMillan. 1–22, 185.
60. Gunn CC. 2002. *The Gunn Approach to the Treatment of Chronic Pain.* New York: Churchill Livingstone. 6–7.
61. Gunn CC. 2002. *The Gunn Approach to the Treatment of Chronic Pain.* New York: Churchill Livingstone. 7.
62. Hubbard DR and Berkoff M. 1993. Myofascial trigger points show spontaneous needle EMG activity. *Spine* 13:1888.
63. Gunn CC and Milbrandt WE. 1976. Tenderness at motor points: a diagnostic and prognostic aid for low back injury. *J. Bone and Joint Surg.* 6:815–825.
64. Simons DG and Travel J. 1981. Letter to editor re: myofascial trigger points, a possible explanation. *Pain* 10:106–109.
65. Sheon RP, Moskowitz RW, and Goldberg VM. 1982. *Soft Tissue Rheumatic Pain: Recognition, Management, Prevention.* Philadelphia: Lea and Febiger.
66. Sola AE. 1984. Treatment of myofascial pain syndromes. In *Advances in Pain Research and Therapy,* vol. 7. Ed. C Benedetti, CR Chapman, and G Morrica. New York: Raven Press. 467–485.
67. Sola AE. 1981. Myofascial trigger point therapy. *Resident and Staff Physician* 27:8, 38–46.
68. Travell J and Simons DG. 1983. *Myofascial Pain and Dysfunction: The Trigger Point Manual.* Baltimore: Williams and Wilkins.
69. McCain G. 1983. Fibromyositis. *Clin. Rev.* 38:197–207.
70. Calliet R. 1977. *Soft Tissue Pain and Disability.* Philadelphia: FA Davis.
71. Ochoa JL, Torebjork E, Marchettini P, Sivak M. 1985. Mechanisms of neuropathic pain: cumulative observations, new experiments, and further speculation. In *Advances in Pain Research and Therapy,* vol. 9. Ed. HL Fields, R Dubner, F Cervero. New York: Raven Press. 431.
72. Gunn CC. 2002. *The Gunn Approach to the Treatment of Chronic Pain.* New York: Churchill Livingstone. 8.
73. Staub NC and Taylor AE. 1984. *Edema.* New York: Raven Press. 273–275, 463–486, 657–675.
74. Gunn CC. 2002. *The Gunn Approach to the Treatment of Chronic Pain.* New York: Churchill Livingstone. 9.
75. Klein L, Dawson MH, and Heiple KG. 1977. Turnover of collagen in the adult rat after denervation. *J. Bone and Joint Surg.* 59A:1065–1067.
76. Gunn CC. 2002. *The Gunn Approach to the Treatment of Chronic Pain.* New York: Churchill Livingstone. 10.
77. Gunn CC. 2002. *The Gunn Approach to the Treatment of Chronic Pain.* New York: Churchill Livingstone, 4.
78. Bergsman O. 1994. Bioelektrische Phänomene und regulation in der Komplementärmedizin. Wien:Facultas.
79. Huneke F and Huneke W. 1928. Unbekannte Fernwirkungen der Lokalanästhesie. *Med. Welt* 27:1013–1014. *Nachdruck Hipppokrates* 28(1957):251–253.
80. Gunn CC. 2002. *The Gunn Approach to the Treatment of Chronic Pain.* New York: Churchill Livingstone. 5.
81. Asbury AK and Fields HL. 1984. Pain due to peripheral nerve damage: An hypothesis. *Neurology* 34:1587–1590.
82. Fields HL. 1987. *Pain.* New York: McGraw-Hill.
83. Bradley WG. 1974. *Disorders of Peripheral Nerves.* Oxford: Blackwell Scientific Publications.
84. Thomas PK. 1984. Symptomatology and differential diagnosis of peripheral neuropathy: clinical features and differential diagnosis. In *Peripheral Neuropathy,* Vol. II. Ed. PJ Dyck, PK Thomas, EH Lambert, and R Bunge. Philadelphia: WB Saunders Co. 1169–1190.
85. Tichauer ER. 1977. The objective corroboration of back pain through thermography. *J. Occup. Med.* 19:727–731.

86. Gunn CC. 1978. Transcutaneous neural stimulation, acupuncture and the current of injury. *American J. Acupunct.* 6(3):191–196.
87. Gunn CC and Milbrandt WE. 1976. Tennis elbow and the cervical spine. *CMAJ* 114:803–809.
88. Gunn CC and Milbrandt WE. 1978. Early and subtle signs in low back sprain. *Spine* 3:267–281.
89. Loh L and Nathan PW. 1978. Painful peripheral states and sympathetic blocks. *J. of Neur. Neurosurg. and Psych.* 41:664–671.
90. Rowe PC and Calkins H. 1998. *Am. Jour. Med.* 105(1):155–215.
91. Davies R, Slater JDH, Forsling ML, and Payne N. 1976. The response or arginine vasopressin and plasma renin to postural change in normal man, with observations on syncope. *Clin. Sci.* 51:267–274.
92. Sneddon JF, Counihan PJ, Bashir Y, et al. 1993. Impaired immediate vasoconstrictor responses in patients with recurrent neurally mediated syncope. *Am. J. Cardiol.* 71:72–76.
93. Thompson HL, Atherton JJ, Khafagi FA, and Frenneaux MP. 1996. Failure of reflex venoconstriction during exercise in patients with vasovagal syncope. *Circulation* 93:953–959.
94. Benditt DG, Chen M-Y, Hansen R, Buetikofer J, and Lurie K. 1995. Characterization of subcutaneous microvascular blood flow during tilt table-induced neurally mediated syncope. *J. Am. Coll. Cardiol.* 25:70–75.
95. Benditt DG, Goldstein M, Adler S, Sakaguchi S, and Lurie K. 1995. Neurally mediated syncopal syndromes: pathophysiology and clinical evaluation. In *Cardiac Arrhythmias*, 3rd ed. Ed. WJ Mandel WJ. Philadelphia: JB Lippincott. 879–906.
96. Kosinski D and Grubb BP. 1994. Neurally mediated syncope with an update on indications of usefulness of head-upright tilt table testing and pharmacologic therapy. *Curr. Opin. Cardiol.* 9:53–64.
97. Van Lieshout JJ, Wieling W, Karemaker JM, and Eckberg DL. 1991. The vasovagal response. *Clin. Sci.* 81:575–586.
98. Waxman MB, Cameron DA, and Wald RW. 1993. Role of ventricular vagal afferents in the vasovagal reaction. *J. Am. Coll. Cardiol.* 21:1138–1141.
99. Fortney SM, Shneider VS, and Greenleaf JE. 1996. The physiology of bed rest. In *Handbook of Physiology,* Vol. II, Section 4: *Environmental Physiology.* Ed. MJ Fregley and CM Blatters. New York: Oxford University Press. 889–939.
100. Manyari DE, Rose S, Tyberg JV, and Sheldon RS. 1996. Abnormal reflex venous function in patients with neuromediated syncope. *J. Am. Coll. Cardiol.* 27:1730–1735.
101. Rowe PC and Calkins H. 1998. A symposium: chronic fatigue syndrome. *Am. Jour. Med.* 105(3A):15S–17S.
102. Robinson SC. 1940. Hypotension: the ideal normal blood pressure. *N. Engl. J. Med.* 223:407–416.
103. Pilgrim JA, Stansfield S, and Marmot M. 1992. Low blood pressure, low mood: *BMJ* 304:75–78.
104. Wessely S, Nickson J, and Cox B. 1990. Symptoms of low blood pressure: a population study. *BMJ* 301:362–365.
105. Streeten DHP. 1987. *Orthostatic Disorders of the Circulation: Mechanism, Manifestations, and Treatment.* New York: Plenum.
106. Low PA, Opfer-Gehrking TL, McPhee BR, et al. 1995. Prospective evaluation of clinical characteristics of orthostatic hypotension. *Mayo. Clin. Proc.* 70:617–622.
107. Schondorf R and Low PA. 1993. Idiopathic postural orthostatic tachycardia syndrome: an attenuated form of acute pandysautonomia: *Neurology* 43:132–137.
108. Hoeldtke RD and Davis KM. 1991. The orthostatic tachycardia syndrome: evaluation of autonomic function and treatment with octreotide and ergot alkaloids. *J. Clin. Endocrinol. Metab.* 73:132–139.
109. Rosen SG and Cryer PE. 1982. Postural tachycardia syndrome: reversal of sympathetic hyperresponsiveness and clinical improvement during sodium loading. *Am. J. Med.* 72:847–850.
110. Low PA, Opfer-Gehrking TL, Taxtor SC, et al. 1995. Postural tachycardia syndrome (POTS), *Neurology* 45(suppl5):S19–S25.
111. Lewis T. 1932. A lecture on vasovagal syncope and the carotid sinus mechanism. *BMJ* 1:873–876.
112. Caulkins H, Shyr Y, Frumin H, Schork A, and Morady F. 1995. The value of the clinical history in the differentiation of syncope due to ventricular tachycardia, atrioventricular block, and neuro-cardiogenic syncope. *Am. J. Med.* 98:365–373.
113. Sheldon R. 1994. Effects of aging on responses to isopoterenol tilt-table testing in patients with syncope. *Am. J. Cardiol.* 74:459–463.
114. Harrison MH, Kravik SE, Geelan G, Keil L, and Greenleaf JE. 1985. Blood pressure and plasma rennin activity as predictors of orthostatic tolerance. *Aviat. Space Environ. Med.* 1059–1064.

115. Camfield PR and Camfield CS. 1990. Syncope in childhood: A case control clinical study of the familial tendency to faint. *Can. J. Neurol. Sci.* 17:306–308.
116. Henderson DA and Shelokov A. 1959. Epidemic neuromyasthenia-clinical syndrome? *N. Engl. J. Med.* 260:757–764.
117. Rowe PC, Bou-Holaigah I, Kan JS, and Calkins H. 1995. Is neurally mediated hypotension an unrecognized cause of chronic fatigue? *Lancet* 345:623–624.
118. Bou-Holaigah I, Rowe PC, Kan JS, and Calkins H. 1995. Relationship between neurally mediated hypotension and the chronic fatigue syndrome. *JAMA* 274:961–967.
119. Rowe PC, Bou-Holaigah I, Kan JS, and Calkins H. 1995. Improvement in symptoms of chronic fatigue syndrome is associated with reversal of neurally mediated hypotension (Abstract). *Pediatr. Res.* 37:33A.
120. Bou-Holaigah I, Calkins H, Flynn JA, et al. 1997. Provocation of hypotension and pain during upright tilt table testing in adults with fibromyalgia. *Clin. Exp. Rheum.* 15:239–240.
121. Almquist A, Goldenberg IF, Milstein S, et al. 1989. Provocation of bradycardia and hypotension by isoproterenol and upright posture in patients with unexplained syncope. *N. Engl. J. Med.* 320: 346–351.
122. Ross BA, Hughes S, Anderson E, and Gillete PC. 1991. Abnormal responses to orthostatic testing in children and adolescents with recurrent unexplained syncope. *Am. Heart. J.* 122:748–754.
123. Calkins H, Shyr Y, Frumin H, Schork A, and Morady F. 1995. The value of the clinical history in the differentiation of syncope due to ventricular tachycardia, artrioventricular block, and neurocardiogenic syncope. *Am. J. Med.* 98:365–373.
124. Thilenius OG, Ryd KJ, and Husayni J. 1992. Variation in expression and treatment of transient neurocardiogenic instability. *Am. J. Cardiol.* 69:1193–1195.
125. Sakaguchi S, Shultz JJ, Remole SC, Adler SW, Lurie KG, and Benditt DG. 1995. Syncope associated with exercise, an manifestation of neurally mediated syncope. *Am. J. Cardiol.* 75:476–481.
126. Sneddon JF, Scalia G, Ward DE, McKenna WJ, Camm AJ, and Frenneaux MP. 1994. Exercise induced vasodepressor syncope. *Br. Heart. J.* 71: 554–557.
127. Sra JS, Jazayeri MR, Dhala A, et al. 1993. Neurocardiogenic syncope: diagnosis, mechanism, and treatment. *Cardiol. Clin.* 11:183–191.
128. Kosinski DJ and Grubb BP. 1994. Neurally mediated syncope with an update on indications and usefulness of head-upright tilt table testing and pharmacologic therapy. *Curr. Opin. Cardiol.* 9:53–64.
129. Streeten DHP and Anderson GH Jr. 1992. Delayed orthostatic intolerance. *Arch. Intern. Med.* 152:1066–1072.
130. Schondorf R and Low PA. 1993. Idiopathic postural orthostatic tachycardia syndrome: an attenuated form of acute pandysautonomia? *Neurology* 43:132–137.
131. Yunus MB, Masi AT, Calabro JJ, Miller KA, and Feigenbaum SL. 1981. Primary fibromyalgia (fibrositis): clinical study of 50 patients with matched normal controls. *Semin. Arthritis. Rheum.* 11:151–171.
132. Buchwald D and Garrity D. 1994. Comparison of patients with chronic fatigue syndrome, fibromyalgia, and multiple chemical sensitivities. *Arch. Intern. Med.* 154:2049–2053.
133. Amadio PC, Mackinnon SE, Merritt WH, et al. 1991. Reflex sympathetic dystrophy syndrome: consensus report of an ad hoc committee of the American Asociation for Hand Surgery on the definition of reflex sympathetic dystrophy syndrome. *Plast. Reconstru. Surg.* 87(2):371–375 [Medline].
134. Campbell JN, Meyer RA, and Raja SN. 1992. Is nociceptor activation by alpha-1 adrenoreceptors the culprit in sympathetically mediated pain? *Am. Pain. Soc. J.* 1:3–11.
135. Cleary AG, Sills JA, and Davidson JE. 2001. Reflex sympathetic dystrophy. *Rheumatology* (Oxford) 40(5): 590–591 [Medline].
136. Dzwierzynski WW and Sanger JR. 1994. Reflex sympathetic dystrophy. *Hand Clin.* 10(1):29–44 [Medline].
137. Kemler MA, Reulen JP, and Barendse GA. Impact of spinal cord stimulation on sensory characteristics in complex regional pain syndrome type 1: A randomized trial.
138. Kemler MA, Rijks CP, DeVet HC. 2001. Which patients with chronic reflex sympathetic dystrophy are most likely to benefit from physical therapy? *J. Manipulative Physiol. Ther.* 24(4):272–278 [Medline].
139. Lankford LL. 1990. Reflex sympathetic dystrophy. In *Rehabilitation of the Hand.* Ed. JM Hunter, et. al. Mosby-Year Book. 763–786.
140. Parrillo, S. August 20, 2001. Reflex sympathetic dystrophy. E-medicine.com
141. Lynch ME. 1992. Psychological aspects of reflex sympathetic dystrophy: A review of the adult and paediatric literature. *Pain* 49(3):337–347 [Medline].

142. Perez RS, Kwakkel G, and Zuurmond WW. 2001. Treatment of reflex sympathetic dystrophy (CRPS type 1): a research synthesis of 21 randomized clinical trials. *J. Pain Symptom. Manage.* 21(6): 511–526 [Medline].
143. Roberts WJ. 1986. A hypothesis on the physiological basis for causalgia and related pains. *Pain* 24(3):297–311 [Medline].
144. Schwartzman RJ. 1993. Reflex sympathetic dystrophy. *Curr. Opin. Neurol. Neurosurg.* 6(4): 531–536 [Medline].
145. Schwartzman RJ and McLellan TL. 1987. Reflex sympathetic dystrophy: A review. *Arch. Neurol.* 44(5): 555–561 [Medline].
146. Schwartzman RJ. 2000. New treatments for reflex sympathetic dystrophy. *N. Engl. J. Med.* 343(9):654–656 [Medline].
147. Stanton-Hicks M, Janig W, Hassenbusch S, et. al. 1995. Reflex sympathetic dystrophy: changing concepts and taxonomy. *Pain* 63(1):127–133 [Medline].
148. Sundaram S and Webster GF. 2001. Vascular diseases are the most common cutaneous manifestation of relex sympathetic dystrophy. *J. Am. Acad. Dermatol.* 44(6):1050–1051 [Medline].
149. Tong HC and Nelson VS. 2000. Recurrent and migratory reflex sympathetic dystrophy in children. *Pediatr. Rehabil.* 4(2): 87–89 [Medline].
150. Van Hilten JJ, van de Beek WJ, and Vein AA. 2001. Clinical aspects of multifocal or generalized tonic dystonia in reflex sympathetic dystrophy. *Neurology* 56(12):1762–1765 [Medline].
151. Wilder RT, Berde CB, Wolohan M, et al. 1992. Reflex sympathetic dystrophy in children: Clinical characteristics and follow-up of seventy patients. *J. Bon. Joint Surg. [Am]* 74(6):910–919 [Medline].
152. Beck RW, Savino PJ, Repka MX, et al. 1984. Optic disc stricture in anterior ischemic optic neuropathy. *Ophthalmology* 91:1334.
153. Schnabel J. 1892. Das glaucomatose Sehnervehnleiden. *Arch. fur. Augen.* 24:273.
154. Johnson BL and Wisotzkey HM. 1977. Neuroretinitis associated with herpes simplex encephalitis in an adult. *Am. J. Opthalmol.* 83:481.
155. Fenton RH and Eason HA. 1884. Beçhet's syndrome: A histopathologic study of the eye. *Arch. Ophthalmol* 72:71
156. Gelwan MJ, Kellen RI, Burde RM, and Kupersmith MF. 1988. Sarcoidosis of the anterior visual pathway: successes and failure. *J. Neurol. Neurosurg. Psychiatr.* 51:1473.
157. Gass JD and Olson CL. 1976. Sarcoidosis with optic nerve and retinal involvement, *Arch. Ophthalmol.* 94:945.
158. Jordan DR, Anderson RL, Nerad JA, et al. 1988. Optic nerve involvement as the initial manifestation of sarcoidosis. *Can. J. Opthalmol.* 23:232.
159. Grant WM. 1986. *Toxicology of the Eye,* 3rd ed. Springfield, IL: Charles C. Thomas.
160. Ruskin, A. 1979. SPG remote effects including psychosomatic symptoms, rage reactions, pain and spasm; *Archives of Physical Med. Rehabil.,* Vol. 60. 142–147 [Klinghardt D, 23 World Conference].
161. Doty RL, Deems DA, Frye RE, Pelberg R, and Shapiro R. 1988. Olfactory sensitivity, nasal resistance, and autonomic function in patients with multiple chemical sensitivities. *Archy Otolaryngeal Head and Neck Surgery* 114(12):1422–1427.
162. Millquist E. 2001. A capsaicin cough test in patients with multiple chemical sensitivity. Presented at 19th Annual International Symposium on Man and His Environment. June 7–10, Dallas, TX.
163. Rea WJ. 1996. *Chemical Sensitivity, Vol. III, Clinical Manifestations of Pollutant Overload.* Boca Raton, FL: CRS Press. 1727.
164. Shirakawa S, Rea WJ, Ishikaw S, and Johnson AR. 1992. Evaluation of the autonomic nervous system response by pupillographical study in the chemically sensitive patient. *Environ. Med.* 8(4):121–127.
165. Payan DG, Levine JD, and Goetzl EJ. 1984. Modulation of immunity and hypersensitivity by sensory neuropeptides. *J. Immunol.* 132(4):1601–1604.
166. Foreman J. and Jordan C. 1983. Histamine release and vascular changes induced by neuropeptides. *Agents Actions* 13:105.
167. Rea WJ. 1992. *Chemical Sensitivity, Vol. I, Mechanisms of Chemical Sensitivity.* Boca Raton, FL: Lewis Publishers. 18.
168. Ader R, ed. 1981. *Psychoneuroimmunology.* New York: Academic Press.
169. Goetzl EJ. 1987. Leukocyte receptor for lipid and peptide mediators. *Fed. Proc.* 46(1):190–191.
170. Payan DG, McGillis JP, and Goetzl EJ. 1986. Neuroimmunology. In *Advances in Immunology.* Ed. FJ Dixon, KF Austen, L Hood, and JW Uhr. 2:199–323. New York: Academic Press.
171. Spector NH, ed. 1985. Neuroimmunomodulation (Proceedings of the First International Workshop on Neuroimmunomodulation). Bethesda, Maryland, International. Workshop on Group Neuroimmunomodulation.

172. Brooks WH, Cross RJ, Roszman TL and Markesbery WR. 1981. Neuroimmunomodulation: neural anatomical basis for impairment and facilitation. *Ann. Neurol.* 12:56–61.
173. Cross RJ, Markesbery WR, Brooks WH, and Roszman TL.1984. Hypothalmic-immune interactions: neuromodulation of natural killer activity by lesioning of the anterior hypothalamus. *Immunology* 51:399–405.
174. Janovic BD and Isakovic K. 1973. Neuro-endocrine correlates of immune response. I. Effects of brain lesions on antibody production, arthus reactivity and delayed hypersensitivity in the rat. *Int. Arch. Allergy* 45:360–372.
175. Bulloch K and Pomerantz W. 1984. Autonomic nervous system innervation of thymic-related lymphoid tissue in wildtype and nude mice. *J. Comp. Neurol.* 228:57–68.
176. Cannon JG, Tatro JB, Reichlin S, and Dinarello CA. 1986. Alphamelanocyte stimulating hormone inhibits immunostimulatory and inflammatory actions of interleukin-1. *J. Immunol.* 137:2232–2236.
177. Fujiwara R and Orita K. 1987. The enhancement of the immune response by pain stimulation in mice: I. The enhancement effect of PFC production via sympathetic nervous system in vivo and in vitro. *J. Immunol.* 138:3699–3703.
178. Goetzl EJ, Chernov-Rogan T, Cooke MP, Renold F, and Payan DG. 1985. Endogenous somatostatin-like peptides of rat basophilic leukemia cells. *J. Immunol.* 135:2707–2712.
179. Weinreich D and Undem BJ. 1987. Immunological regulation of synaptic transmission in isolated guinea pig autonomic ganglia. *J. Clin. Inves.* 79:1529–1532.
180. Said SI. 1984. Vasoactive intestinal polypeptide (VIP): current states. *Peptides* 5:143–150.
181. Said SI. 1982. Vasoactive peptides in the lung, with special reference to vasoactive intestinal peptide. *Exp. Lung Res.* 3:343–348.
182. Stead RH, Tomioka M, Quinonez G, Simon GT, Felten SY, and Bienenstock J. 1987. Intestinal mucosal mast cells in normal and nematode-infected rat intestines are in intimate contact with peptidergic nerves. *Proc. Natl. Acad. Sci. U.S.A.* 84:2975–2979.
183. Selye H. 1946. The general adaptation syndrome and the disease of adaptation. *J. Allergy* 17:23.
184. Pischinger A. 1975. *Matrix and Matrix Regulation: Basis for a Holistic Theory in Medicine.* Ed. H Heine. Eng. tran. N MacLean. Brussels, Belgium: Editions Haug International. 168.
185. Perry VH, Brown MC, and Gordon S. 1987. The macrophage response to central and peripheral nerve injury: A possible role for macrophages in regeneration. *J. Exp. Med.* 165:1281–1223.
186. Bernton EW, Beach JE, Holaday JW, Smallridge RC, and Fein HG. 1987. Release of multiple hormones by a direct action of interleukin-1 on pituitary cells. *Science* 238(4826):519–521.
187. Besedovsky H, Del Rey A, Sorkin E, and Dinarello CA. 1986. Immunoregulatory feedback between interleukin-21 and glucocorticoid hormones. *Science* 233:652–654.
188. Woloski BM, Smith EM, Meyer WJ III, Fuller GM, and Blalock JE.1985. Corticotropin-releasing activity of monokines. *Science* 230(4729):1035–1037.
189. Benveniste EN, Merrill JE, Kaufman SE, and Golde DW. 1985. Purification and characterization of a human T-lymphocyte-derived glial growth-promoting factor. *Proc. Natl. Acad. Sci. U.S.A.* 82:3930–3934.
190. Merrill JE, Kutsunai S, Mohlstrom C, Hofman F, Groopman J, and Golde DW. 1984. Proliferation of astroglia and oligodendroglia in response to human T cell-derived factors. *Science* 224(4656):1428–1430.
191. Dafny N, Prieto-Gomez B, and Reyes-Vasquez C. 1985. Does the immune system communicate with the central nervous system? Interferon modifies central nervous activity. *J. Neuroimmunol.* 9:1–12.
192. Beneviste EN, Kutsunai S, and Merrill JE. 1986. Immunoregulatory molecules modulate glial cell growth. In *Leukocytes and Host Defense.* Ed. JJ Oppenheim and D M Jacobs. New York: Alan R. Liss. 221–226.
193. Williams CA, Schupf N, and Hugli TE. 1985. Anaphylatoxin C5a modulation of an alpha-adrenergic receptor system in the rat hypothalamus. *J. Neuroimmunol.* 9:29–40.
194. Rea WJ. 1996. *Chemical Sensitivity, Vol. III, Clinical Manifestations of Pollutant Overload.* Boca Raton, FL: Lewis Publishers. 1740–1749.
195. Levine JD, Goetzl EJ, and Basbaum AI. 1987. Contribution of the nervous system to the pathophysiology of rheumatoid arthritis and other polyarthritides. *Rheum. Dis. Clin. N. Am.* 13:369–383.
196. Colpaert FC, Donnerer J, and Lembeck F. 1983. Effects of capsaicin on inflammation and on substance P content of nervous tissues in rats with adjuvant arthritis. *Life Sci.* 32:1827.
197. Turnbull JA. 1944. Changes in sensitivity to allergenic foods in arthritis. *J. Dig. Dis.* 15:182–190.
198. Marshall R, Stroud RM, Kroker GF, et al. 1984. Food Challenge effects on fasted rheumatoid arthritis patiens: a multicenter study. *Clin. Ecol.* 2(4):181–190.
199. Haapasaari J, Essen RV, Kahanpää A, Kostiala AA, Holmberg K, and Ahlqvist. 1982. Fungal arthritis stimulating juvenile rheumatoid arthritis. *Br. Med. J.* (Clin. Res. Ed.) 285(6346):923–924.

200. Randolph TG. 1976. Ecologically oriented rheumatoid arthritis. In *Clinical Ecology.* Ed. LD Dickey. 210–212.
201. Lundberg JM, Saria A, Brodin E, Rosell S, and Folers K. 1983. A substance P antagonist inhibits vagally induced inflammation and bronchial smooth muscle contraction in the guinea pig. *Proc. Natl. Acad. Sci. U.S.A.* 80(4):1120–1124.
202. Lundberg JM and Saria A. 1983. Capsaicin-induced desensitization of airway mucosa to cigarette smoke, mechanical and chemical irritants. *Nature* 302: 251–253.
203. Walker KB, Serwonska MH, Valone FH, et al. 1988. Distinctive patterns of release of neuroendodrine peptides after nasal challenge of allergic subjects with ryegrass antigen. *J. Clin. Immunol.* 8(2):108–113.
204. Schiogolev S, Urba W, Longo D, and Goetzl EJ. Unpublished data. Contributions of Neuropeptides to the altered vascular permeability induced by IL-2-LAK cell therapy. *Clin. Res.*
205. Day R, Lemaire S, Nadeau D, Keith I, and Lemaire I. 1987. Changes in autacoid and neuropeptide contents of lung cells in asbestos-induced pulmonary fibrosis. *Am.Rev.Respir.Dis.* 136(4):908–915.
206. Wallengren D, Ekman R, and Moller H. 1986. Substance P and vasoactive intestinal peptide inbullous and inflammatory skin disease. *Acta Derm. Venereol.* (Stockh.) 66:23–28.
207. Foreman JC, Jordan CC, and Piotrowski W. 1982. Interaction of neurotensin with the substance P receptor mediating histamine release from rat mast cells and the flare in human skin. *Br. J. Pharmacol.* 77:531–539.
208. Mann F. 1972. *Acupuncture: The Ancient Art of Chinese Healing and How It Works Scientifically.* Random House. 6.
209. Mann F. 1972. *Acupuncture: The Ancient Art of Chinese Healing and How It Works Scientifically.* Random House. 7.
210. Mann F. 1972. *Acupuncture: The Ancient Art of Chinese Healing and How It Works Scientifically.* Random House. 11.
211. Kellgren JH. 1939–42. On the distribution of pain arising from deep somatic structures with charts of segmental pain. *Clin. Sci.* 4:35–46.
212. Mann F. 1972. *Acupuncture: The Ancient Art of Chinese Healing and How It Works Scientifically.* Random House. 17–22.
213. Mann F. 1972. *Acupuncture: The Ancient Art of Chinese Healing and How It Works Scientifically.* Random House. 23–24.
214. Koblank A. 1958. *Die Naseals Reflexorgan.* Haug, Ulm. Works Scientifcally.
215. Ashkenaz DM. 1937. An experimental analysis of centripetal visceral pathways based upon the viscero-0pannicular reflex. *Am. J. of Phys.* 120:587–595.
216. Norton S. 1986. Toxic responses of the central nervous system. In *Casarett and Doull's Toxicology: The Basic Science of Poisons,* 3rd ed. Ed. CD Klaassen, MD Amdur, and JDoull. New York: Macmillan. 359–386.
217. Bondareff W. 1965. The extracellular compartment of the cerebral cortex. *Anat. Rec.* 152:119–127.
218. Kuhlenbeck H. 1970. *The Central Nervous System of Vertebrates,* Vol.3. New York: Academic Press.
219. Westergaard E, van Deurs B, and Bondsted HE. 1977. Increased vesicular transfer of horseradish peroxidase across cerebral endothelium evoked by acute hypertension. *Acta Neuropathol.* 37:141–152.
220. Cervos-navarro J and Rozas JI. 1978. The arteriole as a site of metabolic change. *Adv. Neurol.* 20:17–24.
221. Welsh FA and O'Connor MJ. 1978. Patterns of microcirculatory failure during incomplete cerebral ischemia. *Adv. Neurol.* 20:133–139.
222. Hazama F, Amano S, and Ozaki T. 1978. Pathological changes of cerebral vessel endothelial cells in spontaneously hypertensive rats with special reference to the role of these cells in the development of hypertensive cerebrovascular lesions. *Adv.Neurol.* 20:359–369.
223. Jacobs JM, MacFarlane RM, and Cavanagh JB. 1976. Vascular leakage in the dorsal root ganglia of the rat, studied with horseradish peroxidase. *J. Neruol. Sci.* 29:95–107.
224. Jacobs JM. 1977. Penetration of systemically injected horseradish peroxidase into gangliaand nerves of the autonomic nervous system. *J. Neurocytol.* 6:607–618.
225. Olney JW, Rhee V, and de Gubareff T. 1977. Neurotoxic effects of glutamate on mouse area postrema. *Brain Res.* 120:151–157.
226. Zimmerberg J and Chernomordik L. 2005. Perspectives *Synaptic Membranes Bend to the Will of a Neurotoxin. Science* 310(December 9):1626–1627.
227. Rigoni M, et al. 2005. *Science* 310:1678.
228. Zimmerberg J. 2000. *Traffic* 1:366.

229. Zimmerberg MM Kozlov, *Nat, Rev. Mol. Cell Biol.* 15 November 2005;10.1038/nrm 1784.
230. Chernomordik V, Frolov VA, Leikina E, Bronk P, and Zimmerberg J. 1998. *J. Cell Biol.* 140:1369.
231. Chernomordik LV, et al., 1993. *FEBS Lett.* 318: 1.
232. Chernomordik LV and Kozlov MM. 2003. *Annu. Rev. Biochem.* 72:175.
233. Giraudo CG, et al. 2005. *J. Cell Biol.* 170:249.
234. Reese C, Heise F, and Mayer A. 2005. *Nature* 436, 410.
235. Xu Y, Zhang YF, Su Z, McNew JA, and Shin YK. 2005. *Nat. Struct. Mol. Biol.* 12:417.
236. Jahn R, Lang T, and Sudhof TC. 2003. *Cell* 112:519.
237. Chanturiya A, Chernomordik LV, and Zimmerberg J. 1997. *Proc.Natl.Acad.Sci.U.S.A.* 94:14423.
238. Brown WJ, Chambers K, and Doody A. 2003. *Traffic* 4:214.
239. Monje, et. al. 2003. *Science* 302(5651):1760.
240. Ullian EM, Christopherson KS, and Barres BA. 2004. Role for glia in synaptogenesis. *Gilia* 47(3):209–216.
241. Winter JO. Coatings to help medical implants connect with neurons. http://www.nanotechwire.com/news.asp?nid=6514&ntid=&pg=9
242. Nedergaard M. 1994. Direct signaling from astrocytes to neurons in cultures of mammalian brain cell. *Science.* 263(5154):1768–1771.
243. Newman EA. 2003 New roles for astrocytes: regulation of synaptic transmission. *Trends in Neurosci.* 26(10):536–542.
244. Czeh B, Michaclis T, Watahabe T, et al. 2001. Stress-induced changes in cerebral metaboloites, hippocampal volume, and cell proliferation are prevented by antidepressant treatment with tianeptine. *Science News.* April 17, 1:159.
245. Fields DR. 2004. The other half of the brain. *Scient. Am.* 290(4):54–61.
246. Kempermann G and Neumann H. 2003. Microglia: The enemy within. *Science.* 302:1689–1691.
247. Heim MH, Kerr IM, Stark GR, and Darnell JE Jr. 1995. Contribution of STAT SH_2 groups to specific interferon signaling by the JAK-STAT pathway. *Science* 267:1347–1349.
248. Friede FL. 1966. *Topographic Brain Chemistry.* New York: Academic Press.
249. Olney JW. 1971. Gluatmate-induced neuronal necrosis in the infant mouse hypothalamus. *J. Neuropathol. Exp. Neurol.* 30:75–90.
250. Rea WJ. 1994. *Chemical Sensitivity, Vol II, Sources of Total Body Load.* BocaRaton, FL: Lewis Publishers. 735.
251. Jacobs JM, Carmichael N, and Cavanagh JB. 1977. Ultrastructural changes in the nervous system of rabbits poisoned with methyl mercury. *Toxicol. Appl. Pharmacol.* 39:249–261.
252. Rapoport S. 1976. *Blood-Brain Barrier in Physiology and Medicine.* 129–152. New York: Raven Press.
253. Hanig JP and Herman EH. 1991. Toxic responses of the heart and vascular systems. In *Casarett and Doull's Toxicology: The Basic Science of Poisons*, 4th ed. Ed. MO Amdur, J Doull, and CD Klaasen. New York: Pergamon Press. 450.
254. Arrighi RE. 1974. Mammalian Chromosomes. In *The Cell Nucleus,* Vol. II. Ed. H Busch. New York: Academic Press. 1–32.
255. National Institute on Drug Abuse. Apr. 1988. Facts about teenagers and drug abuse. NIDA Capsules, C-83-07a.rev.
256. Anthony DC and Graham DG. 1991. Toxic responses of the nervous system. In *Casarett and Doull's Toxicology: The Basic Science of Poisons*, 4th ed. Ed. MO Amdur, J Doull, and CD Klaasen New York: Pergamon Press. 434.
257. Baselt RC, ed. 1982. *Disposition of Toxic Drugs and Chemicals in Man*, 2nd ed. Davis, CA: Biomedical Publications.
258. Monmaney T. 1987. Are we led by the nose: *Discover* 8:48–56.
259. Linnik MD, Zobrist RH, and Hatfield MD. 1993. Evidence supporting a role for programmed cell death in focal cerebral ischemia in rats. *Stroke* 24:2002–2009.
260. Charriaut-Marlangue C, Margaill I, Represa A, Popovici T, Plotkine M, and Ben-Ari Y. 1996. Apoptosis and necrosis after reversible focal ischemia: an in situ DNA fragmentation analysis. *J. Cereb. Blood Flow Metab.* 16:186–194.
261. Li Y, Chopp M, Jiang N, and Zaloga C. 1995. In situ detection of DNA fragmentation after focal cerebral ischemia in mice. *Brain Res. Mol. Brain Res.* 28:164–168.
262. MacManus JP, Buchan AM, Hill IE, Rasquinha I, Preston E. 1993. Global ischemia can cause DNA fragmentation indicative of apoptosis in rat brain. *Neurosci. Lett.* 164:89–92.
263. Emery E, Aldana P, Bunge MB, et al. 1998. Apoptosis after traumatic human spinal cord injury. *J. Neurosurg.* 89:911–920.

264. Crowe MJ, Bresnahan JC, Shuman SL, Masters JN, and Beattie MS. 1997. Apoptosis and delayed degeneration after spinal cord injury in rats and monkeys. *Nat. Med.* 3:73–76. [Erratum, *Nat. Med.* 1997, 3:240.]
265. Li M, Ona VO, Chen M, et al. 2000. Functional role and therapeutic implications of neuronal caspase-1 and -3 in a mouse model of traumatic spinal cord injury. *Neuroscience* 99:333–342.
266. Liu XZ, Xu XM, Hu R, et al. 1997. Neuronal and glial apoptosis after traumatic spinal cord injury. *J. Neurosci.* 17:5395–5406.
267. Rink A, Fung KM, Trojanowski JQ, Lee VM, Neugebauer E, and McIntosh TK. 1995. Evidence of apoptotic cell death after experimental traumatic brain injury in the rat. *Am. J. Pathol.* 147:1575–1583.
268. Thomas LB, Gates DJ, Richfield EK, O'Brien TF, Schweitzer JB, and Steindler DA. 1995. DNA end labeling (TUNEL) in Huntington's disease and other neuropathological conditions. *Exp. Neurol.* 133:265–272.
269. Troost D, Aten J, Morsink F, and de Jong JM. 1995. Apoptosis in amyotrophic lateral sclerosis is notrestricted to motor neurons: Bcl-2 expression is increased in unaffected postcentral gyrus. *Neuropathol. Appl. Neurobiol.* 21:498–504.
270. Smale G, Nichols NR, Brady DR, Finch CE, Horton WE Jr. 1995. Evidence for apoptotic cell death in Alzheimer's disease. *Exp. Neurol.* 133:225–230.
271. Hengartner MO. 2000. The biochemistry of apoptosis. *Nature* 407:770–776.
272. Wyllie AH, Kerr JF, and Currie AR. 1980. Cell death: the significance of apoptosis. *Int. Rev. Cytol* 68:251–306.
273. Kerr JF, Wyllie AH, and Currie AR. 1972. Apoptosis: a basic biological phenomenon with wide-ranging implications in tissue kinetics. *Br. J. Cancer.* 26:239–257.
274. Wyllie AH. 1980. Glucocorticoid-induced thymocyte apoptosis is associated with endogenous endonuclease activation. *Nature* 284:555–556.
275. Liu X, Zou H, Slaughter C, and Wang X. 1997. DFF, a heterodimeric protein that functions downstream of caspase-3 to trigger DNA fragmentation during apoptosis. *Cell* 89:175–184.
276. Horvitz HR. 1999. Genetic control of programmed cell death in the nematode Caenorhabditiselegans. *Cancer Res.* 59:Suppl:1701s–1706s.
277. Yuan J, Shaham S, Ledoux S, Ellis HM, and Horvitz HR. 1993. The C. elegans cell death gene ced-3 encodes a protein similar to mammalian interleukin-1 beta-converting enzyme. *Cell* 75:641–652.
278. Yuan JY and Horvitz HR. 1990. The Caenorhabditis elegans genes ced-3 and ced-4 act cellautonomously to cause programmed cell death. *Dev. Biol.* 138:33–41.
279. Miura M, Zhu H, Rotello R, Hartwieg EA, and Yuan J. 1993. Induction of apoptosis in fibroblasts by IL-1 beta-converting enzyme, a mammalian homolog of the C. elegans cell death gene ced-3. *Cell* 75:653–660.
280. Li P, Nijhawan D, Budihardjo I, et al. 1997. Cytochrome c and dATP-dependent formation ofApaf-1/caspase-9 complex initiates an apoptotic protease cascade. *Cell* 91:479–489.
281. Hengartner MO, Horvitz HR. 1994. C. elegans cell survival gene ced-9 encodes a functional homolog of the mammalian proto-oncogene bcl-2. *Cell* 76:665–676.
282. Conradt B and Horvitz HR. 1998. The C. elegans protein EGL-1 is required for programmed cell death and interacts with the Bcl-2-like protein CED-9. *Cell* 93:519–529.
283. Alnemri ES, Livingston DJ, Nicholson DW, et al. 1996. Human ICE/CED-3 protease nomenclature. *Cell* 87:171.
284. Li H, Zhu H, Xu CJ, and Yuan J. 1998. Cleavage of BID by caspase 8 mediates the mitochondrial damage in the Fas pathway of apoptosis. *Cell* 94:491–501.
285. Wang X. 2001. The expanding role of mitochondria in apoptosis. *Genes. Dev.* 15:2922–2933.
286. Shi Y. 2002. Mechanisms of caspase activation and inhibition during apoptosis. *Mol. Cell.* 9:459–470.
287. Deveraux QL, Schendel SL, and Reed JC. 2001. Antiapoptotic proteins: the bcl-2 and inhibitor of apoptosis protein families. *Cardiol. Clin.* 19:57–74.
288. Xu D, Bureau Y, McIntyre DC, et al. 1999. Attenuation of ischemia-induced cellular andbehavioral deficits by X chromosome-linked inhibitor of apoptosis protein overexpressionin the rat hippocampus. *J. Neurosci.* 19:5026–5033.
289. Yuan J and Yankner BA. 2000. Apoptosis in the nervous system. *Nature* 407:802–809.
290. Friedlander RM and Yuan J. 1998. ICE, neuronal apoptosis and neurodegeneration. *Cell Death Differ.* 5:823–831.
291. Troy CM and Salvesen GS. 2002. Caspases on the brain. *J. Neurosci. Res.* 69:145–150.
292. Fink KB, Andrews LJ, Butler WE, et al. 1999. Reduction of posttraumatic brain injury and free radical production by inhibition of the caspase-1 cascade. *Neuroscience* 94:1213–1218.

293. Yakovlev AG, Knoblach SM, Fan L, Fox GB, Goodnight R, and Faden AI. 1997. Activation of CPP32-like caspases contributes to neuronal apoptosis and neurological dysfunction after traumatic brain injury. *J. Neurosci.* 17:7415–7424.
294. Springer JE, Azbill RD, and Knapp PE. 1999. Activation of the caspase-3 apoptotic cascade in traumatic spinal cord injury. *Nat. Med.* 5:943–946.
295. Li M, Ona VO, Guegan C, et al. 2000. Functional role of caspase-1 and caspase-3 in an ALS transgenic mouse model. *Science* 288:335–339.
296. Ona VO, Li M, Vonsattel JP, et al. 1999. Inhibition of caspase-1 slows disease progression in a mouse model of Huntington's disease. *Nature* 399:263–267.
297. Friedlander RM, Brown RH, Gagliardini V, Wang J, and Yuan J. 1997. Inhibition of ICE slows ALS in mice. *Nature* 388:31–31. [Erratum, *Nature* 1998, 392:560.]
298. Gervais FG, Xu D, Robertson GS, et al. 1999. Involvement of caspases in proteolytic cleavage of Alzheimer's amyloid-beta precursor protein and amyloidogenic A beta peptide formation. *Cell* 97:395–406.
299. Friedlander RM, Gagliardini V, Hara H, et al. 1997. Expression of a dominant negative mutant of interleukin-1 beta converting enzyme in transgenic mice prevents neuronal cell death induced by trophic factor withdrawal and ischemic brain injury. *J. Exp. Med.* 185:933–940.
300. Loddick SA, MacKenzie A, and Rothwell NJ. 1996. An ICE inhibitor, z-VAD-DCB attenuates ischaemic brain damage in the rat. *Neuroreport* 7:1465–1468. [Erratum, *Neuroreport* 1999, 10(9):inside back cover.]
301. Hara H, Friedlander RM, Gagliardini V, et al. 1997. Inhibition of interleukin 1 beta converting enzyme family proteases reduces ischemic and excitotoxic neuronal damage. *Proc. Natl. Acad. Sci. USA* 94:2007–2012.
302. Rabuffetti M, Sciorati C, Tarozzo G, Clementi E, Manfredi AA, and Beltramo M. 2000. Inhibition of caspase-1-like activity by Ac-Tyr-Val-Ala-Asp-chloromethyl ketone induces long-lasting neuroprotection in cerebral ischemia through apoptosis reduction and decrease of proinflammatory cytokines. *J. Neurosci.* 20:4398–4404.
303. Kang SJ, Wang S, Hara H, et al. 2000. Dual role of caspase-11 in mediating activation of caspase-1 and caspase-3 under pathological conditions. *J. Cell Biol.* 149:613–622.
304. Benchoua A, Guegan C, Couriaud C, et al. 2001. Specific caspase pathways are activated in the two stages of cerebral infarction. *J. Neurosci.* 21:7127–7134.
305. Zhu S, Stavrovskaya IG, Drozda M, et al. 2002. Minocycline inhibits cytochrome c release and delays progression of amyotrophic lateral sclerosis in mice. *Nature* 417:74–78.
306. Martinou JC, Dubois-Dauphin M, Staple JK, et al. 1994. Overexpression of BCL-2 in transgenic mice protects neurons from naturally occurring cell death and experimental ischemia. *Neuron* 13:1017–1030.
307. Plesnila N, Zinkel S, Le DA, et al. 2001. BID mediates neuronal cell death after oxygen-glucose deprivation and focal cerebral ischemia. *Proc. Natl. Acad. Sci. USA* 98:15318–15323.
308. Schielke GP, Yang GY, Shivers BD, and Betz AL. 1998. Reduced ischemic brain injury in interleukin-1 beta converting enzyme-deficient mice. *J. Cereb. Blood Flow Metab.* 18:180–185.
309. Fink K, Zhu J, Namura S, et al. 1998. Prolonged therapeutic window for ischemic brain damage caused by delayed caspase activation. *J. Cereb. Blood Flow Metab.* 18:1071–1076.
310. Ader R. 1987. Behavioral Influence on Immunity. Paper presented at the meeting of the Cabinet of Environmental Medicine. Nashville, TN.
311. Butler, JR, Rea WJ, Laseter JL, et al.1986. Unpublished data. EHC-Dallas.
312. Didriksen N and Butler JR. 1977. Unpublished data.
313. Rea WJ. 1980. Personal communication.
314. Rea WJ, Butler JR, Laseter JL, and DeLeon IR. 1984. Pesticides and brain function changes in a controlled environment. *Clin. Ecol.* 2(3):145–150.
315. Rea WJ. 1994. *Chemical Sensitivity, Vol. II, Sources of Total Body Load.* Boca Raton. FL: Lewis Publishers. 638.
316. Rea WJ. 1996. *Chemical Sensitivity, Vol. III, Clinical Manifestations of Pollutant Overload.* Boca Raton. FL: Lewis Publishers. 1760–1761.
317. Delpech A. 1856. Nore sur les accidents que dévelope, chez les ouviers en caoutchouc, l'inhalation du sulfure de carbone envapeur. *Bull. Acad. Imp. Med.* 21:350.
318. Rea WJ. 1994. *Chemical Sensitivity, Vol. II, Sources of Total Body Load.* Boca Raton. FL: Lewis Publishers. 713.
319. Charcot JM. 1874. Des las scherose laterale amyotrophique: symptomatologie. *Prog. Med.* 29:453.

320. Davidson M and Feinleib M. 1972. Carbon disulfide poisoning: a review. *Am. Heart J.* 83:100.
321. Proctor NH and Hughes JP. 1978. *Chemical Hazards of the Workplace.* 149. Philadelphia: J.B. Lippincott.
322. Burnett WW, King EG, Grac M, and Hall WF. 1977. Hydrogen sulfide poisoning: review of 5 years experience. *Canadian Med. Assoc. J.* 117:1277–80.
323. Savage EP, Keefe TJ, Mounce LM, Heaton RK, Lewis JA, and Bursar PJ. 1988. Chronic neurological sequelae of acute organophosphate pesticide poisoning. *Arch. Environ. Health* 43:38–45.
324. Anger WK. 1984. Neurobehavioral testing of chemicals: impact on recommended standards. *Neurobehav. Toxicol. Teratol.* 6:147–153.
325. Rea WJ. 1994. *Chemical Sensitivity, Vol. II, Sources of Total Body Load.* Boca Raton. FL: Lewis Publishers. 837.
326. Kilburn K. 1998. *Chemical Brain Injury.* New York: Van Nostrand Reinhold. 93.
327. Kilburn K. 1998. *Chemical Brain Injury.* New York: Van Nostrand Reinhold. 96 (5.2).
328. Kilburn K. 1998. *Chemical Brain Injury.* New York: Van Nostrand Reinhold. 96–104.
329. Kilburn K. 1998. *Chemical Brain Injury.* New York: Van Nostrand Reinhold. 118.
330. Kilburn K. 1998. *Chemical Brain Injury.* New York: Van Nostrand Reinhold. 114.
331. Kilburn,K. 1998. *Chemical Brain Injury.* New York: Van Nostrand Reinhold. 114–123.
332. Kilburn K. 1998. *Chemical Brain Injury.* New York: Van Nostrand Reinhold. 13.
333. Kilburn K. 1998. *Chemical Brain Injury.* New York: Van Nostrand Reinhold. 157.
334. Kilburn K. 1998. *Chemical Brain Injury.* New York: Van Nostrand Reinhold. 152.
335. Kilburn K. 1998. *Chemical Brain Injury.* New York: Van Nostrand Reinhold. 153.
336. Kilburn K. 1998. *Chemical Brain Injury.* New York: Van Nostrand Reinhold. 159.
337. Kilburn K. 1998. *Chemical Brain Injury.* New York: Van Nostrand Reinhold. 166–179.
338. Kilburn K. 1998. *Chemical Brain Injury.* New York: Van Nostrand Reinhold. 198–251.
339. Kilburn K. 1998. *Chemical Brain Injury.* New York: Van Nostrand Reinhold. 234–283.
340. Kilburn K. 1998. *Chemical Brain Injury.* New York: Van Nostrand Reinhold. 284–275.
341. Kilburn K. 1998. *Chemical Brain Injury.* New York: Van Nostrand Reinhold. 296–305.
342. Anger WK and Johnson BL. 1988. *Chemical Affecting Behavior in Neurotoxicity of Industrial and Commercial Chemicals.* Ed. JL O'Donoghue. Boca Raton, FL: CRC Press. 51–148.
343. Singer R. 1990. *Neurotoxicity Guidebook.* New York: Van Nostrand Reinold. 11–18.
344. Anger WK and Johnson BL. 1988. *Chemical Affecting Behavior in Neurotoxicity of Industrial and Commercial Chemicals.* Ed. JL O'Donoghue. Boca Raton, FL: CRC Press. 8.
345. Bell IR. 2002. Effects of food allergy on the central nervous system. In *Food Allergy and Intolerance,* 2nd ed. Ed. J Brostoff and SJ Challacombe. London: Saunders. 660.
346. Binkley KE and Kutcher S. 1997. Panic response to sodium lactate infusion in patients with multiple chemical sensitivity syndrome. *J. Allergy Clin. Immunol.* 99:570–574.
347. Gorman JM, Papp LA, Coplan JD, et al. 1994. Anxiogenic effects of CO_2 and hyperventilation in patients with panic disorder. *Am. J. Psychiatry.* 151:547–553.
348. Bourin M, Baker GB, and Bradwen J. 1998. Neurobiology of panic disorder. *Psychosam. Res.* 44:163–180.
349. Papp LA, Klein DF, Martinez J, et al. 1993. Diagnostic and substance specificity of carbon-dioxide-induced panic. *Am. J. Psychiatry* 150:250–257.
350. Perna G, Cocchi S, Bertani A, Arancio C, and Bellodi L. 1995. Sensitivity to 35% CO_2 in healthy first-degree relatives of patients with panic disorder. *Am. J. Psychiatry* 152:623–625.
351. Abou-Donia MB. 2003. Organophosphorus Ester-Induced Chronic Neurotoxicity. 58(8):484–497. Department of Pharmacology and Cancer Biology Duke University Medical Center. Durham, North Carolina. *Arch. Env. Health.*
352. Abou-Donia MB. 1994. Organophosphorus pesticides. In *Handbook of Neurotoxicology.* Ed. LW Chang and RS Dyer. New York: Marcel Dekker. 419–47.
353. Koelle GB. 1946. Protection of cholinesterase against inevitable inactivation by diisopropyl fluorophosphate in vitro. *J. Pharmacol. Exp. Ther.* 88:232–37.
354. Abou-Donia MB. 1992. *Neurotoxicology.* Boca Raton, FL: CRC Press. 3–24.
355. Whittaker M. 1994. The pseudocholinesterase variants: esterase levels and increased resistance to fluoride. *Acta. Genet. Basel.* 14:281–85.
356. Lockridge O. 1990. Genetic variants of human serum cholinesterase influence metabolism of the muscle relaxant succinylcholine. *Pharmacol. Ther.* 47:35–60.
357. Mackness B, Mackness MI, Arrol S, et al. 1997. Effect of the molecular polymorphisms of human paraoxonase (PONI) on the rate of hydrolysis of paraoxon. *Br. J. Pharmacol.* 122:265–268.

358. Davies HG, Richter RJ, Keifer M, et al. 1996. The effect of the human serum paraoxonase polymorphism is reversed with diazoxon, soman, and sarin. *Nat. Genet.* 14:334–336.
359. Morita H, Yanagisawa N, Nakajima T, et al. 1995. Sarin poisoning in Matsumoto, Japan. *Lancet* 346:290–293.
360. Okumura T, Takasu N, Ishimatsu S, et al. 1995. Report on 640 victims of the Tokyo subway sarin attack. *Ann. Emerg. Med.* 28:129–135.
361. Yokoyama K, Araki S, Murata K, et al. 1998. Chronic neurobehavioral effects of Tokyo subway sarin poisoning in relation to posttraumatic stress disorder. *Arch. Environ. Health* 53:249–256.
362. Masuda N, Takatsu M, Morinari H. 1995. Sarin poisoning in Tokyo subway. *Lancet* 345:1446–1447.
363. Smith MI, Elvove E, Frazier WH. 1930. The pharmacologicalaction of certain phenol esters withspecial reference to the etiology of so-called ginger paralysis. *Public Health Rep.* 45:2509–2524.
364. Johnson MK. 1975. The delayed neuropathology caused by some organophosphorous esters: mechanism and challenge. *Crit. Rev. Toxicol.* 2:289–316.
365. Abou-Donia MB. 1981. Organophosphorous ester-induced delayed neurotoxicity. *Annu. Rev. Pharmacol. Toxicol.* 21:511–548.
366. Abou-Donia MB and Lapadula LM. 1990. Mechanisms of organophosphorus ester-induced delayed neurotoxicity: type I and type II. *Annu. Rev. Pharmacol. Toxicol.* 30:405–440.
367. Cavanagh JB and Patangia GN. 1995. Changes in the central nervous system in the cat as the result of tri-o-cresylphosphate poisoning. *Brain* 88:165–180.
368. Himuro K, Murayama S, Nishiyama K, et al. 1998. Distal sensory axonopathy after sarinintoxification. *Neurology* 51:1195–1197.
369. Davies OR and Holland PR. 1972. Effect of oximes and atropine upon the development of delayedneurotoxic signs in chickens following poisoning by DFP and sarin. *Biochem. Pharmacol.* 21:3145–3151.
370. Abou-Donia MB. 1992. Triphenyl phosphite: a type II organophosphorus compound-induced delayed neurotoxic agent. In *Organophosphates: Chemistry, Fates, and Effects. Part IV: Toxic Effects—Organismal.* Ed. JE Chambers, PE Levi. San Diego, CA: Academic Press. 327–351.
371. Abou-Donia MB, Wilmarth KR, Jansen KF, et al. 1996. Triphenylphosphine (TPP): a type III organophosphorous compound-induced delayed neurotoxic agent (OPIDN). *Toxicologist* 30:311.
372. Abdel-Rahman AA, Jensen KF, Farr CH, et al. 1997. Daily treatment of triphenylphosphine (TPP) produces organophosphorous-induced delayed neurotoxicity (OPIDN). *Toxicologist* 36:19.
373. Bloch H, Hottinger A. 1943. Uber die spezifitat der cholinesterase-hemmung durch tri-okresyl phosphat. *Int. Z Vitaminforsch* 13:90.
374. Earl CJ and Thompson RHS. 1952. The inhibitory action of triortho-cresyl phosphate and cholinesterases. *Br. J. Pharmacol.* 7:261–69.
375. Aldridge WN and Barnes JM. 1966. Further observations on the neurotoxicity of organophosphorus compounds. *Biochem. Pharmacol.* 15:541–547.
376. Johnson MK. 1969. The delayed neurotoxic effect of some organophosphorus compounds. *Br. Med. Bull.* 114:711–717.
377. Lotti M. 1992. The pathogenesis of organophosphate delayed neuropathy. *Crit. Rev. Toxicol.* 21:465–487.
378. Curtes JP, Develay P, and Hubert JP. 1981. Late peripheral neuropathy due to acute voluntary intoxication by organophosphorous compounds. *Clin. Toxicol.* 18:1453.
379. de Jager AE, van Weerden TW, Houthoff HJ, et al. 1981. Polyneuropathy after massive exposure to parathion. *Neurology* 31:603–605.
380. Stamboulis E, Psimaras A, and Vassilopoulos D. 1991. Neuropathy following acute intoxication, with mercarbam (Opester). *Acta Neurol. Scand.* 83:198.
381. Winrow CJ, Hemming ML, Allen DA, et al. 2003. Loss of neuropathy target esterase in mice links organophosphate exposure to hyperactivity. *Nat. Genet.* 33:477–485.
382. O'Callahan JP. 2003. Neurotoxic esterase: Not so toxic? *Nat. Genet.* 33:1–2.
383. Bus J, Maurissen J, Marable B, et al. 2003. Association between organophosphate exposure and hyperactivity? *Nat. Genet.* 34(3):235.
384. Abou-Donia MB. 1995. Involvement of cytoskeletal proteins in the mechanisms of organophosphorous ester delayed neurotoxicity. *Clin. Exp. Pharmacol. Physiol.* 22:358–359.
385. Schulman H. 1988. *Advances in Second Messenger and Phosphorylation Research.* New York: Raven Press. 39–111.
386. Patton SE, Lapadula DM, and Abou-Donia MB. 1986. Relationship of tri-o-cresyl phosphate-induced delayed neurotoxicity to enhancement of in vitro phosphorylation of hen brain and spinal cord. *J. Pharmacol. Exp. Ther.* 239: 597–605.

387. Suwita E, Lapadula DM, and Abou-Donia MB. 1986. Calcium and calmodulin-enhanced in vitro phosphorylation of hen brain cold-stable microtubules and spinal cord neuro-filament triplet proteins after a single oral dose of tri-ocresyl phosphate. *Proc. Natl. Acad. Sci. USA* 83:6174–6178.
388. Abou-Donia MB, Viana ME, Gupta RP, et al. 1993. Enhanced calmodulin binding concurrent with increased kinase-dependent phosphorylation of cytoskeletal proteins following a single subcutaneous injection of diisopropyl phosphorofluoridate in hens. *Neurochem. Res.* 22:165–173.
389. Gupta RP and Abou-Donia MB. 1998. Tau proteins-enhanced Ca2+/calmodulin (CaM)-dependent phosphorylation by the brain supernatant of diisopropyl phosphorofluo-ridate (DFP)-treated hen: tau mutants indicate phosphorylation of more amino acids in tau by CaM kinase II. *Brain Res.* 813:32–43.
390. Gupta RP and Abou-Donia MB. 1994. In vivo and in vitro effects of diisopropyl phosphorofluoridate (DFP) on the rate of hen brain tubulin polymerization. *Neurochem. Res.* 19:435–444.
391. Gupta RP and Abou-Donia MB. 1995. Neurofilament phosphorylation and [125I] calmodulin binding by Ca2+ /calmodulin-dependent protein kinase in the brain subcellular fractions of diisopropyl phosphorofluoridate (DFP)-treated hen. *Neurochem. Res.* 20(9):1095–1105.
392. Gupta RP and Abou-Donia, MB. 1995. Diisopropyl phosphorofluoridate (DFP) treatment alters calcium-activated proteinase activity and cytoskeletal proteins of the hen sciatic nerve. *Brain Res.* 677:162–166.
393. Gupta RP, Bing G, Hong JS, et al. 1998. cDNA cloning and sequencing of Ca2+ almodulin-dependent protein kinase subunit and its mRNA expression in diisopropylhosphorofluoridate (DFP)-treated hen central nervous system. *Mol. Cell. Biochem.* 181:29–39.
394. Gupta RP and Abou-Donia MB. 1999. Tau phosphorylation by diisopropyl phosphorofluoridate (DFP)-treated hen brain supernatant inhibits its binding with microtubules: role of Ca2+/calmodulin-dependent protein kinase II in tauphosphorylation. *Biochem. Pharmacol.* 53:1799–1806.
395. Gupta RP, Damodaran TV, and Abou-Donia MB. 2000. C-fosmRNA induction in the central and peripheral nervous systems of diisopropyl phosphorofluoridate (DFP)-treated hens. *Neurochem. Res.* 25(3):327–334.
396. Gupta RP, Abdel-Rahman AA, Jensen KF, et al. Forthcoming. Altered expression of neurofilament subunits in diisopropyl phosphorofluoridate (DFP)-treated hen spinal cord and their presence in axonal aggregates.
397. Damodaran TV and Abou-Donia MB. 2000. Alterations in levels of mRNAs coding for glial fibrillary acidic protein (GFAP) and vimentin genes in the central nervous system of hens treats with diisopropyl phosphorofluoridate (DFP). *Neurochem. Res.* 25:809–816.
398. Jensen KF, Lapadula DM, Anderson JK, et al. 1992. Anomalous phosphorylated neurofilament aggregations in central and peripheral axons of hens treated with tri-ortho-cresyl phosphate (TOCP). *J. Neurosci. Res.* 33:455–460.
399. Abou-Donia MB, et al. Unpublished results, 2004. 50. Jamal G. Neurological syndromes of organophosphorus compounds. *Adverse Drug React. Toxicol. Rev.* 1997;16:133–170.
400. Jamal GA, Hansen S, Julu O, and Peter O. 2002. Low level exposures to organophosphorus esters may cause neurotoxicity. *Toxicology.* 181–182(27 December): 23–33.
401. Jamal G. 1997. Neurological syndromes of organophosphorus compounds. *Adverse Drug React. Toxicol. Rev.* 16:133–170.
402. Gershon S and Shaw FB. 1961. Psychiatric sequelae of chronic exposure to organophosphorous insecticides. *Lancet* 1:1371–1374.
403. Dille JR and Smith PW. 1964. Central nervous system effects of chronic exposure to organophosphate insecticides. *Aerosp. Med.* 35:475–478.
404. Metcalf DR and Holmes JH. 1969. EEG, psychological and neurological alterations in humans with organophosphorous exposure. *Ann NY Acad Sci* 160:357–365.
405. Duffy FH, Burchfield JL, Bartels PH, et al. 1979. Long-term effects of an organophosphate upon the human electroencephalogram. *Toxicol. Appl. Pharmacol.* 47: 161–176.
406. Pilkington A, Buchanan D, Jamal GA, et al. 2001. An epidemiological study of the relations between exposure to organophosphate pesticides and indices of chronic peripheral neuropathy and neuropsychological abnormalities in sheep farmers and dippers. *Occup. Environ. Med.* 58(11):702–710.
407. Russell Rand Macri J. 1979 Central cholinergic involvement hyper-reactivity. *Pharmocol. Biochem. Behav.* 10: 43–48.
408. Joubert J and Joubert PH. 1988. Chorea and psychiatric changes in organophosphate poisoning. A report of 2 further studies. *J. Afr. Med. J.* 74:32–34.
409. Rosenstock L, Keifer M, Daniell WE, et al. 1991. Chronic central nervous system effects of acute organophosphate pesticide intoxication. *Lancet* 338:223–227.

410. Steenland K, Jenkins B, Ames RG, et al. 1994. Chronic neurological sequelae to organophosphate pesticide poisoning. *Am. J. Public Health* 84:731–736.
411. Callender TJ, Morrow L, and Subramanian K. 1994. Evaluation of chronic neurological sequelae after acute pesticide exposure using SPECT brain scans. *J. Toxicol. Environ. Health* 41:275–284.
412. Stephens R, Spurgeon A, Calvert IA, et al. 1995. Neuropsychological effects of long-term exposure to organophosphates in sheep dip. *Lancet* 345:1135–1139.
413. Colosio C, Tiramani M, and Maroni M. 2003. Neurobehavioral effects of pesticides: state of the art. *Neurotoxicology* 24:577–591.
414. Scremin O, Shih TM, Huynh L, et al. 2003. Delayed neurologic and behavioral effects of subtoxic doses of cholinesterase inhibitors. *J. Pharmacol. Exp. Ther.* 304:1111–1119.
415. Echbichon DJ and Joy RM. 1995. *Pesticides and Neurological Diseases,* 2nd ed. Boston and London: CRC Press.
416. Savage EP, Keefe TF, Mounce LM, et al. 1988. Chronic neurological sequelae of acute organophosphates pesticide poisoning. *Arch. Environ. Health* 43:38–45.
417. Maroni M, Jarvisalo J, and La Ferla L. 1986. The WHO-UNDP edidemiological study on the health effects of exposure to organophosphorous pesticides. *Toxicol. Lett.* 33:115–123.
418. McConell R, Keifer M, and Rosenstock L. 1994. Elevated quantitative vibrotactile threshold among workers previously poisoned with methamidophos and other organophosphate pesticides. *Am. J. Ind. Med.* 25:325–334.
419. Yokoyama K, Araki K, Murata K, et al. 1998. A preliminary study on delayed vestibulocerebellar effects of Tokyo Subway sarin poisoning in relation to gender difference: frequency analysis of postural sway. *J. Occup. Environ. Med.* 40(1):17–21.
420. Rodnitzky RL, Levin HS, and Mick DL. 1975. Occupational exposure to organophosphate pesticides. *Arch. Environ. Health* 30:98–103.
421. Maizlish N, Schenker M, Weisskopf C, et al. 1987. A behavioral evaluation of pest control workers with short-term, low-level exposure to the organophosphate diazinon. *Am. J. Indust. Med.* 12:153–172.
422. Daniell W, Barnhart S, Demers P, et al. 1992. Neuropsychological performance among agricultural pesticide applicators. *Environ. Res.* 59(1):217–228.
423. Beach JR, Spurgeon A, Stephens R, et al. 1996. Abnormalities on neurological examination among sheep farmers exposed to organophosphate pesticides. *Occup. Environ. Med.* 53(8):520–525.
424. London L, Myers JE, Neil V, et al. 1997. An investigation into neurological and neurobehavioral effects of long-term agrochemical use among deciduous fruit farm workers in the Western Cape, South Africa. *Environ. Res.* 73(1–2):132–145.
425. Bazylewicz-Walckzak B, Majzakova W, and Szymczak M. 1999. Behavioral effects of occupational exposure to organophosphorous pesticides in female greenhouse plantin workers. *Neurotoxicology* 20(5):819–826.
426. Steenland M. 1996. Chronic neurological effects of organophosphate pesticides. *Br. Med. J.* 312:1311–1312.
427. Levin HS, Rodnitzsky RL, Mick DL, et al. 1976. Anxiety associated with exposure to organophosphorous compounds. *Arch. Gen. Psychiatry* 33:225–228.
428. Kaplan JG, Kessler J, Rosenberg N, et al. 1993. Sensory neuropathy associated with Dursban (chlorpyrifos) exposure. *Neurology* 43:2193–96.
429. Stokes L, Stark A, Marshall E, et al. 1995. Neurotoxicity among pesticide applicators exposed to organophosphates. *Occup. Environ. Health* 52:648–653.
430. Fielder N, Feldman RG, Jacobson J, et al. 1996. The assessment of neurobehavioral toxicity: SOGOMSEC joint report. *Environ. Health Perspect* 104(Suppl 2):179–191.
431. Srivastava AK, Gupta BN, Bihari V, et al. 2000. Clinical, biochemical and neurobehavioral studies on workers engaged in the manufacture of quinalphos. *Food Chem. Toxicol.* 38(1):65–69.
432. Nishiwaki Y, Maekawa K, Ogawa Y, et al. 2001. Effects of sarin on the nervous system in rescue team staff members and police officers 3 years after the Tokyo subway sarin attack. *Environ. Health Perspect.* 109:1169–1173.
433. Durham WF, Wolfe HR, and Quinby GE. 1965. Organophosphorus insecticides and mental alertness. *Arch. Environ. Health* 10:55–66.
434. Tabershaw IR and Cooper WC. 1966. Sequelae of acute organic phosphate poisoning. *J. Occup. Med.* 8:5–20.
435. Petras JM. 1981. Soman neurotoxicity. *Fundam. Appl. Toxicol.* 1:242–249.
436. Faro MD and Windle WF. 1969. Transneuronal degeneration in brains of monkeys asphyxiated at birth. *Exp. Neurol.* 24:38–53.

437. McDonough JH, Dochterman LW, Smith CD, et al. 1995. Protection against nerve agent-induced neuropathology, but not cardiac pathology, is associated with the anticonvulsant action of drug treatment. *Neurotoxicology* 15:123–132.
438. Kadar T, Cohen G, Sarah R, et al. 1992. Long-term study of brain lesions following soman, in comparison to DFP and metrazol poisoning. *Hum. Exp. Toxicol.* 11:517–523.
439. Kadar T, Shapira S, Cohen G, et al. 1995. Sarin induced neuropathology in rats. *Hum. Exp. Toxicol.* 14:252–259.
440. Veronesi B, Jones K, and Pope C. 1990. The neurotoxicity of subchronic acetylcholinesterase (AChE) inhibition in rat hippocampus. *Toxicol. Appl. Pharmacol.* 104:440–456.
441. Abdel-Rahman A, Shetty AK, and Abou-Donia MB. 2002. Acute exposure to sarin increases blood brain barrier permeability and induces neuropathological changes in the rat brain: dose-response relationships. *Neuroscience* 113(3):721–741.
442. McDonald BE, Costa LG, and Murphy SD. 1988. Spatial memory impairment and central muscarinic receptor loss following prolonged treatment with organophosphates. *Toxicol. Lett.* 40:47–56.
443. Rafaelle K, Olton D, and Annau Z. 1990. Repeated exposure to diisopropylfluorophosphate (DFP) produces increased sensitivity to cholinergic antagonists in discrimination retention and reversal. *Psychopharmacology (Berl)* 100:267–274.
444. Bushnell PJ, Padilla SS, Ward T, et al. 1991. Behavioral and neurochemical changes in rats dosed repeatedly with diisopropyl fluorophosphate. *J. Pharmacol. Exp. Ther.* 256:741–750.
445. Kassa J, Koupilova M, and Vachek J. 2001. The influence of low level sarin inhalation exposure on spatial memory in rats. *Pharmacol. Biochem. Behav.* 70:175–179.
446. Prendergast MA, Terry AV Jr, and Buccafusco JJ. 1997. Chronic, low-level exposure to diisopropyl fluorophosphates causes protracted impairment of spatial navigation learning. *Psychopharmacology (Berl)* 130:276–284.
447. Prendergast MA, Terry AV Jr, and Buccafusco JJ. 1998. Effects of chronic low-level organophosphate exposure on delayed recall, discrimination and spatial learning in monkeys and rats. *Neurotoxicol. Teratol.* 20:115–122.
448. Shih TM, Duniho SM, and McDonough JH. 2003. Control of nerve agent-induced seizures is critical for neuroprotection and survival. *Toxicol. Appl. Pharmacol.* 188:69–80.
449. Petras JM. 1994. Neurology and neuropathology of soman-induced brain injury: an overview. *J. Exp. Anal. Behav.* 61:319–329.
450. Carpentier P, Delamanche IS, Lebert M, et al. 1990. Seizure-related opening of the blood brain barrier induced by soman: possible correlation with the acute neuropathology observed in poisoned rats. *Neurotoxicology* 11:493–508.
451. Clement JG and Broxup B. 1993. Efficacy of diazepam and avizafone against soman-induced neuropathology in brain of rats. *Neurotoxicology* 14:485–504.
452. Shih TM, Koviak TA, and Capacio BR. 1991. Anticonvulsants for poisoning by the organophosphorous compound Soman: pharmacological mechanisms. *Neurosci. Biobehav. Rev.* 15:349.
453. Pazdernik TL, Emerson MR, Cross R, et al. 2001. Soman-induced seizures: Limbic activity, oxidative stress, and neuroprotective proteins. *J. Appl. Toxicol.* 21:S87–S94.
454. deGroot DMG, Bierman EPB, Bruijnzeel PLB, et al. 2001. Beneficial effects of TCP on soman intoxication in guinea pigs: seizures, brain damage, and learning behavior. *J. Appl. Toxicol.* 21:S57–S65.
455. Taysee L, Calvet JH, Buee J, et al. 2003. Comparative efficacy of diazepam and avizafone against sarin-induced neuropathology and respiratory failure in guinea pigs: Influence of atropine dose. *Toxicology* 188:197–209.
456. Abou-Donia MB, et al. Unpublished results, 2004.
457. Kim YB, Hur GH, Shin S, et al. 1999. Organophosphate-induced brain injuries: delayed apoptosis mediated by nitric oxide. *Environ. Toxicol. Pharmacol.* 7:147–152.
458. Abdel-Rahman AA, Dechkovskaia AM, Goldstein LB, et al. 2004. Neurological deficits induced by malathion, DEET and permethrin, alone or in combination in adult rats. *J. Toxicol. Environ. Health A* 67:331–356.
459. McLeod CG, Singer AW, and Harrington DG. 1984. Acute neuropathology in soman-poisoned rats. *Neurotoxicology* 5:53–58.
460. Churchill L, Pazdernik TL, and Jackson JL. 1985. Soman-induced brain lesions demonstrated by muscarinic receptor autoradiography. *Neurotoxicology* 6:81–90.
461. Baille V, Dorandeu F, Carpentier P, et al. 2001. Acute exposure to a low or mild dose of soman: biochemical, behavioral and histopathological effects. *Pharmacol. Biochem. Behav.* 69:561–569.

462. Sidell FR. 1974. Soman and sarin: Clinical manifestations and treatment of accidental poisoning by organophosphates. *Clin. Toxicol.* 7(1):1–17.
463. West I. 1968. Sequelae of poisoning from phosphate ester pesticides. *Ind. Med. Surg.* 37(11):832.
464. Namba T, Nolte CT, Jackrel J, et al. 1971. Poisoning due to organophosphate insecticides. *Am. J. Med.* 50:475.
465. Kilburn KH. 1999. Evidence for chronic neurobehavioral impairment from chlorpyrifosa, and organophosphate insecticide (Dursban) used indoors. *Environ. Epidemiol. Toxicol.* 1:153–162.
466. Blondell J and Dobozy VA. 14 Jan 1997. *Review of Chlorpyrifos Poisoning Data.* Washington, DC: U.S. Environmental Protection Agency.
467. Abou-Donia MB, Abdel-Rahman AA, Goldstein LB, et al. 2003. Sensorimotor deficits and increased brain nicotinic acetylcholine receptors following exposure to chlorpyrifos and/or nicotine in rats. *Arch. Toxicol.* 77:452–458.
468. Abdel-Rahman AA, Dechkovskaia AM, Mehta-Simmons H, et al. 2003. Increased expression of glial fibrillary acidic protein in cerebellum and hippocampus: Differential effects on neonatal brain regional acetylcholinesterase following maternal exposure to combined chlorpyrifos and nicotine. *J. Toxicol. Environ. Health A* 66:2047–2066.
469. Eng LF and Ghirnikar RS. 1994. GFAP and astrogliosis. *Brain Pathol.* 4:229–237.
470. Olney JW, de Gubareff T, and Labruyere J. 1983. Seizure-related brain damage induced by cholinergic agents. *Nature* 301:520–522.
471. Dawson VL, Dawson TM, Lonedon ED, et al. 1991. Nitric oxide mediates glutamate neurotoxicity in primary cortical cultures. *Proc. Natl. Acad. Sci. USA* 88:6368–6371.
472. Solberg Y and Belkin M. 1997. The role of excitotoxicity in organophosphorus nerve agents central poisoning. *Trends Pharmacol. Sci.* 8:183–185.
473. Raveh L, Chapman S, Cohen G, et al. 1999. The involvement of the NMDA receptor complex in the protective effect of anticholinergic drugs against soman poisoning. *Neurotoxicology* 20:551–560.
474. Dawson TM, Dawson VL, and Snyder SH. 1992. A novel neuronal messenger molecule in brain: the free radical, nitric oxide. *Ann. Neurol.* 32:297–331.
475. Montague PR, Gancayco CD, Winn MJ, et al. 1994. Role of NO production in NMDA receptor-mediated neurotransmitter release in cerebral cortex. *Science* 263(5149):973–977.
476. Bagetta G, Massoud R, Rodino P, et al. 1993. Systematic administration of lithium chloride and tacrine increases nitric oxide synthase activity in the hippocampus of rats. *Eur. J. Pharmacol.* 237:61–64.
477. Kim YB, Hui GH, Lee YS, et al. 1997. A role of nitric acid oxide in organophosphate-induced convulsions. *Environ. Toxicol. Pharmacol.* 1(3):53–56.
478. Thompson CB. 1995. Apoptosis in the pathogenesis and treatment of disease. *Science* 267:1456–1462.
479. Tsujimoto Y. 1997. Apoptosis and necrosis: intracellular ATP level as a determinant for cell death modes. *Cell Death Differ.* 4:429–434.
480. Murphy AN, Fiskum G, and Beal MF. 1999. Mitochondria in neurodegeneration: cell life and death. *J. Cereb. Blood Flow Metab.* 19:231–245.
481. Floyd RA. 1999. Antioxidants, oxidative stress, and degenerative neurological disorders. *Proc. Soc. Exp. Biol. Med.* 222:236–245.
482. Gupta RP, Milatovic D, and Dettbarn WD. 2001. Depletion of energy metabolites following acetylcholinesterase inhibitor-induced status epilepticus: protection by antioxidants. *Neurotoxicology* 22:271–282.
483. Lieberman AD, Craven MR, Lewis HA, et al. 1998. Genotoxicity from domestic use of organophosphate pesticides. *J. Occup. Environ. Med.* 40(11):954–957.
484. Abou-Donia MB, et al. 2004. Unpublished data.
485. Abou-Donia MB, et al. 2004. Unpublished results.
486. Sberna G, Saez-Valero J, Li QX, et al. 1998. Acetylcholinesterase is increased in the brains of transgenic mice expressing the C-terminal fragment (CT100) of the B-amyloid protein precursor of Alzheimer's disease. *J. Neurochem.* 71:723–731.
487. Calderon FH, von Bernhardi R, De Ferrari G, et al. 1998. Toxic effects of acetylcholinesterase on neuronal and glial-like cells in vitro. *Mol. Psychiatry* 3:247–255.
488. Damodaran TV, Jones KH, Patel AG, et al. 2003. Sarin (nerve agent GB)-induced differential expression of mRNA coding for the acetylcholinesterase gene in the rat central nervous system. *Biochem. Pharmacol.* 65:2041–2047.
489. Yang L, Heng-Yi H, and Zhang XJ. 2002. Increased expression of intranuclear AChE involved in apoptosis of SK-N-SH cells. *Neurosci. Res.* 42:261–268.

490. Andres C, Seidman S, Beeri R, et al. 1998. Transgenic acetylcholinesterase induces enlargement of murine neuromuscular junctions but leaves spinal cord synapses intact. *Neurochem. Int.* 32:449–456.
491. Abou-Donia MB, Wilmarth KR, and Jensen KF. 1996. Neurotoxicity resulting from coexposure to pyridostigmine bromide, DEET, and permethrin: implications of Gulf War chemical exposures. *J. Toxicol. Environ. Health* 48:35–56.
492. Rea WJ. 1992. *Chemical Sensitivity, Vol. 1.* Boca Raton, FL:Lewis Publishers. 47–154.
493. Kurt TL. 1998. Epidemiological association in U.S. veterans between Gulf War illness and exposure to anticholinesterases. *Toxicol. Lett.* 11:1–5.
494. Anger WK, Storzbach D, and Binder LM. 1999. Neurobehavioral deficits in Persian Gulf veterans: evidence from a population-based study. *J. Int. Neuropsychol. Soc.* 5:203–212.
495. McCauley LA, Rischitelli G, and Lambert WE. 2001. Symptoms of Gulf War vertrans possibly exposed to organophosphate chemical warfare agents at Khamisiyah, Iraq. *Int J. Occup. Environ. Health* 7:3170–3175.
496. Storzbach D, Campbell KA, and Binder LM. 2000. Psychological differences between veterans with and without Gulf War unexplained symptoms. *Psychosom. Med.* 26: 726–735.
497. White RF, Proctor SP, and Heeren T. 2001. Neuropshchological functions in Gulf War veterans: relationship to self-reported toxicant exposures. *Amer. J. Ind. Med.* 40:42–44.
498. Institute of Medicine of the National Academies. 1995. *Health Consequences of Service during the Persian Gulf War: Initial Findings and Recommendation for Immediate Action.* Washington, DC: National Academies Press.
499. Augerson WS. 2000. *A Review of the Scientific Literature as It Pertains to Gulf War Illnesses, Vol. 5, Chemical and Biological Warfare Agents.* Santa Monica, CA: Rand Corporation.
500. Hyams KC, Wignall FS, and Roswell R. 1995. War syndromes and their evaluation: From the U.S. Civil War to the Persian Gulf War. *Ann. Intern. Med.* 125:398–405.
501. Baker DG, Mendenhall CL, and Simbart LA. 1997. Relationship between posttraumatic stress disorder and self-reported physical symptoms in Persian Gulf War veterans. *Arch. Intern. Med.* 157:2076–2078.
502. Horner RD, Kamins KG, Feussner JR, et al. 2003. Occurrence of amyotrophic lateral sclerosis among Gulf War veterans. *Neurology* 61:742–749.
503. Haley RW. 2003. Excess incidence of ALS in young Gulf War veterans. *Neurology* 61:750–756.
504. Freudenthal RL, Rausch L, Gerhart JK, et al. 1993. Subchronic neurotoxicity of oil formulations containing either tricresyl phosphate or tri-orthocresyl phosphate. *J. Am. Coll. Toxicol.* 12:409–416.
505. Daughtrey W, Biles WR, and Jortner B. 1996. Subchronic delayed neurotoxicity evaluation of jet engine lubricants containing phosphorus additives. *Fundam. Appl. Toxicol.* 32:244–249.
506. Haley, BE. 2001. The relationship of the toxix effects of mercury to exacerabation of the medical contition classified as Alzheimer's disesase. Unpublished Manuscript, Department of Chemistry, University of Kentucky, presented to meetings of International Academy of Oral Medicine and Toxicology, Oak Brook, Illinois, Sept. 7–8.
507. Guyton AC. 1987. *Human Physiology and Mechanisms of Disease.* Philadelphia: WB Saunders Co. 443.
508. Dosch P. 1984. *Manual of Neural Therapy.* 1st English ed (translation of 11th German ed., revised) Lindsay, A. translation Heidelberg, Germany: Karl F. Haug Publishers.
509. Guyton AC and Hall JE (eds.). 2000. *Medical Physiology,* 10th ed. Philadelphia: WB Saunders Co. 698.
510. Gray H. 1918. *Anatomy of the Human Body.* http://www.bartleby.com/107/indexillus.html, Philadelphia: Lea & Febiger. 1825–1861.
511. Guyton AC and Hall JE (eds.). 2000. *Medical Physiology,* 10th ed. Philadelphia: WB Saunders Co. 90. Fig. 8–3.
512. Guyton AC. 1981. *Textbook of Medical Physiology,* 6th ed. Philadelphia: WB Saunders Co. 761.
513. Rea WJ. 1996. *Chemical Sensitivity. Vol. III. Clinical Manifestations of Pollutant Overload.* Boca Raton, Fl: CRS Press. 1889.
514. Rea WJ. 1996. *Chemical Sensitivity. Vol. III. Clinical Manifestations of Pollutant Overload.* Boca Raton, Fl: CRS Press. 1890, Fig. 27.3.
515. Rea WJ. 1996. *Chemical Sensitivity. Vol. III. Clinical Manifestations of Pollutant Overload.* Boca Raton, Fl: CRS Press. 1891, Fig. 27.5.
516. Rea WJ. 1995. *Chemical Sensitivity. Vol. III. Clinical Manifestations of Pollutant Overload.* Boca Raton, Fl: CRS Press. 1741, Fig. 26.5.
517. Guyton AC and Hall JE. (eds.). 2000. *Medical Physiology,* 10th ed. Philadelphia: WB Saunders Co. 93.
518. Rea WJ. 1995. *Chemical Sensitivity, Volume III: Clinical Manifestation of Pollutant Overload.* Boca Raton, Fl: CRC Press. 1750. Fig. 26.7.

519. Rea WJ. 1996. *Chemical Sensitivity. Vol. III. Clinical Manifestations of Pollutant Overload.* Boca Raton, Fl: CRS Press. 1804, 1814.
520. Friedlander RM. 2003. Apoptosis and Caspases in Neurodegenerative Diseases. *N. Engl. J. Med.*, 348;14:1366, Fig. 1.
521. Rea WJ. 1996. *Chemical Sensitivity, Volume III, Clinical Manifestations of Pollutant Overload.* Boca Raton, Fl: CRS Press. 1757. Fig. 26.9.
522. Kilburn K. 1998. *Chemical Brain Injury.* New York: Van Nostrand Reinhold. 184. Fig. 10.2.
523. Waxman S. 1998. Demyelinating Diseases—New Pathological Insights, New Therapeutic Targets. *NEJM.* 338:5 (323).
524. Rea WJ, Pan Y and Fenyves I. *Abstract-Neurotoxicity: The Central Nervous System.*
525. Anger 1986. Patty's Industrial Hygiene and Toxicology, Vol.2, New York: John Wiley & Sons. 1981-1982.

3 Immune System

INTRODUCTION

When studying chemical sensitivity and chronic degenerative disease the dynamics of homeostasis in regard to immunity and autoimmunity are in a very delicate balance. This relationship is evidenced by the number of autoimmune and immune-mediated diseases that are treated by physicians without knowing the etiology. Once local homeostatic control of the connective tissue matrix is lost, the immune system comes into full action acting as an amplification system for responses to noxious stimuli. **This means care must be taken to insure that the immune system is not only robust but is also in balance with itself and the other amplification systems. These systems include the neurovascular, especially the autonomic nervous system, and the endocrine systems. If this balance is not maintained disease will be initiated or propagated**. These endocrine and autonomic nervous systems are the second and third leg of the equilateral triangle of the amplification systems that attempt to maintain the dynamics of homeostasis once local control is lost. The balance of these systems conserves energy, thus reducing nutrient depletion and preventing fatigue and weakness that occur when heroic attempts by the body are made to obtain and maintain homeostasis. The massive contamination of our environment increases the difficulty of maintaining dynamic homeostasis in the immune system because, generally the total body pollutant load, which constantly disrupts the equilibrium mechanism, increases rapidly and chronically. In addition to the nonspecific total body noxious incitant load, a specific body burden of toxic substances like pathogenic *Escheria coli* or *Streptoccocus hemolyticus* or different specific viruses like polio influenza often increases, thereby augmenting the need for more immune output. This increased environmental demand strains the IgA, IgM, and IgG function of immunoglobulins, as well as the innate immune system. Other examples of specific toxic substances that increase body burden include benzene, which may trigger leukemia;[1] formaldehyde, which may trigger chemical sensitivity;[2] or pesticide exposure, which may trigger inflammatory vascular disease and neurological dysfunction.[3] There are now over 60,000 chemicals in our environment and **when significant numbers are taken into the body these have to be eliminated, utilized, or sequestered**. Sequestering of pollutants for survival can chronically disrupt physiology and, thus homeostasis, resulting in chronic metabolic, vascular, musculoskeletal, neurological, endocrine, and immune dysfunction even though this process can be life saving.

   The gastrointestinal tract is the soul of the immune system, since the majority of the immune function (75%) takes place there. A myriad of anatomical connections and functional responses of the immune system exist within the gastrointestinal tract. In not only the gastrointestinal tract, but also other organs and systems, the immune system is intricately connected. These systems include the autonomic nervous system that is important for proper response and regulation and the vascular system in which the response can be devastating with metabolic changes from pollutant triggering with resultant vasospasm and tissue hypoxia. (See Chapters 1,2 and Cardiovascular and Gastrointestinal chapters in Volume 2 - Mechanisms of Chemical Sensitivity and Chronic Degenerative Disease.) This refined response occurs because anatomically the lymph channels and nodes are open for intricate responses between nerve impulses and immune function, and vice versa.

PROPERTIES OF ENTERING NOXIOUS EXCITANTS

The effect on the immune system from entering noxious stimuli into the body must also be taken into consideration when studying the patient with chemical sensitivity and chronic degenerative disease. These responses usually go through the skin, mucosa and connective tissue matrix, which if overwhelmed, generates an immune response. Responses of the immune system may be linear from triggering agents of various doses of the entering noxious substances. The responses can also be biphasic as well. Therefore, we have a potential of four types of responses from an entering noxious stimulus. The responses of the immune system would potentially include stimulation or inhibition by chemicals and/or stimulation or inhibition of the homeostatic response.

In addition to the usual linear dose responses there are many immune related hormetic-like dose response (low-dose stimulation and high-dose inhibition or visa versa) relationships that exist with a myriad of exogenous toxic chemicals, medications, drugs, and even endogenous substances produced in the body. According to Calabrese[4] there are over 100 drugs, 70 endogenous agonists, and 90 environmental contaminates that have hormetic effects. See Tables 3.1, 3.2, and 3.3. These doses are J- or U-shape curves for responses in lieu of linear dose responses. It has now been shown that 50% of the reactions to chemicals are linear and 50% are not. The various dose response curves make it much more difficult to defend against contaminants. This biphasic reaction can thus complicate the diagnosis and treatment of any entering substance. Tables 3.1, 3.2, and 3.3 show many substances that have been shown to have hormetic properties.

These findings demonstrate that within the immune system biphasic dose response relationships are common and highly generalized according to the model, end point, and chemical class. The clinician should be aware of this varied phenomenon of entering noxious substances taking into consideration not only the linear but also the J- or U-shaped effects of the chemical and also the linear and J- or U-shape responses of the immune system. These responses occur with the properties of each specific entering noxious stimuli.

Lymphocyte activation has been widely studied due to its control role in overall immune responsiveness and its association to various diseases such as delayed hypersensitivity, immune deficiency, infectious disease, and cancer. An evaluation of lymphocyte activation by Calabrese has revealed that biphasic dose responses are prominent.[4]

Other immune responses such as antibody-forming response to sheep red blood cells, plaque forming cells, neutrophil production and migration also can be biphasic. The mechanism of this biphasic response is unknown but two separate receptor populations forming an allosteric effect by a single receptor population is known.[5] Another idea of biphasic response mechanism is that 1L-8 stimulation comes from a heterogeneous cell population.

Stem Cells

Stem progenitor cell therapies are emerging in various medical disciplines.[6] Stem cells from the bone marrow originate to supply deficient core cells. Embryonic stem cells' ability to develop into any type of cell–cell pluripotency–is both a benefit and a bane to scientists, **who must keep harvested cells from maturing before they are needed and then mold their identities to suit patients' needs when they need to be activated**. "One of the greatest challenges in this work is to harness and direct cell differentiation." How to tell one stem cell to form blood, another skin, and another liver tissue is the issue. Complex combinations of growth factors and chemical and genetic signals drive the process, which are only in recent years beginning to be understood. Until they do, embryonic stem cell therapies won't make the leap from lab mice to humans. The adult body has a small number of stem cells in many tissues and organs—where they lie low until activated by illness or injury. Unlike embryonic stem cells, adult stem cells haven't proved able to morph into every kind of cell and may be limited to becoming cell types within their tissue of origin. An adult stem cell in the brain, for example, can become a neuron or glial cell, both neural cells, but not a bone or liver cell.

Similarly stem cells from a newborn's cord blood (considered adult cells because they aren't from embryos) produce only blood cells. Recently, though, cord tissue has been found to contain mesenchymal cells capable of generating bone and cartilage.

In general, adult stem cells are scarcer in the body and harder to culture than embryonic cells, yet large numbers are needed for therapies. So far only adult stem cells have been tested in humans.

TABLE 3.1
Pharmacological Agents that Induce Immune-System-Related Hormetic-like Biphasic Dose Responses

11-Desoxycorticqal (compound S)
Acetaminophen
Allicin (main organic allyl sulfur component in garlic)
Amoxicillin
Aqueous mistletoe extract
Atropine
Auranofin
Azathioprine
Bleomycin
Bombesin analog
Bryostatin 1
Bryostatin 2
Bryostatin 5
Budesonide
C5TH-antagonist
Captopril
Captopril disulfide
Carotenoid photodegradation products
Cerebrolysine (amino acid mixture)
Chlorpromazine
Chloroquine
Chloramphenicol
Cianidenol derivatives
Cimetidine
Cinnarizine
cis-(Z)-Clopenthixol
trans-(E)-Clopenthixol
Clofazimine
Clonidine
Cocaine
Colcichine
Coumarin
Cyclophosphyamide (CP) (and some derivatives e.g., 4-CP)
Cyclosporine
Systeamine
DES
Deoxyuridine
Dexamethasone
Diazepam
Dibucaine
Dibutyrly cyclic AMP
Docasahexaneoic acid
Doxycyline

Dynorphin
Enisoprost
Enisoprost free acid
Ethylenimine
Fenfluramine
Fetuin
Fluoxetine
FMet-Leu-Phe (fMLP)
Fofizopan
Gelatin
Gramicidin
Halofantrine hydrochloride
Heparin
Hydrochloraquine
Hydrocortisone
Hydroquione
Hydroxychloroquine
Imidazole
Indomethacin
Isopentenyladenosine
Isophosphamide
Isoprinosine
Leu-enkephalin
Levamine (amino acid mixture)
Levamisole
Lithium
LPS
MDMA (ecstasy)
Met-enkephalin
Methiamide
Methicillin
Methimazole
Methotrexate
Metiamide
Methoxamine
Microcolin A
Misteltoe lectin
Monochlorethyl cyclophosphaminde
Morphiceptin
Morphine
N-Acetylecysteine
Nafcillin
NAN-190
Nitrogranulogen (NGG)

Opioids
DAGO
DALA
DAMEA
Deltorphin
DPDPE
DSLET
DTLET
U-50488
Osbeckia extraqct
Penicillamine
Pentoxifylline
Phosphoramide mustard
Plumbagin
PMA
PolyA:U
Porcine uterine fluid (PUF)
Procainamide
Promethazine
Propranolol
Propylthiouracil
PTU
Pyrithioxine
Resveratrol
Retinoic acid
Sarafotoxin
Sheep RBC's
Sulfated disaccharide
Sulfasalazine
Synthetic prostaglandins
Tamoxifen
Teremifene
Tetramisole
Tetracycline
Thapsigargin
Theophylline
Thimerosal
Thiosalicylic acid
Toremifene
Timethoxybenzoate
Trifluperazine
Tucaresol
UMDH
Verapamil
(W-7) N-(6-Aminohexyl)-5-chloro-1-napththaqlenesulfonamide
Vinblastine
Xylazine

Source: Reprinted from Calabrese, E. J., *Crit Rev in Toxic.,* 35, 89–295, 2005. With permission.

TABLE 3.2
Toxic Substances Inducing Immune-System-Related Hormetic-like Biphasic Immune Responses

1. Alcohol
2. Arsenate
3. Arsenite
4. Asbestos
5. Azide
6. Cadmium chloride
7. Carbon Monoxide
8. Chlorothalonil
9. Chromium (potassium dichromate)
10. Cobalt
11. Copper sulfate
12. Copper(2) acetate (4)
13. *d*-Limonene
14. Diacetoryscirpenol
15. Diesel exhaust particle
16. 1,1-Dimethylhydrazine
17. Dioxane
18. Ethyl carbamate
19. EMT (electromagnetic frequencies) (radio waves)
20. Formaldehyde
21. Hexachlorocyclohexane (α-isomer)
22. Hydoquinone (metabolite of benzene)
23. Lead
24. Mercuric acetate
25. Mercuric chloride
26. Mecuric nitrate
27. Methoxacetic acid (metabolite of 2-methoxyethanol)
28. Methyl mercury
29. Methyl nitrosourea
30. *N*-Methyl-*N'*-nitro-*N*-nitrosoguanidine
31. *Nickel*
32. Paraquat
33. PBB
34. PCB
35. Pentachlorophenol
36. Pesticide mixture (atrazine, metribuzine, endosulfan, lindane, aldicarb, and dieldrin)
37. Phenylarsine oxide
38. Selenium
39. Silver nitrate
40. Sodium azide
41. T-2 toxin
42. TCDD
43. Tributyltin
44. Trichothecenes
    a. Isosatratoxin
    b. Roridin A
    c. Satratoxin G
    d. Satratoxin H
    e. Verrucarin
45. Triphenyl phosphate
46. Vanadium (sodium metaranadate)
47. X-ray
48. Zinc chloride

Source: Reprinted from Calabrese, E. J., *Crit Rev in Toxic.,* 35, 89–295, 2005. With permission.

Stem cells are the primary cells from which immune cells and lymph channels originate. There are many types of stem cells that come from many origins.

Heart disease: Adult bone marrow stem cells injected into heart arteries are believed to improve cardiac function in victims of heart attack or heart failure.[6]

Respiratory system: Here stem cell research has moved more slowly than other areas mainly because of the structural lung complexity, cellular hetrogenicity, and the low turnover rate of the epithelial.[7]

Leukemia and other cancers: In various studies leukemia patients treated with stem cells from bone marrow and umbilical cord blood emerged free of disease; donor blood stem cells have also reduced non-Hodgkin's lymphoma and pancreatic and ovarian cancer in some patients.

Rheumatoid arthritis: Adult stem cells may be helpful in jumpstarting repair of eroded cartilage. In human trials, joint pain lessened temporarily after donor stem cell therapy in some patients, and some then responded better to standard drug therapies.

Parkinson's disease: Since fetal tissue implants had mixed success in reducing neurological symptoms, some researchers say the best hope is that a patient's own neural stem cells may eventually be coaxed to mature into the dopamine-producing cells needed to treat the disease.

TABLE 3.3
Endogenous Agents Inducing Immune-System-Related Hormetic-like Biphasic Dose-Response Relationships

4-oxo-all-*trans*-Retinoic acid	Hydrogen peroxide
13-*cis*-Retinoic acid	IL-1
α-1-Proteinase inhibitor	IL-1β
(formerly called serum α-1- antitryspin)	IL-1α
α-Endorphin	IL-10
ACTH	Isoproterenol
Atrial Natriuretic factor	Linoleic acid
All-*trans*- retinoic- acid	LPS
α-Carotene (photodegrated)	Major secretory protein
α-Endorphin	(Msp)
Bombesin	Melatonin
Bovine interferon alpha-1	Metergoline
C16 Fatty acid	Natural products
cAMP	Asbeskia aspera extracts
Carbon monoxide	Bryostatins 1,2,5-
Cholecystokinin	Garlic aqueous extract
Complement component C5a	Gelatin
Cortisol	Microcolin A
Corticosterone	Neuromedin C
Cortisone	Neuromedin N
cGMP	Neurotensin
Dopamine	Nitric oxide
Endothelin-1	Norepinephrine
Endothelin-3	Oleic acid
Epinephrine	Palmitic acid
Estradiol	PGA analogs (E/EFA)
Estrogen metabolite (2 OH E)	Phospholipase A_2 (PLA_2)
Estrone	Platelet activating factor
Ganglioside mixture	Platelet-derived growth factor
Gastric releasing peptide (GRP)	Progesterone
Granulocyte-macrophage	Prostaglandin E_2 (PGE_2)
colony-stimulating factor (GM-CSF)	Prostaglandin $F_2α$
Hemin	Serotonin (5-HT)
Histamine	Somatostatin (SS)
Human growth hormone	Spiperone
Human neutrophil peptides	Stearic acid
1-3 (HNP 1-3)	Substance P
Hydrocortisol	Testosterone
Hydorcortisone 17-butyrate	TNF-α
	Transforming growth factor-beta
	(TGF-β)
	Vasoactive peptide

Source: Reprinted from Calabrese, E. J., *Crit Rev in Toxic.,* 35, 89–295, 2005. With permission.

Type I diabetes: Basic research is focused on understanding how embryonic stem cells might be trained to become the type of pancreatic islet cells that secrete needed insulin.[6] Recent developments using proteins to spur cell differentiation may speed progress. See Figures 3.1 and 3.2.

Stem cells appear to be in the future for the repair and study of pollutant damaged lungs.

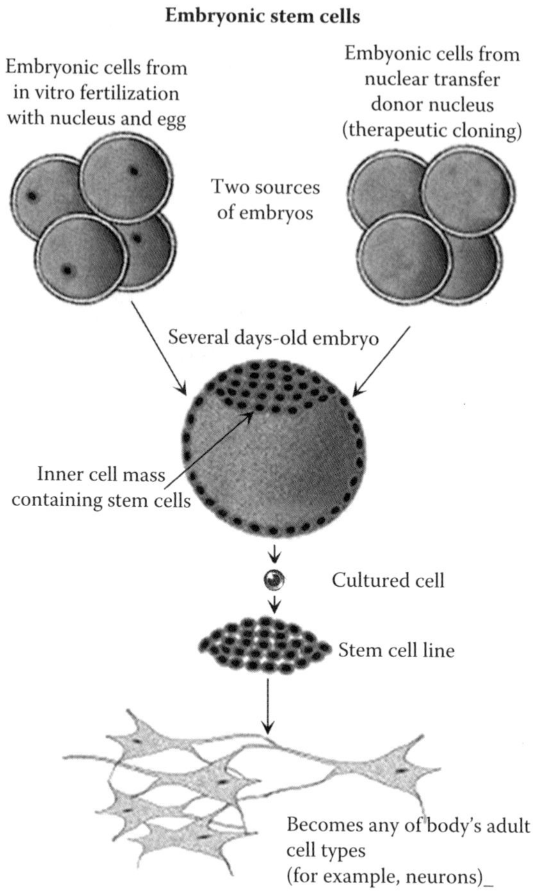

Embryonic stem cells

Embryonic cells from in vitro fertilization with nucleus and egg

Embyonic cells from nuclear transfer donor nucleus (therapeutic cloning)

Two sources of embryos

Several days-old embryo

Inner cell mass containing stem cells

Cultured cell

Stem cell line

Becomes any of body's adult cell types (for example, neurons)

FIGURE 3.1 Embryonic stem cells. (Reproduced from *National Geographic*, 12, July 2005. With permission.)

LYMPH NODES, LYMPHATIC ORGANS, AND THE LYMPHATIC SYSTEM

Development of Lymphatic Channels and Lymphocytes

The origin of the lymphatic channels has been poorly understood, and, thus, **little emphasis has been placed on their anatomy or physiology as not only the conduits of protein but also as a significant part of immune regulation**. The following pages on lymph anatomy and lymphatic innervation will help better define the origin of lymphatics and immune function in the patients with chemical sensitivity and/or chronic degenerative disease. One viewpoint contends the lymphatics originates from the venous endothelium; and another attributes their origin to local mesoderm. The origin may reflect function in some cases. Regardless of where it originates, the lymphatic system first appears as six primary lymph sacs starting late in the sixth week of pregnancy.[8] These sacs form the rudimentary immune system, which is essentially molded into a sophisticated system to combat exposure to foreign stimuli. See Figure 3.3.

Two jugular lymph sacs appear at the angle between the precardinal (future internal jugular) veins and the subclavian veins. In the abdomen, a retroperitoneal lymph sac forms on the posterior body wall at the root of the mesentery during the eighth week of pregnancy. Somewhat later, a cysterna chyli forms at the same level, but dorsal to the aorta. **This organ becomes the main collecting vessel for the lymphatics of the gut, where 75% of the immune function exists**. At about the same time, a pair of posterior lymph sacs arises at the bifurcation of the femoral and sciatic veins. By the end of the ninth week, lymphatic vessels connect these lymph sacs.

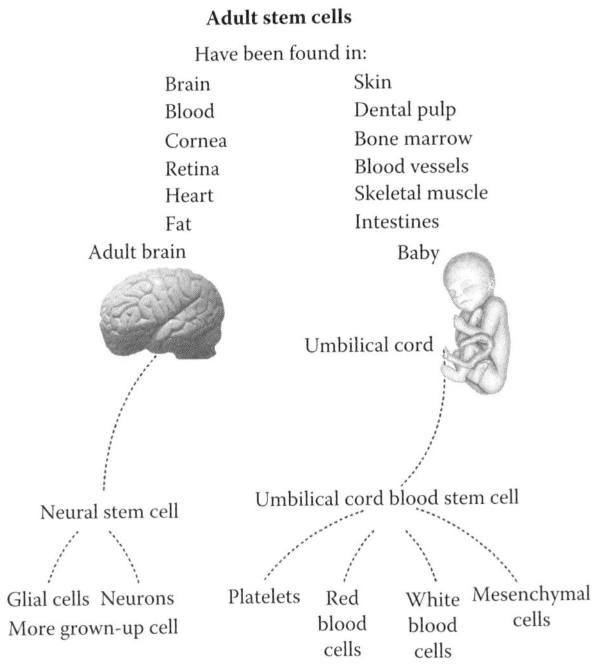

Immune System

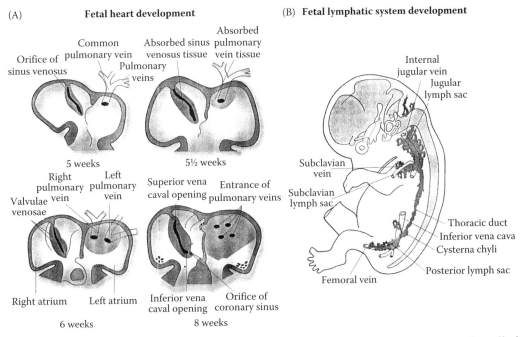

FIGURE 3.2 Adult stem cells are the body's ultimate reserve for repair. At this time, unfortunately, there appears to be a limited number for repair. (*National Geographic*, 13, July 2005. With permission.)

FIGURE 3.3 (A) Stages in the absorption of the common pulmonary vein and its branches into the wall of the left atrium and changes over time in the opening of the sinus venous into the right atrium. (B) Two jugular lymph sacs appear at the angle of the precardinal (future internal jugular) veins and the subclavian veins. (Carlson, B., *Human Embryology and Developmental Biology: Development of Body Systems*, Mosby, 414, 2003. With permission.)

Two major lymphatic vessels connect the cysterna chyli with the jugular lymph sacs. An anastomosis forms between these two channels. A single lymphatic vessel consisting of the caudal part of the right channel, the anastomotic segment, and the cranial part of the left channel ultimately becomes the definitive thoracic duct (TD) of the adult.

The TD is the main collecting vessel of the lymphatic system. Little is known about the intrathoracic tributaries of the TC, which are named intercostal, mediastinal, and bronchomediastinal trunks. Riquet et al.[9] performed studies on 530 adult cadavers. The lympahtics of different organs were catheterized and injected with a dye: lungs ($n = 360$), heart ($n = 90$), esophagus ($n = 50$), and diaphragm ($n = 30$). The lymphatic tributaries draining the lymph from these organs to the TD were dissected along their course to the TD and classified.

The TD tributaries were observed in 147 cases: right lung ($n = 46$), left lung ($n = 69$), heart ($n = 8$), esophagus ($n = 13$), and diaphragm ($n = 11$). Connections with the TD were observed at its origin ($n = 13$), within the mediastinum ($n = 87$), and at the level of the TD arch ($n = 47$). Tributaries from the lung issued from lower paratracheal nodes 4 R ($n = 14$) and 4 L ($n = 31$), subaortic 5 ($n = 4$), subcarinal 7 ($n = 18$), pulmonary ligament 9 ($n = 7$), upper tracheal 2 L ($n = 28$), paraortic 6 ($n = 11$), and celiac nodes ($n = 2$). Tributaries from the heart connected with the TD in the mediastinum in 1 case (4 L) and with the TD arch in seven cases. Tributaries from the esophagus connected with the TD within the mediastinum in 13 cases; anodal routes were frequent ($n = 5$). The TD tributaries from the diaphragm were observed in 11 cases; always connecting with the TD at is origin.

According to Riquest et al.[9] tributaries appear to be located at unchanging levels. Lymph of intrathoracic organs may thus drain into the general circulation through the TD. The tributaries may represent a potential route for tumor cells dissemination. When incompetent, due to valve insufficiency, they permit chylous lymph to backflow into the intrathoracic lymph nodes. Injury at this level may lead to intrathoracic chylous effusions or just congestion of the thoracic lymphatic system.

This variation in anatomy may have significance in the pain patterns of the patient with hypersensitivity and chronic degenerative disease. They may vary with tributary anomalies. Some anomalies can account for the altered physiology as seen in some of these patients with chemical sensitivity and/or chronic degenerative disease as a chain of dysmetabolic events that can occur (Figures 3.4 through 3.9).[10-14] Understanding that there is a myriad of variations in the thoracic lymphatic system allows the clinician the opportunity of explaining the varied symptomology in patients with chemical sensitivity in the chest and those individuals' physical changes which are measured via thermography. As technology becomes more sophisticated we will find more correleation between lymphatic, anatomical, and physiological changes and the clinical courses of patients with chemical sensntivitiy and/or chronic degenerative disease.

The TD drains lymph from most of the body and the left side of the head into the venous system at the junction of the left internal jugular and subclavian veins. The right lymphatic duct, which drains the right side of the head and upper part of the thorax and right arm, also empties into the venous system at the original location of the right jugular sac. At times this regionalization may be important in therapy for various tumors and infections.

REGIONALIZATION OF THE IMMUNE RESPONSE BY LYMPHATIC TISSUE NODES AND CHANNELS AFTER MUCOSAL ENTRY

Mucosa Associated Lymphoid Tissue (MALT); Nasal Associated Lymphoid Tissue (NALT); Bronchial Associated Lymphoid Tissue (BALT); Gut Associated Lymphoid Tissue (GALT)

The lymphatics are the body's access to the immune system and as pointed out in the previous section their anatomy and physiology are very important in understanding immune function and regionalization. The immune cells enter the body from the mucosal surfaces after antigen stimulation from the lumen, and then they go to local, regional, and distal lymph nodes and other lymph organs, like the

Immune System

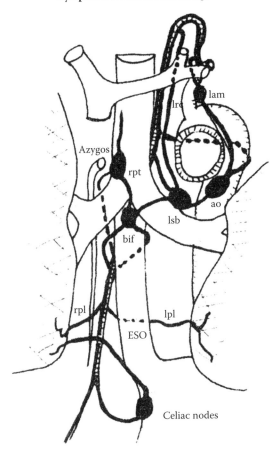

Lymphatic vessels from the lungs

FIGURE 3.4 Lymphatic vessels from the lungs. Connections with the thoracic duct: right paratracheal nodes (rpt) (4 R); left superior bronchial (lsb) 4 L; left recurrent chain (lrc); aortic arch nodes (ao) (aortic subclavian and carotid nodes chains) (5); left anterior mediastinal (lam) (6); right inferior pumonary (rpl)ligament nodes (9); left inferior pulmonary (lph) ligament nodes (9); nodes of tracheal bifurcation (bif) (intertracheobroncial nodes) (7). Numbers in parentheses refer to the 1997 regional lymph nodes classification [7]. (ESO = esophagus.) (Reprinted from Riquet, M., LePimpec Barthes, F., Souilamas, R., and Hidden, G., *Ann Thorac Surg.*, 73, 893, Fig. 1, 2002. With permission.)

thymus gland, spleen, and liver. A great deal of communication exists between the local connective tissue matrix especially the submucosal plexis and in the lamina propria as well as the distal immune, the neurological, and the endocrine systems, thus keeping information constantly flowing, in order to maintain the complete dynamics of homeostasis. **Functional changes often occur in patients with chemical sensitivity and chronic degenerative disease that result in immune dysfunction.**

Once the lymph channels and nodes are completely formed one can access and influence the immune system through either of the nervous or endocrine systems, as well as through the mucosa, submucosa, and/or lamina propria plexi and the connective matrix. The function of these channels and amplification systems can be altered by excess noxious stimuli. They can become hypersensitive or get clogged, thus causing malfunction of the immune system. Fluid enters the lymphatic system through the connective tissue matrix after the fluid is shifted from the end capillary vessels or mucosa. **As shown in Chapter 1 and Cardiovascular chapter in Volume 2 - Mechanisms of Chemical Sensitivity and Chronic Degenerative Disease, the lymphatic system is the principle source of protein return once this substance leaves the blood vessels.** In addition, the proteins

Right lymphatic trunk from the heart

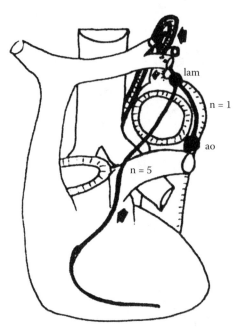

FIGURE 3.5 Right lymphatic trunk from the heart ascending between the aorta and the pulmonary trunk (arrow) before joining the upper part of the left anterior mediastinal node chain (lam) and then emptying into the arch of the thoracic duct (arrow). (ao = aortic arch nodes.) (Riquet, M., LePimpec Barthes, F., Souilamas, R., and Hidden, G., *Ann. Thorac. Surg.*, 73, 896, Fig. 5, 2002. With permission.)

Left lymphtic trunk from the heart

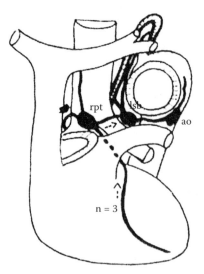

FIGURE 3.6 Left lymphatic trunk from the heart draining into the right paratracheal nodes (rpt). From there the lymph may follow the azygos vein (arrow) or reach the left suprabronchial nodes (lsb) and also the aortic arch nodes (ao). From these lymph node groups, tributaries may further connect with the thoracic duct. (Riquet, M., LePimpec Barthes, F., Souilamas, R., and Hidden, G., *Ann. Thorac. Surg.*, 73, 896, Fig. 6, 2002. With permission.)

Lymphatic vessels from the esophagus

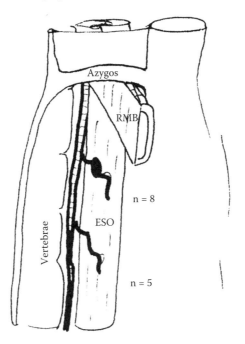

FIGURE 3.7 Lymphatic vessels from the esophagus (EOS) may connect with the thoracic duct directly (anodal route) or after crossing the epiesophageal lymph nodes. (RMB = right main bronchus.) (Riquet, M., LePimpec Barthes, F., Souilamas, R., and Hidden, G., *Ann. Thorac. Surg.*, 73, 897, Fig. 7, 2002. With permission.)

Tributaries from the esophagus

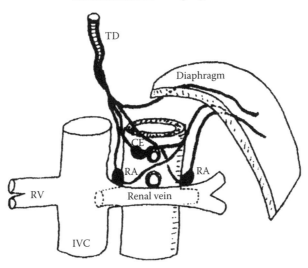

FIGURE 3.8 Tributaries from the diaphragm connecting with the origin of the thoracic duct; anodal route ($n = 4$). (CE = celiac nodes; IVC = inferior vena cava; RA = renal artery nodes; RV = renal vein; TD = thoracic duct). (Riquet, M., LePimpec Barthes, F., Souilamas, R., and Hidden, G., *Ann. Thorac. Surg.*, 73, 898, Fig. 8, 2002. With permission.)

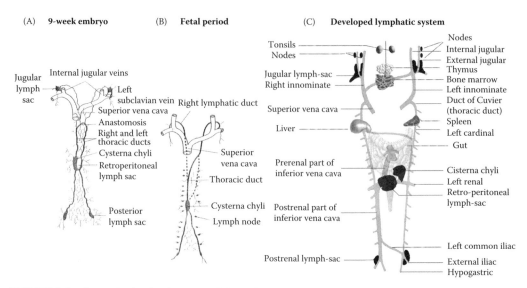

FIGURE 3.9 Stages in the development of the major lymphatic channels. (A) shows 9-week-old embryos. (B) shows the fetal period. Between A and C, the transformation between the reasonably symmetrical disposition on the main lymphatic channels and the asymmetrical condition that is characteristic of the adult can be seen. (C) shows the fully developed lymphatic system. (Modified from Carlson, B., *Human Embryology and Developmental Biology: Development of Body Systems,* Mosby, 414, 2003.)

are one of the principle sources of chemical transport to the tissues. All these systems (immune, endocrine, nervous) exchange information by creating an informational feedback loop for the maintenance of local, regional, and central homeostatic dynamics and the maintenance of acquired immunity. For example, the nasal associated lymphoid tissue (NALT), Figure 3.9 characterized by the tonsils, adenoids, and other lymphoid tissue in the nose and pharynx, initiate acquired lymphatic responses from not only inhaled and ingested substances but also from the gut-associated lymphoid tissue (GALT) and bronchial-associated lymphoid tissue (BALT). Also, these tissues process antigens in the regional lymph nodes of the neck, which then go into the bloodstream via the TD. BALT and GALT absorb antigens from the mucosa; then these antigens go to bronchial and mesenteric lymph nodes, where they are processed and then move to the cysterna chili, which finally goes to the TD into the bloodstream. (For more information, see Gastrointestinal chapter in Volume 2 - Mechanisms of Chemical Sensitivity and Chronic Degenerative Disease.)

Mucosal surfaces and the skin throughout the body are in constant contact with environmental antigens and are in a position to sample antigens for the immune system, which allows a constant production of immunity factors to counteract noxious stimuli. One of the largest complexes of mucosal immune cells (75% of immune system) is in the GALT. **Specialized antigen sampling M (MALT) cells overlaying Peyer's patches process antigen and present it to circulatory naive T- and B-cells, which traffic through the Peyer's patches**. According to the common mucosal immune hypothesis, these T and B-cells enter the TD and systemic circulation and home to various mucosal effector sites in the gastrointestinal tract and the upper and lower respiratory tract. In these sites, the sensitized T- and B-cells protect against potential pathogens by producing a specific IgA against their antigens (i.e., *Pseudomonas, Staphylococci, E. Coli,* foods, chemicals).[15] **Excess noxious stimuli occurring chronically often will disturb this intricate mechanism resulting in the first altered steps of chemical sensitivity and chronic degenerative disease. These changes in the antigen process of the gut often are the origin of disease**.

Several adhesion molecules, including mucosal addressin cellular adhesion molecule-1 (MADACAM-1), L-selectin, 4β-7 integrin, lymphocyte function–associated antigen-1 (LAF-1), and intercellular adhesion molecule-1 (ICAM-1) regulate trafficking of lymphocytes within the mucosal immune system.[16] MADCAM-1 is considered the key molecule for this direction and is expressed on the surface of endothelial cells within the high endothelial venules of Peyer's patches, mesenteric lymph nodes, and venules of the lamina propria.[17] MADCAM-1 has also been observed on splenic sinusoidal cells and within the NALT.[18,19]

Experimentally, lack of enteral stimulation decreases GALT cell mass, reduces intestinal and respiratory tract IgA levels, and impairs established mucosal immunity to a specific antigen like *Pseudomonas* aerugenosa.[20–22] Having fasted over 2,500 hypersensitive patients in the environmental control unit, researchers at the EHC-Dallas noted that other factors must come into play when there is a lack of internal stimulation. **We have observed that the fasting patient with no trauma does not get the pneumonia-like symptoms that the fasting trauma patient recovering by utilizing total parentral nutrition does**. Somehow the hypersensitive food and chemically sensitive patient who is fasting down regulates the immune response just enough to allow better tolerance of the offending substance, but not enough to make the patient susceptible to bacterial infection. However, the effectiveness of the respiratory defenses established against *Pseudomonas* are probably weakened, but not reduced to the same degree that the trauma patient with parenteral nutrition experiences. Kudsk has shown that enteral feeding decreases the incidence of pneumonia compared with a three day starvation protocol with or without total parenteral nutrition[23–25] in rats. However, in all of our years with experience in fasting patients (3–7 days) we have not seen a single respiratory infection.

Clearly, MADCAM-1 expressed within such sites as the nasal passages, mammary glands, and genitourinary system allows for sharing of immunologic protection generated within the gut, according to the common mucosal hypothesis.

Integration of the dynamics of local homeostasis including nonspecific macrophages (which line the lymph node), and leukocyte responses with the lymphatic responses and the other parts of the immune system (Peyer's patches, GALT, BALT, MALT, NALT, etc.) is important in maintaining the dynamics of the homeostatic amplification responses, which encompass neurological, neurotransmitter, and neuropeptide responses. When a constant surge of noxious stimuli enters the body, the overload causes a local disruption in the body. These other systems respond by giving aid immediately.

The immune system plays a major role in the amplification response of homeostasis over the local connective tissue response in most cases, though some immune responses are primary. All lymphocytes go through the thymus gland where the T-cells are activated, and originate from stem cells in the bone marrow where the B-cells are present and where some are processed. **Another very important fact is that 75% of the body's immune response action occurs in the gastrointestinal tract (Peyer's patches)**. (See Gastrointestinal chapter in Volume 2 - Mechanisms of Chemical Sensitivity and Chronic Degenerative Disease for details of the gut lymphocyte system.) The other 25% of the body's immune response action occurs in the other major immune organs, including the lymph nodes, the liver, the spleen, and the lymph channels. The dynamics between these organs, which constantly assure good immunity, are complex. (See Chapter 1.) It is paramount that the clinician understands the dynamics and function of this system in order to help the patient maintain the dynamic equilibrium of homeostasis with sufficient immunity and thus wellness. However, while attaining immunity, autoimmunity must be averted. (See Chapter 1 for the details of the microdynamics of the lymph system.) **In addition to the macrophages, two types of lymphocytes, the T- and B-cells, make up the lymph nodes**. (See Chapter 1.) Both types of lymphocytes are derived originally in the embryo from pluripotent hemopoietic stem cells that differentiate and form lymphocytes. The lymphocytes that are formed eventually end up in the lymphoid tissue. Differentiation or "reprocessing" of lymphocytes can occur.[26]

For example, the lymphocytes that are destined to eventually form activated T-lymphocytes first migrate to and are preprocessed in the thymus gland. Clearly, this process occurs in the young. However, the thymus gland involutes in midlife, yet many people still have a vigorous production of T-lymphocytes. Therefore, other yet to be found mechanisms, which are responsible for cell-mediated immunity, which is frequently disturbed in chronic pollutant overload must come into play in this action at this time of life.[27] (See section of senescence). **Goldstein has shown that both**

types of thymosin can be produced by any cell of the body. Therefore, this fact could be a reason why the inactive thymus does not affect immunity in the long run. Rat studies show that T-cells can be developed in the intestinal epithelium independent of the thymus.[28] These intraepithelial (IE) lymphocytes functions are as follows:

1. Survellince of the interstial epithelial layer for the detection of microbial pathogens
2. Removal of damaged or transformed epithelial cells
3. Mainanence of epithelial integrity via secretion of trophic factors important for epithelial cell growth and differentiation

Most IE lymphocytes are CD_8 suppressor cells which contain granules of substances like perforin, serine esterases, and granzyme. Some type of IE lymphocytes can enhance production of lymphocytes in the gut when they are even suppressed peripherally.

Studies at the EHC-Dallas and Buffalo have shown that chronic toxic total body overload will suppress the cell-mediated immunity as measured by "delayed recall antigens" such as tuberculin, streptococcus, staphylococcus, proteus, candida, tetanus, trichophyton. (For details of the lymphatic organs, the lymph nodes, and lymphatic function, see Chapter 1; for details of the gut immune response to foreign antigens, see Gastrointestinal chapter in Volume 2 - Mechanisms of Chemical Sensitivity and Chronic Degenerative Disease.)

NEUROIMMUNE REGULATION

Neuroimmune regulation of the NALT, BALT, MALT, and GALT is of great importance because many neurotransmitters are released after nerve stimulation especially from the autonomic nervous system. The density of nerves is up to one per $200\mu m^2$ in the gastrointestinal mucosa, which gives the gut and immune system a unique communication network with the nervous system in order to function optimally. Apart from the enteric nervous system, the vast majority of nerve fibers does not reach the interstitial epithelium but terminate at the submucosal or mysenteric plexus (i.e., connective tissue matrix). From the plexus region, nerve varicosities extend to the surrounding area from where the released neurotransmitters and neuropeptides may be expected to diffuse and reach to mucosal targets.

The effect of neural activation is the release of agonist substances (neurotransmitters, neural peptides, neurohormones) that act through neurocrine, paracrine, and endocrine pathways to activate specific receptors on immune cells. (See Chapter 4: Endocrine System.) In the endocrine and paracrine communication between the effector and the commander cell, physical contact on the nerve is not required, but the neural-mediated activation of neurotransmitters at close proximity between the cells is a prerequisite.

Within the lymphoid system, the classical paradigm of neuroimmune communication is represented by norepinephine (NE), the neurotransmitter of most post ganglionic sympathetic fibers and lymphocytes in the spleen.[29] The presence of noradrenergic nerves establishing synapse-like contact between lymphocytes and NE varicosities[30] generates a release of **NE upon splenic nerve stimulation,[31,32] creating action of NE on lymphocytes resulting in altered lymphocytic function.[33,34] In addition NE can also act in a paracrine manner diffusing away from parenchymal nerve ends to be available in a variety of immune cells. The NE also helps regulate immunoinflammatory responses**. When measuring the autonomic nervous system by pupillography and HRV, our studies on thousands of patients with chemical sensitivity and chronic degenerative disease have shown at least 85% with sympathetic responses produce the NE effect. These findings immediately suggest that autonomic dysfunction is an integral part of dyshomeostasis of not only the nervous system but also the immune system in patients with chemical sensitivity and/or chronic degenerative disease.

Neuropeptides released from the central and peripheral nerve varicosities and by the cells of the immune system also act as conventional neurotransmitters, transducing environmental signals into specific chemical messages. **Almost all steps of lymphocyte activation are affected by neuropeptide release including those by neuropeptide Y, vasoactive intestinal peptide (VIP), substance P (5PV), metencephalin, choleocystokinin, calcitonin gene related peptide(CGRP), neurotension, and somatostatin.** Most nerve terminals, which are mostly nonadrenergic, are located in primary and secondary lymphatic organs.[33–38] All primary (bone marrow,[34,39] thymus)[33,36,40] and secondary lymphoid organs (spleen, lymph nodes and MALT [34,39,36,41–43] have been shown to possess varying degrees and patterns of nonadrenergic innervation.

Cholinergic innervation has been found in the bone marrow,[44] spleen,[45] lymph nodes,[46] palentine tonsils,[43] appendix,[39] thymus,[48] and GALT. The existing data suggests that cholinergic innervation of lymphoid tissue is present but sparse and in some organs nonexistent. In the gastrointestinal system neuropeptides are expressed and released from neurons in the enteric nervous system and also from extrinsic afferent varicosities. **Enteric neurons lie in the mesenteric and submucosal plexus while most gastrointestinal peptidergic afferents emanate from the dorsal root ganglia.** Neuropeptides include CGRP, substance P, neurokinin A, encephalin and galantine.

There is a close spatial relationship between peptidergic nerves and mast cells, eosinophils, lymphocytes, plasma cells, macrophages and capsaicin sensitive peptidergic nerves of the spleen, lymph nodes, thymus, tonsils and lymphoid aggregates, nasal, lung, and gastrointestinal mucosa.

In vitro pharmacological studies showed that peripheral lymphocyte proliferation enhancement was mediated by alpha-adrenergic receptors, whereas inhibition was mediated by activation of the β-receptor. It has been shown that **activation of the sympathetic nervous system will increase circulating B- and T-cells, lymph circulation, redistribution of natural killer and T-helper cells, decreased proliferative response for T-dependent mitogens, and modulated antibody response**.

Acetylcholine or cholinergic agonists have been shown to stimulate T-cell secretion of IL-2, IL-4, and interferon 2, and **to inhibit IL-6, to increase T-cell proliferation, enhance cytotoxic activity of T-cells, increase T- and B-cell mobility, modulate antibody secretion, and protection of mature T-cells from apoptotic signals**.

Many effects on mucosal immunity by neuropeptides exist including increase in IgA and IgG in the intestinal fluid, normal development of gut interepithelial lymphocytes, modulation of GALT dysfunction after lack of parenteral nutrition, decrease IgE and increase IgA in Peyer's patches number of B-cells, alter migration of T-cells to Peyer's patches, enhance IgA production in mucosal lymphoid tissues, increases NK and cytotoxicity of interepithelial lymphocytes, and contradicts and decreases IL-2 expression of L-PL lymphocyte. See Figure 3.10.

CLINICAL IMPLICATIONS OF FOOD AND CHEMICAL SENSITIVITY IN RELATION TO THE AUTONOMIC NERVOUS SYSTEM AND IMMUNE SYSTEM

All described (and other potential) interactions occurring among immune-endocrine cells and neural endings in the GALT and BALT may be relevant for the regulation and coordination of the in vitro noninflammatory and inflammatory response against food and chemical derived antigens. However, these interactions must be viewed as a dynamic interplay involving the classic immune and nonimmune cellular network that eventually resides in the mucosal and submucosal tissues and the central (hypothalamus, reticular activating system, limbic system, pineal gland and area postrema of the fourth ventricle), peripheral, and enteric neuroendocrine system. **These dynamic interactions are further represented first by the migratory capacity of activated immunocytes and second, by the fundamental state of the effector cell**. Thus, for instance, activated lymphocytes are able to cross the blood brain barrier allowing direct local cross talk between different microenvironments of the hypothalamus, reticular activating system, the limbic system, the pineal gland, and the area postrema of the fourth ventricle. The effectiveness of the neuroimmune interplay in the control of the immune reactions may largely depend on the functional state of immune cells **since it has been shown that primed cells are more prepared to sense and react against an antihomeostatic message than resting cells**. It is also important to remember that both afferent and efferent

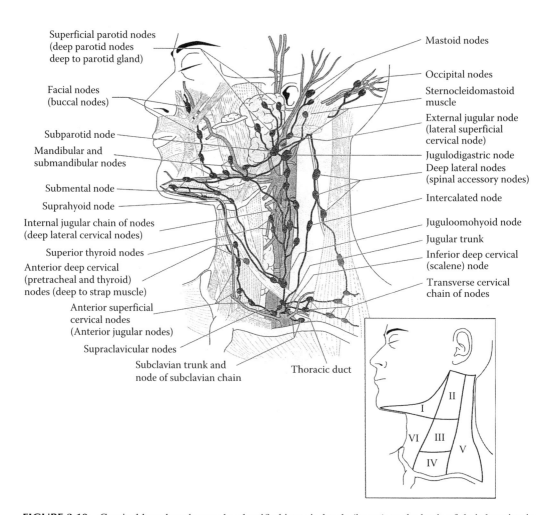

FIGURE 3.10 Cervical lymph nodes can be classified into six levels (insert) on the basis of their location in the neck. Many patients have oral food and chemical reactivity and have swelling in the submandibular nodes. These lymphatics along with the tonsils, adenoids, and other lymphoid tissue of the nasal pharynx make up the NALT (nasal associated lymphoid tissue) which initiate acquired lymphatic responses for the GALT and BALT. (Modified from Souba, W.W., Fink, M.P., Jurkovich, G.J., et al., *Principles & Practice*, WebMD Professional Publishing, New York, 162, 2006.)

neuroendocrine interactions are involved in that interplay. In this circuitry chemokines, neurotransmitters, growth factors, antibodies, neurohormones, and neuropeptides act as the chemical messengers. Food and chemical allergic hypersensitivity reactions are mounted when the offending antigen is able to cross link with surface IgE on mast cells and basophils. Nonallergic hypersensitivity reactions can occur with the non-IgE triggering of the mast cell. These interactions cause release from the mast cells of a broad array of bioactive molecules that alter epithelial and smooth muscle physiology by inducing water and ionic secretion, increase vascular and epithelial permeability to small molecules, and cause gross abnormalities of gastrointestinal motility patterns.[52,53] During the course of these events some of the bioactive molecules released by immunocompetent cells activate other immune and nonimmune cells including nerve terminals, which in turn liberate pro- and counter-regulatory substances that purposely help to maintain the homeostatic balance.

In this sense, cytokines, antibodies, and neuropeptides alter electrical activity and neuropeptide secretions by peripheral, enteric and central neuron[50,51] in order to modulate neurogenesis in lymphoid tissues[52] and to regulate neurogenic inflammation in the gastrointestinal tract.[53] The opposite is also true since nerve stimulation has been shown to inhibit the effects of antigen challenge.[54] The complexity of the response is ultimately represented by the fact that food and chemical antigens themselves **may induce a variety of effects on the nervous system, leading to a lower threshold of discharge to normal stimuli** (hyperexcitability or visceral hyperalgesia), which could be responsible for some of the clinical manifestation of food and chemical sensitivities.[55] This phenomenon may also explain the spreading phenomenon that occurs with pollutant overload in the patients with chemical sensitivity amd/or chronic degenerative disease where they become more sensitive to more and more foods and chemicals until they become universally sensitive.

The law of nerve damage (see Chapter 2) says that when a nerve is damaged or severed all tissue downstream from the damage will become temporarily hypersensitive. However, some nerves become permanently hypersensitive as seen in many patients with chemical sensitivity and/or chronic degenerative disease. If this is the case food and chemical sensitivity will occur via the aforementioned mechanism.

Numerous experimental studies have confirmed that food induced immunological reactions are influenced by neural stimulation as well as neuropeptide and cytokine release. Thus for instance, it has been shown that intestinal secretory responses and permeability in some cases to egg ovaalbumin luminal challenge were inhibited by systemic capsaicin challenge, tetradotoxin, and NP-Y, reduced by interferon YB, but potentiated by substance P and OPIOD antagonists.[48,56-58] Additional evidence supports the involvement of capsacin sensitive fibers as well as VIP, SP, opioids, CGRP, **CRH (corticotropin releasing hormone), and many other peptides in the pathogenesis of several immune-mediated inflammatory disorders such as asthma, nasal allergy, contact and delayed type hypersensitivity reaction, rheumatoid arthritis, systemic lupus erythematosis, thyroiditis, urticaria, chronic arachnoiditis, inflammatory bowel disease, enteric parasitic infestation, and celiac disease**.[57-59, 61-63]

IMMUNITY

When foreign substances enter the body and disturb homeostasis, both types of immunity—innate and acquired—react automatically. With these two types of immunity, the body has the ability to resist almost all types of organisms and toxins or to neutralize noxious stimuli. Thus, innate and acquired immunity keep the body from being overwhelmed (i.e., acutely or chronically), as they maintain homeostasis. Both types of immunity can be changed by chemical sensitivity and/or chronic degenerative disease, creating ill health followed by fixed named disease.

INNATE IMMUNITY

Eighty-five percent of immunity is in the innate immune system. Thus innate immune alterations may be dominant in chemical sensitivity and chronic degenerative disease. **Most of the immune system's mast cells, natural killer lymphocytes, and eosinophils are in the innate immune system in addition to complements, adhesive molecules, leukocytes, and macrophages**. Innate immunity is present at birth. It results from general, nonspecific processes rather than from the reaction directed at a specific disease organism like that seen in humoral immunity. **Innate immunity is triggered by tissue damage**. For example, it has been shown that ozone will not only disrupt mucosal integrity but also alter mucosal clearance of bacteria, which alters and increases the vulnerability to infections.[64] Innate immunity combats the initial nonspecific total body load, and generally, combats the specific load, especially for infections and toxics. It can trigger endothelium support with vasospasm, and also by altering or even triggering the clotting mechanism. Once triggered by local noxious stimuli local mediators make vessels leaky allowing extravasation of complement from blood vessels to tissues.

Surrounding vessels become spastic and blood flow is decreased. Activated vascular endothelium leads to vasospasm at the injury site apparently in order to isolate the incitant; however, subsequent release of local mediators makes the surrounding vessels dilate with an increase in blood flow. Tissue hydrostatic pressure increases, which causes fluid to flow to lymphatic vessels, toward lymph nodes. Endothelial injury and complement activation lead to the recruitment of innate cells. Endothelium up-regulates adhesion molecules that grab immune cells from the circulation. These vessels' adhesion molecules include E-selectra and VCAM-1/ICAM-1. Chemotactic factors cause immune cells to activate matching adhesion molecules (integrins), which become trapped at inflammatory sites. Initial cells responding to pathologic invasions after the PMN's are the macrophages, **which express innate conserved pattern recognition receptors (PRRs). These PRRs recognize pathogens' "molecular motifs" such as lipopolysaccharide (LPS) bacterial glycoprotein or pathogen DNA fragments**. Once these PRRs are activated, macrophages secrete IL-1, TNF_a and ILB/MGP-1 to recruit more phagocytes. Recognition of molecular patterns is the key event in initiating immune responses. Examples of PRRs include: Toll-lite receptors TLR-2, which bind lipoproteins of gram negative bacteria; TLR-3, which recognize viral double stranded RNA; TLR-y, which recognize LPS; TLRq, which recognize unmethylated CpG DNA; and NoD_2, which bind muramyl dipeptide of bacterial cell walls as in Crohn's disease. These PRRs activate a chain of events that result in gene transcription, causing cells to produce more immune mediators. The gene products are inflammatory cytokines, and surface receptors. **PRRs are expressed by dendritic cells, epithelial cells, and macrophages**.

This innate immune process, which maintains the organism in defense of the assault from a hostile environment recruits homeostatic response cells, which activate processes like phagocytosis by leukocytes, macrophages, basophils, and eosinophils. Other innate immunity processes neutralize (described in Chapter 1) and destroy an organism or toxics by acid secretions of the stomach, by resistance of the skin and mucosal barriers, and by the presence of certain chemical compounds (opsonins, lysozymes, proteases, hydrolyases, the antipollutant ezymes [superoxide dismutase, glutathione peroxidase, catalase, etc.,]). These substances attack and destroy foreign invaders (organism or toxins) in the blood, extracellular matrix, and end organ system, and they sequester and isolate pollutants and/or microorganisms in tissue areas as discussed in Chapter 1. **The natural larger multifilter in the connective tissue matrix from the mucosa to the cell is also part of innate immunity**. Innate immunity includes processes that would be overwhelmed if the total body pollutant load were in excess or if nutrient depletion occurred constantly. Chronic excess of total environmental non specific load and specific body load (i.e., *Streptococcus hemolyticus*, organophosphate pesticides, xylene, etc.) acting upon innate immunity will disturb the dynamics of the body's homeostasis, causing the individual to be chronically ill or hypersensitive. In order to maintain the energy-efficient dynamics of homeostasis, and thus optimum health, the total body load must be reduced not only through avoidance with total load reduction but also through optimum nutrition.

Other innate immunity substances include compounds such as lysozymes (mucolytic polysaccharides and proteases), which cause bacteria to dissolve, and polypeptides, which react with and inactivate certain gram positive bacteria. The complement cascade (see later in this chapter), another part of innate immunity, is a system of 20 proteins, which can be activated via the alternate pathway to destroy bacteria and toxic chemicals, but is somewhat limited in capacity. Complement can be found in the arterial lesions of atherosclerosis. Complement function activates proinflammatory mediators such as C-3a or C-5a, which can activate endothelium, and enhance leukocyte recruitment to inflammatory sites. Therefore, the maturation of the atherosclerotic lesions beyond the foam cell stage is strongly dependent on an intact complement system.

Innate immunity also functions by natural killer lymphocytes and by forming immune complexes with gamma globulin to neutralize certain chemicals. These complexes can recognize and destroy foreign cells, tumor cells, and some infected cells, as well as toxics. This innate immunity is in a dynamic homeostatic balance in order for the body not to attack and destroy itself when it is not under attack by foreign invaders. **Innate immunity has an extremely energy-efficient function when its** dynamics are in homeostatic balance. The best method of containing noxious exposure is preventing the immune system from being damaged by the constant bombardment of toxic substances (the total body pollutant load). If in excess, these substances will continue to cause injury to the energy-efficient immune function. This innate immunity is broad in nature and generalized in response, allowing many different substances to be involved in maintaining basic health. Most of the processes in innate immunity can be modified or impaired in chemical sensitivity and chronic disease (see Chapter 1).

Several non-IgE independent mast cell secretagogues have been described as part of innate immunity. These are calcium ionophores, cationic compounds, substance P, neutrophil defensins, bee venom peptides, and complement derived anaphylatoxin C-3a. These compounds activate mast cells independent of Fce-R-1 dependent pathways.[65–68] Calcium ionophore A 23187 increases free intracellular calcium concentration, which circumvents membrane-associated G proteins from directly transporting calcium ions from the extracellular fluid into the cell;[69] cationic secretagogues activate a pertussis toxin-sensitive G protein, which is distinct from the pertussis toxin-insensitive G protein activated by Fce-R-1 stimulation.[70,71] The cationic mast cell secretogogues appear to act by electrostatic and hydrophobic interactions with the cell membrane.

Another innate immunity feature of the mast cell is its production of nitric oxide and superoxide. Superoxide is a strong oxidizing molecule, which causes neutrophil attraction and adhesion to endothelial cells[72–74] and may play a role in inflammation. Superoxide dismutase will neutralize this molecule. Nitric oxide and superoxide can be converted into a highly cytotoxic compound, peroxynitrite, which can produce tissue damage.[66]

Mast cells are involved in innate immunity to bacterial infection. They are activated by lipopolysaccharide. Tumor necrosis factor and complement are released by mast cells. Mast cells also attract tumor necrosis factor from bacteria and neutrophils. Mast cell increases showed less infection in mice.[75]

Transition to Innate and Humoral Immunity in the Newborn and Infant

The complement system and the family of Toll-like receptors (TLRs) are two central arms of innate immunity that are critical to host defense as well as the development of adaptive immunity in the newborn. Most pathogens are presumably allergens activating both complement and TLRs, suggesting that coordinated interactions between these two systems may be important in the development of disease. Indeed, Halisch and Köhl have recently shown that C5a negatively regulates TLR4-induced synthesis of IL-12 family members in macrophages through extracellular signal-regulated kinase and phosphoinositide 3 kinase-dependent pathways.[76] Interactions between PRRs can increase and diversify the recognition and overall handling of pathogens by the innate immune system that are otherwise limited by the genetic bottleneck. Although these interactions have not been formally explored in the pathogenesis of asthma, studies that explore the interaction of the complement system with other innate immune pathways are clearly warranted.

It is becoming increasingly clear that immunoregulatory events occurring at the interface of innate and adaptive immunity play an important role in asthma pathogenesis. Complement stands at this interface. The data reviewed suggest that the complement activation pathway serves as a central regulator of adaptive immune responses to a variety of inhaled substances as well as being part of innate immunity. The emerging paradigm suggests that C3/C3a generation at the airway surface serves as a common pathway for induction of (AHR) airway hyperresponsiveness by a variety of environmental triggers. On the other hand, C5/C5a plays a dual immunoregulatory role by protecting against the initiation of Th2-mediated immune responses during initial contact with allergens through its ability to influence CD/T-cell interactions and a more traditional proinflammatory role once immune responses are established. Thus, **factors regulating the balance between C5a and C3a generation may determine the tendency to develop tolerance or immunity to inhaled allergens, respectively.** We are in the initial stages of understanding the complex interactions between various components of the complement pathway as well as their interactions with other innate immune pathways.

The effects of C3a on processes occurring during allergen sensitization have not been directly evaluated. Studies in C3aR KO mice show the C3a regulates Th2 cytokine production, suggesting that either C3a regulates recruitment and activation of Th2 cells or regulates processes occurring during sensitization or during the effector phase of the allergic response. Although this has not been evaluated directly, studies in which complement activation was blocked at the level of C3 either at the time of sensitization or after sensitization showed that complement blockade effectively suppressed Th2 cytokine production and associated effector functions after priming, but not at the time of antigen sensitization.[77] Although this approach is not specific for C3a, the results suggest that C3a does not influence either the nature of magnitude of the T-cell response at the sensitization phase, but may regulate Th2 cytokine production and/or activate both Th2 effector cells and non-T-cells capable of producing Th2 cytokines. Thus, determination of the exact mechanism(s) by which C3a mediated Th2-driven immune responses in the lung await studies examining the role of C3aR signaling at the time of initial sensitization.

Based on the knowledge at hand, **the balance of C5a–C3a in the airways during early life exposures to allergens may be a major determinant of the development of tolerance or immunity to inhaled antigens**. Although the exact levels of each of these mediators in early life are unknown, it has proposed the following scenario: that high levels of C3a, induced locally through exposure to environmental triggers of asthma such as ozone, cigarette smoke, RSV, or particulate matter, would drive sensitization to inhaled allergens at the airway surface. This level of C3 would presumably set up a Th2 environment in the lungs, which is consistent with data suggesting that the lungs have a Th2 skew at birth and shortly thereafter. If C3 levels persist unopposed by C5a during early allergen exposures, the ensuring Th2 inflammatory response would perpetuate the process by inducing additional C3 production by the airway epithelium. This scenario is consistent with the relatively higher levels of C3a over C5a found in asthmatic airways as compared with those of control subjects after allergen challenge.[78] The conditions under which C5a would dominate are unknown, but may include frequent microbial exposure and/or lack of exposure to environmental triggers such as pollution and smoke. Nonetheless, the balance between these two anaphylatoxins is clearly an important determinant of protection or sensitization to inhaled antigen; thus, additional studies aimed at determining the mechanisms driving differential production of C3a and C5a in the airways are clearly warranted.

After fetal life, the mammary glands are not only anatomically and physiologically adapted to providing sustenance to the offspring in the form of colostrums and milk, they also transmit antibodies evoked by the mother against microorganisms with which she has been infected via mucosal routes. Before birth, human fetuses receive via the placenta maternal IgG, whose specificities reflect the systemic antigenic stimulation to which the mother has been exposed. This IgG protects the fetus in utero and persists in the infant's circulation following normal birth until approximately 6 months of age, when the infant's own immune system matures. There is little IgG in human milk at any time. **IgA, with specificity for microorganisms present in the mother's intestinal tract and to an extent in the respiratory tract, is the principle antibody isotype transmitted in milk throughout lactation**. In this case, passive immunity is passed from the mother via breast milk and placenta and could be considered innate immunity. Others might consider this immunity to be acquired even though the baby did not generate it. Innate immunity certainly persists in the well neonate until the complete immune system matures and has major significance throughout life. Some cases of eczema and other gastrointestinal inflammations have been brought under control by giving mother's milk over a long period of time, suggesting gastrointestinal IgA deficiency.

Galt derived lymphoblasts from mesenteric lymph nodes migrate to the mammary glands of lactating mothers. These cells produce the IgA.[79] (See Figure 3.11.) The function of T-cells in the mammary gland is unknown.

The following is a case report of a boy who had severe eczema and did not have innate or good acquired immunity.

Immune System

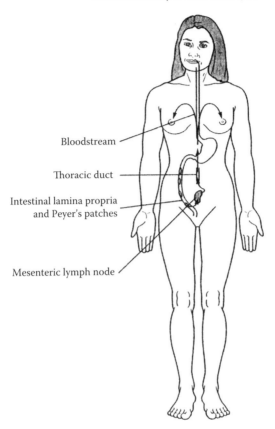

Enteromammary circulation of lymphocytes

- Bloodstream
- Thoracic duct
- Intestinal lamina propria and Peyer's patches
- Mesenteric lymph node

FIGURE 3.11 Enteromammary circulation of lymphocytes. IgA-committed B cells that have responded to antigen in the intestine leave the Peyer's patches in the lymph and migrate into the mesenteric lymph nodes. From there, they pass via the thoracic duct into the bloodstream, which carries them to the lactating mammary glands where they extravasate and become dimeric IgA-secreting plasma cells. (Modified from Brostroff, J., and Gamlin, L., *Food Allergies and Food Intolerance: The Complete Guide to Their Identification and Treatment,* Healing Arts Press, Rochester, VT, 47, Fig. 3.4, 2000. With permission.)

Case study. A nine-month–old white male presented with symptoms of severe eczema, hives, red blotches on his face, neck swelling, and aortic stenosis. He was sick from birth and sensitive to his mother's milk, soy, corn formula, and goat's milk. The mother had severe eczema and showed sensitivity to 23 out of 41 foods tested. She also tested sensitive to 13 out of 18 molds tested. Tests showed IgE – 338 (c-300), T_{11} – 5,224 (c-1,260 – 2,650 mm^3), T_4 helper-3591 (c-657–1770 mm^3), T_8 suppressor- 469 (c325-1,050 mm^3), B– 19 (c82- 477 mm^3); cell cycle-66 - 98%, CMI-3 (c-4-8 positive), and IgA extremely low. This patient was started on pooled mother's milk 4 oz q.i.d. and neutralizing injections for foods to which he was specifically sensitive. His eczema cleared, and he has done well for seven years with slight episodes of exacerbations when he eats food to which he is sensitive or is inadvertently exposed to a chemical that is toxic to him.

ACQUIRED IMMUNITY

Acquired immunity plays an important role as part of total immunity. It occurs in response to exposure and is the major part of the immune system's amplification process although making up only 15% of the body's immune response. Often, acquired immunity only develops after a bacterial or viral infection, parasitic disease, or a toxin enters the body, thus inducing the proliferation of antibodies. The toxics, or toxins, and the exposure for disease development may take several weeks to present. Sometimes symptoms may not appear even months after the first attacks to the body occur (i.e., tuberculosis, histoplasmosis) or until a chronic chemical exposure occurs and the body (in order for acute survival) masks its response clinically. Resistance may peak and then gradually erode over time.

In some instances, acquired immunity can protect either by surviving a disease or by vaccination. For example, an individual who has had measles or polio has permanent acquired immunity as do those who have been vaccinated (although a booster may be necessary in some cases). Alternatively, certain vaccines, like botulism or tetanus toxin, can be protective against doses as high as 100,000 times the amount that would be lethal without immunity. Vaccination protects by boosting a person's immune response without him/her having to have a disease. Substances that normally do not harm the body but can cause disturbances when the body's homeostasis is altered also may respond to vaccination. These substances include mold, pollens, and foods, which, once neutralized by the intradermal provocation and neutralization technique, can prevent reactions against ingestion or inhalation of large volumes of offending substances. Sometimes vaccinations for these aforementioned substances only result in temporary immunity lasting 3–4 days, while at other times and after repeated injections for days, months, or years, permanent immunity may be achieved. **The phenomenon is extremely important in the individual with food and chemical sensitivities where treatment by injections is necessary for long-term survival and well being**.

Clearly, however, excess immune responses may be harmful to certain individuals. Damage also may come from immunizations on isolated tissues. For example, an excess immune response may cause tissue necrosis if there is an overt reaction to a certain bacteria, toxics, pollens, or molds. Often mucosa or skin cuts, parenchymal inflammation, or poison ivy reactions occur due to excess immune responses. Flulike symptoms have been observed after the administration of excess autogenous lymphocytic factor (ALF).

In addition, neutralization of influenza symptoms by the appropriate provocation neutralization dose of influenza vaccine has been observed thousands of times after the onset of influenza. Presumably, the symptoms are due to the acute reaction of an immune response but may also have to do with neural sensitization.

Harm of standard dose vaccination may be due to excess mercury preservative or perhaps excess antibody response to the bacteria or virus, as appears to be the case in some autistic children. A strong association exists between vaccine reactions, especially measles vaccines, and some cases of autism. We have seen some patients develop brain damage from an acute reaction to DPT injections. Some patients with the postpolio syndrome respond to injections of specific low doses of killed polio vaccine, suggesting a long-term adverse effect to either the nonspecific dosed to the stock polio vaccine or the response to overcoming the disease. At the present time, not all is known about the status of vaccines and their aid in maintaining the dynamics of homeostasis. Obviously, vaccines help to protect the general population against certain organisms and substances, but some harm may occur. Caution should be used, as well as constant diagnostic vigilance for the warning signs of early vaccine injury. **If the clinician perceives the early warning signs of vaccine injury he/she can often prevent permanent neurological injury**.

Humoral immunity exists (acquired immunity response) when the B-cell produced antibodies (IgG, IgM), which are immunoglobulin molecules in the blood, attack the invading agent. Activated large lymphocytes specifically designed to destroy foreign substances are also part of acquired immunity. This type of immunity is cell-mediated immunity or **T-cell-mediated immunity** (Table 3.4), **and is usually found to be disturbed in the patient with food and chemical sensitivity or viral infections** like the human immunodeficiency virus.

Both the humoral antibodies and the activated lymphocytes (cell-mediated immunity) are formed in lymphoid tissue of the body, which is able to respond to any acute or chronic stimuli. However, lymphocytes function often become disturbed with chronic environmental overload in a nutrient deficient patient. These deficiencies will further cause dysfunction of the immune system by causing

TABLE 3.4
Cell-Mediated Immunity—Evaluation of T-Cell Functions

Tuberculin	Staphylococcal
Proteus	Candida
Streptococcal	Tetanus
Trichophyton	Diptheria

Source: From EHC-Dallas, 2004.
Note: Delayed recall antigen response at 48 hours.

a vicious downward dyshomeostatic cycle. Due to acquired immunity does not occur until after the first invasion by a foreign organism or toxic substances the body must have some mechanism for recognizing the initial invasion. Each toxic or each type of organism usually contains one or more specific chemical compounds or configurations in its makeup that are different from all other compounds, making recognition and initiation of the immune response possible. In general these compounds are proteins or large polysaccharides,[80] which allow the immune system to identify that they are noxious foreign stimuli, which must be neutralized.

For a substance to be antigenic, the antigen, usually, must have a high molecular weight, 8000–10,000 Daltons although hapten may have higher weights or greater.

In addition to size, the process of antigenicity usually depends on regularly recurring molecular groups (epitopes) on the surface of the large molecule, which allows identification and explains why proteins and large polysaccharides are almost always antigenic. This identification is because they both have a specific type of stereochemical characteristic.[80] Foreign substances often communicate with the body by the presence of lectins and/or electromagnetic frequencies, which allows for rapid and precise identification.

We have included the body's reaction to molds or their mycotoxins here because a severe exposure can cause more than an allergic response. The toxicity of these substances has now been shown to cause immune alteration and even suppress immune function. Mold and mycotoxins can cause inflammation of any target organ, but especially the lungs, bronchioles, brain, and cardiovascular systems.

Although substances with molecular weights of less than 8000 Daltons only seldom act as antigens, immunity can, nevertheless, be developed against substances of low-molecular weight if they form haptens. Many toxic chemicals fit in this category. The chemical first combines with a substance that is antigenic, such as a protein. In one of the body's first responses, proteins are frequently found to neutralize the entering chemical. Chemical neutralization may or may not occur. However, the combination of the chemical and protein will elicit an immune response similar to any other antigen antibody response. The antibodies or activated lymphocytes that develop against the combination will then often react separately against either the protein or the hapten. Therefore, on a second exposure to the hapten, some of the antibodies or lymphocytes will react to it before the antigen can spread through the body causing damage.[80] Others escape the reaction, which can cause havoc in some chemically sensitive and chronic degenerative diseased patients, while others have an accentuated response, which will cause dyshomeostasis. One example of hapten formation is the case of a 40-year-old white male who worked with toluene diisocyanate. He was not affected by exposure until he went on vacation. When he returned, he received a second exposure, and then showed symptoms of asthma and vasculitis, which increased until he became incapacitated. After recovery, he returned to work and, again, had frequent attacks of asthma. Specific serum IgE showed him to be positive to toluene diisocyanate and a plasma protein (i.e., albumin), which in this case is haptogenic, causing his asthma. Other studies might show the toluene dissocyanate to directly act with IgE not forming a hapten, but acting directly through this mechanism.

The haptens that elicit immune responses of this type are usually low-molecular-weight drugs, chemical constituents in dust, breakdown products of animal dander, degenerative products of scaling skin, various industrial chemicals, the toxin of poison ivy, and so forth. Due to the massive environmental contamination of our planet, thousands of toxic chemicals enter the body on a daily basis. These have to be defended against, and all are potential haptens. Once these noxious substances enter the body, **the plasma proteins, including albumin, have a tendency to grab onto these chemicals in order to neutralize them**. However, neutralization is frequently not complete. These chemicals and the haptens also can react in other innate immunity processes, such as directly triggering complement, kinins, prostaglandins, leukotrienes etc., causing the body to destroy the foreign invader. Clearly, the acquired immunity process by hapten formation is a major, though not the total, method for destroying foreign chemicals. All of these immune and nonimmune processes that generate reactions must be in homeostatic balance in order not to harm the body, but to deal with the noxious stimuli. Literally thousands of low-molecular weight chemicals are now directly in the environment. **Many of these chemicals deregulate the suppressor T-cells as observed in 85% of the cases at EHC-Dallas, but some deregulation of immune function may come from hapten antigens**. These environmental chemicals will often sensitize T-cells causing suppression of the suppressor function or causing disruption of the dynamics of homeostasis, which may then lead to the hypersensitivity response. Approximately 15% of environmental pollutants will suppress the helper T-cells, while 5% will suppress B-cell function. Sometimes exposure to environmental toxins can cause low natural killer cells and suppress their activity.

Acquired immunity is the product of the body's lymphocyte system, which must be robust in order to keep the dynamics of homeostasis in an optimum state. By this accepted fact, few totally well people exist today, suggesting at least a mild breakdown in both acquired and innate immunity, which may result in chronic fatigue, fibromyalgia, and brain dysfunction before end-stage disease occurs. In an individual who has a genetic lack of lymphocytes or whose lymphocytes have been destroyed by radiation or chemical exposure, no acquired immunity can develop. Within days after birth, such a person dies of fulminating bacterial infection unless treated by heroic measures. More often, less severe birth injuries occur to the lymphocytes, causing chronic disease. Bubble children were found at one time. They needed controlled environments that proctected them from the bacteria and virus contaminated air in order to reduce their specific infectious total body load. These individuals have low lymphocyte and gamma globulin counts, which are enhanced by the lessened total body load of bacteria and viruses in the controlled environment, which causes and utilizes less stress to the immune system.[80] Clearly, the lymphocytes are essential to survival of the human being. In bubble children, gene therapy was tried and appeared to be somewhat successful in some cases by enhancing the acquired immune response. All the children had SCID-XI, which kills cells crucial to the body's immune system. They were transfused with the corrected gene and did well recovering complete immune function.[81]

It was eventually reported that with time they again lost their immunity.

In the majority of patients who have a less altered immune system than the bubble babies, the controlled less polluted environments, which are designed to decrease particulate and chemical exposure, allow the patient to respond well to the reduction of the total environmental load. These patients usually have hypersensitivity or chronic degenerative disease and have not responded to other forms of medical treatment. (See Treatment chapter in Volume 4 - Mechanisms of Cardiovascular Disease and Chemical Sensitivity.)

LYMPHOCYTES

In the more common immune deregulation the suppression or excess stimulation of helper, suppressor, or killer cell numbers and/or function occurs. Here the patient is alive, but functions suboptimally. Often this suboptimum function occurs for years until the patient develops a fatal disease (e.g., AIDS, tumor, or chronic degenerative disease) or if treated gets well. This immune alteration is one of the ways chemical sensitivity and chronic degenerative disease occurs.

The lymphocytes are located most extensively in the lymph nodes, as shown in Chapter 1. They are also found in special lymphoid tissues, such as in the spleen, submucosal areas of the gastrointestinal tract (GALT, MALT, Peyer's patches), and in the bone marrow. All of these areas must be kept in dynamic homeostatic balance in order to obtain and maintain health.

As is shown in Figure 3.9, the lymphoid tissue is distributed advantageously in the body to intercept the invading organism or toxins before they can spread too widely and damage end-organ tissue. This defense mechanism is especially true of the gastrointestinal tract and other areas open to the environment.[80] Once local control is lost, these areas of defense are often regionalized in order to give several lines of defense against the invading noxious stimulus. This message of foreign substance entry involves the local and regional defense phenomenon (see Chapter 1) for equalizing the dynamics of homeostasis. The intruders trigger an attempt by the body to wall them off. This type of compartmentalization is essential to ward off noxious stimuli after it penetrates the epithelial, mucosal, and immune barriers. In most instances, the invading agent after penetrating the mucosa and skin first enters the connective tissue matrix and the tissue fluids and then is carried by way of lymph vessels to the lymph node or other lymphoid tissue. For instance, the lymphoid tissue of the gastrointestinal tract is exposed immediately and constantly to antigens invading through the gut.

As shown in Chapter 1 and Gastrointestinal and Respiratory chapters in Volume 2 - Mechanisms of Chemical Sensitivity and Chronic Degenerative Disease, a vast network of defense parameters are capable of immune responses that will ward off foreign substances entering the body through each of the barriers. These immune responses are always ready to function to protect against foreign substance invasions, but these functions can become damaged with excess chronic environmental overload, both nonspecific and specific. **Foreign compounds like pesticides on the food, solvents in the drinking water, or excess mold, especially mycotoxins, in air, food, or water, can alter the immune response, triggering cascades of inflammatory cytokines and causing the mucosa to leak**. This leak allows bigger and other foreign substances to enter the body often, impairing phagocytosis. With persistent or recurrent insult by foreign substances, the dynamics of the body's homeostasis will become disturbed, resulting in nutrient depletion. The patient develops dyshomeostasis followed by one or both aspects of chronic disease, either the hypersensitivity response or chronic degenerative disease response.

The lymphoid tissue of the throat and pharynx (the tonsils and adenoids) (NALT) is well located to intercept antigens that enter by way of the upper respiratory tract or the mouth. The lymphoid tissue in the lymph nodes in the initial exposure areas is exposed to antigens that invade the peripheral tissue of the body, again showing an example of regionalization of lymph nodes. Here, more antibody processing occurs. Finally, the lymphoid tissue of the spleen and bone marrow plays the specific role of intercepting antigenic agents that have burst past the local and regional areas and have succeeded in reaching the circulation.[80] In most cases, these organs will rapidly remove or neutralize noxious incitants. However, with a chronic entry of foreign substances, dyshomeostasis occurs with metabolic changes (hypoxia, nonspecific mesenchyme reaction), followed by tissue changes (sols to gels, gels to sols). At this point, the immune system becomes chronically disturbed, and fixed-named irreversible disease usually occurs.

Although most of the lymphocytes in normal lymphoid tissue look alike when studied under the microscope, these cells are distinctly divided into two major populations. One of the cell populations, the T-lymphocytes, is responsible for forming the activated lymphocytes that provide cell-mediated immunity, and the other population, the B-lymphocytes, is responsible for forming the antibodies that provide humoral immunity. Both types of immunity may be disturbed by chronic pollutant overload disrupting the dynamics of homeostasis.[80] We have observed that the cell-mediated immunity is frequently poor in the hypersensitive state especially in individuals with chemical sensitivity. Cell-mediated immunity tests are low indicating poor T-cell function. When the B-cells are overstimulated many types of hypersensitive states and chronic degenerative diseases have been noted with both types of responses occurring during the course of a patient's illness. High antibody titers develop to substances like the Epstein-Barr virus, Cytomegalic virus, *Candida albicans*, and many

TABLE 3.5
EHCD Mold and Mycotoxin Immunity Study Mold & Mycotoxins—T & B—Lymphocytes

($n = 38$)	No.	% Abnormal
WBC	3	8
Lym	13	34
T_{11}	19	50
T_4	14	37
T_8	14	37
B_4	8	21

Mold & Mycotoxins—Cell-Mediated Immunity (CMI)		
($n = 33$) Positive/Tested	No.	%
Abnormal	**14**	**42**
1/7	6	18
2/7	8	24
3/7	5	15
Normal	**19**	**58**
4/7	8	24
5/7	5	15
6/7	1	3

Source: From EHC-Dallas.

Note: T & B-lymphocytes in 38 patients exposed to molds and mycotoxins, (aflatoxin, satratoxins, trichothecene).

kinds of bacteria, which appear to cause the symptoms of an altered B-lymphocyte immune system response rather than due to the primary offenders (Table 3.5).

The B-lymphocytes are destined for formation of antibodies and preprocessed in the liver during mid fetal life and in the bone marrow in late fetal life and after birth. B-lymphocytes are responsible for humoral immunity, which also can be disturbed by chronic pollutant overload causing dyshomeostasis and eventually fixed-named disease. Figure 3.12 shows the two lymphocyte systems for formation, respectively, of the activated T-lymphocytes and antibodies.[80]

T-LYMPHOCYTES

There are generally two types on immune regulation—the central and peripheral. In the central the T-lymphocytes, after their origination in the bone marrow, first migrate to the thymus gland. Here they divide rapidly and at the same time develop extreme diversity for reacting against different specific antigens. That is, one clone of the (thymic) T-lymphocytes develops specific reactivity against one antigen. The next T-lymphocyte develops specificity against another antigen. This process continues until different (thymic) T-lymphocytes with specific reactivities against literally millions of different antigens form. These different types of processed T-lymphocytes then leave the thymus and spread throughout the body to lodge in lymphoid tissue everywhere.[80] Thus, integrity and optimal health of the thymus gland is very important. As one can see, this distribution allows another tier of immunity to be present for protection of the organism in maintaining the dynamics of homeostasis. It is also a set up for malfunction, which could occur anywhere in the body creating local, regional, or distal malfunction in the dynamics of homeostasis. A total body pollutant overload would alter the dynamics of homeostasis, causing nutrient depletion and eventually resulting in disease.

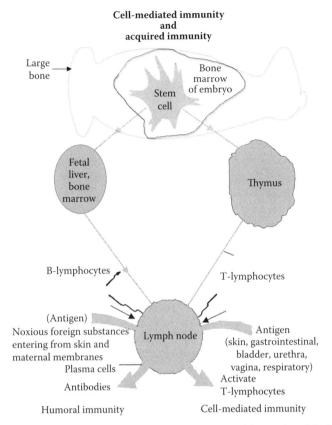

FIGURE 3.12 Diagram shows cell-mediated immunity and humoral immunity. (Modified from Guyton, A. C., and Hall, J. E., *Textbook of Medical Physiology,* 9th ed., p. 447, Philadelphia: WB Saunders Co., 1996. With permission.)

The thymus also makes certain that any T-lymphocytes leaving the thymus will not react against proteins or other antigens that are present or part of the body's own tissues; otherwise, the T-lymphocytes would be lethal to the body in only a few days. The thymus gland selects the T-lymphocytes that will be released first and mixes them with virtually all the specific "self-antigens" from the body's own tissues. If a T-lymphocyte reacts, it is destroyed and phagocytized instead of **being released, which is what happens to as many as 90% of the cells**. Thus, the only cells that are finally released are those that are nonreactive against the body's own antigens. Instead, the body reacts only against antigens from an outside source, such as from bacteria, a toxin, or even a transplanted tissue from a donor.[80] Failure of this process of apoptosis can lead to various types of autoimmune disease and tumors.

Most of the preprocessing of the T-lymphocytes in the thymus occurs shortly before birth of a baby and for a few months after birth. Beyond this time, this periodic removal of the T-lymphocytes from the thymus gland diminishes, but does not eliminate activated lymphocytes. However, **removal of the thymus several months before birth can prevent development of all cell-mediated immunity**. Since this cellular type of immunity is mainly responsible for regulation of transplanted organs, such as hearts and kidneys, one can transplant organs with little likelihood of rejection if the thymus is removed from an animal a reasonable period before birth.[80] Obviously, this does not occur in humans; although radiation of the thymus was commonplace in the early and mid twentieth century. This disturbance would lead to an individual vulnerable to environmental incitants, who would early in life develop asthma, vasculitis, blood clots, or other hypersensitive phenomena due to excessive exposure to environmental incitants.

As stated previously, immature T-cells migrate from the bone marrow to the thymus, where they begin to express receptors for antigen.[82] The majority of these receptors have two chains, α and β, and are called α/β receptors. The receptors recognize an immunogenic peptide held in the cleft of a major-histocompatibility-complex (MHC) molecule—the MHC-antigen-peptide complex. In the thymus, epithelial cells and other antigen-presenting cells display a wide variety of complexes composed of self peptides bound to MHC molecules. T-cells with receptors that bind these complexes with sufficient strength (i.e., avidity) survive (positive selection), whereas T-cells with very-low-avidity interactions with the complexes die. From the outset of the process of creating a repertoire of mature T-cells, therefore, positive selection favors antiself T-cells.[83–88]

But a second process, termed negative selection, causes the death of T-cells with receptors having a high avidity for self peptides.[88–90] Negative selection is the major mechanism of self-tolerance, and autoimmune disease can be the result if it fails.

Only 3% of the T-cell precursors that enter the thymus survive positive and negative selection.[91] This purged population, composed of T-cells with receptors of low and intermediate avidity to self (for brevity, we subsequently will omit mentioning the receptor when referring to avidity), leaves the thymus and inhabits lymphoid organs. Curiously, **the emigrants from the thymus were selected for survival because of their ability to bind to self peptides with low avidity, yet they constitute the population of T-cells that deals with foreign antigens**. An explanation for this paradox is that a given T-cell receptor can cross-react with multiple peptides; this cross-reactivity maintains the flexibility required by the immune system to adapt to a changing environment.[92,93] Moreover, because foreign antigens are not normally present in the thymus, T-cells with the potential for high-avidity binding to foreign peptides evade negative selection and escape into the periphery (i.e., other lymphoid tissues and organs).

Peripheral immunoregulatory mechanisms control the magnitude and class of immune responses and discriminate self from nonself. Intrinsic homeostatic mechanisms that occur during the initial antigen-induced activation of T-cells control the magnitude and class of immune responses, including the emergence of type 1 helper T (Th1) and type 2 helper T (Th2) cells and type 1 regulatory T (Tr1) and type 3 regulatory T (Tr3) cells. These mechanisms entail antigen-induced cell death, anergy, and the secretion of cytokines. However, to deal with the threat of autoimmunity effectively, the immune system must also **discriminate self from nonself, which is carried out in the periphery predominantly by suppressor T-cells**. Decrease in T suppressor cells as seen in individuals with chemical sensitivity will keep the T-cells upregulated (Figure 3.13).[94]

As a consequence of selection within the thymus, some T-cells with intermediate avidity for self antigens enter the periphery,[95–98] where they have the potential to become pathogenic effector cells.[92,98–99] To avoid pathogenic autoimmunity, various peripheral regulatory mechanisms fine-tune the self-reactive T-cell repertoire. These mechanisms suppress the expansion of self-reactive clones with an avidity that is not sufficiently high to eliminate them intrathymically but is high enough to induce pathogenic autoimmunity in the periphery. **Thanks to these peripheral regulatory mechanisms, autoimmune disease does not occur despite the presence of numerous self-reactive clones in the mature T-cell population.**

General Intrinsic Mechanisms

Intrinsic mechanisms induced when the T-cell receptor engages MHC–antigen-peptide complexes are a major point of control over the magnitude and class of immune responses (Figure 3.14). One of these mechanisms entails the avidity and duration of binding of the T-cell receptor with MHC–antigen-peptide complexes—avid binding of relatively long duration favors activation, whereas a weak, brief encounter does not.[100,101] Notably, engagement of the receptors can induce not only activation and differentiation of the T-cell but also apoptosis.[102]

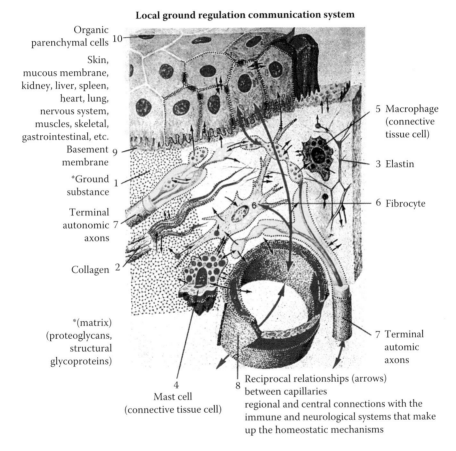

FIGURE 1.3 Reciprocal relationships (arrows) between capillaries (8), ground substance (matrix), (proteoglycans and structural glycoproteins 1, collagen 2, elastin 3), connective tissue cells (mast cell 4, defense cell 5) fibrocyte (6), terminal autonomic axons (7), and organic parenchymal cells (10). Occurs Basement membrane (9), the fibrocyte (6) is the regulatory center of the ground substance. Only this type of cell is able to synthesize extracellular component. The main mediators and filters of information are the proteoglycans, structural glycoproteins, and the cell surface sugar film, glycocalyx (dotted line on all the cells, collagen and elastin). (Modified from Pischinger, A., *Matrix and Matrix Regulation: Basis for a Holistic Theory in Medicine*, Ed. H. Heine, Eng. Trans. N. MacLean, Brussels, Belgium: Editions Haug International, 21, 1991.)

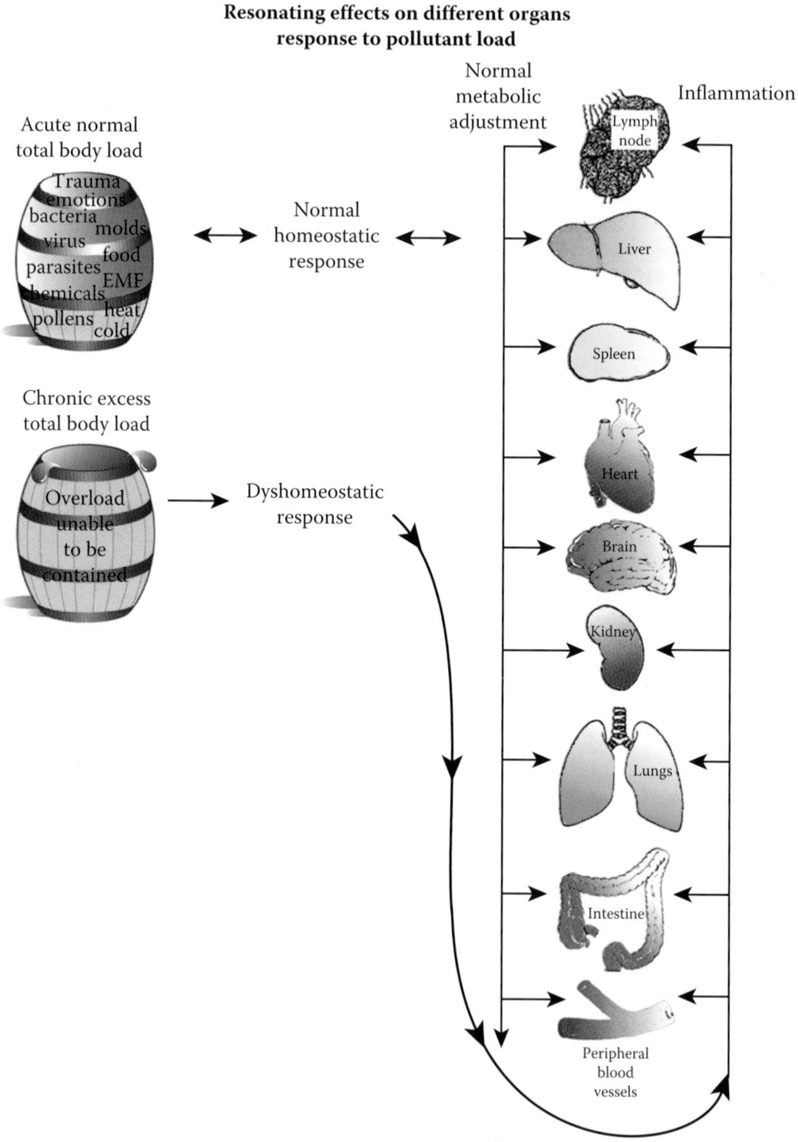

FIGURE 1.11 When the body exceeds its total pollutant load, dyshomeostatic responses occur, finally resulting in inflammation. (From EHC-Dallas.)

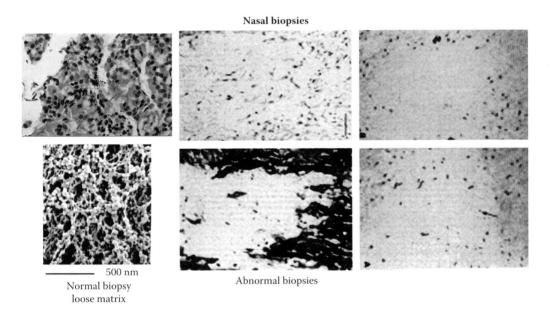

FIGURE 1.17 Different patterns of extracellular matrix in mold sensitive and chemically sensitive patients—nasal biopsies. Note: increased density of matrix in each slide due to inflammation. Biopsies show the ECM sieve as clogged. (Courtesy of Dr. Lindsey Wing, Sydney Australia. Personal communication.)

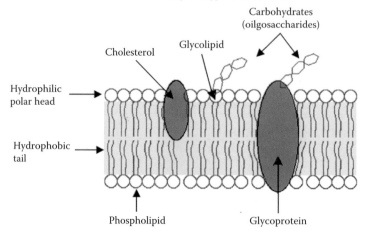

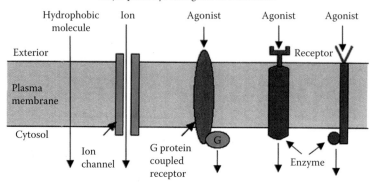

FIGURE 1.23 A cell needs to communicate with its environment so that it can make appropriate responses. The external signal may enter a cell via four major pathways. 1. Hydrophobic molecules, 2. 1- Ion channels, 3. G-protein-coupled receptors, 4. Enzymes. (From Schematic Drawing of a Typical Membrane, http://www.web-books.com/MoBio/Free/Ch1B; Major Pathways for Signals to Enter a Cell, http://www.web-books.com/MoBio/Free/images/Ch6A1.gif. Free images.)

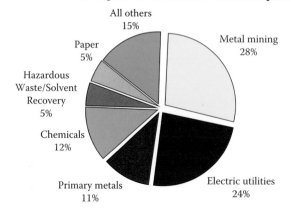

FIGURE 1.44A The public has been exposed to toxic chemicals. (From Toxic Release Inventory. 2003. Total Disposal or Other Releases, EPA 260-R-05-001 May 2005.)

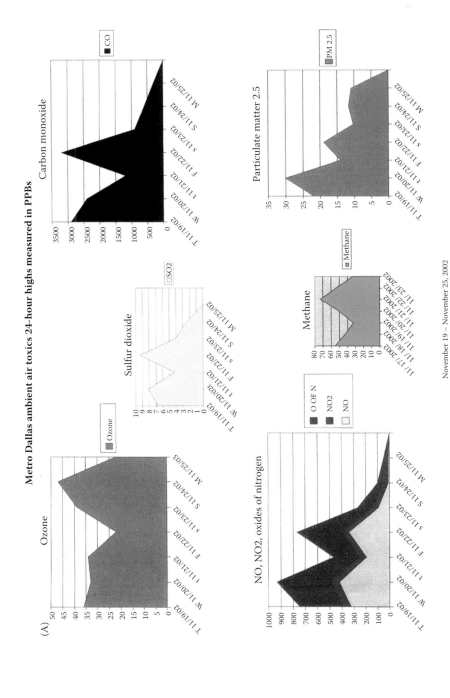

FIGURE 1.45A Metro Dallas ambient air toxics 24-hour highs measured in PPBs. (From EHC-Dallas and local EPA.)

(C)

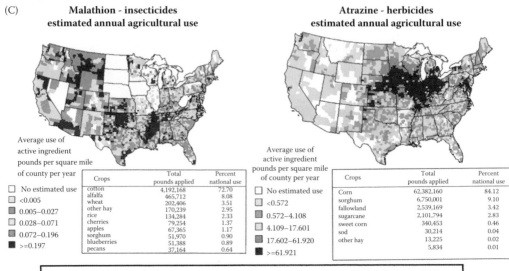

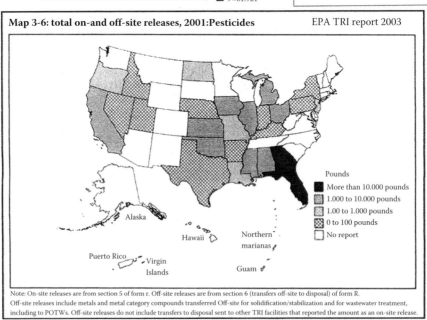

FIGURE 1.45C Pesticide and herbicide U.S. maps. (From EPA TRE Report, 2003.)

(F1)

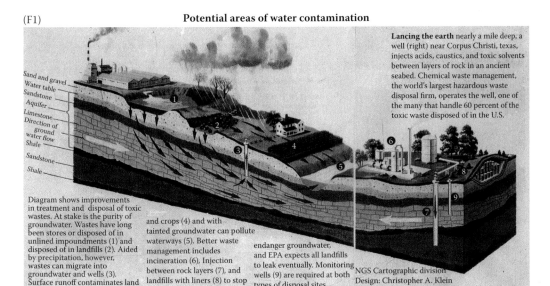

Potential areas of water contamination

Lancing the earth nearly a mile deep, a well (right) near Corpus Christi, texas, injects acids, caustics, and toxic solvents between layers of rock in an ancient seabed. Chemical waste management, the world's largest hazardous waste disposal firm, operates the well, one of the many that handle 60 percent of the toxic waste disposed of in the U.S.

Diagram shows improvements in treatment and disposal of toxic wastes. At stake is the purity of groundwater. Wastes have long been stores or disposed of in unlined impoundments (1) and disposed of in landfills (2). Aided by precipitation, however, wastes can migrate into groundwater and wells (3). Surface runoff contaminates land and crops (4) and with tainted groundwater can pollute waterways (5). Better waste management includes incineration (6), Injection between rock layers (7), and landfills with liners (8) to stop leakage. Injection wells can endanger groundwater, and EPA expects all landfills to leak eventually. Monitoring wells (9) are required at both types of disposal sites.

National geographic March, 1985 "Hazardous waste"

NGS Cartographic division
Design: Christopher A. Klein
Research: William L. Blewett
Painting by: Pierre mion

Chlorination of water

Water	PPM $ChCl_3$	PPM $ChCl_2Br$	PPM $ChClBr$	PPM $ChBr_3$
Spring	26	3	1	10
Well	5	0	0	7
City	29,000	100	12,600	5,700

Plastic components that leach into liquids from the container vs. glass containers

This phthalate extraction study was performed to show the presence of certain phthalate compounds in plastic containers, tygon tubing, and our own phthalate kit glass bottle.

Five different samples were prepared:

1. Tall plastic bottle 100 ml WL-1. Used for drugs of abuse urine collection (1).
2. Short plastic bottle 60 ml graduated, sterile royaline - used for drugs of abuse urine collection (2).
3. Kit plastic bag: speci-grad SPA - R/J - 94. Used to ship our drug and toxic kits. Pat# 4,932,791.
4. Glass bottle (phthalate kit) 30 ml amber glass/green lid. Used for phthalate, organophosphate and herbicides urine collection.
5. Tygon plastic tubing: formulation: R-3603, I.D. = 1/8, 16 GB2 O.D.= 1/4. 14-169- IE lot # LC007U. A 1 1/2 inch piece cut and placed in urine in a glass culture tube and shaken for about 3 hrs.

Contamination by containers

Blank urine was added to samples 1–4 and shaken gently for a period of a period of approximately 18 hr. The samples were then prepared as written in the toxics' SOP. A2 L aliquot was analyzed by GS/MSD #4 in SIM mode. The results are shown below.

Phthalate compound	1. Tall plastic bottle	2. Short plastic bottle	3. Kit plastic bottle	4. Amber glass bottle	5. Tygon plastic tubing
Dimethylphalate	18.8	—	25.6	—	30.9
Diethylphthalate	270.8	—	45.1	—	10.6
Dibutylphthalate	—	74.8	53.5	—	29.3
Butylbenzylphthalate	—	—	—	—	—
Di-2-ethylhexylphthalate	—	—	—	—	162.7
Di - Octylphthalate	—	—	—	—	—

Analyst: MYS   All values in mg/ml (ppb)
Source: Laseter, J.L. 1995. Personal communication

FIGURE 1.45F1 Pollutants of water. (*National Geographic*, "Hazardous Waste," March 1985; From EHC-Dallas, 2002; Laseter, J.L., Plastic versus Glass Containers. Personal communication.)

(G)

The gulf of Mexico's dead zone
Reconstructing an 180-yr record of natural and anthropogenic induced
hypoxia from the sediments of the Louisiana continental shelf
Science Daily
http://www.sciencedaily.com/releases/2008/07/080714160000.htm
University of Michigan (2008, July 17). Record-setting Dead Zones Predicted for Gulf of Mexico, Chesapeake Bay.
Science Daily. Retrieved November 18, 2009, from http://www.sciencedaily.com/releases/2008/07/080714160000.htm

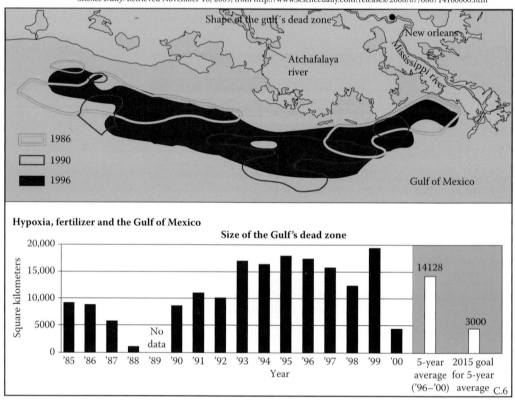

FIGURE 1.45G Toxins in foods. (From Pangborn, J.B. and Smith, B., On Man and His Environment in Health and Disease, Presented at the 13th Annu. Int. Symp., Dallas, Feb. 23, 1995.)

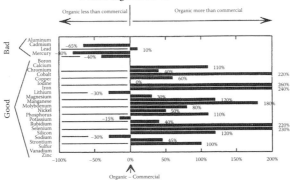

FIGURE 1.45H Toxins in foods.

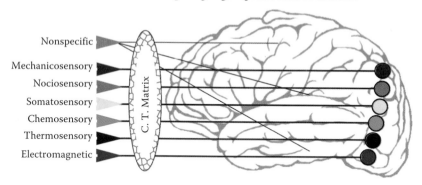

FIGURE 1.46 Receptors to the brain. Most receptors have connective tissue matrix as part of their anatomy. (From EHC-Dallas.)

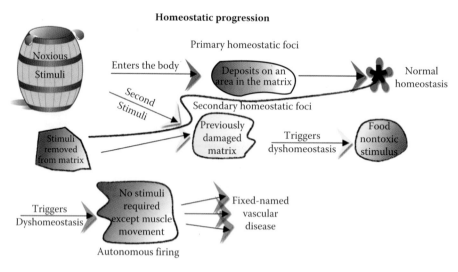

FIGURE 1.64 Primary and secondary foci. Homeostatic triggering homeostasis and then dyshomeostasis. (From EHC-Dallas, 2002.)

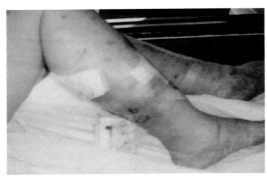
4+ Edema and ulcerated legs before treatment

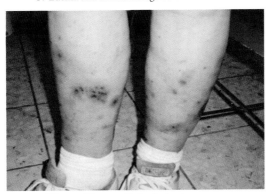
Ulcerated legs totally healed, edema gone
after 3 months of treatment

FIGURE 1.68 Ulcerated legs healed after massive avoidance of pollutants and food. Other treatment included injection therapy with molds, chemicals, and foods to which she was sensitive; intravenous and oral nutrient therapy; sauna/physical therapy; oxygen therapy, and use of autogenous lymphocytic factor. (From EHC-Dallas, 2004.)

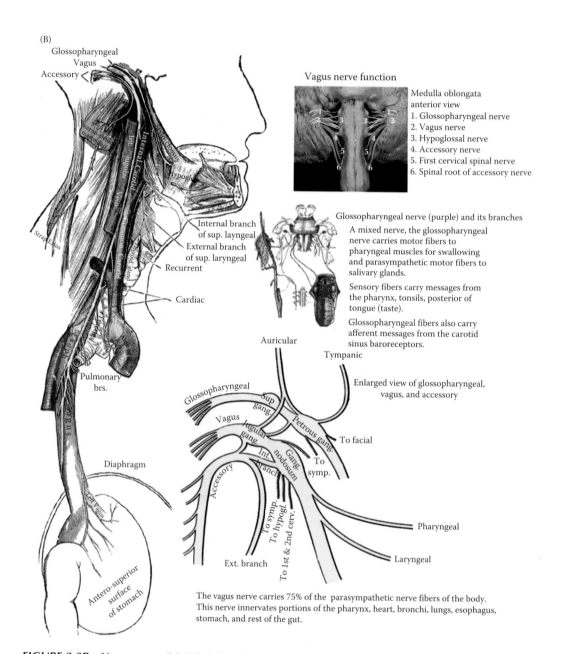

FIGURE 2.3B Vagus nerve. (Modified from Gray, H., *Anatomy of the Human Body*, 1918; Medulla oblongato anterior view from *The Anatomy Project*, Pathenon Publishing Group, 1997.)

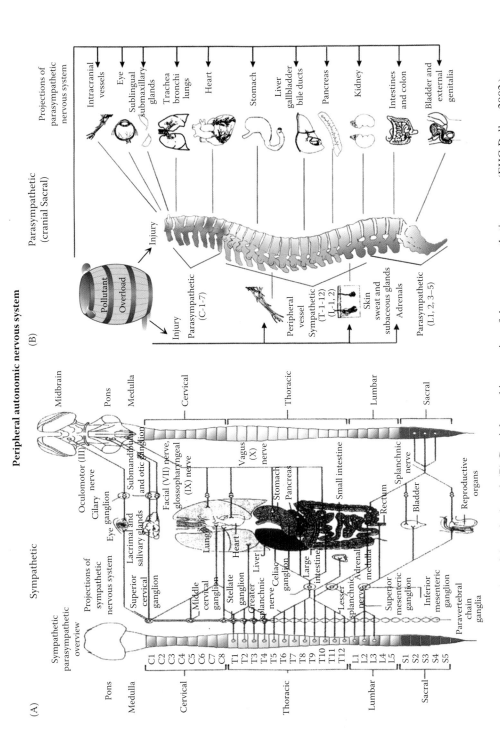

FIGURE 2.6 (A) Sympathetic innervation of the spinal cord. (B) Cranial-sacral innervation of the parasympathetic nerves. (EHC-Dallas, 2002.)

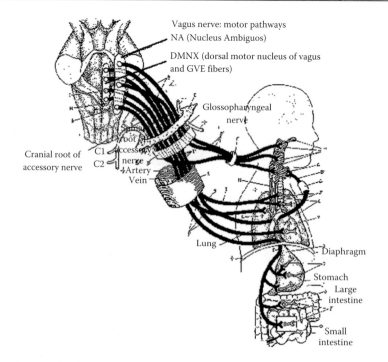

FIGURE 2.7 Routes of reflex mechanisms that may be triggered after pollutant injury in the chemically sensitive and chronic degenerative diseased individual. Responses may be isolated to each single organ, i.e., gallbladder or appendix, or they may be generalized. (EHC-Dallas; Buffalo, 2002; Rea, W.J., *Chemical Sensitivity, Vol. III, Clinical Manifestatons of Pollutant Overload.* Boca Raton, FL: CRC Lewis Publishers. 1735, 1996. With permission.)

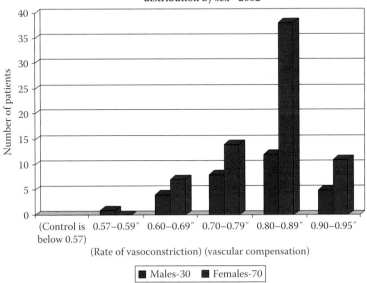

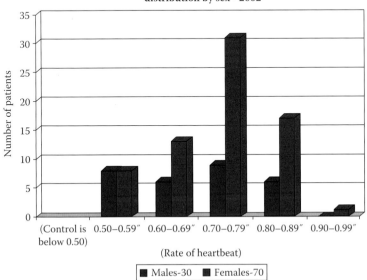

FIGURE 2.14 Patients vascular compensation and chronotrophic responses are lost with severe autonomic stimulation as measured elevation from the supine to the upright position as measured by the heart rate variability machine by the method of Riftine.[16] (EHC-Dallas, 2002.)

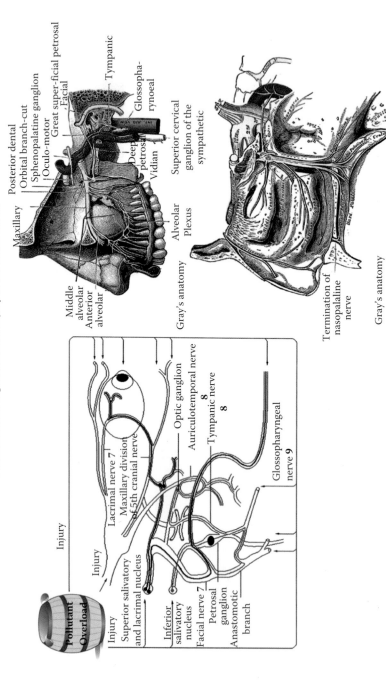

FIGURE 2.16 Potential areas of pollutant injury of the ocular nerves and ocular autonomic nerves showing alveolar branches of superior maxillary nerve and sphenopalatine ganglion of the trigeminal nerve. (Modified from Rea, W.J., *Chemical Sensitivity, Vol. III: Clinical Manifestations of Pollutant Overload*, CRC Press, Boca Raton, FL, 1996; Gray, H., *Anatomy of the Human Body*, Lea & Febiger, Philadelphia, 1918, New York, www.bartleby.com. 2000.)

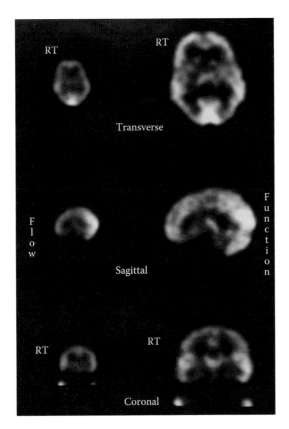

FIGURE 2.23 Triple-headed SPECT camera brain scan of individual with breast implant. (From EHC-Dallas, 2005.)

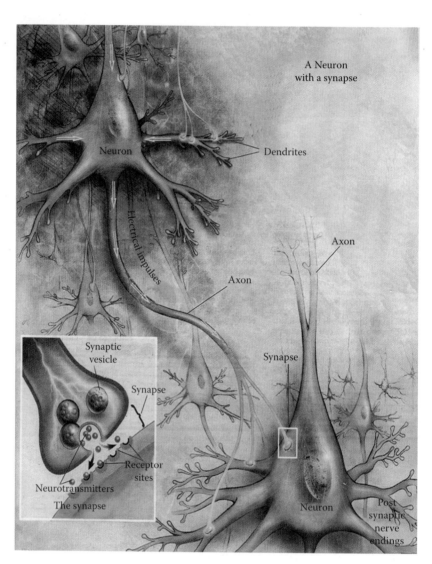

FIGURE 2.36 A neuron with a synapse. (Courtesy of ADEAR.)

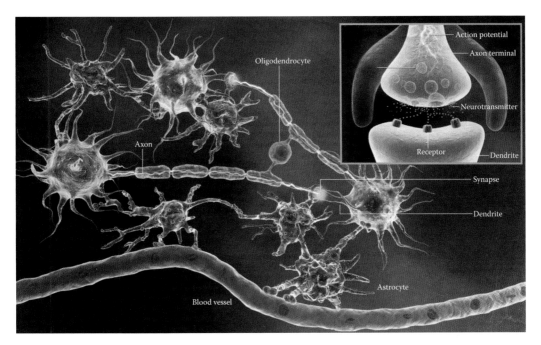

FIGURE 2.39 Glia and neurons work together in the brain and spinal cord. A neuron sends a message down a long axon and across a synaptic gap to a dendrite on another neuron. Astrocyte glia bring nutrients to neurons, as well as, surround and regulate synapses. Oligodendrocyte glia produce myelin that insulates axons. When a neuron's electrical message [action potential] reaches the axon terminal (*insert*), the message induces vesicles to move to the membrane and open, releasing neurotransmitters [signaling molecules] that diffuse across a narrow synaptic cleft to the dendrite's receptors. Similar principles apply in the body's peripheral nervous system, where Schwann cells perform myelination duties. (Fields, D.R., *Scientific American*, 57, 2004. With permission.)

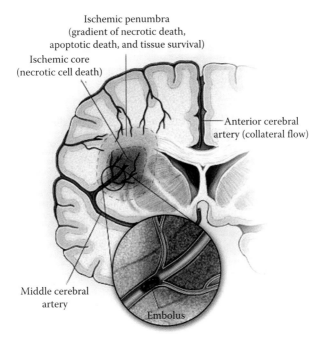

FIGURE 2.41 An embolus in the bifurcation of the middle cerebral artery. The territory perfusion by this artery and areas with little or no collateral flow are subjected to extreme hypoxia and necrotic cell death. In the penumbra, where there is some degree of collateral blood flow, a gradient of tissue perfusion establishes a threshold among necrotic cell death, apoptotic cell death, and tissue survival. (Friedlander, R.M., *N Engl J Med.*, 348(14), 1366, 2003. With permission.)

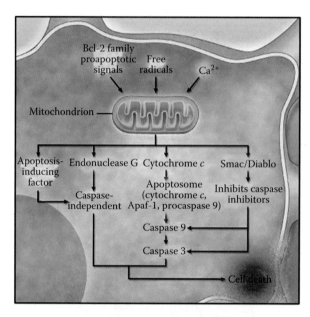

FIGURE 2.42 Key mediators of the caspase pathway in the mitochondria. Three main signals cause the release of apoptogenic mitochondrial mediators: proapoptotic members of the Bcl-2 family, elevated levels of intracellular calcium, and reactive oxygen species. Four mitochondrial molecules mediating downstream cell-death pathways have been identified: cytochrome *c*, Smac/Diablo, apoptosis-inducing factor, and endonuclease G. Cytochrome *c* binds to Apaf-1, which, together with procaspase 9, forms the "apoptosome," which activates caspase 9. In turn, caspase 9 activates caspase3. Smac/Diablo binds to inhibitors of activated caspases and causes further caspase activation. Apoptosis-inducing factor and endonuclease G mediate caspase-independent cell-death pathways. (Friedlander, R.M., *N Engl J Med.*, 348(14), 1368, 2003. With permission.)

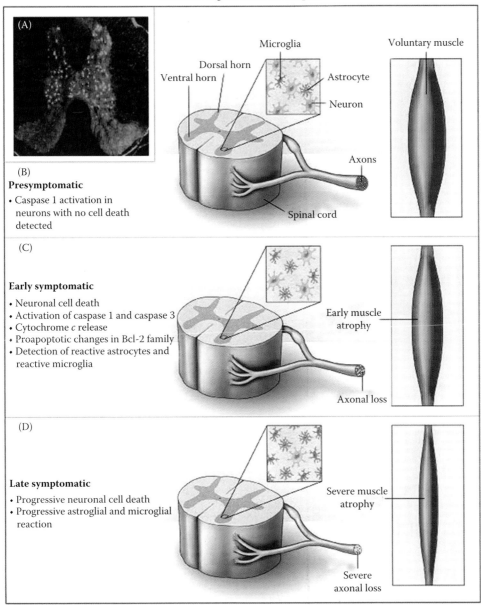

FIGURE 2.43 Neurologic lesions in mice with ALS. Panel A shows activation of neuronal caspase 1 in an axial section of the spinal cord of a 90-day-old mouse with ALS (immunostained with a caspase 1 antibody). At this age, the mouse is at the beginning or the middle of the symptomatic stage. There is no caspase 1 activation in the dorsal horn or in the white matter. In the presymptomatic stage (Panel B), the earliest cell-death signal detected is the activation of neuronal caspase 1. At this stage, there are no overt signs of cell death or strong tissue reaction. In the early symptomatic stage (Panel C), there is widespread activation of caspase 1 and caspase3, release of cytochrome c, and proapoptotic changes in Bcl-2 family members. Ventral motor neurons and axons die, and reactive microgliosis and astrocytosis are present. As the disease advances, the findings described above become more overt (Panel D) and are accompanied by progressive muscle atrophy. (Friedlander, R.M., *N Engl J Med.*, 348(14), 1370, 2003. With permission.)

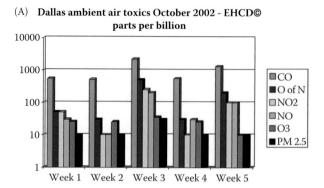

FIGURE 2.45A Dallas Air, October 2002.

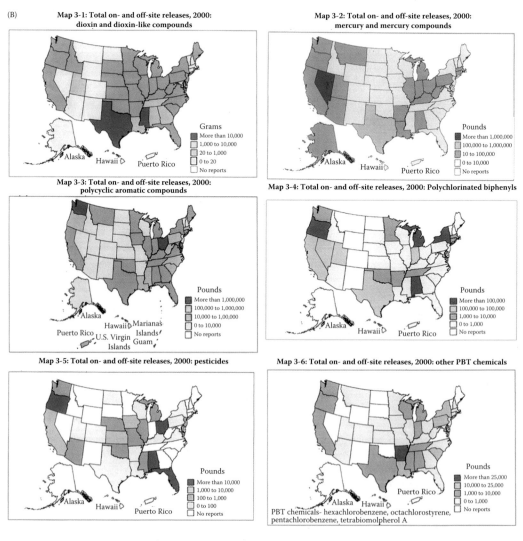

FIGURE 2.45B Total on and off site releases 2000: Map 3-1-Dioxin and Dioxin-like compounds, Map 3-2-mercury and mercury compounds, Map 3-3-polycyclic aromatic compounds, Map 3-4-polychlorinated biphenyls, Map 3-5-pesticides, Map 3-6-other PBT chemicals. (Environmental Protection Agency)

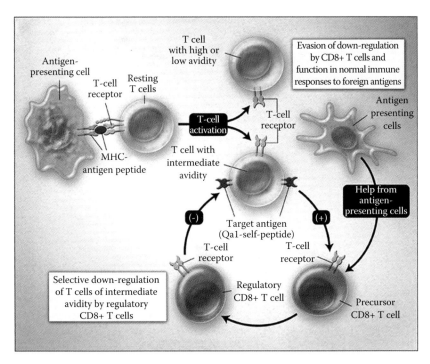

FIGURE 3.13 Regulatory pathway mediated by Qa1-dependent CD8+ T-cells. (Jiang, H., and Chess, L., *NEJM,* 354(11), 1166–1176, 2006. With permission.)

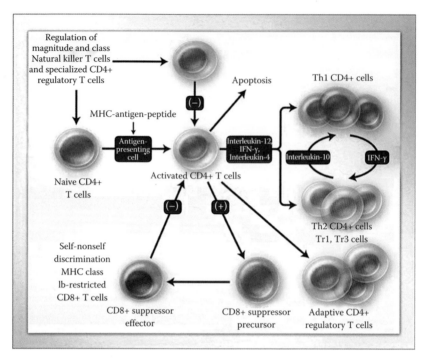

FIGURE 3.14 Control of activated T-cells by suppressor T-cells. (Jiang, H., and Chess, L., *NEJM,* 354(11), 1166–1176, 2006. With permission.)

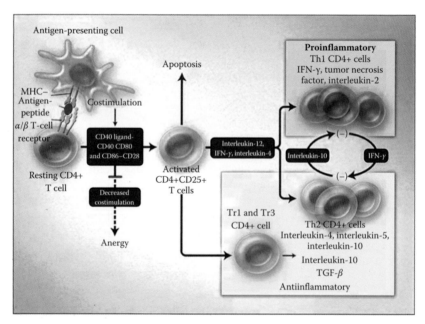

FIGURE 3.15 Control of antigen-induced activation of T-cells by intrinsic mechanisms. (Jiang, H., and Chess, L., *NEJM,* 354(11), 1166–1176, 2006. With permission.)

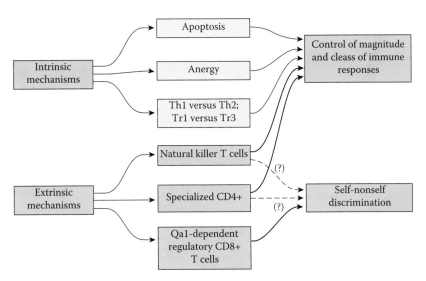

FIGURE 3.16 Peripheral regulatory mechanisms. (Jiang, H., and Chess, L., *NEJM,* 354(11), 1166–1176, 2006. With permission.)

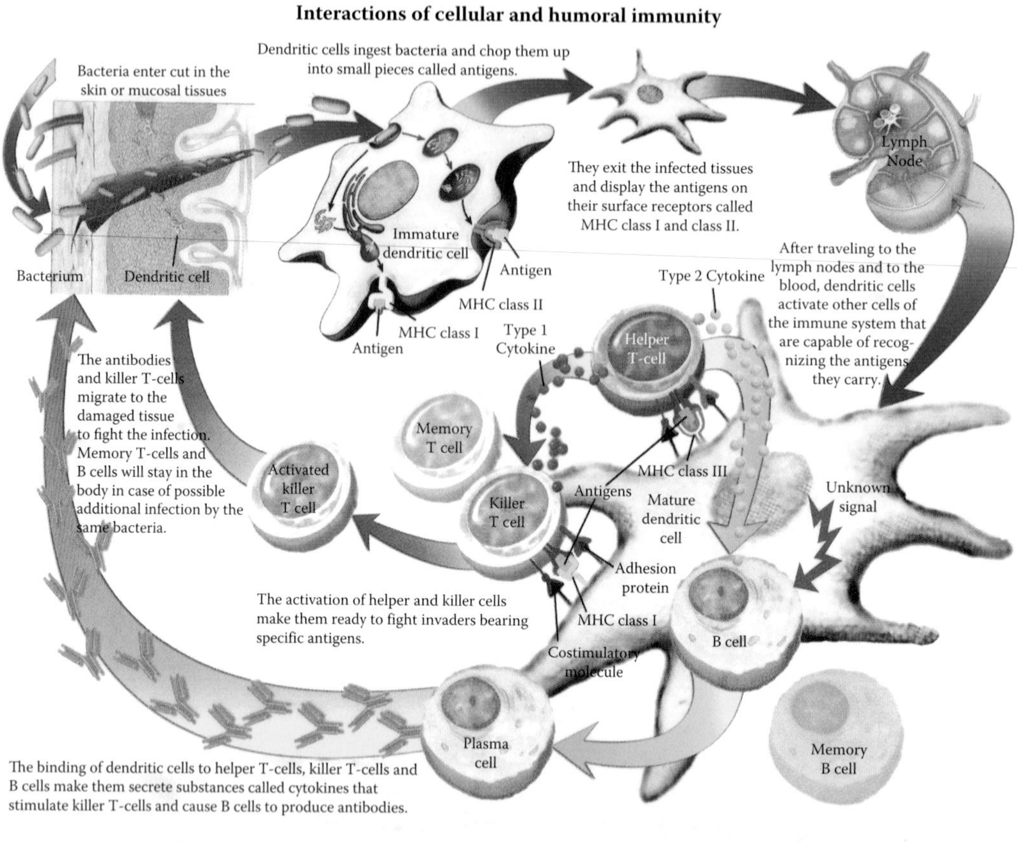

FIGURE 3.21 Interactions of cellular and humoral immunity as defense against invaders. (Immunoscience Lab., Inc. 8693 Wilshire Blvd., Ste. 200, Beverly Hills, CA 90211. http://www.immunoscienceslab.com 2004. With permission.)

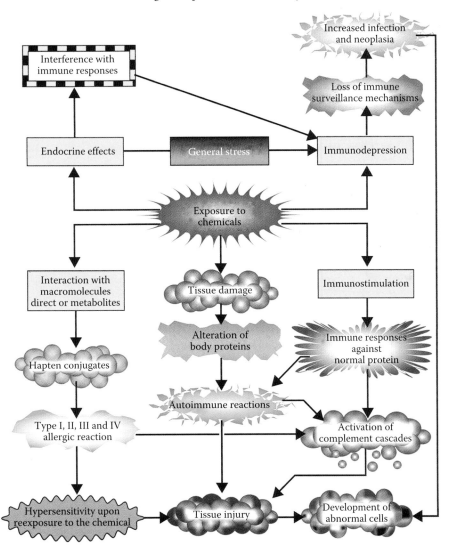

FIGURE 3.22 Toxicological response in the immune system produced by chemicals. (Immunoscience Lab, Inc. Toxicological Response in the Immune System Produced by Chemicals. http://www.immunoscienceslab.com 2004. With permission.)

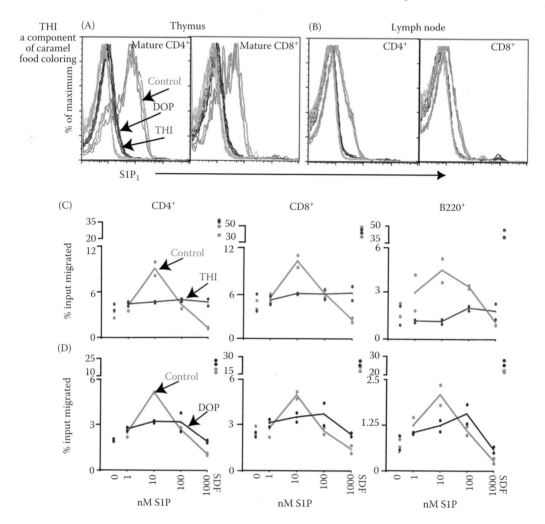

FIGURE 3.28 THI and DOP treatment lead to reduced $S1P_1$ expression and impaired ability to migrate to S1P. (A) Mature single-positive thymocytes and (B) peripheral lymph node cells were stained for $S1P_1$ expression. Green lines show staining of cells from three control-treated mice, blue from three DOP-treated mice, and red from three THI-treated mice. Gray lines show staining of all nine cell types with a control antibody. Staining is representative of at least three experiments. (C and D) Splenocytes from THI- (red), DOP- (blue), or control- (green) treated mice were tested for their ability to migrate to the indicated concentrations of S1P or to 0.3 μg of stromal-cell derived factor (CXCL12) per ml of medium. Individual points show migration in replicate wells. When data from three experiments were compiled, the impaired migration of $CD4^+$ and $CD8^+$ T-cells from DOP- and THI-treated mice to 10 nM S1P was statistically significant ($P < 0.05$). The impaired migration of $B220^+$ cells was significant for THI ($P = 0.05$) but not for DOP. (Schwab, S. R., Pereira, J. P., Matloubian, M., Xu, Y., Huang, Y., and Cyster, J. G., *Science*, 309, 1736, Fig. 2, 2005. With permission.)

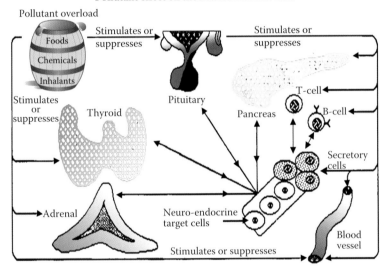

FIGURE 4.5 Pollutant effect on the neuroendocrine cells in the chemically sensitive individual. Note: Pollutants disturb (1) self-regulation; (2) regional regulation; (3) distal regulation; (4) regulation of microcirculation; and (5) regulation of the immune system. (Modified from Rea, W. J., *Chemical Sensitivity, Vol. III: Clinical Manifestations of Pollutant Overload,* CRC Press, Boca Raton, FL, 1607, Fig. 24.2, 1995.)

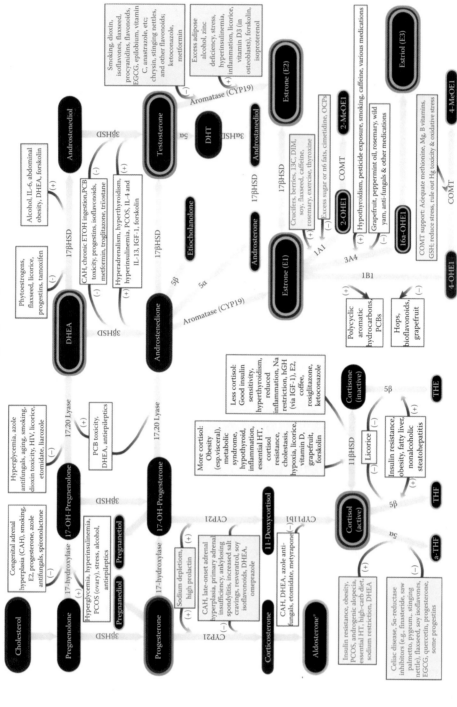

FIGURE 4.12 Steroidogenic pathways. (Courtesy of Genova Diagnostics.)

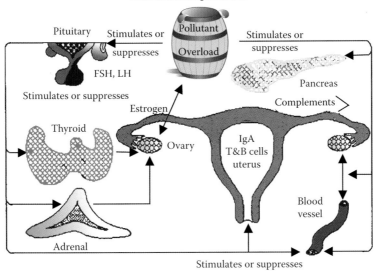

FIGURE 4.16 Immune and endocrine systems' defense of the uterus against environmental pollutants. (Modified from Rea, W. J., *Chemical Sensitivity, Vol III: Clinical Manifestations of Pollutant Overload*, CRC Press, Boca Raton, FL, 1638, Fig. 24.7, 1996.)

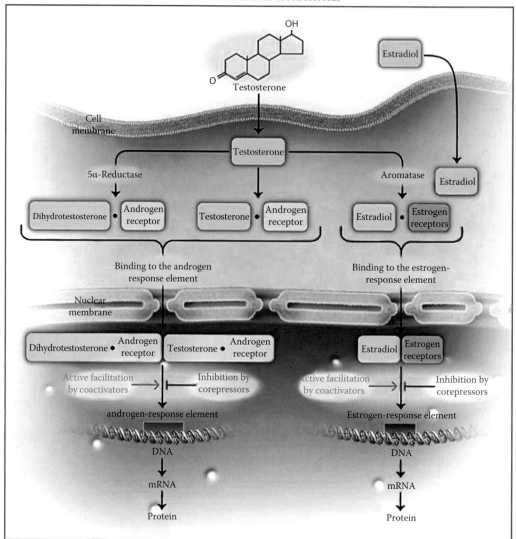

FIGURE 4.17 Intracellular actions of sex steroids. Testosterone enters the cell, some of which is reduced by 5α-reductase to dihydrotestosterone. Both hormones then bind to the androgen receptor, with dihydrotestosterone having 10 times the binding affinity of testosterone. The resulting ligand–receptor complex then attaches to the androgen-response element—a step facilitated by coactivators and inhibited by corepressors—leading to the transcription of messenger RNA (mRNA), protein synthesis, and androgen effects. Some testosterone entering the cell is aromatized to estradiol through the action of aromatase. Estradiol then binds to its specific receptor. The resulting ligand–receptor complex attaches to the estrogen-response element of chromatin—a step facilitated by coactivators and inhibited by corepressors—leading to mRNA transcription, protein synthesis, and estrogen effects. Each step in the process provides opportunities for the introduction of sex-based differences. (Federman, D. D., *NEJM*, 354(14), 1507–1514, Fig. 2, 2006. With permission.)

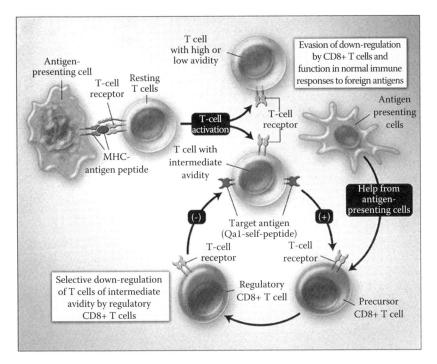

FIGURE 3.13 (See color insert following page 364.) Regulatory pathway mediated by Qa1-dependent CD8+ T-cells. (Jiang, H., and Chess, L., *NEJM*, 354(11), 1166–1176, 2006. With permission.)

Other receptor–ligand interactions are also pivotal. One of them involves CD40 ligand, a cell-surface molecule that makes an early appearance on activated T-cells.[103] It is essential for T-cell–induced antibody formation by B-cells and for causing antigen-presenting cells (e.g., dendritic cells) to trigger cell-mediated immune responses. The interaction of CD40 ligand with CD40, its receptor on B-cells and dendritic cells, causes up-regulation of two surface proteins on these cells, CD80 and CD86.[104,105] **When CD80 and CD86 interact with CD28 on T-cells, the outcome is T-cell activation**; by contrast, an interaction with cytotoxic T-lymphocyte–associated antigen 4 (CTLA-4) on T-cells causes anergy (i.e., a nonspecific state of inactivation) or immune tolerance (i.e., an antigen-specific state of inactivation). **Blockade of the CD80–CD28 or CD40 ligand–CD40 pathways induces anergy**[106–108] **or tolerance, whereas blockade of CTLA-4 enhances the immune response.**[109]

Homeostatic control mechanisms are induced during the initial activation of all T-cells when the T-cell receptor engages complexes of MHC and antigen peptide. T-cell activation depends on the avidity and duration of binding of the receptor to MHC–antigen-peptide complexes,[102,100] as well as on costimulatory molecules (e.g., CD40 ligand, CD28, and cytotoxic T-lymphocyte–associated antigen 4). These initial events influence the outgrowth, function, and death of T-cells, as well as their differentiation into subgroups of helper T (Th) or regulatory T (Tr) cells that secrete distinct arrays of regulatory cytokines. These control mechanisms function in the absence of suppressor cells. IFN denotes interferon, and TGF transforming growth factor.

Superimposed on the intrinsic homeostatic mechanisms are pathways of immunoregulation mediated by subgroups of suppressor T-cells, including natural killer, CD4+, and CD8+ T-cells. Each of the regulatory T-cell subgroups expresses distinct receptors, uses different effector mechanisms, and functions predominantly at different stages during the course of the peripheral immune response. **The natural killer T-cells and specialized CD4+ regulatory T-cells are involved**

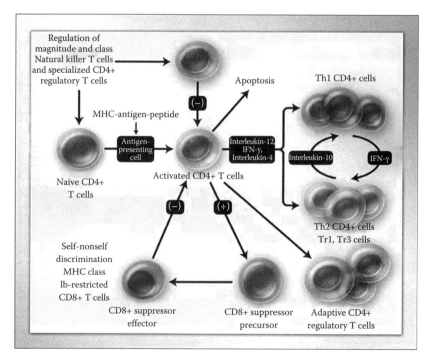

FIGURE 3.14 (See color insert following page 364.) Control of activated T-cells by suppressor T-cells. (Jiang, H., and Chess, L., *NEJM*, 354(11), 1166–1176, 2006. With permission.)

mainly in controlling the magnitude and class of immune responses, whereas CD8+ regulatory T-cells are involved in self–nonself discrimination. MHC denotes major histocompatibility complex, TFN interferon, TGF transforming growth factor, Th helper T-cells, and Tr regulatory T-cells. See Figure 3.15.

The regulatory pathway mediated by Qa-1-dependent CD8+ T-cells is initiated by the antigen-induced activation of T-cells during the primary immune response. One of the consequences of initial T-cell activation is the differential expression of certain Qa-1–self-peptide or HLA-E–self-peptide complexes, expressed as a function of the avidity of the T-cells during initial T-cell activation. These self-peptide complexes, preferentially expressed on the activated T-cells with intermediate avidity, serve as the target structures recognized by regulatory CD8+ T-cells. These target structures trigger the regulatory CD8+ T-cells to differentiate into suppressor cells, which in turn specifically down-regulate activated T-cells with intermediate avidity that express the same target structure during the secondary immune response.[116] See Figure 3.16.

The regulatory pathway mediated by Qa-1-dependent CD_8 and T-cells is initiated by the antigen-induced activation of T-cells during the primary immune response. They target functionally triggering regulating CD_8 and T-cells to differentiate into suppressor cells, which in turn down regulate activated T-cells with intermediate fluidity that express the same target structure decreasing the secondary immune response.

The controls for immunity and prevention of autoimmunity[110] depend upon the shaping of the T-cell repertoire in the thymus and regulation of T-cells in the periphery.

The peripheral regulatory mechanism apparently has a key role in the immunity response and is based on the two concepts of homeostasis and the potential autoimmunity. The immune system is a homeostatic organization that must regulate itself to avert insufficient immunity and suppress

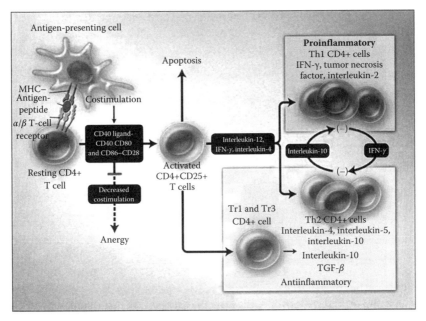

FIGURE 3.15 (See color insert following page 364.) Control of antigen-induced activation of T-cells by intrinsic mechanisms. (Jiang, H., and Chess, L., *NEJM*, 354(11), 1166–1176, 2006. With permission.)

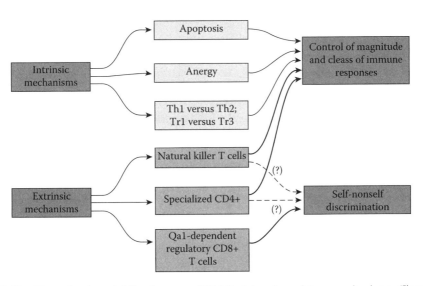

FIGURE 3.16 (See color insert following page 364.) Peripheral regulatory mechanisms. (Jiang, H., and Chess, L., *NEJM*, 354(11), 1166–1176, 2006. With permission.)

excessive immune responses. Protective immunity has a considerable potential for error because it entails the production of potent proinflammatory molecules and killer cells that can destroy not only the invading microorganisms and cancer cells but also normal cells. **To ensure effective immunity to foreign antigens but avert pathogenic autoimmunity in the periphery, the immune system must control the magnitude and class of the immune response but also discriminate self from nonself**. The two immunoregulaory mechanisms have a direct clinical relevance to hypersensitivities and reactions, responses to pathogens, and autoimmunity. One frequently sees responses with suppression of the CD_8 T-cells in the patient with chemical sensitivity. However, one also sees augmentation of the killer cells from 2 to 100% overcontrol levels. Certainly, the magnitude and class of immunity is influenced in these chemically sensitive patients but the control of immunity versus autoimmunity is maintained. Ocassionaly one will see antinuclear antibodies in the range of 1:40–1/640, but autoimmunity never seems to develop in these patients with chemical sensitivity. Frequently low with blood counts do occur in the patients with chemical sensitivity probably through these mechanisms.

B-LYMPHOCYTES

The liver and bone marrow preprocess the B-lymphocytes. However, much less is known about the details for preprocessing B-lymphocytes than those for preprocessing T-lymphocytes. In the human being, B-lymphocytes are known to be preprocessed in the liver during mid fetal life and in the bone marrow during late fetal life or after birth.[111]

B-lymphocytes are different from T-lymphocytes in two ways: First, instead of the whole cell developing reactivity against the antigen, as occurs for the T-lymphocytes, **the B-lymphocytes actively secrete antibodies that are the reactive agents. These agents are large protein molecules that are capable of combining with and destroying the antigenic agent. Second, the B-lymphocytes have even greater diversity than the T-lymphocytes, thus forming many, many millions, and perhaps even billions of types of B-lymphocyte antibodies with different specific reactivities.**[111] This forming of antibodies may occur in excess or be deficient in hypersensitivity or chronic degenerative disease when the individual becomes supersensitive to food, pollens, molds, and many chemicals and/or when the individual is nutrient depleted.

After preprocessing, the B-lymphocytes, like the T-lymphocytes, migrate to the lymphoid tissue throughout the body where they lodge near to, but slightly removed from the T-lymphocyte areas. When a specific antigen encounters the T and B-lymphocytes in the lymphoid tissue, certain of the T-lymphocytes become activated to form activated T-cells and some specific B-lymphocytes form antibodies. The activated T-cells and antibodies, in turn, react highly specifically against the particular type of antigen that initiated their development. The mechanism of this specificity follows.

Millions of specific lymphocytes are preformed in the lymphoid tissue. When stimulated by the appropriate antigens, millions of different types of preformed B-lymphocytes and preformed T-lymphocytes are capable of forming highly specific antibodies or sensitized T-cells. Each one of these preformed lymphocytes is capable of forming only one type of antibody or one type of T-cell with a single type of specificity. Only the specific type of antigen with which it can react can activate it. Once its antigen activates the specific lymphocyte, it reproduces wildly, forming tremendous numbers of cloned lymphocytes. If exposure is chronic, nutrient drain will occur in the T- and B-cells reproduction, often causing nutrient deficiency and forming a poor response to other antigens. **If it is a B-lymphocyte, its progeny will eventually secrete antibodies that then circulate throughout the body. If it is a T-lymphocyte, its progeny are sensitized T-cells that are released into the lymph and carried to the blood and then circulated through all the tissue fluids and back into the lymph**, sometimes circulating around and around in this circuit for months or years.[111] Often the chemically sensitive patient has low suppressor T-cells. In our experience, as high as 90% of the hypersensitive patients have either low suppressor T-cells or low cell-mediated immunity skin tests for T-cell function. This suggests that there is a toxic overload with immune dysfunction that suppresses these sensitized T-cells thus allowing an imbalance in the helper to suppressor cell ratio. Once the imbalance occurs the patient becomes hypersensitive and dysfunctional.

All of those different T-lymphocytes are capable of forming one specific cell type or sensitized T-cell (a clone of lymphocyte), which signifies the lymphocytes in each clone are alike and are derived originally from one or a few early lymphocytes of its specific type.[111]

Only several hundred to a thousand genes code for the different types of antibodies and T-lymphocytes. There are millions of different specificities of antibody molecules or T-cells that can be produced by the lymphoid tissue. The whole gene for forming each type of T- cell or B-cell is never present in the original stem cells from which the functional immune cells are formed; instead, there are only multiple "gene segments"—actually, hundreds of such segments. During preprocessing of the respective T- and B-cell lymphocytes, these gene segments become mixed up with one another in random combination, in this way finally forming the whole genes. Due to the several hundred types of gene segments, as well as the millions of different orders in which the segments can be arranged in single cells, one can understand why millions or even billions of different cell gene types can occur. For each functional T- or B-lymphocyte that is finally formed, the gene structure codes for only a single antigen specificity. These mature cells then become the highly specific T- and B-cells that spread to and populate the lymphoid tissue.[111] **In the chemically sensitive patient many different chemicals may sensitize the T-cells, thus allowing for the rapid and "hair trigger" response one sees**.

Each clone of lymphocytes is responsive to only a single type of antigen (or to similar antigens that have almost exactly the same stereochemical characteristics). The reason for this is the following: In the case of the B-lymphocytes, each one has on the surface of its cell membrane about 100,000 antibody molecules that will react highly specifically with only the one specific type of antigen. Therefore, when the appropriate antigen comes along, it immediately attaches to the cell membrane; this leads to the activation process. In this case of the T-lymphocytes, molecules similar to antibodies, surface receptor proteins (or T-cell markers) are on the surface of the T-cell membrane, and these, too, are highly specific for the one specified activating antigen.[111] **One often sees a spreading phenomena occur when the patient may initially become sensitive to one substance**, i.e., formaldehyde. Later, he becomes sensitive to cousins of formaldehyde, i.e., acetaldyde, paraldehyde, etc. Then, as the load increases, the patient becomes sensitive to breakdown products, i.e., alcohol or other nonrelated substances like phenol, glycol, etc. Eventually, as the sensitization spreads, nonrelated substances like foods and molds become involved. This spreading phenomenon may be due to a loss of recognition of antigens and/or changes in regulatory mechanisms.

Aside from the lymphocytes in lymphoid tissue, literally millions of macrophages are present in the same tissue. These line the sinusoids of the lymph nodes, spleen, and other lymphoid tissue, and they lie in apposition to many of the lymph node lymphocytes. Most invading organisms are first phagocytized and partially digested by the macrophages, and the antigenic product is liberated into the macrophage cytosol. The macrophages then pass these antigens by cell-to-cell contact directly to the lymphocytes, thus leading to activation of the specified clones. The macrophages, in addition, secrete a special activating substance that promotes the growth and reproduction of the specific lymphocytes.[111]

Most antigens activate both T-lymphocytes and B-lymphocytes at the same time. Some of the T-cells that are formed are helper cells; they, in turn, secrete specific substances (lymphokines) that further activate the B-lymphocytes. Indeed, without the help of these T-cells, the quantity of antibodies formed by the B-lymphocytes is usually slight.[111] At times even though there is a large load exposure, as for example molds or chemicals, one sees a limited antibody response even though T-cells are sensitized (Table 3.6).

Before exposure to a specific antigen, the clones of B-lymphocytes remain dormant in the lymphoid tissue. On entry of a foreign antigen, the macrophages in the lymphoid tissue phagocytize the antigen and then present it to the adjacent B-lymphocytes. **In the normal progression of events one often sees a subset of chemically sensitive patients who have a poor ability to phagocytize and destroy the noxious stimuli as shown in Chapter 1**. Either the bacteria or fungus may not be destroyed easily or totally resulting in chronic recurring infections. **Therefore, the inappropriate antigen may be presented to the B-lymphocytes and T-cells, which then result in long-term inappropriate responses**. The antigen is presented to T-cells at the same time, and activated "helper" T-cells then, also, contribute to the activation of the B-lymphocytes. (See Figure 3.10

TABLE 3.6
Blood Mold Antibody Assay (N = 100 Patients)*

Mold or Mycotoxin	IgA Elevated (n)	IgA Tested (n)	IgA %	IgG Elevated (n)	IgG Tested (n)	IgG %	IgM Elevated (n)	IgM Tested (n)	IgM %	IgE Elevated (n)	IgE Tested (n)	IgE %	Nonsensitve	Controls
Aspergillus fumigatus	4	21	19	9	21	43	5	21	24	4	21	19	0	1600
Aspergillus niger	23	82	28	22	82	27	25	82	30	10	32	12	0	1600
Aspergillus veriscolo	6	17	35	6	17	35	2	17	12	2	17	12	0	1600
Alternaria tenuis	4	48	8	19	48	40	14	48	29	6	48	13	0	1600
Chaetomium globosum	11	42	26	16	42	38	15	42	36	7	42	17	0	1600
Chrysagem	0	1	0	0	1	0	0	1	0	0	1	0	0	1600
Cladosporium herbarum	12	67	18	25	67	37	20	67	30	8	67	12	0	1600
Epicoccum nigrum	7	26	27	25	67	37	20	67	30	8	67	12	0	1600
Fumigatus	0	2	0	0	2	0	0	2	0	0	2	0	0	1600
Fusarium nachilform	0	2	0	0	2	0	0	2	0	0	2	0	0	1600
Geotrichum candidiolum	0	6	0	2	6	33	1	6	17	0	6	0	0	1600
Penicillium notatum	10	84	12	35	84	42	19	84	23	9	84	11	0	1600
Rhizopus nigrans	2	9	22	7	9	78	6	9	67	1	9	11	0	1600
Stachybotrys chartarum	16	85	19	35	85	41	28	85	33	13	85	15	0	1600
Aflatoxins	26	92	28	26	92	28	35	92	38	15	92	16	0	1600
Satratoxins	21	88	24	27	88	31	26	88	30	13	88	15	0	1600
Trichothecene	32	88	36	38	88	43	27	88	31	10	88	11	0	1600

Source: From EHC-Dallas, 2003.

Notes: Immunoglobulin (Ig) E was significantly lower than IgG, IgM, IgG, and IgM. Ig levels determined by Vodjani, Immunosciences Laboratores, Beverly Hills, California.

*Not all individuals had the same assay, given the differences in findings in each home.

again.) Those B-lymphocytes specific for the antigen immediately enlarge and take on the appearance of lymphoblasts. Some of the lymphoblasts further differentiate to form plasmablasts, which are the precursors of plasma cells. In these cells, the cytoplasm expands and the rough endoplasmic reticulum vastly proliferates. They then begin to divide at a rate of once every 10 hours for approximately nine divisions, giving in four days a total population of about 500 cells for each original plasmablast. **The mature plasma cell then produces gamma globulin antibodies at an extremely rapid rate—about 2000 molecules per second for each plasma cell**. The antibodies are secreted into the lymph and carried to the circulating blood. This process continues for several days or weeks until final exhaustion and death of the plasma cells occur.[111] In this aforementioned subset of chemically sensitive patients who have altered phagocytosis due to chemical exposure, we have seen low numbers of B-cells or a gammaglobulin deficiency. Though the latter is uncommon, we have supplemented the chemically sensitive patient with intramuscular or intravenous gamma globulin. The following case illustrates this phenomenon.

Case study. A 55-year-old-white female college professor of fine arts presented with the current complaints of vasculitis and left leg ulcer. Five years previous to admission this patient developed vasculitis and osteomyelitis of the left ankle after a sewer flooded her classroom. She was exposed to a large volume of hydrogen sulfide and hexane. Culture of the leg wound revealed Staphylococcal luquanensis species (coagulase negative), Entercoccus fecalis, Acinetobacter lwoffii, and diphtheroids. The patients P_vO_2 was 42mm Hg, indicating poor tissue oxygen extraction. The patient was sensitive to multiple foods, chemicals, and molds, and she was supersensitive to multiple bacteria. Her T_{11} cells were depressed but her helper and suppressor cells were normal. She had low gamma globulin. She was treated with avoidance of pollutants and was housed in a less-polluted environment. She was placed on a rotary diet, avoiding the foods she was sensitive to and drinking less-polluted spring water in glass bottles. Treatment also included intravenous and oral nutrition, specific injection therapy for biological inhalants, foods, chemicals, and bacteria to which she was sensitive. She also had 36 days of oxygen therapy for 2 hrs/day at 6 L/mm. She was given intravenous gamma globulin on a weekly and monthly basis. This treatment finally brought the vasculitis and leg ulcer under control.

In this case the patient apparently did not produce enough specific gamma globulins from her plasma cells. We suspect that the aforementioned patient had enough excess hexane in her body to suppress the gamma globulin generating mechanism and thus, was deficient in gamma globulin. This patient was supersensitive to the bacteria that we cultured from the wound, which caused adverse reactions in the tissue by creating free radicals, hypoxia, and ulcers. These reactions damaged the tissue, setting her body up for the lack of bacterial resistance that had developed since the vasculitis inception five years previously.

Some of the lymphoblasts formed by activation of a clone of B-lymphocytes do not go on to form plasma cells but instead form moderate numbers of new B-lymphocytes similar to those of the original clone. In other words, the B-cell population of the specifically activated clone becomes greatly enhanced. In addition, the new B-lymphocytes are added to the original lymphocytes of the clone. They also circulate throughout the body to repopulate all the lymphoid tissue, but immunologically, they remain dormant until activated once again by a new quantity of the same antigen (memory cells). Subsequent exposure to the same antigen will then cause a much more rapid and much more potent antibody response because there are many more memory cells than there are original B-lymphocytes of the specific clone.[111]

Once a patient reaches the hypersensitive state of chronic disease, memory may not be so beneficial since entry of the minutest amount of a substance may trigger an adverse clinical response, often seen in the odor sensitive chemically wounded individual. These patients will develop symptoms immediately within a few seconds of exposure. Reactions that follow may include arrhythmia, headache, diarrhea, bloating, wheezing, skin ulceration, and urgency to urinate. The clinical response depends on what areas of the body have been previously sensitized. Figure 3.17 shows the difference between the primary response for forming antibodies that occurs on first exposure to a specific antigen and the secondary response that occurs after a second exposure to the same

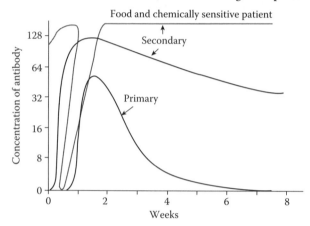

FIGURE 3.17 Time course of the antibody response in the circulating blood to a *primary* injection of antigen and to a *secondary* injection several months later.

antigen. Note the delay in the appearance of the primary response, its weak potency, and its short life. The secondary response, by contrast, begins rapidly after exposure to the antigen (often within hours,) is far more potent, and forms antibodies for many months, rather than for only a few weeks.

The increased potency and duration of the secondary response explains why vaccination is usually accomplished by injecting an antigen in multiple doses with periods of several weeks or several months between injections.[111] It must be emphasized that this long-term response often does not occur clinically in the food and chemically sensitive patient. Often, induced immunity to foods in the chemically sensitive or severely damaged patient only lasts 3–7 days. Most patients eventually respond to injection therapy and avoidance regimen with long-term relief lasting weeks to years while many others achieve permanent immunity. This phenomenon of short-term relief gained from intradermal injections seems to be related to the amount of total body pollutant load and the amount of previous damage that has already occurred. Why the immunity is initially so short-lived is unknown.

ONSET OF ANTIBODIES

The antibodies are gamma globulins (immunoglobulins) that have molecular weights between 160,000 and 970,000. They usually constitute about 20% of all the plasma protein.[111] representing a large portion of immune responses. However, one must keep in mind that other types of responses for neutralizing toxics may occur. For example, albumin, serum, and plasma can "grab on" to toxics and directly neutralize them by chemical reactions. Many patients with chemical exposure will respond to intravenous administration of these substances. Intravenous administration of gamma globulin helps some people temporarily, but rarely permanently.

All the immunoglobulins are composed of combinations of light and heavy polypeptide chains; most are a combination of two light and two heavy chains, as shown in Figure 3.18.[112] Some of the immunoglobulins, though, have combinations of as many as ten heavy and ten light chains, which give rise to the high-molecular-weight immunoglobulins. Yet, in all immunoglobulins, a light chain at one of its ends parallels each heavy chain, thus, forming a heavy-light pair. There are always at least two, and as many as ten, such pairs in each immunoglobulin molecule.[112]

Figure 3.19,[113] binding of antigen molecules to one another by bivalent antibodies, shows the individual antigen by the shaded area on a designated end of each light and heavy chain (the variable portion); the remainder of each chain is the constant portion. The variable portion is different for each specificity of antibody, and it is this portion that attaches specifically to a particular type of

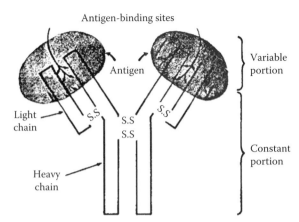

FIGURE 3.18 Structure of the typical IgG antibody, showing it to be composed of two heavy polypeptide chains and two light polypeptide chains. The antigen binds at two different sites on the variable portions of the chains. (Guyton, A. C., and Hall, J. E., *Textbook of Medical Physiology,* 9th ed., Philadelphia: WB Saunders Co., 450, Fig. 34-3 1996. With permission.)

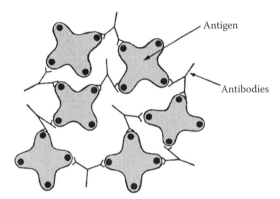

FIGURE 3.19 Binding of antigen molecules to one another by bivalent antibodies. (Guyton, A. C., and Hall, J. E., *Textbook of Medical Physiology,* 9th ed., Philadelphia: WB Saunders Co., 450, Fig. 34-4 1996. With permission.)

antigen. The constant portion of the antibody determines other properties of the antibody, establishing such factors as diffusivity of the antibody in the tissues, adherence of the antibody to specificity structures within the tissues, attachment to the complement complex, the ease with which the antibodies pass through membranes, and other biological properties of the antibody.[111]

Each antibody is specific for a particular antigen; thus, binding is caused by the antibody's unique structural organization of amino acids in the variable portions of both the light and heavy chains. The amino acid organization has a different steric shape for each antigen specificity so that when an antigen comes in contact with it, multiple prosthetic groups of the antigen fit as a mirror image with those of the antibody, thus allowing for rapid bonding between the antibody and the antigen. Then bonding is noncovalent, but when the antibody is highly specific, there are so many bonding sites that the antibody–antigen coupling is, nevertheless, exceedingly strong, and held together by (1) hydrophobic bonding; (2) hydrogen bonding; (3) ionic attractions; and (4) van der Waals forces. It also obeys the thermodynamic mass action law.[112]

Note, especially in Figure 3.19, the two variable sites on the illustrated antibody for attachment of antigens, making this type of antibody bivalent. A small proportion of the antibodies, which consist of combination of up to ten light and ten heavy chains, have as many as ten binding sites.[113]

CLASSES OF ANTIBODIES

There are five general classes of antibodies, respectively named *IgM, IgG, IgA, IgD,* and *IgE.* Ig stands for immunoglobulin.[111]

For the purpose of our present limited discussion, two of these classes of antibodies are of particular importance: **IgG, which is a bivalent antibody and constitutes about 75% of the antibodies of the normal person**, and IgE, which constitutes only a small percentage of the antibodies, but is especially involved in some forms of allergy. The IgM class is also interesting because large shares of the antibodies formed during the primary response to the incitant are of this type. These antibodies have ten binding sites that make them exceedingly effective in protecting the body against invaders, even though IgM antibodies are few.[111] IgA is the surface antibody prevalent on all mucosal surfaces (gastrointestinal, pulmonary, and genitourinary tracts). **IgA antibodies especially, are present in the gastrointestinal tract accounting for 75% of the antibody formation in this organ**. (See Gastrointestinal chapter in Volume 2 - Mechanisms of Chemical Sensitivity and Chronic Degenerative Disease for more information.)

Antibodies act mainly in two ways to protect the body against invading agents: (1) by direct attack on the invader; and (2) by activation of the complement system that then has multiple means of its own for destroying the invader.[111] When this mechanism misfires, antibodies do not seem to protect the chemically sensitive patient. Rather, antibodies seem to make patients worse. Often with chronic pollutant overload one sees depressed and/or elevated total complements, as well as its components, which of course cause local edema and pain if the complements attack the patient. **Also, in 20% of patients with early aperiodic and periodic dysfunction, one sees mildly elevated autoantibodies such as antinuclear antibodies in the range of 1:40–1:160, which usually revert to normal once the pollutant overload is reduced**. These autoantibodies are probably due to the low suppressor cells, which return to normal with treatment, restoring the balance between immunity and autoimmunity.

Figure 3.19 shows antibodies (designated by the Y-shaped bars) reacting with antigens (designated by the shaded dumbbells). Due to the bivalent nature of the antibodies and the multiple antigen sites on most invading agents, the antibodies can inactivate the invading agent by one of several ways, as follows: **Agglutination**, in which multiple large particles with antigens on their surfaces, such as bacteria or red cells, are bound together into a clump; **Precipitation**, in which the molecular complex of soluble antigen (such as tetanus toxin) and antibody become so large that it is rendered insoluble and precipitates; **Neutralization**, in which the antibodies cover the toxic sites of the antigenic agent; **Lysis**, in which some potent antibodies are occasionally capable of directly attacking membranes of cellular agents, thereby causing rupture of the cell. These direct actions of antibodies attacking the antigenic invader probably, under normal conditions, are not strong enough to play a major role in protecting the body against the invader. Most of the protection comes through the amplifying effects of the complement system described below.[111]

As stated earlier, "complement" is a collective term that describes a system of about 20 proteins, many of which are enzyme precursors. The principal actors in this system are 11 proteins designated C1 through C9, B, and D, shown in Figure 3.20.[114] All of these proteins are present normally among the plasma proteins as well as among the plasma proteins that leak out of the capillaries into the tissue spaces. The enzyme precursors are normally inactive, but they can be activated in two ways: (1) the classical pathway, and (2) the alternate pathway.[111]

Thirty-three percent of chronically ill patients treated in the hospital environmental control unit (ECU) had abnormal complement. The majority had abnormally low complements while the others had increased complements suggesting the acute phase reactions (Table 3.7).

The **classical pathway** is activated by antigen–antibody reaction, which sets in motion many steps. Once the noxious stimulus enters the body, an antibody binds with an antigen, which uncovers or activates a specific reactive site on the "constant" portion of the antibody. This process, in turn, binds directly with the C1 molecule of the complement system, setting into motion a "cascade" of sequential reactions, as shown in Figure 3.20, beginning with activation of the proenzyme C1 itself.

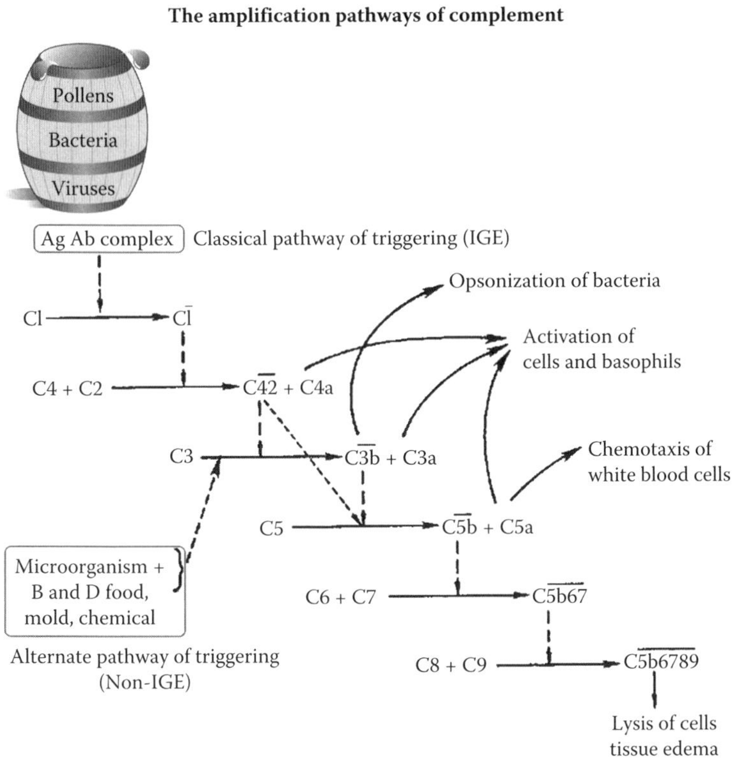

Immune System

The amplification pathways of complement

FIGURE 3.20 Cascade of reactions during activation of the classical pathway of complement. (Guyton, A. C. and Hall, J. E., *Textbook of Medical Physiology*, 9th ed., Philadelphia: WB Saunders Co., 450, Fig. 34-5, 1996. With permission.)

TABLE 3.7
Total Hemolytic Serum Complement in 100 Patients Hospitalized in the ECU

Normal Low	80–120	67%
High	0–80	23%
	120 + −	10%

This route is activated at times in the chemically sensitive patient who has an antigen–antibody reaction such as an individual who is sensitized to toluene diisocynate. If bacteria or virus are not present to attack, activation of this cascade results in tissue edema. The complement attacks the body causing tissue damage both to the cells and the extraclellular matrix. Edema may occur in any or numerous organs, but especially in the bronchial tubes or sinuses, which causes asthma or sinusitis, accompanied by recurrent wheezing, as well as recurrent nasal and sinus congestion.

Only a few antigen–antibody combinations are required to activate many molecules in the first stage of the complement system. The C1 enzymes that are formed then activate successively increasing quantities of enzymes in the later stages of the system, so that from a small beginning, an extremely large "amplified" reaction occurs, thus setting in motion a cascade of responses needed to restore the dynamics of homeostasis. Multiple end-products are formed, as shown in Figure 3.20, and several of these cause important effects that help to prevent damage by the invading organism or toxin. Among the more important actions of the complement cascade performs are the following:

(1) Opsonization and phagocytosis. One of the products of the complement cascade, C3b, strongly activates phagocytosis both in neutrophils and macrophages, causing it (C3b) to engulf either the bacteria or foreign compound to which the antigen-antibody complexes are attached. Then, if the foreign invader is bacteria, holes are punched in the walls, thus, killing the bacteria (opsonization). This process often enhances destruction of the number of bacteria by many hundredfold. **Some toxic chemicals, like hexachlorobenzene, impair phagocytosis, thus allowing infection to occur**. Similarly, if the complement cascade is depleted by excess firing, due to a chronic exposure to noxious stimuli, recurrent infection occurs due to inadequacy of this amplification response. This fact is another reason that reduction in the specific and nonspecific total body load is a significant treatment modality.

(2) **Lysis**. One of the most important of all the products of the complement cascade is the lytic complex, which is a combination of multiple complement factors and is designated C5b6789. This group of substances ruptures the cell membranes of bacteria or other invading organism. When chronic chemical exposure occurs often, innocent normal tissue will be damaged once the cascade is activated, since no microorganisms are present. If the tissue is not destroyed, the area will still be damaged, leaving the tissue vulnerable to the deposition of new entering noxious environmental incitants. With the development of an easier triggering response of the secondary homeostatic foci, the body is rendered more vulnerable to dyshomeostasis.

(3) **Agglutination**. The complement products also change the surfaces of the invading organisms, causing them to adhere to one another, thus promoting agglutination.

(4) **Neutralization of viruses**. The complement enzymes and other complement products can attack the structures of some viruses, thereby rendering them nonvirulent.

(5) **Chemotaxis**. Fragment C5a causes chemotaxis of neutrophils and macrophages, thus causing large numbers of these phagocytes to migrate into the local region of the antigenic agent. At times, this process will rapidly eliminate the foreign substance, thus restoring homeostasis. At other times these cells will release cytokines causing inflammation. When a chronic and continuous amount of noxious stimuli occurs, the patient is primed for long-standing chronic inflammation, which can lead to a sudden health disaster long-term.

(6) **Activation of mast cells and basophils**. As shown earlier in the chapter, fragments C3a, C4a, and C5a all activate mast cells and basophils, causing them to release histamine, heparin, and several other substances into the local fluids. These substances in turn cause increased local blood flow, increased leakage of fluid and plasma protein from the end blood vessels into the tissue, and other local tissue reactions that help inactivate or immobilize the antigenic agent. The same factors play a major role in inflammation. (See Gastrointestinal chapter in Volume 2 - Mechanisms of Chemical Sensitivity and Chronic Degenerative Disease for more information on the basophils.)

INFLAMMATORY EFFECTS

As discussed in Chapter 1, in addition to inflammatory effects caused by activation of the mast cells and basophils, several other complement products contribute to local inflammation. These products cause the already increased blood flow to increase still further, the capillary leakage of proteins to increase, and the proteins to coagulate in the tissue spaces, thus preventing movement of the invading organism or toxic substance through the tissue.[111] This response causes local heat and edema to occur. From studies using specific cutaneous thermography early, ridgid, depressed, poor homeostatic temperature responses occur, initiating dyshomeostatic reactions that lead to hyperextended heat responses. This sequence of events helps eliminate new tissue invading organisms but may become over or underactive when nonbiological noxious stimuli enter the body. When overactivity occurs, one sees local, regional, and sometimes central dyshomeostatic responses with mild to generalized edema. For example, often a patient with an inhaled exposure to ozone develops this mild edema syndrome. Another example is the patient who develops generalized edema after eating a food to which he is sensitive. Though the vessels leak, no inflammation occurs. Therefore, there is no heat. This effect is initially and usually neuropathic in nature.

Sometimes the alternate pathway of the complement system is activated without the intermediation of an antigen–antibody reaction. This activation occurs especially in response to large polysaccharide molecules in the cell membranes of some invading mircoorganism, or by foods, certain chemicals, or mold, and mycotoxins. These substances react with complement factors B and D, forming an activation product that activates factor C3 that then sets off the remainder of the complement cascade beyond the C3 level. Thus, essentially all the same final products of the system are formed as in the classical pathway, and these cause the same effects as those just listed to protect the body against the invader.[111] **Again, chronic activation of the complement system due to chronic pollutant overload puts a strain on the nutrient pool, resulting in a focal site tissue vulnerability to new noxious incitants.**

Since the alternate pathway does not involve an antigen–antibody reaction, it is one of the first lines of defense against invading microorganisms and toxics. It is capable of functioning even before a person becomes immunized against the organism.[111] However, in chronic periodic homeostatic disturbances (the hypersensitive type) the patient often triggers this pathway constantly developing more food and/or chemical sensitivities.

CLINICAL REGULATION OF THE IMMUNE SYSTEM

As shown in Chapter 1 and in the beginning of this chapter, the regulation of the immune system though very complex, is an automatic homeostatic function that is basically influenced by an excess in total environmental pollutant load and the state of nutrition. Regulation of the immune system is very complicated and has many varied responses. See Figure 3.21.

Often responses to incitant overload (either specific or nonspecific) are subtle involving almost imperceptive **clinical responses including transient watering of the eye, runny nose, shortness of breath, very mild abdominal cramps and bloating, transient bladder spasm, vascular deregulation or transient brain dysfunction, etc**. Faulty internal production as well as the body's genetics, abnormal release and altering of information reception mechanisms, and deficient nutrition will change immune regulation. Toxicological responses can occur in the immune system and may be produced by a myriad of toxic chemicals (e.g., polyaromatic hydrocarbons, volatile organic hydrocarbons, pesticides, mycotoxins, etc.). Figure 3.22 shows the various routes that trigger the immune system to give either immune stimulation, resulting in hypersensitivity or abnormal cells, and immune depression.

Therefore, the clinician holds the key to manipulate the immune system in the patient's favor if he understands and applies the knowledge of how to access and influence these mechanisms. For example, when a noxious stimulus enters the body, upon exposure to the specific antigens, as presented by adjacent macrophages, the T-lymphocytes of the specific lymphoid tissue clone proliferate and release a large number of activated T-cells in ways that parallel antibody release by the activated B-cells. The principal difference is that instead of releasing antibodies, whole activated T-cells are formed and released into the lymph. They then pass into the circulation and are distributed throughout the body, passing through the capillary walls into the tissue spaces, back into the lymph and blood once again, and circulating again and again throughout the body, sometimes lasting for months or even years.[111] **These sensitized T-cells often over- or underrespond in the periodic and aperiodic homeostatic disturbances, causing increased or decreased helper and suppressor T-cells**. Once this disturbance occurs, it is difficult to restore homeostasis, even when the total load is reduced and the specific incitants are removed. Often, what appear to be permanent defects in the number and function of the T-cells occurs.

In addition, **T-lymphocyte memory cells are formed in the same way as the B memory cells are formed in the antibody system**. That is, when an antigen activates a clone of T-lymphocytes, many of the newly formed lymphocytes are preserved in the lymphoid tissue to become additional T-lymphocytes of that specific clone; in fact, these memory cells even spread throughout the lymphoid tissue of the entire body. Therefore, on subsequent exposure to the same antigen, the release of activated T-cells occurs far more rapidly and much more powerfully than in the first response.[111] This phenomenon, at times, is very detrimental to the chronically pollutant overloaded patient because

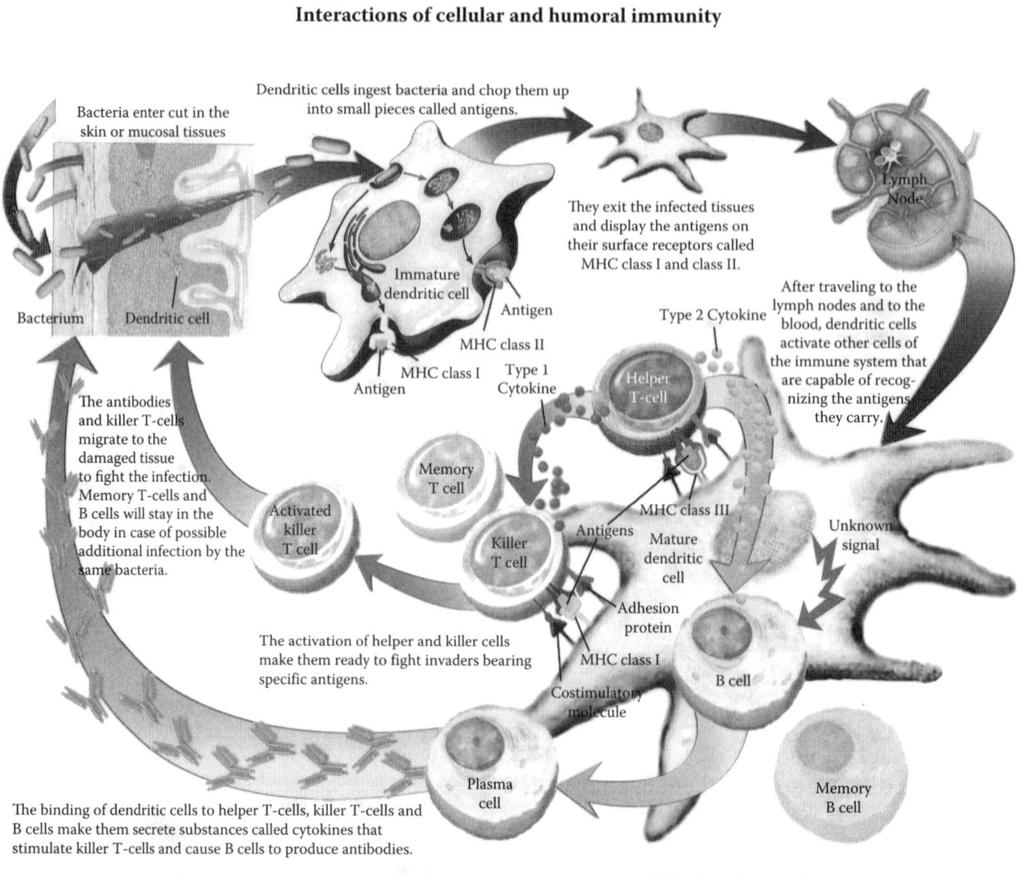

Interactions of cellular and humoral immunity

Dendritic cells ingest bacteria and chop them up into small pieces called antigens.

Bacteria enter cut in the skin or mucosal tissues

They exit the infected tissues and display the antigens on their surface receptors called MHC class I and class II.

Bacterium Dendritic cell

Immature dendritic cell

Antigen

MHC class II

MHC class I   Type 1
Antigen   Cytokine

Type 2 Cytokine

Lymph Node

After traveling to the lymph nodes and to the blood, dendritic cells activate other cells of the immune system that are capable of recognizing the antigens they carry.

The antibodies and killer T-cells migrate to the damaged tissue to fight the infection. Memory T-cells and B cells will stay in the body in case of possible additional infection by the same bacteria.

Activated killer T cell

Memory T cell

Helper T-cell

Killer T cell

Antigens

MHC class III

Mature dendritic cell

Unknown signal

Adhesion protein

The activation of helper and killer cells make them ready to fight invaders bearing specific antigens.

MHC class I

Costimulatory molecule

B cell

Plasma cell

Memory B cell

The binding of dendritic cells to helper T-cells, killer T-cells and B cells make them secrete substances called cytokines that stimulate killer T-cells and cause B cells to produce antibodies.

FIGURE 3.21 (See color insert following page 364.) Interactions of cellular and humoral immunity as defense against invaders. (Immunoscience Lab., Inc. 8693 Wilshire Blvd., Ste. 200, Beverly Hills, CA 90211. http://www.immunoscienceslab.com 2004. With permission.)

with instant recognition of the entry antigen hypersensitivity occurs, which results in patient malfunction. At times, when the overload is reduced, one still sees propagation of the response, which becomes dyshomeostatic, appearing to be caused by excess noxious stimuli, but it may also be the result of a deficiency of the nutrient that fuels the response system. Much research in therapy is being directed at eliminating the excess tissue memory for minute exposure of toxic substances while simultaneously replacing the nutrition.

Antigens bind with receptor molecules on the surfaces of T-cells in the same way that they bind with antibodies. These receptor molecules are composed of a variable unit similar to the variable portion of the humoral antibody, but its stem section is firmly bound to the cell membrane. As many as 100,000 receptor sites are present on a single T-cell.[111] At times these receptor recognitions are disturbed by chronic pollutant overload. This overload will cause the cells not to recognize friendly substances, such as food, vitamins, minerals, amino acids, and lipids or to misinterpret their presence as hostile. **With poor recognition, attacks occur rendering the patient sensitive to a greater number of foods and nutrients**.

Clearly, there are multiple types of T-cells. They are classified into three major groups: **(1) helper T-cells; (2) natural killer T-cells (cytotoxic T-cells)**; and **(3) suppressor T-cells**. The functions of each of these are quite distinct.[111]

The helper T-cells are by far the most abundant of the T-cells, usually constituting more than three-quarters of the total number. As their name implies, they help in the functions of the

Immune System

Toxicological response in the immune system

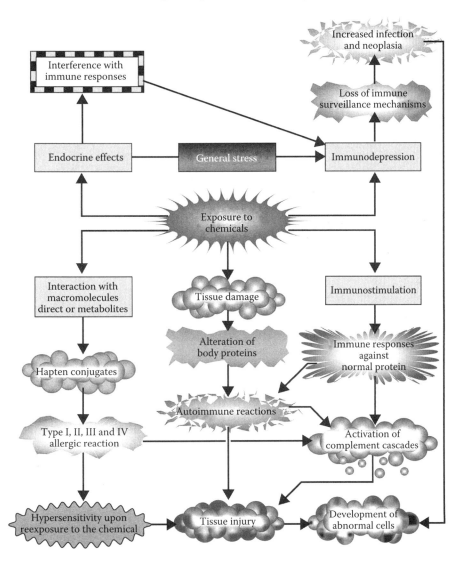

FIGURE 3.22 (See color insert following page 364.) Toxicological response in the immune system produced by chemicals. (Immunoscience Lab, Inc. Toxicological Response in the Immune System Produced by Chemicals. http://www.immunoscienceslab.com 2004. With permission.)

immune system in many ways. In fact, they serve as the major regulator of virtually all immune functions, as shown in Figure 3.23.

The helper T-cells regulate immune function by forming a series of protein mediators (lymphokines), which act on other cells of the immune system, as well as on bone marrow cells. Among the important lymphokines secreted by the helper T-cells are the following: Interleukin-2, 3, 4, 5, 6, Granulocyte-monocyte colony stimulating factor, Interferon-γ. **In the absence of the lymphokines from the helper T-cells, the remainder of the immune system is almost paralyzed**. In fact, the helper T-cells are inactivated or destroyed by the acquired immune deficiency syndrome (AIDS) virus, which leaves the body almost totally unprotected against infectious disease, thereby leading to the now well-known rapid lethal effects of AIDS.[111] Only 15% of chemically sensitive patients have moderately low helper T-cells, while 85% have low suppressor cells. This fact of low suppressor

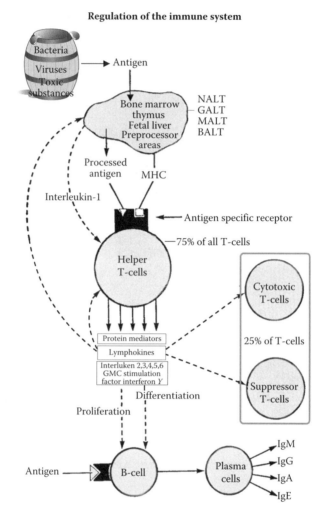

FIGURE 3.23 Regulation of the immune system, emphasizing the pivotal role of the helper T-cells. (Guyton, A. C. and Hall, J. E., *Textbook of Medical Physiology,* 9th ed., Philadelphia: WB Saunders Co., 451, Fig. 34-6, 1996. With permission.)

cells helps explain why these patients live to develop chronic disease. Interestingly, enough of the immune system continues to function in these patients even though they are hypersensitive. Those patients with low helper cells do not seem to have an increased frequency of infections. Therefore, some specific type of helper cells may be damaged, which has nothing to do with infections.

There are two groups of chemically sensitive and chronic degenerative patients: those who have chronic recurring infection and those who do not. In both cases, there appears to be no correlation between the helper T-cells and the number of infections these patients experience.

Some of the specific regulatory functions of the T-cells are the following. In the absence of helper T-cells, the clones for producing NK, natural killer (cytotoxic) T-cells and suppressor T-cells are activated very little by most antigens. The lymphokine interleukin-2 has an especially strong stimulatory effect in causing growth and proliferation of both cytotoxic and suppressor T-cells. IL-2 may not be quantitatively sufficient in the chemically sensitive patient to produce a homeostatic balance. **We see activation of the IL-2 in most cases of chemical sensitivity and some in chronic degenerative disease, but apparently this activation is not sufficient to keep the levels of T_8 suppressor cells in the normal range until the total body load is decreased**. In addition, several of the other lymphokines have less potent effects, especially interleukin-4 and interleukin-5.

Cytotoxic cells seem to get larger in numbers. Much less is known about the suppressor T-cells than about the others, but suppressor T-cells are capable of suppressing the functions of both cytotoxic and helper T-cells. These suppressor functions may serve the purpose of regulating the activities of the other cells, **keeping them from causing excessive immune reactions that might be severely damaging to the body**. This phenomenon is seen in the chemically sensitive patient who has had hypersensitive reactions to many foods and chemicals. These chemically sensitive patients have trophedema and/or chronic inflammation of their small blood vessels, GI tract, sinuses, bronchial tree, or bladder. They become more vulnerable to other environmental incitants, which constantly trigger more inflammatory responses. Due to their ability to alter the immune system, the suppressor cells are classified, along with the helper T-cells, as regulatory T-cells. One scenario for the function of the regulatory suppressor T-cells is the following: The helper T-cells activate the suppressor T-cells; these cells, in turn, act as a negative feedback controller of the helper T-cells; and this automatically sets the level of activity of the helper T-cell system, thus regulating the dynamics of homeostasis. Probably, the **suppressor T-cells play an important role in limiting the ability of the immune system to attack a person's own body tissues**.[111] Suppressor cell number and function is limited in the chemically hypersensitive individual and also in some cases of chronic degeneration. One sees, moreover, excess reactivity of the T-helper cells and thus immune deregulation. Often, one sees tissue edema when the degenerative cascade of hypoxia and the nonspecific mesenchyme reaction are activated. This lack of suppression in the chemically sensitive patient probably explains the multiple aches and pains these patients complain of. Skeletal and smooth muscle may be constantly attacked, resulting in tissue damage with more secondary homeostatic foci (interference fields) occurring.

- The direct actions of antigen to cause B-cell growth, proliferation, formation of plasma cells, and secretion of antibodies are also slight without the "help" of the helper T-cells. Almost all the interleukins participate in the B-cell response, but especially interleukins 4, 5, and 6, which have such potent effect on the B-cells that they have been called B-cell stimulating factors of B-cell growth factors. The B-cells are deranged in 10% of chemically sensitive patients suggesting that the interleukins are not as effective in these patients.
- The lymphokines also affect the macrophages. First, they slow or stop the migration of the macrophages after they have been chemotactically attracted into the inflamed tissue area, thus causing great accumulation of macrophages. Second, they activate the macrophages to cause far more efficient phagocytosis, allowing them to attack and destroy the greatly increasing the number of invading organisms or toxic chemicals. At times, **toxic chemicals such as hexachlorobenzene will damage the macrophage causing impaired phagocytosis and killing capacity**, suggesting that lympokine production is less than optimal or that the macrophage, is unable to respond to the lymphokine stimulation.
- Some of the lymphokines, especially interleukin-2, have a direct positive feedback effect in stimulating activation of the helper T-cells themselves. This substance acts as an amplifier in further enhancing the helper cell response as well as the entire immune response to an invading antigen. Thus, lymphokines like IL-2 equalize the dynamics of homeostasis. Often one sees high IL-2 in the chemically sensitive patient.

The natural killer cell (cytotoxic T-cell) is a direct-attack cell that is capable of killing microorganisms and tumor cells, and, at times, even some of the body's own cells. These cells are large granular lymphocytes. Natural killer cells are distinguished from other T-cells by using monoclonal antibodies CD3−, CD2+ , CD16+, and CD56 (NKH1)+ on the surface markers. See Figure 3.23 again. Killer cells move constantly through the body's tissues, combing the tissues for deviant T-cells to kill them. **Although natural killer cells attack virus or bacteria, they are the principle defense against tumor cells**. In addition, they are the most aggressive fighters in the immune system, being mobilized at the site where foreign substances enter. They are often elevated in chemically sensitive patients frequently causing some tissue damage. Immunosuppressive chemicals (like toxins, benzenes, etc.), chemotherapy, cortisone etc., markedly increase the risk of malignant tumors by destroying or suppressing killer T-cells. The receptor proteins on the surfaces of the NK, natural killer cells (cytotoxic cells), cause them to bind tightly to those organisms or cells that contain their binding-specific antigen (ligand). Interaction of the ligand with the cell surface receptor of the natural killer cell stimulates a Ca^{2+} dependent phosphodiesterase, phospholipase C (PLC), to hydrolyze phosphotidyl-inositol- bis-phosphate (PIP2). The products of hydrolyzed PIP2 are diacylglycerol (DAG) and inositol-1, 4, 5- triphosphate ((IP3). Both of these secondary messengers act by different mechanisms to elicit an appropriate cellular response. DAG activates protein kinase C (PKC), which results in phosphorylation of numerous protein substrates. IP3 releases Ca^{2+} from intracellular stores and leads to elevated intracellular levels with activation of lethal intracellular events[118] Therefore, IP-3 and Ca^{2+} may play the key role in cytolysis mediated by NK cells. PKC represents a partial enzyme and has been implicated in triggering various lymphokines from the killer T-cells such as Interleukin-1 (IL-l), IL-2, and interferon, all of which directly or indirectly participate in the cytotoxic activity of NK cells. See Figure 3.24.

Then, they kill the attacked cell in the manner shown in Figure 3.25.

After binding, the NK cells (cytotoxic T-cell) secrete hole-forming proteins (perforins, cytolysins, serine esterase) that literally punch large round holes in the membrane of the attacked cell. Then fluid flows rapidly into the cell from the interstitial space. In addition, the

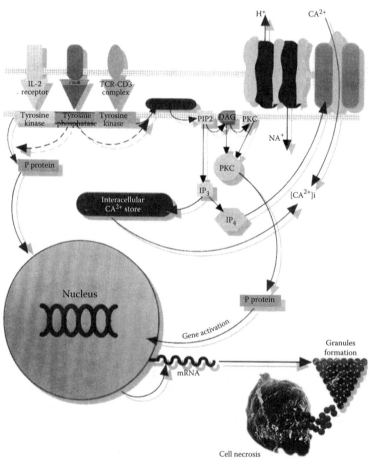

FIGURE 3.24 Mechanism of tumor cell lysis and signal transduction by natural killer cells. (Vojdani, A., *The Role of Natural Killer Cell, Apoptosis and Cell Cycle in Health and Diseases*, 14, 1998. With permission.)

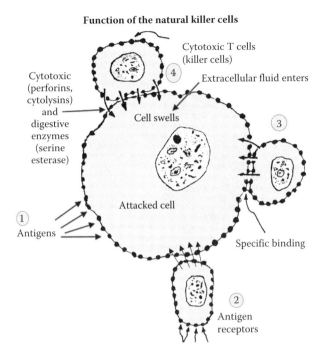

FIGURE 3.25 Direct destruction of an invading cell by sensitized lymphocytes. (Guyton, A. C. and Hall, J. E., *Textbook of Medical Physiology,* 9th ed., Philadelphia: WB Saunders Co., 452, Fig. 34-7, 1996. With permission.)

cytotoxic cell releases cytotoxic substances directly into the attacked cell. Almost immediately, the attacked cell becomes greatly swollen and usually dissolves shortly thereafter.[111]

Especially important is that these NK cells (cytotoxic killer cells) can pull away from the victim cells after they have punched holes and delivered their cytotoxic substances. They move on to kill many more cells. Indeed, even after destruction of all the invaders, many of these cells persist for months thereafter in the tissues.[111] This condition can become dangerous to the chronically compromised pollutant overloaded individual in that their own cells may be attacked even though they are free of microorganisms or tumor cells.

Some of the cytotoxic T-cells are especially lethal to tissue cells that have been invaded by viruses because many virus particles become entrapped in the membranes (filtering effect) of these cells and attract the T-cells in response to the viral antigenicity.[111] These T-cells then attack the damaged tissue causing local inflammation or edema. **The cytotoxic cells also play an important role in destroying cancer cells by forcing them into apoptosis.** The cytotoxic cells additionally attack heart transplant T-cells or other types of cells that are foreign to the person's own body.[111]

Natural killer cell (cytotoxic) activity is reduced in a patient with chronic fatigue, toxic chemical exposure, alcohol intoxication, cancer, or with distal metastatic cancer ($P < 0.0001$).[121] Natural killer cells are involved in other biological functions.

Enhancement of NK cell activity may be achieved by biological response modifiers such as IL-2, interferon, plant extracts, plant lectins, and antioxidants[122] **such as vitamin C, etc.**[123] In one study, vitamin C at 60 mg/Kg body weight enhanced NK cell activity tenfold. It also restores lymphocyte blastogenic responses of T- and B-cell mitogen to normal levels in chronic fatigue and chemical exposure patients.[123]

As mentioned earlier in this section, nutrition and dietary factors have immunological significance. **It is now well established that food sensitivity, natural toxins in foods, and other dietary components can influence the immune system.** Of course, nutritional deficiency will alter the immune system. Clearly, what we eat is linked to how robust our immunity is. Some molecular

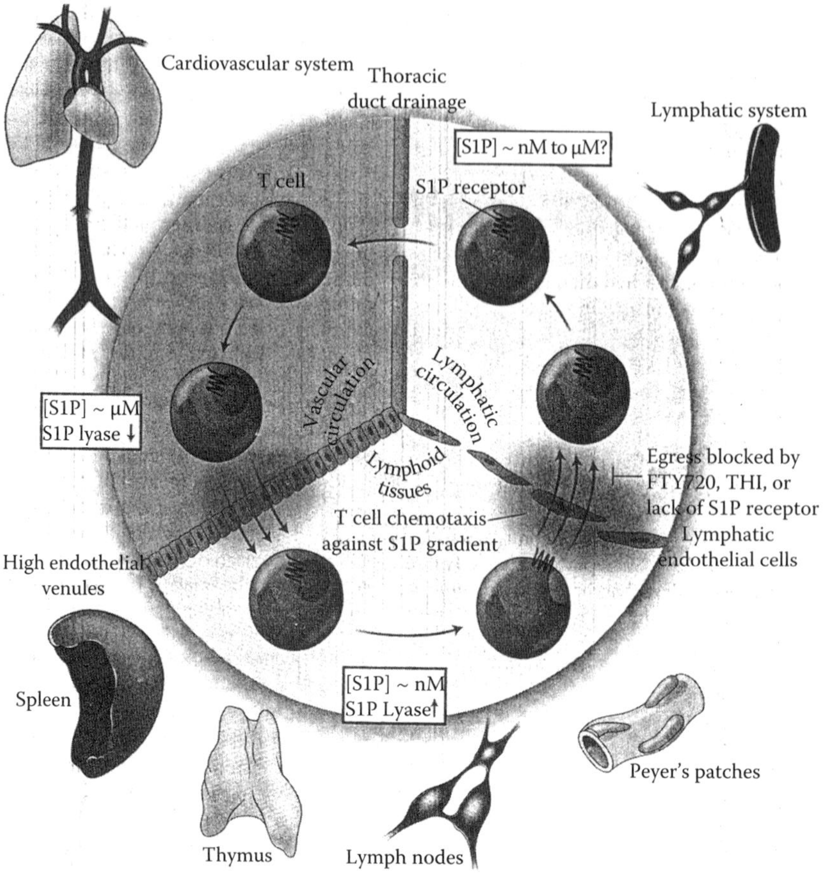

Sphingosine 1-phosphate gradients and T-cell trafficking

FIGURE 3.26 Sphingosine 1-phosphate gradients and T-cell trafficking. In response to chemokine, T-cells in vascular system home into lymphoid tissues by traversing high endothelial venules. Low concentrations of sphingosine 1-phospate ([S1P]) in the lymphoid tissue enables the expression of the sphingosine 1-phosphate receptor on the surface of T-cells, hence allowing their chemotaxis response into the lymphatic system. This steep gradient of sphingosine 1-phosphate is maintained in part by the action of a sphingosine 1-phosphate lyase that degrades this lipid mediator. Inhibition of the lyase by dietary factors perturbs the gradient of sphingosine 1-phosphate and reduces circulating lymphocytes, resulting in immunosuppression. (Bickel, C. and Huey, P., *Science* 309, 1682, 2005. With permission.)

details show how various nutrients and components in the diet can specifically affect the complex immune function. It is well known that **total body pollutant load will affect the immune system since pollutant elevation will cause more food sensitivity and pollutant diminution and/or avoidance of specific foods will also decrease food sensitivity and thus, enhance immunity**.

Schwab et al. has shown that **2-acetyl-4 tetrahydroxybutylimidazole (THI), a component of caramel food colorant III used in food products like beer and barbeque sauce, suppress immunity by increasing the amount of lipid sphingosine-1-phosphate (S1P) in tissues of the immune system**. See Figure 3.26.

Sphingosine has a regular gradient in the immune and cardiovascular system. It modulates the immune function by the regulation of lipid metabolism. The lipid is present in all eukaryotes produced during sphingolipid metabolism. In mammals a steep concentration gradient of sphingosine-1-phosphate exists between the blood vessels and tissues, thereby exposing resident vascular and immune cells bearing a cognate receptor to excess ligand. As lymphocytes traffic in and out of the vascular system, the immune cells must navigate this step of sphingosine 1-phosphate gradient. Schwab et al. discovered that THI (2-acetyl-4 tetrahydroxybutylimidazole) attenuates the activity of sphingosine 1-phosphate lyase in lymphoid tissues. This intracellular enzyme irreversibly degrades sphingosine 1-phosphate into phosphoethanolamine and 2-hexadecanal. **Thus, by neutralizing the concentration gradient of sphingosine 1-phosphate, THI perturbs T-cell egress, causing a logjam in lymphoid tissue. Immunosuppression ensues.** Other additives, preservatives, and color dyes as well as inhaled toxics may well do the same thing.

Naïve lymphocytes circulate through lymphatic and vascular conduits (see Figure 3.16). In addition, they traffic in and out of peripheral tissues by crossing the vascular and lymphatic endothelial cell barriers. Modulation of lymphocyte trafficking is a new approach to control immune pathology. **Recent work with the experimental compound FTY720, which binds to the extracellular domain of sphingosine 1-phosphate receptors, has revealed an unexpected role of sphingosine 1-phosphate in the control of lymphocyte traffic**[125] Expression of sphingosine 1-phosphate receptor $S1P_1$ (formerly known ad EDG-1) in T-cells is critical for their proper egress from lymph nodes, Peyer's patches, and the thymus[126] This receptor is expressed on the surface of T-cells when the extracellular concentration of sphingosine 1-phosphate is low; however, the receptor is rapidly internalized by cells upon exposure to the higher concentrations of the ligand.[127] The concentration of sphingosine 1-phosphate is extremely high in the blood (0.4 to 1.5 µM)—about 25 times that in thymic tissue.[128] Thus, in lymphoid tissues, the sphingosine 1-phosphate receptor is present on the surface of immune cells. But during transit in the blood and lymphatic fluid, the receptor is sequestered in the intracellular endosomal vesicles. The expression of sphingosine 1-phosphate receptors allows the cells to sense the direction of egress by moving against the vascular concentration gradient. **FTY720 is effective in suppressing the immune response (mobilization of T-cells into the vascular and lymphatic systems) because it persistently induces internalization of the sphingosine 1-phosphate receptor and presumably interferes with the compass system of the cell.**

Schwab et al. observed that oral administration of THI to mice increased the concentration of sphingosine 1-phosphate in lymphoid organs. Consequently, this induced internalization of the receptor and accumulation of lymphocytes within the lymph organs resulting in lymphopenia and immunosuppression. Taking a clue from previous findings that the **immunosuppressive effects of THI are reversed by vitamin B_6**, Schwab et al. implicated sphingosine 1-phosphate lyase, which needs vitamin B_6 as a cofactor to degrade sphingosine 1-phosphate. The authors provided pharmacological evidence that THI attenuates enzyme activity in vivo, and further support the finding in mouse hematopoietic cells in which expression of the lyase is knocked down by RNA interference. Another inhibitor of vitamin B_6 dependent enzymes also acted as an immunosuppressant. These findings are of interest as they illuminate the role of sphingosine 1-phosphate lyase in immune regulation.

Sphingosine 1-phosphate lyase was originally discovered in the yeast *Saccharomyces cerevisiae* **because of its ability to confer sphingosine tolerance.**[129] It is involved in many processes including resistance to the chemotherapeutic drug cisplatin, cell survival in mammals, embryonic development in *Glosophila melanogaster* and *Caenorhabditis elegans*, and the regulation of a transcription factor in flies.[127] **As vitamin B_6 deficiency is associated with immune deficiency in humans, these findings suggest that additional dietary factors may modulate immunity by influencing sphingosine 1-phosphate metabolism.**

The findings of Schwab et al. suggest that sphingosine 1-phosphate gradients play a fundamental role in the immune response. How this gradient is generated is not clear. Most mammals appear to expend considerable energy maintaining the vascular gradient of this lipid. For example, vascular endothelial cells secrete sphingosine kinase to generate sphingosine 1-phosphate extracellularly.[130] Other enzymes in the metabolic pathway such as sphingosine 1-phosphate phosphatase, and genetic manipulation of the lyase may be involved. Expression of sphingosine 1-phosphate receptors is inducible by growth factors and cytokines, but receptor internalization may limit the cellular response. The concept of a sphingosine 1-phosphate gradient may also help resolve an apparent discrepancy in the biology of this lipid mediator. It is abundantly present in blood under normal conditions, **yet** is capable of inducing numerous pathologies including inflammation, angiogenesis, and cell proliferation.[131] **The hydrophobic as well as amphipathic physiolochemical property of sphingosine 1-phosphate may afford unique modes of receptor activation**. Indeed, lipid gradients may behave differently than water-soluble polypeptide gradients in the context of signaling within tissues. Of course this lipid gradient will attract lipophilic xenobiotics, which may disturb the metabolism in patients with chemical sensitivity and/or chronic degenerative disease. This immunity then may become dysfunctional.

On a more practical level, modulation of activity and expression of sphingosine 1-phosphate lyase by intrinsic and dietary factors may lead to immunological consequences. It is not yet known how THI inhibits the lyase. Whether sphingosine 1-phosphate lyase is involved in regulation of the vascular homeostasis and/or growth also remains to be determined. But this work provides the molecular basis for the regulation of immunity by dietary factors, lending credence to the saying, we are what we eat. An example of how diet affects immunity is the patient on total parental nutrition who has immune suppression. As the patient replaces the nutrition the immunity returns.

Schwab et al.[132] found that lymphocyte egress from lymphoid organs is critical for immune surveillance and immune effector function and depends on intrinsic expression of sphingosine 1-phosphate receptor-1 (S1P$_1$). We have often seen "constipation" of the lymph nodes where they are large and rubbery after a toxic exposure. Often it takes weeks to months for them to decrease in size presumably through suppression of this SIP mechanism (Figure 3.27).

As S1P$_1$ is essential for lymphocyte egress from the thymus and secondary lymphoid organs, Schwab et al.[133] tested whether THI (food colorant 2-acetyl-4-tetrahydroxybutylimidazole) and DOP (dioctyl phthalate) were acting by altering S1P$_1$ function. Flow cytometric analysis revealed that S1P$_1$ expression was almost undetectable on mature CD4 or CD8 single-positive thymocytes (CD4$^+$ or CD8$^+$) from THI- and DOP-treated mice but was readily detected on control thymocytes (Figure 3.28).[140]

Similarly, THI and DOP treatment caused a loss of surface S1P$_1$ on lymph node CD4 and CD8 T-cells (Figure 3.28, part B). S1P$_1$ mediates chemotaxis of lymphocytes consistent with the loss of surface expression. Neither CD4 nor CD8 splenic T-cells from THI- and DOP-treated mice could migrate in response to S1P, and a similar, although less severe, impairment was seen in follicular B-cells. **This phenomenon may also be the case in some other food additives, inhaled chemicals,**

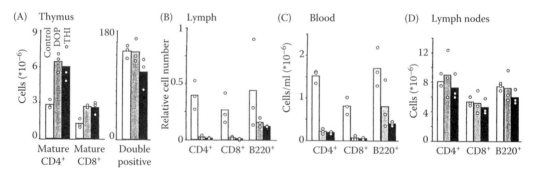

FIGURE 3.27 THI and DOP cause accumulation of mature cells in the thymus and loss of lymphocytes from the lymph and blood. (A to D) Lymphocyte numbers in the indicated tissues from mice given drinking water alone (control, white bard), given water containing 30 mg of DOP per liter (gray bars), or given water containing 50 mg of THI per liter (black bars) for three days. Mature thymocytes ((A) were defined as CD4$^+$ or CD8$^+$ and CD69^{lo}CD62L^{hi}, and double-positive thymocytes were defined as CD4$^+$ and CD8$^+$. Points indicate values from individual mice and bars indicate the average. When data from three experiments were combined, the increase in mature thymocytes and the decrease in blood and lymph lymphocytes were statistically significant ($p < 0.05$), with the exception of the decrease in B220$^+$ cells in the lymph of DOP-treated mice. (Schwab, S. R., Pereira, J. P., Matloubian, M., Xu, Y., Huang, Y., and Cyster, J. G., *Science*, 309, 1736, Fig. 1, 2005. With permission.)

THI and DOP treatment lead to reduced $S1P_1$

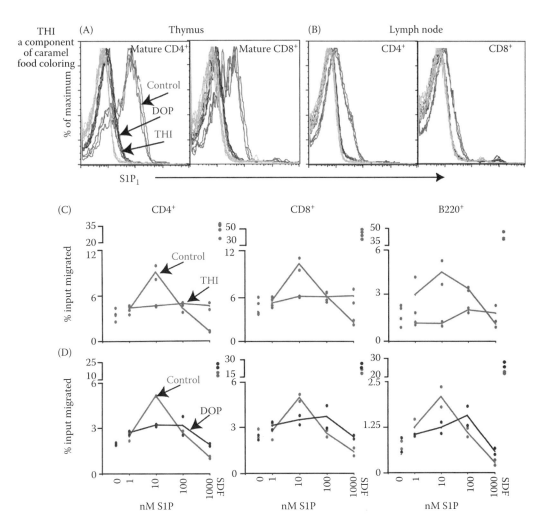

FIGURE 3.28 (See color insert following page 364.) THI and DOP treatment lead to reduced $S1P_1$ expression and impaired ability to migrate to S1P. (A) Mature single-positive thymocytes and (B) peripheral lymph node cells were stained for $S1P_1$ expression. Green lines show staining of cells from three control-treated mice, blue from three DOP-treated mice, and red from three THI-treated mice. Gray lines show staining of all nine cell types with a control antibody. Staining is representative of at least three experiments. (C and D) Splenocytes from THI- (red), DOP- (blue), or control- (green) treated mice were tested for their ability to migrate to the indicated concentrations of S1P or to 0.3 μg of stromal-cell derived factor (CXCL12) per ml of medium. Individual points show migration in replicate wells. When data from three experiments were compiled, the impaired migration of $CD4^+$ and $CD8^+$ T-cells from DOP- and THI-treated mice to 10 nM S1P was statistically significant ($P < 0.05$). The impaired migration of $B220^+$ cells was significant for THI ($P = 0.05$) but not for DOP. (Schwab, S. R., Pereira, J. P., Matloubian, M., Xu, Y., Huang, Y., and Cyster, J. G., *Science*, 309, 1736, Fig. 2, 2005. With permission.)

and/or food sensitivities where T_8 suppressor cells are suppressed and apparently sequestered in the lymph organs. (See Figure 3.28, parts C and D again.)

Cells internalize $S1P_1$ upon exposure to S1P,[134] and incubation of freshly isolated thymocytes with as little as 0.5 nM S1P down-modulate surface $S1P_1$ expression (Figure. 3.29 part A) In contrast, tissue S1P concentrations rose markedly from ~20, 40, and 150 ng/g of tissue in the thymus, lymph nodes, and spleen, respectively, to 5000 to 10,000 ng/g in each of these tissues after treatment with

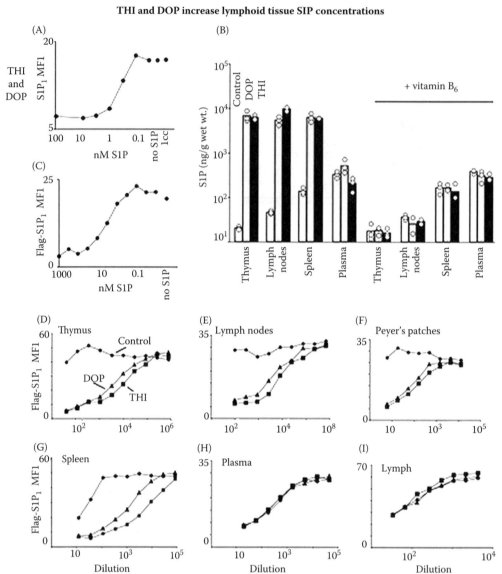

FIGURE 3.29 THI and DOP increase lymphoid tissue S1P concentrations. (A) Mature CD4+ thymocytes from an untreated mouse were incubated in the indicated concentration of S1P, and $S1P_1$ mean fluorescence intensity (MFI) levels were measured. (B) S1P levels in the indicated tissues from control- (open bars), DOP- (gray), or THI- (black) treated mice, measured by LC/MS/MS. Points represent values from individual mice, and bars represent the average. In some cases, the mice also received vitamin B_6 (pyridoxine HCl, 450 mg per liter of water), as indicated. The thymic, lymph node, and spleen S1P concentrations in THI- and DOP-treated mice were significantly different from the control ($P < 0.05$). (C) FLAG-$S1P_1$-expressing WEHI231 cells were incubated in the indicated concentration of S1P, and FLAG-$S1P_1$ MFI levels were measured. (D to G) Extracts enriched in interstitial fluid from the indicated tissues were prepared (14) and titrated onto FLAG-$S1P_1$ WEHI231 cells. The x axis shows dilution of total tissue from which the extract was made, assuming that the tissue density is 1 g/ml. (H and I) As in (C), but x axis shows dilution of plasma or lymph. Maximum FLAG-$S1P_1$ levels vary among experiments because they were performed on different days. All data are representative of at least three experiments, with the exception of MS analysis of vitamin-B_6 treated mice and S1P measurement in Peyer's patches. (Schwab, S. R., Pereira, J. P., Matloubian, M., Xu, Y., Huang, Y., and Cyster, J. G., *Science*, 309, 1737, Fig. 3, 2005. With permission.)

THI and DOP (Figure 3.29, part B). Schwab et al.[135] took two steps to total tissue analysis measuring intracellular and extracellular S1P to selectively determine the receptor-accessible extracellular S1P concentration. First, they developed a bioassay in which S1P-induced down-modulation of FLAG-tagged $S1P_1$ expression in WEHI231 cells could be measured. **When S1P was titrated onto the cells, detectable receptor down-regulation at 1 nM was observed**. (Figure 3.29, part C) Second, they prepared tissue extracts enriched in interstitial fluid by disaggregating lymphoid organs to a single cell suspension in isotonic medium and removing the cells by centrifugation. Since the goal was to gain an indication of the bioavailable S1P, Schawb did not further purify the lipids, and it is possible that their measurements underestimate total extracellular S1P, because some may be sequestered. When they titrated these extracts onto the WEHI231 cells, they found that S1P was present in low to undetectable amounts in extracts from control thymus tissue, lymph nodes, and Peyer's patches (Figure 3.29, D to F).[142] The detection of S1P in the control spleen extract is consistent with this organ being partly composed of plasma-rich red pulp (Figure 3.29, part G). S1P was abundant in plasma, and amounts in lymph were lower but within sixfold of those in plasma (Figure 3.29, parts H and I). Analysis of extracts from THI- and DOP-treated mice revealed that S1P abundance was increased more than 1000-fold in the thymus, lymph node, and Peyer's patch extracts and more than 100-fold in spleen extracts (Figure 3.29, parts D to G). In contrast, there was no difference in plasma and lymph S1P concentrations between the treated and untreated mice (Figure 3.29, parts H and I). This sequestration phenomenon is a favorite way for the body to contain potentially toxic substances and may have further implications on immune function than is the scope of their studies. **This S1P process explains the "constipation of lymph node" phenomenon that is often seen clinically in the patient with chemical sensitivity**. Lymph nodes become large and swollen after an insult from an environmental incitant. Frequently nodes will stay large for days and occasionally weeks before they decrease in size, the symptoms diminish, and the patient gets well.

Of the enzymes involved in determining S1P abundance, S1P lyase, a cytoplasmic enzyme that degrades S1P into phosphoethanolamine and 2-hexadecanal, represented a potential target for THI and DOP, because it is dependent on vitamin B_6.[136–138] **Mice that had received THI or DOP combined with a tenfold molar excess of vitamin B_6 in their drinking water showed no accumulation of mature thymocytes, induction of lymphopenia, down-modulation of $S1P_1$, or increase in lymphoid tissue S1P levels** (Figure 3.29, part B). They therefore tested whether S1P lyase activity in the thymus was affected by THI or DOP treatment. Thymic lysates from treated and untreated mice were incubated with tritiated dhS1P (dihydrosphingosine 1-phosphate), and the products were separated by thin-layer chromatography[139] Lyase activity, measured by the generation of hexadecanal, was severely inhibited in the thymus of DOP- and THI-treated mice.[140] (See Figure 3.30, A and B.)

Excess vitamin B_6 overcame the effect of THI or DOP and restored lyase activity[140] (Figure 3.30, A and B). These findings suggest that THI and DOP cause increased lymphoid tissue S1P abundance because of the inhibition of S1P lyase. **If applicable to humans these findings suggest that flooding the patients' systems with B_6 will help dampen or eliminate adverse lymphocyte responses as seen with the THI and DOP influx**. The B_6 would keep lymphocyte egress from the lymph nodes and other lymph organs flowing normally. Clearly these elegant studies by Schwab show that the intake of **caramel food dye is immunosuppressive. These studies will make the clinician wary of other food additives and again emphasizes why the decrease in total body load through the use of organic foods is essential for the health in the patient with chemical sensitivity and/or chronic degenerative disease.**

Cell Cycle

The cell cycle is the cell's progression from resting to preparation of dividing to mitosis. A scheme of the classical cell cycle is shown in Figure 3.31A and B.[141] The cell cycle compartments are drawn such that their horizontal position reflects their respective DNA content. Cells that contain only one

Phases of the cell cycle

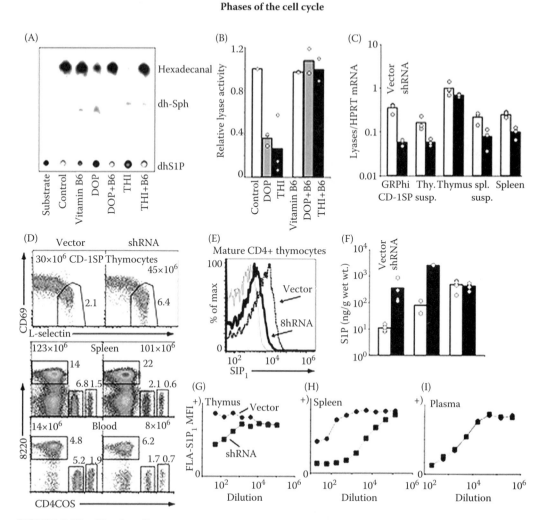

FIGURE 3.30 (Continued)

component of DNA from each parent (2C) are referred to as diploid cells. Cells that have duplicated their genome, and thus have 4C amounts of DNA, are called tetraploid cells.[142]

The cell cycle is classically divided into the following phases: G_0, G_1, S, G_2, and M. The cell cycle phases of G_1 were historically considered to be a time when diploid (2C) cells had little observable activity. Since this time precedes DNA synthesis, the term Gap 1 (G_1) was coined. We now know that there is quite a bit of transcription and protein synthesis during this phase. At a certain point in the cell's life cycle, the DNA synthetic machinery turns on. This phase is labeled S for Synthesis. As the cell proceeds through this phase, its DNA content increases from 2C to 4C. At the end of the S phase, the cell has duplicated its genome and now is in the tetraploid state. After the S phase, the cell again enters a phase that has been historically thought to be quiescent. Since this phase is the second Gap region, it is referred to as G_2. In the G_2 phase, the cell produces the necessary proteins that will play a major role in cytokinesis. After a highly variable amount of time, the cell enters the Mitosis (M phase). The DNA content remains constant at 4C until the cell actually divides at the end of telophase.[142]

Although, at times, the enlarged parent cell finally divides in half to produce two daughter cells, each one is endowed with a complete diploid set of chromosomes. The new daughter cells

FIGURE 3.30 (Opposite) THI and DOP inhibit S1P lyase, and RNAi-mediated S1P lyase knockdown recapitulates the THI- and DOP-induced phenotype. (A and B) Thymic lysates from mice treated as indicated (B6, vitamin B_6) were incubated with tritiated dhS1P. The products were separated by thin-layer chromatography and visualized by autoradiography (A). The lyase product hexadecanal and its metabolites hexadecanol and palmitic acid have a similar R_f (where R_f is the distance moved by the lipid divided by the distance from the origin to the solvent front) (19) and were not distinguished in our assay. dh-Sph, dihydrosphingosine. In (B), quantitative analysis by PhosphorImager (GE Healthcare, Chalfont St. Giles, UK) of thymic lyase activity in three experiments is plotted as percent of control activity. The activity reduction in lysates from THI- and DOP-treated mice was statistically significant ($P < 0.05$). (C) Quantitative real-time PCR analysis of lyase relative to hypoxanthine-guanine phosphoribosyltransferase (HPRT) transcript abundance in the indicated tissues from mice reconstituted with HSC transduced with a retrovirus encoding the shRNA and a GFP reporter (black bars), or the GFP reporter alone (open bars). GFP^{hi} CD4SP indicates GFP-high, $CD4^+$ $CD8^-$ thymocytes; thy. susp., thymic suspension; thymus, total thymus; spl. susp., splenic suspension; spleen, total spleen. Bars show average, points show values from individual animals. The RNA levels in GFP^{hi} thymocytes, thymic suspension, splenic suspension, and total spleen of mice receiving shRNA were significantly different from the control ($P < 0.05$). (D) Flow cytometric analysis of representative thymuses, spleens, and blood, gated by size on lymphocytes. Numbers in the upper corners show the total number of cells in the tissue or cells per milliliter in blood, and the numbers next to gates show the percent of total cells in the gate. (E) Representative flow cytometric analysis of $S1P_1$ expression on mature $CD4^+$ thymocytes. Shaded histograms show staining of both cell types with a control antibody. (F) S1P levels in total thymus, spleen, and plasma measured by LC/MS/MS. The thymic and spleen S1P concentrations in mice that received shRNA were significantly different from the control ($P < 0.05$). (G to I) Representative bioassay measurement of S1P levels in the indicated tissue extracts, as in Figure 3.29. (Schwab, S. R., Pereira, J. P., Matloubian, M., Xu, Y., Huang, Y., and Cyster, J. G., *Science*, 309, 1738, Fig. 4, 2005. With permission.)

immediately enter G_1 and may go through the full cycle again. Alternatively, they may stop cycling temporarily or permanently.[142]

The cell cycle clock programs, and thus sets up, a succession of events by means of a variety of molecules. Its two essential components, cyclins and cyclin-dependent kinases (CDK), associate with one another and initiate entrance into the various stages of the cell cycle. In G_1, for instance, D-type cyclins bind to CDK 4 or CDK 6, and the resulting complexes act on a powerful growth-inhibitory protein known as pRB. This action releases the braking effect of pRB and enables the cell to progress into late G_1 and hence into the S (DNA synthesis) and then to M or mitosis phases. The end result in all these cases is that the cell cycle clock begins to spin out of control, ignoring any external warning to stop. See Figure 3.32.

Progression through the four stages of the cell cycle is also driven by rising levels of the protein cyclins: first the D type, followed by E, A, and then by the B-type cyclin. These act at a crucial step in the cycle late in G_1 at the restriction point (R), when the cell decides whether to commit itself to completing the cycle. For the cell to pass through R and enter S, a molecular "switch" must be flipped from "off" to "on." The switch works as follows: As levels of cyclin D, and later cyclin E, rise, these proteins combine with and activate enzymes the CDK.

- The kinases (acting as part of cyclin-kinase complexes) grab phosphate groups from molecules of adenosine triphosphate (ATP) and transfer them to a protein pRB, the master brake of the cell cycle clock.
- When pRB lacks phosphate, it actively blocks cycling (and keeps the switch in the "off" position) by sequestering other proteins, and the transcription factors.
- But after the cyclin-kinase complexes add enough phosphate groups to pRB, the brake stops working; it releases the factors, freeing them to act on genes and the switch is turned "on."

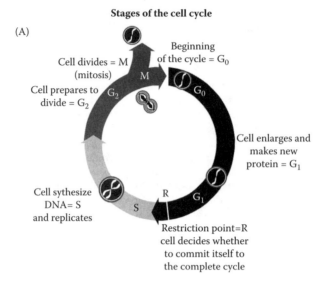

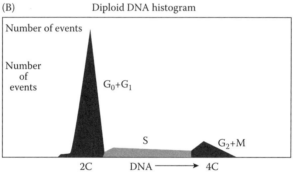

FIGURE 3.31 (A) Stages of the cell cycle. (B) Diploid DNA histogram. Stages of cell cycle (G0, G1, S, G2, and M phases) and DNA histogram generated by flow cytometry. (Vojdani, A., *The Role of Natural Killer Cell, Apoptosis and Cell Cycle in Health and Diseases*, 20, Fig. 11, 1998. With permission.)

- The liberated factors then spur production of various proteins required for continued progression through the cell cycle. See Figure 3.33.

ABNORMAL CELL CYCLE PROGRESSION IN PATIENTS WITH CHRONIC FATIGUE SYNDROME (CFS)

Vojdani tried to determine whether peripheral blood lymphocytes (PBL) isolated from CFS individuals and chemically exposed patients represented a discrete block in cell cycle progression (PBL). He isolated cells from CFS and control individuals; then he cultured, harvested, fixed, and stained them with Propidium iodide. He then analyzed lymphocytes by flow cytometry.

The nonapoptotic cell population in PBL isolated from CFS individuals consisted of peripheral lymphocytes arrested in the late S and G_2/M boundaries, as compared to healthy controls. The arrest was characterized by increased S and G_2/M phases of the cell cycle (from 9 to 33% and from 4 to 21%, respectively).[144] See Table 3.8 and Figure 3.33 again.

Such an abnormality in cell cycle progression is an **indication of abnormal mitotic cell division in patients exposed to chemicals and in patients with chronic fatigue syndrome (CFS)**. From these results Vojdani concluded that PBL from patients with chemical exposure or patients

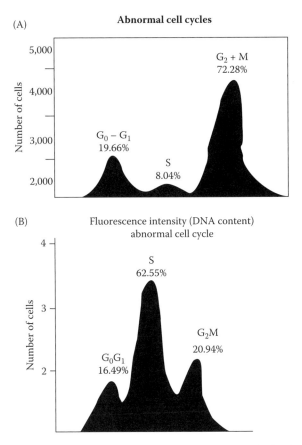

FIGURE 3.32 (A) Cell cycle abnormal G_2+ M phase. (B) Abnormal synthesis phase with a block before mitosis. (From EHC-Dallas, 2002; Griffiths, B.)

with CFS grows inappropriately not only because the signaling pathways in cells are disturbed but because the cell cycle clock becomes **deranged and sends stimulatory messages without the corresponding inhibitory pathways**.

However, in order to hold down the cell proliferation and avoid cancer, the human body equips cells with certain backup systems that guard against runaway division. One such backup system in lymphocytes of CFS patients provokes the cell to undergo apoptosis. This programed cell death occurs if some of the cell's essential components are deregulated or damaged. For example, injury to chromosomal DNA can trigger apoptosis. See Figure 3.34[152] and Table 3.8.[151]

CELL CYCLE AND CANCER

In normal cells, signals consist of growth and controlling messages from the outer surface deep into the nucleus. In the nucleus, the cell cycle clock collects different messages that determine whether the cell should divide. Cancer cells often proliferate excessively because either genetic mutations cause induction of stimulatory pathways that issue too many "go ahead" signals, or the inhibitory pathways can no longer control the stimulatory pathways.[142]

Over the past five years, impressive research has uncovered the destination of stimulatory and inhibitory pathways in the cell. The pathways converge on a molecular apparatus in the cell nucleus, which is often referred to as the "cell cycle clock." This clock is the executive decision-maker of the cell, and it apparently runs rampant in virtually all types of human cancer. In the normal cell, the clock

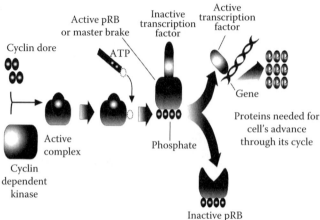

FIGURE 3.33 Molecular switch and synthesis for switch of cell cycle. (Modified from Weinberg, 1996; Vojdani, A., *The Role of Natural Killer Cell, Apoptosis and Cell Cycle in Health and Diseases*, 21, Fig. 11, 1998. With permission.)

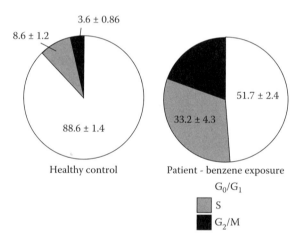

FIGURE 3.34 Cell cycle analysis of peripheral blood lymphocytes from healthy controls and patients exposed to benzene. Note that in patients the majority of cells switched from G_0/G_1 to S and G_2/M phases. (Vojdani, A., *The Role of Natural Killer Cell, Apoptosis and Cell Cycle in Health and Diseases*, 22, Fig. 13, 1998. With permission.)

TABLE 3.8
Percentage of Different Phases of Cell Cycle in Healthy Controls and Patients Exposed to Chemicals: The Difference between Healthy Subjects and Chemically Exposed Individuals Is Statically Significant ($P < 0.0001$)

Phase	Healthy Controls	Chemically Exposed
G_0/G_1	88.6 ± 1.4	51.7 ± 2.4
S	8.6 ± 1.2	33.2 ± 4.3
G_2/M	3.6 ± 0.82	21.0 ± 2.6

Source: Vojdani, A., *The Role of Natural Killer Cell, Apoptosis and Cell Cycle in Health and Diseases*, 23, Table 2, 1998. With permission.

Autoimmunity

If a person should become immune to his or her own tissues, the process of acquired immunity would destroy the individual's own body. The immune mechanism normally "recognizes" a person's own tissue as being distinctive from bacteria or viruses, and the person's immunity system forms few antibodies or activated T-cells against the person's own antigens. Ocassionally, patients with chemical sensitivity do develop mild autoantibody reactions ANA being 1:40–1:62. These levels seem to improve and are eliminated when homeostasis is restored to these patients and the progression is stopped to full blow autoimmune disease. Thus, the dynamics of homeostasis are maintained. This self-tolerance is believed to develop during the preprocessing of the **T-lymphocytes in the thymus and the** preprocessing of **B-lymphocytes in the bone marrow**. The reason for this belief is that injecting a strong antigen into a fetus at the time the lymphocytes are being preprocessed in these two areas prevents the development of clones of lymphocytes in the lymphoid tissue that are specific for the injected antigen. Also, experiments have shown that specific immature lymphocytes in the thymus, when exposed to a strong antigen, become lymphoblastic, proliferate considerably, and then combine with the stimulating antigen—an effect that is believed to cause the cells themselves to be destroyed by the thymic epithelial cells before they can migrate to and colonize the lymphoid tissue.[146] Therefore, it is believed that during the preprocessing of lymphocytes in the thymus and the bone marrow, all or most of those clones of lymphocytes that are specific for the body's own tissue are self-destroyed because of their continual exposure to the body's antigens (apoptosis).

The suppressor T-cells are probably responsible for still another type of "self tolerance." For instance, sometimes an autoimmune reaction occurs acutely against one of the body's tissues, but after a few days or a few weeks, the reaction disappears even though autoimmune antibodies persist in the circulating plasma. What has happened is that the number of suppressor T-cells specifically sensitized to the offending self-antigen has greatly increased. It is believed that these suppressor T-cells function to counteract the effects of the autoimmune antibodies as well as the sensitized helper cells and sensitized cytotoxic T-cells, thus blocking the immune attack on the tissue. However, this is not entirely clear.[146] **Twenty percent of the chemically sensitive and chronically diseased patients we have seen needing hospitalization, have autoantibodies to ANA, nerves, gastrointestinal organs or to some other organs**. These auto antibodies often decrease or disappear when total body pollutant load is decreased. As long as multiple incitants are removed and the total body load is reduced, we have not seen any patient's illness progress to named autoimmune disease such as systemic lupus erythematosis or scleroderma. In fact, some autoimmune diseases like lupus erythematosis and scleroderma improve when the total body pollutant load is reduced and some of the specific antigens, especially molds, are neutralized, as seen in the aforementioned diagram.

FAILURE OF THE TOLERANCE MECHANISM CAUSES AUTOIMMUNE DISEASES

Sometimes people lose some of their immune tolerance to their own tissues. This occurs to a greater extent the older a person becomes. It usually occurs after destruction of some of the body's tissues, which releases considerable quantities of "self-antigens" that circulate in the body and presumably cause acquired immunity in the form of either activated T-cells or antibodies. [146] It also occurs after acute or chronic chemical exposure, possibly due to some tissue destruction because of the inflammation and hypoxia of the nonspecific vasculitis.

Several specific diseases that result from autoimmunity include the following: (1) rheumatic fever, in which the body becomes hypersensitive to tissues in the joints and heart, especially the heart valves, after exposure to a specific type of streptococcal toxin that has an epitope in its molecular structure similar to the structure of some of the body's own self-antigens; often patients with a history of rheumatic fever develop other triggering agents that cause the clinician to misinterpret the patient's heart condition; (2) one type of glomerulonephritis, in which the person becomes hypersensitive to the basement membranes of glomeruli; (3) myasthenia gravis, in which hypersensitivity develops against the acetylcholine receptor proteins of the neuromuscular junction, causing paralysis or usually just weakness; and (4) lupus erythematosis, in which the person becomes hypersensitive to many different body tissues at the same time; this is a disease that causes extensive damage and often rapid death.[146] In the last three conditions, we[147] and others[148] have reported cases in which chemical exposure in the past and or present had occurred. Diseases were triggered by sensitivities to molds, foods, and chemicals. At times, such diseases can be thrown into remission by reducing the total body load, eliminating and neutralizing specific triggering agents, and replacing the nutrition.

AUTOGENOUS LYMPHOCYTIC FACTOR (ALF)

We have developed a technique at the EHC-Dallas to grow the patients' own T-cells in a cell culture medium. When over 30 generations of cell division have occurred, the cells are harvested, sonicated, and processed. The extract is injected two times per week. This injection appears to communicate with the body in a way that stimulates an increase in T-cells and T-cell function. Over 1000 patients have been treated with this autogenuous ALF, with 85% of the patients showing improvement. (See Treatment chapter in Volume 4 - Mechanisms of Cardiovascular Disease and Chemical Sensitivity for more details.)

CELL DEATH

Death along with growth and differentiation is a critical part of the cell cycle in order to maintain the dynamics of homeostasis. As shown in Chapter 1, tissue homeostasis results from the dynamic balance between cell proliferation and cell death. Much is known about cell proliferation in tissue homeostasis and the intricate network of positive and negative signals that regulate cell cycle progression. However, not as much is known about cell death[149]. See Figure 3.35.

Cell death, according to Heine,[150] can either result from necrosis (inflammation) or apoptosis (programmed cell death). Initially, apoptosis was defined on the basis of morphological criteria as a separate mode of cell death, clearly distinct from cell necrosis.[150,151] See Table 3.9.

However, both forms of cell death can occur simultaneously or in a clearly defined temporal or spatial relation within an organ (e.g., in experimental liver damage or cerebrovascular disease/ stroke). Moreover, recent evidence suggests that identical receptors, signal-transduction pathways,

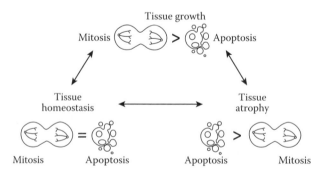

FIGURE 3.35 An organism must establish and maintain equilibrium between cell proliferation and cell death to sustain tissue homeostasis. An imbalance between the rates of these processes can alter the state of tissue growth. (King, K. L., Schwartzman, R. A., and Cidlowski, J. A., *Endrocinology and Metabolism* 3(Supp. A), 93–97, 1996. With permission.)

TABLE 3.9
Cell Death

Necrosis

1. Swelling of cells organelles
2. Rupture of plasma membrane
3. Ion pumping disrupted
4. Increased cellular swelling due to membrane damage
5. Increase in free cyctosolic Ca
6. Membrane phospholipid action

Apoptosis

1. Cell fragments membranes break
2. Cell shrinks but cell membrane intact
3. Nuclease and protease break off chromatin of DNA into fragments
4. Death signals are transduced
5. Fragments broken by proteases, kinases conserve energy

Source: From EHC-Dallas, 2005.

and mechanisms of cytotoxicity can be involved in apoptotic or necrotic cell death. Apoptosis and necrosis seem to represent two extremes of a broad spectrum of cell death modes.[152] Cytotoxic pathways and morphological characteristics may be determined by different intracellular factors and external conditions.[152,154]

In multicellular organisms, the mode of cell death has considerable significance extending beyond the fate of the individual cell. **In apoptosis cells are efficiently and rapidly phagocytized before they lyse and cause inflammation**. This process facilitates tissue reorganization or reconstitution in development and after damage without producing inflammation. Moreover, processes characteristic of apoptosis prevent the spread or release of DNA from transformed or virus-infected cells. One of these processes is the selective fragmentation and packaging of cellular DNA. The genome is usually cleaved into fragments of about 300–500 kbp (so-called high molecular weight [HMW] fragments). Often this action is followed by oligonucleosomal DNA fragmentation (i.e., formation of fragments of about n x 180 bp size). Although DNA fragmentation is not the cause of cell death, it seems to be an important feature of apoptosis in relation to the surrounding tissue, and it is often used as a diagnostic criterion.[153]

As Leist et al.[152] state, apoptosis and necrosis are two forms of cell death that have been defined on the basis of distinguishable, morphological criteria. However, these different types of cell death may involve several common signaling and execution mechanisms. Since various stimuli induce both apoptotic and necrotic death, the mode of cell demise seems to be dependent on intracellular factors. One of these factors is the concentration of ATP. By modulating ATP levels, apoptosis or necrosis can be triggered selectively under otherwise identical conditions. Researchers controlling ATP levels in staurosporine-treated JurkaT-cells detected that apoptotic (but not necrotic) cell death occurred selectively. They quantitated death with high sensitivity by the BM Cell Death Detection ELISA, or by staining with annexin V.[155] Whether to live or die is under control of the cellular sensors. If an adequate amount of superoxide dismutase and other antipollutant enzymes is available, the cell will elect to live. If an excess of reactive oxygen species (ROS) exists, the cell may undergo apoptosis or necrosis. Excess pollutants or nutrient deficiency might alter the decision-making capacity of the cell regarding whether to live or die.

Necrotic cell death results from the following events: trauma, oxidants, toxic excess; hypothermia, hypoxia, ischemia, complement attacks, and metabolic cell poison. Normally, 17% of blood cells across the arterial to the venous circulation are lost, according to Heine.[150] This loss is due to the sequestration of blood cells and destruction for the protein buffering of tissue acidosis. Whether this loss is also due to necrosis or apoptosis is unclear at this time, but apoptosis may be dominant. (See Chapter 1 for more information.)

Necrotic cells exhibit distinctive morphological features and biochemical changes.[156] The earliest of these changes includes swelling of the cell and organelles, such as the mitochondria. The nucleus is usually not affected at this stage. These changes ultimately lead to rupture of the plasma membrane. This rupture causes several events: the cellular contents leak out into the extracellular space, which results in cellular swelling from a loss of selective membrane permeability; and also, ion pumping is disrupted as a result of either membrane damage or depletion of ATP, which is required for ion transport across the membrane.

An influx of water accompanies cation movement across the plasma membrane, which causes more cellular swelling. With an increase of free cytosolic calcium, membrane-bound phospholipids activate and disrupt membrane integrity. Hydrolases are released from the ruptured lysozomes, causing rapid cellular disintegration. During the late stage of necrosis, protein, DNA, and RNA levels rapidly decrease in the cell, but they can still become protein buffers, which are needed to neutralize pollutant-induced acidosis with a drop in pH. DNA is cleared by the lysozomal deoxyribonucleases into fragments displaying a continuing spectrum of sizes, which, when viewed on an argerase gel, appears as a nonspecific smear. **Necrosis typically affects groups of categories of cells, and, if they are in large quantities, an inflammatory reaction develops on the adjacent viable tissue as a response to the released cellular debris, especially in lysozymes, proteases, and hydrolases.**

Apoptosis ("inside out" cell death) gives a morphologically distinct cell death from necrotic cell death. This type of cell death typically affects single cells and other groups of cells scattered throughout a tissue with death coming so rapidly that it is difficult to observe. **Death is not accompanied by inflammation.**

Apoptosis occurs in three stages. In the **first** stage, the cell shrinks and breaks up into a number of membrane-bound fragments that contain structurally intact organelles. These are broken up by nucleases and proteases inside the cell that degrade the high order chromatin structure at the DNA into fragments of 50–300 kilobases and subsequently into smaller DNA pieces, 200 base pair in length (Figure 3.36).[157]

In the **second** stage, death signals are transduced. In the **third** stage, these apoptosis bodies are degraded by serine proteases, cysteine proteases, cycline dependent kinases, granular enzyme, and IL-1 B-converting enzymes. Following this process, they are phagocytized by the neighboring cells—as degeneration of the macrophages is rapid.

It now appears that apoptosis is involved with the net growth of tissues, cellular homeostasis, embryonic development, and the immune response. Inducers of apoptosis include ionizing radiation, Ca^{2+}_2 influx, tumor necrosis factor, viral infection, glucocorticosteroids, embryogenisis, and metamorphosis,[156] T-cell mediated killing of target T-cells[158] and death of auto reactive thymocytes,[159] prooxidants, chemotherapeutic agents, and hydrogen peroxide. **Vitamin E has been shown to prevent apoptosis.**[160]

Changes in carbohydrate moities, phosphotidylserine, and thrombospondin are thought to be included in the recognition of apoptotic bodies.[161] **The expedient clearance of apoptotic bodies before the loss of the cell membrane integrity prevents leakage of cellular components (i.e., lysozmes, proteases, etc.) into the extracellular spaces thereby avoiding the harmful effects of inflammatory response seen in necrotic cell death.** In some instances, apoptosis is not subjected to phagocytosis, but is instead extruded into the adjacent lumen or matrix. The apoptotic bodies show progressive dilution and degeneration of cytoplasmic organelles (secondary necrosis).[162,163] Since cellular fragmentation occurs, the result, within minutes, is a lack of inflammation. Some people think that tissue homeostasis is a balance between tissue proliferation and apoptosis rather than inflammatory necrotic cell death. This view makes good sense because, in some conditions, there is no inflammation, but just cell loss. For instance, hormone-induced tissue atrophy often results in a dramatic increase in apoptosis when dexamethasone is administered to adrenalectomized rats. When apoptosis supercedes proliferation, a reduction in the size of the tissue occurs, i.e., a steroid that induces the death of thymocytes to decrease by 50% within 24 hours.[149] Conversely, organs treated with antigens (i.e., lead nitrate or eyrotenome acetate) produce cell proliferation, which results in increased hepatic volume.

Immune System

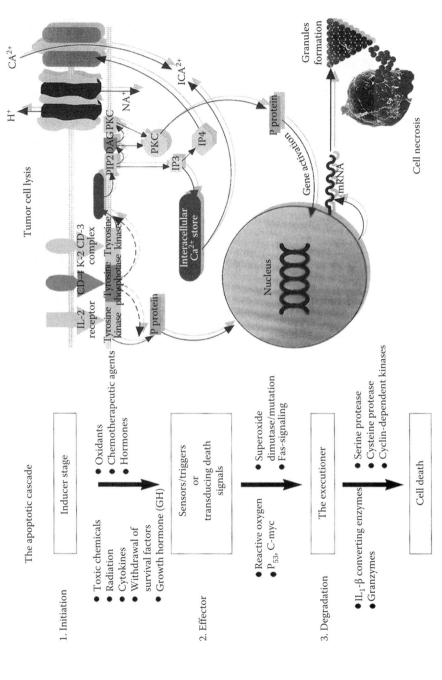

FIGURE 3.36 Cell death. (Modified from Vojdani, A., *The Role of Natural Killer Cell, Apoptosis and Cell Cycle in Health and Diseases*, 24, Tab. 14, 1998. With permission.)

During tumor progression the number of cells undergoing apoptosis increases, possibly to compensate for the cell proliferation of the tumor.[149] Thus, a delicate balance exists between apoptosis and cell division in homeostasis.

Certainly, in the blood, as shown by Heine,[150] there is a dynamic loss of normal cells during normal function. These cells appear to undergo necrotic death according to Heine,[150] but some cell death could be due to apoptosis since no inflammation is present. Apparently, a slow release of mediators occurs, which increases to a point where there are too many inducers causing oxidative stress, and then with the total load increased, inflammation results.

Activation-induced apoptosis is a primary mechanism for down-regulation of an immune response, which leads to immune homeostasis and the selection of T-cells with specificities that could be harmful. This includes deletion of T-cells with self-specificities (autospecific) and T-cells with excessively high affinity for foreign antigens, which may lead to an excessively heightened immune response with septic shock. Surface molecules involved in activation-induced apoptosis involve Fas, Fos, Ligand, and FasL, as well as T-cell receptor (TCR), which modulates the expression and function of these molecules. Fos signaling mechanisms include Nur 77 and fyn kinase and unknown molecules that modulate expressions of FasL. Apoptosis signals are further modulated by monitors or inducers of apoptosis including Bel-2, P53, interleukin-14, converting enzyme, and IGE. In contrast to cell necrosis, apoptosis usually requires energy in the form of ATP as well as a dependent, active process, which triggers the cascade of signal-transducing events.

Some opinion suggests[160] that activation-induced apoptosis is a critical mechanism by which the immune system maintains tolerance to self-antigens by the clonal depletion of autoreactive T- and B-cells[165,166] without the induction of inflammation. Apoptosis is a mechanism for cytotoxic T-cell-mediated target cell lysis and corticosteroid-induced death of immature lymphocytes.[167–179]

Glynn et al. feel that apoptosis accounts for all cell death in human Cd 4 positive T-lymphocytes infected by the HIV virus. Cell suicide (apoptosis) is triggered in response to pathogenic invaders (virus) to halt the spread to neighboring cells.

However, apoptosis also occurs under nonthreatening conditions to replace redundant or unnecessary cells as occurs in tissue morphogenesis and remodeling. Also apoptosis occurs in the normal turn over elimination of cells when they have exhausted their overall usefulness within the organism. In each situation, the dynamics of homeostasis are maintained. Cell suicide has played an undeniably important role in the development, maintenance, and defense of higher organisms, including humans.[170]

Inappropriate apoptosis may contribute to the pathology of several human diseases.[171–176] These can be divided into disorders of excessive apoptosis (re: neuro-degenerating diseases or ischemic changes) and those where insufficient apoptosis occurs (such as autoimmune syndromes, cancer, and sustained pathogenic infections). P53, Bel-2 are now known regulators of apoptosis.

Many reactive chemicals and drugs such as acetoaminophen, diquat, carbon tetrachloride, quinones (i.e., substances with isoprenoic side changes like 1,4-napthoquinone and 1,4-benzoquinone), cyanide polyhydroxyl polyether, methylmercury, organotines etc., have been implicated in apoptosis (programed cell death) as well as necrotic cell death (toxic cell death).[176–185]

In a study of 60 patients exposed to benzene-contaminated water (14 ppb for 2–5 years), Vojdani has shown that 86.6% exhibited an increased rate of apoptosis compared to controls ($p < 0.0001$). There was a discrete block within the cell cycle progression in mitosis. This apoptosis appears to be a protective mechanism against malignancy. In his series of studies, Vojdani found that 10% of his patients had more peripheral lymphocyte mitosis through apoptosis. This high mitosis to apoptosis ratio is predictive of malignancy (leukemia).[186] Therefore, the tendency of normal cells to commit suicide when they are either deprived of their usual growth factors, as seen in this series, or when they are deprived of physical control with their neighbors (gap junction, cellular adhesion molecule) due to chemical exposure is a bold defense against metastasis.

The induction phase of apoptosis is prevented by many antioxidants, for example β-carotene, vitamin C, and vitamin E,[187] **and also by a variety of biological response modifiers such as thymic hormones, cytokinases, leutinans, and viral antigens**.

In autoimmune disorders, the failure to remove autoreactive lymphocytes that arise during the development of or subsequent to an immune response, such as ICE/CED$_3$ (interleutin-1, β-converting enzyme) occurs.

Excessive or failed apoptosis is a prominent morphological feature of several human diseases. Many of the key biochemical players that contribute to the highly ordered process of apoptotic cell death have recently been identified. These include members of the emerging family of cysteine proteases related to mammalian interleukin-1 beta converting enzyme (ICE) and to CED$_3$, the product of a gene that is necessary for programed cell death in the nematode C. elegans.[188] **Among a growing number of potential molecular targets for the control of human diseases where inappropriate apoptosis is prominent, ICE/CED$_3$-like proteases may be an attractive and tangible point for therapeutic intervention**.

ALLERGY AND HYPERSENSITIVITY

An important undesirable side effect of immunity is the development, under some conditions, of allergy or other types of immune hypersensitivity. There are several types of allergy and other hypersensitivities, some of which occur only in people who have a specific allergic tendency but can occur in anybody who becomes overloaded with toxics.[111] Much of the discussion of this book is on the hypersensitivity aspects of chronic disease; therefore, only a brief discussion will be presented here.

These types of hypersensitivity tendencies can cause skin eruptions in response to certain drugs or chemicals, particularly some cosmetics and household chemicals, to which one's skin is often exposed. Another example of such hypersensitivity is the skin eruption caused by exposure to poison ivy.[111]

Delayed-reaction hypersensitivity is caused by activated T-cells and not by antibodies. In the case of poison ivy, the toxin of poison ivy in itself does not cause much harm to the tissues. However, on repeated exposure, it does cause the formation of activated helper and cytotoxic T-cells. Then, after subsequent exposure to the poison ivy toxin, within a day or so the activated T-cells diffuse from the circulating blood in sufficient number into the skin to respond to the poison ivy toxin and elicit a cell-mediated type of immune reaction. Remembering that this type of immunity can cause release of many toxic substances from the activated T-cells as well as extensive invasion of the tissues by macrophages and their subsequent effects, one can well understand that the eventual result of some delayed-reaction allergies can be serious tissue damage. The damage normally occurs in the tissue area where the inciting antigen is present, such as the skin in the case of poison ivy or the lungs when asthma attacks are due to exposure to some air-borne antigens.[111]

Some people have an "allergic" tendency (atopy). It is caused by nonordinary response of the immune system. The allergic tendency is genetically passed on from parent to child and is characterized by the presence of large quantities of IgE antibodies. These antibodies are called reagins or sensitizing antibodies to distinguish them from the more common IgG antibodies. When an allergen enters the body, an allergen-reagin reaction takes place and is followed by a subsequent allergic response.[111] We have observed at the EHC-Dallas and Buffalo that only about 5% of the hypersensitivity reactions are IgE mediated.

A special characteristic of the IgE antibodies (the reagins) is a strong propensity to attach to mast cells and basophils. Indeed, a single mast cell or basophil can bind as many as half a million molecules of IgE antibodies. Then, an antigen (an allergen) that has multiple binding sites binds with several of the IgE antibodies attached to a mast cell or basophil. This causes an immediate change in the cell membrane. This reaction, perhaps, results from a simple physical effect of the antibody molecules being pulled together by the antigen. At any rate, many of the mast cells and basophils rupture; others release their granules without rupturing and secrete additional substances not already preformed in the granules. Some of the many substances that are either released immediately or secreted shortly thereafter include histamine, slow-reacting substance of anaphylaxis (which is a mixture of toxic leukotrienes), eosinophil chemotactic substance, a protease, a neutrophil chemotactic substance, heparin, and platelet-activating factors. These substances cause dilatation of the local blood vessels, attraction of eosinophils and neutrophils to the reactive site, damage to the local tissues by the protease, increased permeability of the capillaries, loss of fluid into the tissue, and contraction of local smooth muscle cells. **Therefore, a number of different types of abnormal tissue responses can occur, depending on the type of tissue in which the allergen-reagin reaction** (the body's state of nutrition and the local tissue resistance) occurs. We want to emphasize that 5–20% of the population have atopic tendencies but that many more people have sensitivities that are not mediated by IgE. Data shows that anyone in society can develop sensitivities. **In a large series of patients with chemical sensitivity (over 20,000), we have seen only 5% to be IgE-mediated**. Among the different types of allergic reactions caused in this manner are anaphylaxis, angioedema, and urticaria.

Since these are all vascular phenomena, they are discussed in the vascular chapter. In addition, since the vascular system is the common end-organ response, other non-IgE mechanisms can come into play, resulting in anaphylaxis, angioedema, and urticaria.

Over 150 cases of recurrent anaphylaxis have been treated and improved at the EHC-Dallas and Buffalo. This excellent result was accomplished using the following treatment: total body pollutant load was reduced and maintained at a low level by performing and maintaining a massive avoidance program of pollutants in the air, food, and water, and a rotary diet of organic foods, eliminating the severe, specific offenders. Injection therapy to neutralize the food, biological inhalants, and chemical sensitivities was instituted. Adequate nutrition in the form of oral and intravenous vitamins, mineral, amino acids, and lipids was maintained, and ALF was administered. (See Vascular chapter in Volume 2 - Mechanisms of Chemical Sensitivity and Chronic Degenerative Disease for more information.)

AGING AND THE IMMUNE SYSTEM

The study of aging and the immune system is extremely important because some changes occur in children, some occur after twenty years, and some occur into the mid thirties. Aging is a complex process that negatively imparts the development of the immune system and its ability to function. **T-cell function is altered in vivo and in vitro in the elderly compared** to the Young.[189] These changes are generously perceived as reflecting a deterioration of the immune system. Most tests for T-cell function are depressed in elderly patients[190] including the rejection phenomena.[191] Several studies have suggested a positive association between good T-cell function in vitro and individual longevity[192–194] and between absolute lymphocyte count and longevity.[195]

Cytotoxic T-cells are compromised in old mice and in humans. It has been found that this compromise is the result of an age-related decrease in the proportion of cells expression of perforin and the amount of perforin per cell.[196] **There is an age associated decline in both the humoral and cellular response.**[197] The former may be the result of the latter, because observed changes in both the B-cell germline encoded repertoire and the age-associated decrease in somatic hypermutation of the B-cell antigen receptor (BCR) are now known to be critically affected by helper T-cell aging.[198]

Deterioration of the immune system with aging is believed to contribute to morbidity and mortality in man due to the greater incidence of infection, autoimmunity, and cancer. The immuno-senescence mechanisms that underlie these age-related defects range from changes in the (a) hematopoietic bone marrow (HBM); b) stem cell defects; c) thymus involution; d) defects in antigen presenting cells (APC); e) aging of the resting immune cells; f) disrupted activation pathways in immune cells; g) replicative senesence of clonally expanding cells. Each of these functions will be discussed separately.

A. Bone Marrow. Hematopoiesis is compromised because of the severely reduced capacity to produce colony stimulating factors[199] and increased production of proinflammatory factors such as IL-6[200] because lower numbers of progenitor cells are present in the bone marrow and because of the age-related decline in proliferative potential of putative haematopoietic stem cells (HSC).[201] There are fewer bone marrow precursors migrating to the thymus.

Hematopoietic stem cells from adult bone marrow were found to have shorter telomeres than fetal liver-derived or umbilical cord-derived stem cells. A survey of 500 bone marrow transplant patients[202] concluded that aging was associated with reduced numbers of progenitor cells.[203] These finding are consistent with the conclusion that reducing the amount of proliferation stress on the HSC (environmnetal control, fasting, and decrease in total body load) results in better telomere length, despite the presence of telomerase and provide evidence for replicate aging of the hematopoietic stem cells. It is also suggested that ways of manipulating telomere length may be applicable therapeutically in this and other contacts.[204,205] **Oxidation-resistant vitamin C formulations may be one possible way to accomplish this**[206] CD34 cells mobilize less effectively in cytokine output in the elderly as compared to young donors.[207]

B. Stem Cell. Embryonic and adult stem cells are poorly studied. It is not known what the number is in aging, and/or whether they are sufficient to reproduce, and/or whether these numbers diminish with age. There appears to be defects with age. It is known however that the proliferation activity of the primitive cells is greatly reduced over the first year of life and that there is a compensatory increase in relative and absolute stem cell number with age to a point. The changes are strain dependent in mice and related to both longevity and the age of the individual mouse.[208] This change may not have anything to do with humans however, it is likely to happen. The aging of stem cells and/or T-cell precursors may directly influence processes of thymic involution.[209,210] There is a decline in the proliferative potential of putative hematopoietic stem cells.[197]

C. Thymus Gland. Chronic involution of the thymus gland is thought to be one of the major contributing factors to aging however, some conditions have reversible thymus functions. Temporary thymic involution occurs with pregnancy and is restored fully at the end of lactation. In animals it can occur during the various seasons. Therefore, dynamic hormonal changes can restore thymic function.

Changes characteristic of thymic involution begin during or shortly after the first year of birth and occur progressively throughout life. The 3–5% annual reduction rate continues until middle age and then slow down to less than 1% per year for the rest of life until 120 years. This **serious reduction of the thymus cellular micro environment is a well controlled physiological process and is presumably under both local and global regulation by the cells of the reticular endothelial system mesh work and the neuroendocrine system**. Serum thymic function levels start to decline at 20 years of age and are completely gone by age 50 or 60, while thymosin and thymopeoesis levels seem to decline earlier starting at the age of 10. However, it has now been shown by Goldstein[211,212] that all cells of the body produce thymosin. Other hormones such as testosterone, estrogen, and hydrocortisone result in marked involution of the thymus. The thymus is the body's clock genetically programed for eventual death. **Only thymosin administration or transplantation will improve the thymus function**. There has been thymic hormone extracted from thymus glands. These have been found to be two different substances with different reactions. Although it has been 30 years of experimentation some of the results look promising.[210]

Infection, pregnancy, stress, drug, or hibernation induced thymic involution are all reversible in younger individuals, which leads to the suggestion that thymic atrophy is an energy-saving process.[213] Since thymic involution occurs around the age of 30 there are several reasons why this may occur. **First**, there is an avoidance of undesired tolerization of newly generated T-cells to pathogens that in later life have entered the thymus;[214] **second**, full T-cell generation was secondary to the generation in early life of a memory cell repertoire for a mostly limited pathogen presence; and **third**, decreased T-cell response in the elderly may be due to decreased thymic-function, decreased accessory cell (monocytes, dendritic cell) function or both.

T-cells require stimulation via the antigen-specific TCR for activation. In addition they also require stimulation via nonpolymorphic antigen nonspecific costimulating receptor (i.e., CD-28) and by molecules expressed on APC. Aberrations of these molecules and their receptors would also lead to compromised T-cell responses. **The degree of phosphorylization decreases in CD_3 with aging**. Cytokine function and response may decrease with aging.

Markers of activated macrophages neopterin, IL-6[+1], IR antagonist, and sTNF-R are elevated in the plasma of the healthy elderly. IL-1, IL-3, IL-4, IL-6, and TNF-alpha are increased with the aged and are known to control isotype switch and immunoglobulin production during B-cell differentiation. In particular, along with B-cell proliferation IL-6 stimulates thymic and peripheral T-cells and, in cooperation with IL-1, induces T-cell differentiation to cytotoxic T-cells and activates NK cells.[215–220] These observations emphasize the importance of IL-6 in both the nonspecific and specific immune responses as well as a variety of other systems and may also be relevant to several aspects of age-associated pathological events including arteriosclerosis, osteoporosis, fibrosis, and dementia.

D. There can be a defective feedback of **aged CD4+ cells** on thymocyte development and differentiation.[221,189]

E. **Immunosenescence** is characterized by the shrinkage of the T-cell repertoire, the accumulation of polyoclonal expansions (mega clones) of memory/effector cells directed toward ubiquitous infectious agents, the involution of the thymus, and the exhaustion of naive T-cells and the chronic inflammatory status (inflammation of aging). Deterioration of the immune system (immunosenescence) is believed to contribute to morbidity and mortality in man due to a greater incidence of infection, autoimmune phenomena, and cancer. **Deregulation of the T-cell function is thought to play a critical role in this process of senescence of aging**. Most tests of T-cell function both in vitro and in vivo are altered in the elderly. Several studies have suggested a positive association with good T-cell function and absolute T-cell counts with longevity.[189,191–194] two year nonsurvival is associated with a cluster of poor T-cell proliferations to mitogen, high CD_8 (cytotoxic/suppressor cell), low CD_4 (helper/delayed type hypersensitivity) and CD19 (B) cell counts.

A decrease of CD_3 cells may occur with age but there appears to be an increase in memory cells. Also, here is an increase in activated T-cells (HLA-DR, CD25) and **killer (NK)** cells. The "homing" environment of lymphocytes to lymph nodes and organs seems to deteriorate with age and thus blood and lymph organ counts may be different. **An increase in nonpathogenic auto antibody occurs with age** probably because some of such antibodies can penetrate living cells and activate them[222] helping to explain deregulated T-cell function in the elderly. Peripheral nonthymic regulatory control may be damaged in the elderly.[222–226] Regulating CD_8 secretion of TGF-B[227] is decreased. See Figure 3.37.

F. Despite declining immune function, aging is associated with increased auto antibodies. The Th1/Th2 paradigm is an important conceptual tool to characterize T-helper cells, since cytokine profiles appear to be skewed toward these two major types and some of the data discussed may be viewed in terms of Th1 and Th2 shifts with age. The Th1 cells are characterized by their ability to secrete IFN-gamma, while the Th2 cells are characterized by IL-4 secretions. **The Th1 cells are considered proinflammatory cells, important against intracellular pathogens requiring cell-mediated immunity where as Th2 cells are anti-inflammatory and mediate humoral responses**.[229,230] However, cytokine regulation may be different with IL-2 gamma, TNF-B, IL-12, and IL-15 supporting primarily stimulatory cellular responses, while IL 4, IL 5, IL 6, IL 10, IL 13 primarily support humoral response. T-cells when stimulated produce IL-2, which is dominant in the elderly. IL-2 is now considered the key factor in the production of the cellular response. It is impaired in the elderly because of reduced efficiency of costimulatory signals. T-cell stimulation results in preferential Th2 responses.

G. Once the T-cells have been successfully stimulated by an antigen, costimulated by cytokine availability, and with utility assured, **the T-cell response requires waves of clonal expansion followed by contraction when the antigen is no longer present and reexpansion on contact with**

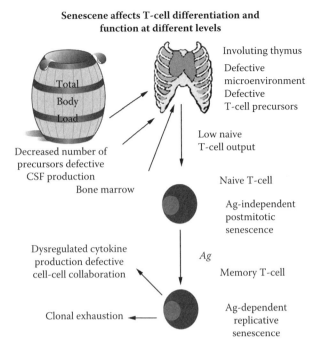

FIGURE 3.37 Senescence affects T-cell differentiation and function at different levels. First, there is a decreased number of bone marrow precursors migrating to the thymus. In the involuting thymus the function of both the thymocytes and the cells from the microenvironment is compromised in the ability to support T-cell differentiation, resulting in a decreased output of new naive T-cells in the elderly. Long-lived naive T-cells suffer antigen-independent postmitotic senescence even in their quiescent state. T-cells which have been activated and become memory cells are maintained in a proliferative state and are, therefore, subjected to replicative senescence and clonal exhaustion. These changes result in a defective capacity of T-cells to collaborate in other aspects of the immune response, hematopoiesis and lymphopoiesis. (Modified from Pawelec, G., Effros, R. B., Caruso, C., Remarque, E., Barnett, Y., and Solana, R., *BioScience,* 53, Fig. 1, update February 1999.)

the antigen again. McCarron et. al.[231] found that T-cell expansion for longevity of clones averaged 52 PD in neonates, adults (20–30 years) 40 PD, and the elderly (70–90 years) 32 PD. Human T-cells infected with HTL-V-1 (Herpes saimiri virus) are immortal. Though this possibility is rare, it may be possible for some other substances (i.e. mycoplasma, foods, molds, chemicals) may activate this propensity to immortality. This phenomenon also appears to be true with fibroblasts.

Most T-cell clones are from the peripheral blood and may not be representative of the T-cell pool. **Major lymphoid organs on which the T-cell can be obtained are skin and gut**. These cells appear to grow indefinitely. The majority of patients had dermatitis and thus due to the clinical condition showed aberrant T-cell immortality.[232–234] However, even these cells showed T-cells decreasing expansion of CD28 expression.

A loss of telomeric DNA and gradual shortening of telomeres has been proposed to result after a certain number of cell divisions that occur with aging. This phenomenon happens in fibroblasts as well as T-cells. **However, inability to repair also occurs with age**. The rate of telomere shrinking may not always be linear and may be hormetic and geometrical depending upon the patient.[235–247]

There is some evidence that memory cells are not quiescent long-lived cells, but represent T-cell clones in a constant state of activation.[248–253] **At one point memory cells may not require an antigen to proliferate and have a lower threshold of triggering as seen in many cases of mold, food and chemical sensitivity**.[252] Even with thymectomy, though naive T-cells disappear, memory cells are long-lived.[253] **Therefore, memory T-cells may have a life expectancy of 15–34 years**.[254] DNA repair is decreased in the elderly.[290–299]

However, antigens may persist at low level such as virus, fungal, bacterial, or chemical. It is clear in patients with chemical sensitivity and chronic degenerative disease that they are constantly being exposed to environmental triggers but usually they are triggered at a lowered threshold.

Alterations of T-cells numbers are not significant in the aged versus the young. However, age-associated changes occur in the T-subsets with a relative loss of CD_4 cells.[260,261] **There are more memory cells and fewer naive T-cells in the elderly**. This fact is coupled with the decreased repression to new antigens and a retained ability to respond to recall antigens. Studies in vivo in the elderly have shown a decrease in CD 45RB expression in CD4 cells.[255,256] **CD 95 also decreases in some patients with old age**.[257] Several studies have shown a decreased CD_3 proliferation in the elderly.[258,259]

Also in humans most T-cells are CD7, but the frequency of CD7 negative cells increases with age.[262] However, **CD28 expression is the closet biomarker of aging as they decrease with age. CD_8 cells also decrease with age**.[263] Thus in CD_8 cells up to 30% of the entire population may consist of oligo or even monoclonal cells expressing the same TCR-VB marker.[264,265] Alterations in CD_8 subsets may be more common than CD_4 in the elderly.[266] Moreover, the realization that mature thymus derived T-cells can reacquire sensitivity to positive and negative selection **outside the thymus, in germinal centers**[267] **indicate that in theory the generation and selection of T-cells may take place even in the absence of a functional thymus**. The generation of functional mature T-cells with diverse TCR2 repertoires from CD34+ human stem cells in the absence of thymic influence in vitro indicates a potential approach to enhance T-cell generation despite compromised thymic function.[271] ALF developed at the EHC-Dallas appears to be one of these substances. (See Treatment chapter Volume 4 - Mechanisms of Cardiovascular Disease and Chemical Sensitivity for more information.)

The incidence and severity of infectious disease increases with the elderly more than twice the young.[269–274] **Actually, the prime cause of death in aged 80 or over is infection**.[275–281] This propensity is associated with increase neutrophils and more IL-8 in the elderly. There is an increased susceptibility to chemical fumes, i.e., Teflon® in aged mice lungs. Thus, there is likely more chemical sensitivity in humans since these reports show the aged have increased sensitivity to more chemical fumes. Vaccination may have a less long effect in the elderly due to less cell mediated immunity.[282–289]

SUMMARY

Both innate and acquired immunity can be and are altered in individuals with chemical sensitivity and chronic degenerative disease. Many different areas are involved in specific patients depending on the hormetic effect of the chemical and biochemical individuality of the patient's homeostatic response. With pollutant overload changes can occur in the lymphatic channels, the lymph nodes and lymph node egress as well as the lymphatic cells. Changes in mucosal function and the effects of the autonomic nervous system are evident with environmental pollutant overload.

REFERENCES

1. Department of Labor. 1977. Occupational exposure to benzene. Federal Register 42:22516.
2. National Institute for Occupational Safety and Health. 1976. U.S. Department of Health, Education, and Welfare. Criteria for a Recommended Standard Occupational Exposure to Formaldehyde. (NIOSH) 77–126 Washington, DC: U.S. Government Printing Office 21–81.
3. Rea WJ. 1994. *Chemical Sensitivity, Vol. II, Sources of Total Body Load*. Boca Raton, FL: Lewis Publishers. 889–893.
4. Calabrese EJ 2005. Hormetic dose-response relationships in immunology: occurrence, quantitative features of the dose response, mechanistic foundations, and clinical implication. *Crit Rev in Toxic*. 35:89–295.
5. Wu J, Cui N, Piao H, et al. 2002. Allosteric modulation of the mouse Kir 6.2 channel by intracellu7lar H+ and ATP. *J. Physiol*. 543(1&2):495–504.

6. *National Geographic.* 2005. What are embryonic stem cells? July:12–13.
7. Bishop AE. Aug. 15, 2008. Chaimanís Summary. Proceedings of the American Thoracic Society (5)6:673.
8. Carlson, B. 2003. *Human Embryology and Developmental Biology: Development of Body Systems.* Mosby. 414.
9. Riquet M, LePimpec Barthes F, Souilamas R, and Hidden, G. 2002. Thoracic Duct Tributaries from Intrathoracic Organs. *Ann Thorac Surg.* 73:892–899.
10. Riquet M, LePimpec Barthes F, Souilamas R, and Hidden, G. 2002. Thoracic Duct Tributaries from Intrathoracic Organs. *Ann Thorac Surg.* 73:893, Fig. 1.
11. Riquet M, LePimpec Barthes F, Souilamas R, and Hidden, G. 2002. Thoracic Duct Tributaries from Intrathoracic Organs. *Ann Thorac Surg.* 73:896, Fig. 5.
12. Riquet M, LePimpec Barthes F, Souilamas R, and Hidden, G. 2002. Thoracic Duct Tributaries from Intrathoracic Organs. *Ann Thorac Surg.* 73:896, Fig. 6.
13. Riquet M, LePimpec Barthes F, Souilamas R, and Hidden, G. 2002. Thoracic Duct Tributaries from Intrathoracic Organs. *Ann Thorac Surg.* 73:897, Fig. 7.
14. Riquet M, LePimpec Barthes F, Souilamas R, and Hidden, G. 2002. Thoracic Duct Tributaries from Intrathoracic Organs. *Ann Thorac Surg.* 73:898, Fig. 8.
15. Ikeda S, Kudsk KA, Fukatsu K, et al. 2003. Enteral feeding preserves mucosal immunity despite in vivo MadCAM-1 blockade of lymphocyte homing. *Ann. Surg.* 237(5):677–685.
16. Connor EM, Eppihimer MJ, Morise Z, et al. 1999. Expression of mucosal adressin cell adhesion molecule-1 (MAdCAM-1) in acute and chronic inflammation. *J. Leukoc. Biol.* 65:349–355.
17. Briskin M, Winsor-Hines D, Shyjan A, et al. 1997. Human mucosal adressin adhesion molecule-1 is preferentially expressend in intestinal tract and associated lymphoid tissue. *Am. J. Path.* 151:97–110.
18. Kraal G, Schornagel K, Streeter PR, et al. 1995. Expression of the mucosal vascular adressin, MAdCAM-1, on sinus-lining cells in the spleen. *Am. J. Path.* 147:763–771.
19. Csencsits KL, Jutila MA, and Pausal DW. 1999. Nasal-associated lymphoid tissue: Phenotypic and functional evidence for the primary role of peripheral node adressin in naïve lymphocyte adhesion to high endothelial venules in a mucosal site. *J. Immunol.* 163:1382–1389.
20. Li J, Kudsk KA, Gocinski B, et al. 1995. Effects of parenteral nutrition on gut-associated lymphoid tissue. *J. Trauma.* 39:44–52.
21. Kudsk KA, Ki J, and Renegar KB. 1996. Loss of upper respiratory tract immunity with parenteral feeding. *Ann. Surg.* 223:629–638.
22. King BK, Kudsk KA, Li J, et al. 1999. Route and type of nutrition influence mucosal immunity to bacterial pneumonia. *Ann. Surg.* 229:272–278.
23. Kudsk KA, Croce MA, Fabian TC, et al. 1992. Enteral versus parenteral feeding: Effects on septic morbidity after blunt and penetrating abdominal trauma. *Ann. Surg.* 215:503–513.
24. Moore FA, McCroskey BL, Haddis T., et al. 1988. Total enteral nutrition vs total parenteral nutrition after major torso injury: Attenuation of hepatic protein reprioritization. *Surgery* 104:199–207.
25. Moore EE and Jones TN. 1986. Benefits of immediate jejunostomy feeding after major abdominal trauma: A prospective, randomized study. *J. Trauma.* 26:874–879.
26. Guyton AC and Hall JE. 1996. *Textbook of Medical Physiology,* 9th ed. Philadelphia: WB Saunders Co. 193–194.
27. Guyton AC and Hall JE. 1996. *Textbook of Medical Physiology,* 9th ed. Philadelphia: WB Saunders Co. 446.
28. Thymosins MJ: 2007. Clinical promise after a decades-long search. *Science* 316:682–683.
29. Felten SY and Olschowka JA. 1987. Noradrenergic sympathetic innervation of the spleen: II. Tyrosine hydroxylase (TH)-positive terminal from synaptic-like contacts on lymphocytes in the splenic white pulp. *J. Neurosci. Res.* 18:37–48.
30. Elenkow IJ and Vizi ES. 1991. Presynaptic modulation of release of noradrenaline from the sympathetic nerve terminal in the rat spleen. *Neuropharmacology* 30:1319–1324.
31. Friedman EM and Irwin MR. 1997. Modulation of immune cell function by autonomic nervous system. *Parmacol. Ther.* 74:27–38.
32. Sanders VM and Munson AE. 1984. β-Adrenoceptor mediation of the enhancing effect of norephinephrine on the murine primary antibody response in vitro. *J. Paharmacol. Exp. Ther.* 230:183–192.
33. Bulloch K and Pomerantz W. 1984. Autonomic nervous system innervation of thymic-related lymphoid tissue in wild-type and nude mice. *J. Comp. Neurol.* 228:57–68.
34. Calvo W. 1968. The innervation of bone marrow in laboratory animals. *Am. J. Anat.* 123:315–328.

35. Felten DL, Ackerman KD, Wiegand SF, and Felten SY. 1987. Noradrenergic sympathetic innervation of the spleen. I. Nerve fibers associated with lymphocytes and macrophages in specific compartments of the splenic white pulp. *J. Neurosci. Res.* 18:28–36.
36. Felten SY and Felten DL. 1991. Innervation of lymphoid tissue. In *Psychoneuroimmunology,* 2nd ed. Ed. R Ader, DI Felten, and N. Cohen. San Diego, CA: Academic Press. 27–69.
37. Nohr D and Weihe E. 1991. The neuroimmune link in the bronchus-associated lymphoid tissue (BALT) of cat and rat: Peptides and neural markers. *Brain. Behav. Immun.* 5:84–101.
38. Weihe E and Krekel J. 1991. The neuroimmune connection in human tonsils. *Brain. Behav. Immun.* 5:41–54.
39. Felten DL, Overhage JM, Felten SY, and Schmedtje JF. 1981. Noradrenergic sympathetic innervation of lymphoid tissue in the rabbit apprendix: Further evidence for a link between the nervous and immune systems. *Brain. Res. Bull.* 7:595–612.
40. Felten DL, Felten SY, Carlson SL, Olschowska JA, and Livant S. 1985. Innervation of lymphoid tissue. *J. Immunol.* 135:755–765s.
41. Giron LT, Crutcher KA, and Davis JN. 1980. Lymph nodes: A possible site for sympathetic neuronal regulation of immune responses. *Ann. Neurol.* 8:520–522.
42. Jesseph JM and Felten DL. 1984. Noradrenergic innervation of the gut-associated lymphoid tissues (GALT) in the rabbit. *Anat. Rec.* (Abstract) 208:A81.
43. Yamashita T, Kumazawa H, Kozuki K, Amano H, Tomado K, and Kumazawa T. 1984. Autonomic nervous system in human palatine tonsil. *Acta. Otolaryngol.* Suppl 416:63–71.
44. DePace DM and Webber RH. 1975. Electrostimulation and morphologic study of the nerves to the bone marrow of the albino rat. *Acta. Anatomic.* 93:1–18.
45. Kudoh G, Hoshi K, and Murakami T. 1979. Fluorescence micropscopic and enzyme histochemical studies of the innervation of the human spleen. *Arch. Histol. Jap.* 42:169–180.
46. Ottaway CA, Lewis DL, and Asa SL. 1987. Vasoactive intestinal peptide-containing nerves in Peyerís patches. *Brain. Behav. Immun.* 1:148–158.
47. Fatani JA, Quayyum MA, Mehta I, and Singh U. 1986. Parasympathetic innervation of the thymus: A hitochemical and immunocytochemical study. *J. Anat.* 147:115–119.
48. Crowe SE and Perdue MH. 1992. Gastrointestinal food hypersensitivity: Basic mechanism of pathophysiology. *Gastroenterology* 103:1075–1095.
49. Shanahan F, Denburg JA, Fox J, Bienenstock J, and Befus AD. 1985. Mast cell heterogeneity. Effects of neurocenteric peptides on histamine release. *J. Immunol.* 135:1331–1337.
50. Besedowsky HO, Sorkin E, Felix E, and Hass H. 1977. Hypothatlmic changes during the immune response. *Eur. J. Immunol.* 7:323–325.
51. Hölt V, Garzon J, Schulz R, and Herz A. 1984. Corticotropin-releasing factor is excitatory in the guinea-pig ileum and activates an opioid mechanism in their tissue. *Eur. J. Pharmacol.* 101:165–166.
52. Kannan Y, Bienenstock J, Ohta M, Stanisz AM, and Stead RH. 1996. Nerve growth factor and cytokine mediate lymphoid tissue-induced neurite out-growth from mouse superior cervical ganglia *in vitro*. *J. Immunol.* 156:313–320.
53. Karalis K, Muglia LJ, Bae D, Hilderbrand H, and Majzoub JA. 1997. CRH and the immune system. *J. Neuroimmunol.* 72:131–136.
54. Miura M, Inoue H, Ichinose M, Kimura K, Katsumata U, and Takishima T. 1990. Nonadrenergic noncholinergic inhibitory nerve stimulation on the allergic reaction in cat airways. *Am. Rev. Respir. Dis.* 141:29–32.
55. Weinreich D, Unolem BJ, and Leal-Caroloso JM. 1992. Functional effects of mast cell activation in sympathetic ganglia. *Ann. NY. Acad. Sci.* 664:293–308.
56. Djuric VJ, Wang L, Bienenstock J, and Perdue MH. 1995. Naloxone exacerbates intestinal and systemic anaphylaxis in the rat. *Brain. Behav. Immun.* 9:87–100.
57. McKay DM, Berin MC, Fondacaro JD, and Perdue MH. 1996. Effects of neuropeptide Y and substance P on antigen-induced ion secretion in rat jejunum. *Am. J. Physiol.* 271:G987–992.
58. McKay DM, Bienenstock J, and Perdue MH. 1993. Inhibition of antigen-induced secretion in the rat jejunum by interferon α/β. *Reg. Immunol.* 5:43–59.
59. Bucala R. 1998. Neuroimmunomodulation by macrophage migration inhibitory factor (MIF). *Ann NY Acad Sci* 840:74–82.
60. Holzer P. 1988. Implications of tachykinins and calcitonin gene-related peptide in inflammatory bowel disease. *Digestion* 59:269–283.
61. Manzini S, Maggi CA, Geppetti P, and Bacciarelli C. 1987. Capsaicin desensitization protects from antigen induced bronchospasms in conscious guinea-pigs. *Eur. J. Pharmacol.* 138:307–308.

62. Minault M, Lecron JC, Labrouche S, Simonnet G, and Gombert J. 1995. Characterization of binding sites for neuropeptide FF on T lymphocytes of the Jurkat cell line. *Peptides* 16:105–111.
63. Stanisz AM. 1994. Neuronal factors modulating immunity. *Neuroimmuno-modulation* 1:217–230.
64. Hollingsworth JW, Kleeberger SR, and Foster WM. Ozone and pulmonary innate immunity. *Proc. Am. Thorac. Soc.* 4:240–246.
65. Befus AD, Mowat C, Hu J, et al. 1999. Neutrophil defensins induce histamine secretion from mast cells: Mechanism of action. *J. Immunol.* 163:947–953.
66. Foreman JC. 1993. Non-immunological stimuli of mast cells and basophil leukocytes. In *Immunopharmacology of Mast Cells and Basophils* Ed. JC Foreman. San Diego, CA: Academic Press. 57–69.
67. Leung KBP, Flint KC, Hudspith BN, et al. 1987. Some further properties of human pulmonary mast cells recovered by bronchoalveolar lavage and enzymic dispersion of lung tissue. *Agent Actions* 20:213–215.
68. Mousli M, Hugli TE, Landry Y, and Bronner C. 1994. Peptidergic pathway in human skin and rat peritoneal mast cell activation. *Immunopharmacology* 27:1–11.
69. Foreman JC, Mongar JL, and Gomperts BD. 1973. Calcium ionophores and movement of calcium ions following the physiological stimulus to a secretory process. *Nature* 245–251.
70. Nakamura T, Ui M. 1984. Islet-activating protein, pertussis toxin, inhibits Ca^{2+}-induced and guanine nucleotide-dependent release of histamine and arachindonic acid from rat mast cells. FEBS *Lett* 261:171–174.
71. Saito H, Okajima F, Molski TF, et al. 1987. Effect of a ADP-ribosylation of GTP-binding protein by pertussis toxin on immunoglobulin E-dependent histamine release from msat cells and basophils. *J. Immunol.* 138:3927–3934.
72. Henderson WR and Kaliner M. 1978. Immunologic and non-immunologic generation of superoxide from mast cells and basophils. *J. Clin. Invest.* 61:187–196.
73. Henderson WR and Kaliner M. 1979. Mast cell granule peroxidase: Location secretion and SRS-A inactivation. *J. Immunol.* 122:1322–1328.
74. Nie XF, Ibbotson G, and Kubes P. 1996. A balance between nitric oxide and oxidants regulates mast cell-dependent neutrophil-endothelial cell interaction. *Circ. Res.* 79:992–999.
75. Galli SJ, Maurer M, and Lantz CS. 1999. Mast cells as sentinels of innate immunity. *Curr Opin Immunol* 11:53–59.
76. Hawlisch H, Köhl J. 2006. Complement and toll-like receptors: Key regulators of adaptive immune responses. *Mol. Immunol.* 43:13–21.
77. Traube C, Rha YH, Takeda K. Park JW, Joetham A, Balhorn A, Dakhama A, Giclas PC, Holers VM, and Gelfand EW. 2003. Inhibition of complement activation decreases airway inflammation and hyper-responsiveness. *Am. J. Respire. Crit. Care. Med.* 168:1333–1341.
78. Krug N, Tschernig T, Erpenbeck VJ, Hohlfeld JM, and Köhl J. 2001. Complement factors C3a and C5a are increased in bronchoalveolar lavage fluid after segmental allergen provocation in subjects with asthma. *Am. J. Respir. Crit. Care. Med.* 164:1841–1843.
79. Brostroff J and Gamlin L.2000. *Food Allergies and Food Intolerance: The Complete Guide to Their Identification and Treatment.* Rochester, VT: Healing Arts Press. 47, Fig. 3.4.
80. Guyton AC and Hall JE. 1996. *Textbook of Medical Physiology,* 9th ed. Philadelphia: WB Saunders Co. 445–453.
81. WebMD Medical News. April 27, 2000. "Gene Therapy Frees 'Bubble Children.'" Marina Cavazzana-Calvo, MD, lead researcher on gene therapy team.
82. Jiang H and Chess L, 2006. Mechanisms of disease: Regulation of immune responses by T cells. *NEJM* 354(11):1166–1176.
83. Bevan MJ. 1977. In a radiation chimaera, host H-2 antigens determine immune responsiveness of donor cytotoxic cells. *Nature* 269:417–418.
84. Bevan MJ and Fink PJ. 1978. The influence of thymus H-2 antigens on the specificity of maturing killer and helper cells. *Immunol. Rev.* 42:3–19.
85. Waldmann H. 1978. The influence of the major histocompatibility complex on the function of T-helper cells in antibody formation. *Immunol. Rev.* 42:202–223.
86. Zinkernagel RM. 1978. Thymus and lymphohemopoietic cells: Their role in T cell maturation in selection of T cellsí H-2-restriction-specificity and in H-2 linked Ir gene control. *Immunol. Rev.* 42:224–270.
87. Hengartner H, Odermatt B, Schneider R, et al. 1988. Deletion of self-reactive T cells before entry into the thymus medulla. *Nature* 336:388–390.
88. von Boehmer H and Kisielow P. 1990. Self-nonself discrimination by T cells. *Science* 248:1369–1373.

89. Nossal GJ. 1994. Negative selection of lymphocytes. *Cell* 76:229–239.
90. Sprent J and Webb SR. 1995. Intrathymic and extrathymic clonal deletion of T cells. *Curr. Opin. Immunol.* 7:196–205.
91. Shortman K, Egerton M, Spangrude GJ, and Scollay R. 1990. The generation and fate of thymocytes. *Semin. Immunol.* 2:3–12.
92. Mason D. 1998. A very high level of crossreactivity is an essential feature of the T-cell receptor. *Immunol. Today* 19:395–404.
93. Garcia KC, Degano M, Pease LR, et al. 1998. Structural basis of plasticity in T cell receptor recognition of a self peptide-MHC antigen. *Science* 279:1166–1172.
94. Jiang H and Chess L. 2006. Mechanisms of Disease:Regulation of Immune Responses by T Cells. *NEJM* 354(11):1166–1176.
95. Bouneaud C, Kourilsky P, and Bousso P. 2000. Impact of negative selection on the T cell repertoire reactive to a self-peptide: a large fraction of T cell clones escapes clonal deletion. *Immunity* 13:829–840.
96. Sandberg JK, Franksson L, Sundback J, et al. 2000. T cell tolerance based on avidity thresholds rather than complete deletion allows maintenance of maximal repertoire diversity. *J. Immunol.* 165:25–33.
97. Jiang H, Curran S, Ruiz-Vazquez E, Liang B, Winchester R, and Chess L. 2003. Regulatory CD8+ T cells fine-tune the myelin basic protein-reactive T cell receptor V beta repertoire during experimental autoimmune encephalomyelitis. *Proc. Natl. Acad. Sci. USA* 100:8378–8383.
98. Kuchroo VK, Anderson AC, Waldner H, Munder M, Bettelli E, and Nicholson LB. 2002. T cell response in experimental autoimmune encephalomyelitis (EAE): role of self and cross-reactive antigens in shaping, tuning, and regulating the autopathogenic T cell repertoire. *Annu. Rev. Immunol.* 20:101–123.
99. Jiang H, Wu Y, Liang B, et al. 2005. An affinity/avidity model of peripheral T cell regulation. *J. Clin. Invest.* 115:302–312.
100. Savage PA, Boniface JJ, Davis MM. 1999. A kinetic basis for T cell receptor repertoire selection during an immune response. *Immunity* 10:485–492.
101. Davis MM, Boniface JJ, Reich Z, et al. 1998. Ligand recognition by alpha beta T cell receptors. *Annu. Rev. Immunol.* 16:523–544.
102. Lenardo M, Chan KM, Hornung F, et al. 1999. Mature T lymphocyte apoptosis-immune regulation in a dynamic and unpredictable antigenic environment. *Annu. Rev. Immunol.* 17:221–253.
103. Lederman S, Yellin MJ, Krichevsky A, Belko J, Lee JJ, and Chess L. 1992. Identification of a novel surface protein on activated CD4+ T cells that induces contact-dependent B cell differentiation (help). *J. Exp. Med.* 175:1091–1101.
104. Caux C, Massacrier C, Vanbervliet B, et al. 1994. Activation of human dendritic cells through CD40 cross-linking. *J. Exp. Med.* 180:1263–1272.
105. Klaus SJ, Pinchuk LM, Ochs HD, et al. 1994. Costimulation through CD28 enhances T cell-dependent B cell activation via CD40-CD40L interaction. *J. Immunol.* 152:5643–5652.
106. Boussiotis VA, Freeman GJ, Gribben JG, Nadler LM. 1996. The role of B7-1/B7-2:CD28/CLTA-4 pathways in the prevention of anergy, induction of productive immunity and down-regulation of the immune response. *Immunol. Rev.* 153:5–26.
107. Jenkins MK. 1994. The ups and downs of T cell costimulation. *Immunity* 1:443–446.
108. Lenschow DJ, Walunas TL, Bluestone JA. 1996. CD28/B7 system of T cell costimulation. *Annu. Rev. Immunol.* 14:233–258.
109. Chambers CA, Kuhns MS, Egen JG, Allison JP. 2001. CTLA-4-mediated inhibition in regulation of T cell responses: Mechanisms and manipulation in tumor immunotherapy. *Annu. Rev. Immunol.* 19:565–594.
110. Cohn M. 2004. Whither T-suppressors: If they didnít exist would we have to invent them? *Cell. Immunol.* 227:81–92.
111. Guyton AC and Hall JE. 1996. *Textbook of Medical Physiology,* 9th ed. Philadelphia: WB Saunders Co. 453–473.
112. Guyton AC and Hall JE. 1996. *Textbook of Medical Physiology,* 9th ed. Philadelphia: WB Saunders Co. 450, Fig. 34–3.
113. Guyton AC and Hall JE. 1996. *Textbook of Medical Physiology,* 9th ed. Philadelphia: WB Saunders Co. 450, Fig. 34–4.
114. Guyton AC and Hall JE. 1996. *Textbook of Medical Physiology,* 9th ed. Philadelphia: WB Saunders Co. 450, Fig. 34–5.
115. Immunosciences Lab., Inc. 2004. 8693 Wilshire Blvd., Ste. 200, Beverly Hills, CA 90211. http://www.immunoscienceslab.com
116. Immunosciencelab, Inc. 2004. Toxicological Response in the Immune System Produced by Chemicals. http://www.immunoscienceslab.com

117. Guyton AC and Hall JE. 1996. *Textbook of Medical Physiology,* 9th ed. Philadelphia: WB Saunders Co. 451, Fig. 34–6.
118. Aristo Vojdani MT. 1998. *The Role of Natural Killer Cell, Apoptosis and Cell Cycle in Health and Diseases.* 13.
119. Vojdani A. 1998. *The Role of Natural Killer Cell, Apoptosis and Cell Cycle in Health and Diseases.* 14.
120. Guyton AC and Hall JE. 1996. *Textbook of Medical Physiology,* 9th ed. Philadelphia: WB Saunders Co. 452, Fig. 34–7.
121. Vojdani A.. 1998. *The Role of Natural Killer Cell, Apoptosis and Cell Cycle in Health and Diseases.* 6.
122. Vojdani A. 1998. *The Role of Natural Killer Cell, Apoptosis and Cell Cycle in Health and Diseases.* 15.
123. Vojdani A. 1998. *The Role of Natural Killer Cell, Apoptosis and Cell Cycle in Health and Diseases.* 48–50.
124. Bickel C and Huey P. 2005. *Science* 309: 1682.
125. Brinkmann, V, Cyster JG, and Hla T. 2004. *Am. J. Transplant* 4:1019.
126. Allende ML, Dreier JL, Mandala S, and Proia RL. 2004. *J. Biol. Chem.* 279:15396.
127. Saba JD and Hla T. 2004. *Circ. Res.* 94:724
128. Schwab SR, Pereira JP, Matloubian M, Xu Y, Huang Y, and Cyster JG. 2005. Lymphocyte sequestration through SIP lyase inhibition and disruption of SIP gradients. *Science* 309:1735–1739.
129. Saba JD, Nara R, Bielawska A, Garrett S, and Hannun YA. 1997. *J. Biol. Chem.* 272:26087.
130. Ancellin N, et al. 2002. *J. Biol. Chem.* 277:6667.
131. Hia, T. 2004. *Semin. Cell Dev. Biol.* 15:513.
132. Schwab SR, Pereira JP, Matloubian M, Xu Y, Huang Y, and Cyster JG. 2005. Lymphocyte sequestration through SIP lyase inhibition and disruption of SIP gradients. *Science* 309:1736, Fig. 1.
133. Schwab SR, Pereira JP, Matloubian M, Xu Y, Huang Y, andCyster JG. 2005. Lymphocyte sequestration through SIP lyase inhibition and disruption of SIP gradients. *Science* 309:1736, Fig. 2.
134. Lee MJ, et al. 1998. *Science* 279:1552.
135. Schwab SR, Pereira JP, Matloubian M, Xu Y, Huang Y, and Cyster JG. Lymphocyte sequestration through SIP lyase inhibition and disruption of SIP gradients. *Science* 309:1737, Fig. 3.
136. Le Stunff H, Milstien S. and Spiegel S. 2004. *J Cell Biochem* 92:882.
137. Saba JD and Hla T. 2004. *Circ. Res.* 94:724.
138. Van Veldhoven PP. 2000. *Methods Enzymol.* 311:244.
139. Materials and methods are available as supporting material on Science Online.
140. Schwab SR, Pereira JP, Matloubian M, Xu Y, Huang Y, and Cyster JG. 2005. Lymphocyte sequestration through SIP lyase inhibition and disruption of SIP gradients. *Science* 309:1738, Fig. 4.
141. Vojdani A. 1998. *The Role of Natural Killer Cell, Apoptosis and Cell Cycle in Health and Diseases.* 20, Fig. 11.
142. Vojdani A. 1998. *The Role of Natural Killer Cell, Apoptosis and Cell Cycle in Health and Diseases.*
143. Vojdani A. 1998. *The Role of Natural Killer Cell, Apoptosis and Cell Cycle in Health and Diseases.* 21, Fig. 11.
144. Vojdani A. 1998. *The Role of Natural Killer Cell, Apoptosis and Cell Cycle in Health and Diseases.* 23, Tab. 2.
145. Vojdani A. 1998. *The Role of Natural Killer Cell, Apoptosis and Cell Cycle in Health and Diseases.* 22, Fig. 13.
146. Guyton AC and Hall JE. 1996. *Textbook of Medical Physiology,* 9th ed. Philadelphia: WB Saunders Co. 452–453.
147. Rea WJ. 1996. *Chemical Sensitivity, Vol. III, Clinical Manifestation of Pollutant Overload.* Boca Raton, FL: Lewis Publishers.
148. Anthony H and Maberly J. 2000. Multiple chemical sensitivity. *Jr. Soc. Med.* 93(3):160–161.
149. King KL, Schwartzman RA, and Cidlowski JA. 1996. Apoptosis in life, death and the cell cycle. *Endrocinology and Metabolism* 3(Supp. A):93–97.
150. Heine H. 1999. *Homotoxicology: A Synthesis of Medical Schools of Thought on a Scientific Basis.* Institut fur Antihomotoxische Medizin und Grundregulation Storschung Bahnackystrabe 16 D-76532 Baden-Baden, Germany. 5.
151. Vojdani A, Vojdani E, and Cooper E. 2003. Antibodies to myelin basic protein, myelin oligodendrocytes peptides, alpha-beta-crystalline lymphocyte activation and cytokine production in patients with multiple sclerosis. *J. Intern. Med.* 254(4):363–374.
152. Leist M, Hühnle S, Single B, and Nicotera P. 1998. Differentiation between apoptotic and necrotic cell death by means of the BM cell death detection ELISA or Annexin V staining. *Biochemica* 2:25.

153. Leist M and Nicotera P. 1997. The shape of cell death. *Biochem. Biophys. Res. Commun.* 236:1–9.
154. Nicotera P and Leist M. 1997. Energy supply and the shape of death in neurons and lymphoid cells. *Cell Death Differ.* 14:435–442.
155. Leist M, Hühnle S, Single B, and Nicotera P. 1998. Differentiation between apoptotic and necrotic cell death by means of the BM cell death detection ELISA or Annexin V staining. *Biochemica.* 2:25.
156. Walker NI, Narmon BV, Gobe GE, and Kerr JF. 1988. Patterns of Cell Death. *Methods Achiev Exp Pathol.* 13:18–54.
157. Vojdani A. 1998. *The Role of Natural Killer Cell, Apoptosis and Cell Cycle in Health and Diseases.* 24, Tab. 14.
158. Lin L, Chatiroudi A, Silverster G, Wernett ME, et al. 2002. Visualization and quantification of T cell-mediated cytotoxicity using cell-permeable fluorogenic caspase substrates. *Nature Medicine* 8:185–189.
159. Bunin A. 2006. Mechanism of apoptosis induced by TCR signals in thymocytes. Emory University Proquest® Dissertations & Theses. Publication number: 3189862. 5309.
160. De Forrest V, et al. *Free Radical Biology and Medicine* 16(6): 675–684.
161. Ponner BB, Stach C, Zoller O, et al. 1998. Induction of apoptosis reduces immunogenicity of human T-cell lines in mice. *Scandinavian Journal of Immunology* 47(4):343–347.
162. Lanzon RJ, Patton CW, and Weissman IL. 1993. A morphological and immunohistochemical study of programmed cell death in botrylus schlosseri (tunicate ascidiacea). *Cell Tissue Res.* 272:115–127.
163. Takase KI, Ishikawa M, and Hoshiai H. 1995. Apoptosis in the regeneration process of unfertilized mouse ova. *Tohoku Journal of Experimental Medicine* 175(1):69–76.
164. Monte JD. 1996. The role of programmed cell death as an emerging new concept for the pathogenesis of autoimmune tissue.*Clinical Immunology and Pathology* 80(3 Sept.):, 82–84, Art. 36.
165. Maher S, Toomey D, Condron C, and Bouchier-Hayes D. 2002. Activation-induced cell death: the controversial role of fas and fas ligand in immune privilege and tumor counterattack. *Immunology and Cell Biology* 80:131–137.
166. Marleau AM and Sarvetnick N. 2005. T cell homeostasis in tolerance and immunity. *Journal of Leukocyte Biology* 78:575–584.
167. Crowston JG, Akbar AN, Constable PH, et al. 1998. Antimetabolite-induced apoptosis in Tenonís capsule fibroblasts. *Investigative Opthalmology and Visual Science* 39:449–454.
168. Sanderson CJ. 1976. The mechanism of T-cell mediated cytotoxicity II: Morphological studies of cell death by time-lapse microcinematography. *Proc. Roy. Soc. Lond. B* 192:241–255.
169. Matter A. 1979. Microcinematographic and electron microscopic analysis of target cell lysis induced by cytotoxic T lymphocytes. *Immunology* 26:179–190.
170. Nicholson DW. 1996. ICE-CED$_3$-like Proteases as Target for the Control of Inappropriate Apoptosis. *Nature Biotechnology* 14:297–301.
171. Bursch W, Oberhammer F, and Schulte-Hermann R. 1992. Cell death by apoptosis and its protective role against disease. *Trends. Pharm. Sci.*13:245–251.
172. Carson DA and Ribiero JM. 1993. Apoptosis and disease. *Lancet* 341:1251–1254.
173. Barr PJ and Tomei LD. 1994. Apoptosis and its role in human disease. *BioTechnology* 12:487–493.
174. Häcker G and Vaux DL. 1995. The medical significance of physiological cell death. *Medicinal Res. Rev.* 15:299–311.
175. Thompson CB. 1995. Apoptosis in the pathogenesis and treatment of disease. *Science* 267:1456–1462.
176. Walker PR, Smith C, Youdale T, et al. 1991. Topoisomerase II-reactive chemotherapeutic drugs induce apoptosis in thymocytes. *Cancer Res* 51:1078.
177. Brown DB, Sun XM, and Cohen GM. 1993. Dexamethasone-induced apoptosis involves cleavage of DVA to large fragments prior to internucleosomal freagmentation. *J Biol Chem* 268:3037.
178. Vivian B, Rossi AD, Chow SC, Nicotera P. 1995. Organotin compounds induce calcium overload and apoptosis in PC12 cells. *Neurotoxicoloogy* 16:19.
179. Kunimoto M. 1994. Methyl mercury induces apoptosis of rat cerebellar neurons in primary culture. *Biochem. Biophys. Res. Commun.* 204:310.
180. Reynolds ES, Kanz MF, Chicco P, Moslen MT. 1984. 1,1-Dichloroethylene: an apoptotichepatotoxin? *Environ. Health Perspect.* 57:313.
181. Ledda-Columbano GM, Coni P, Curto M, et al. 1991. Induction of two different modes of cell death, apoptosis and necrosis in rat liver after a single dose of thioacetamide. *Am. J. Pathol.* 139:1099.
182. Cohen JJ and Duke RS. 1984. Glucocorticoid activation of a calcium-dependent endonuclease in thymocyte nuclei leads to cell death. *J. Immunol.* 132:38.
183. Rossi AD, Larsson O, Manzo L, et al. 1993. Modification of Ca^{2+} signaling by inorganic mercury in PC12 cells. *FASEB7,* 1507.

184. Aw, TY, Nicotera P, Manzo L, and Orrenius S. 1990. Tributyltin stimulates apoptosis in rat thymocytes. *Arch. Biochem. Biophys.* 283:46.
185. Raffray M, McCarthy D, Snowden RT, Cohen GM. 1993. Apoptosis as a mechanism of tributyltin cytotoxicity to thymocytes: relationship of apoptotic markers to biochemical and cellular effects. *Toxicol App Pharmacol* 119:122.
186. Vojdani A. 1998. *The Role of Natural Killer Cell, Apoptosis and Cell Cycle in Health and Diseases.* 27.
187. Vojdani A. 1998. *The Role of Natural Killer Cell, Apoptosis and Cell Cycle in Health and Diseases.* 25.
188. Yuan J, Shaham S, Ledoux S, Ellis HM, and Horvitz R. 1993. The C-elegans cell death gene CED-3 encodes a protein similar to mammalian interleukin-1B-converting enzyme. *Cell.* 75:641–652.
189. Pawelec G, Effros RB, Caruso C, Remarque E, Barnett Y, and Solana R. 1999. T-cells and aging. *BioScience* (update February).
190. Thoman ML and Weigle WO. 1989. The cellular and subcellular bases of immunosenescence. *Adv Immunol* 46:221–262.
191. Tielen FJ, Vanvliet ACM, Degeus B, Nagelkerken L, and Rozing J. 1993. Age-related changes in CD4+ T-cell subsets associated with prolonged skin graft survival in aging rats. *Transplant. Proc.* 25:2872–2874.
192. Roberts-Thomson IC, Whittingham S, Youngchaiyud U, and Mackay IR. 1974. Ageing, immune response and mortality. *Lancet* 2:368–370.
193. Murasko DM, Weiner P, and Kaye D. 1988. Association of lack of mitogen induced lymphocyte proliferation with increased mortality in the elderly. *Aging: Immunology and Infectious Disease* 1:1–23.
194. Wayne SJ, Rhyne RL, Garry PF, and Goodwin JS. 1990. Cell-mediated immunity as a predictor of morbidity and mortality in subjects over 60. *J. Gerontol.* 45:45–48.
195. Bender BS, Nagel JE, Adler WH, and Andres R. 1986. Absolute peripheral blood lymphocyte count and subsequent mortality of elderly men: The Baltimore Longitudinal Study of Aging. *J. Am. Geriatr. Soc.* 34:649–654.
196. Rukavina D, Laskarin G, Rubesa G, et al. 1998. Age-related decline of perforin expression in human cytotoxic T-lymphocytes and natural killer cells. *Blood* 92:2410–2420.
197. Greeley EH, Kealy RD, Ballam JM, Lawler DF, and Segre M. 1996. The influence of age on the canine immune system. *Vet. Immunol. Immunopathol.* 55:1–10.
198. Yang XH, Stedra J, and Cerny J. 1996. Relative contribution of T and B cells to hypermutation and selection of the antibody repertoire in germinal centers of aged mice. *J. Exp. Med.* 183:959–970.
199. Buchanan JP, Peters CA, Rasmussen CJ, and Rothstein G. 1996. Impaired expression of hematopoietic growth factors: A candidate mechanism for the hematopoietic defect of aging. *Exp. Gerontol.* 31:135–144.
200. Cheleuitte D, Mizuno S, and Glowacki J. 1998. In vitro secretion of cytokines by human bone marrow: Effects of age and estrogen status. *J. Clin. Endocrinol. Metab.* 83:2043–2051.
201. Lansdorp PM, Dragowska W, Thomas TE, Little MT, and Mayani H. 1994. Age-related decline in proliferative potential of purified stem cell candidates. *Blood Cells* 20:376–381.
202. Kendall MD, Johnson HRM, and Singh J. 1980. The weight of the human thymus gland at necropsy. *Journal of Anatomy* 131:485–499.
203. Wynn RF, Cross MA, Hatton C, et al. 1998. Accelerated telomere shortening in young recipients of allogeneic bone-marrow transplants. *Lancet* 351:178–181.
204. Wright WE, Brasiskyte D, Piatyszek MA, and Shay JW. 1996. Experimental elongation of telomeres extends the lifespan of immortal x normal cell hybrids. *EMBO J.* 15:1734–1741.
205. Van Steensel B and De Lange T. 1997. Control of telomere length by the human telomeric protein TRF1. *Nature* 385:740–743.
206. Furumoto K, Inoue E, Nagao N, Hiyama E, and Miwa N. 1998. Age-dependent telomere shortening is slowed down by enrichment of intracellular vitamin C via suppression of oxidative stress. *Life Sci.* 63:935–948.
207. Anderlini P, Przepiorka D, Lauppe J, et al. 1997. Collection of peripheral blood stem cells from normal donors 60 years of age or older. *Br. J. Haematol.* 97:485–487.
208. De Haan G, Nijhof W, and Van Zant G. 1997. Mouse strain-dependent changes in frequency and proliferation of hematopoietic stem cells during aging: Correlation between lifespan and cycling activity. *Blood* 89:1543–1550.
209. Thoman ML. 1997. Effects of the aged microenvironment on CD4(+) T cell maturation. *Mech Ageing Dev* 96:75–88.

210. Pawelec G and Solana R. 1997. Immunosenescence. *Immunol. Today* 18:514–516.
211. Naz RK, Kaplan P, Badamchian M, and Goldstein AL. 1995. Effects of synthetic thymosin-alpha 1 and its analogs on fertilizability of human sperm: Search for a biologically active, stable epitope. *Arch. Androl.* 35(1):63–69.
212. Philip D, Badamchian M. Scheremeta B. Nguyen M, Goldstein AL, and Kleinman HK. 2003. Thymosin beta 4 and a synthetic peptide containing its actin-binding domain promote dermal wound repair in db.db diabetic mice and aged mice. *Wound Repair Regen* 11(1):19–24.
213. George AJT and Ritter MA. 1996. Thymic involution with ageing: Obsolescence or good housekeeping? *Immunol. Today* 17:267–272.
214. Turke PW. 1997. Thymic involution. *Immunol. Today* 18:407.
215. Catania A, Airaghi L, Motta P, et al. 1997. Cytokine antagonists in aged subjects and their relation with cellular immunity. *J. Gerontol. Ser. A-Biol. Sci. Med.* 52:B93–97.
216. Mysliwska J, Bryl E, Foerster J, and Mysliwski A. 1998. Increase of interleukin 6 and decrease of interleukin 2 production during the ageing process are influenced by the health status. *Mech. Ageing. Dev.* 100:313–328.
217. Cohen HJ, Pieper CF, Harris T, Rao KMK, and Currie MS. 1997. The association of plasma IL-6 levels with functional disability in community-dwelling elderly. *J. Gerontol. Ser. A-Biol. Sci. Med.* 52:M201–208.
218. Fagiolo U, Cossarizza A, Scala E, et al. 1993. Increased cytokine production in mononuclear cells of healthy elderly people. *Eur. J. Immunol.* 23:2375–2378.
219. Riancho JA, Zarrabeitia MT, Amado JA, Olmos JM, and Gonzalezmacias J. 1994. Age-related differences in cytokine secretion. *Gerontology* 40:8–12.
220. Wiedmeier SE, Mu HH, Araneo BA, and Daynes RA. 1994. Age- and microenvironment-associated influences by platelet-derived growth factor on T cell function. *J. Immunol.* 152:3417–3426.
221. Mehr R, Perelson AS, Fridkishareli M, Globerson A. 1996. Feedback regulation of T cell development: Manifestations in aging. *Mech. Ageing Dev.* 91:195–210.
222. Portales Perez D, Alarcon Segovia D, Llorente L, et al. 1998. Penetrating anti-DNA monoclonal antibodies induce activation of human peripheral blood mononuclear cells. *J. Autoimmun.* 11:563–571.
223. Candore G, Dilorenzo G, Mansueto P, et al. 1997. Prevalence of organ-specific and non organ-specific autoantibodies in healthy centenarians. *Mech. Ageing Dev.* 94:183–190.
224. Mariotti S, Chiovato L, Franceschi C, and Pinchera A. 1998. Thyroid autoimmunity and aging. *Exp Gerontol* 33:535–541.
225. Nobrega A, Haury M, Gueret R, Coutinho A, and Weksler ME. 1996. The age-associated increase in autoreactive immunoglobulins reflects a quantitative increase in specificities detectable at lower concentrations in young mice. *Scand. J. Immunol.* 44:437–443.
226. Ray SK, Putterman C, and Diamond B. 1996. Pathogenic autoantibodies are routinely generated during the response to foreign antigen: A paradigm for autoimmune disease. *Proc. Natl. Acad. Sci. USA* 93:2019–2024.
227. Crisi GM, Chen LZ, Huang C, and Thorbecke CJ. 1998. Age-related loss of immunoregulatory function in peripheral blood CD8 T cells. *Mech Ageing Dev* 103:235–254.
228. Pawelec G, Effros RB, CarusoC, Remarque E, BarnettY, Solana R. 1999. T cells and aging. *BioScience* (update February): 53, Fig. 1. www.bioscience.org
229. Mosmann TR and Sad S. 1996. The expanding universe of T-cell subsets: Th1, Th2 and more. *Immunol. Today* 17:138–146.
230. Romagnani P, Parronchi M, Delios M, et al. 1997. An update on human Th1 and Th2 cells. *Int. Arch. Allergy Immunol.* 113:153–156.
231. McCarron M, Osborne Y, C. Story C, Dempsey JL, Turner R, and Morley A. 1987. Effect of age on lymphocyte proliferation. *Mech. Ageing Dev.* 41:211–218.
232. Reinhold U, Pawelec G, Fratila A, Leippold S, Bauer R, and Kreysel H-W. 1990. Phenotypic and functional characterization of tumor infiltrating lymphocytes in mycosis fungoides: Continuous growth of CD4+CD45R+ T-cell clones with suppressor-inducer activity. *J. Invest. Dermatol.* 94:304–309.
233. Kaltoft K, Pedersen CB, Hansen BH, Lemonidis AS, Frydenberg J, and Thestrup-Pedersen K. 1994. In vitro genetically aberrant T cell clones with continuous growth are associated with atopic dermatitis. *Arch. Dermatol. Res.* 287:42–47.
234. Kaltoft K, Pedersen CB, Hansen BH, and Thestruppedersen K. 1995. Appearance of isochromosome 18q can be associated with in vitro immortalization of human T lymphocytes. *Cancer Genet Cytogenet* 81:13–16.
235. Harley CB, Futcher AB, and Greider CW. 1990. Telomeres shorten during ageing of human fibroblasts. *Nature* 345:458–460.

236. Lindsay J, McGill NI, Lindsay LA, Green DK, and Cooke HJ. 1991. In vivo loss of telomeric repeats with age in humans. *Mutation Research* 256:45–48.
237. Allsopp RC and Harley CB. 1995. Evidence for a critical telomere length in senescent human fibroblasts. *Exp. Cell. Res.* 219:130–136.
238. Slagboom PE, Droog S, and Boomsma DI. 1994. Genetic determination of telomere size in humans: A twin study of three age groups. *Am. J. Hum. Genet.* 55:876–882.
239. Satoh H, Hiyama K, Takeda M, et al. 1996. Telomere shortening in peripheral blood cells was related with aging but not with white blood cell count. *Jpn. J. Hum. Genet.* 41:413–417.
240. Allsopp RC, Vaziri H, Patterson C, et al. 1992. Telomere length predicts replicative capacity of human fibroblasts. *Proc. Natl. Acad. Sci. USA* 89:10114–10118.
241. Vaziri H, Schachter F, Uchida I, et al. 1993. Loss of telomeric DNA during aging of normal and trisomy-21 human lymphocytes. *Am. J. Hum. Genet.* 52:661–667.
242. Frenck RW, Blackburn EH, and Shannon KM. 1998. The rate of telomere sequence loss in human leukocytes varies with age. *Proc. Natl. Acad. Sci. USA* 95: 5607–5610.
243. Iwama H, Ohyashiki K, Ohyashiki JH, et al. 1998. Telomeric length and telomerase activity vary with age in peripheral blood cells obtained from normal individuals. *Hum. Genet.* 102:397–402.
244. Hiyama K, Hirai Y, Kyoizumi S, et al. 1995. Activation of telomerase in human lymphocytes and hematopoietic progenitor cells. *J. Immunol.* 155:3711–3715.
245. Kosciolek BA and Rowley PT. 1998. Human lymphocyte telomerase is genetically regulated. *Gene Chromosome Cancer* 21:124–130.
246. Kruk PA, Rampino NJ, and Bohr VA. DNA damage and repair in telomeres: Relation to aging. *Proc. Natl. Acad. Sci. USA* 92:258–262.
247. Martens UM, Zijlmans JM, Poon SS, et al. 1998. Short telomeres on human chromosome 17p. *Nat. Genet.* 18:76–80.
248. Michie CA, McLean A, Alcock C, and Beverley PCL. 1992. Lifespan of human lymphocyte subsets defined by CD45 isoforms. *Nature* 360:264–265.
249. McLean AR and Michie CA. 1995. In vivo estimates of division and death rates of human T lymphocytes. *Proc. Natl. Acad. Sci. USA* 92:3707–3711.
250. Beverley PCL, Michie CA, and Young JL. 1993. Memory and the lifespan of human T-lymphocytes. *Leukemia* 7:S50–54.
251. Bruno L, Von Boehmer H, and Kirberg J. 1996. Cell division in the compartment of naive and memory T lymphocytes. *Eur. J. Immunol.* 26:3179–3184.
252. Tanchot C, Lemonnier FA, Perarnau B, Freitas AA, and Rocha B. 1997. Differential requirements for survival and proliferation of CD8 naive or memory T-cells. *Science* 276:2057–2062.
253. Swain SL, Bradley LM, Croft M, et al. 1991. Helper T-cell subsets: Phenotype, function and the role of lymphokines in regulating their development. *Immunol. Rev.* 123:115–144.
254. Effros RB and Pawelec G. 1997. Replicative senescence of T-lymphocytes: Does the Hayflick Limit lead to immune exhaustion? *Immunol. Today* 18:450–454.
255. Shinohara S, Sawada T, Nishioka Y, et al. 1995. Differential expression of Fas antigen and bcl-2 protein on CD4(+) T cells, CD8(+) T cells, and monocytes. *Cell Immunol.* 163:303–308.
256. Kudlacek S, Jahandideh-Kazempour S, Graninger W, Willvonseder R, and Pietschmann P. 1995. Differential expression of various T cell surface markers in young and elderly subjects. *Immunobiology* 192:198–204.
257. Aspinall R, Carroll J, and Jiang SS. 1998. Age-related changes in the absolute number of CD95 positive cells in T cell subsets in the blood. *Exp. Gerontol.* 33:581–591.
258. Phelouzat MA, Arbogast A, Laforge TS, Quadri RA, and Proust JJ. 1996. Excessive apoptosis of mature T lymphocytes is a characteristic feature of human immune senescence. *Mech. Ageing Dev.* 88:25–38.
259. Potestio M, Caruso C, Gervasi F, et al. 1998. Apoptosis and ageing. *Mech. Ageing Dev.* 102:221–237.
260. Miller RA. 1997. Age-related changes in T cell surface markers: A longitudinal analysis in genetically heterogeneous mice. *Mech. Ageing Dev.* 96:181–196.
261. Vaziri H and Benchimol S. 1998. Reconstitution of telomerase activity in normal human cells leads to elongation of telomeres and extended replicative life span. *Curr. Biol.* 8:279–282.
262. Kukel S, Reinhold U, Oltermann I, and Kreysel HW. 1994. Progressive increase of CD7(-) T cells in human blood lymphocytes with ageing. *Clin. Exp. Immunol.* 98:163–168.
263. Callahan JE, Kappler JW, and Marrack P. 1993. Unexpected Expansions of CD8-Bearing Cells in Old Mice. *J. Immunol.* 151:6657–6669.
264. Posnett DN, Sinha R, Kabak S, and Russo C. 1994. Clonal Populations of T Cells in Normal Elderly Humans: The T Cell Equivalent to Benign Monoclonal Gammapathy. *J. Exp. Med.* 179:609–618.

265. Ricalton NS, Roberton C, Norris JM, Rewers M, Hamman RF, and Kotzin BL. 1998. Prevalence of CD8(+) T-cell expansions in relation to age in healthy individuals. *J. Gerontol. Ser. A Biol. Sci. Med.* 53:B196–203.
266. Clegg CH, Rulffes JT, Wallace PM, and Haugen HS. 1996. Regulation of an extrathymic T-cell development pathway by oncostatin M. *Nature* 384:261–263.
267. Zheng B, Han SH, Zhu Q, Goldsby R, and Kelsoe G. 1996. Alternative pathways for the selection of antigen-specific peripheral T cells. *Nature* 384:263–266.
268. Pawelec G, Müller R, Rehbein A, Hähnel K, and Ziegler BL. Extrathymic T cell differentiation in vitro from CD34+ stem cells. *J Leukocyte Biol* 64:733–739.
269. La Croix AZ, Lipson S, Miles TP, and Whilte L. 1989. Prospective study of pneumonia hospitalization and mortality of US older people: The role of chronic conditions, health behaviors and nutritional status. *Public Health Rep* 104:350–360.
270. Ackermann RJ and Munroe PW. 1996. Bacteremic urinary tract infection in older people. *J. Am. Geriatr. Soc.* 44:927–933.
271. Chattopadhyay B and Al-Zahawi M. 1983. Septicemia and its unacceptably high mortality in the elderly. *Journal of Infection* 7:134–138.
272. Gorse GJ, Thrupp LD, Nudleman KL, Wyle FA, Hawkins B, and Cesario TC. 1984. Bacterial meningitis in the elderly. *Arch. Intern. Med.* 144:1603–1607.
273. Barker WH and Mullooly JP. 1980. Impact of epidemic type A influenza in a defined adult population. *Am. J. Epidemiol.* 112:798–813.
274. Sprenger MJW, Mulder PGH, Beyer WEP, Van Strik R, and Masurel N. 1993. Impact of influenza on mortality in relation to age and underlying disease. *Int. J. Epidemiol.* 22:334–340.
275. Yoshikawa TT. 1997. Perspective: Aging and infectious diseases: Past, present, and future. *J. Infect. Dis.* 176:1053–1057.
276. Horiuchi S and Wilmoth JR. 1997. Age patterns of the life table aging rate for major causes of death in Japan, 1951–1990. *J. Gerontol. Ser. A-Biol. Sci. Med.* 52:B67–77.
277. Ljungquist B, Berg S, and Steen B. 1996. Determinants survival: An analysis of the effects of age at observation and length of the predictive period. *Aging-Clin. Exp. Res.* 8:22–31.
278. Meyer KC, Ershler W, Rosenthal NS, Lu XC, and Peterson K. 1996. Immune dysregulation in the aging human lung. *Amer. J. Respir. Crit. Care. Med.* 153:1072–1079.
279. Meyer KC, Rosenthal NS, Soergel P, and Peterson P. 1998. Neutrophils and low-grade inflammation in the seemingly normal aging human lung. *Mech. Ageing. Dev.* 104:169–181.
280. Johnston CJ, Finkelstein JN, Gelein R, and Oberdorster G. 1998. Pulmonary inflammatory responses and cytokine and antioxidant mRNA levels in the lungs of young and old C57BL/6 mice after exposure to teflon fumes. *Inhal. Toxicol.* 10:931–953.
281. Yagi T, Sato A, Hayakawa H, and Ide K. 1997. Failure of aged rats to accumulate eosinophils in allergic inflammation of the airway. *J. Allerg. Clin. Immunol.* 99:38–47.
282. Degreef GE, Vantol MJD, Kallenberg CGM, et al. 1992. Influence of ageing on antibody formation invivo after immunisation with the primary T-cell dependent antigen helix-pomatia haemocyanin. *Mech Ageing Dev.* 66:15–28.
283. Lesourd B. 1995. Protein undernutrition as the major cause of decreased immune function in the elderly: Clinical and functional implications. *Nutr. Rev.* 53:S86–94.
284. Ruben FL, Nagel, and Fireman PP. 1978. Antitoxin responses in the elderly to tetanus-diptheria in the (Td) immunisations. *Am. J. Epidemiol.* 108:145–155.
285. Armitage KB, Duffy EG, Mincek MA, et al. 1993. Transient normalization of lymphocyte blastogenic and specific antibody responses following boosting of healthy elderly subjects with tetanus toxoid. *J. Gerontol.* 48:M19–25.
286. Steger MM, Maczek C, Berger P, and Grubeck-Loebenstein B. 1996. Vaccination against tetanus in the elderly: do recommended vaccination strategies give sufficient protection? *Lancet* 348:762.
287. Schatz D, Ellis T, Ottendorfer E, Jodoin E, Barrett D, and Atkinson M. 1998. Aging and the immune response to tetanus toxoid: Diminished frequency and level of cellular immune reactivity to antigenic stimulation. *Clin. Diagn. Lab. Immunol.* 5:894–896.
288. Powers DC. 1994. Effect of age on serum immunoglobulin G subclass antibody responses to inactivated influenza virus vaccine. *J. Med. Virol.* 43:57–61.
289. Provinciali M, Distefano G, Colombo M, et al. 1994. Adjuvant effect of low-dose interleukin-2 on antibody response to influenza virus vaccination in healthy elderly subjects. *Mech. Ageing Dev.* 77:75–82.
290. Miller SD, Crouch EA, and Busbee DL. 1997. An accessory protein of DNA polymerase alpha declines in function with increasing age. *Mutat. Res-Fundam. Mol. Mech. Mut.* 374:125–138.

291. Turner DR, Morley AA, Seshadri RS, Sorrell JR. 1981. Age-related variations in human lymphocyte DNA. *Mech Ageing Dev* 17:305–309.
292. Hartwig M and Korner IJ. 1987. Age-related changes in DNA unwinding and repair in human peripheral lymphocytes. *Mech. Ageing Dev.* 39:73–78.
293. Jacobs PA, Brunton M, and Court Brown WM. 1964. Cytogenetic studies in leucocytes on the general population: subjects of ages 65 years and more. *Ann. Hum. Genet.* 27:353–362.
294. Rattan SI. 1989. DNA damage and repair during cellular ageing. *Int. Rev. Cytol.* 116:47–88.
295. Boerrigter METI, Wei JY, and Vijg J. 1995. Induction and repair of benzo[a]pyrene-DNA adducts in C57BL/6 and BALB/c mice: Association with aging and longevity. *Mech Ageing Dev* 82:31–50.
296. Hart RW and Setlow RB. 1974. Correlation between desoxyribonucleic acid excision repair and lifespan in a number of mammalian species. *Proc. Natl. Acad. Sci. USA* 71:2169–2173.
297. Cortopassi GA and Wang E. 1996. There is substantial agreement among interspecies estimates of DNA repair activity. *Mech. Ageing Dev.* 91:211–218.
298. Moriwaki SI, Ray S, Tarone RE, Kraemer KH and Grossman L. 1996. The effect of donor age on the processing of UV-damaged DNA by cultured human cells: Reduced DNA repair capacity and increased DNA mutability. *Mutat Res-DNA Repair* 364:117–123.
299. Lee SW, Fukunaga N, Rigney DR, Shin DY, and Wei JY. 1997. Downregulation of DNA topoisomerase I in old versus young human diploid fibroblasts. *Mutat. Res-Fundam. Mol. Mech. Mut.* 373, 179–184.
300. Souba WW, Fink MP, Jurkovich GJ, et al. 2006. American college of surgeons ACS surgery. *Principles & Practice*. New York, NY: WebMD Professional Publishing. 162.

4 Endocrine System

INTRODUCTION

In previous chapters, we began the detailed discussion of the three limbs of homeostatic physiology for amplification of the dynamics of the homeostatic response. It was noted that once local homeostatic control is exceeded, this response might be disrupted or imbalanced by noxious stimuli in the patients with chemical sensitivity and/or chronic degenerative disease. Our first focus was the sensory spinal nerves, autonomic and central nervous system limb discussed in Chapter 2. The next limb of amplification of the homeostatic response, the immune system, we discussed in Chapter 3.[1] The final limb of anatomic physiology of the amplification systems that is altered in chemical sensitivity and/or chronic degenerative disease is the endocrine system, which is the subject of this chapter. **The dynamic function of these three limbs creates all of the homeostatic amplification systems, which are the major subunits of the central homeostatic mechanism**. The dynamics of the amplification systems can be activated after a localized homeostatic subunit response in the connective tissue matrix is not contained, triggering the secondary amplification system of the homeostatic response in the patient with chemical sensitivity and/or chronic degenerative disease. The triad of the neurological, immune, and endocrine systems, as an integral part of the ground regulation system (GRS), culminates at their proximal end in the hypothalamus, the limbic, the reticular activating system, the area postrema of the fourth ventricle, pituitary gland, and pineal gland of the brain. Distal connections to the end organs such as the thyroid, parathyroid, adrenal glands, pancreas, ovaries, testicles, and thymus gland occur. In addition, these circuits find distal connections to the end organs, neuroendocrine cells, and receptors via the autonomic nervous system (ANS) and the connective tissue matrix. When the functions of any of these organs or systems are out of balance, the dynamics of homeostasis are also out of balance, placing a strain on its energy efficiency that occurs in the patient with chemical sensitivity and/or chronic degenerative disease resulting in weakness and frequently pain. Certainly, a demand will be placed on the nutrient pool resulting in the potential for more altered dynamics of homeostasis, which is seen in the patient with chemical sensitivity and/or chronic degenerative disease.

INTEGRATED PHYSIOLOGY

In general, **the hormonal system is concerned principally with the dynamics of the homeostatic control of the various metabolic functions of the body, such as the rates of chemical reactions in the cells, and the transport of substances through cell membranes, or other aspects of cellular metabolism like growth, synthesis, and secretion**. These functions are very essential and often disturbed in the patient with chemical sensitivity and/or chronic degenerative disease (Table 4.1). Some hormonal effects occur in seconds, while other effects require several days simply to start and then continue for weeks or even months. The hormonal systems are essential in vascular regulation and response as well as immune functions. Their absence or insufficiencies often create dyshomeostasis or at least strain homeostasis, resulting in inefficient energy consumption.

Since both the hormonal and nervous systems are a part of the GRS, many interrelations exist between them. One should realistically look at these systems as an integrated neuroendocrine system that often is malfunctioning in the patient with chemical sensnitivity and/or chronic degenerative disease. For example, the adrenal medulla and pituitary glands secrete their hormones almost entirely in response to appropriate neural stimuli. In other interrelations, the different pituitary hormones control secretion of the majority of the other endocrine glands, which results in

TABLE 4.1
Hormonal System Concerned with Various Metabolic Functions

1. Rates of chemical reaction in cells
2. Transport of substances through cell membranes
3. Aspects of cellular metabolism—growth, synthesis, secretion

Source: From EHC-Dallas, 2002.

homeostasis. The autonomic nerves secrete norepinephrine and acetylcholine locally, etc. and are often working overtime in the pollutant-injured patient with chemical sensitivity and/or chronic degenerative disease.

Dynamic homeostatic regulation of hormones has two types of effects. One is a local effect and the other is distal. Local hormones are those like acetylcholine, which is released from parasympathetic and peripheral skeletal nerve endings, and norepinephrine, which is released from the sympathetic nerve endings. These particular hormones have an integral role in chemical sensitivity and chronic degenerative disease as the human biological system produces neurohormones.

Regional and distal hormones like secretin are released from the mucosa of the duodenal wall and transported in the blood to the pancreas where they cause the release of a watery bicarbonate rich secretion. This buffering solution is often necessary and its need exceeded in the chronically ill patient who is usually acidic. Cholecystokinin is released from the small intestine, and then transported to the gallbladder to cause contraction and to the pancreas to cause digestive enzyme secretion. Excess cholecystokinin has been observed to cause severe hypersensitive reactions including laryngeal edema in the patient with chemical sensitivity.

The usual sequence of cholecystokinin is to stimulate secretion of bile; the pancreas then down regulates the activity of secretin and releases peptides. This activity forms a negative efferent feedback loop to inhibit secretin release. The release of secretin from the duodenal cells is mediated through this mechanism including cAMP, Ca^{++}, and protein kinase C.

There are two types of secretin receptors: (1) a high-affinity secretin receptor, which has a higher affinity for vasoactive intestinal polypeptide (VIP), which has a high affinity for vascular walls and thus a regulator of intestinal oxygen supply; and (2) a lesser-affinity receptor.

Secretin inhibits gastric acid, serotonin, gastric release, and gastric emptying and stimulates bicarbonate production from the pancreas. This is a negative feedback loop. Secretin release is stimulated by bile acids and folic acids and is inhibited by somatostatin and met-encephalin and a variety of chemical pollutants. **The metabolic impact of secretin is that it stimulates the production of arachidonic EPA (eicosapentaenoic acid), augments FA (fatty acids) increase and metabolism to bile acids, and also stimulates CCK-B mucin, which plays a neuromodulatory role in the regulation of GABAergic neuronal activity, which also stimulates speech**. We often see delayed gastric emptying in the patient with chemical sensitivity and/or chronic degenerative disease, which occurs probably through these mechanisms. The delayed emptying can cause distal gastrointestinal upset and bloating.

All these hormonal substances have local, regional, and distal effects on the dynamics of the homeostatic control mechanism, and they must all be in balance to assure adequate function. Most of the general hormones involved in the dynamics of homeostasis are secreted by the specialized endocrine glands. The individual types will be discussed in this chapter (Figure 4.1).

In general, the control of the hormone secretion rate depends on the negative feedback characteristic of the homeostatic mechanism. The rate of secretion of every hormone is controlled by some internal control homeostatic subsystem. In most instances, this control is exerted through a negative homeostatic feedback mechanism and works as follows: **The endocrine gland has a natural tendency to oversecrete the hormone**. Due to this tendency, the hormone exerts more and more

Endocrine System

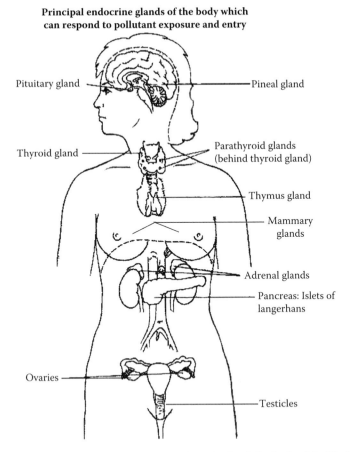

FIGURE 4.1 Anatomical loci of the principal endocrine glands of the body. (Modified from Guyton, A. C. and Hall, J. E., *Textbook of Medical Physiology,* 9th ed., Philadelphia: WB Saunders Co., 926, Fig. 74-1, 1996.)

of its control on the target organ. The target organ, in turn, performs its function. When there is an excess of hormone formation, usually some factor about the organ's function then feeds back to the endocrine gland and causes a negative effect on the gland by decreasing its rate of secretion. Usually, **the important factor to be controlled is not the secretory rate of the hormone itself, but the degree of activity of the target organ although both activities are important**. Therefore, only when the target organ's activity rises to an appropriate level will the feedback to the gland be powerful enough to slow further secretion of the hormone. If the target organ responds poorly to the hormone, the endocrine gland will almost always secrete a continuous amount of its hormone until the target organ eventually reaches the appropriate level of activity, but at the expense of excessive secretion of the controlling hormone, which, if continued secretion occurs, eventually causes altered dynamics of homeostasis because of nutrient depletion and energy loss. A clinical example of this phenomenon is the patient who has a chronic chemical exposure such as 2,4 dinitrophenol, phthalate esters, thiocyanates, which challenges the thyroid to produce the thyroid hormone for a longer time and in greater quantity than normal. The thyroid strains to keep up with the demand because the liver has upregulated the cytochrome p-450 enzymes to degrade the toxic chemical but in parallel degrades the thyroid hormone. Due to the increased liver metabolism, the thyroid hormone is removed at a faster rate. Therefore, the thyroid function almost becomes hyper to keep up with the demand. Then, after a period of straining and stress, the thyroid totally depletes itself resulting in a hypothyroid condition and thyroiditis with all the clinical manifestations.

Several general patterns exist in which the endocrine glands store and secrete their hormones for dynamic homeostatic balance. **For instance, all the protein hormones such as thyroid, adrenal, pineal, and neuroendocrine cells are formed by the granular endoplasmic reticulum of the glandular cells in the same manner secretory proteins are formed**. This reticulum is dispersed throughout the cell. After several cleavages, the hormone is sent to the golgi apparatus where another part of the hormone is cleaved to shape the final product. The golgi apparatus compacts the hormone into secretory vesicles or granules that are stored in the cytoplasm until the appropriate signal for release occurs. When they are released, the dynamics of homeostasis can be restored and/or maintained.

The two groups of hormones derived from tyrosine, the thyroid, and adrenal medullary hormones are both formed by actions of enzymes in the cytoplasm, i.e., compartments of the glandular cells. Epinephrine and norepinephrine are absorbed into vesicles and stored and increased production and release occur with acute stress. Appropriate release of these hormones is important in the maintenance of the dynamics of homeostasis. Chronic pollutant or psychological stress can occur that will then disturb normal homeostatic function often causing symptoms of pain, fatigue, and anxiety.

The thyroid hormones, thyroxin and triiodothyronine, are first formed as component parts of a large protein molecule, thyroglobulin, which is stored in large follicles of the thyroid gland. When thyroid hormones are to be secreted, specific enzymes[3] cleave the thyroglobulin molecules allowing release of the thyroid hormone. For the hormones of the adrenal cortex, ovaries, and testes, the amounts stored are small, but present in the cells are large amounts of precursor molecules, especially cholesterol and its hormone intermediates. Upon the signal for release, the necessary chemical conversion to the final hormones occurs with the appropriate secretion. Pollutant presence and injury can inhibit or trigger the final hormone output causing dyshomeostasis.The amount of hormones required to control metabolic and endocrine function for appropriate homeostasis ranges from 1 picogram to a few micrograms per milliliter of blood. Also, the rates of secretion are very small. **The potency of hormones helps to exert powerful control over some aspects of homeostatic control**. Without exception, the rate of secretion of every hormone is itself controlled by the dynamics of some internal homeostatic subunit control mechanism. This control generally is exerted by the negative feedback homeostatic characteristic. Clinicians see alternatives of this feedback phenomenon in their chemically sensitive and chronic degenerative diseased patients with chronic noxious stimuli exposure. It has been calculated that at times with 0.02–0.56 µg/ml amount of chlorinated pesticides in the blood, 10–30% of the cell blood volume is displaced with these toxics.[3] If this is the case a contant stimulatory or inhibitory effect would be available for disturbing cell function. See Tables 4.2 through 4.5.

Once hormonal reserve is lost, homeostatic function becomes fragile, and the chronically ill patient physically weakens, tiring easily. This patient then has no tolerance for any stressor including food, nutrient supplementation or even water, minerals or condiments as we have seen in the severely ill patient with chemical sensitivity and/or chronic degenerative disease. **The endocrine hormones most often do not react directly on the intracellular machinery to control the final cellular chemical reactions, but act as an information sharer and stimulator**. These hormones usually combine with hormone receptors on the surfaces or inside of cells, which makes responses extremely vulnerable to connective tissue contamination and dysfunction. The combination of hormone and receptor usually initiates a cascade of reactions in the cell with each stage of reaction in the cascade becoming more powerfully activated than the previous stage, so that even a small initiating stimulus leads to a large final effect. Thus, homeostatic amplification occurs. This amplification of response in the patient with chemical sensitivity and/or chronic degenerative disease is often seen when chronic pollutant exposure (e.g. insecticides, mycotoxins) occurs resulting in a series of hypersensitivity hair-trigger-like responses.

Each cell has between 2000 and 100,000 receptors, and each receptor is highly specific for a single hormone like norepinephrine, epinephrine, dopamine, methacholine, urocholine, etc.

TABLE 4.2
Blood Levels of Organochlorine Pesticides in 200 Chemically Sensitive Patients

Pesticide	LD_{50} (mg/kg) Orl-Rat*	Average Serum Level (ng/ml) in Chemically Sensitive Patients	Molecules/ Erythrocyte in Chemically Sensitivie Patients	Molecules/ Leukocyte in Chemically Sensitive Patients
Endosulfan	18	0.08	24	16.10^3
Dieldrin	46	0.13	41	27.10^3
Heptaclor Epoxide	62	0.56	173	115.10^3
DDT and DDE	110–880	5.35	1,910	1274.10^3
Gamma-Chlordane		0.02	5.9	$3.9.10^3$
Beta- BHC	6000	0.42	173	115.10^3
Hexachlorobenzene	10,000	0.30	126	84.10^3

Source: Rea, W. J., Fenyves, E. F., Seba, D., and Pan, Y., *J. Environ. Bio.,* 22(3), 163–169, 2001.

Note: Variability factor calculated from 1 ppb of the analyte is 18.6%.

*administered orally

TABLE 4.3
Blood Levels of Chlorinated Hydrocarbons in 114 Chemically Sensitive Patients

Pesticide	LD_{50} (mg/kg) Orl-Rat*	Average Serum Level (ng/ml) in Chemically Sensitive Patients	Molecules/ Erythrocyte in Chemically Sensitivie Patients	Molecules/ Leukocyte in Chemically Sensitive Patients
Dichlorobenzenes	500	0.19	155	103.10^3
Chloroform	1194	0.19	192	128.10^3
Dichloromethane	2524	0.62	875	584.10^3
Trichoroethylene	4920	0.09	82	55.10^3
Tetrachloroethylene	8850	0.93	672	448.10^3
1,1,1-Trichloroethane	10,300	0.45	406	271.10^3

Source: Rea, W. J., Fenyves, E. F., Seba, D., and Pan, Y., *J. Environ. Bio.,* 22(3), 163–169, 2001.

Note: Variability factor calculated from 1 ppb of the analyte is 12.6%.

*administered orally

Specific hormones affect specific receptors for orderly dynamics of homeostatic function. For example, peptide and catecholamines are found in decreased numbers on the cytoplasm organelles for steroid hormones and in or on the cell surface protein, while thyroid hormones are usually found in the nucleus. One can see that with pollutant exposure, injury would result in a myriad of responses depending on what specific receptors were stimulated or inhibited.

The dynamics of the homeostatic control of the number of receptors is very important. **The number of receptors in a target cell usually does not remain constant from day to day or even minute to minute because the receptor proteins themselves are often inactivated or destroyed during the course of their function, and at other times, they are either reactivated or new ones are manufactured by the protein manufacturing mechanism of the cell.** For instance, binding of the hormone with its target cell receptors often causes the number of active

TABLE 4.4
Blood Concentrations of Drugs after Therapeutically Effective Dosage in Humans

Drug	LD$_{50}$ (mg/kg) Orl-Rat*	Therapeutic Steady Stae Average (ng/ml) in Humans	Toxic Level (ng/ml) in Humans	Molecules/Erythrocyte (Therapeutic) In Humans	Molecules/Leukocyte (Therapeutic) In Humans
Digoxin		08.–2.0	2.5	130–310	(90–210).10^3
Haloperidol	128	4–2.5		1280–7980	(850–5320).10^3
Clonidine	150	1.0–2.0		530–1040	(350–700).10^3
Amphetamin	50	20–30	>200	17,800–26,600	(11,900–17,700).10^3
Verapamil	114	100–500		26,400–132,000	(17,6000–88,000).10^3
Cocaine		100–500	>1,000	39,600–198,000	(24,400–132,000).10^3
Propoxyphene	135	100–400	>500	36,000–144,000	(24,000–96,000).10^3
Chlorpromazin	141	50–300	>750	18,8000–113,000	(12,600–75,400).10^3
Meperidine (Demerol)	162	400–700	>1,000	194,400–339,600	(129,600–226,400).10^3
Imipramin	250	125–250	>500	53,500–107,200	(35,700–71,400).10^3
Amitriptyline	320	120–250	>500	52,000–108,400	(34,600–72,200).10^3
Morphine	335	10–80	>200	4,200–33,600	(2,800–22,400).10^3
Desipramin	375	75–300	>400	33,700–135,000	(22,500–90,000).10^3
Nortriptyline	502	50–150	>500	22,800–68,400	(15,200–45,600).10^3
Propranolol	515	50–100		23,200–46,300	(15,400–30,800).10^3

Source: Rea, W. J., Fenyves, E. F., Seba, D., and Pan, Y., *J. Environ. Bio.,* 22(3), 163–169, 2001.
*administered orally.

receptors to decrease because of the decreased production of the receptor molecules, which become inactivated by the hormone's presence. In either event, down-regulation of the receptors occurs. This down-regulation decreases the responsiveness of the target tissue for the hormone as the number of active receptor(s) decreases. An example of this down-regulation is found in the male who is prepubertal and has a hypothalamic pituitary axis that is supersensitive to testosterone. After puberty, the supersensitivity is lost, but the male develops many effects, such as muscle hypertrophy, deepened voice, and hair growth, etc. The clinician sees the variability of symptoms such as fatigue, weakness, anxiety, and imbalance with pollutant exposure from pesticides, solvents, and mycotoxins, which may be explained partially by this phenomenon of receptor variability.

In a few instances, hormones cause up-regulation of receptors; that is, the stimulating hormone induces the formation of more receptor molecules than does the normal protein manufacturing output of the target cell. In this instance, the target tissue becomes, progressively, more sensitive to the stimulating effects of the hormone. **In cases of hypoadrenalism, there is an up-regulation of the sensitivity of thyroid receptors and the patient becomes more sensitive to thyroid hormone, which often results in his inability to tolerate the administration of exogenous thyroid hormone**. At times, cortisone supplementation is needed to allow the thyroid hormone to be tolerated. If the chronic noxious stimuli (i.e., polychloro biphenyl (PCB), PBB, isocyanates) are not withdrawn, the thyroid hormone will continue its output until exhausted. The target organ's dynamics of homeostatic physiology is altered until tissue changes occur. Eventually, fixed-named autonomous disease will occur. For example, hypothyroidism may occur when PCB's or thiocyanates stimulate the thyroid causing it to continuously produce hormones until it finally burns out from exhaustion. Then hypothyroidism occurs.

Endocrine System

TABLE 4.5
Reference Levels of Highly Potent Chemicals in Blood (Hormones, Vitamins, etc.) in Humans

Chemical	Specimen	Reference Level (ng/ml)	Molecules/ Erythrocyte	Molecules/ Leukocyte
Adrenocorticotropic hormone (ACTH)	Plasma	0.0007–0.079		
Angiotensin I and II	Plasma	0.010–0.0888		
Estrogens (total)	Serum	0.020–0.080 (M)	8.1–32 (M)	$(5.4–22).10^3$
		0.060–0.400 (F)	24–162 (F)	$(16–108).10^3$
Testosterone, free	Serum	0.052–0.280 (M)	22–116 (M)	$(4.2–22).10^3$
		0.002–0.006 (F)	0.8–2.5 (F)	$(0.6–1.7).10^3$
Thyrotropin-releasing hormone	Plasma	0.005–0.060	1.7–20	$(1.1–13).10^3$
Vitamin D_3, 1,25- dihydroxy	Serum	0.025–0.045	7.2–13	$(4.8–8.6).10^3$
Vitamin B_{12}	Serum	0.1–0.7	32–223	$(21–149).10^3$
Progesterone	Serum	0.13–0.97 (M)	50–310 (M)	$(37–206).10^3$
		0.15–25 (F)	57–9,520 (F)	
Prolactin (h PRL)	Serum	0–20		$(38–6,350).10^3$
Tri-iodothyronine, free	Serum	0.26–0.48	48–88	$(32–59).10^3$
Thyroxine, free	Serum	0.8–2.4	124–371	$(82–247).10^3$
17-Hydroxyprogesterone (17-OHP)	Serum	0.5–2.5 (M)	181–906 (M)	$(121–604).10^3$(M)
		0.2–5.0 (F)	73–1,813 (F)	$(48–1,209).10^3$(F)
Growth hormone (HGH)	Serum or Plasma	0.4–10 (M)		
		1.0–14 (F)		

Source: Rea, W. J., Fenyves, E. F., Seba, D., and Pan, Y., *J. Environ. Bio.*, 22(3), 163–169, 2001.

The initial mechanisms of the hormone receptor action play a major role in hormonal function for regulation of the dynamics of homeostasis of chemical sensitivity and/or chronic degenerative disease as shown in the following examples. First, the combination of neurotransmitters and the receptors almost always causes a conformational change in the protein structure of the receptor, which then causes changes in the membrane permeability of local hormones. Usually, this conformational change opens or closes a channel for one of the ions. For instance, **norepinephrine and epinephrine have especially strong effects on sodium and potassium channels, which will change the action potential of the membrane of the smooth muscle, causing relaxation during bronchospasm. In this way, bronchial homeostasis may occur. Norepinephrine and epinephrine substances can also cause excitation in some areas and inhibition in others. These substances, when secreted from the reticular activating system of the brain, also keep the level of consciousness high.** Since the epinephrine and nonepinephrine secretion is abnormally high in 85% of the patients with chemical sensitivity and/or chronic degenerative disease, dyshomeostasis of the hormonal regulation, of the sodium and potassium channels occurs. This deregulation causes jitteriness, chronic fatigue, edema, and weakness in patients with chemical sensitivity and/or chronic degenerative disease.

Another example of the hormone receptor action is noted when hormones help obtain homeostasis by activating an intracellular enzyme, which then combines with a receptor. For instance, insulin binds with a portion of its membrane receptor that protrudes to the exterior of the cell. This binding then produces a structural change in the receptor molecule itself causing the portion of the molecule that protrudes to the inside of the cell to become an activated kinase. Kinase then promotes phosphorylation of several different substances in the cell. Most of the actions of insulin result secondarily from the phosphorylation process to achieve glucose homeostasis. A second example of the membrane receptor binding to the enzyme is the hormone binding to a special transmembrane receptor, which then becomes the activated enzyme, adenyl cyclase, at its end, and protrudes to the interior of the cell. This cyclase, in turn, causes the formation of the cyclic AMP (cAMP), which has a multitude of effects inside the cell. These effects control cell activity and maintain the dynamics of homeostasis. The cAMP is a secondary messenger and not the hormone itself, but it directly institutes the intracellular changes. Cyclic guanosine monophosphate (cGMP) is a similar secondary messenger.

The hormone receptor action is also seen when some steroid and thyroid hormones bind with intracellular receptors to activate specific portions of DNA. This activation, in turn initiates transcription of specific parts of the gene's DNA to form messenger RNA. **Many toxic chlorinated benzenes, such as hexaclorobenzene, actinomycin D, and irradiations activate and initiate changes in DNA and RNA**. Therefore, minutes to hours to days after an exposure, newly formed proteins appear in the cell and become controllers of new and increased cellular function. Growth hormone (GH) and insulin appear to enhance the translation of messenger RNA in the cytoplasm and promote the formation of polypeptides and proteins (Figure 4.2).[4]

Secondary messengers, such as **cyclic 3, 5-AMP + PPi, cause the following responses: they activate enzymes, alter cell permeability, cause muscle contraction or relaxation, and produce secretion**. Also, the breakdown products of membrane phospholipids can act as secondary messengers. When phospholipase C attaches to a receptor and breaks down some cell membrane lipids, inositol triphosphate and diacylglycerol may be triggered. **As a secondary messenger, inositol triphosphate, mobilizes calcium ions** in both the **endoplasmic reticulum and mitochondria**. It also promotes changes in the secretion of the secretory cells and in the ciliary action of smooth muscle contractions. **The diacylglycerol is the other secondary messenger that activates the enzyme, protein kinase**, which along with increased Ca^{++} aids in cell division and proliferation. In addition, the lipid portion of the diacylglycerol is arachidonic acid, which is the precursor for prostaglandins and other local hormones. Moreover, any noxious stimuli such as toxic chemicals (insecticides, solvents, mycotoxins) can disrupt the internal and external receptors and alter the

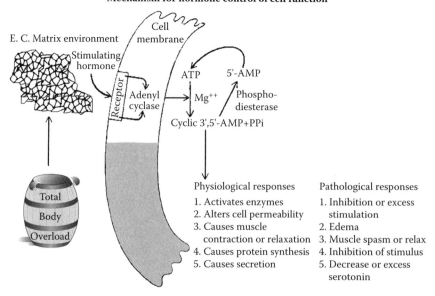

FIGURE 4.2 Diagram shows cyclic 3′,5′- adenosine monophosphate (cAMP) mechanism by which many hormones exert their control of cell function. (Modified from Guyton, A. C., and Hall, J. E., *Textbook of Medical Physiology*, 9th ed., Philadelphia: WB Saunders Co., 930, Fig. 74-2, 1996.)

messenger cascade. Accordingly, the protein kinases and arachidonic acid are altered, causing lipid metabolism dysfunction and altered leukotriene and prostaglandin formation.

Other secondary messengers trigger calmodulin, a protein that binds calcium; calmodulin activates myosin to cause smooth muscle contraction. Many patients with chemical sensitivity and/or chronic degenerative disease have problems with weakness and difficulty moving while others have a large amount of muscle spasms. Either response could be a secondary messenger malfunction.

Endocrine malfunction as seen in chemically sensitive and chronic degenerative diseased patients causes not only local, regional, and distal dyshomeostasis, but also may very strongly affect the neuro-immuno axis of the GRS. Likewise, damage to the neuro-immuno axis can affect the endocrine system. This damage to the endocrine system can have either adverse or positive effects on the chemically sensitive and/or chronically diseased individual, depending on whether the whole system is well orchestrated through the GRS or disrupted by pollutant overload. Pollutants (toluene, organophosphate insecticides, mycotoxins) and other noxious stimuli have been shown to affect endocrine metabolism in a myriad of ways, stimulating or disrupting homeostasis. Often, for example, pollutant entry and injury first manifests in dysfunction of the pineal, ovarian, testicular, adrenal, and thyroid metabolism (Figure 4.3)[5] However, it appears as if the pituitary/hypophysial areas, which are the integrators of the dynamics of homeostatic subunits for central homeostasis, are affected more often since this area appears to be involved in most distal homeostatic reactions. Certainly, the severely damaged catabolic states seen in some extremely chemically sensitive or chronically diseased patients are the result of or influenced by panhypopituitarism or its equivalent.

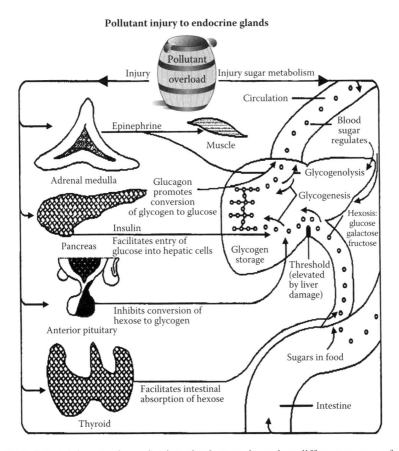

FIGURE 4.3 Pollutant injury to the endocrine glands may deregulate different aspects of glucose and glycogen metabolism. (Modified from Rea, W. J., *Chemical Sensitivity, Vol. III: Clinical Manifestations of Pollutant Overload,* CRC Press, Boca Raton, FL, 1598, Figure 24.1, 1995.)

Consequently, polyendocrinopathy is now commonplace in the treatment of neglected chemically sensitive and chronically disabled individuals. At least 10%, and perhaps as many as 20%, of chemically sensitive individuals have polyendocrinopathy. The following case represents this growing problem of severe homeostatic dysfunction.

Case study. A 40-year-old white female entered the ECU complaining of bone pain, anxiety attacks, and sensitivities to foods and chemicals. For at least the 10 years previous to this admission, she had experienced intolerance to various foods and chemicals. Initially, she was rigid on her food rotation and environmental control, but as she improved, she became careless and was exposed to many chemicals. She developed premenstrual tension accompanied by irregular periods and antiovarian antibodies. She eventually underwent a bilateral oopherectomy, which reduced the autoantibodies. Five years prior to admission, she developed mild adrenal insufficiency, which occasionally required cortisone supplementation. One year prior to admission to the ECU, she developed antithyroid antibodies. Once placed on thyroid extract, antibody titres improved. Three months prior to admission, she developed hypercalcemia and was found to have abnormally high calcium of 12.5 mg/dL (C = 8.5–11.0 mg/dL). This reading dropped to normal after resection of a parathyroid adenoma and partial thyroidectomy. The patient presented with hypomagnesemia, which responded to intravenous magnesium sulfate supplementation and heat depuration/physical therapy. This woman had retained a large number of toxic chemicals (hexane, toluene, trichloroethane) in her blood that appeared to be triggering her problems and disrupting her homeostasis, especially endocrine homeostasis. Oral, intradermal, and inhaled challenges confirmed her sensitivity to biological inhalants, foods, and chemicals. She gradually improved with institution of environmental control and diet.

This chapter discusses in depth each endocrine gland separately as well as the neuroendocrine cells and the entire receptor system, all or any of which may become imbalanced through injury and/or through chronic exposure to noxious stimuli. Following chronic exposure to noxious stimuli, this local imbalance disturbs efficient homeostasis, resulting in a triggering of the GRS and the central homeostatic mechanism. **In the chemically sensitive or chronic degenerative diseased patient, disturbed physiology of the dynamics of homeostasis of the endocrine system frequently results in amplification of regional and distal responses until some part of the homeostatic system is depleted**. Namely, information is changed, metabolism is altered, and fixed-named autonomous disease occurs.

PITUITARY GLAND

Pollutant injury to the pituitary gland has been poorly studied. Nontheless, clinical evidence suggests that injury to this organ occurs more often than the clinician has appreciated. Clearly, homeostasis can be disrupted by entry of noxious stimuli into this gland. This disruption may occur gradually, over a long period of time, like that which occurs in the aging process, or the degeneration occasionally occurs rapidly.

The hypothalmic-pituitary axis controls much of the endocrine system through distal homeostatic reactions orchestrated through the central area. Since the pituitary is the master control area for endocrine function and it is also intimately connected to the hypothalamus, integration of the central homeostatic mechanism through the ground regulation system (GRS), of which it is a part, is highly significant. Some chemicals are known to alter the pituitary, for example, suppression of prolactin and corticosterone. ACTH in rats exposed to organochlorine compounds such as pesticides and solvents has been shown to alter the dynamics of homeostasis. Similarly, in humans, some **individuals exposed to Endosulfan, an organochlorine pesticide that alters the dynamics of the homeostatic regulatory mechanism, have experienced abnormal reproduction**.[6] Endosulfan pesticide is found in a small percentage of chemically sensitive patients, along with another pesticide, which is almost always, DDT. A plasticizer bisphenol A, which is a breakdown product of polycarbonate plastics and dental sealer, has shown to have an adverse effect on the pituitary. The first one is an estrogenic effect, which when produced, will alter estrogen receptors.[6] The other is in the follicle-stimulating hormone (FSH) mechanism. A toxic effect, resulting in

Endocrine System

toxic damage to the membranes and free radical damage resulting in inflammation of the matrix, as well as abnormal endocrine influence on the GRS and the dynamics of homeostatic function. Undoubtedly, the pituitary gland is one of the central mechanisms of homeostatic amplification. These observations set the stage for the kind of endocrine homeostatic function, dysfunction, and amplification of responses to be further discussed in this chapter.

POLLUTANT EFFECTS ON THE NERVOUS SYSTEM OF THE PITUITARY GLAND

The anterior pituitary originates from the Rathke's pouch, which is an embryonic invagination of pharyngeal epithelium. The posterior pituitary comes from part of the hypothalamus being neural in origin and having a large number of glial cells in the gland. This part of the gland will then be prone to any pollutant neurotoxin. The hormone output from the anterior and frontal posterior pituitary has much to do with regulating the dynamics of homeostatic function.

At least two glands (adrenal medulla and posterior pituitary glands) secrete their hormones in response to direct nerve stimuli, which are triggered by stimuli that at times can be noxious. Therefore, if there were a pollutant stimulus to their particular innervation, it would enter directly or through the peripheral autonomic nerves. Hormones would be released and symptoms would occur, often to varied degrees in the chemically sensitive or chronic degenerative diseased individual, resulting in disordered homeostasis. Similarly, if the posterior pituitary released the hormone ADH, this would result in excess water retention, which would then raise the blood pressure of the individual causing disordered homeostasis. **We often see this water retention in both the chronic degenerative diseased and the hypersensitive patient who receives an inadvertent exposure to toxics**. Sometimes the edema is subtle in the patients with early onset of chemical sensitivity and/or chronic degenerative disease but can be massive in those who are chronically exposed. A great deal of evidence shows that nerve influences affect the homeostatic function of the anterior hypophysis, although this part of the pituitary gland probably has no direct innervation. The homeostatic regulatory neurohormones appear to reach the pituitary from the hypothalamus via the hypophysial portal circulation using neurosecretory substances that act on glandular cells to release anterior pituitary hormones. **This physiological fact suggests that the anterior pituitary is governed by a dynamic homeostatic reflex arc comprised of a neural segment ending in the median eminence and a vascular segment formed by the hypophysial portal system**. The relationship between the primary capillary plexus of the portal system and the nerve endings suggests that the chemotransmitters, passing from nerve to capillary, are small peptides in the order similar to the well-characterized posterior lobe hormones. Noxious stimuli to the nerves, secretory cells, or blood vessels may alter the dynamics of homeostatic functions because they are part of the GRS, creating generalized homeostatic body dysfunction as is often seen in chemically sensitive or chronic degenerative diseased patients. **The toxic chemicals can act as false neurotransmitters (neurotransmitter mimics) and include alcohol-mimics GABA, LSD-mimics, serotonin, THC-mimics anandamide**. They also can overstimulate or suppress the available transmitters or receptors. Evidence for these actions has been found at the EHC-Dallas by using intradermal injection challenges with various neurotransmitters such as serotonin, methacholine, epinephrine, dopamine, acetylcholine, histamine, etc., which reproduce or ameliorate symptoms in chemically sensitive and/or chronic degenerated diseased individuals by stimulation or neutralization of the cells and neurovascular system and PG/GAG's of the connective tissue matrix. Often, the injections help the patient return to homeostasis by dampening or correcting the electrical lability mechanism of the connective tissue matrix and the nervous tissue.

POLLUTANT EFFECTS ON THE PHYSIOLOGY OF THE PITUITARY

In the pituitary, there is usually one cell type for each hormone produced. At times, this fact allows the clinician to pinpoint more precisely the origin of homeostatic dysfunction when the pituitary is involved. The anterior pituitary acidophilic cells produce prolactin (luteotropic hormone, LTH) and somatotropic hormone (growth, STH). The anterior pituitary basophil cells produce thyrotropin (TSH), adrenocorticotropin (ACTH), FSH, lutenizing hormone (LH), or interstitial-cell-stimulating hormone (ICSH), and melanocyte-stimulating hormone (MSH). Stimulation of preoptic and anterior hypothalamic nuclei controls secretions of thyroid stimulatory hormone (TSH), while stimulation of the posterior hypothalamus and medial eminence of the infundibulum causes secretion of ADH. Stimulation of the median eminence of the hypothalamus triggers FSH, and stimulation of the anterior hypothalamus gives release of LH. Other nuclei of the hypothalamus control secretion of GH and prolactin by the anterior pituitary gland. **Direct chronic pollutant stimulation of these various parts of the hypothalamus and pituitary could account for many of the various types of homeostatic or altered homeostatic responses. These responses include weakness, fatigue, myalgia, arthralgia, etc. as observed in some chemically sensitive and chronic degenerative diseased patients**. Constant stimulation or suppression of the pituitary causes sustained dyshomeostatic responses that alter the whole neuro-immune-endocrine axis, causing exacerbation or reduction in chemical sensitivity and/or chronic disease. Although we have had several pituitary tumors in patients with chemical sensitivity (one from molybdenum sensitivity and the other from pesticides) there is is sparse animal data. However, one study using hexadimethrine bromide induced pitutitary necrosis. One study by Werley et al. using Webster mice and a chemical, ASTM E981-84, showed lesions in the pituitary.[7] Futhermore, prolactin levels are definitely altered in some patients exposed to toluene. Some drugs stimulate pituitary hypoplasia and others inhibit suggesting altered homeostasis of the anterior acidophilic pituitary cells.[8,9] The following case is an illustration.

Case study. A 45-year-old white chairwoman of a department of physiology at a university medical school in the United States was admitted to the EHC-Dallas with the chief complaints of brain dysfunction, malaise, fatigue, edema, hormonal dysfunction, odor sensitivity, cyanosis, gastrointestinal upset, and obesity. Her illness started years previously, coinciding with a toluene leak discovered in the area of the department of physiology. Prior to collapse, this patient had worked continually in a medical school environment for 23 years, during which time she had been healthy and highly productive. A few months before she collapsed, her family physician of many years detected elevated blood pressure, gastritis, and extreme fatigue. These conditions were followed by metrorrhagia of three weeks' duration. Her fatigue worsened until virallike symptoms began early in January 1986. Many other individuals in her immediate work environment complained of similar symptoms and reported a heavy odor of toluene in the hallways. Her office was located above a passageway for automobile traffic. In addition, the positron emission tomography facility immediately adjacent to her building was being repaired, and heavy black smoke was released from the lines into the air.

Due to her history of exposure to toluene, the presence of toluene in her blood 1.4 ng/mL (control 1.0 ng/mL), and the known cerebral and endocrinological effects of toluene exposure, she underwent double-blind inhalation challenge testing in a chemically free booth to assess sensitivity to toluene. She was tested with two controls and toluene. No reaction to the controls was observed. On exposure to toluene, she experienced facial flushing, dizziness, and sensations of flushing in her arms. She appeared very distant and foggy in her thinking. Independent neurological assessment by a psychologist before and after each inhalation challenge test showed significant worsening on neurological screening and on visuoperceptual motor functioning with exposure to toluene. She also was perceived by the objective tester to be affected adversely by toluene exposure, with symptoms of clumsiness, tremulousness, and unsteady movements of hands and legs. These responses confirmed sensitivity to toluene. These effects were not seen on the control challenges. Nine months later, upon reexposure to toluene (1989), she deteriorated rapidly and again went on an amino acid liquid diet (70 g/day). After three weeks on this program the patient showed improvement; the toluene in her blood was 0.8 ng/mL, which was decreased from the original level of 1.4 ng/mL. She is currently working on developing an uncontaminated, preservative- and additive-free diet plan for the future maintenance of her health. Long-term follow up after 19 years showed her to be stabilized but unable to work in the city around the medical center and still sensitive to chemicals.

At the EHC-Dallas, we have seen alterations in the other aforementioned pituitary hormones after injury from noxious stimuli in the chemically sensitive and chronic degenerative diseased individuals. We have frequently observed these hormones return to normal levels as the pollutant is eliminated and the dynamics of orderly homeostasis is restored.

In addition, secretion of the posterior pituitary involves oxytocin and vasopressin (antidiuretic hormone, ADH) and can occur by direct neurostimulation or stimulation from the hypothalamus, since ADH is generated in the supraoptic nucleus of the hypothalamus. Therefore, noxious stimuli can disturb homeostasis in these areas as well. We have seen literally thousands of environmentally sensitive patients at the EHC-Dallas and Buffalo who have peripheral and periorbital edema, presumably resulting from pollutant-stimulated hypothalamic generated pituitary released ADH. This inappropriate ADH syndrome may be cleared with noxious stimuli avoidance with reduction of the total body pollutant load. We have then triggered a recurrence with inhaled or oral challenge using the incitant to which the individual was sensitive.

The hypothalamus receives signals from almost all possible sources of the nervous system. Thus, when a person experiences pain, a portion of the pain signal is transmitted back to the hypothalamus resulting in production of more ADH and its release from the posterior pituitary causing peripheral edema. Likewise, when a person experiences some powerful depressing or exciting news, a portion of the signal is transmitted to the hypothalamus, where the patient may or may not respond adversely or positively. Also, olfactory stimuli discern pleasant or unpleasant smells and transmit strong signal components directly to and through the amygdaloid nuclei into the hypothalamus, which then generates ADH in order to obtain and maintain the dynamics of homeostasis. Even the concentration of nutrients, electrolytes, water, and various hormones in the blood excite or inhibit various portions of the hypothalamus in order to maintain homeostasis. Hence, **the hypothalmus is a major control system for both the input information concerned with homeostasis, as well as the output information used to control secretions of many globally important pituitary hormones** (Figure 4.4).[10]

One of the most significant pituitary hormones is the antidiuretic hormone, ADH, that is antidiuretic and hypertensive by favoring arteriolar constriction and fluid retention. An increase in the osmotic pressure of the plasma stimulates osmoreceptors in the anterior hypothalamus. From such osmoreceptors, whose membranes may be damaged by chronic exposure to noxious stimuli or by other cells stimulated by pollutants, trauma, emotions, or drugs (e.g., tranquilizers, nicotine, morphine), stimuli pass to the cells and fibers of the **supraopticohypophysial nucleus**. This area **appears to be the central area for regulatory homeostasis**. ADH appears to be formed in the neural cell bodies of the supraoptic nuclei of the hypothalamus and to be passed along the length of axons in combination with a carrier substance. It is released at, or near, the nerve ending in the neural lobe of the posterior pituitary. ADH then enters the bloodstream and goes to the kidney to work on the renal tubules, with subsequent disproportionate reabsorption of water. **Fifteen percent of the water reabsorbed in the kidney is under the influence of ADH**. This volume is enough to cause altered dynamics in homeostasis if the total 15% of the reabsorbed water is retained. Some foods and toxic chemicals can have either a direct or sometimes indirect influence on this hormone, causing excess production and leading to peripheral edema. When the chemically sensitive or chronic degenerative diseased individual fasts in a controlled environment and achieves a significant reduction in his or her total body load, a significant reduction of edema, is seen.[11] Frequently, in the first 48 hours of a fast, patients will lose 8–10 pounds, which is almost entirely the result of loss of edema. At times, we have seen 25 pounds of edema lost in four days of fasting without the use of diuretics.[11] This finding suggests alteration of the ADH effect. In addition, when a chemically sensitive patient is challenged with sensitivity to specific foods, at times swelling will occur. Furthermore, when the patient avoids these foods, his or her edema will decrease as the normal dynamics of homeostasis are restored.[11] This same phenomenon occurs in patients who are sensitive to a specific chemical or who are overloaded with many noxious stimuli.[11]

Chronic overload by noxious stimuli may cause homeostatic deregulation by alteration of the chemical reactions within the cells, by changes in the permeability of membranes, or by activation of some other specific cellular substances or changes in the connective tissue matrix. **Hormones**

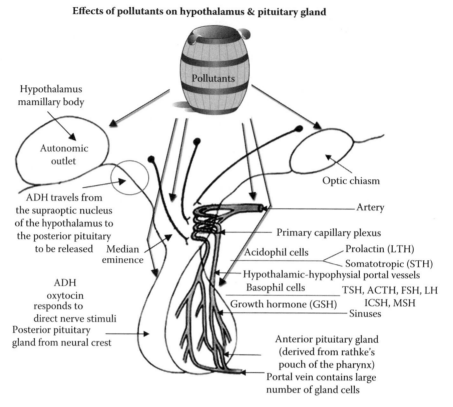

FIGURE 4.4 Hypothalamic-hypophysial portal system. Potential areas of pollutant injury may be isolated or generalized. They also may involve vascular injury, neurological injury, or direct endocrine injury. Note: ADH made in the supraoptic nucleus of the hypothalamus travels by neural routes to the posterior pituitary to be released. (Modified from Guyton, A. C. and Hall, J. E., *Textbook of Medical Physiology*, 9th ed., Philadelphia: WB Saunders Co., 935, Fig. 74-4, 1996.)

achieve their effects by activation of the cyclic AMP of cells, which in turn elicits specific cellular functions, and by activation of the genes of the cells, which causes the formation of intracellular proteins that initiate specific cellular function. Exposure to noxious stimuli can cause injury to both types of activation effects, causing dysfunction with exacerbation of the chemical sensitivity and with altered dynamics of homeostasis.

Noxious stimuli may injure the cyclic AMP mechanism in a variety of ways including damaging the hormone receptor site on the cell membrane, the adenyl cyclase enzyme, or the ATP conversion to cyclic 3,5AMP.[12] The physiological homeostatic parameters that may be altered are enzyme activation, cell permeability, muscle contraction or relaxation, protein synthesis, and secretions.[12] All of these altered parameters are seen in some chemically sensitive and chronic degenerative diseased patients where the dynamics of homeostasis are altered.

Pollutant or other noxious stimuli injury to the steroid hormone action on the genes that cause protein synthesis can alter the protein production that acts as carriers or enzymes, thus altering homeostasis. **Pollutant or noxious stimuli injury may occur to the receptor site within the cytoplasm, in the transport process of the hormone to the nucleus, by the alteration of the protein size, in the combination of small protein and hormone, by activating specific genes to form messenger RNA, or cause diffusion of messenger RNA back into the cytoplasm**, where it promotes the translation process of the ribosomes to form new proteins.[12] Many of these alterations are thought to occur in some chemically sensitive and chronic degenerative diseased patients and result in dyshomeostasis.

Pollutant or other noxious stimuli injury of the output of other hormones of the pituitary has also been suggested. For example, clinicians see apparent pituitary ovarian hormone deregulation in some chemically sensitive and chronic degenerative diseased patients with premenstrual tension syndrome (PMS). In this case dyshomeostasis has occurred. Though often the dysfunction is due to pollutant damage to the ovary, correction of the syndrome with administration of extremely dilute doses of LH, FSH, estrogen, or progesterone may be possible and has been effective at the EHC-Dallas and Buffalo. This correction suggests pituitary deregulation, and both pituitary and ovarian deregulation have been observed in over 1000 cases of PMS treated at the EHC-Dallas and Buffalo.

Some pesticides have been shown to alter the hypothalamic-pituitary axis homeostatic control on the endocrine system. **Suppression of prolactin and corticosterone due to organochlorine compounds has been shown in rats**[13] **and appears to occur in some chemically sensitive and chronic degenerative diseased humans observed at the EHC-Dallas**. Research has shown how endosulphan (an organochlorine pesticide) alters the dynamics of homeostatic regulatory mechanism, resulting in abnormal reproduction.[12] As shown previously, chlordane reduces pituitary hormonal content and produces a dyshomeostatic, estrogenic effect.[12] These changes are frequently seen in the chemically sensitive and chronic degenerative diseased patient.

Growth Hormone

Aside from GHs, all major anterior pituitary hormones exert their effects on specific target organs. We will discuss pollutant or other noxious stimuli injury to these other hormones further on in the chapter as it relates to the specific target organ. **GH does not function through a specific target organ but instead exerts its effects on almost all tissues of the body**. This effect will be discussed separately. GH greatly affects the dynamics of homeostasis and is gradually depleted with age. It is known to have an indirect or direct effect on cartilage and bone growth as well as on other tissues. We have observed and it has been reported that certain pesticides and other toxic chemicals can stimulate, and more often suppress this hormone,[14] resulting in severely altered dynamics of homeostasis. Findings also include herbicides that can cause stunted growth of some young, exposed offspring.[15] These substances can accelerate aging by decreasing production of GH.

GH has many generalized effects, which may be altered by pollutant or other noxious stimuli exposure, resulting in the altered dynamics of homeostasis. These alterations include: a change in the rate of protein synthesis in all cells of the body, which results in malnutrition and, thus, dyshomeostasis as seen in some chemically sensitive and chronic degenerative diseased individuals; a change in the mobilization of fatty acids from adipose tissue, leading to obesity or hypercholesterolemia as is often seen after pollutant exposure; a change in the use of fatty acids for energy, causing weakness; and a change in the rate of glucose utilization throughout the body, resulting in blood sugar swings, which is one of the most clinically obvious manifestations of altered homeostasis. These sugar swings often contribute to chronic fatigue. All of these alterations have been seen at one time or another in the chemically sensitive or chronic degenerative diseased patient with resultant decrease in protein and fat stores and excess utilization of carbohydrates. Damage to GH may alter its ability to enhance the transport of amino acids through cell membranes, causing loss of protein synthesis, and body wasting, as seen in some of the severely injured chemically sensitive patients. Often, the patients with chemical sensitivity and/or chronic degenerative disease present at weights of 60–80 pounds and are unable to tolerate all foods. In addition to pollutant injury to transport defects, injury may also occur to protein synthesis by the ribosomes on which GH has a direct effect. Decreased quantities of mRNA will also occur if pollutant injury decreases GH. **GH also decreases catabolism; therefore, pollutant overload may accelerate catabolism, which then negates this protein-sparing effect of GH and results in a debilitated chemically sensitive or chronic degenerative diseased patient**.

Pollutant or other noxious stimuli injury to GH may decrease the release of fatty acids from adipose tissue. When GH inhibition is present in a chemically sensitive or chronic degenerative diseased individual a decrease of the protein-sparing effect by pollutant or other noxious stimuli injury may decrease the conversion of fatty acids to acetyl Co-A. Subsequently a decrease in energy production occurs as well as decreased acetylation conjugation.

Pollutant or other noxious stimuli injury to GH can also influence carbohydrate metabolism, altering in turn the cellular metabolism of glucose, thus, altering homeostasis. Pollutant injury to GH can cause an increased utilization of glucose for energy, a prevention of glycogen deposition in cells, and an increased uptake of glucose by the cells. This phenomenon results in dramatic blood sugar swings, often seen in a subset of chemically sensitive and chronic degenerative diseased patients whose homeostasis has been disrupted.

Dwarfism and gigantism can occur with pollutant injury to GH. A decrease in growth hormone can also change the output of various glands, such as the adrenal, thyroid, etc., which results in less than optimum function because of inadequate output from these glands, thus causing dyshomeostasis; a vicious downhill cycle then occurs and aging accelerates.

PINEAL GLAND

The pineal gland plays an important role in sexual[16] and reproductive cycles,[16] as well as in homeostatic wake-sleep cycles. Moreover, it contributes to many of the body's immune regulating and anticancer functions.[16] The pineal gland is a vestigial of the third eye, has nerve attachments, and can function as an endocrine organ. It is also one of three areas of the brain that has a 4, rather than 6 cellular blood brain barriers. The other areas with four cellular blood brain barriers are the hypothalamus and area postrema of the fourth ventricle. Since the pineal gland is a neuroendocrine organ like the adrenal medulla and pituitary gland, the pineal gland is dicussed in this endocrine chapter because of its unique endocrine effects. The functional pineal organ clearly again emphasizes the integration of the endocrine and neurological systems' role in the dynamics of normal homeostatic function. Among other endocrine effects, **the pineal gland has the ability to generate energy during efficient sleep**. It is without a doubt one of the most important energy generators, since without sleep the human body does not function.[17] **Initially, sleep cycles are often affected by the early stages of chemical sensitivity and/or chronic degenerative disease and these patients often develop insomnia and thus, fatigue upon excess chemical pollutant, mold, and mycotoxin exposure**.

The pineal gland is controlled by the amount of light seen through the eyes each day and therefore is more directly influenced by environmental factors than most endocrine glands. The pineal gland will activate even with a certain amount of darkness, which helps to regulate homeostasis. The pineal gland participates actively **in regulating the endocrine system[18] and the immune system, resulting in orderly homeostasis. It can restore immune function eightfold during a good night's sleep.[19] It plays a key role in modulating circadian rhythm,[20] along with regulating the gonadal steroids,[21] corticosterone,[22] and thyroid hormones**.[23] Circadian rhythms are a part of normal GRS homeostatic function. (See Chapter 1.)

The pineal gland is also one of the circumventricular organs situated in the caudal epithalamus just above the midbrain tegmentum.[24] It has lost all direct connections with the CNS, but it is connected to the ANS, and is able to participate in local, regional, and distal homeostasis. Nerve impulses due to light pass from the eye to the supraoptic chiasm of the brainstem to the lateral hypothalamus. The impulses proceed from the sympathetic nerve tracts in the spinal cord to the cervical ganglia, move along the sympathetics and blood vessels, and then terminate directly in the parenchymal cells of the pineal gland. **Morphologically, the predominant parenchymal cell is the pinealcyte, which has secretory properties of hypothalamic peptides (LH- RH and somatostatin) and a high concentration of biogenic amines (norepinephrine, serotonin, melatonin, and other indole alkyl amines)**.[24] The other cell types are astrocytes and they are intertwined with pinealcytes. The pineal gland appears to be the only gland that secretes melatonin, which when secreted, goes to the anterior pituitary gland to control and suppress gonadotropin secretion, and thus it plays a very important role in the dynamics of endocrine homeostasis. In animals this suppression happens in the winter. In humans the cycle is not quite known. However, in some chemically sensitive people and chronic degenerative diseased people, this function has been observed to improve during the spring and summer months of increased daylight.[24] If chemically sensitive individuals are exposed to more light during this time, their endocrine system seems to be in balance and to function better. **Conversely, homeostatic balance appears to be more fragile in the fall and winter months.** When the endocrine system of chemically sensitive individuals is more balanced, they possess greater ability and more energy to resist damage from toxic chemicals.[25-27]

In the pineal gland the formation of melatonin from serotonin through N-acetyl serotonin and subsequent O-methylation takes place in the presence of hydroxyindole-O-methyl transferase (HIOMT). This enzyme differs from catechol-O-methyl transferase (COMT) because it is found only in or near the pineal gland. The enzyme, HIOMT, catalyzes the methylation of hydroxyindoles, and COMT, in contrast is present in most cells to bring about methylation of epinephrine and norepinephrine. The presence of melatonin in peripheral nerves, where HIOMT is not found, suggests that melatonin is a neurohormone released by the pineal gland into the general circulation and subsequently taken up by nerves. Epinephrine, norepinephrine, serotonin, 5-methoxyindole acetic acid, and 5-hyroxyindole acetic acid are also found in the pineal gland. Abnormal release of these substances due to stimulation by a pollutant or other noxious substance alters both pineal and peripheral functions, causing dyshomeostasis and exacerbation of chemical sensitivity.

Another function of this gland relates to the amount of HIOMT present in the pineal gland, which is second only to the kidney on a weight-for-weight measurement. Therefore, an injury to this gland may explain the disturbed sleep-wake cycles seen in some chemically sensitive and chronic degenerative diseased individuals.

Many functions besides the action of HIOMT, are characteristic of the pineal gland. **For example, the blood flow through the pineal gland is extremely high.** This increased blood flow function might allow the pineal to be susceptible to more pollutant or noxious stimuli exposure and injury since blood transports the toxic chemicals or noxious stimuli to the gland more frequently than to other areas of the body. In addition, the blood brain barrier is less dense than most other areas of the brain and thus more pollutants can cross over easily. **The amount of iodine uptake is second to the thyroid uptake and creates yet another unknown function for the pineal gland.**

The effects of light on the human pineal gland may differ, but it is evident that absence of natural light is important. Studies show that light causes a profound change in the pineal gland in rats.[28] Similar studies show changes in humans.[29] Since the interior of the human body is dark, the released melatonin acts as a messenger from the light to each end organ. When stimulated, the gland decreases its weight, production of HIOMT, serotonin, mRNA synthesis, glycogen, and succinic dehydrogenase activity, and it increases 5-hydroxytryptophan decarboxylase activity. In our current lifestyle, humans spend 95% of their time indoors, which creates a dyshomeostatic environment in which to observe and record the many effects of the lack of natural light on the pineal gland. For example, Ott[30] reported that lack of light in some cases causes weakness, inability to learn and function, and depression. In contrast, Parry et al.[31] reported that the administration of bright light decreased premenstrual depression. Also, seasonal affective disorders may be dependent upon light. When seen in some chemically sensitive individuals these disorders are considered an example of dyshomeostasis. Many chemically sensitive people have been observed to have severe affective depression that responds well to morning light. This light inhibits their production of melatonin, which has made them depressed, tired, and sleepy. The 509 nm (blue light) shuts off melatonin production and can be used in a treatment capacity.

Another function of the pineal gland relates to the earth's rotation that produces a natural biorhythm yearly (365.25 days) and 24-hour cycles. The pineal gland receives the yearly

and daily impulses and transmits them to the endocrine system. This gland appears to be the alarm clock for all biorhythms.[32–41] Many individuals with chemical sensitivity and/or chronic degenerative disease experience change in their 24-hour clocks and, therefore, they do not function efficiently. This reduced biorhythm function can be due to lack of exposure to natural sunlight, or due to work habits or even unknown reasons, some of which are probably related to individual pollutant overload. Severe change in biorhythms stresses the pineal gland.[42–44] When an individual takes an air flight from the west to the east, dyshomeostasis develops. Also, inappropriate pineal gland function may result from a lack of nutrients for methyl conjugation (methylation) due to xenobiotic excess. Faulty methylation of xenobiotics is the primary measurable defect in conjugation seen in chemically sensitive patients, and it always results in dyshomeostasis.

Melatonin can be collected in the saliva as well as the blood. When metabolized, 75% becomes sulfates, 15% becomes acetylates, and 5% becomes glucuronides.[45–47] Pollutant or other noxious stimuli injury may alter the conjugation systems, resulting in inappropriate pollutant clearing in some chemically sensitive patients, resulting in altered homeostasis. **As a result of faulty clearing, excess melatonin and, thus, altered sleep-wake cycles and immune dysfunction may occur**.

Further influences of the pineal gland show increased melatonin production both decrease sexual maturity in anovulatory females and is found at high levels nightly.[48] Melatonin contains high levels of zinc, and, therefore, pollutant-induced imbalance of zinc and copper imbalance may trigger melatonin alterations.[49–77] Melatonin may lower cholesterol, especially low-density lipoprotein (LDL).[78–84] Its immunoregulatory effects increase natural killer cell (NK) activity. Melatonin increases immune function during a sleep cycle eight times. Melatonin counteracts stress by being an epinephrine antagonist.[85–91]

Clearly, the pineal gland with its production of melatonin is highly significant in regulating the dynamics of homeostasis in many subsystems of the body. Its function can no longer be ignored and medical thinking must take the pineal gland into account when considering homeostatic balance and prevention of disease especially chemical sensitivity and/or chronic degenerative disease. Supplementation (1–20 mg) for sleep as well as immune function modulates can be taken from 1–20 years without any side effects.

NEUROENDOCRINE SYSTEM (PARAGANGLIA, PARANEURON)

The neuroendocrine system is comprised of the adrenal medulla, the pineal gland, and the pituitary gland together with a system of dispersed cell groups that are embryologically coderived from the neural crest and endocrine cells. The neuroendocrine cells have characteristics shared by neurons and endocrine cells and appear to have a common origin with neurons, thus being one of the areas for neuroendocrine homeostatic integration. These cells are able to produce both substance(s) identical to or related to neurotransmitters and protein/polypeptide substance(s) that possess hormonic actions. Neuroendocrine substances appear to influence the dynamics of homeostasis. These cells also possess synaptic vesicle-like and/or neurolike (transmitter) granules that have a receptosecretory function, which results in stimuli acting upon the receptor site on the cell membrane.

Extra-adrenal neuroendocrine cells are distributed throughout the visceral ANS and are allied closely with sympathochromaffin (autonomic) ganglia.[92–94] With this integration of cells, neuroendocrine function for homeostasis occurs and appears to be finely tuned. **These cells include SIF cells, carotid and aortic body glomeruli, parafollicular cells, adrenohypophysial cells, pancreatic islet cells, gastroenteric endocrine cells, parathyroid cells, gastric cells, merkel cells, olfactory cells, melanocytes, mast cells, bronchial endocrine cells, hair cells of the inner ear, and amacrine cells**. The variety of cells and functions in this system of neuroendocrine cells constitutes a diffuse network stretching throughout the body. This network then becomes part of the integrated GRS for local, regional, and global homeostatic function. The unique, diffuse, and varied characteristics of the secretions produced upon chronic pollutant or noxious substance stimulation of this system help explain how the chemically sensitive and chronic degenerative diseased individual might manifest a wide variety of clinical responses including varied symptoms related to many anatomical areas. The neuroendocrine cells' secretions are in contact with and are part of the GRS. In fact, at times the clinical responses resulting from neuroendocrine cells secretion are a final expression of the peripheral GRS. The importance of understanding this system in relation to chemical sensitivity and/or chronic disease cannot be overemphasized. Many of the widespread signs and symptoms of some of the chemically sensitive and chronic degenerative diseased patients can be explained by understanding the pathophysiology of this system. **Chronic pollutant or other noxious stimuli triggering of this system may cause appropriate responses at inappropriate times**, or it may stimulate the system to be overactive or inhibitory. For example, when a noxious stimulus enters the body, and the olfactory nerve is discharged, the sensitive individual may have a sudden urge to urinate or defecate because of neuroendocrine discharge, or he/she may develop a cloudy brain, with brain dysfunction resulting in unclear thinking. Chronic pollutant or other noxious stimuli of the neuroendocrine system can also cause nervous and immune system suppression or triggering and bring about symptoms that might previously have been considered psychosomatic in origin. However, these symptoms are the result of pollutant or other noxious incitant injury to the neuroendocrine cells.

We see in this neuroendocrine system how autonomic nerve fibers richly innervate paraganglia that act as the distal part of the GRS for homeostatic control. For example, the parenchymal chief cells synthesize catecholamines from tyrosine[95] and are supported by sustentacular elements analogous to peripheral nerve Schwann cells.[93,96] This sustentacular network can be detected immunocytochemically.[97,98] In extra-adrenal neuroendocrine cells, the sustentacular cells delineate a characteristic cell nest pattern of the chief cells observed both in normal and neoplastic tissues.[94,98,100]

NEUROENDOCRINE PHENOTYPE

Biologically, the paraganglia constitute one of the major functional subsets among the multiple groups of topographically dispersed neuroendocrine cells, and are considered functionally to be a part of the dynamics of homeostatic regulation through the GRS. The definition of a "neuroendocrine" phenotype is based upon a conjunction of cytoarchitectural, biosynthetic, and cytofunctional criteria.[101] These criteria include polyprotein gene expression with an overlapping biosynthetic profile,[102–105] product storage in argyrophilic dense-core secretory granules,[106,107] physiologic activity involving monoamine precursor uptake and decarboxylation,[108] and **functional capacities to amplify or generate neural signals either by means of humoral stimulation**[95,108] **or by biochemical transduction of microenvironmental perturbations**.[109] Often, they are responsible for the dynamics of local homeostatic amplification.

The oligopeptide or monoamine molecules secreted by paraganglion and other neuroendocrine cells can affect a multiplicity of homeostatic bioregulatory actions (Figure 4.5)[110] **both locally and distally as part of the GRS**.

These actions can be systemic (endocrine function), localized and adjacent to neuroendocrine cells (paracrine function), or self-regulatory (autocrine function). These neuroendocrine cell groups all take part in the maintenance of the dynamics of homeostasis. As an integral part of the GRS such physiologic variation among neuroendocrine cell groups provides a sophisticated, multilevel, homeostatic feedback mechanism for modulating microcirculatory dynamics and hormonal responses in target organs.[104,111,112] For example, **paracrine oligopeptide secretions of some neuroendocrine cells (i.e., choleocystokinin, peptide yy, vasoactive intestinal peptide) regulate local biosynthetic activity in the gastrointestinal tract,**[113] **while paracrine or autocrine interactions of monoamines are thought to underlie the mechanism of chemoreception in certain extra-adrenal paraganglia**.[114] Due to this neuroendocrine system's complexity, pollutant or other noxious stimuli entry and injury to this system may cause deregulation of the dynamics of homeostatic at any level, thereby causing a chain reaction of global homeostatic dysfunction. These dyshomeostatic reactions may occur only to adjacent or local cells, or it may travel to a distant site, where it produces local, regional, or generalized symptoms, depending on the level of local or regional

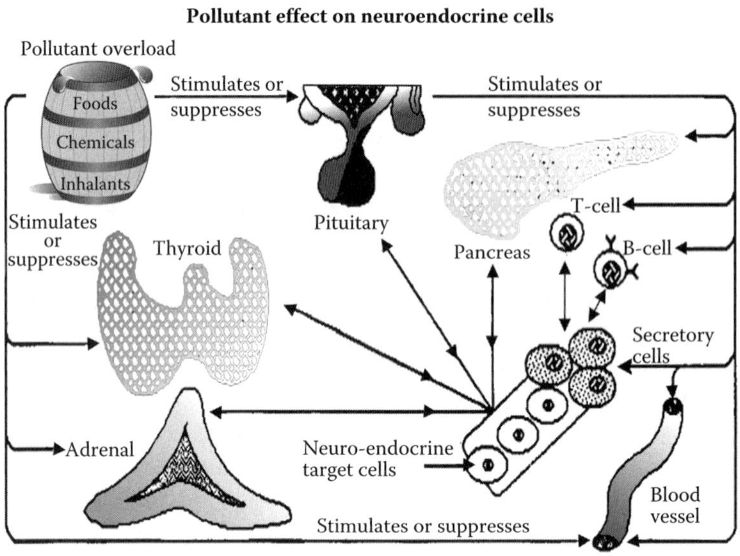

Pollutant effect on neuroendocrine cells

FIGURE 4.5 (See color insert following page 364.) Pollutant effect on the neuroendocrine cells in the chemically sensitive individual. Note: Pollutants disturb (1) self-regulation; (2) regional regulation; (3) distal regulation; (4) regulation of microcirculation; and (5) regulation of the immune system. (Modified from Rea, W. J., *Chemical Sensitivity, Vol. III: Clinical Manifestations of Pollutant Overload*, CRC Press, Boca Raton, FL, 1607, Fig. 24.2, 1995.)

symptoms such as release of endocrine hormones. A chemically sensitive individual might, for example, develop mild nausea, or gut cramps, or generalized symptoms such as anxiety. This clinical picture does occur in some cases of chemical sensitivity.

One role suggested for the function of the neuroendocrine cells is related to their secretions appearing to influence immune function[115] **and thus the dynamics of homeostasis**. (See Gastrointestinal chapter in Volume 2 - Mechanisms of Chemical Sensitivity and Chronic Degenerative Disease) Since this influence appears to be the case, the potential for additional local or widespread effects exists, for once pollutant or other noxious stimuli alter the dynamics of homeostasis, homeostatic deregulation occurs. This wide range of effects is often seen in the food sensitive and chemically sensitive individual who may have mild to severe symptoms from a pollutant or noxious stimulus. For example, phenol exposure may trigger malaise, headache, rhinorrhea, and diarrhea in a chemically sensitive patient, while another chemical such as petroleum-derived alcohol might trigger only localized gut pain. Both responses are examples of disordered dynamics of homeostasis.

Often a small, local homeostatic reaction triggers the spread of symptoms to many other organs, and dyshomeostasis occurs. For example, chemically sensitive individuals are observed clinically to have hypoglycemic effects, even though they have normal blood sugar levels. From this particular neuroendocrine effect, they often become weak, jittery, and develop fuzziness of the head. **In our ECU studies, we measured blood sugars in hundreds of patients who had symptoms of hypoglycemia, only to find that blood sugars measured were normal.** We then injected the patients with different neuro hormones including serotonin, norepinephrine, epinephrine, methcholine, somatostatin, and substance P, and were able to **reproduce the patients' symptoms**. See Table 4.6 and Figure 4.6.

Biosynthetic Profile

The biosynthetic products of neuroendocrine cells may be similar or diverse. See Table 4.7.

Biosynthetic products include biologically active monoamines and regulatory oligopeptides as well as a host of common enzymes, structural proteins (such as neurofilaments[116] **or**

Endocrine System

TABLE 4.6
Neuroendocrine Effects Seen in 100 Chemically Sensitive Patients with Normal Blood Sugars upon Chemical Inhaled Challenges Provoked and Neutralized by Intradermal Injections Using Diluted Solutions

	Dilution	Symptoms and Signs
Serotonin	1/5–1/125	Swelling and headache, vascular regulation and deregulation
Norepinephrine	1/5–1/625	Agitated, jittery, sleepy, depressed, alert, sharp
Epinephrine	1/5–1/3125	Agitated, jittery, sleepy, depressed, depression elimination
Methacholine	1/5–1/125	Agitated, jittery, sleepy, depressed
Substance P	1/5–1/125	Myalgia, headache, bloating, cramps
Somatostatin	1/5–1/125	Myalgia, headache, bloating, cramps
Vasoactive intestinal peptide	1/5–1/3125	Cramps, muscle spasm, headache
Bombesin	1/5–1/625	Laryngeal edema, headache, dizziness, infection
Cholecystokinin	1/5–1/625	Laryngeal edema, gastrointestinal upset
Neuropeptide Y	1/5–1/625	Burping, gas
Dopamine	1/625–1/5	Headache, eyes crossing, chest pain

Source: From EHC-Dallas, 1994.

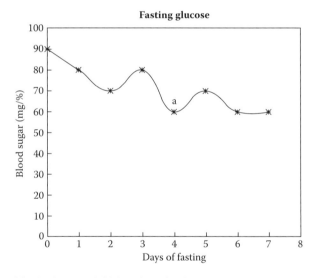

FIGURE 4.6 Fasting blood glucose—2,000 patients in the ECU. Note that two patients needed glucose supplementation. (From EHC-Dallas, 1988.)

intermediate filaments),[117] and membrane markers.[95,118] These cellular products may be stored for later secretion, and with our advanced technology many can now be specifically identified in tissue sections with monoclonal antibodies.[116,117,119–122] Very recently technology also gave us the ability to identify peptide messenger RNAs or neuroendocrine cells in situ, by using labeled nucleic acid probes.[105] Some chemical pollutants resemble these vasoactive substances and signal inappropriate release or direct triggering of target organs. These malfunctioning cycles easily triggered from a stimulation or inhibition reaction create a vicious downhill homeostatic function that for a subset of chemically sensitive patients leads to wave after continuous wave of symptoms including muscle aches, anxiety, jitteriness, fatigue, fuzzy thinking, and many more symptoms.

All neuroendocrine cells contain high levels of an enolase enzyme, which is characteristic of neural tissues.[123] Enolase isoenzymes are products of three independent gene loci: alpha, beta,

TABLE 4.7
Biosynthetic Products of Neuroendocrine Cells

Monoamines	Structural Proteins	Glutamate
Acetylcholine	Cytochrome B (561)	γ-aminobutyric acid (GABA)
Dopamine[a]	Chromagranins A and B[b]	Glycine
Norepinephrine[a]	Intermediate filaments[b]	Aspartate
Epinephrine[a]	Neurofilaments[b]	Adenosine
Serotonin	**Others**	Dynorphin
Histamine	Vasoactive intestinal peptide	Substance P
Oligopeptides	Peptide yy	Neurotensin
Adrenocorticotropin (ACTH)	Cholecystokinin	Oxytocin
Opioid peptides (enkephalins)	Big gastrin	Vasopressin
Calcitonin	Bombesin	Angiotensin II
Somatostatin	Neuropeptide	Conticotropin releasing hormone (CRH)
Enzymes	Secretin	Growth hormone releasing factor (GRF)
Neuron specific enolase (NSE)[b]	β-endorphin	Thyrotropin releasing hormone (TRH)
Tyrosine hydroxylase	Pancreatic polypeptide	Gonadetropin releasing hormone (GuRH)
DOPA decarboxylase	Gastric inhibitory peptide	
Dopamine beta hydroxylase	Leucine encephalin	
Phenylethanolamine N-methyltransferase	Motilin	
	Serpins	

Source: EHC-Dallas, 2009.
[a] Secretory products
[b] Immunoreactive substances

and gamma. The gamma gene is expressed almost exclusively by brain neurons and neuroendocrine cells.[124] This "neuron specific" enolase of gamma subtype can be detected immunocytochemically within the cytoplasmic compartment of all neuroendocrine cells using monoclonal antibodies.[124,125] The other two types, alpha and beta, may not be as organ specific.

Secretory Granules

The biosynthetic products of neuroendocrine cells are stored within relatively homogeneous cytoplasmic granules ranging in size close to the limits of conventional light microscopic resolution (0.5–2 μM). These granules are highlighted by nonspecific chromaffin, argyophil, or argentaffin reactions.[96,126,127] Ultrastructurally, the granules are membrane limited, and their cores are usually electron dense in omiscated tissues.[96,106]

This core phenomenon is probably due to the internal concentration of monoamines and associated nucleotides in addition to enkephalins[128] and other oligopeptides. The granule matrix contains a large proportion of acid soluble proteins categorized as chromaffins.[129] These can be identified immunocytochemically in normal or neoplastic neuroendocrine cells with specific polyclonal or monoclonal antibodies.[107,129–131] Pollutant injury to these subcellular membranes may cause deregulation of storage, which keeps the system in either a depleted or overactive state causing the patient to be sluggish, overactive, or to have other symptoms.

SPECIALIZATION OF NEUROENDOCRINE CELLS

Biochemical studies, including peptide analyses and molecular genetic studies, indicate that intracellular control of polyprotein gene expression in neuroendocrine cells can be exerted at several levels: transcription, posttranscription, translation, or posttranslation.[104] As defined previously, all of the neuroendocrine cells are endowed with the potential to synthesize or store a remarkably broad spectrum of biosynthetic products, some of which resemble neurotransmitter substances of the CNS.[101,105,111,112,132–134]

A consistent range of biosynthetic expression and functional activity is characteristic of neuroendocrine cells: the parenchymal cells produce, store, or secrete catecholamines.[92,94] This presence of catecholamines is associated with the presence of catecholamine-related hydroxylase and decarboxylase biosynthetic enzymes.[95,115] As pointed out earlier, **85% of the patients with chemical sensitivity and/or chronic degenerative disease as measured by HRV show a sympathomentic effect, which can be released at times of trauma**. They also appear to be released with chronic pollutant exposure in a subset of patients with chemical sensitivity and/or chronic degenerative disease that have cerebral involvement resulting in brain fog. Typically, the storage granules also contain abundant opiate peptides.[120,128,131,134] One subset of chemically sensitive patients has excess catecholamine responses characterized by shakiness, stiffness, tachycardia and fatigue etc., and, in fact, some patients even have increased catecholamines levels in their blood. Upon pollutant challenge, another subset of patients develops reactions similar to opiate effects where they become sluggish and very sleepy. Abnormal release of these substances or failure to clear them could account for the increased or prolonged reactions seen in some chemically sensitive patients.

A particular physiologic response of neuroendocrine parenchymal cells is their exquisite sensitivity to local hypoxia. When blood flow or oxygen tension is reduced, a consequent rapid release of catecholamines follows.[95] **Of course, pollutant or other noxious stimuli overload may lead to vasospasm with ensuing local hypoxia to the neuroendocrine cells that then produce the catecholamine effect during altered dynamics of homeostasis**. Frequently this response is misinterpreted as the aforementioned hypoglycemic response. The clinician can now measure this catecholamine effect by using the heart rate variability machine. This machine measures not only the sympathetic (85%) and parasympathetic effect but also the vascular composition responses, which reflect tissue hypoxia. Tissue hypoxia can then be measured by the antecubital venous O_2 levels.

Two major groups of neuroendocrine cells that have been accessible to direct experimental investigations (intra-adrenal and extra-adrenal paraganglia) exhibit further subspecialization. First, the intra-adrenal paraganglia (adrenal medulla) respond to efferent autonomic stimuli by liberating increased quantities of catecholamines into the venous circulation. In this respect, these parenchymal cells resemble sympathetic neurons and contribute to the mobilization of cardiac, hepatic, and other systemic organ responses to somatic injury or stress, as well as to pollutant injury in order to maintain the dynamics of homeostasis. Indeed, the adrenal medulla is viewed as an integral functional element of the sympathetic nervous system.[95] Also, the intercarotid and aorticopulmonary neuroendocrine cells exemplify extra-adrenal paraganglia that transduce chemical changes in the partial pressure of arterial oxygen, carbon dioxide, or pH into efferent nervous signals, which then stimulate ventilation. The exact mechanisms of such chemosensory transduction remain controversial, but they probably involve complementary activities of one or more catecholamines in homeostatic feedback loops that modulate both internal capillary blood flow and thresholds for afferent neural stimulation.[109,114] One frequently sees chemically sensitive and/or chronic degenerative diseased patients develop breathlessness after pollutant or other noxious stimuli exposure (see Neuromuscular chapter in Volume 2 - Mechanisms of Chemical Sensitivity and Chronic Degenerative Disease). All pulmonary parameters including auscultation and pulmonary function tests are normal in these patients. It appears that they have a homeostatic deregulation of the respiratory center and/or the neuroendocrine chemoreceptors.

Functions of other extra-adrenal paraganglia presumably imitate those of the intercarotid bodies or of the adrenal medulla and peripheral sympathetic neurons, but have not been directly probed. Curiously, many extra-adrenal paraganglia are more conspicuous during fetal life or infancy than in the normal adult.[92,93]

TABLE 4.8
Major Families of Extra-Adrenal Neuroendocrine Cells

Family	Sites
Branchiomeric	Jugulotympanic (middle ear)
	Intracarotid (carotid bodies)
	Intrathyroid
	Laryngeal
	Aortico-pulmonary (superior medastinum)
	Coronary-interatrial (heart base)
Intravagal	Ganglion nodosom
	Nasopharynx
	Angle of mandible
	Inferior vagus nerve distribution
Aortico-sympathetic	Paravertebral
	Intrathoracic
	Retroperitoneal
	Organ of Zuckerkandl
Visceral-autonomic	Gastroduodenal region
	Porta hepatis
	Genital tract
	Urinary bladder
	Cauda equina

Source: From EHC-Dallas, 2002.

TOPOGRAPHY OF NEUROENDOCRINE CELL SUBSETS

The topography of neuroendocrine cell subsets is important in understanding regional symptomatology in relationship to the dynamics of homeostasis and dyshomeostasis, as seen in some chemically sensitive and chronic degenerative diseased patients.

Paraganglia and other major subsets of the neuroendocrine cells are distinguished empirically, based upon both topography and histogenesis.

The paraganglion cells occur as macroscopically cohesive units in the adrenal medullae and in some major extra-adrenal locations[92] (Table 4.8).

During embryonic development, the ancestral stem cells of paraganglia migrate from the ventral neural crest.[135–137] Topographically, they remain associated principally with cervical and mediastinal tissues of ontogenetic gill arch derivation with the paravertebral sympathochromaffin ganglia or with peripheral elements of the ortho- and parasympathetic nervous systems located in the viscera of cerebrospinal axis.[94,138] Anatomically constant groups of extra-adrenal paraganglia thus can be grouped into several broad "families" including the branchiomeric, intravagal, aorticosympathetic, and visceral autonomic.[93,94] Each group will be discussed in detail.

BRANCHIOMERIC GROUP

The branchiomeric group includes the jugulotympanic group in the middle ear, the intracarotid bodies, laryngeal group, the aortic-pulmonary bodies in the superior mediastinum, and the coronary-interatrial bodies at the base of the heart. The jugulotympanic group may be affected by pollutant overload locally or distally, resulting in a similar response of pain or, if triggered too long, inflammation in the middle ear area. The intracarotid bodies are chemoreceptors and help regulate carbon dioxide and O_2 content to the brain. Pollutant or other noxious stimuli overload can cause homeostatic deregulation of these bodies with resultant homeostatic vascular deregulation and deregulation of the respiratory and vasomotor areas of the brain.

Laryngeal cell homeostatic deregulation may result in laryngeal spasm and spastic type dysphonias. (See Respiratory chapter in Volume 2 - Mechanisms of Chemical Sensitivity and Chronic Degenerative Disease.) Often one finds, that pesticide or solvent exposure causes this problem of spastic dysphonia. In our experience at the EHC-Dallas once the triggering agent is found and avoided or neutralized the problem is relieved. The injury to the aortic-pulmonary bodies may change perception in the baroreceptors, with resultant mild imperception of pressure change. This imperception may account for some environmentally triggered essential hypertension or hypotension both of which can be devastating in the patient with chemical sensnitivity and/or chronic degenerative disease. **Increased sensitivity may result in symptoms induced by changes in the weather, as seen in a large segment of chemically sensitive and some chronic degenerative diseased individuals**.

Bronchopulmonary neuroendocrine cells are widely distributed within the bronchopulmonary system, both singly and as tight clusters, referred to as "neuroepithelial bodies."[139] They appear during the first trimester differentiation of the fetal lung.[140] Origins of pulmonary neuroendocrine cell groups from both neuroectoderm and gut endoderm have been suggested by immunohistochemical studies as well as patterns of neoplasia.[141–143] Homeostatic dysfunction may lead to hyperplasia or tumors, both of which have been seen in some chemically sensitive and/or chronic degenerative diseased individuals.

Further evidence of neuroendocrine homeostatic regulation of the bronchopulmonary systems comes from studies with recipients of heart–lung transplants. These recipients have an appropriate level of ventilation during exercise as a result of a disproportionate increase in tidal volume with a reduced respiratory rate.[144] It appears that neurally mediated homeostatic feedback from the receptors modulates the pattern of ventilation response during exercise. When the exercise response was studied extensively with denervated heart subjects,[145–148] they exhibited limitations in maximal heart rate and stroke volume.[149] Abnormalities in the pattern of cardiac-rate response with such subjects included: resting tachycardia, a slow heart rate response during mild to moderate exercise, a more rapid heart rate response during strenuous exercise, and a prolonged recovery time for the heart after exercise.[146] The greatest heart rate response occurs at individual points during exercise,[150] suggesting that circulatory catecholamines rather than autonomic innervation are the predominant influence on the cardiac rate[146] and homeostasis. The patients may repeat the typical pattern of nerve severance (patients with distal hypersensitivity of their organs). At the EHC-Dallas and Buffalo, we have seen heart rate variability alterations similar to those just described in some of our chemically sensitive patients who develop increased altered dynamics of homeostasis. **They have developed either a block in the autonomic nervous system, which can occur for instance, with phenol administration, or they have developed an autosympathectomy**.

In addition, the neoplasms of the bronchopulmonary neuroendocrine cell usually present as carcinoid tumors or small cell anaplastic ("oat cell") tumors.[142] These patients have flushing, hypertension and other systemic responses. True intrapulmonary paragangliomas are extremely rare.[151]

Multicentric "minute paragangliomas," which are incidental autopsy or surgical findings, lack apparent clinical significance[152] and actually may be of pleural origin.[153] However, these minute tumors may explain aberrant function in more people than originally thought. Perhaps, in the future, with some concentrated effort and sophisticated technology a correlation of minute paragangliomas' presence with some chemically sensitive patients' dysfunction will be proven. In our series at the EHC-Dallas there appears to be an above normal incidence of neuroendocrine tumors in the chemically sensitive population. The following case report illustrates the problem:

Case report. A 50-year-old white female presented with severe chemical sensitivity accompanied by weakness, fatigue, fibromyalgia, and GI upset for approximately 10-year duration. The patient had a diagnostic workup and was shown to be sensitive to many chemicals and foods. Treatment included massive avoidance of pollutants in air, food, and water; injections for biological inhalants, food, and some chemicals; a rotary diet; and nutrient supplementation to which she responded well. Over the next 2 years, a lung mass developed, which was removed by one of the authors (WJR). The tumor was an oat cell carcinoma (the patient had never been a smoker). Knowing that 2-year survival was less than 1%, her doctors implemented a vigorous program of chemotherapy and intravenous nutritional therapy. The patient is well and tumor free seven years after removal. Some chemical sensitivity is still present, but it is less severe.

Laboratory evidence in a subset of pollutant-injured chemically sensitive individuals suggests that responses similar to those seen with bronchopulmonary endocrine tumors such as flushing and hypertension may be partially mediated through the paraganglion ANS.

In addition, some of the cardiac and especially the noncardiac chest pain experienced in the chemically sensitive and chronic degenerative diseased patient can be attributed to the coronary and interatrial neuroendocrine cells located at the base of the heart. Often this pain is confused with myocardial infarction, which must always be ruled out.

INTRAVAGAL NEUROENDOCRINE CELLS

The intravagal neuroendocrine cells include the ganglion nodosom of the nasopharynx, the angle of the mandible, and the inferior vagus nerve distribution. No one knows the effects of pollutant injury on these areas, but many patients do have trouble with their temporal mandibular joint either as an end result from trauma or sometimes as a trigger from pollutant overload. Possibly the anatomical location of intravagal neuroendocrine cells is a part of the mechanism of triggering. In addition, some chemically sensitive and chronic degenerative diseased patients present with a feeling of being strangulated in the upper larynx and pharynx. This feeling also occurs after pollutant challenge and could be the result of injury or temporary block to the ganglion nodosoms. Again, the dynamics of homeostasis are altered.

Other neuroendocrine cells are found on and in the thyroid gland. Thyroid gland C-cells, characterized by production of calcitonin, are found within the follicles of the thyroid gland. Output from these cells will influence homeostasis. Embryogenetic studies indicate an origin of C-cells from the neuroectoderm or neural crest,[154,155] but some developmental and oncological observations have raised the possibility of an endodermal or a combined endodermal and neural crest origin.[156,157] Calcitonin effects may be abnormal in the chemically sensitive or chronic degenerative diseased individual. These effects may be especially apparent in chemically sensitive patients who have muscle spasm or tetany and show deregulation of their calcium and magnesium balance with normal parathyroid function.

The aortic-autonomic neuroendocrine cells include those of the paravertebral intrathoracic and retroperitoneal areas as well as the organ of Zuckerkandl at the bifurcation of the abdominal aorta. Pollutant or other noxious stimuli injury, which alters the dynamics of homeostasis, is probably always secondary to the triggering of this organ through ANS deregulation, except in those chemically sensitive or chronic disease patients who have aortic bifurcation surgery such as in cases of arteriosclerotic aortic endarterectomy or aneurysm graft replacement.

VISCERAL-AUTONOMIC PARAGANGLION CELLS

The visceral-autonomic paraganglion cells include those in the gastroduodenal tract, parahepatic system, genital tract, bladder, and cauda equina.

The gastrointestinal and pancreatic neuroendocrine cells localize within the mucosa of the stomach, the intestine, or the pancreas. In the pancreas, they form both the classical islets and extrainsular collections of hormonally active cells. **Subtle differences in the spectrum of regulatory peptide biosynthesis may be related to specific anatomic distribution within the gastrointestinal tract**.[141,158] The output of these cells will influence the dynamics of homeostasis as has been seen in many cases of gastritis, ulcers, and upper gastrointestinal distress. Neuroendocrine cells appear during early differentiation of the gastroenteric pancreatic axis both in normal embryos[155,159] and in teratomas.[141] Other pathologic evidence also supports an origin from pluripotential foregut precursors.[158,160] Many pollutant sensitive patients exhibit gastrointestinal upset, which was previously

TABLE 4.9
Peptide Testing in 50 Chemically Sensitive Patients with Gastrointestinal Symptoms Reproduced

Antigen	Positive/Tested	(%)
Vasoactive intestinal peptide	40/40	100
Secretin	39/39	100
β- Endorphin	26/30	87
Pancreastatin frag. 37-52	26/30	87
Cholecystokinin frag. 1-21[a]	27/32	84
Bombesin[a]	29/34	85
Neuropeptide Y	28/36	78
Gastric inhibitory polypeptide	26/30	87
Pancreatic polypeptide	27/32	69
Leucine enkephalin	24/37	65
Peptide YY[a]	16/32	50
Big gastrin I	26/32	87
Motilin	23/27	85
Cholecystokinin frag 25-33	23/28	82

Source: From EHC-Dallas, 1994.
[a] Laryngeal edema also reproduced.

unexplained, but now appears to be the result of stimulatory secretions or inhibition of these neuroendocrine cells. This system is clearly imbalanced in these patients, resulting in altered dynamics of homeostasis. Vasoactive intestinal peptide and somatostatin can be released, causing gastrointestinal upset. We now have a series of 12 patients who have elevated amylase levels from pollutant triggering. **A subset of chemically sensitive patients exists with complete or near total food sensitivity, which appears to be due to gut neuoendocrine dysfunctions**. Treatment of these food sensitivities by neutralization therapy is shown to aid in recovery. See Table 4.9.

Genitourinary Paraganglion Cells

Genitourinary tract neuroendocrine cells are associated with intramural neural plexi in the bladder or vagina, probably representing visceral-autonomic paraganglia. These cells clearly influence local and regional homeostasis in the chemically sensitive individual. Plausibly, these have been related to the sacral-parasympathetic limb of the ANS.[94] In a subset of chemically sensitive individuals, pollutant stimulation **of these cells results in a sudden urge to urinate**. At the EHC-Dallas, this phenomenon has been demonstrated upon inhaled or injected pollutant challenge and subsequently with injected neurohormones, which suggests neuroendocrine homeostatic deregulation of the genitourinary neuroendocrine cells. **Vaginodynia and vulvodynia also occur with pollutant-triggered trauma**. Interstitial cystitis would also be another example of the abnormal matrix trigger occurring through the GRS.

Argyrophilic cells, which are immunoreactive for serotonin, somatostatin, and nonspecific enolase, have been demonstrated in the prostatic urethra, prostatic ductules, and prostate acini.[162] These cells are possibly of endodermal (perhaps hindgut) origin.[162] After an inhaled pollutant or other noxious stimuli exposure, some chemically sensitive and chronic degenerative diseased patients with recurrent prostatitis report prostate pain and eventually inflammation, which is later followed by bacterial infection. The incitant disrupts the dynamics of homeostasis, which results in release of procytokines causing inflammatory damage followed by upregulation of complements 3 and 4, lymphocytes and phagocytes and, thus, inflammation. Many cases of chemically induced (i.e., PCB) prostatitis appear to occur in this manner.

Neuroendocrine cells are found in the normal female genital tract and in some ovarian neoplasms;[163] pathways of normal development indicate that some of the latter neoplasms could arise from coelomic epithelium.[163] Intrauterine paragangliomas have been reported very rarely.[164] Consequently, the presence of these cells within the uterus is open to question.

The nonasthmatic chemically sensitive patient frequently complains of bladder spasm, lower, and upper gastrointestinal spasm, vaginal bleeding, vaginal spasm and pain, peripheral vascular constriction, and a myriad of respiratory signs and symptoms not intimately related to the lung. When **neuroendocrine discharge after pollutant exposure occurs, the chemically sensitive patient may have to urinate and defecate suddenly because of spasm in those areas related to neuroendocrine functions**. The patient may also experience cold hands and feet with accompanying chest pain and breathlessness. This urgency is compatible with homeostatic dysfunction of the lower paraganglion system.

Paragangliomas

Paraganglia tumors may give us some insight into chemical release and the abnormal responses seen in some chemically sensitive and chronic degenerative diseased individuals.

Solitary or multicentric paragangliomas may arise in relationship to any of the topographic families of paraganglia listed in Table 4.8. Overall, approximately 90% of paragangliomas (paraganglion tumors) arise within the adrenal gland and are alternately classified as "pheochromocytomas" in this location. Knowledge of the origin of these and other locations is valuable since they give insight into the extent of the hormone release from the neuroendocrine systems, which causes some chemically sensitive patients to react to pollutant exposure.

Extra-adrenal paragangliomas are most common in jugulotympanic, intercarotid, superior mediastinal, and retroperitoneal paravertebral sites. These areas rich in neuroendocrine cells are either more sensitive to, or more prone to pollutant damage. The majority of these kinds of paragangliomas arise in the head and neck,[94,165-167] where they are allied to anatomic structures of branchial arch ontogeny (branchiomeric paraganglia), or are associated with the vagus nerve. Within the branchiomeric family, intercarotid and jugulotympanic paragangliomas are most common. Laryngeal paragangliomas are relatively uncommon, and intrathyroid paragangliomas are very rare.[94,168-170]

Vagal paraganglioma may arise at the level of the ganglion nodosum or in more inferior portions of the vagus nerve. They typically present in the angle of the mandible or near the fossa of Rosenmueller in the posterior pharynx.[171-174] In the thorax, paragangliomas represented 4% of cases in a large series of neural tumors.[175] Lack et al.[176] reviewed 36 cases of intrathoracic paragangliomas. Mediastinal supra aortic, aorticopulmonary, or coronary-interatrial paragangliomas may impinge upon the atrial walls or upon the base of the heart and the great vessels. Posterior mediastinal tumors may arise from the paravertebral sympathetic trunk.[107,177,178] Paragangliomas are probably more common in the retroperitoneum than in the mediastinum.[94,179,180] They occur less commonly in visceral locations supplied by orthosympathetic or cranial-sacral parasympathetic elements: the orbit,[181] duodenum,[182] hepatic ducts,[183] bladder wall,[184] genitourinary tract, or cauda equina.[185] Urinary bladder and cauda equina tumors are probably the most frequent of the craniosacral locations. The urinary bladder paragangliomas are often functional and resemble adrenal medullary pheochromocytomas both histologically and cytochemically. Often a positive tissue chromaffin reaction occurs,[186] which elicits and defines that all tumors are important to diagnose. All of these are immediate examples of autonomic nervous system dysfunction, connective tissue matrix disruption, and the result of disordered dynamics of homeostasis.

Clinical Manifestations of Neuroendocrine Stimulation

In the chemically sensitive individual the most commonly recognized clinical manifestations of neuroendocrine system stimulation by pollutants are anxiety attacks due to paraganglia surges

followed by chronic fatigue. **In most cases a symptom complex is characterized by headaches, palpitations or tachycardia, and sometimes excessive diaphoresis but usually lack of sweating**.

These neuroendocrine (paraganglia) surges are paroxysmal and often leave the patients with a jittery, shaky feeling accompanied by ravenous hunger, which lasts from a few minutes to several hours until a new level of homeostasis is obtained. Once the paroxysmal nature of these attacks is recognized in the pollutant-injured chemically sensitive patient, they can be cleared under environmentally controlled conditions to avoid the triggering by pollutants or noxious stimuli, thus restoring the dynamics of homeostasis. In addition, alkalinization of the body with sodium bicarbonate, treatment with vitamin C, or oxygen supplementation may stop the reaction. Once the patient is in a basal homeostatic state, signs and symptoms can be reproduced by controlled, measured challenges of biological inhalants (pollen, dust, mold etc.), foods, and toxic chemicals, then these pollutant triggers can be defined. As discussed previously, **the chemically sensitive patient commonly becomes ravenous upon pollutant stimulation**. For example, many of these patients will experience this phenomenon in areas of high pollution such as in airports. With this pollutant stimulation, they will eat any food in sight, even when they know it may be harmful to them. Often, symptoms of pollutant overload can be eliminated by giving the individual one or combinations of some of the following: fresh air, oxygen, magnesium sulfate, ascorbic acid, L-glutathione, multiple B vitamins, or trisalts.

Other functional manifestations of paraganglia pollutant stimulation may rarely be related to abnormal secretion of neuroendocrine peptides. Reports could include abnormal serum levels of biologically active calcitonin, ACTH, vasoactive intestinal peptide, somatostatin, and possibly serotonin. These responses have been affirmed by reports of tumors found in these areas.[121,187–190] Blood levels of serotonin may be increased or decreased.[191] Immunoreactive oligopeptides such as calcitonin,[119] somatostatin,[192] or VIP may be detected in paraganglion tumor cells[193] without evidence of serum elevations or clinical activity. Therefore, at present, whether stimulation of the paraganglion cells by pollutants will give measurable serum elevations in the peripheral blood is uncertain. Clearly, after pollutant injury occurs the effects of both somatostatin and vasoactive intestinal peptides are seen clinically. (See W.J. Rea, *Chemical Sensitivity*, Volume III, Chapter 26 for more information.)

Our observations, along with findings of others investigations indicate the significant role pollutant injury plays in deregulation of neuroendocrine cells and systems in the chemically sensitive patient.

ADRENAL GLANDS

Some physiology of the adrenal glands (Figure 4.7) is described in our discussion of the paraganglia.

The physiology of the dynamics of homeostasis is important in understanding adrenal function in the chemically sensitive and/or chronic degenerative diseased patient, since so many of both types of patients are thought to be adrenal insufficient. Although most chemically sensitive or chronic degenerative diseased patients really are not, in the classical sense of the word, adrenal insufficient, many severe cases show adrenal deregulation and appear to have poor adrenal reserves, which disrupts the dynamics of homeostasis. Adrenal insufficiency therefore, will be given further consideration here. Specifically, the adrenal gland will now be considered in preference to the neuroendocrine system as a whole.

Physiology of the Dynamics of Homeostasis and Dyshomeostasis in the Adrenal Glands

The nerve supply of the adrenal glands is through the ANS. The sympathetic preganglionic fibers for these glands are the axons of cells located in the intermediate lateral columns of the lowest two or three thoracic segments and highest two lumbar segments of the spinal cord. These fibers descend through the greater, the lesser, and the least thoracic and first lumbar splanchnic nerves to the celiac, aorticorenal, and renal ganglia. Most fibers go to the celiac plexus and then to the adrenals. In the adrenal glands these fibers become part of the GRS of the adrenals thus making a direct connection with the hypothalamus and the central homeostatic control center. In addition, the adrenals are

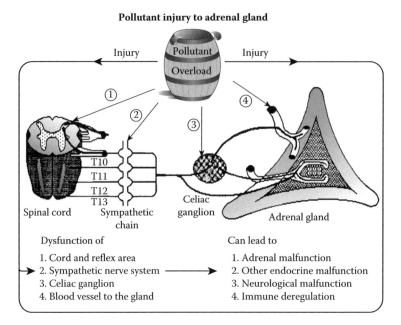

FIGURE 4.7 Pollutant injury to the adrenal gland. (Modified from Rea, W. J., *Chemical Sensitivity, Vol. III: Clinical Manifestations of Pollutant Overload,* CRC Press, Boca Raton, FL, 1619, Fig. 24.3, 1996.)

covered and cemented with connective tissue, which completes the master homeostatic feedback loops. The adrenals represent a classic example of anatomy for the GRS to do its job. Parts of the autonomic nerves also communicate with the renal and phrenic nerves.

Parasympathetic fibers are conveyed to the celiac plexus in the celiac branch of the vagus nerve to the adrenals. **The majority of the fibers of the suprarenal plexus enter through or near the hilus through the adrenal cortex to the medullary cells** (see Figure 4.7). **The fibers ramify profusely and terminate around the medullary chromaffin cells. Other fibers invaginate the adrenal blood vessels, which results in vascular regulation of much of the gland.** Pollutant or other noxious stimuli injury may occur to the olfactory nerve, hypothalamus, spinal cord, sympathetics, celiac ganglion, the spinal autonomic nerves, and the blood vessels, as well as the connective tissue matrix, resulting in a myriad of stimulations or suppressions with the resultant dynamics of the homeostatic function of the adrenal gland, the rest of endocrine and neuroendocrine systems, the neurological system, and the immune system, being disrupted or altered. The law of nerve injury may come into play with all organs distal to the injury becoming hypersensitive and injury to individual areas may alter the whole system causing it to by hyperactive. These systems are all in the integrated communication network of the GRS. These ramifications may reveal why chronic pollutant or other noxious stimuli overload causes many problems in the chemically sensitive and/or chronic degenerative diseased individual, resulting in some cases with much diffuse symptomatology, such as weakness, fatigue, etc. from the subunit dyshomeostasis.

The adrenal medulla contains columnar cells, which secrete catecholamines, epinephrine, and norepinephrine. Excess or chronic stimuli will cause overproduction of those substances creating altered homeostasis. **Some islets of these chromaffin cells secrete mostly epinephrine, while others secrete mostly norepinephrine.** Preganglion fibers enter the medulla and terminate directly on the parenchymal cells or on scattered sympathetic ganglion cells (Figure 4.8). Eighty-five percent of the patients with chemical sensitivity and/or chronic degenerative disease who have autonomic function measured by HRV show a sympathetic effect suggesting high output of these two hormones.

The renin-angiotensin system (RAS) or the renin-angiotensin-aldosterone system (RAAS) is a hormone system that helps regulate long-term blood pressure and blood volume in the

Endocrine System

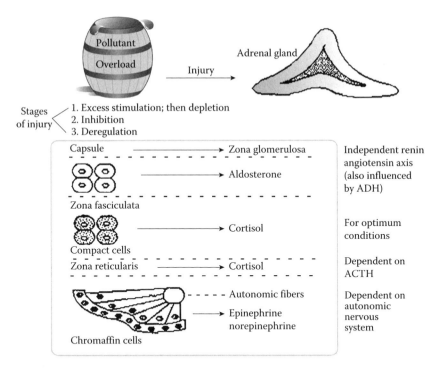

FIGURE 4.8 Pollutant injury to the adrenal gland. Note the preganglionic fibers of the sympathetic nervous system enter the medulla and terminate in the chromaffin cells. (Modified from Rea, W. J., *Chemical Sensitivity, Vol. III: Clinical Manifestations of Pollutant Overload,* CRC Press, Boca Raton, FL, 1620, Fig. 24.4, 1996.)

body. It also affects the heart in heart failure and myocardial infarct conditions. (See Vascular chapter in Volume 2 - Mechanisms of Chemical Sensitivity and Chronic Degenerative Disease.) Homeostatic deregulation may occur at any specific site or on all sites with pollutant or other noxious substance injury. This homeostatic dysfunction may be limited or diffuse, and it is one of the reasons for the **pollution-triggered jitteriness, shakiness, subsequent fatigue, and other symptoms of nor-epinephrine release frequently seen in some chemically sensitive patients**. Often, chemically sensitive patients present with either an excess of epinephrine or with more insufficient levels of other secretions, mainly norepinephrine, epinephrine, or dopamine, as measured in their urine and blood profiles. These abnormal levels indicate injury to their orderly dynamics of the homeostatic metabolic sequences. Those patients with insufficient epinephrine and norepinephrine appear to be mentally depressed, while those with excess levels show symptoms of jitteriness, weakness, and fatigue.

In the adrenal gland, the adrenal cortical cells are divided into three layers. The first layer, the outer zona glomerulosa, produces aldosterone relatively independently, though the angiotensin mechanism is influenced by release of ADH from the pituitary gland (Figure 4.8). Control of aldosterone production is more through the renin angiotensin axis, which starts at the juxta glomerular apparatus of the kidney and helps regulate the dynamics of fluid and blood pressure homeostasis. Pollutant injury to this outer cortical layer may result in edema and hypertension. There is a group of patients with chemical sensitivity and/or chronic degenereative disease who have hypotension with blood pressure ranging from 70/50 to 90/60. These patients exhibit orthostatic characteristics with difficulty on arising from a sitting position and an inability to get proper endpoints for starting injection immunotherapy or neutralization therapy. They are very fragile in triggering a stimulus response, as they can trigger a stimulus response quickly; they have difficulty in the sauna, and show an inability to sweat readily. They find it difficult to maintain homeostasis. Sometimes they need supplementation with florinef saline and albumin to stabilize their blood pressure and homeostasis.

The second and third layers, the zona fasciculata and zona reticularis, produce corticosteroids and are under control of ACTH (from the hypophysis of the pituitary gland). Therefore, a feedback loop for homeostasis exists beneath the adrenal gland and hypophysis. **The adrenal gland receives blood from 30 to 50 small arteries.** Obviously, chronic homeostatic dysfunction of these blood vessels, after direct pollutant or other noxious stimuli damage to any of them, to their ANS attachments, or to the pituitary hypophyseal axis may result in the development of subtle adrenal dysfunction. Early pollutant injury may then develop into more homeostatic deregulation of the endocrine, the neurological, and immune system with a vicious downward metabolic cycle. Then tissue changes occur until fixed-named disease results. **The different types of corticoids are derived from cholesterol; therefore, pollutant or other noxious incitant injury to cholesterol metabolism might cause inappropriate production with dyshomeostasis.**

Pollutant or other noxious incitant overload to the adrenal cortex with subsequent effects on cortisol production may have widespread effects in the chemically sensitive or chronic degenerative diseased individual because of the known homeostatic functions of cortisol. **Gluconeogenesis may be impaired, where the patient is not able to generate the 6–15-fold glucose increase that is often needed with pollutant or other noxious incitant exposure or stress.** Once the dynamics of homeostasis are altered, transport of amino acids from the extracellular fluid into the liver cells may be changed and the amino acids are not easily converted to glucose. DNA transcription and messenger RNA may be impaired by pollutant or other noxious incitant injury. Easy mobilization of amino acids from muscles is under the control of the adrenal cortex and may, therefore be impaired in the patient with disordered homeostasis. The rate of glucose utilization, which is usually decreased by cortical cells, may be increased by pollutant or other noxious incitant injury. NADH oxidation may be accelerated by pollutant stimulation, with rapid and excessive glycolysis causing swings in blood glucose levels (Figure 4.9). This chain of events is frequently seen in some chemically sensitive and/ or chronic degenerative diseased patients. See Figures 4.10, 4.11, and 4.12.

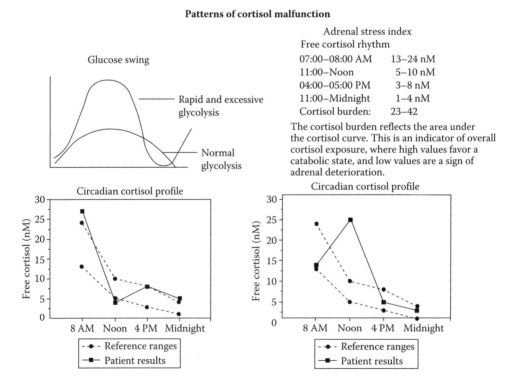

FIGURE 4.9 Patterns of cortisol malfunction. (From EHC-Dallas, 2006.)

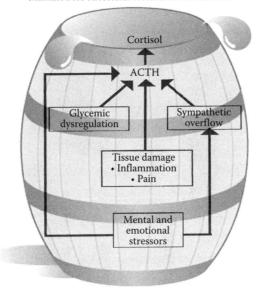

FIGURE 4.10 Inducers of cortisol release. (EHC-Dallas, 2006.)

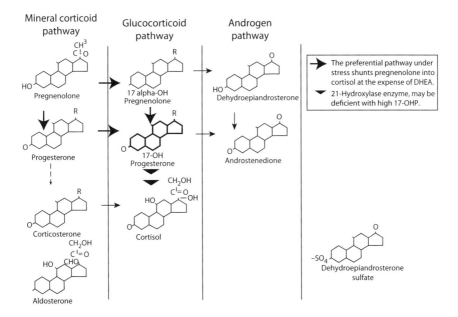

FIGURE 4.11 Adrenal steroid synthesis pathway. (EHC-Dallas, 2006.)

Homeostatic dysfunction resulting in excess cortisone can cause muscle weakness and immune dysfunction due to alteration of protein metabolism. Frequently the patient with chemcial sensitivity and/or chronic degenerative disease has a production of excess or depleted cortisol in one or more quadrants of the 24-hour cycle. These levels are measeured by direct measurement. Although protein metabolism in the periphery may decrease, it will increase in the liver cells. As observed in 1,000 chemically sensitive patients, protein metabolism is disturbed. **Usually with altered homeostasis the**

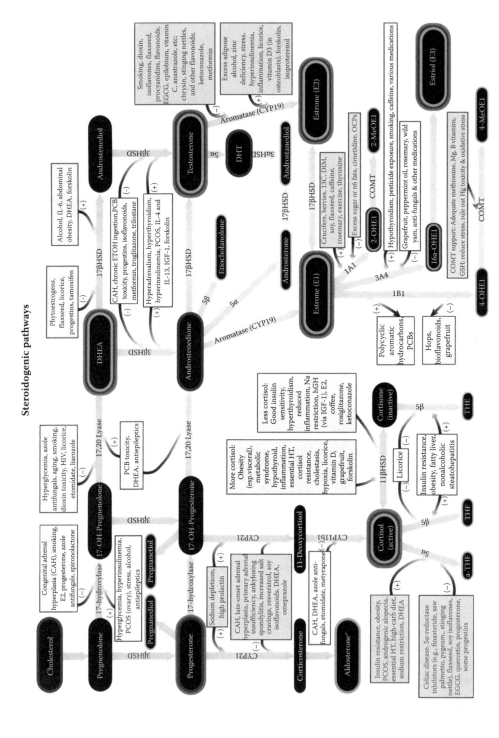

FIGURE 4.12 (See color insert following page 364.) Steroidogenic pathways. (Courtesy of Genova Diagnostics.)

visible effects of protein metabolism are the opposite of those produced by cortisol and results in a decreased rate of deamination of amino acids by the liver, decreased protein synthesis, decreased formation of plasma proteins by the liver, and the aforementioned decreased conversion of amino acids to glucose. Clinically, these patients may smell of ammonia, experience weight loss, and feel fatigued. **The presence of excess ammonia may cause burning of tissues, especially the mucous membranes, sleep disturbances, and mental instability.**

The dynamics of homeostasis may be altered by the mobilization of fatty acids from adipose tissue, which is influenced by the cortical area and may also be altered with pollutant exposure. When a chemically sensitive and/or chronic degenerative diseased individual fasts in a controlled environment, a decreased transport of glucose into the cells may occur, along with an increased mobilization of fats and increased beta oxidation of fat in the cells. This sequence allows for a more efficient functioning of all systems, resulting in normalization of the dynamics of homeostasis after the pollutant or other noxious incitant load has been reduced. **Mobilization of fat in times of stress in order to produce glucose will also mobilize sequestered toxics in the fat**. These toxics can then cause symptoms of their own, which often further stimulate the liver and may overload/impair its ability to detoxify.

Pollutants or other noxious stimuli can up regulate ACTH, which stimulates the cortex and may eventually deplete it of adequate cortisol output. Once tissues are damaged by pollutant exposure and the dynamics of homeostasis are disturbed (see Chapter 1), **inflammation may occur due to insufficient local amounts of cortisol, which can then cause release of lysozymes due to lack of stability of the pollutant-injured lysosomal membranes locally. Inflammation will then result, as seen in the chemically sensitive and/or chronic degenerative diseased patients**.

With disordered homeostasis, plasma proteins may be lost from the capillaries, causing intracellular and extracellular edema, which is seen in the chemically sensitive and/or chronic degenerative diseased individual. With altered dynamics of homeostasis, change in cortisol output by pollutant or other noxious stimuli overload will also affect the immune system with resultant eosinophilia or esoinopenia as seen in some cases of chemical sensitivity or lymphocytopenia may occur as seen in most cases. **Lymphocytopenia is due to excess cortisol at the cellular level, but also may be due to enhanced lympholysis**. This chronic pollutant-derived homeostatic deregulation of cortisol found in the chemically sensitive patient may result in stimulation, inhibition, or atrophy of the lymphoid tissue, which in turn increases or decreases the output of sensitized lymphocytes and antibodies from the lymphoid tissue. Therefore, depending on the type of pollutant stimulation, the chemically sensitive or chronic degenerative diseased individual may experience increased or decreased immunity to infections through the malfunction of the adrenal axis. At the EHC-Dallas and Buffalo, we not only treat more chemically sensitive and chronic degenerative diseased patients who are resistant to infections, but also those in a large subset who are not actively resistant. Decreasing the total body load often clears the recurrent infections due to factors like improved phagocytosis. **One of the causes of the mild anemia frequently seen in the chemically sensitive and chronic degenerative diseased patient may be caused by decreased cortisol production**.

When ACTH release is triggered from the pituitary gland, μ- and B-melanocyte stimulating hormone, B-lipotropin, and B-endorphin are also secreted. The opiate-like effects of B-endorphins are often seen in a subset of chemically sensitive patients who have altered dynamics of homeostasis, frequently accounting for their sluggishness, brain fogginess, and dysfunction. Thousands of challenges with foods and chemicals at the EHC-Dallas and Buffalo under environmentally controlled conditions that have shown opiate effects can occur in the chemically sensitive individual. Lipotropin effects by the data or statistics from pollutant stimulation are unclear as to their status in the chemically sensitive individual. A small subset of individuals with chemical sensitivity has extreme, adverse reactions to sunlight on their skin. These reactions may be affected by inhibition of melanocyte stimulating hormone.

A whole spectrum of adrenal dysfunction may occur after pollutant injury with altered homeostasis. At first, this homeostatic dysfunction may be mild, but if the stimulation is too long or virulent, gross pathology may be seen. Early diagnosis and treatment is the aim of medical intervention, with the hope of eliminating changes and restoring the dynamics of homeostasis before they become irreversible.

CLINICAL PICTURE OF POLLUTANT INJURY AND ADRENAL DYSFUNCTION

The clinical entities seen in homeostatic dysfunction of the adrenal gland due to pollutant or other noxious incitant overload are either hyper- or hypofunction. Most commonly, however, chemically sensitive patients do not present with either of these entities, per se. Rather, because they are **fragile to most exposures, they demonstrate a generalized adrenal instability, which can be** demonstrated on objective autonomic function tests (iris corder [pupillography], heart rate variability analysis), by specific thermography for local end-organ homeostais, low white blood cell counts, elevated venous oxygen levels (indicating poor oxygen extraction), and inability to tolerate 500,000 killed streptococcal or staphylococcal vaccine, which suggests an adrenal homeostatic regulatory problem. Hyper- and hypofunction are discussed independently. However, in the author's experience, primary treatment of hypoadrenal function with steroids is often not beneficial over the long-term. It appears that long-term reduction of total body pollutant load allows for restoration of adrenal reserve, negating the need for cortisol supplementation.

HYPERADRENALISM

The adrenal genital syndrome contains an excess of 17 ketosteroids produced by toxics, which down regulate the 21 hydroxyalase enzyme. The syndrome occurs because of the excess virilizing hormone. In this syndrome, cortisol is underproduced, which stimulates ACTH release and then increases adrenal dysfunction, with the production of excess androgens and disruption of the dynamics of homeostasis.

Some patients have an increase in cortisol production through a hyperplasia of the zona fasciculata and zona reticularis. Environmental pollution may cause Cushing's disease in some patients. Other cases of adrenal virilization have been reported due to pollutant injury or due to injury of the 21-hydroxylase enzyme especially from toxic pesticides.[194]

Still other patients may have hyperplasia or a small tumor of the zona glomerulosa resulting in hyperaldosteronism. The cause of increased aldosterone secretion in idiopathic cyclic edema is not clear. Often an excessive secretion of pituitary antidiuretic hormone under the influence of environmental stimuli is present. The exact pathophysiologic mechanisms are complex and aldosterone secretion is not always elevated. When aldosterone levels are raised, it is often not responsive to stimuli that normally cause its levels to decrease. Food and chemicals have been seen to trigger this edema syndrome as a reaction to challenges. In addition, previously diagnosed idiopathic edema resulting from a variety of environmental causes has been cleared in patients who completed a stay in the ECU. Many cases of pollutant injury causing adrenal virilism have been reported.[194] A large concentration of these occurred in Puerto Rico due to the island's high pollutant levels.[194,195]

ADRENAL INSUFFICIENCY/HYPOADRENALISM

Many stages of adrenal insufficiency are seen in the chemically sensitive and/or chronic degenerative diseased patient. The most common cause currently documented may be from autoimmune disease, which is usually due to multiple pollutant overload or direct chemical inhibition. Many mild types of focal adrenal insufficiency exist in which the whole clinical picture of adrenal insufficiency does not occur in the chemically sensitive and some chronic degenerative diseased patients. The milder form of low adrenal reserve occurs in these patients, in which the adrenals are capable of producing enough hormone to maintain an apparently normal chronic state of health in the absence of stress. This state may be due to chronic malnutrition, chronic excess stimulation of the adrenal with resultant depletion or sudden, life-threatening stress as seen in massive trauma. With the increased demand on adrenal cortical hormones, mild to moderate symptoms or even circulatory collapse may occur. Usually, this chronic type of chemically sensitive patient first presents with fatigue and/or recurrent infections, an increase in allergic diathesis, a mild autoimmune-like disease, vasculitis, or nausea and vomiting. The ACTH test is then used. As shown in Chapter 1,[197] the process of adaptation is important in understanding pollutant injury with altered dynamics of homeostatic dysfunction. Only the adrenal response to adaptation will be discussed here since the clinical features and other ramifications are discussed in Chapter 1.[197]

The alarm reaction in pollutant or other noxious incitant injury is divided into an initial stage of shock and a secondary stage of countershock. If these are balanced, the normal dynamics of homeostasis occurs. **Evidence indicates that during the initial stage of pollutant entry and injury, effects characteristic of an acute release of epinephrine are followed by changes characteristic of relative adrenocortical insufficiency (a fall in blood pressure, decreased blood sugar, decreased blood sodium, and increased blood potassium), which lasts from a few minutes to hours. The stage of countershock then occurs, with effects characteristic of increased adrenocorticoid activity (return of blood pressure to normal or above normal, increased blood sugar, restoration of normal or increased level of blood sodium and normal or decreased level of blood potassium, and enlargement of the adrenal cortex).** This is the process of adrenal adaptation due to pollutant or other noxious stimuli entry with subsequent injury. These effects return to baseline during the first stage of resistance as the normal dynamics of homeostasis occur (Chapter 1), but reappear during the second stage of exhaustion, where altered dynamics of homeostasis occur. **In hypophysectomized animals these changes do not occur, indicating that they are mediated by the hypothalamus and pituitary gland through their production of corticotropin releasing factor (CRF) and ACTH, respectively.**[197] During the recovery phase and as the production of hydrocortisone returns to normal, the production of androgens increases and apparently is related to the healing process, since these hormones that were secreted are known to be anabolic. Therefore, adrenal cortical hormones, especially hydrocortisone, evidently play an important role in the normal dynamics of the homeostatic response to pollutant or other noxious stimuli stress.

The critical role of the adrenal cortex in response to chronic pollutant or other noxious stimuli stress in human subjects manifests in a number of ways. A patient with untreated mild adrenal insufficiency or low adrenal reserve may function reasonably well when environmental conditions are optimum, but he or she tends to tire more easily as the dynamics of homeostasis are altered. If strenuous physical exercise is undertaken or if a meal is skipped, hypoglycemic symptoms as severe as convulsions may develop. Pollutant or other noxious stimuli stress will also accentuate this problem. If an infection such as a common cold develops, acute adrenal insufficiency with nausea, vomiting, falling blood pressure, and collapse may occur in the severe cases of chemical sensitivity and/or chronic degenerative disease. These cases are rare. More commonly, apathy may occur in the milder cases. **The apathetic response is the most common endocrine response seen in chemical sensitivity or chronic degenerative disease or after chronic pollutant entry. Apathy can not only be generated by adrenal insufficiency but may also be generated by dyshomeostasis of the limbic system.** To a point, all of these undesirable developments may be prevented by administration of suitable dosages of hydrocortisone, which will help restore the dynamics of homeostasis. However, reduction of total body load and replacement of nutrition is a much more efficacious course in the long-treatment plan.

Normal adrenocortical function has been demonstrated to be essential for an ability to withstand infections[198] and to achieve a normal homeostatic response. Several studies have indicated that either too little or too much glucocorticoid can impair resistance to infection, resulting in altered dynamics of homeostasis, whereas an optimum level of glucocorticoid enhances resistance to infection. However, prolonged use of extra hydrocortisone will damage multiple systems, resulting in much irreversible damage.

In over 40,000 patients studied at the EHC-Dallas and Buffalo, only three have had overt adrenal insufficiency and they both responded to proper steroid replacement. However, it is becoming evident that many chemically sensitive and/or chronic degenerative diseased patients may develop selective adrenal deficiencies as their dynamics of homeostasis are altered. Studies measuring 11-OH and 17 ketosteroids before and after ACTH stimulation have revealed several patterns of isolated adrenal stepwise alterations. Occasionally, supplementation (e.g., dehydroepiandrosterone) with one type of steroid synthesis pathway deficiency substance will help the patient recover more rapidly. Again, reduction of the total body load and replacement of nutrition is much more efficacious in restoring robust adrenal function. Five to twenty mgm of cortisone acetate may be used temporarily without the complication of obesity, hypertensions, cataracts etc. Often this is necessary for only a short period of time.

PARATHYROID GLAND

The two to six parathyroid glands secrete parathormone, which regulates serum calcium and is stimulated by low calcium and inhibited by high calcium. This homeostatic balance of calcium in the body is crucial for cellular function. Thyrocalcitonin secreted by the thyroid neurochemical C-cells appears to be a second hormone that influences calcium by lowering it in the blood and inhibit reabsorption of bone. Calcium, which is the opposite of parathormone, increases. The dynamics of homeostasis occur between the secretions of these two hormones in normal functioning humans.

Four principle actions of the parathormones occur, which make up the dynamics of homeostasis in several areas. First, parathormone inhibits phosphate reabsorption (or enhances phosphate secretion) by the renal tubule. Second, parathormone reabsorbs calcium and phosphate from the bone. Third, parathormone increases calcium absorption from the GI tract. Fourth and finally, parathormone enhances calcium reabsorption in the renal tubules. Activity of the hormone parathormone also enhances magnesium excretion. Parathormone can interact with vitamin D, which is altered in 30% (15% excessive, 15% deficient) of our chemically sensitive patients. Chronic pollutant or other noxious stimuli overload may alter absorption of calcium and incorporation of it into bones and cells. Increased phosphate has been observed in the serum of some chemically sensitive individuals, but more often it is depleted. This elevation may be due to a mild inhibition of the parathormone output or excess intake of phosphate. **Pollutant injury apparently can occur with excess phosphate intake from soft drinks or organophosphate insecticides.** Chronic fatigue may result from damage to the phosphate energy mechanism (ATP production), causing depletion of phosphate. In patients with renal calcium leaks due to pesticide or solvent exposure, a secondary hyperparathyroidism may occur associated with moderately reduced bone calcium content and density in women and associated with moderately increased blood pressure.[199b] Pesticide injury to the parathyroid gland that results in a magnesium leak may eventually become a calcium leak causing altered dynamics of homeostasis, as we have seen in patients with severe exposure.

The dynamics of homeostasis can be disturbed resulting in an excess production of parathyroid hormone, which leads to hypercalcemia by increased stimulation of the osteoclastic activity of the bone with a release of calcium and phosphate. Absorption of calcium from the gut and reabsorption of calcium from the renal tubule results in inhibition of the tubular reabsorption of phosphate, causing an excessive loss of phosphate in the urine. Most of our severely ill chemically sensitivie patients requiring intravenous hyperalimentation have extremely low phosphorus, partially due to the aforementioned mechanism. An example of pollutant injury causing a parathyroid tumor follows.

Case study. A 55-year-old-white female was exposed to an inordinate amount of pesticide and herbicide surrounding her house in the Corn Belt. She developed severe malnutrition and malabsorption with intolerance of all foods and nutrients. She lost weight to 80 pounds and had to have total intravenous nutrition for 2 years. During this time it was noticed that her serum calcium was slightly elevated 10.8 -11- (c-10.5 mg/dl). She had slightly elevated parathorome levels. She had received no calcium in her intravenous. Parathyroid serum showed equivicable dense area behind the right lower lobe of her thyroid gland. With the intravenous nutrition she gained weight and was eventually able to tolerate food. Her intravenous line was discontinued and the patient did well for 4 years. She then developed nausea, apathy, short term memory loss and fatigue. Her serum Ca was 11.5 – (c-10.5 mg/dl). Parathormone level was high at 73- (c-10-55 pg/ml). Parathyroid scan showed a definite tumor below the right lower lobe of the thyroid. Parathyroidectory was successfully performed. The patient did well.

OVARY

In chemically sensitive and or chronic degenerative diseased women, the hormonal system, consisting of three different hierarchies of hormones, has complex dynamic homeostatic responses to pollutant or other noxious stimuli. Due to the high metabolism, the ovaries are particularly susceptible to early pollutant injury. The dynamics of homeostatic imbalance can occur, resulting in an over- or understimulation of the luteinizing (LH) hormones, FSH, estrogen, and progesterone. Testosterone can also become imbalanced. Estrogen and progesterone imbalance is the most frequent endocrine abnormality observed in the chemically sensitive and/or chronic degenerative diseased female and typically results in altered dynamics of homeostasis.

For all of the sex hormones, cholesterol is the precursor, and pollutant injury to the liver triggers and propigates damaged cholesterol metabolism can cause a change in hormone output, altering the dynamics of homeostasis. This liver damage is a common occurrence in the pollutant overloaded patient. In all steroid-secreting glands, the side chain of cholesterol is degraded to pregnenolone and dehyroepiandosterone (Figure 4.13).[199]

Pregnenolone is transformed into progesterone, which, by degradation of the side chain, becomes testosterone, settling in the vulnerable areas. Once sequestered in the ovaries toxics have been shown to cause great damage.[200]

Estrogen receptors are important in maintaining local and regional control of the dynamics of homeostasis as estrogen circulates throughout the body. The receptors are universally distributed throughout the body but have an unequal distribution. E-R α (estrogen-receptors) are especially found in the adrenal glands, pituitary, uterus, testis, and kidney, while ER-ß are found in the ovary, testis, prostate, and thymus. Both types of estrogen receptors are present in bone, breast, and brain.

When estrogen molecules bind to their receptors, some impulses go inside the cells to activate genes turning some on, and some are turned off. The shape of estrogen receptors plus the estrogen molecule determines the activity. In the cells, the protein regulatory molecules enhance or suppress the effectiveness of receptors bound to various estrogens. There are at least 30 different estrogen receptor enhancing or suppressing proteins that can reduce or strengthen the aperiodic homeostatic function (chemical sensitivity). Estrogen mimics or designer estrogens used to suppress breast cancer recurrence, like tamoxifen or raloxifen, have now been shown to alter ovarian physiology.

These regulatory proteins are found in different amounts in various tissues. This uneven distribution can explain the tissue specific effects of estrogen mimics and designer estrogens. Different shapes of the receptor estrogen complex attract different regulatory molecules that themselves activate different genes. This may explain why the same hormones affect different people differently. Each person has slightly different properties of co activators and co suppressors. Estrogen does not exist and act in just one natural form, but has several forms, which include estrone, estradiol, and estriol. Studies show that each of these forms may have both different and similar effects.

One natural variant of estrogen, dehydroesterone, is effective in preventing hot flashes. It does not lower the level of fatty acids in a woman's blood, suggesting that the only benefit of this estrogen is in limiting hot flashes. Naturally occurring estrogens work selectivley. There is no perfect designer estrogen or selective estrogen modulation yet available. Altered connective tissue matrix results in the modification of cellular messengers and affects the way cells communicate with each other. **These functions seem to work better with natural hormones**; however, the long-term effects of natural estrogens are not known. Estrogen has been known to have a dampening

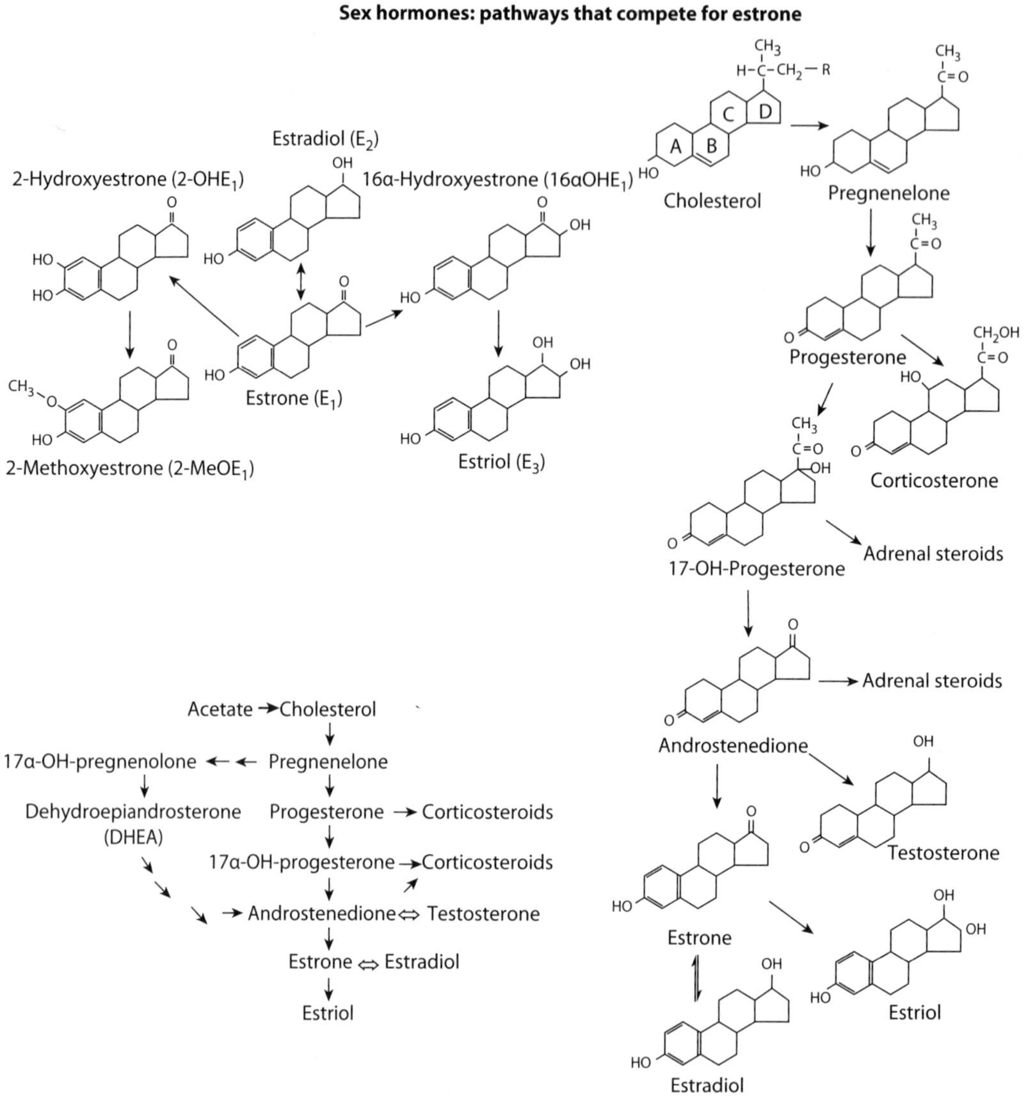

FIGURE 4.13 Sex hormones. Pathways that compete for estrone. (Modified from Bralley, J. A. and Lord, R. S., *Laboratory Evaluations in Molecular Medicine: Nutrients, Toxicants, and Cell Regulators*, Institute of Molecular Medicine, Norcross, GA, 300, Fig. 10-2, 2001.)

effect on the cardiovascular system with a lack of estrogen causing altered dynamics of homeostasis. The late onset of arteriosclerosis in females following menopause is common knowledge. A study by Couch and Wortman[201] supports this observation of the dampening effect, which delays homeostasis from being altered. They found a significant occurrence of migraine in pathologically anovulatory females (polycystic ovary, galactorrhea, amenorrhea) versus in pregnant women or those taking birth control pills when the dynamics of homeostasis were altered. It was suggested that this condition might also be a hypothalamic problem. In addition, estrogen was found to favorably influence cardiac disease and myocardial infarction.[202,203] Floroxymestine, a progesterone compound, also has been shown to help prevent angioedema.[201] Excess estrogen will also alter the dynamics of homeostasis, as it has been shown to have an adverse effect on vessel walls, causing venous inflammation that results in thrombophlebitis, pulmonary emboli, and myocardial infarction.[201]

Androgenic hormones are known to influence the dynamics of homeostasis of the vascular tree in the prevention of some forms of angioedema. Danazole® (Danocrine)[204] has been shown to help prevent angioedema in some cases of hereditary angioedema.

Certain toxic chemicals, especially pesticides have been shown to have a predilection for the ovaries in animals and humans.[205] Lindane, a known carcinogen in animals, has induced a marked disturbance of the estrous cycle in some animals, prolonging the proestrus phase five to seven times and thereby delaying ovulation.[206] Used in humans to treat scabies and lice, Lindane disrupts the dynamics of homeostasis causing mental, cardiac, and liver dysfunction.[206] **Lindane is used as a delousing medication for children and also for scabies**.

> **Case report**. A 35-year-old male, osteopathic physician was exposed to high levels of lindane when he was serving in the military.
>
> The patient developed brain dysfunction and altered electrolyte imablance with low potassium and magnesium. He also developed a bradycardia, tachycardia syndrome accompanied by an episode of gastrointestinal upset with gas and bloating, bouts of mild hypotension and moderate brain dysfunction with loss of judgment. This syndrome progressed over a 2-year period until he was not only unable to heed the advice of his fellow physicians, but also unable to use his basic knowledge as a physician. He continued to abuse the use of potassium. One day after taking excess potassium, he had a cardiac arrest and could not be resuscitated.

In vitro studies measured effects of the pesticide o,p'-DDT and its isomer p,p'-DDT on progesterone, showing alterations in production as well as high levels of Lindane.[207]

The earliest reports of disturbed dynamics of homeostasis with adverse reproductive effects from pollutants began when Hamilton and her counterparts reported spontaneous abortions after our Allied troops were gassed in Europe during World War I.[208] These early reports described women exposed to very high toxic levels of benzene. These women had severely disturbed dynamics of homeostasis and were unable to bear normal live children, or they bore children who failed to thrive. Secondary amenorrhea is likely to result from the effects of toxic substances on the ovaries.[209] More effects of toxic substances on the ovaries are germ-cell destruction and receptor modifications, which then result in altered dynamics of homeostasis. Early onset of menopause has been directly related to cigarette smoking,[210] which has been shown to disturb the dynamics of homeostasis, and it may well be that other toxic chemical exposures may result in the same process. Premature ovarian failure has been experimentally induced using polycyclic hydrocarbons and alkylating agents. A higher risk of cancer of the ovary in cosmetologists and hairdressers may evidence toxic insults.[211] The dyes and solvents used to enhance hair beauty tend to disrupt the dynamics of homeostasis creating a basis for chronic disease. We have seen a high incidence of ovarian difficulties in hairdressers due to these substances causing alterations in the dynamics of homeostais. Data on early-age use of oral contraceptives altering the dynamics of homeostasis also often reveals menstrual irregularities[212] as well as breast tumors[213] and venous inflammation.[214]

Homeostatic impairments induced in the ovaries by sublethal doses of carbaryl and endosulfan have been studied.[215] Exposures produced a reduction in the number of oocytes, a reduction in size and deformity in different stages of oocytes, damage to yolk vesicles in maturing and mature oocytes, an increase in the number of atretic oocytes, development of interfollicular spaces, an increase in the connective tissue of tunica albuginea, and dilation of blood vessels. These substances obviously damaged the connective tissue matrix and the GRS, disrupting the dynamics of homeostasis. Effects were dependent on the dose, duration of exposure, and type of pesticide. Doses of endosulfan were found to be more toxic than carbaryl. Apes fed pesticide (arochlor-1254) in their diet developed altered homeostasis with abnormal estrogen and progesterone levels followed by menstrual irregularities. Dioxin alters the action of estrogen in reproductive organs in a manner that is both age dependent and target organ specific.[145,146]

At the EHC-Dallas, our groups of PMS and endometriosis patients with disordered dynamics of homeostasis resulted in severe menstrual disturbances, and they often had blood levels of toxic organic chemicals, but failed to show big changes in estrogen and progesterone levels. Many of these toxics were estrogen mimics like DDT, DDE, chlordane, etc. Therefore, these products had both the excess estrogen effect as well as toxic effects. Clinically, these patients clearly had disordered dynamics of homeostasis. However, they often responded to minidose hormonal therapy, as well as to massive avoidance and heat depuration programs, which restored the altered dynamics of homeostasis. Detectable levels of organochlorine pesticides were always present in their blood before treatment with the avoidance and depuration programs. After treatment, however, the pesticide level was not detectable. Clearly, these substances altered the dynamics of homeostasis. Patients responded much better as they reduced their total load of pesticides and toxic organic hydrocarbons. The dynamics of homeostasis could be restored and normal function returned.

During sexual differentiation, a number of critical periods exist when the reproductive system is uniquely susceptible to chemical or other noxious stimuli induced changes. At these times, an inappropriate chemical signal can result in severely altered dynamics of homeostasis with resultant irreversible lesions that often result in infertility. However, similarly exposed young adults who are only transiently affected can easily restore homeostasis. The serious reproductive abnormalities that resulted from human fetal exposure to DES, synthetic hormones, and other drugs provide grim examples of the types of lesions that can be produced by interfering with this process. Furthermore, it is of concern that many of the abnormalities are not expressed during fetal life and only became apparent during the stage when presenting alternatives in sex differentiation occurs.[218]

In humans, *in utero* exposure to a hormonally active chemical such as DES, androgen, 2 zearenone mycotoxin or progestin results in physiological disruptions of the dynamics of homeostasis with eventual morphological and pathological alterations in reproductive functions. For example, DES causes cancer, infertility, and severe abnormalities of the cervix, uterus, and Fallopian tubes as well as alterations of reproductive and sex-linked nonreproductive behaviors.[219]

Meyer-Bahlburg et al.[220] and Gray[221] reported that women exposed to DES *in utero* have altered dynamics of homeostasis. We have frequently seen these types of patients and they are clinically fragile. They have less well-established sex-partner relationships, lower sexual desire and enjoyment, decreased sexual excitability, and diminished cortical function. Hines and Shipley[222] found that **women exposed to DES *in utero* showed a more masculine pattern of cerebral lateralization on verbal tests than their sisters**. Such sexual differences in specialization of the two hemispheres of the brain for different types of cognitive processing are well-documented in humans with men tending toward greater left hemisphere specialization for verbal stimuli than women.[223] At the EHC-Dallas we have seen some chemically sensitive patients with altered dynamics of homeostasis exhibit characteristics similar to those described by Meyer-Bahlburg et al.[220] and Gray.[221] Cognitive ability may be different in DES- or other hormonally treated females whose psychopathology includes a tendency toward depression, and/or anxiety.[224,225] An increased incidence occurs in DES-exposed women of immunologic hyperactivity, rheumatic fever, recurrent strep throat, and autoimmune disease.[226,227]

We have seen many cases of chemical sensitivity and chronic disease in children of women who have had DES treatment. As a group these patients are much more difficult to treat.

Swaab and Fliers[228] found that the preoptic area of the hypothalamus is 2.5 times larger in men than women and contains 2.2 times as many cells. A recent report suggested that homosexual men have femalelike INAH-3 (interstitial nuclear anterior hypothalamus) brain structures, implying a biological basis for homosexuality.[229]

http://www.ncbi.nlm.nih.gov/books/bv.fcgi?rid = dbio.box.4131

Organization/Activation Hypothesis

Does prenatal (or neonatal) exposure to particular steroid hormones impose permanent sex-specific changes on the central nervous system? Such sex-specific neural changes have been shown in regions of the brain that regulate "involuntary" sexual physiology. The cyclic secretion of luteinizing hormone by the adult female rat pituitary, for example, is dependent on the lack of testosterone during the first week of the animal's life. The luteinizing hormone secretion of female rats can be made noncyclic by giving them testosterone four days after birth; conversely, the luteinizing hormone secretion of males can be made cyclical by removing their testes within a day of birth.[230] It is thought that sex hormones may act during the fetal or neonatal stage of a mammal's life to organize the nervous system in a sex-specific manner, and that during adult life the same hormones may have transitory, activational effects. This idea is the organization/activation hypothesis.

Interestingly, **the hormone chiefly responsible for determining the male brain pattern is estradiol**. Testosterone in fetal or neonatal blood can be converted into estradiol by the enzyme P450 aromatase. This conversion occurs in the hypothalamus and limbic system the two areas of the brain known to regulate hormone secretion and reproductive behavior.[231,232] Thus, testosterone exerts its effects on the nervous system by being converted into estradiol. But the fetal environment is rich in estrogens from the gonads and placenta. What stops these estrogens from masculinizing the nervous system of a female fetus? Fetal estrogen (in both males and females) is bound by a-fetoprotein. This protein is made in the fetal liver and becomes a major component of the fetal blood and cerebrospinal fluid. It will bind and inactivate estrogen, but not testosterone.

Attempts to extend the organization/activation hypothesis to "voluntary" sexual behaviors are more controversial because there is no truly sex-specific behavior that distinguishes the two sexes of many mammals, and because hormonal treatment has multiple effects on the developing mammal. For instance, injecting testosterone into a week-old female rat will increase pelvic thrusting behavior and diminish lordosis a posture that stimulates mounting behavior in the male when she reaches adulthood[233,234] These behavioral changes can be ascribed to testosterone-mediated changes in the central nervous system, but they could also be due to hormonal effects on other tissues. Testosterone enables the growth of the muscles that allow pelvic thrusting. And since testosterone causes females to grow larger and to close their vaginal orifices, one cannot conclude that the lack of lordosis is due solely to testosterone-mediated changes in the neural circuitry[234–239]

In addition, the effects of sex steroids on the brain are very complicated, and the steroids may be metabolized differently in different regions of the brain. Male mice lacking the testosterone receptor still retain a male-specific preoptic morphology in the brain, and male mice lacking the aromatase enzyme are capable of breeding.[240,241] **These studies show that there is more to sex-specific morphology and behavior than steroid hormones**. Despite best-selling books that pretend to know the answers, there is much more to learn regarding the relationship between development, steroids, and behavior. Moreover, extrapolating from rats to humans is a very risky business, as no sex-specific behavior has yet been identified in humans, and what is "masculine" in one culture may be considered "feminine" in another.[242–244] Kandel et al.[234] concluded that there is ample evidence that the neural organization of reproductive behaviors, while importantly influenced by hormonal events during a critical prenatal period, does not exert an immutable influence over adult sexual behavior or even over an individual's sexual orientation. Within the life of an individual, religious, social, or psychological motives can prompt biologically similar persons to diverge widely in their sexual activities.

MALE HOMOSEXUALITY

Certain behaviors are often said to be part of the "complete" male or female phenotype. The brain of a mature man is said to be formed such that it causes him to desire mating with a mature woman, and the brain of a mature woman causes her to desire to mate with a mature man. However, as important as desires are in our lives, they cannot be detected by in situ hybridization or isolated by monoclonal antibodies. We do not yet know if sexual desires are primarily instilled in us by our social education or are fundamentally "hardwired" into our brains by genes or hormones during our intrauterine development or by other means. **It is clear however, that males sensitive to perfume may not be able to get an erection when exposed**.

In 1991, Simon LeVay[229] proposed that part of the anterior hypothalamus of homosexual men has the anatomical form typical of women rather than of heterosexual men. The hypothalamus is thought to be the source of our sexual urges, and rats have a sexually dimorphic area in their anterior hypothalamus that appears to regulate their sexual behavior. Thus, this study generated a great deal of publicity and discussion.

The interstitial nuclei (neuron clusters) of the anterior hypothalamus (INAH) were divided into four regions. Three of them showed no signs of sexual dimorphism. However, one of them, INAH3, showed a statistically significant difference in volume between males and females; it was claimed that the male INAH3 is, on average, more than twice as large as the female INAH3. Moreover, LeVay's data suggested that the INAH3 of homosexual men was similar in volume to that of women and less than half the size of heterosexual men's INAH3. This finding, LeVay claimed, "suggests that sexual orientation has a biological substrate."

There have been several criticisms of LeVay's interpretation of the data. First, the data are from populations, not individuals. One can also say that there is a statistical range and that men and women have the same general range. Indeed, one of the INAH3 from a homosexual male was larger than all but one of those from the 16 "heterosexual males" in the study. Second, the "heterosexual men" were not necessarily heterosexual, nor were the "homosexual men" necessarily homosexual; the brains came from corpses of people whose sexual preferences were not known. This brings up another issue: homosexuality has many forms, and is probably not a single phenotype. Third, the brains of the "homosexual men" were taken from patients who had died of AIDS. AIDS affects the brain, and its effect on the hypothalamic neurons is not known. Fourth, because the study was done on the brains of dead subjects, one cannot infer cause and effect. Such data show only correlations, not causation. It is as likely that behaviors can affect regional neuronal density as it is that regional neuronal density can affect behaviors. If one interprets the data as indicating that the INAH3 of male homosexuals is smaller than that of male heterosexuals, one still does not know whether that is a cause of homosexuality or a result of it.

Indeed,[240] has shown that the density and size of certain neurons in rat spinal ganglia depend on the frequency of sexual intercourse. In this case, the behavior was affecting the neurons. Fifth, even if a difference in INAH3 does exist, there is no evidence that the difference has anything to do with sexuality. Sixth, these studies do not indicate when such differences (if they exist) emerge. The question of whether differences among the heterosexual male, female, and homosexual male INAH3 occur during embryonic development, shortly after birth, during the first few years of life, during adolescence, or at some other time was not addressed.

In 1993, a correlation was made between a particular DNA sequence on the X chromosome and a particular subgroup of male homosexuals: homosexual men who had a homosexual brother. Out of 40 pairs of homosexual brothers wherein one brother had inherited a particular region of the X chromosome from his mother, the other brother had also inherited this region in 33 cases.[246] One would have expected both brothers to have done so in only 20 cases, on average. Again, this is only a statistical concordance, and one that could be coincidental. Moreover, the control (the incidence of the same marker in the "nonhomosexual" males of these families) was not reported, and the statistical bias of the observations has been called into question, especially since other laboratories have not been able to repeat the result.[247,248] More recent studies[249,250] found little or no increase in the incidence of this DNA sequence when homosexual men were compared to their nonhomosexual brothers. Hu and colleagues concluded that this sequence is "neither necessary nor sufficient for a homosexual orientation." Thus, despite the reports of these studies in the public media, no "gay gene" has been found.

Genes encode RNAs and proteins, not behaviors. While genes may bias behavioral outcomes, we have no evidence for their "controlling" them. The observance of people with schizophrenia, or people whose personalities change radically after a religious conversion or a traumatic experience, indicates that a single genotype can support a wide range of personalities. This is certainly a problem with any definition of a "homosexual phenotype," since people can alternate between homosexual and heterosexual behavior, and the definition of what is homosexual behavior differs between cultures.[180] Thus, whether homosexual desires are formed by genes within the nucleus, by sex hormones during fetal development, or by experiences after birth is still an open question.

The presence of excess estrogens in the environment is also a cause for concern in sexual orientation since these hormones can and have altered the dynamics of homeostasis in males and females. In animals, excess estrogen or toxic estrogen mimics with prenatal or neonatal exposures have a dramatic and permanent influence on brain structure and behavior.[181]

Soto et al.[182] tested xenobiotic compounds that were reported to have estrogenic activity[182–184] or were suspected to be estrogens because of their molecular structure. Among them, the mycotoxins zearalenol and zearnalenone have been used as anabolic food additives for cattle and sheep. These toxics are known to alter the dynamics of homeostasis activity as estrogens. The phytoestrogen coumestrol and the pesticides DDT (p,p' and o,p'), chlordecone (Kepone®), and 1-hydroxychlordane

TABLE 4.10
Estrogenic Response of MCF7 Cells to Insecticides, Phytoestrogens, and Phytohormones

Compound	Concentration[a]	PE[c]	RPE(%)[d]	RPP(%)[e]
Estradiol	30 pM	6.7	100	100
Kepone	30 μM	5.6	81	0.0001
Mirex	30 μM	0.6	—	—
DDT, p,p'	30 μM	5.0	70	0.0001
DDD, o,p'	30 μM	5.8	84	0.0001
1-hydroxychlordene	30 μM[b]	3.1	37	0.0001
Chlorodene	30 μM[b]	1.2	4	—
Heptachlor	30 μM[b]	1.5	8	—
Arochlor 1221	30 μM[b]	1.4	7	—
Giberellic acid	30 μM[b]	1.2	4	—
Chlordane	30 μM[b]	1.3	5	—
Zearalenone	3 nM	6.0	88	1
Zearalenol	3 nM	6.3	93	1
Coumestrol	3 μM	6.3	93	0.001

Response of MCF7 Cells to Natural and Synthetic Estrogens

Compound	Concentration[a]	PE[b]	RPE(%)[c]	RPP[d]
Estradiol	100 pM	4.7	100	100
Diethylstilbestrol	10 pM	5.1	112	1000
1,1-β-Chloromethyl-stradiol	10 pM	5.1	110	1000
Dehydrodoisynolic acid	10 nM	4.8	103	1
Dichloro-doisynolic acid	100 nM	4.6	97	0.1
Allenolic acid	10 nM	4.9	105	1
Hydroxyphenyl-cychlhexanoic acid	10 nM	4.9	105	1

Source: Modified from Soto, A.M., Lin, T.-M., Justicia, H., Silvia, R.M., and Sonnenschein, C., *Advances in Modern Environmental Toxicology, Vol. XXI. Checmically-Induced Alterations in Sexual and Functional Development: The Wildlife/Human Connection*, eds. T, Colborn and C. Clement, Princeton, NJ: Princeton Scientific, 302–303. With Permission.

[a] Indicates the lowest concentration needed for maximal cell yield.
[b] Indicates the highest concentration tested in culture.
[c] Proliferative effect is expressed as the ratio between the highest cell yield obtained with the test chemical and the hormone-free control
[d] Relative proliferative effect, which is calculated as 100 × (PE-1) of the test compound/(PE-1) of E_2. A value of 100 indicates that the compound tested is a full agonist; a value of 0 indicates that the compound lacks estrogenicity at the does tested, and intermediate values suggest that the xenobiotic is a partial agonist.
[e] Relative proliferative potency is the ratio between the dose of E_2 (estrogen/estradial) and that of the xenobiotic needed to produce maximal cell yields × 100.

were all found to be estrogenic (Table 4.10). Alkyl phenols such as p-nonylphenol (present in modified polystyrene) and pentyl phenols induced estrogen responses.[186] 6-Bromo-2-naphthal and allenoic acid, also have estrogen activity.[186]

Endocrine and reproductive changes have been associated with all pesticide types, insecticides (carbaryl, DDT, methoxychlor, aldrin, chlordane, dieldrin, and Kepone®), herbicides (2,4-DNP, 2,45-T), and fungicides (thiocarbamates such as zineb and maneb).[258] Several authors have claimed that once PCBs have passed through the liver, they become estrogenic,[259,260] altering the dynamics of homeostasis.

Besides the synthetics with estrogenic activating properties, a number of natural products have steroidlike action. These include some phenolic plant products,[261,262] fusarium-producing resorcylic acid lactones (mycotoxins in stored grains), and isoflavanoids from legumes. These chemicals can alter the dynamics of homeostasis and sexual differentiation of gonadotropin secretion[261] and the sexual behavior of rodents. The flavonoids are inhibitors of steroid biosynthetics.[263] Of course, the flavonoids are distributed in vegetable products, soy products being the most common ones used. They exhibit many enzymatic activities similar to those of steroids. Nutritionists have touted these substances for their antioxidant properties. Clearly, they are double-edged swords due to these inhibition qualities. **It has been our experience at the EHC-Dallas and Buffalo that many of the supplemental flavonoids** are detrimental to the chemically sensitive and chronic degenerative diseased patients. These substances should be used carefully.

All of these substances should be considered when studying homeostasis, since they may significantly influence it. **Western dietary changes of this century have resulted in higher estrogen intake**. High fat diets result in high intestinal absorption of estrogens, while high fiber diets result in low estrogen levels.[264] Increasing meat intake and reducing vegetable consumption increases estrogens that are more available to steroid-responsive tissues, while concentration of less potent and perhaps estrogen inhibitory isoflavanoid declines. These dietary shifts are associated with increases in the incidence of estrogen-dependent western disease.[265] Estrogen advances menarche in rodents.[266-271] It appears to do the same in humans. One can clearly see where the dynamics of homeostasis and the GRS can be altered or at least influenced by the estrogenic effects, which then result in metabolic, followed by tissue changes, followed by the possibility of fixed-named autonomous disease.

Frem-Titulaer[272] reported that Sáenz de Rodriguez and Pérez-Comas have identified over 1,000 children with adrenogenital syndrome from Puerto Rico. This condition appeared to result from ingestion of chicken containing residual toxic chemicals and estrogen compounds that had originally been used to treat the chicken feed. Dramatic suppression with equal potency to estradiol was found, which altered the dynamics of homeostasis in these children.

Pérez-Comas[273] has shown abnormal adrenal and ovarian development in Puerto Rican babies. He has evaluated 1,053 patients with abnormal sexual development over a period of 14 years. These patients were from Puerto Rico, the mainland United States, Latin America, and Europe. The most frequent initial diagnoses were premature thelarche,[273] gynecomastia,[274] precocious puberty,[273] and premature pubarche.[273] Other conditions associated with increased estrogen levels were asymmetry of breasts, virginal hyperplasia of breasts, and hirsutism. Five patients were diagnosed with pseudoprecocious puberty associated with hypothyroidism and Down's syndrome.

Females were affected by contaminated foods more frequently than males. Total serum estrogen levels were generally increased in 85% of males and 86% of females studied. Prolactin levels were abnormal in 28% of females and 16% of males. FSH levels were increased in 32% of males and 40% of females. LH levels were high in 33% of males and 18% of females.

The years of highest incidence of the adrenogenital syndrome were 1982, 1983, and 1984. Cases have diminished dramatically, with a 50% drop in 1985 when a federal investigation of hormones in meat was being carried out. From May 1 to October 31, 1986, 53 new cases were evaluated. Starting in 1986, 45 of them presented with symptomatology of excess hormones. Overall review of remission data of these patients revealed that a limited diet produced remission in 58% of males and 51% of females. Without diet, remission was observed in only 6% of males and 11% of females. The diet was free of poultry, eggs, and meat until estrogen levels became normal. Afterwards, grain-fed poultry and birds were permitted.

Pelvic sonographic abnormalities were observed in 62% of the females affected by hormone-contaminated foods. Ovarian cysts were directly related to increased estrogen levels in 88% of patients with sonographic abnormalities.

Hormonal and clinical remission of Pérez-Comas's patients usually took 3–6 months on diet-treated patients. Improvement of ovarian and uterine abnormalities that can be seen by sonogram usually takes from 6 to 12 months. However, recurrence of this syndrome is frequently due to non-observance of a diet free of estrogens or hormonelike toxic chemicals, which alter homeostasis and allows for disease processes to be reinstituted. No adequate governmental action has been taken up to the present time.

One case that exemplifies pollutant-triggered, severe ovarian dysfunction is presented.

Case study. A 35-year-old P-4, G-4 white female entered the ECU with the chief complaints of nausea, vomiting, and weight loss for the previous six months. Her environmental and past history were significant in that she had grown up in farm country, had married a farmer, and had lived on a farm all her life. After the birth of each of her children, she noticed a loss of strength and an increase in her premenstrual syndrome. These symptoms were so severe that she would become nonfunctional several days before her period. She also noticed a lot of irregular and excessive uterine bleeding. She then developed severe nausea, vomiting, and weight loss that first occurred only premenstrually but then became constant during the 6 months prior to her admission to the ECU. She became totally food intolerant and lost 30 pounds due to an inability to hold down food. When she was admitted to the ECU, she was emaciated and feeble. Other organ systems were normal, except for areas of excess skin from where the recent weight loss had been.

Significant laboratory tests revealed chlordane, 5 ppm; heptachlor, 0.3 ppm; DDT, 0.8 ppm; dieldrin, 0.6 ppm; lindane, 3 ppm; hexachlorobenzene, 2 ppm; T lymphocytes, 500/mm^3 (C = 1400 to 2200/mm^3); WBC, 30001mm^3; B lymphocytes, 4%; and total hemolytic complement levels = 50% of control.

This patient was placed on intravenous fluids and given nothing by mouth for 1 week. Nausea and vomiting gradually subsided, and she was able to eat an occasional chemically less-contaminated food without vomiting. Treatment with food neutralization injections allowed her to expand her food repertoire gradually, until she could eat a wholesome diet. She was given 15 gms of intravenous, preservative-free vitamin C daily along with multiminerals. Her strength gradually returned until she could be discharged to her home, where a safe oasis had been built. Over the last 4 years, she found that she had to leave her area of farm county when spraying season started due to an exacerbation of her symptoms following exposure to the chemicals in the various sprays used. In the fifth year post diagnosis, she was inadvertently exposed to pesticide and developed uterine bleeding that required a hysterectomy. Since then she has done well. After 15 years, this patient is still clinically fragile but much improved and carries out an active life. She just completed building a new, less-polluted house.

A broad spectrum of the dynamics of ovarian homeostatic dysfunction, from failure to conceive to spontaneous abortion to endometriosis to premenstrual syndrome, has been observed in the chemically sensitive female patient. Each condition is discussed in more detail on the following pages.

Premenstrual syndrome is seen frequently in chemically sensitive and some chronic degenerative diseased patients in whom oral contraceptives act as pollutants. Also, PMS is seen in chemically sensitive patients who have been exposed to other external toxic chemicals. Other side effects of oral contraceptives in the chemically sensitive woman are depression, headaches, and anxiety.

Grant[273] has shown when the hormone balance of a birth control pill is varied with homeostasis being lost and the progesterone strength increased, the side effects (ranging from irregular bleeding, distended veins, arteriole changes, irritability, weight gain, depression, and loss of libido) change.

Administration of high progestin and low estrogen causes regular, scanty withdrawal bleeding. Low estrogen and low progesterone also triggers altered dynamics of homeostasis with irregular bleeding. Low progesterone and high estrogen levels cause venous changes.

Antithrombin III, which prevents clotting, is decreased by estrogen and progesterone administration. After administration, sticky platelets appear to occur.[276]

Vessey[277] demonstrated that women on the bith control pill developed disordered homeostasis having a risk of venous thromboembolisms nine times that of women not on the pill. The increased risk for thrombolic stroke was six times normal. Increased risks for a variety of other conditions included hemorrhagic stroke, two times normal; subarachnoid hemorrhage, six times normal; and myocardial infarction, four times normal.[278] (See Figures 4.14 and 4.15 and Table 4.11.)

Grant[279] showed that with altered homeostasis of progesterone that there is an increased occurrence of gallbladder disease, diabetes, and heart disease in women who take the pill. In this population there also appears to be a change in metabolism resulting in more infections, more food and chemical sensitivity, and more cancer.

English and Welsh studies also reported an increased risk of cancer in young women using the birth control pill.[278]

Grant's research[280] further discovered that the most common foods triggering PMS headaches were wheat, oranges, eggs, tea, coffee, chocolate, milk, beans, corn, cane sugar, mushrooms, and peas. These findings are similar to our study results at the EHC-Dallas with the addition of beef and pork, which were high on the list.

Cigarette smoking clearly alters the dynamics of homeostasis leading to the distribution of nicotine and its major metabolites, nicotine-1'-N-oxide and cotinine, in the urine, amniotic fluid, breast milk, cord blood, and maternal venous blood of pregnant smokers, their fetuses, and neonates. These nicotine products may produce adverse effects on the health of both mother and baby,

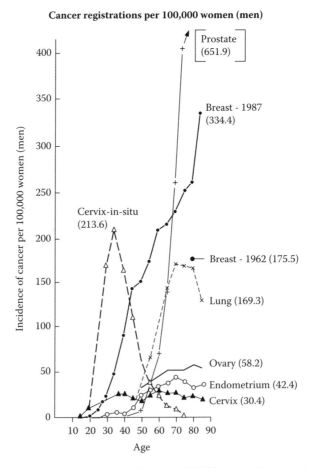

FIGURE 4.14 Percent increase in cancer registrations per 100,000 men and women between 1965 and 1987. Bracketed figures are the highest rate. (From Grant, E., *The Bitter Pill:How Safe Is the Perfect Contraceptive?* Corgi Books, London, 119, 1986. With permission.)

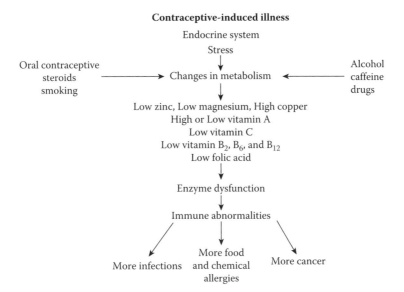

FIGURE 4.15 Grant's concept of the origin of contraceptive-induced illness. (Rea, W. J., *Chemical Sensitivity, Vol. III: Clinical Manifestations of Pollutant Overload,* 1633, Fig. 24.6, 1996; Courtesy of Grant E., *The Bitter Pill:How Safe Is the Perfect Contraceptive?* Corgi Books, London, 119, 1986.)

TABLE 4.11
Deaths Per 100,000 Women of Those in the Pill-Taking Age Group

	England and Wales (1975)	Oxford/FPA OC Users at Entry (1968–1979)	R.C.G.P. Ever-Users (1968–1979)	Walnut Creek all Women % (1968–1977)
All deaths	108.5	52.9	87.2	100
Neoplasms	44.3	25.3	30	45
Cirulatory	21.9	12.3	29.9	15
Accident/violence	17.1	4.6	18.2	19
Follow up			42.0%	86.2%

Source: Rea, W. J., *Chemical Sensitivity, Vol. III: Clinical Manifestations of Pollutant Overload,* 1633, Tab. 24.6, 1996; Courtesy of Grant, E., *The Bitter Pill:How Safe Is the Perfect Contraceptive?* Corgi Books, London, 119, 1986.

resulting in altered dynamics of homeostasis. Additionally, between the 29th and 34th weeks of gestation, a significant increase ($p < 0.05$) in the urinary metabolite ratio, cotinine to nicotine-l′-N-oxide, occurs as compared with controls. Studies on female smokers taking the birth contol pill also have shown a significant increase in the cotinine -to nicotine-l′-N-oxide ratio ($p < 0.02$), as compared with controls. High ratio values (>3) were shown to be due to cotinine excretion. Because this ratio may reflect the relative importance of the μ-C and aliphatic N-oxidative pathways, it could have toxicological importance, as well, and certainly alters homeostasis relating to xenobiotics for the mother and fetus.[282,283]

Fetal liver has been shown to be capable of metabolizing (*in vitro*) various tobacco alkaloids via a variety of C- and N-oxidative reactions, which may be of toxicological significance linked not only to tobacco alkaloids, but also to other xenobiotics.[282,283]

The potential procarcinogenic effects of estrogen have slightly dampened the enthusiasm of hormone replacement therapy. Some studies have focused on problems with estrogen replacement leading to concerns about the role of aberrant estrogen metabolism in premenopausal women.[284] The role played by estrogen oxidative metabolism in hormone sensitive disease has come from studies of estrogen-dependent neoplasms of reproductive organs. With respect to breast cell malignancy, estradiol is believed to act primarily as a promotional factor causing increased growth rates in breast cells already transformed to a cancerous state. However, estradiol is not the only active estrogen. Other metabolites formed from estradiol have the capacity to act as estrogen and in some cases antiestrogens. Studies have shown that **the major metabolites of estradiol and estrone are those hydroxylated at either the C-2 or the C-16 positions**. 2-hydroxyestrone (2-OHE$_1$) is essentially devoid of peripheral biological activity as shown in studies of uterine weight, gonadotropin secretion, and cell proliferation.[285]

2-hydroxyestrone has even been found to exert a modest antiestrogenic effect.[286,287,288] The 16 α-hydroxylated metabolites, 16α-hydroxyestrone (C) and estriol (E$_3$) are estrogen agonists.[289] Women with breast and endometrial cancers have marked elevation of 16α-hydroxyestrone, which is a significant risk factor for such estrogen dependent tumors.[290] Tumors in other estrogen sensitive tissues are also promotated by 16 OHE$_1$. **As the ratio of 2- OHE$_1$ to 16 OHE$_1$ decreases, the severity of recurrent respiratory papillomatosis increases**.[291] Testing for urinary levels of 2-hydroxyestrone and 16 α-hydroxyestrone provides valuable insights regarding risk and dietary patient counseling.[285,287,289] The ratio of 2- OHE$_1$ to 16 OHE [EMT] should be greater than 2.0 + values in the upper normal range. Any woman using hormone replacement therapy who has a low estrogen metabolite index should be monitored closely for improvements in urinary metabolites.

The increased risk from estrogen and estrogen metabolites has led to a search for compounds that safely produce estrogen-like effects and those that decrease the production of 16-OHE$_1$. Soy isoflavones (e.g., daidzein and genistein) are natural compounds that come the closest to meeting these criteria.[292] **Soy isoflavones have little or no effect on induction of tumors, and they have many estrogen-like properties even though they operate through different receptor sites on target tissues**.[293–295] Dietary intake of soy products[296] and flax oil[297] have been shown to favorably modulate the rates of 2- versus 16-hydroxyestrone production.

Diet also can promote chemopreventive activity against breast cancer by increasing the removal of circulating estrogen. Estrogens are metabolized by cytochrome P-450 enzymes that are inducible by compounds found in the Brassilica family like cabbage, Brussels sprouts, and broccoli.[291] Two phytochemicals contained in these foods, indole-3 carbinol (3C) and diindolylmethane (DIM), have been identified as active inducers of certain P-450 – isozymes.[299] The reaction catalyzed by these P-450-isoenzymes produces 2-hydroxylation of estradiol. Induction of P-450 by I3C results in decreased concentrations of several metabolites known to activate the estrogen receptor.[300] If the cytochrome P-450 is sufficiently active, a competing pathway causes 16-hydroxylation. Flaxseed supplementation at 10-gm/day significantly increases urinary 2/16-hydroxyestrone ratio, suggesting a breast cancer chemoprotective effect in premenopauseal women.[301] Women who have low EMI (2/16-hydroxyestrone) may be treated with I3C and DIM supplements, as well as with ground flaxseed.

Sexual Dysfunction

Sexual dysfunction is often seen in female patients with chronic pollutant or other noxious stimuli overload who develop disordered dynamics of homeostasis. These patients often have little or no sex drive. They fail to be stimulated when manual manipulation of the clitoris is attempted. They fail to lubricate or achieve orgasm. Conversely, in some females with dyshomeostasis due to pollutant or other noxious stimuli overload, the opposite, hypersexualism, occurs. Since most sexual function takes place through a series of complex interactions between the neurological and hormonal systems when the dynamics of homeostasis are altered, it is at times difficult to pinpoint one particular pollutant or other noxious stimuli stressor in these patients. However, we have treated enough patients with dysfunction due to specific environmental pollutants and metabolic dysfunction to realize that with the removal of these incitants, resulting in a reduced total body load, sexual homeostasis can then be maintained.

Many chemically sensitive women report that their mates emanate an odor that probably was acquired at work due to contact with such substances as gasoline, pesticides, and phenols. These odors alter the dynamics of homeostasis and immediately impede their desire for sex. Once the odors are removed by showering, changing clothes, or taking a sauna, the desire often returns. These odors from toxic substances clearly mask the natural sexual phermones. Some patients require their husbands to be away from the place of work for two days in order to outgas toxic substances before they have sexual intercourse.

Other chemically sensitive patients have been observed to have the urge for sex, but are so dry vaginally that they need lubrication. If they detoxify, vaginal lubrication returns. If they still do not lubricate vaginally, usually saliva works best for lubrication, since artificial and natural oils and jellies can cause reactions, altering the dynamics of homeostasis and stopping the sexual drive in these severely hypersensitive patients.

Other women are hyperstimulated by odors. They have an increased sex drive and at times will be somewhat indiscriminate with whom they have sex. The parasympathetic fibers are involved in many of these processes, and stimulation or malfunction of these fibers has often been observed and measured in the chemically sensitive patient at the EHC-Dallas. Although infertility is a problem in some chemically sensitive patients exposed to toxic pesticides, the majority of chemically sensitive patients are able to conceive.

UTERUS AND TUBES

Altered homeostasis from chronic pollutant or other noxious stimuli injury to the uterus and fallopian tubes is common. PMS, which is partially ovarian and partially uterine in etiology, is discussed in this section. During PMS the tubes can become inflamed and cause pain. The next section will discuss the environmental aspects of PMS and endometriosis, which results from alteration of the GRS and thus homeostasis.

PREMENSTRUAL SYNDROME

Premenstrual syndrome is a complex disorder of altered homeostasis with a wide range of symptoms that occur regularly before menstruation. PMS is a classic demonstration of how the integrated connective tissue matrix, and hormonal, neurological, and immune systems are imbalanced. The integrated homeostasis of these systems in PMS functions as a result of subunit GRS malfunction. Central homeostatic harmonic function is altered, resulting in PMS. Most women find that PMS typically has three symptom patterns. First, symptoms may start 1–10 days prior to menstruation and continue until the menstrual period begins; these symptoms may begin as a result of sensitivity to progesterone. Second, symptoms may start at ovulation and disappear in a day or two, reappearing later in the premenstrual phase; this symptom pattern may be due to sensitivity to LH. Finally, PMS symptoms may begin at ovulation, continue steadily into the menstrual period, and disappear only toward the end of menses. This latter type may be due to LH, estrogen, and/or progesterone homeostatic deregulation. Since symptoms of PMS are so varied and extreme in some cases, it can be difficult to identify and treat. However, there are special characteristics that distinguish PMS symptoms from those of other disorders.

The key characteristics in identifying PMS include those presented in Table 4.12.

1. Time of onset or increased severity of PMS—PMS tends to begin at puberty, becomes problematic after pregnancy, after discontinuing the birth-control pill, after amenorrhea, or after tubal ligation because efficient homeostasis is altered.
2. Painful menstruation—pain during menstruation is NOT related to PMS; however, it is still related to hormone dysfunction with altered homeostasis.
3. Weight fluctuations—temporary monthly gain of 3–9 pounds is common.

TABLE 4.12
Symptoms of PMS

Feeling "bloated"	Irritability
Feelings of weight increase	Aggression
Breast pain/tenderness	Tension
Skin disorders	Anxiety
Hot flashes	Depression
Headache	Lethargy
Pelvic pain	Insomnia
Change in bowel habits	Change in appetite
Thirst	Crying
Abdominal swelling	Change in sexual desire
Rhinitis	Loss of concentration
Fatigue	Mental confusion
Swelling of hands and feet	Memory loss
	Poor coordination/clumsiness/accidents

Source: Rea, W. J., *Chemical Sensitivity, Vol III: Clinical Manifestations of Pollutant Overload,* CRC Press, Boca Raton, FL, 1635, Tab. 24.7, 1996.

4. Food cravings—eating binges, especially involving sweet or salty foods; after not eating for 4–6 hours, there may be an onset of acute symptoms, panic attacks, or migraine headaches.
5. Lowered tolerance for alcohol—decreased tolerance of alcohol just before onset of the period may be accompanied by alcohol craving;
6. Inability to tolerate birth control pills—women with PMS may experience exacerbation of PMS symptoms when they take oral contraceptives.
7. History of threatened abortion—bleeding in early months of pregnancy followed by successful delivery is common in PMS patients.
8. History of hypertension or toxemia during pregnancy—women who experience a pregnancy complicated by hypertension, toxemia, severe headaches, disturbed vision, or rapidly developing fluid retention, often develop PMS.
9. Increased sex drive—women with PMS often experience an increase in sex drive during the premenstrual phase.

Premenstrual syndrome is not dysmenorrhea, which is pelvic pain during menses. This condition is usually treated with magnesium supplementation, oral contraceptives, and prostaglandin inhibitors, which keep the uterus from cramping. PMS is also not endometriosis, which is a condition in which some cells from the lining of the inside of the uterus, instead of passing out of the body during menstrual flow, actually work their way up through the Fallopian tubes into the abdominal cavity, where they cause pelvic pain. Finally, PMS is not pelvic inflammatory disease, which is infection of the uterus, fallopian tubes, and ovaries with accompanying pain in the pelvis and lower abdomen, as well as nausea, high fever, and/or rapid pulse.

Premenstrual syndrome is a multifaceted disorder that involves a homeostatic dysfunction of the patient's immune, enzyme detoxification, neurological, vascular, and hormonal systems, as well as possibly the nutrient supply. PMS represents local, regional, and central GRS dysfunction. Previous theories have related the problem to a deficiency in progesterone levels or an imbalance in estrogen and progesterone levels, with treatment aimed toward restoring that balance. Laboratory analysis of PMS patients does not always support these theories, but clearly there is altered homeostasis.

Over 50% of premenopausal patients treated for food and chemical sensitivity in the ECU have PMS symptoms. Laboratory tests for the majority of these patients show normal hormone levels, though this does not mean that the hormones are well balanced at any given time in the cycle. As a result of these findings, we believe that PMS patients have altered homeostasis with many patients being overly sensitive to their own hormones. Others have hypersensitivity to other xenobiotics and noxious stimuli, which cause the altered homeostasis with resultant metabolic dysfunction. During the time of their cycle, when certain hormone levels are normally high, patients react with physical and emotional symptoms. The reactions appear to be not highly specific to each hormone. The same symptom caused by progesterone in one patient can be related to estrogen or LH in another. Often PMS is due to imprecise levels of coordinating hormones or altered outputs, which result in a fluctuating level further altering the dynamics of homeostasis. As shown in Neuromuscular chapter in Volume 2 - Mechanisms of Chemical Sensitivity and Chronic Degenerative Disease, Clinical Homeostasis, once an anatomical area is damaged it is then preconditioned to further alter homeostasis with each subsequent exposure. Thus, once initial damage occurs in the ovary and uterus, i.e., disordered connective tissue matrix and estrogen receptors and cell membrane, these organs become conditioned to respond with more sensitivity to the hormones resulting in worse PMS). **This phenomenon is emphasized by the observation that some women will have severe PMS when they ovulate from one ovary and have very mild PMS when they ovulate from the other**. Also, when the uterine target organ is removed and the ovaries are left in, many patients' cases of PMS disappear.

Many other internal and external factors, including nutritional deficiencies, contribute to the PMS disorder in each individual. At the EHC-Dallas and Buffalo, we have observed that many women with PMS are deficient in certain nutrients like magnesium or vitamin B_6. These deficiencies only show up premenstrually when the regional reproductive system is stressed. Once the deficient nutrient is replaced, the dynamics of homeostasis are restored, and these patients are free of the PMS syndrome. Therefore, the total load of stress on the body must be reduced for effective treatment, to restore homeostasis.

As she interacts with her environment, the chemically sensitive patient with PMS increases her total body load as many stressors are continually added. These stressors include such things as outdoor and indoor air pollution, water and food contamination, and emotional and physical stress. Any stress to the immune and enzyme detoxification systems including infection, pregnancy, prior hormone administration, or toxic chemical exposure can upset immune control mechanisms and local, regional, and systemic homeostasis, thereby allowing the inappropriate development of antibodies against any organ in the body and resulting in altered dynamics of homeostasis.

The EHC-Dallas and Buffalo have treated over 4000 women with some form of PMS. In one consecutive series of 30 patients, the mean age was 34 years with a range of 30 to 39 years in 46% of the patients (Table 4.13); 50% of patients had onset of PMS at age 29 years.

However, a smaller portion of patients' PMS started at or shortly after the onset of menstruation. Usually, the mother had a similar history, clearly having altered homeostasis as well. Another interesting facet of these chemically sensitive patients with PMS was that 90% of them had neurological/psychological symptoms during their premenstrual period, and 66% had chlorinated pesticides in their blood, which may well account for their severe ovarian and brain dysfunction. Certainly, their integrated central homeostatic mechanism was not functioning harmoniously, resulting not only in brain dysfunction at the disturbed membrane level but also autonomic dyshomeostasis. Of these patients, 76% were sensitive secondarily to trees, and 80% were sensitive to grass. They noted that their PMS was worse during the spring and fall months, suggesting an environmental influence on the homeostatic mechanism. Thirty-three percent of the patients were sensitive to terpenes and 86% to toxic chemicals such as phenols, formaldehyde, petroleum alcohols, pesticides, and chlorine.

It was noteworthy that, when exposed to a series of toxic chemicals by inhalation, the chemically sensitive PMS patient was usually sensitive to just one chemical, rather than a multiple series, as seen in the more advanced chemically sensitive patient. However, it did appear that multifactoral incitants were present in the form of individual foods, molds, and other stressors that altered homeostasis.

TABLE 4.13
Thirty Chemically Sensitive Patients with PMS Studied in the ECU after 4 Days Deadaptation with Reduction of Total Body Load

Challenge Studies	Percent Positive	Challenge
Trees	70	ID[a]
Grasses	80	ID
Molds	100	ID
Terpenes	33	ID
Foods	100	ID and oral confirmation
Cane sugar	100	
Corn	76	
Wheat	63	
Potato	63	
Tomato	63	
Toxic Chemicals	86	ID
	49	IHDB[b]

Source: From EHC-Dallas. 1986.

Note: Ages: 30–39 years; mean: 34 years. Average age of onset: 29 years. Blood chlorinated pestides: 100%.

[a] ID = Intradermal challenge
[b] Inhaled double-blind challenges—phenol (< 0.005 ppm); chlorine (< 0.33 ppm); formaldehyde (< 0.2 ppm); pesticide (< 0.0034 ppm); saline placebos.

Clearly, the environmental receptor system of the connective tissue matrix is altered in women suffering from PMS. This, in turn, alters the GRS and hormone integrated homeostasis. One hundred percent of the chemically sensitive PMS patients we observed at the EHC were sensitive to one or more foods. The main offenders that triggered or exacerbated their syndrome were cane sugar—100%; corn—76%; wheat, potatoes, tomatoes, brewer's yeast—63%; eggs and bananas—60%; and chicken—56%. We found that 95% of the chemically sensitive PMS patients do well with hormone neutralization, a rotary diet with elimination of the worst offenders, nutritional supplementation, and chemical avoidance. These treatments reduce the patient's total body load, helping to restore orderly efficient homeostasis through a more stable GRS.

The uterus and fallopian tubes are particularly susceptible to pollutant injury due to their internal vascular anatomy, smooth muscle content, and their direct contact with the external environment. The uterus has a rich supply of connective tissue matrix full of immune defense cells, which helps fend off external exposure by local neutralization (Figure 4.16). However, noxious substances can penetrate it and cause severe dyshomeostatic problems as was shown in the previous case study of the woman with pesticide overexposure.

VASCULITIS OF THE REPRODUCTIVE SYSTEM

Human studies involving altered homeostasis in three cases of cervical vascular malformation as a cause of antepartum and intrapartum bleeding in DES exposed females have been reported.[302] Often, abnormal uterine bleeding may be triggered by other environmental incitants, which alter homeostasis, resulting in metabolic changes followed by tissue changes. Here, the blood vessels including the connective tissue matrix become damaged thus altering the GRS. The vessel walls become fragile due to the exposure of triggering agents. Then tissue changes occur, resulting in vessel wall leakeage of blood. Often the patients will have cutaneous bruises that correlate with the uterine bleeding. We have seen numerous cases of uterine pain and bleeding secondary to blood vessel dysfunction

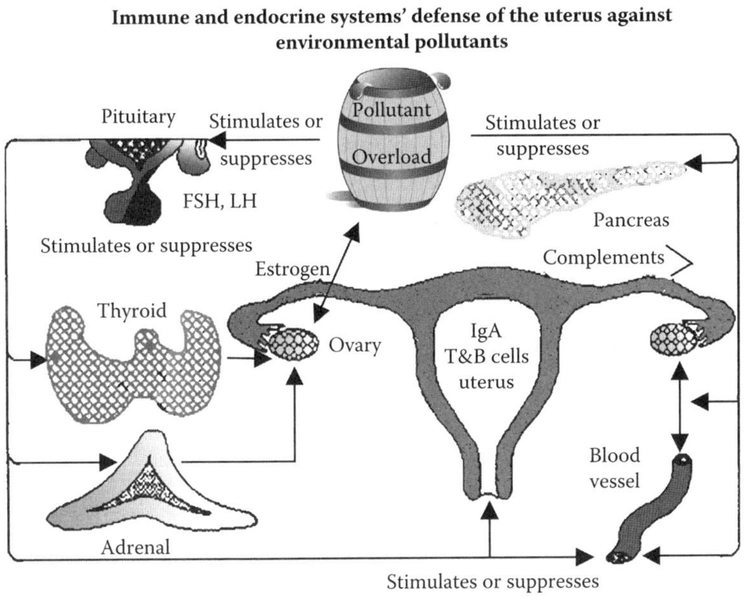

FIGURE 4.16 (See color insert following page 364.) Immune and endocrine systems' defense of the uterus against environmental pollutants. (Modified from Rea, W. J., *Chemical Sensitivity, Vol III: Clinical Manifestations of Pollutant Overload*, CRC Press, Boca Raton, FL, 1638, Fig. 24.7, 1996.)

from toxic exposures. The uterus may be the sole target, or one part of the large, target organ. The following case illustrates the environmentally sensitive patient's problems with uterine bleeding.

> **Case study.** A 36-year-old white female had a severe vasculitis syndrome characterized by spontaneous bruising, peripheral and periorbital edema, and cerebral dysfunction marked by memory loss. In addition, she had slow mentation. Testing revealed she had multiple food and chemical sensitivities. Biopsy of her spontaneous bruises revealed perivascular lymphocytic infiltrates. This patient developed spontaneous uterine bleeding, which could only be controlled by a hysterectomy. Lesions in the uterus and fallopian tubes revealed perivascular lymphocytic infiltrates similar to those seen on the bruised skin. This patient has done well over the past 10 years after moving to a less-polluted area of the United States.

ENDOMETRIOSIS

Endometriosis is characterized by the slow growth of tissue from the endometrium in an inappropriate site such as the peritoneum, myometrium, or ovary. It is one of the most commonly diagnosed causes of dysmenorrhea and infertility once altered homeostasis occurs. A fairly common gynecologic disorder, endometriosis appears to be on the rise and may be due to increased pollutant exposure. While there were only 20 cases reported before 1921,[303] this disease today affects 5 million women per year in the United States, and is more common in industrialized nations and around industrialized centers.[303] In 1992, German researchers reported that women with endometriosis had significantly higher levels of polychlorinated biphenyls in their blood.[303] These are known disrupters of the dynamics of homeostasis. A Canadian study reported to the Ontario Association of Pathologists in 1985 also showed a high rate of PCBs in a study of monkeys with endometriosis.[303]

Rier et al.[304] showed that the incidence of the reproductive disease, endometriosis, was determined in a colony of rhesus monkeys chronically exposed to O-2, 3,7,8-tetrachlorodibenzo-*p*-dioxin (TCDD or dioxin) for a period of 4 years. Ten years after termination of dioxin treatment, the presence of endometriosis was documented by surgical laparoscopy, and the severity of disease was assessed. The incidence of endometriosis was directly correlated with dioxin exposure, and the severity of disease was dependent upon the dose administered ($0 < 0.001$). Three of seven animals exposed to 5 ppt dioxin (43%) and five of seven animals exposed to 25 ppt dioxin (71%) had moderate to severe endometriosis. In contrast, the frequency of disease in the control group was 33%, similar to an overall prevalence of 30% in 304 rhesus monkeys housed at The Harlow Primate Center with no dioxin exposure. This 15-year study indicates that latent female reproductive abnormalities may be associated with dioxin exposure in the rhesus monkey. Therefore, the effects of this toxic chemical may be more diverse than previously reported.

Previous work has described an association of endometriosis in rhesus monkeys following exposure to PCB compounds.[305] Rier et al.'s[304] results support and extend these findings, since dioxin is used as a reference compound for halogenated aromatic hydrocarbons, including PCBs.[306] Extensive literature has documented the incidence of endometriosis in the rhesus model following exposure to single-energy proton irradiation, mixed-energy proton irradiation, and X-rays.[307-309] In Rier et al.'s study,[304] **endometriosis in dioxin-exposed monkeys was first documented seven years following the termination of dioxin exposures**. Immune system defects are a common probable factor that may contribute to the development of endometriosis in each of these animal models. Once immune defects occur the dynamics of homeostasis will usually be altered. Indeed, this observation is consistent with human studies, suggesting that immune mechanisms may contribute to the disease process.[310,311] Dioxin has immunosuppressive activities and is a potent inhibitor of T lymphocyte functions,[312-314] thus being a disrupter of homeostasis. In addition, this toxic chemical modulates steroid receptor expression resulting in altered tissue-specific responses to hormones[217] that also will adversely change homeostasis. Chronic immunosuppression in combination with hormonal deregulation may have facilitated the aberrant growth of endometrial tissue within the peritoneum of dioxin-treated animals.[314] Many theories to explain the pathogenesis of endometriosis exist, and it is not necessary to discuss them in this book. Tsukino et al. found many organochlorine compounds such as polycholorinated dioxins biphenyls and chlorinated pesticides in Japanese women with endometriosis.[315] Heiller et al. found a similar example.[316]

Comparative studies have demonstrated that endometrium and endometrial implants differ in their responses to hormonal fluctuations during the menstrual cycle. Concentrations of estrogen and progesterone cytosol receptors are lower in endometriosis than in eutopic endometrium, and no variations occur during the menstrual cycle or hormone therapy. The response to the hormones appears to be related to the site and type of endometrial lesion. Ovarian endometriosis is less responsive than peritoneal implants, particularly the vesicular type, and is more responsive than fibrotic, nodular implants.

In order to prevent continuous dysmenorrhea in a woman of reproductive age, early diagnosis, and treatment of the progression of disease leading to infertility and advanced pathology should be done.

A diagnosis may be entertained when there is dysmenorrhea. Most women expect to experience some pain as a part of their normal monthly period, but women with endometriosis regularly experience five days of dysmenorrhea each month, a total of five years of pain during their reproductive lifetime. Such women often suffer from dyspareunia and chronic abdominal pain in addition to dysmenorrhea. **A large proportion of patients with endometriosis, approximately 65%, experience painful symptoms, whereas 35% do not**. In over 60% of endometriosis patients, classic bimanual palpation is negative.

In experiments at the EHC-Dallas, steroid hormones play a central role in the hormonal management of endometriosis, as does massive reduction of the toal body pollutant load, and restoration of nutrition. These patients continue to have debilitating symptoms even after surgical removal of the uterus, adnexa, ovaries, and patches of endometriosis in adjacent areas. These symptoms are wide-ranging, from chronic lower abdominal pain, backache, and urinary problems to dizziness, hot and cold sensations, and an inability to tolerate odors. This group of patients respond to a comprehensive treatment approach consisting of environmental control; avoidance of chemicals; rotation of chemically less-contaminated food; and appropriate injection therapy for environmental, food, and chemical incitants; as well as hormone neutralization treatment (Table 4.14). This regimen frequently restores homeostasis allowing the disease to come under control.

TABLE 4.14
Data of 22 Patients with Endometriosis

Associated Disease		Intradermal Skin Testing (Positive)				High Level of Toxic Chemical in Blood	
Disease	No.	Antigen	No.	Antigen	No.	Compound	No.
Fatigue		Food (range: 2–43 kinds; mean:17 kinds)	16	Ethanol	4	Pentane	1
Depression	1	Molds	14	Formaldehyde	7	2-Methylpentane	6
Toxic brain syndrome	2	Dust	14	Women's cologne	5	3-Methylpentane	6
Ovarian cyst	1	Mite	14	Men's cologne	4	Cyclopentane	2
Infertility	1	Danders	123	Orris root	8	n-Hexane	4
Cephalgia	3	Candida	11	Newsprint	3	β-BHC	2
Chest wall syndrome	1	T.O.E.	11	Phenol	1	DDE	3
Collagen vascular disease	1	Terpenes	11	Diesel	1	Heptachlor epoxide	1
Autoimmune thyroid	1	Smuts	10	Fluogen	11	Trans-nonachlor	2
Irritable bowel syndrome	2	Tree pollen	14	CMV	1	Trimethylbenzenes	2
Arthralgia	2	Grass pollen	14	EBV	5	Toluene	2
Myalgia	2	Weed pollen	12	Bacteria	1	Ethylbenzene	1
PMS	2	Insecticides	2	Rayon	1	Styrene	1
Engioedema	1	Estrone	10	Polyester	1	Chloroform	1
Allergic rhinosinusitis	7	Leutinizing hormone	6	Wool	2	1,1,1-Trichloroethane	1
Hypothyroidism	1	Progesterone	10	Cotton	7	Trichloroethylene	1
GI upset	1	Natural gas	3	Silk	2	Tetrachloroethylene	1
EMF sensitivity	1	Cigarette smoke	8	MRV	9	HCB	2
Hypoglycemia	1	Conjugated estrogen	1	Algae	4	Oxychlordane	1
Connective tissue disease	1			Perfume	1		
				Liner RV	1		

(Continued)

TABLE 4.14
Data of 22 Patients with Endometriosis (Continued)

	Inhaled Testing		Immune Parameters			Immune Parameters	
	Positive		Positive				
Compound	Parameter	No.	Parameter	No.	Parameter	No.	
Chlorine	CMI—3 positive reactions	1	Low total lymphocytes	4		High antimicrosomal AB	2
Phenol	CMI—5 positive reactions	1	Low T_4	1		High antinuclear AB	3
Pesticide		1	Low T_8	3		High antithyroglobulin	1
Formaldehyde		1	High T_{11}	1		High antireticulin AB	1
			High T_4/T_8	1		High thyroglobuline AB	1
			Low CH_{50}	2		Postiive hepatitis B surface Ag	1
			Low C_4	1			
			IgG—low	1			
			IgG –high	1			
			IgE—high	1			
			High FBV –early antigen IFA	1			
			High EBV-VCA-IgG	2			
			High EBNA IgG	4			
			High EBNA IgM	3			
				2			

Endocrine System

	No.		No.
Zinc—low	1	Chronic cystic cervicitis, squamous metaplasia, cervix uteri benign	1
Sodium—low	2		
Iron—low	3		
Iron—high	1		
Phosphorus—low	1		
Phosphorus—high	2		
Sulfur—high	1		
Strontium—high	1		
Potassium—high	1		
Calcium—high	1		
Tin—low	1		
Molybdenum—low	1		

Outcome Measurements	No.	Treatment	No.
Abnormal iris corder	4	Sauna	2
Abnormal Triple-headed SPECT Brain Scan	1	Rotation diet	11
Surgery for endometriosis	3	Elimination diet	12
Hysterectomy	4	Injection therapy	19
Abnormal equilibrium	1		

Source: From EHC-Dallas. 2005.
Note: Age range = 12–59 years. Average age = 35.

SPONTANEOUS ABORTION

Studies of various industries have suggested links between occupational exposure to chemicals and reproductive failure. For example, among **dentists and dental assistants, there is a significant relationship between total mercury levels in the hair of exposed women and histories of reproductive failure and menstrual cycle disorders**.[317] In a review article, Landrigan et al.[318] cited human and animal studies that showed the reproductive toxicity effects of ethylene oxide, a biocide used in sterilization of hospital equipment. Nurses handling antineoplastic drugs during the first trimester of pregnancy have higher rate of fetal loss.[319] In the Danish county of Funen, Heidam[320] showed odds ratios for spontaneous abortions were significantly increased among factory workers and painters exposed to such chemicals as nitrous oxide, inorganic mercury, organic solvents, and pesticides. Among women exposed to organic solvents during laboratory work, there was an increased (though not significant) tendency toward miscarriage.[321] In a Swedish study of women working in a rubber plant, unfavorable pregnancy outcomes (threatened abortion, spontaneous abortion, or malformation) were associated with exposure to the tire building process.[322] A study in the plastics industry showed some conflicting results; however, the odds ratio for spontaneous abortions in workers in polyurethane processing factories was increased.[323]

Although there are inherent problems in epidemiological studies such as these cited, laboratory animal research tends to support their conclusions. Solvents such as benzene and its derivatives, including toluene, xylene, etc., which are frequently identified in chemically sensitive individuals and are known as disrupters of homeostasis, cause spontaneous abortion in rabbits. Some solvents cause a dose-dependent increase in postimplantation loss in rats and mice.[324] One study found pregnant mice exposed to methylmercuric chloride had an increase in fetal toxicity. This study also showed chromosome stickiness and clumping in fetal tissues, leading to reduced mitotic divisions.[325]

Data from industrial accidents (e.g., the leakage of methyl isocyanate in Bhopal, India, in 1984) also show fetal toxicity. Varma[326] reported that a survey of 3270 families nine months after the accident indicated that 43% of pregnancies did not result in live births. Studies in mice corroborated these findings.[323]

One of the most widely studied prescription drugs in regard to fetotoxicity is DES.[327–329] Research suggests that other relatively common prescription drugs are also fetotoxic. Spontaneous abortions have been reported with the use of oral anticoagulants for patients with cardiac valve prostheses.[330] The use of the anticoagulant warfarin probably is related to early abortions.[331] In a study of renal transplant patients, successful outcome of pregnancy was associated with reduced exposure to warfarin.[332]

In an investigation of fetal exposure to isotretinoin (retinoic acid), Lammer et al.[333] found an unusually high relative risk of major fetal malformations. These authors suggested that the effect is on cephalic neural crest cell activity, since the malformations tend to be craniofacial, cardiac, and thymic in nature. Other studies have shown similar results associated with spontaneous abortion.[334,335] GRS is formed when organ developments starts and is more susceptible to damage as many adverse actions are dose dependent and have linear or bell shaped toxicity curves.

Other prescription drugs known or suggested to cause spontaneous abortions include imidazole agents, which are used to treat vaginitis;[336] tedral, which has been prescribed for asthmatics with upper respiratory infection;[337] and lithium, which is used to control manic-depression.[338] Animal studies suggest that the glucocorticosteroids budesonide and fluocinolone acetonide, may be fetotoxic,[339] as may be calcium channel blockers used to prevent premature labor,[340] methyldopa (an antihypertensive),[341] and certain antibiotics.[342] Particularly interesting is a study showing that female mice mated to males that had been treated with methyldopa had a higher incidence of abortion and lower incidences of total implantations, even though mating capacity and fertility were not affected.[338] It is clear from all the studies that when normal homeostasis is altered, metabolic and, at times, tissue changes produce adverse outcomes.

Hormones

Progesterone support for *in vitro* fertilization increases the rate of spontaneous abortion.[343] However, in some cases **once abortion has started progesterone administration seems to stop the abortion**. Women who conceive with human gonadotropins have a high rate of spontaneous abortions.[344,345] Animal studies show certain sex steroidal hormones used clinically for detection of early pregnancy and for supportive therapy of pregnancy are embryolethal.[346,347]

Influence of Toxics on Hormonal Homeostasis

There is a realization that environmental contaminants interact with hormone receptors and mimic or antagonize the action of endogenous hormones.[348]

PCBs

PCB's can trigger hormonal dyshomeostasis. PCB's will not only be trapped in fat cells, but also in extracellular lipids, and can be trapped as well in the C.T. matrix resulting in inflammation and eventually uterine vasculitis. Epidemiologic evidence suggests that PCBs are toxic to the reproductive processes.[349] In comparison to a control group, women with missed abortions have shown higher PCB serum levels with increases in the penta- and hexachlorobiphenyls.[350] Animal studies corroborate these findings.[351]

Pesticides/Herbicides

Workers exposed to pesticides in the grape gardens of Andhra Pradesh show an increase in spontaneous abortions, with cytogenetic studies indicating an increase in chromatid breaks and gaps in chromosomes of peripheral blood.[352] Here with an overload of pesticides, as occurs with PCBs, the neuro-endocrine system and connective tissue matrix become damaged resulting in a disinformation response through the GRS and resultant dyshomeostasis resulting.

Results of studies of phenoxy herbicide (Agent Orange) exposure in Vietnam suggests a link between the herbicide and unfavorable outcomes of pregnancy.[353] An increase in chromosome breakage has been observed in males exposed to the herbicide.[354]

Human studies have shown increased incidence of spontaneous abortion related to consumption of alcohol during pregnancy.[355,356] Laboratory studies show nondysjunction chromosomal defects in fertilized eggs of ethanol-exposed female mice[357] and increased rates of spontaneous abortion in pregnant macaques with ethanol administration.[358]

Tetrahydrocannabinol (THC), the principal psychoactive component in marijuana, produced spontaneous abortions when administered to monkeys early in pregnancy;[359] a rapid decrease in chorionic gonadotropin levels and a decrease in progesterone also occurred. When compared to former drug abusers maintained on methadone during pregnancy and to drug-free women, cocaine-using women had a significantly higher rate of spontaneous abortion.[360] At least one study[361] has shown a relationship between caffeine consumption and spontaneous abortion. For more information, see *Chemical Sensitivity,* Volume IV, Chapter 30.

Menopause Dysfunction

Early menopause, excessive bleeding during menopause and exacerbation of peripheral symptoms have been seen in chemically sensitive women. Ovarian function will suddenly cease, resulting in amenorrhea, dry vaginal mucosa, and early onset of osteoporosis and arteriosclerosis. This dyshomeostatic reponse has been seen in some chemically sensitive and many chronic degenerative diseased individuals. A more common syndrome has been the erratic output of estrogen and progesterone that results in a myriad of symptoms including hot flashes, sweating, anxiety, and an increased appetite. Hot flashes due to hormone irregularity are often confused with those due to chemical and food triggering. We have seen many women with chemical sensitivity develop hot flashes that were easily eliminated after the offending foods and chemicals were removed.

VAGINA AND VULVA

Though technically not a part of the endocrine system the vagina and vulva can be affected by the various activities of the endocrine system. For completeness, therefore, these organs are discussed here.

Suckling neonatal mice that ingested milk from chlordecone-treated dams (250, 500, or 1000 μg daily) exhibited dose-dependent changes in the vaginal canals similar to those exposed to estradiol (10, 20, or 40 μg). Changes included mucification, keratinization, and desquamation.[362] o,p-Dichlorodiphenyltrichloroethane (o,p'-DDT)-stimulated DNA synthesis and cell division in the uterine luminal epithelium, stroma, and myometrium resulted in uterine hyperplasia.[363]

Environmental incitants can produce observable vaginal irritation, inflammation, and even infection. Spermicidal foams and jellies can alter homeostasis, triggering vaginal inflammation. Inhaled volatile organic chemicals have been seen to affect the vagina as a target organ, resulting in an increase in inflammation and recurrent infection with less virulant opportunistic invading microbes. It has been observed that laboratory workers who worked with dioxin laden test tubes had recurrent vaginal infections, which were reported after a latency period of 2–3 months.[364] Overuse of antibiotics may allow chronic *Candida* infestation to occur.[365] In utero exposure to DES is the best known cause of vaginal cancer in human offspring.

Once altered homeostasis occurs and hypersensitivity state follows, food sensitivity frequently triggers a vaginal discharge that is often correlated with nasal discharge. However, other factors can influence problems in the vagina. Severe vaginal pain, for example, can often occur with or without sexual intercourse, when homeostasis is altered. Sometimes a patient's vaginal discomfort may be a systemic reaction accompanied by rhinitis, asthma, anxiety, depression, or edema after ingestion of foods or chemicals. At other times, vaginal pain with intercourse may result from a woman's sensitivity to her partner's sperm. With such a patient, **we have successfully used the partner's sperm in an intradermal injection of an autogenous extract to neutralize the sensitivity**, and the woman has then been able to have pain-free intercourse.

Vulvitis also has been seen with severe chemical overload. It can be excrutiatingly painful to the point that patients are unable to eat. Objective evaluation of vulvar burning is difficult, except for occasional mucosal cuts and swelling. Some, though not all, cut and swollen areas are usually tender to touch. As a rule, pain represents a neuropathy, and occasionally rectal manipulation of the muscles of the pelvis will help reduce discomfort. Sometimes these patients also have a burning tongue. Usually, these patients are triggered by chronic environmental pollutant and food exposures that alter homeostasis, and they respond to evironmental manipulation. McKay[366] has described 20 patients with vulvodysnia. She treated them with 10–60 mg/daily of amitriptyline with partial success. We have had no long-term success treating environmentally sensitive patients with this drug, but removal of the noxious stimuli usually restores homeostasis, and the patient improves.

TESTES AND SEMINAL VESICLES

The hypothalamus is responsible for stimulating the anterior pituitary to release gonadotropin hormones that bring on puberty. Before puberty, the hypothalamus is so extremely sensitive to testosterone that if an individual's testicles are chemically stimulated, releasing even a minute amount of testosterone, inhibition of the entire system occurs. Thus, this is an example of how sensitive the GRS is in maintaining homeostasis. In contrast, chronic pollutant or noxious incitant stimulation may imbalance the prepubertal individual, causing a myriad of endocrine and related dysfunctions of the dynamics of homeostasis. After puberty, the hypothalamus loses this extreme sensitivity to testosterone, which allows the secretory mechanisms of the testes to develop full activity. A new level of homeostasis is achieved. This homeostatic process may be altered in some chemically sensitive males who have been seen with sex organ dysfunction. At times, pollutant overload may cause suppression of sperm. For example, the plasticizer, bisphenol, used in polycarbonate bottles will act as an estrogen mimic and cause sex drive suppression.

The sex hormones have an integral part in cell function. Testosterone enters the cell, some of which is reduced by 5 α-reductase to dihydrotestosterone. Both hormones bind to the androgen receptor with dihydrotestosterone having ten times the binding affinity of testosterone. The resulting ligand-receptor complex then attaches to the androgen-response element—a step facilitated by coactivators and inhibited by corepressors—leading to the transcription of messenger mRNA protein synthesis and androgen effects. Some testosterone entering the cell is aromatized to estradiol through the action of aromatase. Testosterone then has not only a role in sexual function but also an anabolic sterod effect influencing positive protein metabolism. See Figure 4.17.

One study showed these toxics including phthalates, DDT, DDE, PCB in human amniotic fluid. Sexual attitudes and preferences appear to be affected by many synthetic chemicals some of which are found in the blood of patients with chemical sensitivity and/or chronic degenerative disease. These include 209 PCBs, 75 dioxins, and 135 tumors.[367] The literature is legion on the effects of toxics on animals. Alligators,[368] mice,[369] rats,[370] many fish,[371] including salmon[372] guinea pigs,[373] panthas,[374] monkeys,[375] dogs,[376] rabbits,[377] gulls,[378] bald eagles,[379] other birds.[380] Some of these effects included small testicles, small penises, and feminization of the species. The chemicals ranged from dioxins to chlorinated pesticedes, atrazine, diazanon, methoprene, dithane, and temephos.

In the past forty years there has been an increase of 30–40% hypospdias, undescended testicles, bisexuality, and feminization of males[381] where 40% showed low sperm counts when they reached adulthood.[382] **PBP exposure in women showed a higher incidence of testicular abnormaities and smaller penises in male babies**.

American women have double the miscarriages as those in HongKong.[383] Pentachlonphenol is associated with habitual abortion, unexplained infertility, menstrual disorders, and premature menopause.[384] Tris fire retardant (2,3-dibromopropylphosphate) is associated with testicular abnomalities, smaller penises, cancer increase, and genetic defects.[385,386]

Some chemically sensitive and/or chronically diseased men begin to exhibit decreasing sexual function in their early 30s and 40s suggesting early climatric. The total body pollutant load is increased and informational changes are sent through the body. Metabolic changes follow. This homeostatic dysfunction may trigger mood swings and depression as well as other physiologic diminution in energy. Tissue changes occur, resulting in the onset of "pot bellies," and apathy as the estrogen pathway becomes dominant.

The male sexual act may be thwarted in many ways by pollutant overload that disturbs homeostasis. **Sexual dysfunction from olfactory stimulation by perfumes and toxic chemicals, which send informational changes throughout the body, has been reported by many chemically sensitive males**. These changes in information cause dyshomeostasis, prevents an erection and may even stop the desire for sex. Others have noted that even though they could achieve an erection, they often could not maintain it if intercourse was occurring in a polluted environment. They recognized that chronic subtle exposures to environmental pollutants disrupted normal sexual function. However, they reported that they could maintain an erection for an extended period of time in a less-polluted environment. This aneccdotal evidence demonstrates that when the noxious stimuli are withdrawn or not present, the dynamics of homeostasis needed for normal sexual function are restored. Apparently, parasympathetic impulses are easily blocked or dampened by environmental pollution in the sensitive individual. **Down-regulation of nitric oxide synthase and formation of nitric oxide occurs**, preventing vasodilation and increased blood supply of the corpus spongiosum. Absence of nitric oxide prevents the erection from being maintained. Pollutant-triggered homeostatic dysfunction occurs where the veins dilate, and the arteries constrict, causing a loss of erection.

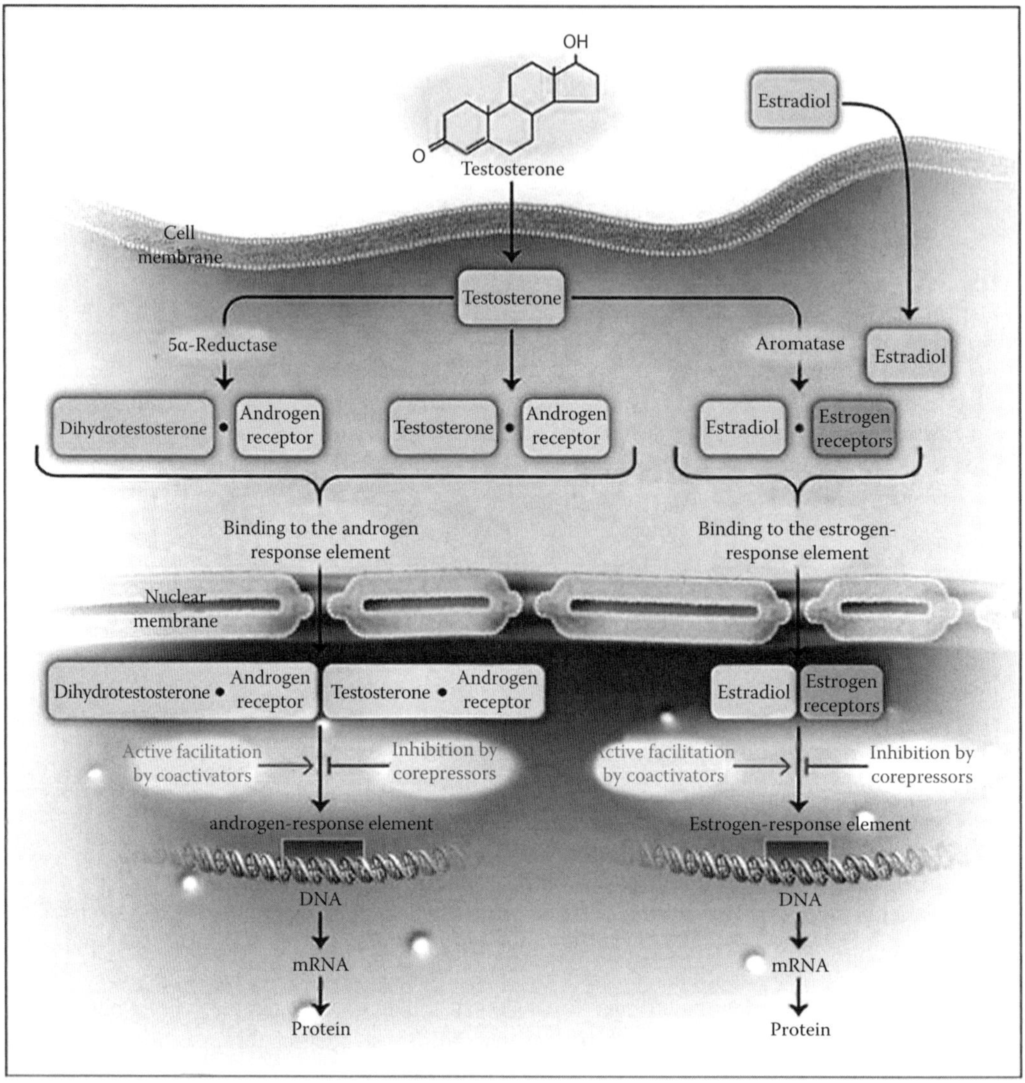

FIGURE 4.17 (See color insert following page 364.) Intracellular actions of sex steroids. Testosterone enters the cell, some of which is reduced by 5α-reductase to dihydrotestosterone. Both hormones then bind to the androgen receptor, with dihydrotestosterone having 10 times the binding affinity of testosterone. The resulting ligand–receptor complex then attaches to the androgen-response element—a step facilitated by coactivators and inhibited by corepressors—leading to the transcription of messenger RNA (mRNA), protein synthesis, and androgen effects. Some testosterone entering the cell is aromatized to estradiol through the action of aromatase. Estradiol then binds to its specific receptor. The resulting ligand–receptor complex attaches to the estrogen-response element of chromatin—a step facilitated by coactivators and inhibited by corepressors—leading to mRNA transcription, protein synthesis, and estrogen effects. Each step in the process provides opportunities for the introduction of sex-based differences. (Federman, D. D., *NEJM,* 354(14), 1507–1514, Fig. 2, 2006. With permission.)

Other chemically sensitive patients have reported an inability to ejaculate, indicating probable chronic pollutant damage to the nerves that leave the cord at L-1 and L-2 and pass through the hypogastric plexus. This damage would result in GRS dysfunction and disorderly efficient homeostasis. We have some cases where the male was continuously stimulated by chronic exposures to certain toxic chemicals or foods, sending informational changes that acted as aphrodisiacs, causing an increase in sexual desire and a pathologic appetite for indiscriminate sex several times a day.

The understanding of the physiology of testosterone in relation to homeostasis is very important in the chemically sensitive male, because of not only its sexual function but also its anabolic steroid effect. Interstitial Leydig cells produce testosterone and are influenced by environmental pollutants such as x-rays, heat, pesticides, and a myriad of other pollutants[387,388,391] that will alter the dynamics of homeostasis. In addition to its role in sexual function, testosterone causes hair growth. Loss of testosterone due to pollutant injury will cause patchy hair loss as seen in many chemically sensitive males. Excess testosterone alters homeostasis, causing acne. The effect of testosterone on protein formation and muscle development may be minimized in some chemically sensitive patients. As a result of homeostatic dysfunction, muscle wasting and weakness may develop. **Pollutant damage may cause a loss of testosterone, causing the metabolic rate to decrease 5–15%, and it may be one of the causes of the cold sensitivity that is seen in most chemically sensitive and chronically diseased patients**. Homeostatic dysfunction resulting in low testosterone may adversely affect red blood cells, and it can decrease the reabsorption of sodium. Many pollutants have been reported to alter testosterone production.

The effects of toxic chemicals on the homeostasis of reproduction are generally more appreciated than their effects on the homeostatic functions of other organs. Studies by Frem-Titulaer et al.[272] discussed in the section on ovaries show the occurrence of precocious puberty in Puerto Rican males due to contaminants in the chicken feed. Other contaminants show a similar picture of altering homeostasis. For example, inhaled or ingested methylchloride resulted in a broad spectrum of homeostatic dysfunction by the presence of uni- or bilateral sperm granulomas in the cauda epididymis, significantly depressed testes weights, significantly lowered testicular spermatid head counts, delayed spermiation, chromatin margination in round spermatids, epithelial vacuolation, luminal exfoliation of spermatogenic cells, and multinucleated giant cells.[366]

Additionally, disturbances have been reported with spermatogenous quantity and quality of sperm as well as other functions of morphology and motility. Benomyl, a systemic fungicide, produced age-related decreases in testicular or epididymal weights, decreased epididymal sperm counts, decreased vas deferens sperm concentrations and/or testicular lesions, and increased incidence of diffuse hypospermatocytogenesis.[387,385,386,388] Dibromochloropropane exposure (initial and residual) brought about reduced body and gonadal weight gains and caused hypospermatogenesis or seminiferous tubular atrophy.[387] Acute exposure may produce irreversible injury.[391]

DDT [1,1-*bis*(*p*-chlorophenyl)-2,2,2-trichloroethane]-exposed embryos demonstrated significant alterations in both testes and ovaries. The testes consisted of mostly stroma with fewer seminiferous cords, while ovaries contained a larger number of distended nodular cords and differences in the distribution of these cords.[392] **Thus, the connective tissue matrix and GRS receive misinformation from these toxics, resulting first, in metabolic and then tissue changes, which then results in autonomous fixed-name disease**.

Disorders of development and function of the male reproductive tract have increased in incidence over the past 30–50 years, apparently, because of the massive contamination of our air, food, and water.[393] Disordered homeostasis appears to be widespread. Examples of such disorders include testicular cancer,[394] maldescent (cryptorchidism),[393,395] and urethral abnormalities (hypospadias),[293] and there has been a striking drop in semen volume and sperm counts in normal adult men[396] as homeostasis becomes altered. Since these changes are recent and appear to have occurred in many countries, we presume that they represent alteration of the dynamics of homeostasis and reflect adverse effects of environmental or lifestyle factors on the male rather than, for example, genetic changes in susceptibility.[393] Since testicular cancer, cryptorchidism, and hypospadias are all errors that probably arise during fetal development,[397–399] these abnormalities and reduced sperm counts may have a common etiology.

Treatment of several million pregnant women between 1945 and 1971 with a synthetic estrogen, DES (diethylstilbesterol), is now recognized to have led to substantial increases in the incidence of cryptorchidism and hypospadias, and decreased semen volume and sperm counts in the sons of these women.[400] DES exposure may also increase the incidence of testicular cancer,[401] and other evidence also points to a link between maternal estrogen concentrations and the frequency of testicular cancer[402] and cryptorchidism.[403] The similarity between these effects and the adverse changes in male reproductive development and function over the past 40 to 50 years raises the question of whether the adverse changes are attributable to altered exposure to estrogens during fetal development. This possibility is not unlikely given the view that humans now live in an environment that can be viewed as a virtual sea of estrogens,[404] whose excess will alter normal homeostatic function.

During the past 50 years, tens of thousands of chemicals have been synthesized and released into the general environment. Some of these chemicals inadvertently interfere with hormone function in animals and, in some cases, humans. The public health implications of these so-called endocrine disruptors have been the subject of scientific debate, media interest, and policy attention over the past several years.[405] The current scientific debate centres on whether there is evidence of significant risks to the general human population.[406]

Mechanisms of Action and Fetal Vulnerability

Numerous assays have reproducibly shown that **some pesticides and other industrial chemicals can directly bind to, or block hormone receptors, thereby initiating or blocking receptor-activated gene transcription**.[407] Other exogenous chemicals act indirectly on hormonal homeostasis by altering steroidogenesis, hormone transport on binding proteins, receptor numbers on target organs, or hormone metabolism.[408] For example, polychlorinated biphenyls (PCBs) interfere with thyroid function through a variety of mechanisms, including increased metabolism of T4 (thyroxine), interference with T4 delivery to the developing brain by displacement from the carrier protein, and interference with the conversion of T4 to T3 (triiodothyronine).[409]

During development the fetus is particularly sensitive to hormonal fluctuations. Exposures to low levels of exogenous hormones or toxicants may result in permanent physiologic changes that are not seen in adults exposed at similar levels.[410] For example, mild hypothyroidism in an adult is not expected to have long-term effects on the brain. In contrast, subtle hypothyroidism during fetal and neonatal life causes disruption of neurotransmitters, neurotropins, axonal growth, and normal mitochondrial function in the developing brain, resulting in retarded cognitive and neuromotor development.[411]

Potential health implications

Reported abnormalities in laboratory animals and wildlife exposed to endocrine-disrupting chemicals include feminization of males, abnormal sexual behaviour, birth defects, altered sex ratios, decreased sperm density, decreased size of testes, breast cancer, testicular cancer, reproductive failure and thyroid dysfunction[412,413] (Table 4.15).

Epidemiologic studies involving workers have found associations between exposure to specific pesticides or industrial chemicals and levels of TSH, testosterone, and prolactin in adults.[425–427] Some of these studies have also found significant associations with other relevant end points, including diminished sperm quality, impaired sexual function and testicular cancer.[428,429] Numerous studies have found associations between occupational exposure to solvents or pesticides and subfertility or adverse effects on offspring such as hypospadias or cryptorchidism, but it is unclear whether these effects are due to endocrine mechanisms[430,431] (Table 4.16).

Hormones and Neurobehavioral Effect

Prenatal or early postnatal exposure to certain environmental pollutants has been associated with learning and behavioural abnormalities. In some cases, there is evidence that these neurologic abnormalities may be due to an endocrine mechanism. For example, a single low dose of dioxin during the development of the hypothalamic-pituitary-gonadal axis in the rat has been shown to produce a feminizing effect on the behaviour of male offspring, reflecting altered sexual differentiation of the brain.[446]

TABLE 4.15
Examples of Mechanisms of Actions of Endocrine-Disrupting Chemicals

Chemical/Reference	Use	Health Effect	Mechanism
DES[414]	Synthetic estrogen	Humans (prenatal exposure; vaginal cancer, reproductive tract abnormalities (females); cryptorchidism, hypospadias, semen abnormalities (males)	Estrogen receptor agonist
Methoxychlor[408, 413]	Insecticide	Rodents: accelerated puberty, abnormal ovarian cycling (females); aggressive behavior (males) receptor agonistovarian cycling (females); aggressive behaviour (males)	Metabolite is an estrogen
DDT[416]	Insectidice	Rodents (males): delayed puberty, reduced sex accessory gland size, altered sex differentiationaltered sex differentiation	Metabolite(DDE)is an androgen receptor antagonist
Vinclozolin[417]	Fungicide	Rodents (males): feminization, nipple development, hypospadias	Androgen receptor antagonist
PCBs[418,419]	No longer manufactured; still in electrical transformers, capacitors, toxic waste sites, food chain	Humans (in utero exposure): delayed neurological development; IQ deficits	Accelerated T4 metabolism, decreased T4 levels, elevated TSH levels (high doses: thyromimetic)
Atrazine[420–424]	Herbicide	Rodents (females) mammary tumors abnormal ovarian cycling. Humans: some evidence of breast and ovarian tumors	Reduces gonadotropin-releasing hormone from hypothalamus, reduces pituitary LH levels, interferes with metabolism of estradiol, blocks estrogen receptor binding
Dioxin[414]	By-product of industrial processes including waste incineration; food contaminant	Rodents (in utero exposure): delayed puberty, increased susceptibility to mammary cancer (females); decreased testosterone, hypospadias, hypospermia, delayed testicular descent, feminized sexual behaviour (males). Humans: decreased T3 and T4 levels, decreased testosterone levels,* cancer*	Aryl hydrocarbon receptor agonist; increases estrogen metabolism, decreases estrogen-mediated transcription, decreases gene estrogen levels, decreases testosterone levels by interfering with HPG axis

Source: EHC-Dallas, 2002; Modified from Solomon, G. M. and Schettler, T., *CMAJ* 163(11).

Note: DES = diethylstilbestrol, DDT = dichlorodiphenyltrichloroethane, PCBs = polychlorinated biphenyls, T4 = thyroxine, TSH = thyroid stimulating hormone, IQ = intelligence quotient, LH = luteinizing hormone, HPG axis = hypothalamic–pituitary–gonadal axis, T3 = triiodothyronine.

Exposures in adults.

Thyroid hormone is known to affect development of the fetal brain.[411] Thyroid disruption, including goiter and neurobehavioural abnormalities, has been found in wildlife and laboratory animal populations feeding on the organochlorine-contaminated food chain of the Great Lakes.[447–449] In utero and lactational exposure of nonhuman primates to environmentally relevant levels of PCBs has been shown to cause impaired learning.[450] In humans, higher levels of PCBs in breast milk have correlated

TABLE 4.16
Trends in Human Health Effects Potentially Related to Endocrine Function

End Point/Reference	Degree of Change	Region	Trend
Hypospadias[434]	4.3% per year	Canada	Increasing incidence
	3.3% per year	US	
Cryptorchidism[434]	3.5% per year	Canada	Increasing incidence
	1.6% per year	US	
Sperm count[435,436]	–0.7%/mL per year*	Canada	Decreasing
	–3%/mL per year	US	
	–5.3%/mL per year	Europe	
Testicular cancer[437–439]	2.1% per year	Canada	Increasing incidence
	2.3% per year	US	
	2.3–5.2% per year†	Europe	
Prosatate cancer[440,441]	3% per year	Canada	Increasing incidence‡
	5.3% per year	US	
Breast cancer[442,443]	3.3% per year	Saskatchewan	Increasing incidence
	1.9% per year	US	
Sex ratio[444]	–1.0 males/10,000 per year	Canada	Shift toward females
	–0.5 males/10,000 per year	US	
Age at breast Development[445]	11.2–9.96 years in white population	US	Shifting earlier

Source: EHC-Dallas, 2002; Modified from Solomon, G. M. and Schettler, T., *CMAJ* 163(11).

* This trend disappears when data from before 1984 are included.

† Range is dependent on country, with Sweden at the lower and the former East Germany at the upper end of the range.

‡ International trends in prostate cancer are complicated by the introduction of the prostate specific antigen screen, but prostate cancer mortality also increased (by about 1% per year through 1995 in the US and Canada), implying that improved diagnosis may not fully explain the rising incidence trends.

---

with higher TSH levels in nursing infants.[451] Blood levels of certain PCBs have positively correlated with TSH levels and negatively correlated with free T4 levels in children aged 7–10.[452] In addition to the antithyroid effects of PCBs, animal studies have revealed evidence of altered neurotransmitter and neuroreceptor levels, which may be primary or secondary to the thyroid effects.[453]

Cohort studies involving children environmentally exposed to PCBs in utero through maternal consumption of Great Lakes fish have revealed delayed psychomotor development and increased distractibility in those most highly exposed.[454,455] In one study, at age 11, the most exposed children were more than three times as likely to perform poorly on intelligence quotient (IQ) tests and more than twice as likely to be at least two years behind in reading comprehension as the least exposed children in the study.[454] **Some entire population groups, such as the Inuit in Canada, currently have body burdens of PCBs that exceed levels known to affect cognitive functioning.**[456]

Beyond Endocrine Disruption

Although much attention has focused on the endocrine system, disruption of other biological signalling pathways is an important related issue. The structural and functional development of the brain is dependent on the integration of hormones, neurotransmitters, neurotropins and locally produced steroids.[363]

Chemicals that interfere with neurotransmitters, such as the organophosphate pesticides, have many similarities to chemicals that interfere with hormones. Toxic effects are generally reversible in adults. In the developing brain, however, effects may be permanent and result in functional deficits. For example, rodents exposed to a single low dose of an organophosphate pesticide in a critical period of neonatal life have been found to have permanently decreased brain density of muscarinic receptors and hyperactive behaviour when tested as adults.[458] Recent research has demonstrated that, in the developing brain, neurotransmitters perform growth regulatory and morphogenic functions.[459] For example, inhibition of acetylcholinesterase results in reduced axonal outgrowth and accumulation of neurofilaments in vitro.[460] It appears that immature neurologic systems, like immature endocrine systems, are sensitive to low doses of exogenous agents that have no apparent effect on adults.

Implications and Ongoing Activities

Hormones act at extremely low levels (parts per trillion); therefore, in theory, even exposures to low levels of hormonally active agents may be of concern, particularly during sensitive periods of fetal development. Furthermore, endocrine-mediated effects may be subtle and manifest primarily in populations rather than in individuals. For example, slight overall declines in sperm density or IQ may have little relevance for individual but important adverse implications for the population.[461]

Low-level exposures to endocrine-disrupting chemicals are ubiquitous in today's environment. Persistent chemicals such as DDT, PCBs and dioxins are detectable in nearly 100% of human blood samples, and even some of the shorter-lived potential endocrine disruptors are frequently detected in general population surveys of residues in blood or urine.[462,463] The ubiquitous nature of the exposures combined with the nontrivial potential health effects justifies further research, education, and preventive action to reduce human exposures to endocrine disruptors.

Humans can be exposed to estrogens or estrogen mimics via inhalation, ingestion, or injection, although few concrete facts exist that enable the impact of each of these routes to be evaluated. Variations in diet may have the greatest impact in human absorption of estrogen, which may alter the dynamics of homeostasis. It has been argued that the low incidence of breast cancer in Japan and China is related to their high-fiber, low-fat diet,[464] rich in soy proteins. The key argument is that endogenous estrogens, which are implicated in the etiology of breast cancer and excreted via the bile, are more readily metabolized and reabsorbed from the gut when the gut contains low amounts of fiber. This so-called enterohepatic recirculation of estrogens means effectively that a **woman on a low-fiber diet would be exposed to more of her endogenous estrogen than a woman on a high-fiber diet**. The relative consumption of fats (especially animal fats), proteins, and refined carbohydrates can also substantially affect estrogen excretion and metabolism. The overall effect of eating a modern western diet is certain to increase exposure to endogenous estrogens,[464] which may result in altered dynamics of homeostasis. Whether such a diet in a pregnant woman could expose the male fetus to sufficient estrogen to induce reproductive-tract abnormalities is a matter for speculation. However, Evidence that links occurrence of cryptorchidism and testicular cancer to endogenous estrogen concentrations in the mother,[402] particularly in first pregnancies when concentrations of bioavailable estrogen can be high.[465] Other studies indicate that breast cancer in the mother is a significant risk factor for testicular cancer in male offspring.[466]

EXOGENOUS ESTROGENS

Synthetic estrogens pose a hazard to humans because they alter homeostasis by the negative feedback and less the 100% gain principle. Many are manufactured to be orally active and resistant to degradation. DES and other synthetic estrogens were used widely in the livestock industry for 20 to 30 years, and, for at least the first 20 years of their use, it was not recognized that they might pose a risk to humans.[467] In 1970, as a result of abnormalities in children born to DES-treated women,[400] procedural changes in the use of DES were introduced to minimize the risk to humans.[468] It remains a matter of conjecture as to how stringently these procedures were adhered to and what the level of human exposure to anabolic estrogens was between the 1950s and 1970s. Orally active anabolic estrogens were banned in Europe in 1981, and many of the anabolic estrogens used now in the livestock industry are not orally active.[481] However, one sees an inordinate amount of potbellies and other signs of excess estrogen and lack of testosterone in males over 40 years old in the U.S. **The increasing evidence of obesity in the American female poplulation appears to be directly related to the hormone-contaminated food supply.**

The other use of synthetic estrogens that has increased in the past 20–40 years is in the oral contraceptive pill (e.g., ethinyl estradiol). There are reports that ethinyl estradiol is detectable in water sources, but there is little data on concentration levels of this hormone in drinking water. However, like DES,[469] ethinyl estradiol does not bind to sex-hormone-binding globulin (SHBG) in the liver unlike most estrogen in the blood is normally bound to SHBG. This means that ethinyl estradiol has a very high-potency if ingested. The negative feedback and <100% gain characteristic is, again, applied here. Interaction is altered and dyshomeostasis occurs, resulting in the metabolic tissue changes.

Many chlorinated pesticides have now been found to have feminizing effects. DDT appears to damage DNA and suppresses sperm counts. **Most estrogenic compounds have more potent 1/50 to 1/10,000th than those of DES or the natural estrogen 17B-estradiol**. Kepone and Mirex, chlorinated pesticides, were also found to give estrogen effects and suppress the sperm count. See Tables 4. 17 and 4.18.[470,471]

Bisphenol-A, a plasticizer in polycarbonate containers and bottles and also a dental processor, and octyphenol are also synthetic estrogen disrupters. Many environmental estrogens bind estrogen receptors. Yet, a comparison of the structures of estradiol and those of environmental estrogens show how dissimilar they are.[376-384] See Figure 4.18.

As stated in the section on endometriosis, dioxin is a synthetic estrogen as are some of the plasticizers such as phthalates.

Phytoestrogens

Changes in diet in many industrialized countries over the past 40 years may have increased the amount of estrogens in the environment because many plants and fungi contain so-called

TABLE 4.17
Toxic Estrogens or Mimics

Chemical	β-Gal EC$_{50}$(μM)
17 β-Estradiol	0.0001
Endosulfan	< 33
Dieldrin	< 33
Toxaphene	< 33
Chlordane	ND*
Endosulfan + deildrin	0.092
Endosulfan + toxaphene	0.121
Endosulfan + chlordane	0.189
Dieldrin + toxaphene	0.210
Dieldrin + chlordane	0.286
Toxaphene + chlordane	0.306

Source: EHC-Dallas, 2002.

TABLE 4.18
Toxic Estrogens or Mimics

Chemical	h-ER Binding IC_{50} (μM)
Endosulfan	< 50
Dieldrin	< 50
Toxaphene	< 50
Chlordane	ND*
17 β-Estradiol	0.001
Endosulfan + deildrin	0.324
Endosulfan + toxaphene	0.339
Endosulfan + chlordane	0.363
Dieldrin + toxaphene	0.198
Dieldrin + chlordane	0.514
Toxaphene + chlordane	0.533

Source: EHC-Dallas, 2002.

* The IC_{50} value was not determined because chlordane did not appear to demonstrate competitive binding activity at any concentration tested. It has been reported that the IC_{50} values for endosulfane and tosaphene are 631 and 470 μM, respectively.

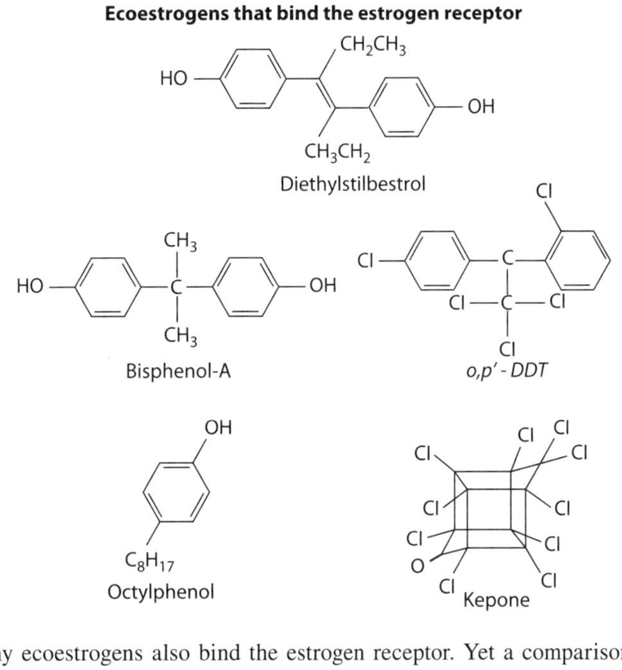

Ecoestrogens that bind the estrogen receptor

Diethylstilbestrol

Bisphenol-A

o,p'-DDT

Octylphenol

Kepone

FIGURE 4.18 Many ecoestrogens also bind the estrogen receptor. Yet a comparison of the structures of estradiol and those of the ecoestrogens shows how physically dissimilar they are. It is difficult to understand how so many differently shared structural keys can fit the same lock. Standard receptor-binding assays, therefore, may not always give accurate information about the potential strength of an environmental estrogen. (EHC-Dallas, 2002.)

phytoestrogens.[475,478] In animals it is well recognized that ingestion of such plants or fungi results in phytoestrogen-induced abnormalities in normal reproductive function.[475] Soya is the richest source of phytoestrogens[478] and its consumption, especially as a substitute for meat protein, has increased enormously in the past two or more decades. Attempts have been made to estimate the biological effects of phytoestrogens in humans,[487] and, at face value the conclusion is that exposure to phytoestrogens alone would probably be insufficient to induce major direct estrogenic effects in most people. Phytoestrogens may reduce exposure to endogenous estrogens by inhibiting production of SHBG by the liver and, thus, decreasing the ratio of bioavailable endogenous estrogen.[464]

Dietary fats differentially regulate serum cholesterol and both the amount and type of dietary fat are clinically important.[472–477] Formal recommendations on dietary fat consumption remain the principal means of reducing moderately elevated LDL cholesterol levels in the general population. Nevertheless, the mechanisms by which fatty foods modulate serum lipoproteins are not completely known, and optimization of dietary recommendations will require a more thorough understanding of those mechanisms. Triglyceride-derived fatty acids are generally accepted as the active agents regulating serum cholesterol. This review examines the evidence supporting this assumption and proposes a new working hypothesis, namely that the effect of dietary fats is due not only to fatty acids but also to trace levels of sterols and related compounds previously thought to be biologically inactive. The following two hypotheses are examined:

- *The fatty acid hypothesis*: **the ability of food fats to alter serum cholesterol is due to the structure of the fatty acids they contain**. Important elements of fatty acid structure include the presence, number, location, and *cis/tran* isomerism of double bonds and fatty acid chain length. This hypothesis is almost universally accepted because fatty acids are by far the most abundant component of food fats. Trans fatty acid seems detrimental to the system.
- *The sterol hypothesis*: **the ability of food fats to alter serum cholesterol is due in part to the small quantities of sterols and related compounds they contain**
- In animal fats, the sterol is cholesterol, whereas in vegetable fats there are several sterols collectively called phytosterols, which are sutructurally related to cholesterol. Sterols, oxidized sterols, and the cholesterol precursor, squalene, are the principal materials considered.

These hypotheses are not mutually exclusive, and it is likely that both contribute significantly to variance in lipoprotein levels.

The effect of dietary fats on serum cholesterol is widely assumed to be due solely to the fatty acids and cholesterol they contain. Phytosterols, sterol oxidation products, and sterol precursors such as squalene, however, are often present in dietary fats. Little is known of the physiology of these substances in natural foods, and most published diet studies do not consider them at all. **Supplementaion of the diet with high-dose phytosterols is now recommended for prevention of heart disease, but both recent and old data strongly suggest that the lower levels of** phytosterols naturally present in vegetable fats may also reduce cholesterol absorption and serum cholesterol substantially. Moreover, unmeasured phytosterols may confound otherwise well-controlled diet studies because there is an inverse correlation between phytosterol and saturated fatty acid content of vegetable fats. Sterol oxidation products, many of which are found in foods, are potent regulators of lipoprotein and cholesterol transport pathways in vitro. Squalene is a phytosterol precursor abundant in olive oil that is at least partly absorbed and then quantitatively converted to cholesterol. The effects of dietary triglyceride-derived fatty acids have not been experimentally separated from the effects of trace fat components in most clinical studies. A better understanding of the activity of sterol-related dietary components is needed to reduce variability in diet studies, to accurately

TABLE 4.19
Summary of Dietary Fat Trials (Unsaturated Versus Saturated) in Which Phytosterol Content Was Controlled

Dietary PS mg/day	Unsaturated Fat Type	Serum Cholesterol mg/dL	Dietary PS mg/day	Saturated Fat Type	Cholesterol Mg/dL	Reduction %	Ref Below
0	Trilinolein	286	0	Palmitic/Oleic	340	15.9	1
0	Corn Oil	122	0	Coconut Oil	165	26.1	1
0	Corn Oil	62	0	Coconut Oil	615	24.9	1
345	Corn Oil	47	398	Cocoa Butter	184	21.1	2
345	Corn Oil	196	398	Cocoa Butter	234	16.2	2
345	Corn Oil	162	398	Cocoa Butter	218	25.7	2
345	Corn Oil	140	398	Cocoa Butter	195	28.2	2
345	Corn Oil	223	398	Cocoa Butter	260	14.2	2
345	Corn Oil	194	398	Cocoa Butter	251	22.7	2
41	Corn Oil	381	47	Butter Oil	485	21.4	3
41	Corn Oil	293	47	Butter Oil	421	30.4	3
41	Corn Oil	340	47	Butter Oil	450	24.4	3
41	Corn Oil	256	47	Butter Oil	321	20.2	3
41	Corn Oil	176	47	Butter Oil	242	27.3	3
41	Corn Oil	176	47	Butter Oil	291	39.5	3
70	Safflower Oil	255	62	Butter Oil	320	20.3	3
42	Safflower Oil	184	36	Butter Oil	289	36.3	3
39	Corn Oil	496	34	Butter Oil	549	14.8	3
39	Corn Oil	450	34	Butter Oil	460	2.2	3
71	Safflower Oil	139	57	Butter Oil	233	40.3	3
364	Safflower Oil	178	346	Lard	210	15.2	4
403	Safflower Oil	352	362	Lard	418	15.8	4
Mean		**254 ± 24**			**325 ± 27**	**22.8 ± 1.9**	

Source: Modified from Ostlund, R.E. Jr., Racette, S.B., and Stenson, W.F., Ann. Rev. Nutr., 478, 2002.

Each line represents a single subject with the left three columns depicting a diet period with unsaturated fat and the next three columns a period with saturated fat.

PS = measured dietary phytosterols, reduction = percent reduction of serum cholesterol with unsaturated fat compared with saturated fat, ref = reference work below. The mean values for all 22 subjects ± SEM are at the bottom of the table.

assess the effects of dietary fatty acids and to maximize the effectiveness of dietary treatment for hypercholesterolemia. See Table 4.19.[478]

Estrogens in Milk

It is widely accepted that in developed countries too great a consumption of dairy products occurs, which is a trend that probably started in the 1940s and 1950s. This consumption may lead to changes in estrogen metabolism due to excess intake of estrogens from the dairy product, as described above, but may also have other implications. In modern farming practice, most dairy cows are pregnant; however, unlike women, they continue to lactate during pregnancy. **Cows' milk therefore contains substantial amounts of estrogens (mainly estrone sulfate)**.[481] Fortunately, estrogen in pasteurized milk is "lost" during formulation into baby-milk powder,[481] although it is unclear why this occurs or whether the estrogen might appear in other dairy products. It is uncertain to what extent estrone sulfate in cows' milk is absorbed from the gut of an adult or child or whether dietary constituents such as fiber might alter absorption.

ESTROGENIC CHEMICALS

Many of the chemicals with which we have contaminated our environment in the past 50 years are weakly estrogenic.[479,480] These chemicals are remarkably resistant to biodegradation, are present in our food-chain, and accumulate in our bodies,[481] allowing for an increase in total body load which will alter the dynamics of homeostasis. In various animals, high-concentrations of these chemicals have been associated with reproductive abnormalities,[481] which include changes in the semen quality of adult rats exposed neonatally to polychlorinated biphenyls via their mother's milk.[482] Even more disturbing is evidence that a single maternal exposure of rats (on day 15 of pregnancy) to nannogram amounts of the chlorinated hydrocarbon TCDD (dioxin) has no effect on the mother, but increases the frequency of cryptorchidism in male offspring and causes dose-dependent reductions in their testicular weight and sperm count in adulthood.[483] The latter effects appear to be a consequence of decreased Sertoli-cell numbers. It is not known, whether exposure of humans to these chemicals is sufficient to induce "estrogenic effects" directly, or via any of the mechanisms described above, but similar effects are widespread in wild animals.[480]

Changes in human exposure to estrogens are difficult to quantify, especially when the suspected alterations are in the metabolism and bioavailability of endogenous estrogens during pregnancy. The most reasonable and safest assumption is that pregnant women and **humans in general are exposed to more, rather than less estrogens than was the case 50 years ago**. The extent of this increase, its source(s), and its consequences are likely to differ among countries and among individuals if the routes of exposure listed in Table 4.19 are all valid. Whether increased human exposure to estrogens could account for the increased incidence of abnormalities in male reproductive development and function is unknown, but "weak estrogens" may be more potent in the fetus and neonate than in the adult.[480] Certainly an increase in the total body load would place a strain upon, if not alter, the dynamics of homeostasis. Moreover, many (and perhaps all) of the reproductive-tract abnormalities can be brought about by estrogen-induced changes during fetal development, thus providing strong circumstantial support for a role for environmental and dietary estrogens in the development of these abnormalities.[484]

SERTOLI CELL NUMBER AND SPERM OUTPUT

Although there are physiological pathways via which estrogens can impair development of the male reproductive tract, how can impairment of development be linked to falling sperm counts in adult men? The answer, according to studies in animals, is by altering homeostasis while changing the rates of multiplication of Sertoli cells. Sertoli-cell multiplication occurs during fetal, neonatal, and prepubertal life and is controlled to a large extent by FSH.[485,486] Inhibition of FSH secretion reduces Sertoli-cell multiplication,[485] **and in neonates FSH secretion is exquisitely sensitive to inhibition by exogenously administered estrogens**,[485] including estrogen mimics. At a fixed time in postnatal life, Sertoli-cell multiplication ceases, coincident with "maturation" of these cells.[485,487] Importantly, this maturation also coincides with dramatic decreases in the secretion of estrogens and MIS (mullerian).[485,487] The "fixing" of Sertoli-cell number during prepubertal development has important consequences in adult life because Sertoli cells are responsible for orchestrating and regulating spermatogenesis. Each Sertoli cell can only support a fixed number of germ cells during development into spermatozoa.[488,489] Therefore, the lower the number of Sertoli cells the lower will be the "ceiling" for sperm output. Studies in animals have all shown that alteration of Sertoli-cell number (up[490] or down[489]) in early life determines testicular size and sperm output in adulthood. In such studies the quality of the spermatozoa is not affected, just the quantity. Of particular significance is the fact that estrogen administration to animals in fetal and early neonatal life results in smaller testes and reduced sperm counts in adult life.[485] Moreover, the sons of women exposed to DES during pregnancy show an increased incidence of low sperm counts,[407] consistent with what would be predicted from animal data.[485]

Semen quality data collected systematically from reports published worldwide indicate that sperm density has declined appreciably between 1938 and 1990, although it cannot be concluded whether or not this decline is continuing. Concomitantly, the incidences of some genitourinary abnormalities, including testicular cancer and possibly maldescence and hypospadias, have increased. Such remarkable changes in semen quality and the occurrence of genitourinary abnormalities over a relatively short period is probably due more to environmental rather than genetic factors.[485]

Effects of pollutants reported in men were impotency and sterility.[491] Impotency occurs after chronic exposure to nitrous oxide anesthetics. Since the 1970s, individuals have linked adverse reproductive effects with anesthetic gases and dibromochloropropane (DBCP).[492] Chemical agents can interfere with the process by which testicular products are formed and, therefore, interfere with functions of a variety of processes throughout the body.[491] Additional complaints associated with the genitourinary system include back pain secondary to broad ligament swelling, vaginal discharge, vaginitis, urinary frequency, and urgency. According to Hunt,[491] the reproductive system in men and women has an important characteristic. Cyclic cellular growth and differentiation occurs in each of the tissues, which provides a special susceptibility to toxic effects and may affect the capacity of individuals to develop mature ova and sperm, to provide a suitable environment for the fertilized ovum and to allow for the normal growth and development of the embryo and fetus.

Rosenblum and Rosenblum[493] reported methyltestosterone is effective in preventing episodes of hereditary angioneurotoxic edema (for further information, see W.J. Rea, *Chemical Sensitivity*, Volume III, Chapter 28). At the EHC-Dallas, we have identified many men and women with chlorinated pesticides in their blood (see W.J. Rea, *Chemical Sensitivity*, Volume II, Chapter 13[231] and Chapter 28). Some of these chemicals have appeared to cause congenital abnormalities, especially of the metabolic type. Many have sexual dysfunction relating to their chemical sensitivity. Diagnosis and treatment are similar to that used with any patient with chemical sensitivity.

THYROID

In the complex patient with dyshomeostasis who has multiple sensitivities to food and chemicals or chronic disease with endocrine involvement, the thyroid gland, has to be considered. A correlation between Hashimoto's lymphocytic thyroiditis and allergic or autoimmune disease has been identified.[494,495] **Evidently, some chemically sensitive patients, especially women, have an autoimmune endocrinopathy with associated ovarian, thyroid, and adrenal involvement**. Periods of high-hormone stress such as puberty, childbirth, and menopause partially account for this endocrinopathy, but evidence for environmental chemical triggers, which alter homeostasis, has also been accumulating in recent years. For example, Bahn et al.[496] noted an increased prevalence of thyroid antibodies among workers exposed to polybrominated biphenyls (PBBs). Gaitan et al.[496] have shown increased incidences of thyroiditis and hypoactive thyroid in people exposed to water drawn from oil shale areas. He has traced this incidence across both the North and South American Continents. Bastomsky has induced thyroid disease in rats treated with polyhalogenated aromatic hydrocarbons and substances such as PCBs and PBBs.[498] Saiffer[499] has reported thyroid dysfunction in many chemically sensitive patients in the San Franciso area. At the EHC-Dallas, we have also seen numerous cases of thyroiditis. In fact, 25% of the chemically sensitive and chronically diseased patients we see at the EHC-Dallas have thyroiditis.

Agents known to have goitrogenic and/or antithyroid effects on the thyroid of humans and other animal species are widespread and have often been found in the chemically sensitive and chronic degenerative diseased individual. These agents are listed in Table 4.20.

Physiology and Pathophysiology

Environmental compounds may cause goiter not only by acting directly on the thyroid gland causing altered dynamics of homeostasis but also by acting indirectly, altering the thyroid's homeostatic regulatory character of negative feedback and < 100% gain, the excretion of thyroid hormones,

TABLE 4.20
Environmental Agents Producing Goitrogenic and/or Antithyroid Effects

	Goitrogenic/Antithyroid Effects		
	In Vivo		In Vitro
	Humans	Animals	Systems
Sulfurated organics			
Thiocyanate	+[a]	+	+
Isothiocyanates	NT[b]	+	+
Thioglycosides (goitrin)	+	+	+
Aliphatic disulfides	NT	+	NT
Polyphenols			
Bioflavonoids	NT	+	+
Phenolic and phenolic-carboxylic derivatives			
Resorcinol (1,3-dihydroxybenzene)	+	+	+
Orcinal (5-mehtylresorcinal)	NT	+	+
2-Methylresorcinal	NT	+	+
4-Chlororesorcinal	NT	+	+
Phloroglucinal	NT	+	+
Pryogallol	NT	+	+
3,4-Dihydroxbenzoic acid	NT	NT	+
3,5-Dihydroxybenzoic acid	NT	NT	+
3-Chloro-4-hydroxybenzoic acid	NT	NT	+
2,4-Dinitrophenol	NT	+	O[c]
Phthalate esters and phthalic acid derivatives			
Dihydroxybenzoic acids	NT	NT	+
PCB's[d] and Pbbs[e]			
PCBs (Aroclor)	NT	+	NT
PBB oxides	+	+	NT
Other organochlorines			
p,p'-DDT[f], p,p'-DDE[g] and dieldrin	NT	+	NT
2,3,7,8-Tetrachlorodibenzo–p-dioxin	NT	+	NT
Polycyclic aromatic hydrocarbons			
3,4-Benzapyrene	NT	+	NT
3-Methylcholanthrene	NT	+	NT
7,12-Dimethylbenzanthracene	NT	+	NT
Inorganics			
Excess iodine	+	+	+
Lithium	+	+	+

Source: Gaitan, B., *The Thyroid Gland: A Practical Clinical Treatise.* Ed. L. Van Middlesworth. Chicago: Year Book Medical Publishers, 263–280, 1986.

[a] Active
[b] Nontested.
[c] Inactive.
[d] Polychlorinated biphenyls
[e] Polybrominated biphenyls
[f] Dichlorodiphenyltrichloroethane
[g] Dichlorodiphenyldichloroethylene

Endocrine System

and/or the peripheral metabolism. Price et al.[500] have shown that if rats are challenged with PCBs or other chlorinated compounds, an increase in the response of the cytochrome detoxification mechanism occurs in the liver. Thus, these compounds moderately accelerate removal of thyroid hormone from the blood, which initiates the positive feedback characteristic with a vicious downward cycle. The increase in metabolism of these toxics removes thyroid hormone from the blood, causes thyroid production to increase, and then eventually depletion occurs as the thyroid gland wears out.

AGENTS ACTING DIRECTLY ON THE THYROID

Gaitan[501] has reported the various environmental goitrogenic and antithyroid compounds and their sites of action in the thyroid gland. These are shown in Figure 4.19. Goitrogens act in the thyroid gland to interfere with the dynamics of homeostasis and the process of hormonal synthesis yet the mechanism that induces the trophic changes leading to goiter formation is not yet well understood, but it involves altered homeostasis. Antithyroid compounds can be divided into three categories according to the way they act on iodine metabolism in the thyroid.

Class I
The thiocyanates or thiocyanate-like compounds appear primarily to inhibit the active concentrating mechanism of iodine, and their goitrogenic activity can be overcome by iodine administration. These ions have a molecular volume and charge similar to those of iodine. This fact is the reason that they compete with iodine for transport in the follicular cell. **Cyanogenic glycosides in a variety of staple foods such as cassava, almonds, apples, walnuts, and sorghum may exert a goitrogenic effect when converted to thiocyanate in the living organism.**[503–505]

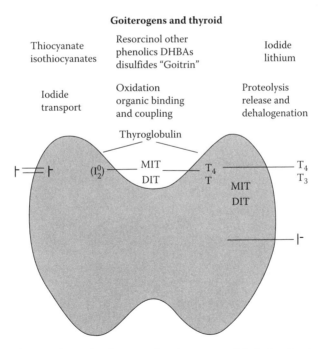

FIGURE 4.19 Naturally occurring and anthropogenic goiterogens and their site of action in the thyroid gland. DHBAs = dihydrobenzoic acids; I^- = iodide; MIT = monoiodotyrosine; DIT = diiodotyrosine; T_4 = thyroxine; T_3 = triiodothyronine. (Modified from Rea, W. J., *Chemical Sensitivity, Vol III: Clinical Manifestations of Pollutant Overload,* CRC Press, Boca Raton, FL, 1657, Fig. 24.10, 1996; Courtesy of Gaitan.)

Many chemically sensitive individuals react to these foods upon oral challenge, and some of these foods have been seen to exacerbate thyroiditis in these individuals.

The isothiocyanates act on the thyroid mainly by rapidly converting to thiocyanate. The naturally occurring butyl, allyl, and methyl isothiocyanates do not inhibit the thyroidal peroxidase enzyme. However, isothiocyanates not only use the thiocyanate-metabolic pathway, but also react spontaneously with amino groups, forming disubstituted thiourea derivatives, which produce a thiourea-like antithyroid effect. Additive antithyroidal effects of thiocyanate, isothiocyanate, and the thioglycoside, "goitrin," occur with combinations of these naturally occurring goitrogens.

Class II

According to Gaitan,[501] the thiourea or thionamide-like goitrogens interfere with the processes of organification of iodine to form the active thyroid hormones, and their action usually cannot be antagonized by iodine. These goitrogens prevent the oxidation of iodine by thyroid peroxidase and impair covalent binding of iodine to thyroglobulin. In small amounts, they inhibit the formation of active thyroid hormones from iodotyrosine precursors. In larger quantities, they impair formation of monoiodotyrosine (MIT) and diiodotyrosine (DIT). These types of compounds do not prevent transport of iodine into the thyroid gland, and pharmacologic doses would be required to suppress iodine uptake. The naturally occurring thioglycoside L-5-vinyl-2-thioxazolidone, "goitrin," is a main representative in this category. Goitrin is unique in that it does not degrade like other thioglycosides. Goitrin acts *in vitro* on the thyroidal peroxidase (Table 4.21), on I_{50},[226] on organification,[506] and *in vivo* it exerts a thionamide-like effect.[503]

Class III

Naturally occurring bioflavonoids are polyhydroxylated C_6–C_3–C_6 structures. These polyphenols possess intrinsic antithyroid activity. Except for the glycoside hesperidin, its aglycon **hesperidin, catechin, and phytoretin are potent inhibitors of the thyroid peroxidase enzyme**.

Class IV

The antithyroid activity appears to be exerted, as in the case of resorcinol, by the presence of two free metapositioned OH-groups in the benzene ring of the benzypyran unit.

According to Gaitan,[501] resorcinol and its phenolic and phenolic-carboxylic dihydrobenzoic acids (DHBAs) parent compounds,[507,508] and the aliphatic disulfides[398,507] also act on this step of thyroid hormone synthesis. The goitrogenic and antithyroid effects of resorcinol are enhanced when its conjugation with glucuronic acid is affected, which often occurs in the chemically sensitive population. These phenolic compounds apparently can cause or exacerbate chemical sensitivity. Throughout this book, numerous phenolic challenges have been shown to trigger problems in the chemically sensitive individual. Phenols are degraded by the conjunction of the large sugar molecule, glucaric acid. As shown previously, large sugar molecules like glucuronic acid are part of the CT matrix and GRS as well as the intracellular endoplasmic reticulum. Therefore, these phenoloic substances, if chronically present can easily overwhelm the body's homeostatic mechanism.

Class V

Agents in this group interfere with the processes of proteolysis and release of thyroid hormones. The most important representative of this group is iodine.[509–511] An excess intake of iodine, arbitrarily defined as 2 mg or more per day, inhibits the synthesis and release of thyroidal hormones and eventually produces "iodine goiter" and hypothyroidism. Lithium has also been shown to belong to this category.[512–514] **Brominated, fluorinated, and chlorinated compounds also can give a pseudohalogen effect and trigger hypothyroidism** (see *Chemical Sensitivity*, Volume I, Chapter 4).[1] This pseudohalogen effect seems to be relevant in many chemically sensitive patients, who have a tendency toward thyroid dysfunction.

TABLE 4.21
Effect of the Concentration of Naturally Occurring Substances Producing I_{50}[a] on Thyroid Peroxidase[b]: Comparison with the Antithyroid Drugs Propylthiouracil and Methimazole

Compound	I_{50} (mmoles/mL)	Potency (Propylthiouracil/Inhibitor)
6-Propylthiouracil	7.2	1.0
Methimazole		
L-5-vinyl-2-thiooxazolidone	4.2	1.7
"goitrin"	4.0	1.8
Isothiocyanates		
Butyl	NI[c](1×10^2)	0
Allyl	NC(1×10^2)	0
Methyl	NC(1×10^2)	0
Bioflavonoids		
Catechin	0.72	10.0
Phloretin	0.54	13.3
Hesperitin	3.9	1.8
Hesperidin	61.9	0.1
Phenols		
Resorcinol	0.3	26.7
Phloroglucinal	0.2	37.9
Pyrogallol	3.8	1.9
Orcinol	1.0	7.2
3,4-Dihydroxybenzoic acid	7.0	1.0
3,5-Dihydroxybenzoic acid	2.0	3.6
Phthalates		
Diiobutyl phthalate	NC(1×10^4)	0
Dioctyl phthalate	NC(1×10^4)	0
o-Phthalic acid	NC(1×10^4)	0
m-Phthalic acid	NC(1×10^2)	0

Source: Gaitan, B., *The Thyroid Gland: A Practical Clinical Treatise.* Ed. L. Van Middlesworth. Chicago: Year Book Medical Publishers, 263–280, 1986.

[a] 50% inhibition.
[b] A glucose-glucose oxidase system was used for H2O2 generation.
[c] No inhibition

AGENTS ACTING INDIRECTLY ON THE THYROID

According to Gaitan,[515] the antithyroid effect of 2,4-dinitrophenol (DNP) is complex. First, it is due in part to an inhibition of the pituitary TSH mechanism.[516] In addition, DNP interferes with T_4 binding,[517–519] further decreasing serum T_4 concentration. Polychlorinated biphenyls exert a similar effect.[520,522] Finally, DNP accelerates the disappearance of T_4 from the circulation, exaggerating even more the reduction of serum T_4 concentration.[517] At the EHC-Dallas, over 2,000 chemically sensitive patients have been challenged with the ambient dose of 2,4-DNP (< 0.0034 ppm) and experienced the reproduction of their symptoms and signs.

Thyroid hormones are excreted into the intestine in both free and conjugated forms, along with small amounts of their deiodinated metabolites. **Glucuronide conjugation occurs mainly in the liver by the action of a UDP-glucuronyltransferase, and sulfate conjugation occurs mainly in the kidney by the action of a sulfate transferase**. However, under normal circumstances little T_4

and T_3 are excreted in conjugated form. As has been shown in many chapters in this book, especially on discussion of mechanisms, Chapter 1, gluconronidation is often disturbed in the chemically sensitive individual. This disturbance may account for the occurrence of thyroiditis and hypothyroidism in many chemically sensitive patients.

Potent hepatic microsomal enzyme inducers, PCBs have properties of both phenobarbital and the polycyclic hydrocarbon type of inducer. They greatly enhance the biliary excretion of circulating T_4 as T_4-glucuronide, which then is lost in stools, at least in the rat. This excretion is probably secondary to induction of hepatic microsomal T_4-uridine diphosphate-glucuronyl transferase. Enhanced peripheral metabolism and reduced binding of T_4 to serum proteins in PCB-treated animals result in markedly decreased serum T_4 concentrations, activation of the pituitary-thyrotropin-thyroid axis, and eventually in goiter formation.[522,520] Furthermore, PCB-treated animals exhibit decreased serum T_4, but unchanged T_3 levels. This observation may be explained by increased peripheral deiodination of T_4–T_3 and increased thyroidal T_3 secretion that results from the state of relative iodine deficiency induced by accelerated metabolism of T_4.

PBBs,[521] dioxin (TCDD),[522,523] and the polycyclic aromatic hydrocarbons (PAH), 3-methylcholanthrene (MCA) and 3,4-benzpyrene (BaP),[524-526] appear to act similarly to PCBs, but there is some indication that PBBs and MCA also interfere with homeostasis, and thus the process of hormonal synthesis in the thyroid gland.

The bioflavonoid phloretin not only interferes with homeostasis but also inhibits thyroidal peroxidase. In addition it has been shown to affect the peripheral metabolism of thyroid hormones, and phloretin polymers interact with TSH, preventing its action at the follicular thyroid cell,[527,528] indicating that this class of substances can alter thyroid hormone economy in a complex manner.

GENERAL PROPERTIES, DISTRIBUTION, AND EPIDEMIOLOGY

Environmental antithyroid and goitrogenic compounds that interrupt homeostasis are naturally occurring or anthropogenic. Gaitan[501] has shown that they can be present in foodstuffs, in contaminated water supplies, or in wastewater effluents, or they can be airborne or occur as waste products of industrial processes. Most of these contaminants cause homeostatic dysfunction in the chemically sensitive and chronic degenerative diseased individual.

Sulfurated Organics Thiocyanate (SCN), Isothiocyanates, and Thioglycosides (Goitrin)

Extensive reviews on sources, metabolic pathways, and action of cyanogenic glycosides, thioglycosides, isothiocyanates, and thiocyanates that interfere with thyroid homeostasis have been published.[1,501,504,505]

Thiocyanate (SCN) and isothiocyanates have been demonstrated as goitrogenic principles in *Cruciferae* (i.e., the food cabbage). The potent antithyroid compound "goitrin," a thioglycoside, was isolated from yellow turnips and from *Brassica* seeds. Cyanogenic glucosides (thiocyanate precursors) have also been found in several staple foods (cassava, maize, bamboo shoots, sweet potatoes, lima beans). After ingestion these glucosides can be readily converted to SCN, which can alter the dynamics of homeostasis by causing damage to a widespread tissue enzyme.

Isothiocyanates, as previously mentioned, not only use the thiocyanate-metabolic pathways, but also react spontaneously with amino groups forming disubstituted thiourea derivatives, which produce a thiourea-like antithyroid effect. Thus, the actual concentration of thiocyanates or isothiocyanates in a given foodstuff may not represent its true goitrogenic potential, nor does the absence of these compounds negate a possible antithyroid effect because inactive precursors can alter the dynamics of homeostasis when converted into goitrogenic agents either in the plant itself or in the animal after its ingestion.

Thioglycosides undergo a Lossen arrangement to form isothiocyanate derivatives and in some instances, thiocyanate. Therefore, **the amount of thiocyanate in the urine is a good indicator** of the presence of thioglycosides in food. It has been demonstrated that a mustard oil glucoside, glucobrassicin, yields SCN under the action of thioglucosidase, "myrosinase," an enzyme present in plants. Ingestion of pure progoitrin, a naturally occurring thioglycoside, disturbs the dynamics of homeostasis, eliciting antithyroid activity in rats and humans in the absence of myrosinase. The antithyroid activity of progoitrin is due to its partial conversion into the more potent goitrogen, 1,5-vinyl-2-thiooxazolidone or goitrin in the animal. This ability of plants and animals to readily convert inactive precursors into goitrogenic agents must be considered when investigating the possible etiologic role of dietary elements in sporadic or endemic goiter. Of course, this fact is probably one reason that the chemically sensitive and/or chronic degenerative diseased individual has trouble tolerating some foods. Local thyroid and homeostatic units become dysfunctional by the goitergens, resulting in informational change, metabolic change (metabolism), and then tissue change in the thyroid gland.

Anthropogenically, thiocyanate is found in high concentrations (1 g/L) in wastewater effluents of coal-conversion processes and in body fluids as a metabolite of hydrogen cyanide gas consumed while smoking.[507] Several goiter endemics have been attributed to the presence of these sulfurated organics in foodstuffs. Two goiter endemics have been ascribed to the presence of these goitrogenic substances in milk.[501] One case was in Tasmania, where a seasonal variation in goiter prevalence in school children was noted in spite of adequate iodine intake and in which an isothiocyanate, cheirolin, was suspected as the principal goitrogen. The other occurred in Finland, where goitrin present in cows' milk from the region of endemic goiter was considered the causative factor. Thiocyanate from a cyanogenic glucoside (linamarin) in cassava, a staple food, acting in the presence of extreme iodine deficiency, is thought to be the cause of endemic goiter and cretinism in central Africa.[529] Many chemically sensitive patients and/or chronic degenerative diseased individuals have transient thyroid pain and tenderness in the neck, which can occur from food and contaminant exposure similar to these reports of periodic goiter. The pain subsides after a day or two once the pollutant reaction has taken its course.

Sporadic goiter and hypothyroidism were also documented in patients on long-continued administration of thiocyanate for treatment of hypertension.[501,530]

Aliphatic Disulfides

The major volatile components of onions and garlic have been identified as small aliphatic disulfides that are known to alter thyroid homeostasis and cause marked antithyroid activity in rats.[501,507] Gaitan[501] has shown that organic disulfides have also been identified as water contaminants in the United States and in water supplying a Colombian district with endemic goiter.[503,531] The most frequently isolated compounds in the United States are dimethyl, diethyl, and diphenyl disulfides, but dimethyl trisulfide, dimethyl sulfoxide, and diphenylene sulfide have also been isolated. Organic sulfide pollutants are also present in high concentration in wastewater effluents of coal-treatment plants.[507] Many people eat a lot of garlic for treatment of various maladies such as chronic candida infections. Although we have not associated goiter with the ingestion of garlic, we recommend using caution in the chronic use.

Polyphenols

Bioflavonoids are C_6–C_3–C_6 aromatic phenolic compounds widely distributed in nature. They are important stable organic constituents of a wide variety of plants. Bioflavonoids in high concentrations are present in polymeric (tannins) and oligomeric (pigments) forms in various staple foods (i.e., millet, sorghum, beans, ground nuts, etc.) of the Third World.[505,532]

This type of polyphenol, which has been shown to be goitrogenic in rats, is known to alter the dynamics of homeostasis[505] Furthermore, according to Gaitan,[501] they are potential immediate precursors of potent phenolic antithyroid monomers. Actually, cyanidin, a naturally occurring compound used as the model subunit of flavonoid-type humic substances (HS), yields the antithyroidmonomers resorcinol, phloroglucinol, and orcinol by reductive degradation.[533] It is clear that polyphenols can make chronically ill patients worse as they tend to alter homeostasis. Many natural polyphenols have been touted as beneficial antioxidants. However, we have noticed that **most chemically sensitive patients do not tolerate them very well and usually develop a specific phenol dyshomeostatic reaction.**

Decaying organic matter (plants and animals) rich in these phenolic materials becomes the substrate of flavonoid-type HS during the process of fossilization. HS are, high molecular-weight complex polymeric compounds and are the principal organic components of soils and waters.[534,535] They are also present in coals, shales, and possibly other carbonaceous sedimentary rocks. Thus, **bioflavonoid structures may be the link between phenolic goitrogens in foodstuffs and those present in rocks, soils, and water.** Both types alter the dynamics of homeostasis and can trigger chemical sensitivity.

In western Sudan,[536] the prevalence of goiter is higher in remote rural villages, where the staple diet consists of only millet, rather than in small towns of the same area, where the diet includes a combination of millet, dura, and wheat. Rich in bioflavonoids,[532] **millet has been shown experimentally to disturb the dynamics of homeostasis and to be goitrogenic.**

Phenolic and Phenolic-Carboxylic Derivatives

As stated previously in this chapter, phenols will alter more general as well as local dynamics of thyroid homeostasis. Phenols can damage the thyroid in experimental studies and have been found to do so in the chemically sensitive and chronic degenerative diseased individual. These substances will be discussed in more detail here.

Resorcinol (1,3 Dihydroxybenzene)

The prototype of this group of compounds alters the dynamics of homeostasis by their antithyroid and goitrogenic properties both in humans and in experimental animals.[501,507] Resorcinol and other parent antithyroid phenolic and phenolic-carboxylic compounds,[501,507] phloroglucinol, pyrogallol, 5-methylresorcinol [orcinol], 3,4-and 3,5-DHBA and the ortho-[o] and meta-[m] phthalic acids) are monomeric byproducts of reduction, oxidation, and microbial degradation of HS,[471,533] which can alter the dynamics of homeostasis. At the heart of the "humification" process are the chemical and microbial-mediated production and polymerization of phenolic and carboxylic benzene-rings, which are known to alter both local and generalized dynamics of homeostasis of the thyroid and liver. Up to 70% of flavonoid HS may be made up of these subunits.[534,535] As much as 8% of shale bitumen is constituted by phenols.[507] Phenols are also the major organic pollutants in aqueous effluents from coal-conversion processes.[507] Resorcinol and other antithyroid phenolic pollutants comprise as much as 4 g/L in the aqueous effluent from coal liquefaction units.[537] These pollutants may enter community water supplies, constituting a potential environmental goitrogenic factor in humans and other animal species as well as a local and generalized disrupter of the dynamics of homeostasis. Coal-conversion waste-waters contain, in addition to phenolics, thiocyanate, and sulfides (S2-),[538,539] all of which possess antithyroid and goitrogenic properties and will potentially increase the total body load, straining or even altering the dynamics of orderly homeostasis.

Resorcinol is used industrially in the production of pharmaceuticals, dyes, plasticizers, textiles, resins, and adhesives for wood, plastics, and rubber products.[540]

Resorcinol has been identified as a water contaminant in the United States.[541] **Resorcinol and a substituted resorcinol have also been isolated from water supplies of endemic goiter districts in western Colombia**[546] **and the coal-rich Appalachian area of eastern Kentucky.**[501] As early as the 1950s, the goitrogenic effect of resorcinol was demonstrated when patients applying resorcinol

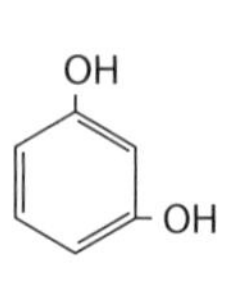

FIGURE 4.20 Resorcinol. (EHC-Dallas 2004)

Endocrine System

ointments for the treatment of varicose ulcers, developed goiter and hypothyroidism.[543,544] Several observations suggest that resorcinol crosses the human placenta and may cause both goiter and neonatal hypothyroidism.[545] The presence of halogenated organic compounds with known or potential harmful effects that alter the dynamics of homeostasis has awakened public health and environmental concerns. These compounds are produced by the chlorination of water supplies, sewage, and power plant cooling waters.[546–548] Present at µg/L concentrations (parts per billion) in treated domestic sewage and cooling waters, 4-chlororesorcinol and 3-chloro-4-hydroxybenzoic acid possess antithyroid activities. Whether these pollutants exert additive or synergistic antithyroid effects and/or act as "triggers" of autoimmune thyroiditis requires investigation, particularly **because more than 60 soluble chloro-organics have been identified in the primary and secondary effluents of typical domestic sewage treatment plants**. Clearly, these substances alter the thyroid homeostatic subunit causing many symptoms in the chemically sensitive and chronic degenerative diseased individual and apparently do influence the function and balance of their thyroid hormone.

2,4-DNP

Derivatives of DNP are widely used in agriculture and industry. An insecticide, herbicide, and fungicide,[507] **DNP is also used in the manufacturing of dyes, to preserve timber, and as an indicator; it is also a byproduct of ozonization of parathion**.

Administration of 2,4-DNP to human volunteers resulted in rapid and pronounced decline of circulating thyroid hormones.[549,550] The biological significance of this observation and the public health impact of this pollutant on the thyroid are still unknown. **We, at the EHC-Dallas, have challenged by inhalation over 2000 patients with 2,4-DNP, which has triggered a virtual textbook of medical symptoms and altered their homeostasis**. See Figure 4.21.

Phthalate Esters and Phthalic Acid Derivatives: DHBAs

Phthalates are ubiquitous in their distribution and have been frequently identified as water pollutants.[507] Most commonly, they result from **industrial pollution or artificial contamination, including plastic liquid and food containers**, but phthalates are also reported to occur naturally in shale, crude oil, petroleum, plants, and as fungal metabolites.[540,551] Although phthalates and phthalic acids do not possess intrinsic goitrogenic activity,[508] they undergo biodegradation by gram-

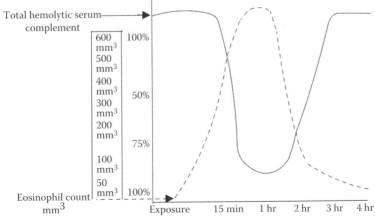

FIGURE 4.21 Symptoms and sign reproduced of rhinitis, sneezing, dizziness, and fatigue after double-blind inhaled challenge of 2,4-DNP after four days of deadaptation with the total pollutant load reduced. (Modified from Rea, W. J., *Chemical Sensitivity, Vol III: Clinical Manifestations of Pollutant Overload*. CRC Press, Boca Raton, FL, 1665, Fig. 24.11; Courtesy B. Chen at Robens Institute Surrey Guildford, England. 1998.)

negative bacteria with production of intermediate metabolites, such as DHBA,[507,552] known to alter homeostasis and possess antithyroid properties.[508] Thus, phthalates, with bacterial intermediates, may become a source of environmental goitrogenic compounds. These abundant pollutants exert deleterious effects on the thyroid of humans and other animal species as evidenced by Price's[500] studies on high-doses of phthalates and Chen's[500] study of low-dose exposure and in combination with PCB exposure (Figure 4.22). Electron microscopic study shows altered homeostasis with extremely enlarged endoplasmic reticulum and vacuolization of the mitochondria even in animals that were dosed with low-dose DEHP alone (Figure 4. 23).

PCBs and PBBs

PCBs and PBBs are aromatic compounds containing two benzene nuclei with two or more substituent chlorine or bromine atoms.[540] Evidence is mounting that shows dietary PCBs and PBBs are disrupters of homeostasis and can have deleterious effects on health. There is both growing concern and uncertainty about the long-range effects of bioaccumulation and contamination of our ecosystem with these chemicals. That uncertainty extends to the potentially harmful effect of these pollutants on the thyroid.[540] However, we have seen many nonthyroid effects with the dynamics of homeostatis altered in our chemically sensitive and chronic degenerative diseased patients and also in some of our patients with thyroiditis and thyroid tumors who have PCBs in their blood. Clearly, the PCBs not only alter local homeostatic functions but also affect the generalized integrated control function of the central homeostatic mechanism.

Bahn reported an increased prevalence of primary hypothyroidism (11%), which was documented among workers from a plant that manufactured PBBs and PBB oxides.[553] These subjects had

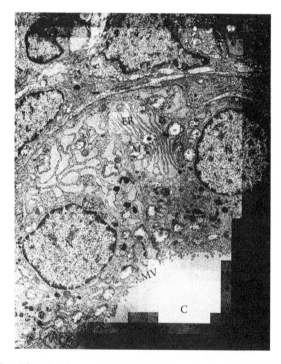

FIGURE 4.22 Thyroid malfunction in a rat fed DEHP (50 mg/kg) and BHT (25 mg/kg)—low dose—for 2 weeks. Note enlarged endoplasmic reticulum (ER), mitochondria (M), and disturbed nuclei (N); vaculated microvilli (MV). Colloid (C) are normal. (Modified from Rea, W. J., *Chemical Sensitivity, Vol III: Clinical Manifestations of Pollutant Overload*, CRC Press, Boca Raton, FL, 1665, Fig. 24.11; Courtesy B. Chen at Robens Institute, Surrey Guildford, England, 1998.)

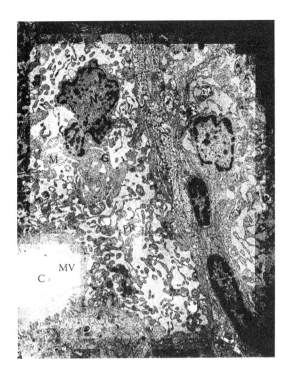

FIGURE 4.23 Thyroid changes in rats fed DEHP (50 mg/kg) for two weeks; endoplasmic reticulum (ER) markedly enlarged; mitochondria (M) vacuolization; Golgi (G) disturbance; nuclear (N) disruption. Normal: colloid (C) and microvilli (MV). (Modified from Rea, W. J., *Chemical Sensitivity, Vol III: Clinical Manifestations of Pollutant Overload,* CRC Press, Boca Raton, FL, 1666, Fig. 24.12; Courtesy B. Chen at Robens Institute, Surrey Guildford, England, 1998.)

elevated titers of antithyroid microsomal antibodies, indicating that hypothyroidism was probably a manifestation of lymphocytic autoimmune thyroiditis or perhaps a PBB-induced pathogenic autoimmune response, or an exacerbation of underlying subclinical disease. Further, environmental pollutants operating in genetically predisposed individuals may trigger the pathogenic mechanisms that lead to goiter formation and autoimmune thyroiditis. The presence of organic goitrogens (resorcinol, substituted resorcinols, thiocyanates, disulfides) and potential "triggers" of the autoimmune response (PAH and halogenated hydrocarbons) in coal[554–559] and water supplies and endemic areas underlies the need to test this hypothesis.[560–562] Goitrous autoimmune thyroiditis has been observed after administration of carbon tetrachloride and the PAH (polyaromatic hydrocarbon), methylcholanthrene, and DMBA (dimethylbenzanthracene) to the BUF (Buffalo) inbred strain of rats.[564,565] Similarly, injection of mouse thyroglobulin with bacterial lipopolysaccharide induces autoimmune thyroiditis in "good-responder" mice, whereas "poor-responder" strains develop little pathologic response. "Good" and "poor" responders differ in their H-2 haplotype.[566] A relation between loci in the major histocompatibility complex and susceptibility to autoimmune thyroid disease has also been demonstrated in humans. For instance, **the histocompatibility HLA DR⁵ antigen is seen with increased frequency in patients with goitrous thyroiditis, whereas atrophic thyroiditis is associated with the HLA-DR3.**[564] Thus, environmental pollutants operating in genetically predisposed individuals may also trigger the homeostatic mechanism and then the pathogenic mechanisms that lead to goiter formation and autoimmune thyroiditis. The presence of organic goitrogens (resorcinol, substituted resorcinols, thiocyanates, disulfides) and potential "triggers" of the autoimmune response (PAH and halogenated hydrocarbons) in coal[507,537,539,559,567] and water supplies of endemic areas[501,531,568] underlie the need to test this hypothesis. **The incidence rate of this disorder has steadily increased in the United States during the past five decades.**[510,563]

Known to cause marked alterations in homeostasis, metabolism, and in thyroid gland structure and function of birds,[507] DDT, DDE, and dieldrin induce microsomal-enzyme activity[569] that may affect thyroid hormone metabolism in a way similar to that of the polyhalogenated biphenyls and PAH. The impact of these pollutants on the human thyroid is unknown, but it is well known that they can alter homeostasis.

TCDD, one of the most toxic small organic molecules, is a contaminant in the manufacturing process of several pesticides and herbicides, including Agent Orange. Also, a potent inducer of hepatic microsomal enzymes, TCDD markedly enhances the metabolism and biliary excretion of T_4-glucuronide in excess and clearly alters the dynamics of homeostasis.[522,521] Rats treated with TCDD concomitantly develop hypothyroxinemia, increased serum TSH concentrations, and goiter, probably as a result of T_4 loss in the bile.[522] The impact on the thyroid to humans exposed to this agent is unknown, and new studies of thyroid function and thyroid hormone metabolism in those cases are necessary. However, this toxic substance has been proven to be a disregulator of the central homeostatic mechanism and, therefore, may well trigger local thyroid dysfunction.

PAH, 3,4,-BaP, MCA, and 7,12-DMBA

Found repeatedly in food and domestic water supplies, polyaromatic hydrocarbons (PAHs)[549,570,573] (see *Chemical Sensitivity* Volume II, p.539[574] and p.583[575]) are present in industrial and municipal waste effluents. They also occur naturally in coal, soils, ground water, surface water, and in their sediments and biota. One of the most potent of the homeostatic disruptures and producers **of carcinogenic PAH compounds, 3,4-BaP is widely distributed and, as in the case of other PAHs, is not efficiently removed by conventional water treatment processes**. Therefore, many chemically sensitive and chronic degenerative diseased individuals have deregulation of the dynamics of homeostasis frequently in their body due to PAH exposure. With treatment of these patients, thyroid involvement is best documented and understood.

Genetically predisposed individuals (i.e., HLA DR^5 antigens) appear to have greater susceptibility to organic PAH and microbial water pollutants that trigger goiter. **Autoimmune thyroiditis does develop after administration of PAH and carbon tetrachloride in the rat.**[543,576] Workers exposed to PCBs manifested low thyroxin levels with increased thyrotropin, and individuals experience repeated exacerbation of thyroiditis with organophosphate exposure. The following is an illustrative case.

> **Case study**. A 58-year-old black male electrical worker in the field for 19 years developed a cold thyroid nodule, which was removed by one of the authors (WJR), and found to be a benign thyroid adenoma. He was also diagnosed with fibromyalgia and hypertension. His blood level of PCB was 10 ppb. His fibromyalgia and hypertension disappeared after the PCB levels became nondetectable following a course of heat depuration treatment. However, subsequent inhaled challenge with PCB's reproduced his symptoms and signs and he showed sensitivity to chlorine and phenol. This patient is well after 10 years of post treatment that included a program of avoidance of pollutants in air, food and water.

A defect in T-suppressor cells in patients with thyroiditis is postulated by Farid[577] and Scherbaum.[578] The EHC-Dallas and Buffalo have seen suppression of the suppressor cells in many of our chemically sensitive patients with endocrine dysfunction. People with pollutant-triggered thyroid problems frequently complain of fatigue, transient sore throats, coldness with subnormal body temperatures, and dry skin. They often have other features of chemical sensitivity including odor sensitivity. Alternately, they may just have pain and fatigue as seen in many patients with chronic degenerative disease. They also have had many viral infections and possibly recurrent *Candida* problems. As discussed previously, changes occur in autoantibodies to the thyroid and the microsomes. In addition, the patient may have an abnormally low thyroid profile.

In summary, many toxic agents in the environment can alter the dynamics of homeostasis both locally in the thyroid, adrenal, genital organs, and neuroendocrine cells and centrally in the hypothalamus, pineal gland, reticular activating system, area postrema of the fourth ventricle, and pituitary gland. These toxics have been shown to cause an imbalance in the endocrine amplification system. Once chronic metabolic changes occur, followed by tissue changes, autogenous conditions then come into play. Because of these tissue changes, fixed-named disease occurs, resulting in thyroiditis and hypothyroidism and other endocrine organ failure. Diagnosis and treatment of pollutant-triggered conditions that alter the endocrine system are consistent with those methods used with other pollutant-related illnesses and are, therefore, not given further discussion here.

SUMMARY

Single and multiple chemicals in various doses either individually and/or in combinations can cause individual or multiorgan dysfunction of the endocrince system. The astute clinician must be aware of these factors in order to help the patient with hypersensitivity and/or chronic degenerative disease.

REFERENCES

1. Rea WJ. 1992. *Chemical Sensitivity, Vol. I, Mechanisms of Chemical Sensitivity.* Boca Raton, FL: Lewis Publishers.
2. Guyton AC and Hall JE. 1996. *Textbook of Medical Physiology,* 9th ed. Philadelphia: WB Saunders Co. 926, Fig. 74-1.
3. Rea WJ, Fenyves EF, Seba D, and Pan Y. 2001. Organochlorine pesticides and chlorinated hydrocarbon solvents in the blood of chemically sensitive patients: A statistical comparison with therapeutic medication and natural hormones. *J.Environ Bio.* 22(3):163–169.
4. Guyton AC and Hall JE. 1996. *Textbook of Medical Physiology,* 9th ed. Philadelphia: WB Saunders Co. 930, Fig. 74-2.
5. Rea, WJ. 1995. *Chemical Sensitivity, Vol. III, Clinical Manifestations of Pollutant Overload.* Boca Raton, FL: CRC Press. 1598, Fig. 24.1.
6. Rea, WJ. 1995. *Chemical Sensitivity, Vol. III, Clinical Manifestations of Pollutant Overload.* Boca Raton, FL: CRC Press. 1599.
7. Werley MS, Burleigh-Flayer HD, Fowler EH, Rybka ML, and Ader AW. 1996. Development of pituitary lesions in ND4 Swiss Webster mice when estimating the sensory irritancy of airborne chemicals using ASTM method E981–84. *Am. Ind. Hyg. Assoc. J.* 57(8):712–716.
8. Wallace LA. 1987. The total exposure assessment methodology (TEAM) study: Summary and analysis, Vol. 1. EPA/600/6-87-002A, Washington, DC: U.S. Environmental Protection Agency.
9. Williams GM and Weisburger JH. 1986. Chemical carcinogens. In *Casarrett and Doull's Toxicology: The Basic Science of Poisons,* 3rd ed. Ed. CD Klaasen, MO Amdur, and J Doull. New York: Macmillan. 107.
10. Guyton AC and Hall JE. 1996. *Textbook of Medical Physiology,* 9th ed. Philadelphia: WB Saunders Co. 935, Fig. 74-4.
11. Rea, WJ. 1995. *Chemical Sensitivity, Vol. III, Clinical Manifestations of Pollutant Overload.* Boca Raton, FL: CRC Press. 1600.
12. Rea, WJ. 1995. *Chemical Sensitivity, Vol. III, Clinical Manifestations of Pollutant Overload.* Boca Raton, FL: CRC Press. 1601.
13. Cooper RL, Chadwick RW, Rehnberg GL, Goldman, Booth KC, et al. 1989. Effect of lindane on hormonal control of reproductive function in the female rat. *Toxicol. Appl. Pharmacol.* 99:384–394.
14. Woodward JC and Jee, WSS. 1991. Skeletal system. In *Handbook of Toxicologyic Pathology.* Ed. WM Haschek and CG Rousseaux. San Diego, CA: Academic Press. 511–517.
15. The damage is strikingly different from species to species. *Ronment.* Y B7.
16. Rea, WJ. 1995. *Chemical Sensitivity, Vol. III, Clinical Manifestations of Pollutant Overload.* Boca Raton, FL: CRC Press. 1602.
17. Bartsch C, Bartsch H, and Peschke E. 2009. Light, Melatonin and Cancer: Current results and future perspectives. *Biological Rhythm Research* 4(1):17–35.
18. Hoffman RA and Reiter RJ. 1965. Influence of compensatory mechanisms and the pineal gland on dark-induced gonadal atrophy in male hamsters. *Nature* 207:658–659. doi:10.1038/207658a0.
19. Reiter RJ and Ellison NM. 1970. Delayed puberty in blinded anosmic female rats: Role of the pineal gland. *Biology of Reproduction.* Society for the Study of Reproduction, Vol. II. 216–222.

20. Reiter RJ, Paredes SD, Manchester LC, and Tan D-X. 2009. Reducing oxidative stress: a newly-discovered genre for melatonin. *Critical Reviews in Biochemistry and Molecular Biology* 44(4):175–200.
21. Reiter RJ. The pineal gland: An important link to the environment. 1986. *News. Physiol. Sci.* 1:202–205.
22. Champney TH, Webb SM, Richardson BA, and Reiter RJ. 1985. Hormonal modulation of cyclic melatonin production in the pineal gland of rats and syrian hamsters: Effects of thyroidectomy or thyroxine implant. *Chonobiology International* 2(3):177–183.
23. Reiter RJ, Hoffmann JC, and Rubin PH. 1986. Pineal gland: Influence on gonads of male rats treated with androgen three days after birth. *Science* 160(3826):420–421.
24. Reiter RJ, Hoffman RA, and Hester RJ. 2005. The effects of thiourea, photoperiod and the pineal gland on the thyroid, adrenal and reproductive organs of female hamsters. *Published Online* 162(3):263–268.
25. Hayes DK, Pauly JE, and Reiter RJ. 1990. *Chronobiology: Its Role in Clinical Medicine, General Biology, and Agriculture. Part B.* New York: John Wiley & Sons.
26. Reiter RJ. 1982. *The Pineal and Its Hormones: Proceedings of an International Symposium. Jan. 2–9, 1982.* New York: Alan R. Liss.
27. Reiter RJ. 1982. *The Pineal,* Vol. 7. Montreal: Eden Press.
28. Reiter RJ and Tan DX. 2002. Melatonin: An antioxidant in edible plants. *Annals of the New York Academy of Sciences* 957:341–344.
29. Reiter RJ. Pineal Research Review, Vol. I. *The Quarterly Review of Biology* 60(1):110–111.
30. Ott JN. 1985. Color and light: their effects on plants, animals and people. *Int. J. Biosoc.Res.Spec. Subj.* 8:1–35.
31. Parry BL, Rosenthal NE, Tamarkin L, and Wehr TA. 1987. Treatment of a patient with seasonal premenstrual syndrome. *Am. J. Psychiatry* 144:762–766.
32. Reiter RJ, et al. 1995. A review of the evidence supporting melatonin's role as an antioxidant. *J. Pineal. Res.* 18(1):1–11.
33. Reiter RF, et al. 1994. Melatonin as a free radical scavenger: Implications for aging and age-related diseases. *Ann. N Y Acad. Sci.* 719:1–12.
34. Poon AM, et al. 1994. Evidence for a direct action of melatonin on the immune system. *Biol. Signals* 3(2):107–117.
35. Herbal Roulete. 1995. *Consumer Reports* Nov.:698–705.
36. Cowley G. Melatonin. 1995. *Newsweek* Aug. 7:46–49.
37. Hearn W. 1995. Melatonin caution: Latest magic bullet may not hit target. *American Medical News* Nov. 6:20–30.
38. Reiter RJ, et al. 1993. Melatonin, hydroxyl radical-mediated oxidative damage, and aging: A hypothesis. *J. Pineal. Res.* 14(4):151–168.
39. Maestroni GJ. 1995. T-helper-2 lymphocytes as a peripheral target of melatonin. *J. Pineal. Res.* 18(2):84–89.
40. Kloeden PE, et al. 1993. Timekeeping in genetically programmed aging. *Exp. Gerontol.* 28(2):109–118.
41. Kloeden PE, et al. 1994. Artificial life extension: The epigenetic approach. *Ann. NY Acad. Sci.* May 31:719–82.
42. Blask DE, Sauer LA, and Dauchy RT. 2002. Melatonin as a chronobiotic/anticancer agent: cellular, biochemical, and molecular mechanisms of action and their implications for circadian-based cancer therapy. *Current Topics in Medicinal Chemistry* 2(2):113–132.
43. Castroviejo DA, Escames G, Carazo A, Leon J, Khaldy H, and Reiter RJ. 2002. Melatonin, mitochondrial homeostasis and mitochondrial-related diseases. *Current Topics in Medicinal Chemistry* 2(2):133–151.
44. Cuzzocrea S and Reiter RJ. 2002. Pharmacological actions of melatonin in acute and chronic inflammation. *Current Topics in Medicinal Chemistry* 2(2):153–165.
45. Guerrero JM and Reiter RJ. 2002. Melatonin–immune system relationships. *Current Topics in Medicinal Chemistry* 2(2):167–179.
46. Tan D, Reiter RJ, Manchester LC, et al. 2002. Chemical and physical properties and potential mechanisms: Melatonin as a broad spectrum antioxidant and free radical scavenger. *Current Topics in Medicinal Chemistry* 2(2):181–197.
47. Kennaway DJ and Wright H. 2002. Melatonin and circadian rhythms. *Current Topics in Medicinal Chemistry,* 2(2):199–209.
48. Lerchla A, Zachmann A, Alib MA, and Reiter RJ. 1998. The effects of pulsing magnetic fields on pineal melatonin synthesis in a teleost fish (brook trout *Salvelinus fontinalis*). *Neuroscience Letters* 256(3):171–173.
49. Pappolla M, et al. 1997. Melatonin prevents death of neuroblastoma cells exposed to the Alzheimer amyloid peptide. *Journal of Neuroscience* 17(5):1683–1690.
50. Reiter R, et al. 1997. Prophylactic actions of melatonin in oxidative neurotoxicity. *Ann. N Y Acad. Sci.* 825:70–78.

51. Arendt, et al. 1987. Some effects of jet-lag and their alleviation by melatonin. *Ergonomics* 30(9):1379–1393.
52. Bendheim, et al. 1992. Nearly ubiquitous tissue distribution of the scrapie agent precursor protein. *Neurology* 42:149–156.
53. Bendheim, et al. 1984. Antibodies to a scrapie prion protein. *Nature* 310:418–421.
54. Benitez-King, et al. 1993. Calmodulin mediates melatonin cytoskeletal effects. *Experientia* 49(8):635–641.
55. Bolton, et al. 1987. Isolation and structural studies of the intact scrapie agent protein. *Arch. Biochem. Biophys.* 258(2):579–590.
56. Cai, et al. 1993. Release of excess amyloid β protein precursor. *Science* 259:514–516.
57. DeArmond, et al. 1984. Identification of prion amyloid filaments in scrapie-infected brain. *Cell* 41:221–235.
58. Dyrks, et al. 1992. Amyloidogenicity of and βA4-bearing amyloid protein precursor fragments by metal-catalized oxidation. *J. Biol. Chem.* 267(25):18210–18217.
59. Ghetti, et al. Gerstmann-staussier-Scheinker disease: II. Neurofibrillary tangles and plaques with PrP-amyloid coexist in an affected family. *Neurology* 39:1453–1461.
60. Glenner, et al. 1984. Alzheimer's disease: initial report of the purification and characterization of a novel cerebrovascular amyloid protein. *Biochem. Biophys. Res. Commun.* 120(3):885–890.
61. Huerto-Delgadillo, et al. 1994. Effects of melatonin on microtubule assembly depend on hormone concentration: Role of melatonin as a calmodulin antagonist. *J. Pineal. Res.* 17:55–62.
62. Kitamoto, et al. 1986. Amyloid plaques in Creutzfeldt-Jakob disease stain with prion protein antibodies. *Ann. Neurol.* 20:204–208.
63. Kitamoto, et al. 1991. A prion protein missense variant in kuru plaque cores in patients with Gerstmann-Straussler syndrome. *Neurology* 41:306–310.
64. LeBars, et al. 1991. PET and plasma phaarmococokinetic studies after bolus intravenous administration of (11 C) melatonin in humans. *Nucl. ed. Biol.* 18(3):357–361.
65. Mishima, et al. 1994. Morning bright light therapy for sleep and behavior disorders in elderly patients with dementia. *Acta. Psychiatr. Scan.* 89:1–7.
66. Oesch, et al. 1985. A cellular gene encodes scrapie PrP 27–30 protein. *Cell* 40:735–746.
67. Pappolla, et al. 1992. Immunohistochemical evidence of antioxidant stress in Alzheimer's disease. *Am. J. Pathol.* 140(3):621–628.
68. Pappolla, et al. 1995.Neuropathology and molecular biology of Alzeheimer's disease. In *Perspectives on Behavioral Medicine Alzeheimer's Disease.* Ed. Stein et al. Sand Diego, CA: Adademic Press. 3–20.
69. Pappolla, et al. 2001. The heat shock/oxidative stress connection: relevance to Alzeheimer disease. *Mol. and Chem. Neuropath.* 28:21–34.
70. Pappolla, et al. 1997. Disruption of β-Fibrillogenesis by the pineal hormone melatonin. *Soc. for Neurosci. Abstr.* 23(1–2):1881.
71. Pappolla, et al. 1997. Melatonin prevents death of neuroblastoma cells exposed to the Alzheimer amyloid peptide. *J. Neurosci.* 17(5):1683–1690.
72. Pappolla, et al. 1988. Inhibition of Alzheimer β-Fibrillogenesis by melatonin. *J. Biol. Chem.* 273(13):7185–7188.
73. Pappolla, et al. 1991. The pineal gland: A circadian or seasonal aging clock? *Aging* 3(2):99–101.
74. Pappolla, et al. 1991. The pineal control of aging: The effects of melatonin and pineal grafting on the survival of older mice. *Ann. NY Acad. Sci.* 621:291–313.
75. Prusiner. 1991. Molecular biology of prion diseases. *Science* 252:1515–1522.
76. Puckett, et al. 1991. Genomic structure of the human prion protein gene. *Am. J. Hum. Genet.* 49:320–329.
77. Reiter, et al. 1993. Antioxidant capacity of melatonin: A novel action not requiring a receptor. *Neuroendoc. Lett.* 15(1–2):103–116.
78. Reiter. 1995. The pineal gland and melatonin in relation to aging: A summary of the heories and of the Data. *Exp. Gerontol.* 30(3/4):1992–2212.
79. Reiter, et al. 1999. Oxidative toxicity in models of neurodeneration: Responses to melatonin. *Rest. Neurol. Neurosci.* 12(2–3):135–142.
80. Robakis, et al. 1994. Involvement of amyloid as a central step in the development of Alzheimer's disease. *Neurobiol. Aging.* 15(supp. 2):S127–129.
81. Safar, et al. 1990. Molecular mass biochemical composition, and physiocochemical behavior of the infectious form of the scrapie precursor protein monomer. *Proc. Natl. Acad. Sci. USA.* 87(16):6373–6377.
82. Selkoe. 1997. Amyloid β protein and the genetics of Alzheimer's disease. *J. Bio. Chem.* 27(31): 182958–18298.
83. Skene, et al. 1990. Daily variation in the concentration of melatonin and 5-Methoxypsoralen in the human pineal gland: Effect of age and Alzheimer's disease. *Brain Research* 528:170–174.

84. Synder, et al. 1994. Amyloid &#x0b32 aggregation in mixtures of amyloid with different chain length. *Biophys. J.* 67:1216–1228.
85. Souetre, et al. 1989. Abnormal melatonin response to 5-methoxypsoralen in dementia. *Am. J. Psychiatry.* 146(8):1037–1040.
86. Tagliavini, et al. 1991. Amyloid protein of German-Straussler-Schneinker disease (Indiana kindred) is an 11 kd fragment of prion protein with an N-terminal glycine at codon58. *EMBO J.* 10(3):513–519.
87. Turk, et al. 1987. Purification and properties of the cellular and scrapie hamster prion proteins. *Eur. J. Biochem.* 176:21–30.
88. van Renberg, et al. 1999. A new model for the pathophysiology of Alzheimer's disease. *SAMJ.* 87(9):1111–1115.
89. Weidemann, et al. 1989. Identification, biogensis, and localization of precursors of Alzheimer's disease A4 amyloid protein. *Cell* 57:115–126.
90. Wisniewski, et al. 1991. Peptides homologous to the amyloid protein of Alzheimer's disease containing a glutamine for glutamine acid substation have accelerated amyloid fibril formation. *Biophys. Res. Commun.* 17(3):1247–1254c.
91. Wisniewski, et al. 1994. Acceleration of Alzheimer's fibril formation by apolipoprotein E in vitro. *Am. J. Pathol.* 145(5):1030–1035.
92. Couplan RE. 1965. *The Natural History of the Chromaffin Cell.* London: Longmans, Gree & Co. 1–279.
93. Kjaergaard J. 1973. *Anatomy of the Carotid Glomus and Carotid Glomus-Like Bodies (Non-Chromaffin Paraganglia): With Electron Microscopy and Comparison of Human Foetal Carotid, Aorticopulmonary, Subclavian, Tempanojugular, and Vagal Glomera.* Trans. A LaCour. Copenhagen: F.A.D. L.'s Forlag. 328.
94. Gleener GG and Grimley PM. 1974. *Tumors of the Extra-Adrenal Paraganglion System (Including Chemoreceptors), Fascicle 9, Atlas of Tumor Pathology,* (2nd Series. Washington, DC: Armed Forces Institute of Pathology.
95. Carmichael S and Winkler H. 1985. The adrenal chromaffin cell. *Sci. Am.* 253(2):40–49.
96. Grimley PM and Glenner GG. 1968. Ultrastructure of the human carotid body: A perspective on the mode of chemoreception. *Circulation* 37:648–665.
97. Ferri GL, Probert D, Cocchia F, et al. 1982. Evidence for the presence of S-100 protein in the glial component of the human enteric nervous system. *Nature* 297:409–410.
98. Lloyd RV, Blaivas M, and Wilson BS. 1985. Distribution of chromogranin and S100 protein in normal and abnormal adrenal medullary tissues. *Arch. Pathol. Lab. Med.* 109:633–635.
99. Chaudhry AP, Haar JG, Koul A, and Nickerson PA. 1979. A nonfunctioning paraganglioma of vagus nerve. *Cancer* 432:1689–1701.
100. Robertson DI and Cooney TP. 1980. Malignant carotid body paraganglioma: Light and electron microscopic study of the tumor and its metastases. *Cancer* 46:2623–2633.
101. Grimley PM and DeLellis RA. 1986. Multisystem neuroendocrine neo-plasms. In *Pathology of Incipient Neoplasia.* Ed. J Albores-Savedra and D Henson. Philadelphia: WB Saunders. 425.
102. DeLellis RA and Wolfe HG. 1981. The polypeptide hormone-producing eneuroendocrine cells and their tumors. *Meth. Achiev. Exp. Pathol.* 10:190–220.
103. Rosenfeld MG, Amara SG, Birnberg NC, et al. 1983. Calcitonin, prolactin and growth hormone gene expression as model systems for the characterization of neuroendocrine regulation. *Recent Prog. Horm. Res.* 39:305–351.
104. Douglass J, Civelli O, and Herbert E. 1984. Polyprotein gene expression: generation of diversity of neuroendocrine peptide. *Ann. Rev. Biochem.* 53:665–715.
105. Wolfe H, Childers H, Montminy M, Goodman R, et al. 1985. Use of antisense RNA probes for morphologic detection of peptide-producing cells by in situ hybridization. *Lab. Invest.* 52:77A.
106. Payne CM, Nagle RB, and Borduin V. 1984. Methods in laboratory investigation. An ultrastructural cytochemical stain specific for neuroendrocrine neoplasm. *Lab. Invest.* 51:350–365.
107. Balsera E, Lloyd RV, Livingston SK, Lavallee M, and Azar HA. 1986. Immunohistochemistry and electron microscopy of neuroendocrine neoplasms. *Lab. Invest.* 54:4A.
108. Pearse AGE. 1969. The cytochemistry and ultrastructure of polypeptide hormone-producing cells of the APUD series and the embryologic, physiologic, and pathologic implications of the concept. *J. Histochem. Cytochem.* 17:303–313.
109. Wasserman K. 1980. Recent advances in carotid body physiology. *Fed. Proc.* 39:2626.
110. Rea WJ. 1995. *Chemical Sensitivity, Vol. III, Clinical Manifestations of Pollutant Overload.* Boca Raton, FL: CRC Press. 1607, Fig. 24.2.
111. Polak JM and Bloom SR. 1979. The diffuse neuroendocrine system: Studies of this newly discovered controlling system in health and disease. *J. Histochem. Cytochem.* 27:1398–1400.

112. Snyder SH. 1980. Brain peptides as neurotransmitters. *Science* 209:976–983.
113. Larsson LI, Goltermann N, Rehfeld JF, and Schwartz TW. 1979. Somatostatin cell processes as pathways for paracrine secretion. *Science* 205:1393–1395.
114. Krammer EB. 1978. Carotid body chemoreceptor function: Hypothesis based on a new circuit model. *Proc. Natl. Acad. Sci.* 75:2507–2511.
115. Angeletti RH, Nolan JA, and Zaremba S. 1985. Catecholamine storage vesicles: Topolgraphy and function. *Trends Biochem. Sci.* 10:240–243.
116. Trojanowski JQ and Lee VMY. 1985. Expression of neurofilament antigens by normal and neoplastic human adrenal chromaffin cells. *New Engl J. Med.* 313:101–104.
117. Sibley RK. 1985. The intermediate filament profile of neuro and neuroendocrine neoplasms. *Lab. Invest.* 52:62A.
118. Angeletti RH, Nolan JA, and Zaremba S. 1985. Catecholamine storage vesicles: topolgraphy and function. *Trends Biochem. Sci.* 10:240–243.
119. Hassoun J, Monges G, Giraud P, et al. 1984. Immunohitochemical study of pheochromocytomas: An investigation of methionine-enkephalin, vasoactive intestinal peptide, somatostatin, corticotrophin, endorphin, and calcitonin in 16 tumors. *Am. J. Patho.* 114:56–63.
120. Lloyd RV, Shapiro B, Sisson JC, et al. 1984. An immunohistochemical study of pheochromocytomas. *Arch. Pathol. Lab. Med.* 108:541–544.
121. Sano T, Saito H, Inaba H, et al. 1983. Immunoreactive somatostatin and vasoactive intestinal polypeptide in adrenal pheochromocytoma. *Cancer* 52:282–289.
122. Verhofstad AA, Steinbusch HWM, Joosten HWJ, et al. 1983. Immunocytochemical localization of nonadrenaline, adrenaline and serotonin. In *Immunocytochemicsty: Practical Applications in Pathology and Biology*. Bristol: Wright PSG. 143–168.
123. Schmechel DE. 1985. Subunit of the glycolytic enzyme enolase: Nonspecific or neuron specific: *Lab Invest.* 52:239–242.
124. Lloyd RV and Warner TF. 1984. Immunohistochemistry of neuron-specific enolase. In *Advance in Immunochistochemistry* Ed. RA DeLellis. New York: Masson Publishing. 127–140.
125. Thomas P, Battifora H, Maderino G. 1986. Is neuron-specific endlase specific? An immunohistochemical comparison of a monoclonal and polyclonal antibody against neuron-specific enolase. *Lab Invest.* 54:63A.
126. Chambers RC, Bowling MC, and Grimley PM. 1968. Glutaraldehyde fixation in routine histopatholgy. *Arc. Path.* 85:18–30.
127. Smith DM and Haggitt RC. 1983. A comparative study of generic stains for carcinoid secretory granules. *Am. J. Surg. Pathol.* 7:61–68.
128. Lewis RV, Stern AS, Kimura S, et al. 1980. An about 50,000 dalton protein in adrenal medulla: a common precursor of [met] and [leu]-enkephalin. *Science* 200:1450–1461.
129. O'Connor DT, Burton D, and Deftos LJ. 1983. Chromogranin A: Immunohistology reveals its universal occurrence in normal polyprptide hormone producing endocrine glands. *Life Sci.* 33:1657–1663.
130. Wilson BS and Lloyd RV. 1984. Detection of chromogranin in neuroendocrine cells with a monoclonal antibody. *Am. J. Pathol.* 115:458–468.
131. Johnson TL, Lloyd RV, Shapiro B, et al. 1985 Cardiac paragangliomas: A clinicopathologic study of four cases. *Lab. Invest.* 52:31A.
132. Tischler AS, Dichter MA, Biales B, and Greene LA. 1977. Neuroendocrine neoplasms and their cells of origin. *N. Engl. J. Med.* 296:919–925.
133. Roth J, Leroith D, Shiloach, et al. 1982. Thye evolutionary origins of hormones, neurotransmitters, and other extracellular chemical messengers. *N. Engl. J. Med.* 306:523–526.
134. DeLellis RA, Tischler AS, Lee AK, Blount M, and Wolfe HF. 1983. Leuenkephalin-like immunoreactivity in proliferative lesions of the human adrenal medulla and extra-adrenal paraganglia. *Am J. Surg. Pathol.* 7:29–37.
135. Schimke RN. 1980. The neurocristopathy concept: Fact or fiction. In *Advances in Neuroblastoma Research: Proceedings of the 2nd Symposium on Advances in Neuroblastoma Research, Philadelphia, 1979*. Ed. AE Evans. New York: Alan R. Liss.
136. Kissel P, Andre JM, and Jacquier A. 1981. *The Neurocristopathies*. New York: Masson Publishing. 262.
137. Le Dourain NM. 1982. *The Neural Crest*. Cambridge, MA: Cambridge University Press. 259.
138. Llena JF. 1983. Paraganlioma in the cerebrospinal axis. *Prog. NeuropathoL* 5:261–276.
139. Stahlman M and Gray ME. 1984. Ontogeny of neuroendocrine cells in human fetal lung: I. An electron microscopic study. *Lab. Invest.* 51:449–463.
140. Stahlman MT, Kasselberg AG, Orth DN, and Gray ME. 1985. Ontogeny of neuroendocrine cells in human fetal lung: II. An immunohistochemical study. *Lab. Invest.* 52:52–60.

141. Bosman FT and Louwerens JWK. 1981. APUD cells in teratomas. *Am. J. Pathol.* l04:174–180.
142. Gould VB, Linnoila RI, Memoli VA, Warren WH. 1983. Biology of disease. Neuroendocrine components of the bronchopulmonary tract: Hyperplasias, dysplasias, and neoplasms. *Lab. Invest.* 5:519–537.
143. Manning JT, Ordonez NG, Rosenberg HS, and Walker WE. 1985. Pulmonary endodermal tumor resembling fetal lung: Report of a case with immunohistochemical studies. *Arch. Pathol. Lab. Med.* 109:48–50.
144. Sciurba FC, Owens GR, Sanders MH, et al. 1988. Evidence of an altered pattern of breathing during exercise in recipients of heart-lung transplants. *N. Engl. J. Med.* 319:1186–1192.
145. Donald DE and Shepherd JT. 1963. Response to exercise in dogs with cardiac denervation. *Am. J. Physiol.* 205:393–400.
146. Savin WM, Haskell WL, Schroeder JS, and Stinson EB. 1980. Cardiorespiratory responses of cardiac transplant patients to graded, symptom-limited exercise. *Circulation* 62:55–60.
147. Pope SE, Stinson EB, Daughters GT, et al. 1980. Exercise response of the denervated heart in long-term cardiac transplant recipients. *Am. J. Cardiol.* 46:213–218.
148. Campeau L, Pospisil L, Grondin P, Dyrda I, and Lepage G. 1970. Cardiac catheterization findings at rest and after exercise in patients following cardiac transplantation. *Am. J. Cardiol.* 25:523–528.
149. Clark DA, Schroeder JS, Griepp RB, et al. 1973. Cardiac transplantation in man: Review of first three years' experience. *Am. J. Med.* 54:563–576.
150. Davies CT, Few J, Foster KG, and Sargeant AJ. 1974. Plasma catecholamine concentration during dynamic exercise involving different muscle groups. *Eur. J. Appl. Physiol.* 32:195–206.
151. Carter D and Eggleston JC. 1979. *Tumors of the Lower Respiratory Tract: Fascicle 17, Atlas of Tumor Pathology,* 2nd Series. Washington, DC: Armed Forces Institute of Pathology.
152. Ichinose HR, Hewitt RL, and Drapanas T. 1971. Minute pulmonary chemodectoma. *Cancer* 28:692–700.
153. Costero I, Barroso-Moguel R, and Martinez-Palomo A. 1972. Pleural origin of some of the supposed chemodectoid structures of the lung. *Beitr. Pathol.* 146:351–365.
154. Pearse AGE and Polak IM. 1978. The diffuse neuroendocrine system and the APUD concept. In *Gut Hormones.* Ed. SR Bloom. Edinburgh: Churchill Livingstone. 33.
155. Andrew A. 1976. APUD cells, apudomas and the neural crest. *S. Afr. Med. J.* 50:890–898.
156. Jubb KV and McEntee K. 1959. The relationship of ultimobranchial remnants and derivatives to tumor of the thyroid gland in cattle. *Cornell Veterinarian* 49:41–69.
157. Fernandes BI, Bedard YC, and Rosen I. 1982. Mucus-producing medullary cell carcinoma of the thyroid gland. *J. Clin. Pathol.* 78:536–540.
158. Dayal Y. 1983. Endocrine cells of the gut and their neoplasms. In *Pathology of the Colon, Small Intestine and Anus.* Ed. HT Norris. New York: Churchill Livingstone. 267–302.
159. Sidhu GS. 1979. The endodermal origin of digestive and respiratory tract APUD cells: Histopathologic evidence and a review of the literature. *Am. J. Pathol.* 96:5–20.
160. Cox WF and Pierce GB. 1982. The enodermal origin of the endocrine cells of an adenocarcinoma of the rat. *Cancer* 50:1530.
161. Di Sant'Agnese PA, DeMesy Jensen KL, Churukian CJ, and Agarwal MM. 1985. Human prostatic endocrine-paracrine (APUD) cells. *Arch. Pathol. Lab. Med.* 109:607–612.
162. Di Sant'Agnese PA and DeMesy Jensen KL. 1984. Somatostain andlor somatostatin-like immunoreactive endocrine-paracrine cells in the human prostate gland. *Arch. Pathol. Lab. Med.* 108:693–696.
163. Scully R B, Aguirre P, and DeLellis RA. 1984. Argyrophilia, scrotonin and peptide hormones in the female genital tract and its tumors. *Int. J. Gynecol. Pathol.* 3:51–70.
164. Young TW and Thrasher TV. 1982. Nonchromaffin paraganglioma of the uterus. *Arch. Pathol. Lab. Med.* l06:608–609.
165. Oberman HA, Holtz F, Sheffer LA, and Magielski IB. 1968. Chemodectomas (nonchromaffin paragangliomas) of the head and neck. *Cancer* 21:838–851.
166. Lack BB. 1978. Hyperplasia of vagal and carotid body paraganglia in patients with chronic hypoxemia. *Am. J. Pathol.* 91:497–516.
167. Batsakis, JD. 1980. Paragangliomas of the head and neck. In *Tumors of the Head and Neck.* Ed. JD Batsakis. Baltimore: Williams & Wilkins. 369–380.
168. Gallivan MVB, Chun B, Rowden G, and Lack BB. 1979. Laryngeal paraganglioma: Case report with ultrastructzzral analysis and literature review. *Am. J. Surg. Pathol.* 3:85–92.
169. Lack, BB, Cubilla AL, and Woodruff JM. 1979. Paragangliomas of the head and neck region. *Hum. Pathol.* 10:191–218.
170. Buss DH, Marshall RB, Baird FG, and Myers RT. 1980. Paraganglioma of the thyroid gland. *Am. J. Surg. Pathol.* 4:589–593.

171. House JM, Goodman ML, Gacek RR, and Green GL. 1972. Chemodectomas of the nasopharynx. *Arch. Otolaryngol.* 96:38–141.
172. Kahn LB. 1976. Vagal body tumor (nonchromaffin paraganglioma, chemodectoma, and carotid body-like tumor) with cervial node metastasis and familial association. *Cancer* 38:2367–2377.
173. Tannir NM, Cortas N, and Allam C. 1983. A functioning catecholamine-secreting vagal body tumor: A case report and review of the literature. *Cancer* 52:932–935.
174. Persson AV, Frusha JD, Dial PF, and Jewell BR. 1985. Vagal body tumor: Paraganglioma of the head and neck. *Calif. J. Clinicians* 35:232–237.
175. Reed JC, Hallet KK, and Felgin DS. 1978. Neural tumors of the thorax: Subject review from the AFIP. *Radiology* 126:9–17.
176. Lack BB, Stillinger RA, Colvin DB, Groves RM, and Burnette DG. 1979. Aortic-pulmonary paraganglioma: Report of a case with ultrastructural study and review of the literature. *Cancer* 43:269–278.
177. Olson JL and Salyer WR. 1978. Mediastinal paragangliomas (aortic body tumor): A report of four cases and a review of the literature. *Cancer* 41:2405–2412.
178. Gallivan MVB, Chun B, Rowden G, and Lack BB. 1980. Intrathoracic paravertebral malignant paraganglioma. *Arch. Pathol. Lab. Med.* l04:46–51.
179. Melicow MM. 1977. One hundred cases of pheochromocytoma (107 tumors) at the Columbia-Presbyterian Medical Center, 1926–1976: A clinicopathological analysis. *Cancer* 40:1987–2004.
180. Lack BB, Cubilla AL, Woodruff JM, and Lieberman PH. 1980. Extra-adrenal paragangliomas of the retroperitoneum: A clinicopathologic study of 12 tumors. *Am. J. Surg. Pathol.* 4:109–129.
181. Thacker WC and Duckworth JK. 1969. Chemodectoma of the orbit. *Cancer* 23:1233–1238.
182. Perrone T, Sibley RK, and Rosai I. 1985. Duodenal gangliocytic paraganglioma: An immunohistochemical and ultrastructural study and a hypothesis concerning its origin. *Am. J. Surg. Pathol.* 9:31–411.
183. Miller TA, Weber TR, and Appelman HD. 1972. Paraganglioma of the gallbladder. *Arch. Surg.* 105:637–639.
184. Leestma JB and Price BB Jr. 1971. Paraganglioma of the urinary bladder. *Cancer* 28:1063–1073.
185. Lipper S and Decker RB. 1984. Paraganglioma of the cauda equina: A histologic, immunohistochemical, and ultrastructural study and review of the literature. *Surg. Neurol.* 22:415–420.
186. Albores-Saavedra J, Maldonado MB, Ibarra J, and Rodriguez HA. 1969. Pheochromocytoma of the urinary bladder. *Cancer* 23:1110–1118.
187. White MC and Hickson BR. 1979. Multiple paragangliomata secreting catecholamines and calcitonin with intermittent hypercalcemia. *J. R. Soc. Med.* 72(7):532–535.
188. Apple D and Kreines K. 1982. Cushing's syndrome due to ectopic ActH production by a nasal paraganglioma. *Am. J. Med. Sci.* 283:32–35.
189. Grizzle WE, Tolbert L, Pittman CS, Siegel AL, and Aldrete JS. 1983. Corticotropin production by tumors of the autonomic nervous system. *Arch. Pathol. Lab. Med.* 108:545–550.
190. Farrior JB III, Hyams VJ, Benke RH, and Farrior RH. 1980. Carcinoid apudoma arising in a glomus jugulare tumor: Review of endocrine activity in glomus jugulare tumors. *Laryngoscope* 90:110–119.
191. Pfeiffer CC. 1975. *Mental and Elemental Nutrients.* New Canaan, CT: Keats Publishing.
192. Saito H, Saito S, Sano T, Kagawa N, Hizawa K, and Tatara K. 1982. Immunoreactive somatostatin in catecholarmine-producing extra-adrenal paraganglioma. *Cancer* 50:560–565.
193. Tischler AS, Lee AK, Nunnemacher G, Said SI, DeLellis RA, Morse GM, and Wolfe HJ. 1981. Spontaneous neurite outgrowth and vasoactive intestinal peptide-like immunoreactivity of cultures of human paraganglioma cells from the glomus jugulare. *Cell Tissue Res.* 219:543–555.
194. Luoma JR. May 15, 1990. Scientists are unlocking secrets of dioxins devastating power. *New York Times.*
195. Louis GMB, Gray LEF, Marcus M, et al. 2008. Environmental factors and puberty timing: Expert panel research needs. *Pediatrics* 121(Suppl.):S192–S207.
196. Guyton AC and Hall JE. 1996. *Textbook of Medical Physiology,* 9th ed. Philadelphia: WB Saunders Co. 965.
197. Rea WJ. 1992. *Chemical Sensitivity, Vol I, Mechanisms of Chemical Sensitivity.* Boca Raton, FL: Lewis Publishers. 17.
198. Jeffries, WM. 1981. *Safe Uses of Cortisone.* Springfield, IL: Charles C. Thomas.
199. Bralley JA and Lord RS. 2001. *Laboratory Evaluations in Molecular Medicine: Nutrients, Toxicants, and Cell Regulators.* Norcross, GA: Institute of Molecular Medicine. 300, Fig. 10-2.
199b. Jorde R, J Sundsfjord J, Haug E, Bønaa KH. 2000. Relation Between Low Calcium Intake, Parathyroid Hormone, and Blood Pressure. *Hypertension* 35:1154–1159.
200. Rea WJ. 1996. *Chemical Sensitivity, Vol. III, Clinical Manifestations and Pollutant Overload.* Boca Raton, FL: CRC Press. 1624.
201. Couch JR and Wortman L. 1984. Anovulatory states as a factor in occurrence of migraine. Paper presented at Migraine Trust, Fifth International Symposium.

202. Matthews KA, Flory JD, Owens JF, Harrison KT, and Berga SL. 2003. Influence of estrogen replacement therapy on cardiovascular responses to stress of healthy post menopausal women. *Psychophysiology* 38(3):391–398.
203. Schuit SCE, Oei H-HS, Witteman JCM, et al. 2004. Estrogen receptor and gene polymorphisms and risk of myocardial infarction. *JAMA* 291(24):2969–2977.
204. Gelfand JA, Sherms RJ, Ailing DW, and Frank MM. 1976. Treatment of hereditary angio-edema with Danazol: Reversal of clinical and biochemical abnormalities. *N.Engl. J. Med.* 295:1444.
205. Rea WJ. 1994. *Chemical Sensitivity, Vol. II, Sources of Total Body Load.* Boca Raton, FL: Lewis Publishers. 837.
206. Uphouse L. 1987. Decreased rodent sexual receptivity after lindane. *Toxicol. Lett.* 39(1):7–14.
207. Thomas KB and Colborn T. 1992. Organochlorine endocrine disruptors in human tissue. In *Advances in Modern Environmental Toxicology, Vol. XXI, Chemically-Induced Alterations in Sexual and Functional Development: The Wildlife/Human Connection.* Ed. T Colborn and C Clement. Princeton, NJ: Princeton Scientific. 379.
208. Hamilton A and Hardy HL. 1974. *Industrial Toxicology,* 3rd ed. Acton, England: Publishing Sciences.
209. Mattison DR. 1980. Morphology of oocyte and follicle destruction by polycyclic aromatic hydrn carbons in mice. *Toxicol. Appl. Pharmacol.* 53(2):249–259.
210. Everson RB, Sandier DP, Wilcox AJ, et al. 1986. Effect of passive exposure to smoking on age at natural menopause. *Br. Med. J. [Clin. Res.]* 293(6550):792.
211. De Coufle P, Lloyd JW, and Salvin LG. 1977. Causes of death among construction machinery operators. *J. Occup. Med.* 19(2):123–128.
212. Pérez-Comas, A. 1982. Precocious sexual development in Puerto Rico. *Lancet* 1:1299–1300.
213. Meirik, O.1986. Oral contraceptives and breast cancer in young women. Some notes on a current controversy. *Acta Obstet. Gynecol. Scand. [SuppLi]*134:5–7.
214. Muntean, W. 1987. Spontaneous deep vein thrombosis in children and adolescents [letter]. *J. Pediatr. Surg.* 22(2):188.
215. Hayes WJ. 1982. Baltimore: Williams & Wilkins. 253, 444.
216. DeVito MJ, Thomas T, Martin B, Umbreit TH, and Gallo MA. 1992. Antiestrogenic action of 2,3,7,8-tetra-chlorodibenzo-p-dioxin: tissue-specific regulation of estrogen receptor in CDI mice. *Toxicol. Appl. Pharmacol.* 113:284–292.
217. Safe S, Astroff B, Harris M, et al. 1991. 2,3,7,8-tetrachlorodibenzo-p-dioxin (TCDD) and related compounds as antiestrogens: Characterization and mechanisms of action. *Pharmacol. Toxicol.* 69:400–409.
218. Gray LB. 1992. Chemical-induced alterations of sexual differentiation: a review of effects in humans and rodents. In *Advances in Modern Environmental Toxicology, Vol. XXI, Chemically-Induced Alterations in Sexual and Functional Development: The Wildlife/Human Connection.* Ed. T Colborn and C Clement. Princeton, NJ: Princeton Scientific. 203.
219. Gray LB. 1992. Chemical-induced alterations of sexual differentiation: a review of effects in humans and rodents. In *Advances in Modern Environmental Toxicology, Vol. XXI, Chemically-Induced Alterations in Sexual and Functional Development: The Wildlife/Human Connection.* Ed. T Colborn and C Clement. Princeton, NJ: Princeton Scientific. 206.
220. Meyer-Bahlburg HFL, Ehrhardt AA, Feldman JF, et al. 1985. Sexual activity level and sexual functioning in women prenatally exposed to diethylstibestrol. *Psychosomatic Med.* 47(6):497–511.
221. Gray LB. 1992. Chemical-induced alterations of sexual differentiation: A review of effects in humans and rodents. In *Advances in Modern Environmental Toxicology, Vol. XXI, Chemically-Induced Alterations in Sexual and Functional Development: The Wildlife/Human Connection.* Ed. T Colborn and C Clement. Princeton, NJ: Princeton Scientific. 207.
222. Hines M and Shipley C. 1984. Prenatal exposure to diethyistilbestrol (DES) and the development of sexually dimorphic cognitive abilities and cerebral lateralization. *Develop. Psych.* 20(1):81–94.
223. McGlone J. 1980. Sex differences in human brain asymmetry: A critical survey. *Behav. Brain Sci.* 3:215–263.
224. Hines, M. 1992. Surrounded by estrogens? Considerations for neurobehavioral development in human beings. In *Advances in Modern Environmental Toxicology, Vol. XXI, Chemically-Induced Alterations in Sexual and Functional Development: The Wildlife/Human Connection.* Ed. T Colborn and C Clement. Princeton, NJ: Princeton Scientific. 269.
225. Hines M. 1992. Surrounded by estrogens? Considerations for neurobehavioral development in human beings. In *Advances in Modern Environmental Toxicology, Vol. XXI, Chemically-Induced Alterations in Sexual and Functional Development: The Wildlife/Human Connection.* Ed. T Colborn and C Clement. Princeton, NJ: Princeton Scientific. 274.

226. Blair PB. 1992. Immunologic studies of women exposed in utero to diethyl stilbestrol. In *Advances in Modern Environmental Toxicology, Vol. XXI, Chemically-Induced Alterations in Sexual and Functional Development: The Wildlife/Human Connection.* Ed. T Colborn and C Clement. Princeton, NJ: Princeton Scientific. 289.
227. Noller KL, Blair PB, O'Brien PC, and Mellon LJ. 1988. Increased occurrence of autoimmune disease among women exposed in utero to diethyl stilbestrol. *Fertil. Steril.* 49:1080–1082.
228. Swaab DF and Fliers BA. 1985. Sexually dimorphic nucleus in the human brain. *Science* 228:1112–1114.
229. LeVay S. 1991. A difference in hypothalamic structure between heterosexual and homosexual men. *Science* 253:1030–1037.
230. Barraclough CA and Gorski RA. 1962. Studies on mating behavior in the androgen-sterilized female rat in relation to the hypothalamic regulation of sexual behavior *J. Endocrinol.* 25:175–182.
231. Reddy RV, Naftolin F, and Ryan KJ. 1974. Conversion of androstenedione to estrone by neural tissues from fetal and neonatal rats. *Endocrinology* 94:117–121.
232. McEwen BS, Leiberburg L, Chaptal C, and Krey LC. 1977. Aromatization: Important for sexual differentiation of the neonatal rat brain. *Horm. Behav.* 9:249–263.
233. Phoenix CH, Goy RW, Gerall AA, and Young WC. 1959. Organizing action of prenatally administered testosterone propionate on the tissues mediating mating behavior in the female guinea pig. *Endocrinology* 65:369–382.
234. Kandel ER, Schwartz JH, and Jessell TM. 1995. *Essentials of Neural Science and Behavior.* Norwalk, CT: Appleton and Lange.
235. Harris GW and Levine S. 1965. Sexual physiology of the brain and its experimental control *J. Physiol.* 181:379–400.
236. De Jonge FH, Muntjewerff JW, Louwerse AL, and Van de Poll NE. 1988. Sexual behavior and sexual orientation of the female rat after hormonal treatment during various stages of development. *Horm. Behav.* 22:100–115.
237. Moore CL. 1990. Comparative development of vertebrate sexual behavior: Levels, cascades, and webs. In *Contemporary Issues in Comparative Psychology.* Ed. DA Dewsbury. Sunderland, MA: Sinauer Associates. 278–299.
238. Moore CL, Dou H, and Juraska JM. 1992. Maternal stimulation affects the number of motor neurons in a sexually dimorphic nucleus of the lumbar spinal cord. *Brain Res.* 572:52–56.
239. Fausto-Sterling A. 1995. Animal models for the development of human sexuality: A critical evaluation. *J. Homosexuality* 28:217–236.
240. Breedlove SM. 1992. Sexual dimorphism in the vertebrate nervous system. *J. Neurosci.* 12:4133–4142.
241. Fisher CR, Graves KH, Parlow AF, Simpson R and E. 1998. Characterization of mice deficient in aromatase (ArKO) because of a targeted disruption of the *cyp19* gene. *Proc. Natl. Acad. Sci. USA* 95:6965–6970.
242. Jacklin D. 1981. Methodological issues in the study of sex-related differences. *Dev. Rev.* 1:266–273.
243. Bleier R. 1984. *Science and Gender.* New York: Pergamon. 80–114.
244. Fausto-Sterling A. 1992. *Myths of Gender.* New York: Basic Books.
245. Breedlove SM. 1997. Sex on the brain. *Nature* 389:801.
246. Hamer DH, Hu S, Magnuson VL, Hu N, and Pattatucci AML. 1993. A linkage between DNA markers on the X chromosome and male sexual orientation. *Science* 261:321–327.
247. Risch N, Squires-Wheeler E, and Keats BJB. 1993. Male sexual orientation and genetic evidence. *Science* 262:2063–2065.
248. Marshall E. 1995. NIH's "gay gene" study questioned. *Science* 268:1841.
249. Hu S, et al. 1995. Linkage between sexual orientation and chromosome Xq28 in males but not in females. *Nature Genet.* 11:248–256.
250. Rice G, Anderson C, Risch N, and Ebers G. 1999. Male homosexuality: Absence of linkage to microsatellite markers at Xq28. *Science* 284:665–667.
251. Carroll J and Wolpe PR. 1996. *Sexuality and Gender in Society.* HarperCollins, New York.
252. Hines M. 1992. Surrounded by estrogens? Considerations for neurobehavioral development in human beings. In *Advances in Modern Environmental Toxicology, Vol. XXI, Chemically-Induced Alterations in Sexual and Functional Development: The Wildlife/Human Connection.* Ed. T Colborn and C Clement. Princeton, NJ: Princeton Scientific. 261.
253. Soto AM, Lin TM, Justicia H, Silvia RM, and Sonnenschein C. 1992. An "in culture" bioassay to assess the estrogenicity of xenobiotics (B-screen). In *Advances in Modern Environmental Toxicology, Vol. XXI, Chemically-Induced Alterations in Sexual and Functional Development: The Wildlife/Human Connection.* Ed. T Colborn and C Clement. Princeton, NJ: Princeton Scientific. 302.

254. Meyers CY, Matthews WS, Roll LL, Koib VM, and Parady TE. 1977. Carboxylic acid formation from kepone. In *Catalysis in Organic Synthesis.* Ed. GW Smith. New York: Academic Press. 213–215, 253–255.
255. Palmiter RD and Mulvihill BR. 1978. Estrogenic activity of the insecticide kepone on the chicken oviduct. *Science* 201:356–358.
256. Hammond B, Katzenollenbogen BS, Kranthammer N, and McConnell J. 1979. Estrogenic activity of the insecticide chiordecone (kepone) and interaction with uterine estrogen receptors. *Proc. Natl. Acad. Sci. U.S.A.* 76:6641–6645.
257. Soto AM, Lin TM, Justicia H, Silvia RM, and Sonnenschein C. 1992. An "in culture" bioassay to assess the estrogenicity of xenobiotics (B-screen). In *Advances in Modern Environmental Toxicology, Vol. XXI, Chemically-Induced Alterations in Sexual and Functional Development: The Wildlife/Human Connection.* Ed. T Colborn and C Clement. Princeton, NJ: Princeton Scientific. 305.
258. Moses M. 1993. Pesticides. In *Occupational and Environmental Reproductive Hazards: A Guide for Clinicians.* Ed. M. Paul. Baltimore: Williams & Williams. 296.
259. Metzler M. 1985. Role of metabolism in determination of hormonal activity of estrogens: Introductory remarks. In *Estrogens in the Environment: Influences on Development.* Ed. JA McLachlan. New York: Elsevier. 187–189.
260. Blaich G, Pfaff B, and Metzler M. 1987. Metabolism of diethylstibestrol in hamster hepatocytes. *Biochem. Pharmacol.* 36:3135–3140.
261. Whitten P and Naftolin F. 1991. Dietary plant estrogens: a biologically active background for estrogen action. In *The New Biology of Steroid Hormones.* Ed. R Hochberg and F Naftolin. New York: Raven Press. 155–167.
262. Whitten PL. 1992. Chemical revolution to sexual revolution: historical changes in human reproductive development. In *Advances in Modern Environmental Toxicology, Vol. XXI, Chemically-Induced Alterations in Sexual and Functional Development: The Wildlife/Human Connection.* Ed. T Colborn and C Clement. Princeton, NJ: Princeton Scientific. 313.
263. Kellis JT and Vickery LB. 1984. Inhibition of human estrogen synthetase (aromatase) by flavones. *Science* 225:1032–1034.
264. Adlercreutz H. 1990. Western diet and western diseases: some hormonal and biochemical mechanisms and associations. *Scand. J. Clin. Lob. Invest.* 50 Suppl. 201:3–23.
265. Trowell HC and Burkitt DB. 1983. *Western Diseases: Their Emergence and Prevention.* London: Edward Arnold.
266. Gorski, RA. 1968. Influence of age on the response to perinatal administration of a low dose of androgen. *Endocrinology* 82:1001–1004.
267. Hines M, Alsum P, Roy M, Gorski RA, and Goy RW. 1987. Estrogenic contributions to sexual differentiation in the female guinea pig: Influences of diethylstilbestrol and tamoxifen on neural, behavioral, and ovarian development. *Horm. Behav.* 21:402–17.
268. Ramirez VD and Sawyer CH. 1965. Advancement of puberty in the female rat by estrogen. *Endocrinology* 76:1158–1168.
269. Matsumo A and Arain Y. 1977. Precocious puberty and synaptogenesis in the hypothalamic arcuate nucleus in pregnant mare serum gonadotropin (PMSG) treated immature female rats. *Brain Res.* 129:375–378.
270. Matsumo A and Arain Y. 1980. Sexual dimorphism in "wiring pattern" in the hypothalamic arcuate nucleus and its modifications by neonatal hormonal environment. *Brain Res.* 190:238–242.
271. Lephart BD, Mathews D, Noble JF, and Ojeda SR. 1989. The vaginal epithelium of immature rats metabolizes androgens through an aromatase-like reaction: Changes through the time of puberty. *Biol. Reprod.* 40:259–267.
272. Frem-Titulaer LW, Cordero JF, Haddock L, et al. 1986. Premature theiarche in Puerto Rico: A search for environmental factors. *AJDC* 140:1263–1267.
273. Grant B. 1985. *The Bitter Pill: How Safe Is the "Perfect Contraceptive"?* London: Transworld Publishers. 38.
274. Kimball AM., Hamadeb R, Mahmood RAH, Khalfan S, Mubsin A, Ghabrial F, and Armenian HK. 1981. Gynaecomastia among children in Bahrain. *Lancet* 1:671–672.
275. Grant B. 1985. *The Bitter Pill: How Safe Is the "Perfect Contraceptive"?* London: Transworld Publishers.
276. Wright HP. 1960. General properties of blood: the formed elements. In *Medical Physiology and Biophysics.* Ed. TC Ruch and JF Fulton. Philadelphia: WB Saunders. 502–528.
277. Vessey MP. 1980. Female hormones and vascular disease: An epidemiological overview. *Br. J. Fam. Plann.* 6(3):1.
278. Grant B. 1985. *The Bitter Pill: How Safe Is the "Perfect Contraceptive"?* London: Transworld Publishers. 118.

279. Grant B. 1985. *The Bitter Pill: How Safe Is the "Perfect Contraceptive"?* London: Transworld Publishers. 61–64.
280. Grant, B. 1985. *The Bitter Pill: How Safe Is the "Perfect Contraceptive"?* London: Transworld Publishers. 89.
281. Hibberd AR, O'Connor V, and Gorrod JW. 1978. Detection of nicotine, nicotine-i-N-oxide and cotinine in maternal and foetal body fluids. In *Biological Oxidation of Nitrogen.* Ed. JW Gorrod. New York: Elsevier North Holland Biomedical. 353–361.
282. Hibberd, AR. 1979. *Studies on the Metabolism and Excretion of Nicotine and Some Related Compounds.* PhD thesis. University of London, UK
283. Hibberd AR, Abrahams Y, and Gorrod JW. 1980. Metabolism of nicotine, cotinine and nicotine$\Delta^{1'(5')}$ iminium ion by human foetal liver, in vitro. In *Clinical Pharmacy III.* Ed. H Turakka and B Van der Kleign. New York: Elsevier North Holland Biomedical. 79–88.
284. Jernström H, Klug TL, Sepkovic W, Bradlow H, and Nared SA. 2003. Predictors of the plasma ratio of 2-hydroxyestrone to 16α-hydroxytroe among pre-menopausal-nulliparo women from four ethnic groups. *Carcinogenesis* 24(5):991–1010.
285. Auborn K, Abramson A, Bradlow HL, et al. 1998. Estrogen metabolism and laryngeal papillomatosis: A pilot study on dietary prevention. *Auticancer Res.* 18(6):4569–4573.
286. Gupta M, McDougal A, Safe S. 1998. Estrogenic and antiestrogenic activities of 16 α- and 2-hydroxymetabolite of 17β-estradiol in MCF-7 and T47D human breast cancer cells. *The Journal of Steroid Biochemistry and Molecular Biology.* 67(5):413–419.
287. Bradlow HL, Davis L, Lin G, Sepkovic D, and Tiwari R. 1995. Effects of pesticides on the ratio of 16 α- and 2-hydroxyestrone: A biologic market of breast cancer risk. *Environ Health Perspect.* 103(suppl 7):147–150.
288. McDougal A and Safe S. 1998. Induction of 16 α-/2- hydroxyestrone metabolite ration in MDT-7 cells by pesticides, carcinogens and antiestrogens does not predict mammary carcinogens. *Environ Health Perspect.* 106(4):203–206.
289. Fishman J and Martucci C. 1980. Biological properties of 16 α-hydroxyestrone: Implcation in estrogen physiology and pathophysiology. *J. Clin. Endocrinol. Metab.* 51:611–615.
290. Wright J. 2008. Reducing the hormone related cancer risk international anti-aging system. ias@smart-drugs.com
291. Linus Pauling Institute at Oregon State University. lpi.oregonstate.eduinfocenterphytochemicvals/:3c/-42k-
292. www.chiro.org/nutritionFULL/soy-isoflavones-for-womens-health.shtml-29k. Soy isoflavones for women's health. Is soy a viable alternative to traditional estrogen hormone replacement?
293. Lephart ED, West TW, Weber KS, et al. 2002. Neurobehavioral effects of dietary soy phytoestrogens: Effects of estrogen-like endocrine disrupters on development of brain and behavior. *Neurotoxicol. Teratol.* 24(1):5–16.
294. Messina MJ and Wood E. 2008. Soy isoflavones, estrogen therapy, and breast cancer risk: Analysis and commentary. *Nutri. J.* 7:17.
295. Wood CD, Register T, Franke AA, Anthony MS, and Cline JM. 2006. Dietary soy isoflavones inhibit estrogen effects in the postmenopausal breast. *Cancer Res.* 66:1241–1249.
296. McCann SE, Moysich KB, Freudenheim JL, Ambroson CB, and Shields PG. 2002. The risk of breast cancer associated with dietary lignans differs by CYP17 genotype in women. *J. Nuor.* 132:3036.
297. Thompson DS, Kirshner MA, Klug TL, Kastango KB, and Polleck BK. 2003. A preliminary study of the effect of fluoxetine treatment on the 2:16-alpha-hydroxyestrone ratio in young women. *Ther. Drug Monit.* 25(1):125–128.
298. Frambeise and the Surgical Menopause Survivors: A Survivor's Guide to Surgical Menopause; estrogen interactions. March 22, 2006. Surmeno.blogspot.com/2006/03/estrogen-interactions.htm-88k.
299. Crowell JA, Page JG, Levine BS, Tomlinson MJ, and Hebert CD. 2006. Indole-3-carbinal. But not it major digestive product 3,3-diindolylmethane, induces reversible hypatocyte hypertrophy and cytochrome P450. *Toxicol. App. Pharmacol.* 211(2):115–123.
300. www.rrpf.org/therapies/13c-dim.html-75k-:13c/dim
301. McCann SE, Wactawski-Wende J, Kufel K, et al. 2007. Changes in 2-hydroxyestrone and 16 α- hydroxyestrone metabolism with flaxseed consumption: Modification by COMT and CYP181 genotype. *Cancer Epidemiol. Biomarkers Prev.* 16:256–262.
302. Follen MM, Tox HB, and Levine RU. 1985. Cervical vascular malformation as a cause of antepartum and intrapartum bleeding in three diethylstilbestrol-exposed progeny. *Am. J. Obstet. Gynecol* 153(8):890–891.
303. Witze, A. Feb.14, 1994. Chemicals suspected in endometriosis: Possible role of dioxin, other environmental factors studied. *Dallas Morning News.*

304. Rier SB, Martin DC, Bowman RB, Dmowski WP, and Becker JL. 1993. Endometriosis in rhesus monkeys *(Macaca mulatta)* following chronic exposure to 2,3,7,8-tetrachlorodibenzop-dioxin. *Fundam. Appl. Toxicol.* 21:433–441.
305. Campbell JS, Wong J, Tryphonas L, et al. 1985. Is simian endometriosis an effect of immunotoxicity? Presented at the Ontario Association of Pathologists 48th Annual Meeting, London, Ontario.
306. Poland A and Knutson JC. 1982. 2,3,7,8-tetrachlorodibenzop-dioxin and related halogenated aromatic hydrocarbons: Examination of the mechanism of toxicity. *Ann. Rev. Pharmacol. Toxicol.* 22:517–554.
307. Wood DH, Yochmowitz MG, Salmon YL, Eason RI, and Boster RA. 1983. Proton irradiation and endometriosis. *Aviat. Space Environ. Med.* 54:718–724.
308. Fanton JW and Golden JG. 1991. Radiation-induced endometriosis: In *Macaca mulatta*. *Radiat. Res.* 126:132–140.
309. Wood, DH. 1991. Long term mortality and cancer risk in irradiated rhesus monkeys. *Radiat. Res.* 126:132–140.
310. Dmowski WP, Braun D, and Gebel H. 1991. The immune system in endometriosis. In *Modern Approaches to Endometriosis* Ed. BJ Thomas and JA Rock. Boston: Kluwer Academic. 97–111.
311. Hill JA. 1992. Immunologic factors in endometriosis and endometriosis-associated reproductive failure. *Infertil. Reprod. Med. Clin. N. Am.* 3:583–596.
312. Holsapple MP, Snyder NK, Wood SC, and Morris DL. 1991. A review of 2,3,7,8-tetrachlorodibenzo-p-dioxi n-induced changes in immunocompetence: 1991 update. *Toxicology* 69:219–255.
313. Neubert R, Jacob-Muller U, Stahlmann R, Helge H, and Neubert D. 1991. Polyhalogenated dibenzo-p-dioxins and dibenzofurans and the immune system. *Arch. Toxicol.* 65:213–219.
314. Tomar RS and Kerkvliet NI. 1991. Reduced T-helper cell function in mice exposed to 2,3,7,8-tetrachlorodibenzop-dioxin (TCDD). *Toxicol. Lett.* 57:55–64.
315. Tsukino H, Hanaoka T, Sasaki H, et al. 2005. Associations between serum levels of selected organochloine compounds and endometriosis in infertile Japanese women. *Environ Res.* Sep 99(1):118–125.
316. Heiller JF, Donnez J, Lison D. 2008. Organochlorines and endometriosis: A mini-review. *Chemosphere* 71(2):203–210. Epub 2007 Nov 19.
317. Sikorski R, Juszkiewicz T, Paszkowski T, and Szprengier-Juszkiewicz T. 1987. Women in dental surgeries: Reproductive hazard in occupational exposure to metallic mercury. *Int. Arch. Occup. Environ. Health* 59(6):551–557.
318. Landrigan P J, Meinhardt TJ, Gordon J, Lipscomb JA, et al. 1984. Ethylene oxide: An overview of toxicologic and epidemiologic research. *Am. J. Ind. Med.* 6(2):103–115.
319. Selevan SG, Lindbohm ML, Hornung RW, and Heniminki K. 1985. A study of occupational exposure to anti neoplastic drugs and fetal loss in nurses. *N. Engl. J. Med.* 313(19):1173–1178.
320. Heidam LZ. 1984. Spontaneous abortions among dental assistants, factory workers, painters, and gardening workers: a follow-up study. *J. Epidemiol. Community Health* 38(2):149–155.
321. Axelson O, Johansson B, and Flodin U. 1983. Unidentified risk factor [letter]. *J. Occup. Med.* 25(3):181.
322. Axelson O, Edling C, and Andersson L. 1983. Pregnancy outcome among women in a Swedish rubber plant. *Scand. J. Work Environ, Health* 2(Suppl.):79–83.
323. Lindbohm ML, Hemminki K, Kyyrönen P. 1985. Spontaneous abortion among women employed in the plastics industry. *Am. J. Ind. Med.* 8(6):579–586.
324. Ungváry G and Tárai B. 1985. On the embryotoxic effects of benzene and its alkyl derivatives in mice, rats and rabbits. *Arch. Toxicol.* [Suppl.] 8:425–430.
325. Curle DC, Ray M, and Persaud TV. 1983. Methylmercury toxicity: In vivo evaluation of teratogenesis and cytogenetic changes. *Anat. Anz.* 153(1):69–82.
326. Varma DR. 1987. Epidemiological and experimental studies on the effects of methyl isocyanate on the course of pregnancy. *Environ. Health Perspect.*72:153–157.
327. Ludmir J, Landon MB, Gabbe SG, Samuels P, and Mennuti MT. 1987. Management of the diethylstibestrol-exposed pregnant patient: A prospective study. *Am. J. Obstet. Gynecol.* 157(3):665–669.
328. Menczer J, Dulitzky M, Ben-Baruch G, and Modan M. 1986. Primary infertility in women exposed to diethylstiboestrol in utero. *Br. J. Obstet. Gynaecol.* 93(5):503–507.
329. Henderson L and Regan T. 1985. Effects of diethylstiboestrol-dipropionate on SCEs, micronuclei, cytotoxicity, aneuploidy and cell proliferation in maternal and foetal mouse cell treated in vivo. *Mutat. Res.* 144(1):27–31.
330. Ben Ismail M, Abid F, Trablsi S, Taktak M, and Fekih M. 1986. Cardiac valve prostheses, anticoagulation, and pregnancy. *Br. Heart J.* 55(1):101–105.
331. Lee PK, Wang RY, Chow JS, et al. 1986. Combined use of warfarin and adjusted subcutaneous heparin during pregnancy in patients with an artificial heart value. *J. Am. Coll. Cardiol.* 8(1):221–224.

Endocrine System

332. O'Donnell D, Sevitz H, Seggie JL, Meyers AM, Botha JR, and Myburgh JA. 1985. Pregnancy after renal transplantation. *Aust. N.Z. J. Med.* l5(3):320–325.
333. Lammer BJ, Chen DT, Hoar RM, Agnish ND, et al. 1985. Retinoic acid embryopathy. *N. Engl. J. Med.* 313(14):837–841.
334. Stern RS, Rosa F, and Baum C. 1984. Isotretinoin and pregnancy. *J. Am. Acad. Dermatol.* 10(5 Pt l): 851–854.
335. Rosa FW, Wilk AL, and Kelsey FO. 1986. Teratogen update: Vitamin A congeners. *Teratology* 33:355–364.
336. Rosa FW, Baum C, and Shaw M. 1987. Pregnancy outcomes after first-trimester vaginitis drug therapy. *Obstet. Gynecol.* 69(5):75 1–755.
337. Matsuoka R, Gilbert BF, Bruyers H Jr., and Optiz JM. 1985. An aborted human fetus with truncus arteriosus communis possible teratogenic effect of Tedral. *Heart Vessels* 1(3):176–178.
338. Källen B and Tandberg A. 1983. Lithium and pregnancy: a cohort study on manic depressive women. *Acta Psychiatr. Scand.* 68(2):134–139.
339. Kihlstroöm I and Lundberg C. 1987. Teratogenicity study of the new glucocorticosteroid budesonide in rabbits. *Arzneimittel Forsch.* 37(1):43–46.
340. Olab MB and Rahwan RG. 1986. Evaluation of the antiabortifacient and embryotoxic effects of methylenedioxyindene and methylenedioxyindan calcium antagonists. *Gen. Pharmacol.* 17(5):549–552.
341. Kar RN, Khan K, and Mukherjee SK. 1984. In vivo mutagenic effect of methyldopa: I. Dominant lethal test in male mice. *Cytobios* 41:151–159.
342. Lynch PJ. 1984. Abortion in sows after injection of a suspension of penicillin and streptomycin. *Aust. Vet. J.* 61(1):29.
343. Nikkanen V, Katainen P, and Puroinen O. 1987. Progesterone support of the luteal phase in in vitro fertilization program: A hazard? *Ann. Chir. GynacoL* [Suppl.] 202:42–44.
344. Bohrer M and Kemmann B. 1987. Risk factors for spontaneous abortion in menotropin-treated women. *Fertil. Steril.* 48(4):571–575.
345. Rock JA, Wentz AC, Cole KA, et al. 1985. Fetal malformations following progesterone therapy during pregnancy: A preliminary report. *Fertil. Steril.* 44(1):17–19.
346. Hendrickx AG, Korte R, Leuschner F, et al. 1987. Embryotoxicity of sex steroidal hormone combinations in nonhuman primates: I. Norethisterone acetate + ethinyl estradiol and progesterone + estradiol benzoate *(Macaca mulatta, Macaca fascicularis,* and *Papio cynocephalus). Teratology* 35(1):119–127.
347. Hendrickx AG, Korte R, Leuschner F, et al. 1987. Embryotoxicity of sex steroidal hormones in nonhuman primates: 11. Hydroxyprogesterone caproate, estradiol valerate. *Teratology* 35(1):129–136.
348. Foster WG, Agzarian J. 2008. Toward less confusing terminology in endocrine disruptor research. *J Toxicol. Environ. Health B Crit. Rev.* 11(3–4):152–161.
349. Rogan WJ, Gladen BC, and Wilcox AJ. 1985. Potential reproductive and postnatal morbidity from exposure to polychiorinated biphenyls: epidemiologic considerations. *Environ. Health Perspect.* 60:233–239.
350. Bercovici B, Wassermann M, Cucos S, Ron M, Wassermann D, and Pines A. 1983. Serum levels of polychlorinated biphenyls and some organochlorine insecticides in women with recent and former missed abortions. *Environ. Res.* 30(l):169–174.
351. McNulty WP.1985. Toxicity and fetotoxicity of TCDD, TCDF and PCB isomers in rhesus macaques (Macaca mulatta). *Environ. Health Perspect.* 60:77–88.
352. Rita P, Reddy PP, and Reddy SV. 1987. Monitoring of workers occupationally exposed to pesticides in grape gardens of Andhra Pradesh. *Environ. Res.* 44(1):l–5.
353. Sterling TD and Arundel AV. 1986. Review of recent Vietnamese studies on the carcinogenic and teratogenic effects of phenoxy herbicide exposure. *Int. J. Health Serv.* 16(2):265–278.
354. Kaye CI, Rao S, Simpson SJ, Rosenthal FS, and Cohen MM. 1985. Evaluation of chromosomal damage in males exposed to Agent Orange and their families. *J. Craniofac. Genet. Dev. Biol.* [Suppl.] 1:259–265.
355. Blume SB. 1986. Women and alcohol: a review. *JAMA* 256(11):1467–1470.
356. Abel BT. 1984. Prenatal effects of alcohol. *Drug Alcohol Depend* 14(1):1–10.
357. Kaufman MH. 1983. Ethanol-induced chromosomal abnormalities at conception. *Nature* 302(5905): 258–260.
358. Clarren S K, Bowden DM, and Astley SJ. 1987. Pregnancy outcomes after weekly oral administration of ethanol during gestation in the pig-tailed macaque *(Macaca nemestrina). Teratology* 35(3):345–354.
359. Asch RH and Smith CG. 1986. Effects of delta 9-THC, the principal psychoactive component of marUuana, during pregnancy in the rhesus monkey. *J. Reprod. Med.* 31(12):1071–1081.

360. Chasnoff IJ, Burns WJ, Schnoll SH, and Burns KA. 1985. Cocaine use in pregnancy. *N. Engl. J. Med.* 313(11):666–669.
361. Srisuphan W and Bracken MR. 1986. Caffeine consumption during pregnancy and association with late spontaneous abortion. *Am. J. Obstet. Gynecol.*154(1):14–20.
362. Eroschenko VP and Osman F. 1986. Scanning electron microscopic changes in vaginal epithelium of suckling neonatal mice in response to estradiol or insecticide chlorodecone (kepone) passages in milk. *Toxicology* 38(2):175–185.
363. Rolirson AK, Schmidt WA, and Stancel GM. 1985. Estrogenic activity of DDT: estrogen-receptor for profiles and the responses of individual uterine cell types following o,p-DDT administration. *Toxicol. Environ. Health* 16(3–4):493–508.
364. EHC-Dallas. 1994. Unpublished data.
365. Graham B, Chignell AH, and Eykyn S. 1986. Candida endophthalmitis: a complication of prolonged intravenous therapy and antibiotic treatment. *J. Infect.*13(2):167–173.
366. McKay M. 1993. Dysesthetic ("essential") vulvodynia: treatment with amitriptyline. *J. Reprod. Med.* 38(1):9–13.
367. Colborn T, et al. 1996. *Our Stolen Future: Are We Threatening Our Fertility, Intelligence and Survival? A Scientific Detective Story.* New York: Penguin Books USA.
368. Ohanjanyan, O. 1999. Health and Environment officer, WECF, "Persistent Organic Pollutants and Reproductive Health, 'Hormorne Disrupters.'" Sept. 5. http://www.earthsummit2002.org/wcaucus/Caucus%Position%20Papers/agriculture/pest.
369. Colborn T, et al. 1996. *Our Stolen Future: Are We Threatening Our Fertility, Intelligence and Survival? A Scientific Detective Story.* New York: Penguin Books USA. 145–159.
370. "Mickey/Minnie." 1999. *Earth Island Journal.* 14(3). Gary Santolo of the Sacramento, California, consulting firm of C2HMHill headed the Kesterson filed study.
371. Jett DA, et al. 2001. Cognitive function and cholinergic neurochemistry in weanling rats exposed to chlorpyifos. *Toxicol. Appli. Pharm.* 174(2):89–98.
372. Colborn T, et al. 1996. *Our Stolen Future: Are We Threatening Our Fertility, Intelligence and Survival? A Scientific Detective Story.* New York: Penguin Books USA. 192–195.
373. Daly H. 1993. Laboratory rat experiments show consumption of Lake Ontario salmon causes behavioral changes: support for wildlife and human research results. *J. Gr. Lakes Res.* 19(4):784–788.
374. Luoma JR. 1995. Havoc in the Hormones. *Audubon*, July/August. www.magazine.audubon.org.
375. Facemire C, et al. 1995. Reproductive impairment in the Florida panther: Nature or nurture? *Envir. Health Persp. Suppl.* 103(4):79–86.
376. Why do you think they call them DIE-oxins? *Endometriosis Association Newsletter,* Spring 2002.
377. Hayes H, et al. 1991. Case control study of canine malignant lymphoma: positive association with dog owner's use of 2,4 Dichlorophenoxyacetic acid herbicides. *J. Nat. Cancer Instit.* 83(17).
378. Cox C. 1996. Nonyl phenol and related compounds. *J Pesticide Reform.* 16(1):15–20.
379. Johnson C. Endocrine disrupting chemicals with transexualism. www.transadvocate.org/news/htm.
380. Giesy J, et al. 1994. Deformities in birds of the Great Lakes Region:Assigning causality. *Environ Sci & Techn* 28(3):128–135.
381. Skakkebaek N. 1993. *Sexual Abnormalities: Assault on the Male.* Horizon video BBC.
382. Male infertility and other reproductive problems in men. *Environ Health.* March 17, 2002. http://eces.org/ec/health/malereproductiveproblems.shtml.
383. Rapp DJ. 2001. *How to Keep Yourself and Your Loved Ones Out of Harm's Way: Our Toxic World,* 5th ed. Buffalo, NY: Environmental Medical Research Foundation. 171.
384. Gehard I. 1991. Prolonged exposure to wood preservatives induces endocrine and immunologic disorders in women. *Am. J. Obstet. & Gyn.* 165(2):487–88.
385. Sherman, JM. *Chemical Exposure and Disease: Diagnostic and Investigative Techniques.* Princeton, NY: Princeton Scientific 94.
386. IARC Monographs on the Evaluation of the Carcinogenic Risk of Chemical to Humans: Some Halogenated Hydrocarbons. 1979. International Agency for Research on Cancer, Lyon, France. 20:575–585.
387. Ball HS. 1984. Effect of methoxychlor on reproduction systems of the rat. *Proc. Soc. Exp. Biol. Med.* 176(2):187–196.
388. Carter SD, Hem JF, Rehnberg GL, and Laskey JW. 1984. Effect of benomyl on the reproductive development of male rats. *J. Toxicol. Environ. Health* 13(1):53–68.
389. Common Chemical Exposure May Affect Male Reproductive Development. Sudy finds June 01, 2005, http://www.usmc.rochester.edu/news/story/index.cfm?id=807

390. Melcarek H. 1999. Chemicals Found to Affect Male Reproductive System in New Way: Pesticide and You Beyond Pesticides/National Coalition Against the Misuse of Pesticides. 19(1):18–19.
391. Saridifer SH, Wilkins RT, Loadholt CB, Lane LG, and Eldridge JC. 1979. Spermatogenesis in agricultural workers exposed to dibromochloropropane (DBCP). *Bull. Environ. Contam. Toxicol.* 23(4–5):703–710.
392. Swartz WJ. 1984. Effects of 1,1-bis-[p-chlorophenyl]-2,2,2-trichloroethane (DDT) on gonadal development in the chick embryo: a histological and histochemical study. *Environ. Res.* 35(2):333–345.
393. Giwercman A and Skaltkeback NB. 1992. The human testis: an organ at risk? *Int. J. Androl.* 15:373–375.
394. Osterlind A. 1986. Diverging trends in incidence and mortality of testicular cancer in Denmark, 1943–1982. *Br. J. Cancer* 53:501–505.
395. Jackson MB. John Radcliffe Hospital Cryptochidism Research Group. 1988. The epidemiology of cryptorchidism. *Horm. Res.* 30:153–156.
396. Carlsen B, Giwercman A, Keiding N, and Skakkeback, NB. 1992. Evidence for decreasing quality of semen during past 50 years. *Br. Med. J.* 305:609–613.
397. Forest MG. 1982. Development of the male reproductive tract. In *Aspects of Male Infertility* Ed. R de vere White. Baltimore: Williams & Wilkins. 1–60.
398. Hutson JM, Williams MPL, Fallat MB, and Arrah A. 1990. Testicular descent: New insights into its hormonal control. In *Oxford Reviews of Reproducuve Biology.* Ed. SR Milligan. Oxford: Oxford University Press. 1–56.
399. Skakkeback N B. 1987. Carcinoma: In-situ and cancer of the testis. *Int. J. Androl.* 10:1–40.
400. Stillman RJ. 1982. In utero exposure to diethylstilbestrol: adverse effects on the reproductive tract and reproductive performance in male and female offspring. *Am. J. Obstet. Gynecol.* 142:905–921.
401. Arai Y, Mon T, Suzuki Y, and Bern HA. 1983. Long-term effects of perinatal exposure to sex steroids and diethylstilbestrol on the reproductive system of male mammals. *Int. Rev. Cytol.* 84:235–268.
402. Prener A, Hsieh CC, Engholm G, Trichopoulos GD, and Jensen OM. 1992. Birth order and risk of testicular cancer. *Cancer Causes Control* 3:265–272.
403. Depue RH. 1984. Maternal and gestational factors affecting the risk of cryptochidism and inguinal hernia. *Int. J. Epidemiol.* 13:311–318.
404. Field B, Selub M, Hughes CL. 1990. Reproductive effects of environmental agents. *Semin. Reprod. Endocrinol* 8:44–54.
405. Solomon GM and Schettler T. Environment and health: 6. Endocrine disruption and potential human health implications. *CMAJ* 163(11).
406. Committee on Hormonally Active Agents in the Environment, National Research Council. Hormonally Active Agents in the Environment. 1999. Washington, DC: National Academy Press.
407. Cooper RL and Kavlock RJ. 1997. Endocrine disruptors and reproductive development: A weight-of-evidence overview. *J. Endocrinol.* 152:159–166.
408. Swartz WJ and Corkern M. 1992. Effects of methoxychlor treatment of pregnant mice on female offspring of the treated and subsequent pregnancies. *Reprod. Toxicol.* 6(5):431–437.
409. Brouwer A, Morse DC, Lans MC, Schuur AG, Murk AJ, Klasson-Wehler E, et al. 1998. Interactions of persistent environmental organohalogens with the thyroid hormone system: mechanisms and possible consequences for animal and human health. *Toxicol. Ind. Health* 14:59–84.
410. Bigsby R, Chapin RE, Daston GP, Davis BJ, Gorski J, Gray LE, et al. 1999. Evaluating the effects of endocrine disruptors on endocrine function during development. *Environ. Health Perspect.* 107(Suppl 4): 613–618.
411. Haddow JE, Palomaki GE, Allan WC, Williams JR, Knight GJ, Gagnon J, et al. 1999. Maternal thyroid deficiency during pregnancy and subsequent neuropsychological development of the child. *N. Engl. J. Med.* 341(8):549–555.
412. Crisp TM, Clegg ED, Cooper RL, Wood WP, Anderson DG, Baetcke KP, et al. 1998. Environmental endocrine disruption: an effects assessment and analysis. *Environ. Health Perspect.* 106(Suppl 1):11–56.
413. Vom Saal FS, Cooke PS, Buchanan DL, Palanza P, Thayer KA, Nagel SC, et al. 1998. A physiologically based approach to the study of bisphenol A and other estrogenic chemicals on the size of reproductive organs, daily sperm production, and behavior. *Toxicol. Ind. Health* 14:239–260.
414. Giusti RM, Iwamoto K, Hatch EE. 1995. Diethylstilbestrol revisited: a review of the long-term health effects. *Ann. Intern. Med.* 122:778–788.
415. Vom Saal FS, Nagel SC, Palanza P, Boechler M, Parmigiani S, Welshons WV. 1995. Estrogenic pesticides: binding relative to estradiol in MCF-7 cells and effects of exposure during fetal life on subsequent territorial behavior in male mice. *Toxicol. Lett.* 77(1–3):343–350.
416. Gray LE. 1998. Xenoendocrine disruptors: laboratory studies on male reproductive effects. *Toxicol. Lett.* 102–103:331–335.

417. Gray LE, Ostby J, Monosson E, Kelce WR. 1999. Environmental antiandrogens: low doses of the fungicide vinclozolin alter sexual differentiation of the male rat. *Toxicol. Ind. Health* 15(1–2):48–64.
418. Jacobson JL and Jacobson SW. 1996. Intellectual impairment in children exposed to polychlorinated biphenyls in utero. *N. Engl. J. Med.* 335:783–789.
419. Zoeller RT, Dowling A, and Vas AA. 2000. Developmental exposure to polychlorinated biphenyls exerts thyroid-like effects on the expression of RC3/neurogranin and myelin basic protein messenger ribonucleic acids in the developing rat brain. *Endocrinology* 141:181–189.
420. Bradlow HL, Davis DL, Lin G, Sepkovic D, Tiwari R. 1995. Effects of pesticides on the ratio of 16/2-hydroxyestrone: a biologic marker of breast cancer risk. *Environ. Health Perspect.* 103(Suppl 7):147–150.
421. Tran DQ, Kow KY, McLachlan JA, Arnold SF. 1996. The inhibition of estrogen receptor-mediated responses by chloro-s-triazine-derived compounds is dependent on estradiol concentration in yeast. *Biochem. Biophys. Res. Commun.* 227:140–146.
422. Cooper RL, Goldman JM, Stoker TE. 1999. Neuroendocrine and reproductive effects of contemporary-use pesticides. *Toxicol. Ind. Health* 15:26–36.
423. Donna A, Crosignani P, Robutti F, Betta PG, Bocca R, Mariani N, et al. 1989. Triazine herbicides and ovarian epithelial neoplasms. *Scand. J. Work. Environ. Health* 15:47–53.
424. Kettles MA, Browning SR, Prince TS, Horstman SW. 1996. Triazine herbicide exposure and breast cancer incidence: an ecologic study of Kentucky counties. *Environ. Health Perspect.* 105:1222–1227.
425. Steenland K, Cedillo L, Tucker J, Hines C, Sorensen K, Deddens J, et al. 1997. Thyroid hormones and cytogenetic outcomes in backpack sprayers using ethylenebis(dithiocarbamate) (EBDC) fungicides in Mexico. *Environ. Health Perspect.* 105:1126–1130.
426. Sweeney MH, Calvert GM, Egeland GA, Fingerhut MA, Halperin WE, and Piacitelli LA. 1997–98. Review and update of the results of the NIOSH medical study of workers exposed to chemicals contaminated with 2,3,7,8-tetrachlorodibenzodioxin. *Teratog. carcinog. mutagen.* 17(4–5):241–247.
427. Manzo L, Artigas F, Martinez E, Mutti A, Bergamaschi E, Nicotera P, et al. 1996. Biochemical markers of neurotoxicity: a review of mechanistic studies and applications. *Hum Exp Toxicol* 15(Suppl. 1):S20–35.
428. Hardell L, Ohlson CG, and Fredrikson M. 1997. Occupational exposure to polyvinyl chloride as a risk factor for testicular cancer evaluated in a case-control study. *Int. J. Cancer.* 73:828–830.
429. Whelan EA, Grajewski B, Wild DK, Schnorr TM, and Alderfer R. 1996. Evaluation of reproductive function among men occupationally exposed to a stilbene derivative: II. Perceived libido and potency. *Am. J. Ind. Med.* 29:59–65.
430. Gold EB and Tomich E. 1994. Occupational hazards to fertility and pregnancy outcome. *Occup. Med.* 9:435–469.
431. Weidner IS, Moller H, Jensen TK, and Skakkebaek NE. 1998. Cryptorchidism and hypospadias in sons of gardeners and farmers. *Environ. Health Perspect.* 106: 793–796.
432. Hunter DJ, Hankinson SE, Laden F, Colditz GA, Colditz GA, Manson JE, Willett WC, et al. 1997. Plasma organochlorine levels and the risk of breast cancer. *N. Engl. J. Med.* 337:1253–1258.
433. Hoyer AP, Grandjean P, Jorgensen T, Brock JW, and Hartvig HB. 1998. Organochlorine exposure and risk of breast cancer. *Lancet* 352:1816–1820.
434. Paulozzi LJ. 1999. International trends in rates of hypospadias and cryptorchidism. *Environ. Health Perspect.* 107:297–302.
435. Younglai EV, Collins JA, and Foster WG. 1998. Canadian semen quality: an analysis of sperm density among eleven academic fertility centers. *Fertil. Steril.* 70:76–80.
436. Swan S, Elkin EP, and Fenster L. 1997. Have sperm densities declined? A reanalysis of global trend data. *Environ. Health Perspect.* 105:1228–1232.
437. Liu S, Wen SW, Mao Y, Mery L, and Rouleau J. 1999. Birth cohort effects underlying the increasing testicular cancer incidence in Canada. *Can. J. Public Health* 90(3):176–180.
438. McKiernan JM, Goluboff ET, Liberson GL, Golden R, and Fisch H. 1999. Rising risk of testicular cancer by birth cohort in the United States from 1973–1995. *J. Urol.* 162(2):361–363.
439. Bergstrom R, Adami HO, Mohner M, et al. 1996. Increase in testicular cancer incidence in six European countries: A birth cohort phenomenon. *J. Natl. Cancer Inst.* 88:727–733 [Abstract].
440. Levy IG, Iscoe NA, Klotz LH. 1998. Prostate cancer: 1. The descriptive epidemiology in Canada. *CMAJ* 159(5):509–513. www.cma.ca/cmaj/vol-159/issue-5/0509.htm [Abstract].
441. Haas GP and Sakr WA. 1997. Epidemiology of prostate cancer. *CA Cancer J Clin* 47:273–287.
442. Wang PP and Cao Y. 1996. Incidence trends of female breast cancer in Saskatchewan, 1932–1990. *Breast Cancer Res. Treat.* 37(3):197–207.
443. Wolff MS, Collman GW, Barrett JC, and Huff J. 1996. Breast cancer and environmental risk factors: epidemiological and experimental findings. *Annu. Rev. Pharmacol. Toxicol.* 36:573–596 [Abstract].

444. Allan BB, Brant R, Seidel JE, and Jarrell JF. 1997. Declining sex ratios in Canada. *CMAJ* 156(1):37–41. www.cma.ca/cmaj/vol-156/issue-1/0037.htm [Abstract].
445. Herman-Giddens ME, Slora EJ, Wasserman RC, Bourdony CJ, Bhapkar MV, Koch GG, et al. 1997. Secondary sexual characteristics and menses in young girls seen in office practice: a study from the Pediatric Research in Office Settings network. *Pediatrics* 99:505–512.
446. Mably TA, Moore RW, Goy RW, Peterson RE. 1992. In utero and lactational exposure of male rats to 2,3,7,8-tetrachlordibenzo-p-dioxin: 2. Effects on sexual behavior and the regulation of LH secretion in adulthood. *Toxicol Appl Pharmacol* 114:108–117.
447. Leatherland JF. 1998. Changes in thyroid hormone economy following consumption of environmentally contaminated Great Lakes fish. *Toxicol. Ind. Health* 14:41–57.
448. Daly HB, Stewart PW, Lunkenheimer L, and Sargent D. 1998. Maternal consumption of Lake Ontario salmon in rats produces behavioral changes in the offspring. *Toxicol. Ind. Health* 14:25–39.
449. Moccia RD, Fox GA, and Britton A. 1986. A quantitative assessment of thyroid histopathology of herring gulls from the Great Lakes and a hypothesis on the causal role of environmental contaminants. *J. Wildl. Dis.* 22:60–70.
450. Rice DC. 1999. Behavioral impairment produced by low-level postnatal PCB exposure in monkeys. *Environ. Res.* 80(2 pt 2):S113-121.
451. Koopman-Esseboom C, Morse DC, Weisglas-Kuperus N, Lutkeschipholt IJ, Van der Paauw CG, Tuinstra LG, et al. 1994. Effects of dioxins and polychlorinated biphenyls on thyroid hormone status of pregnant women and their infants. *Pediatr Res* 36:468–473 [Abstract].
452. Osius N, Karmaus W, Kruse H, and Witten J. 1999. Exposure to polychlorinated biphenyls and levels of thyroid hormones in children. *Environ Health Perspect* 107:843–849.
453. Tilson HA and Kodavanti PR. 1997. Neurochemical effects of polychlorinated biphenyls: An overview and identification of research needs. *Neurotoxicology* 18:727–743.
454. Jacobson JL and Jacobson SW. 1990. Effects of in utero exposure to PCBs and related contaminants on cognitive functioning in young children. *J. Pediatr.* 116:38–45.
455. Lonky E, Reihman J, Darvill T, Mather J, and Daly H. 1996. Neonatal behavioral assessment scale performance in humans influenced by maternal consumption of environmentally contaminated Lake Ontario fish. *J. Great Lakes Res.* 22:198–212.
456. Muckle G, Dewailly E, and Ayotte P. 1998. Prenatal exposure of Canadian children to polychlorinated biphenyls and mercury. *Can. J. Public Health* 89(Suppl. 1):S20-5, 22-7.
457. Lauder JM. 1988. Neurotransmitters as morphogens. *Prog. Brain Res.* 73:365–387.
458. Ahlbom J, Fredriksson A, and Eriksson. 1995. Exposure to an organophosphate (DFP) during a defined period in neonatal life induces permanent changes in brain muscarinic receptors and behavior in adult mice. *Brain Res.* 677:13–19.
459. Lauder JM and Schambra UB. 1999. Morphogenic roles of acetylcholine. *Environ. Health Perspect.* 107(Suppl. 1):65–69.
460. Bigbee JW, Sharma KV, Gupta JJ, and Dupree JL. 1999. Morphogenic role for acetylcholinesterase in axonal outgrowth during neural development. *Environ. Health Perspect* 107(Suppl. 1):81–87.
461. Rice DC. 1998. Issues in developmental neurotoxicology: Interpretation and implications of the data. *Can. J. Public Health* 89(Suppl. 1):S31-6, S34-40.
462. Hill RH Jr, Head SL, Baker S, Gregg M, Shealy DB, Bailey SL, et al. 1995. Pesticide residues in urine of adults living in the United States: reference range concentrations. *Environ. Res.* 71:99–108.
463. Archibeque-Engle SL, Tessari JD, Winn DT, Keefe TJ, Nett TM, and Zheng T. 1997. Comparison of organochlorine pesticide and polychlorinated biphenyl residues in human breast adipose tissue and serum. *J. Toxicol. Environ. Health* 52:285–293.
464. Adlercreutz H. 1990. Diet, breast cancer and sex hormone metabolism. *Ann. N.Y. Acad. Sci.* 595:281–290.
465. Bernstein L, Depue RH, Ross PK, et al. 1986. Higher maternal levels of free estradiol in first compared to second pregnancy: A study of early gestational differences. *J. Natl. Cancer Inst.* 76:1035–1044.
466. Moss AR, Osmond D, Rachetti P, Torti FM, and Gurgin V. 1986. Hormonal risk factors in testicular cancer: A case control study. *Am. J. Epidemiol.* 124:39–52.
467. McLachlan JA., ed. 1985. *Estrogens in the Environment.* Amsterdam: Elsevier.
468. Lamming GE. 1985. Report of the scientific group on anabolic agents in animal production. *Comm. Eur. Commun. Rep. Eur.* 8913:4–25.
469. Sheehan DM and Young M. 1979. Dielthystilbestrol and estradiol binding to serum albumin and pregnancy plasma of rat and human. *Endocrinology* 104:1442–1446.
470. See source of table. Arnold SF, et al. 1996. *Science.* Tab. 17.

471. See source of table. Arnold SF, et al. 1996. *Science.* Tab. 18.
472. Hegsted DM, McGandy RG, Myers ML, Stare SM, and Stare FJ. 1965 Quantitative effects of dietary fat on serum cholesterol in man. *Am. J. Clin. Nutr.* 17:281–295.
473. Keys A, Anderson JT, and Grand F. 1965. Serum cholesterol response to changes in the diet: IV. Particular saturated fatty acids in the diet. *Metabolism* 14:776–787.
474. Grundy SM and Denke MA. 1990. Dietary influences on serum lipids and lipoproteins. *J. Lipid. Res.* 31:1149–1172.
475. Howell WH, McNamara DJ, Tosca MA, Smith BT, and Gaines JA. 1997. Plasma lipid and liprprotein responses to dietary fat and cholesterol: A meta-analysis. *Am. J. Clin. Nutr.* 65:1747–1764.
476. Khosla P and Sundram K. 1996. Effects of dietary fatty acid composition on plasma cholesterol. *Prog. Lipid. Res.* 35:93–132.
477. Yu S, Derr J, Etherton TD, and Kris-Etherton PM. 1995. Plasma cholesterol-predictive equations demonstate that steraic acid is neutral and monounsaturated fatty acids are hypocholesterolemic. *Am. J. Clin. Nutr.* 61:1129–1139.
478. Ostlund RE Jr., Racette SB, and Stenson WF. 2002. Effects of trace components of dietary fat on cholesterol metabolism: phytosterols, oxysterols, and squalene. *Ann. Rev. Nutr.* 6(12):349–359.
479. Field B, Selub M, and Hughes, CL. 1990. Reproductive effects of environmental agents. *Semin. Reprod. Endocrinol.* 8:44–54.
480. Colborn T and Clement C, eds. 1992. *Chemically-Induced Alteration in Sexual and Functional Development: The Wildlife/Human Connection.* Princeton, NJ: Princeton Scientific Publishing.
481. Hamon M, Fleet IR, and Heap RB. 1990. Comparison of oestrone sulphate concentrations in mammary secretions during lactogenesis and lactation in dairy ruminants. *J. Dairy Res.* 57:419–422.
482. Sager D, Girar D, and Nelson D. 1991. Early postnatal exposure to PCBs: spasm function in rats. *Environ. Toxicol. Chem.* 10:737–746.
483. Mably TA, Bjerke DL, Moore RW, Gendron-Fitzpatrick A, and Peterson, RB. 1992. In utero and lactational exposure of male rats to 2,3,7,8-tetrachlorodibenzo-p-dioxin 3: effects on spermatogenesis and reproductive capability. *Toxicol. Appl. Pharmacol.* 114:118–126.
484. Sharpe RM and Skakkeback NB. 1993. Are oestrogens involved in falling sperm counts and disorders of the male reproductive tract? *Lancet* 341:1392–1395.
485. Sharpe RM. 1993. Falling sperm counts in men: is there an endocrine cause? *J. Endocrinol.* 137:357–360.
486. Cortes D, Muller J, and Skakkeback NE. 1987. Proliferation of Sertoli cells during development of the human testis assessed by sterological methods. *Int. J. Androl.* 10:589–596.
487. Hirobe S, He W-W, Lee MM, and Donahoe PK. 1992. Mollerian inhibiting substance messenger ribonucleic acid expression in granulosa and Sertoli cells coincides with their mitotic activity. *Endocrinology* 131:854–862.
488. Russell LD and Peterson RN. 1984. Determination of the elongate spermatid sertoli cell ratio in various mammals. *J. Reprod. Fertil.* 70:635–641.
489. Orth JM, Gunsalus GM, and Lamperti AA. 1988. Evidence from sertoli cell-depleted rats indicates that spermatid numbers in adults depend on number of serttoi cells produced during prenatal development. *Endocrinology* 122:787–794.
490. Cooke S, Porcelli J, and Hess RA. 1992. Induction of increased testis growth and sperm production in adult rats by neonatal administration of the goitrogen prophylthiouracil (PTU): the critical period. *Biol. Reprod.* 46: 146–154.
491. Hunt VR. 1982. The reproductive system: sensitivity through the life cycle. *Ann. Am. Conf. Ind. Hyg.* 3:53–59.
492. Hunt, VR. 1979. *Work and the Health of Women.* Boca Raton, FL: CRC Press.
493. Rosenblum AH and Rosenblum P. 1952. Gastrointestinal allergy in infancy: significance of eosinophiles in the stools. *Pediatrics* 9:311.
494. Eishi Y and McCullagh P. 1988. PVG rats, resistant to experimental allergic thyroiditis, develop high serum levels of thyroidglobulin after sensitization. *Clin. Immunol. Immunopathol.* 49(1):101–106.
495. Mitsunaya M. 1987. Cytophilic anti-thyroglobulin antibody and antibody-dependent macrophage-mediated cytotoxicity in Hashimoto's thyroiditis. *Acta Med. Okayama* 41(5):205–214.
496. Bahn AK, Mills JL, Snyder PJ, Gann PH, Houten L, Bialik O, Hollmann L, and Utiger, RD. 1980. Hypothyroidism in workers exposed to polybrominated biphenyls. *N. Engl. J. Med.* 302:31–33.
497. Gaitan B, Island DP, and Liddle GW. 1969. Identification of a naturally occurring goitrogen in water. *Trans. Assoc. Am. Physicians* 132:141–152.
498. Bastomsky CH. 1977. Goiters in rats fed polychlorinated biphenyls. *Can. J. Physiol. Pharmacol.* 55:288–292.

499. Saifer P. 1988. Personal communication.
500. Price SC, Ozalp S, Weaver R, Chescoe D, Mullervy D, and Hinton RH. 1988. Thyroid hyperactivity caused by hypolipodaemic compounds and polychlorinated biphenyls: The effect of coadministration in the liver and thyroid. *Arch. Toxicol.* [Suppl] 12:85–92.
501. Gaitan B. 1986. Environmental goitrogens. In *The Thyroid Gland: A Practical Clinical Treatise.* Ed. L Van Middlesworth. Chicago: Year Book Medical Publishers. 263–280.
502. Gaitan B. 1980. Goitrogens in the etiology of endemic goiter. In *Endemic Goiter and Endemic Cretinism.* Ed. JB Stanbury and B Hetzel. New York: John Wiley & Sons.
503. Langer P and Greer MA, eds. 1977. *Antithyroid Substances and Naturally Occurring Goitrogens.* Basel, S. Karger, A.G.
504. Ermans AM, Mbulamoko NB, Delange F, Ahiuwalia R, eds. 1980. *Role of Cassava in the Etiology of Endemic Goiter and Cretinism.* IDRC- 1 36e. Ottawa: International Development Research Centre.
505. Van Etten CH. 1969. Gotrogens. In *Toxic Constituents of Plant Foodstuffs.* Ed. IB Liener. New York: Academic Press.
506. Gaitan B, Cooksey RC, Matthew D, and Presson R. 1983. In vitro measurement of antithyroid compounds and environmental goitrogens. *J. Clin. Endocrinol. Metab.* 56(4):767–773.
507. Gaitan, B. 1992. Adverse effects of environmental pollutant exposure on the thyroid. In *Principles and Practice of Environmental Medicine.* Ed. AB Tarcher. New York: Plenum Book.
508. Cooksey RC, Gaitan B, Lindsay RJ, Hill J, and Kelly K. 1985. Humic substances: a possible source of environmental goitrogens. *Organ. Geochem.* 8:77–80.
509. Wolff J. 1969. Iodide goiter and the pharmacologic effects of excess iodide. *Am. J. Med.* 47:101.
510. Gaitan B. 1975. Iodine deficiency and toxicity. In *Proceedings of the Western Hemisphere Nutrition Congress IV.* Ed. PL White and N Selvey. Acton, MA: Publishing Sciences Group.
511. Suzuki H. 1980. Etiology of endemic goiter and iodide excess. In *Endemic Goiter and Endemic Cretinism* Ed. JB Stanbury and B Hetzel. New York: John Wiley & Sons.
512. Lazarus JH and Bennie BH. 1972. Effect of lithium on thyroid function in man. *Acta Endocrinol.* 70:166.
513. Emerson CH, Dyson WL, and Utiger RD. 1973. Serum thyrotropin and thyroxine concentrations in patients receiving lithium carbonate. *J. Clin. Endocrinol. Metab.* 36:338.
514. Child C, Nolan G, and Jubiz W. 1976. Changes in serum thyroxine, triiodothyronine and thyrotropin induced by lithium in normal subjects and in rats. *Clin. Pharmacol. Ther.* 20:715.
515. Gaitan B. 1988. Goitrogens. *Bailliere's Clin. Endocrinol. Metab.* 2(3):683–702.
516. Goldberg RC, Wolff J, and Greep RO. 1957. Studies on the nature of the thyroid-pituitary interrelationship. *Endocrinology* 60:38.
517. Goldberg RC, Wolff J, and Greep RO. 1955. The mechanism of depression of plasma protein bound iodine by 2,4-dinitrophenol. *Endocrinology* 56:560.
518. Wayne BJ, Koutras DA, and Alexander WD, eds. 1964. *Clinical Aspects of Iodine Metabolism.* Oxford: Blackwell Scientific Publications.
519. Wolff J, Standaert MB, and Rall, J. 1961. Thyroxine displacement from the serum and depression of serum protein-bound iodine by certain drugs. *J. Clin. Invest.* 40:1373.
520. Bastomsky CH, Murphy PVN, and Banovac K. 1976. Alterations in thyroxine metabolism produced by cutaneous application of microscope immersion oil: effects due to polychlorinated biphenyls. *Endocrinology* 98:1309.
521. Allen-Rowlands CF, Casracane VD, Hamilton MF, and Seifter J. 1981. Effect of polybrominated biphenyls (PBB) on the pituitary-thyroid axis of the rat (41099). *Proc. Soc. Exp. Biol. Med.* 166:506.
522. Bastomsky CH. 1977. Enhanced thyroxine metabolism and high uptake goiters in rats after a single dose of 2,3,7,8-tetrachlorodibenzo-p-dioxin. *Endocrinology* 101:292–296.
523. Potter CL, Sipes IG, and Russell DH. 1983. Hypothyroxinemia and hypothermia in rats in response to 2,3,7,8-tetrachlorodibenzo-p-dioxin administration. *Toxicol. Appl. Pharmacol.* 69:89.
524. Newman WC, Fernandez RC, Slayden RM, and Moon, RC. 1971. Accelerated biliary thyroxine excretion in rats treated with 3~methyl-cholanthrene. *Proc. Soc. Exp. BioL Med.* 138:899–900.
525. Bastomsky CH and Papapetrou PD. 1973. Effect of methylcholanthrene on biliary thyroxine excretion in normal and Gunn rats. *J. Endocrinol.* 56:267.
526. Goldstein JA and Taurog A. 1968. Enhanced biliary excretion of thyroxine glucuronide in rats pretreated with benzopyrene. *Biochem. Pharmacol.* 17:1049.
527. Melander A, Sundler F, and Ingbar SH. 1973. Effect of polyphloretin phosphate on the induction of thyroid hormone secretion by various thyroid stimulators. *Endocrinology* 92:1269.
528. Toccafondi RS, Brandi ML, and Melander A. 1984. Vasoactive intestinal peptides stimulation of human thyroid cell function. *J. Clin. Endocrinol. Metab.* 58:157.

529. Delange F and Ahluwalia R, eds. 1983. *Cassava Toxicity and Thyroid: Research and Public Health Issues.* IDRC-207e. Ottawa: International Development Research Centre.
530. Roti B, Grundi A, and Braverman LB. 1983. The placental transport, synthesis and metabolism of hormones and drugs which affect thyroid function. *Endocrine Rev.* 4:11.
531. Gaitan B. 1973. Water-borne goitrogens and their role in the etiology of endemic goiter. *World Rev. Nutr. Diet* 17:53.
532. van Hulse JH, ed. 1980. *Polyphenols in Cereals and Legumes.* IDRC-145e. Ottawa: International Development Research Centre.
533. Burges NA, Hurst HM, and Walkden, B. 1964. The phenolic constituents of humic acid and their relationship to the lignin of the plant cover. *Geochem. Cosmo Chim. Acta* 1547:1554.
534. Choudhry GG. 1981. Humic substances: I. Structural aspects. *Toxicol. Environ. Chem.* 4:209.
535. Schnitzer M and Khan S. 1972. *Humic Substances in the Environment.* New York: Marcel Dekker.
536. Osman AK and Fatah AA. 1981. Factors other than iodine deficiency contributing to the endemicity of goitre in Darfur Province (Sudan). *J. Hum Nutr.* 35:302.
537. Pitt WW, Jolley RL, and Jones O. 1979. Characterization of organics in aqueous effluents of coal-conversion plants. *Environ. Int.* 2:167.
538. Jahnig CB and Bertrand RR. 1976. Aqueous effluents from coal-conversion processes. *Chem. Eng. Prog.* 72:51.
539. Klibanov AM, Tu TM, and Scott KP. 1983. Peroxidase-catalyzed removal of phenols from coal-conversion waste waters. *Science* 221:259.
540. Rea WJ. 1994. *Chemical Sensitivity, Vol. II, Sources of Total Body Load.* Boca Raton, FL: Lewis Publishers. 765.
541. Shackelford WM and Keith LH. 1976. *Frequency of Organic Compounds Identified in Water.* Athens, GA: Environmental Research Laboratory.
542. Jolley RL, Gaitan B, Douglas BC, and Felker LK. 1986. Identification of organic pollutants in drinking waters from areas with endemic thyroid disorders and potential pollutants of drinking water sources associated with coal processing areas. *Am. Chem. Soc. Environ. Chem.* 26:59–62.
543. Bull GM and Fraser R. 1950. Myxoedema from resorcinol ointment applied to leg ulcers. *Lancet* 6:851.
544. Quentin JO and Hobson BM. 1961. Varicose ulceration of the legs and myxoedema and goiter following application of resorcinol ointment. *Proc. R. Soc. Med.* 44:164.
545. Walfish PO. 1983. Drug and environmentally induced neonatal hypothyroidism. In *Congenital Hypothyroidism.* Ed. JH Dussault and P Walker. New York: Marcel Dekker.
546. Jolley RL, Gorchev H, and Hamilton, DH Jr., eds. 1978. *Water Chlorination: Environmental Impact and Health Effects, Vol.II.* Ann Arbor, MI: Ann Arbor Science Publishers.
547. Jolley RL, Pitt WW Jr., Scott CD, Jones O Jr., and Thompson JB. 1975. Analysis of soluble organic constituents in natural and process waters by high-pressure liquid chromatography. In *Trace Substance in Environment Health, Vol. IX.* Ed. DD Hemphill. Columbia, MO: The University of Missouri. 247–253.
548. Rea WJ. 1994. *Chemical Sensitivity, Vol. II, Sources of Total Body Load.* Boca Raton, FL: Lewis Publishers. 535.
549. Castor CW and Beierwaltes WH. 1955. Depression of serum protein-bound iodine levels in man with dinitrophenol. *J. Clin. Endocrinol. Metab.* 15:862.
550. Nemeth S. 1958. Short-term decrease of serum protein-bound iodine concentration after administration of 2,4-dinitro-phenol in man. *J. Clin. Endocrinol. Metab.* 18:225.
551. Peakall DB. 1975. Phthalate esters: Occurrence and biological effects. *Residue Rev.* 54:1.
552. Keyser P, Pujar BO, Baton RW, and Ribbons DW. 1976. Biodegradation of the phthalates and their esters by bacteria. *Environ. Health Perspect.* 18:159–166.
553. Bahn AK, Mills JL, Synder PJ, Gann PH, Houten L, Bialik O, Holimann L, and Utiger RD. 1980. Hypothyroidism in workers exposed to polybrominated biphenyls. *N. Engl. J. Med.* 302:31.
554. Gaitan B. 1992. Adverse effects of environmental pollutant exposure on the thyroid. In *Principles and Practice of Environmental Medicine.* Ed. AB Tarcher. New York: Plenum Publishing. 371–387.
555. Pitt WW, Jolley RL, and Jones O. 1979. Characterization of organics in aqueous effluents of coal-conversion plants. *Environ. Int.* 2:167.
556. Jahnig CB and Bertran RR. 1976. Aqueous effluents from coal-conversion processes. *Chem. Eng. Prog.* 72:51.
557. Klibanov AM, Tu TM, and Scott KP. 1983. Peroxidase-catalyzed removal of phenols from coal-conversion waste waters. *Science* 221:259.
558. Meuzelaar HL, Windig W, Harper AM, Huff SM, McClennch WH, and Richards M. 1984. Pyrolysis mass spectrometry of complex organic materials. *Science* 226:268–274.

559. Morris SC, Moskwitz PD, Sevian WA, Silberstein S, and Hamilton LD. 1979. Coal-conversion technologies: some health and environmental effects. *Science* 206:654–662.
560. Gaitan B. 1973. Water-borne goitrogens and their role in the etiology of endemic goiter. *World Rev. Nutr. Diet.* 17:53.
561. Gaitan B. 1983. Endemic goiter in western Colombia. *Ecol. Dis.* 2:295.
562. Gaitan B. 1993. Anfithyroid compounds. In *Thyroid Diseases: Clinical Fundamentals and Therapy.* Ed. F Monaco, MA Sata, B Shapiro, and L Troncone. Boca Raton, FL: CRC Press. 615–625.
563. Matovinovic J and Trowbridge FL. 1980. North America. In *Endemic Goiter and Endemic Cretinism.* Eds JB Stanbury and B Hetzel. New York: John Wiley & Sons.
564. Weetman AP and McGregor AM. 1984. Autoimmune thyroid disease: developments in our understanding. *Endocrine Rev.* 5:309.
565. Biggazi PB and Rose NR. 1975. Spontaneous autoimmune thyroiditis in animals as a model of human disease. *Prog. Allergy* 19:245.
566. Esquivel PS, Kong YM, and Rose NR. 1978. Evidence for thyroglobulin-reactive T cells in good responder mice. *Cell Immunol.* 37:14.
567. Meuzelaar HLC, Windig W, Harper AM, Huff SM, McClennen WH, and Richards JM. 1984. Pyrolysis mass spectrometry of complex organic materials. *Science* 226:268.
568. Gaitan B. 1983. Endemic goiter in western Colombia. *Ecol. Dis.* 2:295.
569. Rogan WJ, Bagniewska A, and Damstra T. 1980. Pollutants in breast milk. *Engl. J. Med.* 30:1450.
570. *Safe Drinking Water Committee: Drinking Water and Health,* 1977. Washington, DC: National Academy of Sciences.
571. Andelman JB and Sness MJ. 1970. Pc lynuclear aromatic hydrocarbons in the water environment. *Bull. WHO* 43:479.
572. Pelkonen O and Nebert DW. 1982. Metabolism of polycyclic aromatic hydrocarbons: etiologic role in carcinogenesis. *Pharmacol. Rev.* 34:189.
573. Lo MT and Sandi B. 1978. Polycyclic aromatic hydrocarbons (polynuclears) in foods. In *Residue Reviews,* Vol. 69. Ed. FA Gunther. New York: Springer Verlag.
574. Rea WJ. 1994. *Chemical Sensitivity,. Vol. II, Sources of Total Body Load.* Boca Raton, FL: Lewis Publishers. 539.
575. Rea WJ. 1994. *Chemical Sensitivity, Vol. II, Sources of Total Body Load.* Boca Raton, FL: Lewis Publishers. 583.
576. Biggazi PB and Rose NR. 1975. Spontaneous autoimmune thyroiditis in animals as a model of human disease. *Prog. Allergy* 19:245.
577. Farid NR. 1987. Immunogenetics of autoimmune thyroid disorders. *Endocrinol. Metab. Clin. N. Am.* 16:229–245.
578. Scherbaum WA. 1993. Pathogenesis of autoimmune thyroiditis. *Nuklearmediziner* 16:241–24.
579. Rea WJ. 1996. *Chemical Sensitivity, Vol III: Clinical Manifestations of Pollutant Overload.* Boca Raton, FL: CRC Press. 1628–1629. Tab.24.4 & 24.5.
580. Rea WJ. 1996. *Chemical Sensistiviy. Vol. III: Clinical Manifestations of Pollutant Overload.* Boca Raton, FL: CRC Press. 1633. Tab. 24.6.
581. Grant E. 1986. *The Bitter Pill: How Safe Is the Perfect Contraceptive?* London: Corgi Books. 119.
582. Rea WJ. 1996. *Chemical Sensitivity, Vol III: Clinical Manifestations of Pollutant Overload.* Boca Raton, FL: CRC Press. 1635. Tab. 24.7.
583. Rea WJ. 1996. *Chemical Sensitivity, Vol III: Clinical Manifestations of Pollutant Overload.* Boca Raton, FL: CRC Press. 1655. Tab. 24.10.
584. Rea WJ. 1996. *Chemical Sensitivity, Vol III: Clinical Manifestations of Pollutant Overload.* Boca Raton, FL: CRC Press. 1658, Tab. 24.11.
585. Rea WJ. 1996. *Chemical Sensitivity, Vol III: Clinical Manifestations of Pollutant Overload.* Boca Raton, FL: CRC Press. 1619. Fig. 24.3.
586. Rea WJ. 1996. *Chemical Sensitivity, Vol III: Clinical Manifestations of Pollutant Overload.* Boca Raton, FL: CRC Press. 1620. Fig. 24.4.
587. Rea WJ. 1996. *Chemical Sensitivity, Vol III: Clinical Manifestations of Pollutant Overload.* Boca Raton, FL: CRC Press. 1632. Fig. 24.5.
588. Rea WJ. 1996. *Chemical Sensitivity, Vol III: Clinical Manifestations of Pollutant Overload.* Boca Raton, FL: CRC Press. 1638. Fig. 24.7.
589. Federman DD. 2006. The biology of human sex differences. *NEJM.* 354(14):1507–1514. Fig. 2.
590. Rea WJ. 1996. *Chemical Sensitivity, Vol III: Clinical Manifestations of Pollutant Overload.* Boca Raton, FL: CRC Press. 1657. Fig. 24.10.

591. Rea WJ. *Chemical Sensitivity, Vol III: Clinical Manifestations of Pollutant Overload*. Boca Raton, FL: CRC Press. 1665. Fig. 24.11.
592. Rea WJ. *Chemical Sensitivity, Vol III: Clinical Manifestations of Pollutant Overload*. Boca Raton, FL: CRC Press.1666. Fig. 24.12.
593. Grande F, Anderson JT, and Keys A.1958. Serum cholesterol in man and the unsaponifiable fraction of corn oil in the diet. *Proc. Soc. Exp. Biol. Med.* 98:436–440.
594. Spritz N, Ahrens EH. Jr, and Grundy SM. 1965. Sterol balance in man as plasma cholesterol concentrations are altered by exchanges of dietary fats. *J. Clin. Invest.* 44:148–1493.
595. Connor WE, Witiak DT, Stone DB, and Armstrong ML. 1969. Cholesterol balance and fecal neutral steroid and bile acid excretion in normal men fed dietary fats of different fatty acid compositon. *J. Clin. Invest.* 48:1363–1375.
596. Grundy SM, Aherns EH, Jr. 1970. The effects of unsaturated dietary fats on absorption, excretion, synthesis, and distribution of cholesterol in man. *J. Clin. Invest.* 49:1135–1152.
597. Ostlund RE Jr., Racette SB, and Stenson WF. 2002. Effects of trace components of dietary fat on cholesterol metabolism: Phytosterols, oxysterols, and squalene. *Ann. Rev. Nutr.* 478.

Index

A

Aanatomical categories of homeostatic mechanism, 12
Abortions, spontaneous, 478
Acetylation, 142
Acetylcholine, 208, 420
    accumulation, 304
    action, 190
    concentrations, 277
    initiated by, 236
    neurotoxic dose of, 213
    neutralizing dose of, 214
    release, 208
    supersensitivity to, 213
Acetylcholinesterase (AChE), 302
    inhibition of, 487
    synthetic OP compounds and, 304
Acquired immunity, 357–360
Activated T-cells by suppressor T-cells, 366
Activation-induced apoptosis, 400
Actively mobile phagocytic leukocytes, 71
Acupressure, 209
Acupuncture, 16
    acupuncture energy flow system (AES), 251
        dermatomes, 252
        hyperaesthetic zones, 254
        meridians, 255, 258–259
        points on body, 253–257
        segmental reference of deep pain, 254
        small and large stimulus, 257
Acute hypersensitivity, 1
Acute nerve injury, 219
Acute pollutant-induced response, 195
Acute respiratory distress syndrome, 70
Acute thrombosis with rupture, 153
Acylation, 142
Adaptation
    principle, 118
    response, 119
Adaptive control, 128
ADH, *see* Antidiuretic hormone (ADH)
Adjustment mechanism, 22
Adjustment responses, 117
    local cell and matrix reactions for, 107–108
        biological rhythms and periodic signals, 109
        principles and facts about, 117–123
Adrenal genital syndrome, 454, 464
Adrenal glands
    homeostatic regulatory dysfunction, symptoms of, 454
    hyperadrenalism, 454
    hypoadrenalism, 454–456
    physiology of homeostasis and dyshomeostasis in, 447, 453–454
        abnormal levels of epinephrine and norepinephrine, 449
        aldosterone, production and control of, 449
        corticosteroids, production and control of, 450
        cortisol malfunction, patterns of, 450
        epinephrine and norepinephrine, secretion of, 448
        inducers of cortisol release, 451
        inflammation, sign of, 453
        mobilization of fatty acids, alteration in, 453
        nerve supply of adrenal, 447–448
        pollutant injury, effect of, 448
        protein metabolism, disturbance in, 451, 453
        RAS, function of, 448–449
    pollutant injury, 447–448
Adrenal steroid synthesis pathway, 451
Adrenocortical glucocortical hormone, 82
Adrenocorticotrophic hormone, 247
Advanced coronary artery disease
    microscopic studies, 63
Agglutination, 374, 376
Aging; *see also* Immune system
    bone marrow, 403
    CD4 + cells, 404
    immunosenescence, 404
    stem cell, 403
    T-cell differentiation and, 405
    thymus gland, 403
Agonist, 102
Air toxics, 91
Albumin
    chemical sensitivity, 83
    colloid osmotic pressure in plasma, 82
    detoxification mechanisms, 82
ALF, *see* Autogenous lymphocytic factor (ALF)
Aliphatic hydrocarbons in blood, 297
Allergic rhinitis, 212
Allergy
    and hypersensitivity, 401–402
    mast cells and basophils, 74
Alpha melanocyte stimulating hormone, 248
Amino acid
    concentration in blood, 82
    dietary alterations and, 82
Ammonia effect, 453
Amplification system, 47
Anaphylaxis, 74
Aneurysm, 54
Angioedema, 458–459
Angiogensis, 148
Antagonist, 102
Antibody
    classes of, 374
    onset of, 372–373
    response in circulating blood, 372
Antidiuretic hormone (ADH), 431

Antigen–antibody reaction
    classical pathway, 374
        cascade of reactions, 375
Antihumoral antishock conditions, 136
Antioxidant therapy, 64–65
Antipollutant enzymes, 129
    production, 53
Aortic wall
    stimulation of mesenchyme in, 54
Aperiodic homeostatic disturbance, 16
Aphrodisiacs, 482
Apoptosis, 398
    rate, neutrophil-mediated inflammatory
        response, 70–71
Argyrophilic cells, 445
Arrhythmias, microscopic studies, 63
Arterial stenosis, 54
Arteriosclerosis, 54
Arthralgia, 217
Astrocytes, 273
Atrazine-herbicides estimated annual agricultural use, 93
Atrial fibrillation
    sympathetic and parasympathetic impulses, 131
Autogenous lymphocytic factor (ALF), 16, 396
Autoimmune thyroiditis, 503–504
Autoimmunity, 366
    diseases, tolerance mechanism, 395–396
Automaticity of body for survival, 8
Autonomic nerves, 46
    of eye, 235
Autonomic nervous system (ANS), 185
    anatomy and physiology, 190–198
    case study, 198
    changes in
        chemical sensitivity, 194
        chronic degenerative disease, 194
    dysfunction
        breakdown of functions, 210–211
        case studies, 191, 205–206, 213–214
        denervation supersensitivity, 212–215
        ectodermal structures dilated, 209
        fascial tightness and spasms, 209
        hands and feet, vasospasm of, 207
        parasympathetic nerve stimulation, 210
        sweating, 210
        symptom-driven cholinergic responses, 208
        wakefulness and sleep, 212
        weakness and shakiness, 205
    hypothalamus, central connection with, 198–199
    and immune system, 351–353
    nerve measurement, heart rate variability by method of
        Riftine, 194
    reflex pathways, 202
        regionalization, 205
        routes of, 204
    triggering, 191–192
    vascular and neuroendocrine response, 218
Axonal degeneration, 220
Axon reflex, 166

B

Bacterial and fungal infection, 70
BALT, *see* Bronchial-associated lymphoid tissue (BALT)

Baroreceptor system, 127
Basement membranes, 39
    alteration in, 40–41
    and tumors, 41
Basic fibroblast growth factor (bFGF), 147
Basophils
    in circulating blood, 74
    responsiveness of, 74
B-cells, renewal mechanism, 76
BDNF, *see* Brain-derived neurophic factor (BDNF)
Beçhet's disease, 233
    sural nerves in, 94
Benomyl effect, 483
Beta-adrenergic effects on T-cells, 244
bFGF, *see* Basic fibroblast growth factor (bFGF)
Bicepital tendonitis, 217
Biochemical individuality, 38
    of response, 121–122
Bioflavonoids, 499–500
Biological rhythms and periodic signals of normal
        homeostasis, 109
Biorhythm, 109
Biphasic model
    and chronic degenerative disease, 5–6
    dose response challenges, 5
    response, 5
Bipolarity, 120–121
Birth control pill, and disordered homeostasis, 466–467
Bisphenol-A, 488
Block water-soluble toxic agents, 273
Blood
    carbon dioxide concentration in, 51
    mineral levels in, 52
    mold antibody assay, 370
    oxygen concentration in, 51
    pressure and heart rate responses to upright tilt table
        testing, 226–227
Blood-brain barrier
    chemically sensitive and/or chronic degenerative
        disease, 262
    extracellular basement membrane, 264
    function of, 263
    glial cells wrapping around capillaries, 263
    noxious injury after penetration of, 264–266
    pollutant injury to, 260–261
    sensitivities to toxicants, 265
Body's communication system
    CTM, 4
        amorphous ground substance of, 15
        canuliculi of, 31
        communication system properties, ECM and, 13
        dynamic properties and physiology, 19–20
        electrical reorientation of, 28
        and electromagnetic energy, 25–26
        end blood vessels embedded in, 17
        energy dissemination systems in, 22–23
        holistic homeostatic control system, 8
        instability, 24
        integration of communication, 10
        lability of, 23–25
        lymph channel in, 59
        mechanical coherence, 27–28
        mechanical support system, 12–13
        molecular sieve, 34–37, 46–48

Index

multiple episodes of skeletal injuries, 12–13
open-ended, instability and lability after pollutant overload, 24
as open system, 21, 23
organ cells and, 7
receptors, 101
remodeling, macrophages role in, 33
sensitivities to foods, molds and chemicals, 12–13
solid structures of, 13–14
three-dimensional architecture, 14
tissue physical properties, 14
traumatic injury, 13
GRS, homeostatic mechanism, 12
intravascular content, 62–66
cells, 67–81
neurological aspects of homeostasis, 85
plasma lipids, 83–85
plasma proteins, 82–83
skin and mucous membranes, 12
vascular homeostasis, 67
diffusion, 48–49
intravascular content, 62–67
lymphatics, 56–59, 62
microcirculation, 52–55
vascular sieve, 49–52
vascular tone, 55–56
Brain
and causing damage, 273
chemicals effect on, 486–487
receptors, 100
response from toxicity, causes of variation, 267
Brain-derived neurophic factor (BDNF), 273
Bronchial-associated lymphoid tissue (BALT), 344–350
Bronchopulmonary neuroendocrine cells, 443
Bronchospasm, 249
Brush pile effect, 28
Butyrylcholinesterase (BChE), 305

C

Calciphylaxis, 63
Calcitonin, 444
Calcitonin-gene-related-peptide (CGRP), 166
Calcium, homeostasis in human, 456
Capillary endothelial swelling, 53
Capillary sphincter response, 54
Carbohydrates as information communication system, 35
Carbon dioxide concentration in blood, 51
Carcinogens, risk assessments, 5
Carcinoma elimination and ECM, 13
Cardiac beat, biorhythm, 109
Cardiac myocytolysis in cardiomyopathy, 63
Carnitine, 85
Caspases, 280
activation, 281
neuronal degeneration and, 283
pathway in mitochondria, 281
Catecholamines, 74
Catechol-O-methyl transferase (COMT), 435
Causlgia, *see* Reflex sympathetic dystrophy (RSD) syndrome
Cell cycle, 389
abnormal, 393
analysis of peripheral blood lymphocytes, 394

and cancer, 393
clock programs, 391
in healthy controls and patients, 394
molecular switch of, 394
phases of, 390
progression in patients with CFS, 392
stages, 392
Cell death, 397, 399
equilibrium between cell proliferation and, 396
Cell-mediated immunity, 359, 363
Cell reaction and recovery, 284
Cellular and humoral immunity
interactions of, 378
Cellular homeostatic control mechanism
stages, 135
Cellular repair process, 140
Central and obstructive apnea, 212
Central homeostasic adjustment, 112
autonomic and somatic integration, 113
awareness, 113
in brainstem, 113
hypothalamus control, 114–115
limbic system control, 115
reticular formation, 113
Central nervous system (CNS), 185
Central neuropathy
associated diagnosis in patients with, 296
intradermal skin testing
for biological inhalants, 300
for chemicals, 299
for metals, 300
organochlorine pesticides in blood with, 298
posturography, 301
pupillograpaphy, 302
Central neurotoxicity
patients with, 296
Triple Headed SPECT Brain scan measurements for, 301
Cervical lymph nodes, 352
CGRP, *see* Calcitonin-gene-related-peptide (CGRP)
Chemically sensitive patients
biopsies of fat and blood in, 50
constant tachypnea or underbreathing, 130
dyshomeostatic patients, 262
effects of autonomic system stimulation from pollutant injury, 211
and healthy volunteer controls with pupillography, 241
HRV-chronotrophic in, 221
immunological data, 58
inhaled double-blind challenges, 292
intradermal peptide testing in, 246
neurotransmitters hypersensitivity by intradermal skin testing, 206
phagocytic index, recurrent infections, 72
pollutant overload
nucleophils and electrophils, noncatalytic reactions between, 142
WBC counts in, 70
Chemical pollutants, 215–216
Chemicals released into Dallas air, 289–290
Chemoreceptors, 188
Chemotaxis, 376
Chloracne, 166

Chlorinated hydrocarbons, in chemically senstitive patients, 423
Cholecystokinin, 420
Chondroitin 4-and 5-sulfate, 15
Chronic degenerative disease, 1
   and/or hypersensitivity, 7, 9
   chemically sensitive and oxygenation at tissue level, 54
   energy drain and, 23
   environmental aspects of, 2
   lymphocytes and, 4
   physiologic changes, 156
   response, 119–120
Chronic fatigue syndrome, 22, 223–224, 433, 447
   fibromyalgia and, 226
Chronic hypersensitivity, 1
Chronic mucosal swelling, 212
Chronic noxious incitant entry
   clinical signs and symptoms, 9–10
   holistic response of body homeostatic mechanism, 9
   kinetics of, 9
   prodromes, 9
   total body pollutant load, 9
Chronic organophosphate induced neuropsychiatric disorder (COPIND), 309–310
Cigarette smoking, impact on mother and child health, 466
Clotting-unclotting equilibrium (CUE), 63–64
Coagulopathy, 65
Collagen
   for communication, 15
   and PG/GAGs, wound heal, 4
   as supporting structure, 14
   types, 14
Colloid in matrix in injured patients and weather changes, 13
Communication network
   pollutant-nutrient balance and pollutant overload, 10–11
   for regulating of homeostatic mechanism, anatomical categories used in, 12
Communicators, monosaccharides and polymers, 36
Compartmentalization of homeostatic dysfunction, 112
COMT, see Catechol-O-methyl transferase (COMT)
Concomanant etiological diagnosis, 297
Congealed plasma, 65
Congestive heart failure
   microscopic studies, 63
Conjugation detoxification process, 53
Conjugation reactions, 142–143
Connective tissue matrix (CTM), 4
   amorphous ground substance of, 15
   canuliculi of, 31
   communication system properties ECM and, 13
   dynamic properties and physiology, 19–20
   electrical reorientation of, 28
   and electromagnetic energy, 25–26
   end blood vessels embedded in, 17
   energy dissemination systems in, 22–23
   holistic homeostatic control system, 8
   instability, 24
   integration of communication, 10
   lability of, 23–25
   lymph channel in, 59
   mechanical coherence, 27–28
   mechanical support system, 12–13
   molecular sieve
      autonomic control, 46–48
      biochemical components, 35
      control of individual local homeostatic response, 36–37
      extracellular fluid, 34–35
      storage capacity, 45
      vesicles on ground regulation, 45–46
   multiple episodes of skeletal injuries, 12
      homeostasis and, 13
      internal joint pathology, 13
   open-ended, instability and lability after pollutant overload, 24
   as open system, 21, 23
   organ cells and, 7
   receptors, 101
   remodeling, macrophages role in, 33
   sensitivities to foods, molds and chemicals, 12–13
   solid structures of, 13–14
   three-dimensional architecture, 14
   tissue physical properties, 14
   traumatic injury, 13
Connon's law, 216
Contraceptive-induced illness origin, see Birth control pill, and disordered homeostasis
COPIND, see Chronic organophosphate induced neuropsychiatric disorder (COPIND)
Coronary artery
   aberrant nonspecific mesenchyme reaction in, 153
Corticotropin releasing factor (CRF), 455
Cortisol production, pollutants effect on, 450–453
Cranial manipulation, 16
CREST syndrome, 159
CRF, see Corticotropin releasing factor (CRF)
CUE, see Clotting-unclotting equilibrium (CUE)
Cupping, 232
Curb lysis in pathological surplus, 137
Cushing's disease, 454
Cyclic AMP (cAMP)
   mechanism for hormone control of cell function, 426
   noxious stimuli effect on, 432
Cyrptococosis, 233
Cytochrome P-450 enzymes, 468
Cytochrome system, 141–142
Cytokines, 33
   cell sources of, 150
   role in wound healing, 149–150, 152–153

D

Danocrine, 459
DDT pesticide, effect on sperm count, 488
Defense system
   early defense mechanism, 133–135
   redox system latency, 132–133
   tissue response to local entry
      case study, 135–136
      lymphocytic phase, 136–137
      macrophage phase, 136
      microphage phase, 136
      tissue bound histiocytic defense phase, 135–136
Degenerative and hypersensitivity response, 121

Index

Degranulating, 75
Dehydration, 28
Dehydroesterone, 457
Delayed-reaction hypersensitivity, 401
Denervation
    law of, 215
    supersensitivity, 212–215
Depression, 243–244
Dermatin sulfate, 15
DES, see Diethylstilbesterol (DES)
Detoxification processes, 1
    reactions, 141
Devic's disease, 215–216
Diabetes
    basement membrane thickening in, 54
    microscopic studies, 63
Diablo, apoptosis-inducing factor, 281
Diarrhea, 23
Diethylstilbesterol (DES), 460, 478, 480, 483, 487
Diffuse myofascial pain syndrome, 218
Dihydrotestosterone, 481
2,4-Dinitrophenol (DNP), 497
Dioxin exposure, and endometriosis, 473–474
Dioxins, halflife, 9
Disease
    defined, 9
    holistic treatment program, 10
    individuals, symptoms in, 9
    origin of, 22
    prevention and reversibility, 10
DNP, see 2,4-Dinitrophenol (DNP)
Dopamine, 188
Dose-response observation, 2
Driving detoxification, 49
Drugs effective dosage in humans, blood concentrations, 424
Drusen of optic nerve, 231–232
Dynamics of homeostasis, alteration in, 419
Dyshomeostasis, 2–3
    responses, 1
Dystrophy syndrome, 18

E

Early defense mechanism, 133–135
ECM, see Extracellular matrix (ECM)
Ecoestrogens, 489
Eczema, 67
Edema, 145
Elastica fragmentation, 232
Electromagnetic energy and CTM, 25–26
Electromagnetic receptors, 188
Electrostatic tone, 43–44
ELF, see Extremely low frequencies (ELF)
Embryonic stem cells, 338, 342
Emotional behavior and motivation over cingulate gyrus, 116
End blood vessels embedded in C.T. matrix, 17
Endocrine-disrupting chemicals, action mechanisms, 485
Endocrine glands
    location in body, 421
    pollutant injury to, 427
    role in homeostatic control mechanism, 420–428
Endometriosis, 470, 473
    dioxin exposure and, 473–474
    and dysmenorrhea, 474
    hormonal management of, 474
    patient data on, 475–477
    responses to hormonal fluctuations, 474
End organ failure
    optimal health and, 6
Endorphins hormone, 247
Endostatin, 41
Endosulfan pesticide, 428
Endothelial swelling, 53
    and oxygen deficiency, 80
End-stage fixed disease, nutrition, 1
Energy
    drain, 22
    efficient regulation of body, 9
    metabolism for homeostatic process, 8
Enolase isoenzymes, 439–440
Environmental load, 2
Environmentally triggered thrombocytopenia
    after deadaptation in ECU with total load decreased, 81
Environmental pollutants, 92
Environmental receptor system (ERS), 20
Environmental switches and gene function, 35
Enzyme and nonenzyme detoxification after pollutant overload, 143
Enzyme rate and nutrient deficiency, 155–156
Eosinophils, 72–73
Epigenes
    epigenetic switches
        and markers, 38
        phenomenon, 37
Epigenome, DNA with, 39
Epinephrine, secretion from adrenal medulla, 448
Epithelialization of skin, 148
Erythrocyte homeostasis
    abnormalities from oxidative injury, 80
Eserine insecticides, 237
Estradiol, role in male brain pattern, 461
Estrogen
    contaminated foods and abnormalities, 464–466
    effect of, 457–458
    endogenous estrogens, 487
    estrogenic chemicals, 492
    exogenous estrogens, 487–488
    human exposure to, 487
    MCF7 cells response to, 463
    in milk, 491
    mimics, 488–489
    natural and synthetic products, 463–464
    phytoestrogens, 488–491
    procarcinogenic effects of, 467–468
    receptors, 457
    Sertoli-cell number, fixing of, 492
    sexual orientation, role in, 463–464
    western diet and, 464, 487
Ethinyl estradiol, 488
Ethylene oxide, toxicity effects, 478
Everchanging dynamic process, 37
Ever-decreasing doses, 243

Exercise-induced T-cell depression, 244
Extra-adrenal paragangliomas, 446
Extracellular fluid, 15
    transport system, 34
Extracellular matrix (ECM)
    electrical potential, 7
    environmental information
        receptor system, 23
    enzymatic reactions in, 35
    instability
        open-ended system, 20
    makeup of, 106
    mechanical fluctuations, 29
    molecular sieve of, 29
    patterns of, 31
    receptors, 103
    reorganization of tissue, 22
Extremely low frequencies (ELF), 27
Extrinsic coagulation cascade, 147
Eye, pollutant injury in
    abducens, 234
    autonomic nerves, 234–236
    case study, 236
    cranial nerves, 228
    nervous system, 228
    oculo-motor nerve, 233
    optic nerve, 231–233
    parasympathetic nervous system, 237
    potential areas, 238
    sphenopalatine ganglion, 233–234
    trigeminal nerve, 233
    trochlear nerve, 233

F

False neurotransmitters, 429
Fat and blood
    chemically sensitive patients, biopsies in, 50
Fatty acids
    degradation and oxidation, 85
    energy requirements and oxidation, 85
Feedback information, 22
Feedback mechanism of dynamics of homeostasis, 9
Feed forward control, 128
Fibroblasts
    fibroblast growth factor 2 (FGF2), 147
    PG/GAGs and, 76
    proteins secreted, 77
Fibromyalgia, 22
Fibrositis, fibromyalgia, fibromyositis, 218
Filtering mechanism of body, 134
Fixed-named disease, 24
    phagocytic dysfunction, 71
Fixed-named end-stage disease, 1
Floroxzymestine, 458
Fluid exchange, 49
Fluoxetine drug, 96
Food and mold injection therapy, 82
Food pollutants, 90

G

GAGs, see Glucose aminoglycans (GAGs)
Gain-phase response, 126
GALT, see Gut-associated lymphoid tissue (GALT)
Gamma globulin injection, 16
Gastrointestinal tract
    and immune system, 337
G-CSM, see Granulocyte colony stimulating
        factor (G-CSM)
Generalization, 205
Genes
    epigenetic switches and, 37
    expression, linear concept of, 36
    and genetic defect, 37–38
    genetics and environmental responses, 39
    with long-term memory, macroorganic structures
        production, 35
Genetically dictated response system, 134
Genitourinary paraganglion cells, 445–446
Giant cell capsule, 78
Gilbert's disease, 38
Glia in brain and spinal cord, 274
Gluconation/glucuronidation, 142
Glucose aminoglycans (GAGs), 11, 74
    environmental receptor and communication vehicle, 15
    PG/GAGs
        and electrical orientation, 28
        heat of redox reaction, 23
        on hyaluronic acid tree, 14
        in mechanical coherence of tissue, 28
        output of, 54
        piezoelectric phenomena of, 28–29
    types of, 15
Glutathione replenishing pathway, 65
Glycoproteins for cell adhesion, 15
GM-CSF, see Granulocyte-monocyte colony stimulating
        factor (GM-CSF)
Goitrin, 496, 498–499
Goitrogens, 495
Gonadotropin hormones, 480
Granulocyte colony stimulating factor (G-CSM), 164
Granulocyte-monocyte colony stimulating factor
    (GM-CSF), 164
Granulocytes and inflammation, 164
Ground regulation system (GRS), 3, 419, 427–429, 437,
    469, 472, 483
    communication
        cell-connective tissue-environmental system
            and, 104
        function of, 44–45
        matrix-cell-reception feeding, 105
        receptors, 101
        system with functional unit, 11
        trophic neurogenic changes, 47
Growth hormone, pollutant effects on, 433–434
Growth rhythm, 109
GRS, see Ground regulation system (GRS)
Gulf of Mexico's dead zone, 97
Gut-associated lymphoid tissue (GALT), 139, 344–350

H

Hair-trigger response, 243
Hard connective tissue (HCT)
    types, 16
HCT, see Hard connective tissue (HCT)
Head's zones, 253

Healing and regeneration rhythm, 109
Heine cylinders
  anatomical configuration, 16
Helper T-cells, 378–379
Hemifusion intermediate, 269–270
Hemolytic serum complement, 375
Hemorrhage into meningeal sheaths, 232
Heparin
  and anaphylaxis, 74
  heparin sulfate, 15
Herbicides use, 90, 93
Herpes simplex, recurrent mouth ulcers, 68
Hexachlorobenzene toxics, 136
High-dose effect
  inhibition, 5–6
HIOMT, *see* Hydroxyindole-O-methyl transferase (HIOMT)
Hippocampus task, 272
Histamine, 147
Histocytic phase, 135
Hoarseness, 23
Holistic treatment program, 10; *see also* Disease
Homeodynamic equilibrium
  fibroblast vs. macrophage vs. mast cell, 154
Homeodynamics, 3
Homeostasis, 1
  chemically sensitive or chronic degenerative disease deregulation, 62–63
  communication network and anatomical categories used in, 12
  control mechanisms
    characteristics, 126
  dynamics of, 3
    blood vessels, 48
    ebb and flow, 24
    energy system, 23
    intracapillary matrix, 48
    normal function facts, 123–125
    orderly process, 22
    physical and mental energy, 6
    within physiologic boundaries, 6
    sequence, 8
    and survival, 7–8
  dysfunction, 2
  and dyshomeostasis in adrenal glands, 447, 453–454
    abnormal levels of epinephrine and norepinephrine, 449
    aldosterone, production and control of, 449
    corticosteroids, production and control of, 450
    cortisol malfunction, patterns of, 450
    epinephrine and norepinephrine, secretion of, 448
    inducers of cortisol release, 451
    inflammation, sign of, 453
    mobilization of fatty acids, alteration in, 453
    nerve supply of adrenal, 447–448
    pollutant injury, effect of, 448
    protein metabolism, disturbance in, 451, 453
    RAS, function of, 448–449
  initiation, 100
  integrated central mechanism, 100
  of isotonia, isoionia and isoosmia, 41–42
  of lymph flow, 61
  neurological aspects of, 85
  organization factor, 25
  and patient management, 6
  process
    energy metabolism for, 8
    holistic, 6
  regional and central, 15
  repair mechanism, 6
  restoration of, 71–72
  triggering, 23, 126
    agents, 3
  worsening and secondary foci, 129–130
Homeostatic control mechanisms, 365
Homeostatic dysfunction, disordered homeostatic response
  aperiodic response, 158–159
  case study, 158–159
  periodic response, 156–158
Homeostatic feedback mechanism, 242
Homosexual men, brain structure in, 460
Hormesis
  hormetic concept, 2–3
    dose-response model, 5
    evaluation, 6
Hormonal system; *see also specific hormone*
  control of hormone secretion rate, 420–421
  effects of dynamic homeostatic regulation of
    distal effect, 420
    local effect, 420
  function of, 419–420
  and nervous systems, relationship between, 419–420
Hormones like growth hormone, 82
Hot flashes, 457, 479–480
Human body
  anatomical loci of endocrine glands, 421
  autonomic effects on organs of, 201
  communication system
    concept and facts, 10
    environment knowledge and, 10
  dysfunction, 1
  and energy automaticity, 8–9
  energy efficient regulation, 9
  energy requirements and oxidation of free fatty acids, 84
  exposure, 305
  filtering
    effect, 80
    mechanism of, 134
  frequent complaints, 293
  functions
    alterations in, 7
    homeodynamic regulation, 9
    organ cells and CTM, 8–9
    redox processes, 8
    state of nutrition and, 8
  health effects related to endocrine function, trends in, 486
  homeostatic mechanism
    dynamics of, 4
    response nature, 6
  normal fluctuating physiology, 3
    periodic activity of, 29
    physical exercise, 29
  open receptor systems, 7
  organs, resonating effects
    response to pollutant load, 26

total pollutant load, 26
triglycerides, role in, 84
type of sensor in, 7
Humoral immunity, 363
Humoral shock phase, 136
Hyaluronic acid, 15
Hydrocortisone role, 455
2-Hydroxyestrone (2-OHE$_1$), 468
Hydroxyindole-O-methyl transferase (HIOMT), 435
Hyperaesthetic zones in internal disease, 254
Hypersensitivity
   and chronic degenerative disease, 9
   phase in treatment, 2
Hypoadrenalism, 424, 454–456
Hypotension, 222
Hypothalamic-hypophysial portal system, 432
Hypothalamus
   ANS, central connection with, 198–199
   autonomic and somatic integration, 115
   lesions and overload of noxious stimuli, 200
Hypothyroidism, 21
Hypoxia
   pathological changes with, 63
   reversible phase of, 80
   tissue, failure of extraction of oxygen, 130
   vicious down hill cycle of, 53

I

Idiopathic cyclic edema, 454
IE, *see* Intraepithelial (IE) lymphocytes
Immune system
   aging
      bone marrow, 403
      CD4 + cells, 404
      immunosenescence, 404
      stem cell, 403
      T-cell differentiation and, 405
      thymus gland, 403
   and ANS, clinical implications of, 351
   clinical regulation, 377
   diaphragm connecting with, 347
   dysfunction
      functional changes, 345
      gastrointestinal tract, 337
      left lymphatic trunk from heart, 346
   lymphatic channels
      stages in, 348
   lymphatic vessels
      from esophagus, 347
      from lungs, 345
   noxious excitants properties of entering
      antibody-forming response, 338
      biphasic dose response relationships, 338
      hormetic-like dose response, 338
      lymphocyte activation, 338
      responses of, 338
      stem cell, 338, 340–341
   regulation of, 380
   response, 137
   right lymphatic trunk from heart, 346
   toxicological response in, 379
Immunity
   acquired, 357–360
   innate, 353–357
Immuno-privileged tissue, 276
Immunosuppression, 385
IMS, *see* Integrated muscle stimulation (IMS)
Inappropriate apoptosis, 400
Incitant injury, 20
Incitant stimulus
   fibroblasts, 22
   nature of, 4–6
Indoor air pollution, 90
   sources of, 94
Inertia, 22
Inflammation, 20, 453
   antioxidant nutrients and, 63
   bacteria or toxic products spread, 162
   effects, 376
   granulocytes and monocytes, 164
   local tissue destruction, 162
   macrophage invasion, 163–164
   macrophage-neutrophil response, 164
   margination, 163
   mechanisms, 159, 162
   neurological system connection with CTM, 166–167
   and noxious environmental stimuli, 22
   oxidative and prolonged response of homeostatic response, 162
   signs, 147
   tissue damage, 147
Information intake and distribution
   nature, 7–8
Injury, 145
   of autonomic nervous system, 74–75
Innate immunity
   C5a–C3a balance, 356
   case study, 357
   chemical sensitivity and chronic degenerative disease, 353
   complement activation, 354
   endothelial injury, 354
   mast cell and, 355
   natural killer lymphocytes, 354
   in newborn and infant, 355
   processes, 354
   system response, 72
   tissue damage, 353
   triggering, 353
Insulin, 82
Integrated muscle stimulation (IMS), 209
Integrins, 147
Interleukin-1 (IL-1), 164
Intermittent flow phenomenon, 52
Interstitial cystitis, 445
Interstitial fluid (IF), 15
Interstitial nuclear anterior hypothalamus (INAH), 460, 462
Intestinal mucosa, 138
Intestinal peristalsis, biorhythm, 109
Intracellular redox equilibrium, glutathione role in, 71
Intradermal injection provocation-neutralization technique, 3
Intradermal neutralization therapy, 17
Intradermal treatment, 3

Intraepithelial (IE) lymphocytes functions, 350
Intraocular inflammation, 233
Intravagal neuroendocrine cells, 444
Intravascular components, 62–63
Intrinsic and extrinsic coagulation cascades, 146
Intrinsic mechanisms, 364
  antigen-induced activation of T-cells, 367
Iris corder, autonomic changes measured by, 239
Ischemic heart disease, microscopic
      findings in, 63
Ischemic stroke, 281
Isoionia, 43
Isoosmia, 41
Isoosmosis, 41
Isotonia, 41–42
Isotretinoin (retinoic acid), 478

J

JAK-STAT signaling pathway, 276
Jittery anxiety attack, 195
J-shaped dose response, 5

K

Keratin sulfate, 15
Kinins, 147
Krebs cycle function and energy generators, 141
Kupffer cells, *see* Macrophages

L

Lability characteristics, 24
Laminin, 45–46
Laser beams, 16
Lassitude, 22
Latent free radical response, 32–33
Lateral epicondylitis, 217
LCT, *see* Loose connective tissue (LCT)
Leukocyte
  balance, 69
  function characteristic, 137
  homeodynamics of, 69
  homeostasis
    topical toxicant load in respiratory tract, 68
  and leukocytolysis
    rate, 136
  and leukolysis, 68–69
  self-destruction, 137
Limbic system triggering, 240
Lindane, 459
Linear and biphasic (U-or J-like) dose effects, 1
Linear concept of gene expression, 36
Link proteins, 14
Lipid soluble substances, 49–50
Lipophilic toxics distribution, 51
Lipophilic xenobiotics, 85
Liver
  capillary sinusoids, 49
  damage, 457
Local anesthesia, 16
Local cell and matrix reactions for adjustment responses,
      107–108
  biological rhythms and periodic signals, 109

Local homeostatic responses and information
      reception, 99–103
Loose areolar tissue, 17
Loose connective tissue (LCT), 16
  reconstruction of, 18
Low-dose effect
  stimulation, 5–6
Lyme disease, 233
Lymphatics, 56
  adult stem cells, 343
  arterial capillaries, 58
  communication in dynamics of holistic
      homeostasis, 57
  dynamic homeostasis and, 76
  enteromammary circulation, 357
  life spans, 76
  lymphatic channels and lymphocytes, development
    anatomy/physiology, 342
      hypersensitivity and chronic degenerative disease,
        pain patterns, 344
      lymph sacs, 342, 344
      TD, 344
  lymph drainage, 57
  lymph flow, 58–59
  lymph node
    afferent denervation of, 245
    nerve ending in, 244
    organization, 78
  lymphocytes, 360
    B-lymphocytes, 368–369
    foreign compounds and, 361
    T-lymphocytes, 362–364
  lymphocytic channels, 62
  lymphocytic phase, 136
    and immune processes, 137
  overflow mechanism, 62
  populations, 76
Lymphocytopenia, 453
Lysis, 374, 376
Lysosomes developing in cardiac tissue, 67

M

Macrophages
  homeodynamic balance with fibroblasts, 77
  liver lining, 79
  and proteolytic enzymes, 4
  tissue in
    classification, 77–78
    integral components of alveolar walls, 78
Magnets application, 16
Malathion insecticides estimated annual agricultural
      use, 93
Male homosexuality, 461–462
MALT, *see* Mucosa associated lymphoid tissue (MALT)
Mammary carcinoma regulator, 45
Margination, 163
Masking phenomenon, 189
Massage, 16
Mass discharge of sympathetic nerves, 196–197
Mast cells
  activation, 376
  degranulating, 75
  factor, apoptosis, 74

Fce-RI dependent activation, 74
human peritoneal adhesions, 75
IgE and non-IgE dependent substances, 73
intraepithelial T-lymphocytes in, 73
nitric oxide, regulated by, 74
types of, 73
Matchstick test, 218
Matrix metalloproteinases (MMPs), 145
Matrix receptor, 103–107
Mechanoreceptor-mediated bradycardia, 222
Mechanoreceptors, 188
Mecholyl insecticides, 237
Meckel's ganglion, 233
Mediterranean anemia, 38
Medusa like appearance, 80
Meissner's corpuscles, 19
Melatonin, 435–436
Membrane permeability, 189
Membrane resting potential (MRP), 32
Mental function
    environmental treatment, effects of, 287
Metallo proteinase, 33
    over expression, 41
Metals and inorganic neurotoxic compounds, 303
Methylation, 142
Methyltestosterone, 493
Microcirculation, 52
    toxic exposures and, 53
Microclot and microplaque formation, 65
Microglia, 276
Minute paragangliomas, 443
Misinformation triggers, 154
MMPs, see Matrix metalloproteinases (MMPs)
Modern medicine and holistic approach, 9–10
Mold and mycotoxin immunity study, 362
Molecular cellular injury in disease, 63
Molecular sieve, 29–30
    basement membranes, 39–41
    control, 31–32
    and CTM
        autonomic control, 46–48
        biochemical components, 35
        control of individual local homeostatic
            response, 36–37
        extracellular fluid, 34–35
        storage capacity, 45
        vesicles on ground regulation, 45–46
    dynamic, 30
    electrostatic tone, 43–45
    experimental stress and
        latent free radical response, 32–33
    genetic and environmental control, 37–39
    kaleidoscope effect, 31
    regulation of and by isoosmia, isotonia and
        isoionia, 41–43
Molecules
    distribution, 25
    random distribution, entropy and enthalpy, 25
Monocytes
    and inflammation, 164
    macrophage system, 77
    monocyte colony stimulating factor (M-CSF), 164
Monosaccharides and polymers as communicators, 36
Mood states scores profile, 295

MRP, see Membrane resting potential (MRP)
Mucosa associated lymphoid tissue (MALT), 133, 344
    adhesion molecules, 348–349
    antigen sampling, 348
    chemical sensitivity and chronic degenerative
        disease, 345
    Goldstein
        thymosin types of, 349–350
    intraepithelial (IE) lymphocytes
        functions, 350
    lymphatic vessels from lung, 345
    lymph nodes, 349
    noxious stimuli, 348
    Peyer's patches, 348–349
    pneumonia-like symptoms, 349
    principle source of protein, 345–346
Mucous membrane
    barrier and sieve effect, 12
    electric and magnetic tone, 7
    integration of communication, 10
Multiple sclerosis
    demyelination and axonal degeneration, 308
Muscarinic effect, 190
Muscle shortening, 216
    bands, 217
    electromyography examination, 217
    inflammatory pain, 218
    multiple tender trigger points, 217–218
    self-perpetuating cycle, 217
Myelinated–unmyelinated fibers, 193
Myocardial fibrosis, 63
Myofascial pain syndrome, 218

N

NAET, see Nambudripod Allergy Elimination Technique
    (NAET)
NALT, see Nasal associated lymphoid tissue (NALT)
Nambudripod Allergy Elimination Technique
    (NAET), 260
Nasal associated lymphoid tissue (NALT), 344–349
    neuroimmune regulation
        acetylcholine/cholinergic
            agonists, 351
        cholinergic innervation, 351
        enteric neurons, 351
        neuropeptides, 350–351
        norepinephine, 350
        sympathetic nervous system,
            activation, 351
National Institute for Occupational Safety and Health
    (NIOSH), 286
Natural killer (NK), 243
    enhancement activity, 383–384
    function of, 383
    and specialized CD4 + regulatory
        T-cells, 365–366
    tumor cell lysis and signal
        transduction, 382
Naturally occurring toxins in food, 98
Necrotic cell death, 397–398
Needle acupuncture, 209
Needle modalities, 16
Negative feedback process, 126

Index

Neovascularization, 148
Nerve ending in lymph nodes, 244
Nerve-mast cell, 247
Nervous system, 185
Neurally mediated hypotension (NMH), 220–222
    case study, 222
    fibromyalgia and, 226
Neuroendocrine cells, 436–437
    biosynthetic products of, 438–440
    clinical manifestations of pollutant stimulation, 446–447
    extra-adrenal neuroendocrine cells, 436
        major families of, 442
    neuroendocrine phenotype, 437–438
    noxious stimulus, effect of, 437
    and paragangliomas, 446
    secretory granules in, 440
    specialization of, 440–441
    topography of subsets of, 442
        branchiomeric group, 442–444
        intravagal cells, 444
        visceral autonomic, 444–445
Neuroepithelial bodies, 443
Neurogenesis, 276
    edema, 218
    inflammation, 166
Neurogenic vascular responses to pollutant stimuli
    homeostatic response, 237
Neurohormonal response, 114
Neuroimmunological mediators
    pathogenesis, 243–245
Neurological system connection with CTM and inflammation, 166–167
Neuromuscular hypotension, 223–224
Neuromyelitis optica (NMO), see Devic's disease
Neurons
    caspases and, 283
    electrical message, 274
    function, 271
    neural dysfunction
        mechanism of action, 286
    neuronal pool, 271
    physiology, principles of, 270
    with synapse, 268
        lipids and, 269
        long-term sensitivities, 272
Neuropathy, 220
Neurotherapy and related therapies, 19
Neurotoxic chemicals
    chemical groupings in, 302
    clinical manifestation of, 288
    organic solvents, 305
    organic substances, 304
    pesticides, 306
Neurotoxicity, 19
    neurotoxic effect on balance, 292
Neurotoxicity target esterase (NTE), 308
Neurotransmitter, 190
Neutralization, 374
Neutrophils
    homeostasis, 71
    inflammatory cells, 146
    mediated inflammatory response, 70
Nicotinic effect, 190
NIOSH, see National Institute for Occupational Safety and Health (NIOSH)
Nitric oxide synthase, 481
NMH, see Neurally mediated hypotension (NMH)
Nocioreceptors, 188
Nodular necrosis of muscularis, 232
Noncarcinogens, risk assessments, 5
Nonimmune detoxification, homeostasis and defense mechanism, 139
Nonimmune enzyme detoxification systems, 154
Nonlinearity of response, 22
Nonrapid eye movement (NREM) sleep, 212
Nonspecific defense mechanism of homeostasis, 160–161
Nonspecific mesenchyme reaction (NSMR), 32; see also Repair mechanism
    angiogensis, 148
    in arteriosclerosis, 54
    case study, 144
    chemical sensitivity and/or chronic degenerative disease, 54
    connective tissue and vascular inflammation, 54
    epithelialization of skin, 148, 150–151
    fibrosis, 54
    hemostasis, 147
    hypoxia/pollutant entry, 53
    inflammation, 147–148
    migratory phase, 148
    proliferative phase and collagen synthesis, 151
    scar remodeling, 152
    wound
        contraction, 151
        healing process, 146
Nonspecific receptors, 188
Norepinephrine, 208, 420
    secretion from adrenal medulla, 448
Noxious excitants, properties of entering, 337
    antibody-forming response, 338
    biphasic
        dose-response relationships, 341, 348
        dose-responses, 339
        immune responses, 340
    hormetic-like dose response, 338
    lymphocyte activation, 338
    responses of immune system, 338
    stem cell
        cord tissue, 338
        heart disease, 340
        leukemia, 340
        Parkinson's disease, 340
        respiratory system, 340
        rheumatoid arthritis, 340
        type I diabetes, 341
Noxious stimuli, 1
    accentuations of physiologic processes by, 64
    communication with, 10
    dissemination of, 16
    dynamics, 3–4
    effect on electrical membrane potential, 189
    entry, linear and biphasic effects, 187–188
    environmental incitants

high-dose inhibition, 3
low-dose stimulation, 3
hormesis, 5
hypothalamus lesion in, 200
inflammation, 22
myriad of, 33
neutralization, 8
onset of symptoms, 9
pharmacological and pathological response, 16
pollutant load, 4
regionalization of, 110
repair damage, 8
swings in his/her physiology, 21
tissue macrophages, 4
NREM sleep, *see* Nonrapid eye movement (NREM) sleep
NSMR, *see* Nonspecific mesenchyme reaction (NSMR)
NTE, *see* Neurotoxicity target esterase (NTE)
Nutrients
antioxidative, 67
enzyme processes and, 132
basement membranes and, 29
and cellular replacement, 51
deficient individual, 2
depletion levels of, 140
and enzymes, 155–156
as fuels
for detoxification, 155
for physiologic adjustment processes, 155
for wound healing, 155
immune function and, 384
inflammation and clotting, 63
injection of, 16
intactness of NSMR and, 143–144
from lungs and gastrointestinal tract, 29
malabsorption, 155
and noxious stimuli, 158
plasma and intracellular, 51
sequestration process in, 140
in serum, 51
total body pollutant overload, 155

O

Occlusive coronary artery disease, 63
Occulomotor nerve, 195
Oligodendrocyte in CNS, 275
Open informational feedback loop response system, 22
Open system for CTM, 21–22
OPICN, *see* Organophosphorus ester-induced chronic neurotoxicity (OPICN)
OPIDN, *see* Organophosphorus ester-induced delayed neurotoxicity (OPIDN)
Opsonization, 375–376
Optic nerve injuries
chemical, 233
radiation, 233
Organisms, energetically open information collecting and response systems, 21
Organization/activation hypothesis, 460–461
Organochlorine pesticides, in chemically sensitive patients, 423
Organophosphate
carbamate pesticide exposure, 141
insecticides, 210, 237

Organophosphorus ester-induced chronic neurotoxicity (OPICN), 303
characteristics of, 310–311
epidemiological studies, 310
mechanisms of, 314
apoptosis, 315
necrosis, 315
occurrence and severity, 316
neurological and neurobehavioral alterations, 312–313
neuropathological
alterations, 312–313
lesions, 313
subclinical exposures, 311–312
toxic exposure, 311
Organophosphorus ester-induced delayed neurotoxicity (OPIDN), 302, 307
mechanisms of, 308–309
protein kinases as targets for, 309
types, 308
Osmosis, 41
Ossified sulci of dura matter, 17–18
Osteoarthritis, 217
Osteopathic manipulation, 16, 209
Ovary
estrogen (*see* Estrogen)
and organization/activation hypothesis, 460–461
pollutants, adverse effects of, 458–460
sex hormones in, 457–458
Overall hypersensitivity and chronic degenerative response, 121
Oxidation process, 63, 65
Oxidative stress on CS and CDD blood, 66
Oxidosis, 8
Oxygen
concentration
in blood, 51
in IF, 51
deficiency, 80
reducing electrical potential, 8
therapy and dietary alteration, 54

P

Pain, 47
reflex, 112
Papilledema, 231
Paragangliomas, 446
Paraoxonase, 305
Parasympathetic nervous system, 195
pollutant chemicals and, 239
Parathormone, 456
Parathyroid glands, 456
P450 aromatase, 461
Patient management, 6
Pattern recognition receptors (PRRs), 354
PCBs, *see* Polychlorinated biphenyls (PCBs)
PD-ECGF, *see* Platelet-derived endothelial cell growth factor (PD-ECGF)
PDGF, *see* Platelet-derived growth factor (PDGF)
Pelvic inflammatory disease, 470
Pentachlonphenol, 481
Peptide histidine isoleucine (PHI), 246
Peptide histidine methionine (PHM), 246

Index

Periodic focal homeostatic disturbances, 16
Periodic homeostatic disturbances, 131
Periodic or aperiodic homeostatic disturbances, 126
Peripheral and periorbital edema, 63
Peripheral autonomic nervous system, 191
    cranial-sacral innervation of parasympathetic nerves, 203
    spinal cord, sympathetic innervation of, 203
Peripheral immunoregulatory mechanisms, 364
Peripheral nervous system (PNS), 185
Peripheral neuritis
    arsenic-exposed and unexposed subjects
        ANOVA analysis, 294
Peripheral regulatory immune mechanism, 367
Peristalsis, 195
Pervasive information-gathering matrix receptors, 47
Pesticides use, 90, 93
Petechiae, 63
Peyer's patches, 244
Phagocytosis, 70, 375–376
    phagocytic index, 72
    toxics, effect on, 133
Phenol
    challenge, abnormal SPECT scan after, 20
    exposure, 131
Pheochromocytomas, 446
PHI, see Peptide histidine isoleucine (PHI)
PHM, see Peptide histidine methionine (PHM)
Phthalate extraction study, 95
Phytoestrogens, 488, 490–491
Phytosterols, 490–491
Piezoelectric effect of mechanical compression, 28
Pilocarpine insecticides, 237
Pinealcytes, 434
Pineal gland, 434
    as alarm clock for biorhythms, 435–436
    environmental factors, influence of, 434
    light, effects of, 435
    location and cell types, 434
    melatonin secretion by, 435–436
    pollutant-induced imbalance of zinc and copper in, 436
    role in homeostatic balance, 435
    wake-sleep cycles, role in, 434
Piperine drug, 96
Pituitary gland
    pollutant effects on, 428
        growth hormone, 433–434
        nervous system of pituitary, 429
        physiology of pituitary, 429–433
Plasma proteins, 49, 82–83
Platelet-derived endothelial cell growth factor (PD-ECGF), 147
Platelet-derived growth factor (PDGF), 147
Platelet homeostasis
    case study, 81
    plasma
        lipids, 83–85
        proteins, 82–83
Pollutant exposure
    damage, 49
    reversibility and irreversibility of cell damage after, 278
        acute central nervous system injury, 279
Pollutant injury
    adrenal glands, 447–448
    and autonomic nerve, 209
    to blood-brain barrier, 261
    of eye
        abducens, 234
        autonomic nerves, 234–236
        case study, 236
        cranial nerves, 228
        nervous system, 228
        oculo-motor nerve, 233
        optic nerve, 231–233
        parasympathetic nervous system, 237
        potential areas, 238
        sphenopalatine ganglion, 233–234
        trigeminal nerve, 233
        trochlear nerve, 233
Pollutants entry and homeostatic response to and fate of noxious stimuli
    adjustment responses, 107
    defense system, 132–143
    local receptors, 99–107
    repair mechanism, 143–156
    total body load, 98–99
    total environmental load, 85–86, 90, 93
Pollutant-stimulated assault in tissues, 71
Polychlorinated biphenyls (PCBs), 484, 486, 498
Polymorphonuclear leukocytes, 70
Positive feedback mechanism, 127–128
Postganglionic response to pollutants, 208
Precipitation, 374
Precipitory injury, 220
Pregnenolone, 457
Premenstrual syndrome (PMS), 465, 469
    altered homeostasis in, 470–471
    characteristics of symptoms of, 469–470
    chemically sensitive patient and, 471–472
    nutritional deficiencies and, 471
    symptom patterns, 469
Primary and secondary foci, 129
Progesterone, and spontaneous abortion, 479
Proinflammatory cytokines, 165
Pro-oxidant pollutants, 65
Prostaglandins, 147
Protein
    permeability
        susceptibility to allergic responses, 50
        vascular and lymphatic channels, 50
    phosphate buffers, 68
    PMNs and lymphocytes, 71
Proteoglycans (PGs), 11
    PG/GAGs
        and electrical orientation, 28
        heat of redox reaction, 23
        on hyaluronic acid tree, 14
        in mechanical coherence of tissue, 28
        output of, 54
        piezoelectric phenomena of, 28–29
PRRs, see Pattern recognition receptors (PRRs)
Psoriasis, 67
Pulsed random control system theory, 52

Pupillography
  autonomic nerve disturbance, 242
  chemically sensitive patients
    and healthy volunteer controls with, 241
  pupilograph, 21
Purpura, 63

R

Radical progenitors in living systems, 33
Radiculopathic pain conditions, 218
Rapid distribution phenomenon, 34
Rapid eye movement (REM), 212
RAS, *see* Renin-angiotensin system (RAS)
Raynaud's phenomenon, 63
Receptors
  characteristics, 102
  chronic stimulation, 102
  expression, 102
  number and activity, 103
  systems, biphasic dose–response relationships, 187
  types, 102
Redosis, 8
Redox processes
  and body functions, 8
  redox potentials using Nehm equations, 8
Redox system latency, 132–133
Reflex pathways, 202; *see also* Autonomic nervous system (ANS)
  regionalization, 205
  routes of, 204
Reflex sympathetic dystrophy (RSD) syndrome
  after head injury and stroke, 227
  demineralization and osteoporosis, 227
  patient study, 229–230
  peripheral nerve injuries, 227
  signs, 227
  stages of, 227–228
Regional homeostatic adjustment, 110–111
Regionalization, 4, 205
Regulation thermography scanning
  chart, 60
  measurement points and index, 59
Regulatory pathway
  antigen-induced activation, 366
  mediated by Qa1-dependent CD8 + T-cells, 365
REM, *see* Rapid eye movement (REM)
Renin-angiotensin system (RAS), 448
Repair mechanism, 1
  healing, nonspecific mesenchyme reaction
    angiogensis, 148
    case study, 144
    epithelialization of skin, 148, 150–151
    hemostasis, 147
    inflammation, 147–148
    migratory phase, 148
    proliferative phase and collagen synthesis, 151
    scar remodeling, 152
    wound contraction, 151
    wound-healing process, 146
Resorcinol, 496, 500–501
Respiratory mucosal edema, 249
Respiratory tract
  topical toxicant load in, 68
Response-ready triggering, 243
Resultant adverse body response, 16
Reticular activating system in brain stem, 114
Reticuloendothelial system, 77
Revascularization of injured area, 148
Rhinorrhea, 23
Rhythmical signal output, 272
Rhythmic periododicity, 29
Riftine method for nerve measurement by heart rate variability, 194
Robust dynamic intestinal health, 138
RSD syndrome, *see* Reflex sympathetic dystrophy (RSD) syndrome

S

Schwann cells, 274
SCT, *see* Soft connective tissue (SCT)
Secretin, 420
  delayed gastric emptying and, 420
  metabolic impact of, 420
  receptors of, 420
Segmental demyelination, 220
Segmental reactions, 111–112
Self-perpetuating cycle, 217
Selye's alarm reaction and response, 157
Sensory neuropeptides
  modulation of immunological responses by, 242
Sensory receptors, 188
Sequestration, 139–140
Serotonin, 147, 189
Sertoli cell multiplication, 492
Sexual dysfunction, in females, 468–469
Sickle cell anemia, 38
Skeletal injuries
  multiple episodes of, 12
    homeostasis and, 13
    internal joint pathology, 13
Skin
  barrier and sieve effect, 12
  galvanometer, 21
  integration of communication, 10
  membrane, electric and magnetic tone, 7
  testing for chemicals, 297
Sleep apnea disturbance, 212
Sleep-wake
  cycles, 434–435
  rhythm, 109
Smokers, microscopic studies, 63
Smooth muscle innervation, 208
SNAP (soluble NSF attachment protein) receptors (SNARE) proteins, 266
Soft connective tissue (SCT), 13
Somatostatin (SOM), 242
Sound sensitivity, 2
Soy isoflavones, 468
Spasm, 217
Spastic dysphonia, 443
Specific incitant response
  cigarette smoke, 120
  contact lens, 120
Sphenomaxillary fossa, 233

Sphingosine 1-phosphate gradients and T-cell
    trafficking, 384
  THI and DOP treatment, 387–389
Spleen functional structure, 79
Spondylosis, 219
  evolution of chronic degenerative disease
    and, 219
Spontaneous bruising, 63
Spontaneous oxidation, 24
Spreading phenomenon, 122–123, 243
Spring effect, 28
Squalene, 490
Stable dynamic function, 3
Staphylococcal infection, 162
Steroidogenic pathways, 452
Stimulatory phase, 120
Stress-inducing mental states, 137
Subcutaneous tissue edema, 218
Sublingual therapy, 3
Subsequential healing process, 22
Subtle brain dysfunction with mental fogginess, 22
Sudett's atrophy, 18
Sugar swings, 433–434
Sulfur conjugation, 142
Superoxide equation, 64
Supersensitivity, 212–215
  and normal response, 216
  smooth and striated muscle, 216
Sural nerves in Behcet's Disease, 94
Surgical and medical sympathectomy, 194–195
Switch phenomenon, 123
Sympathetic nervous system
  mass discharge, 196
  pollutant chemicals and, 240
Symptomatology, 49
Synchronization of biorhythms, 109

T

Tachycardia, 220–222
T-cells
  mediated immunity, 358
  renewal mechanism, 76
Temporary vessel occlusion, 53
Testes
  chemicals as endocrine disruptors, 483–484
    mechanisms of action of, 484–485
  disorders due to contamination, 483
  homeostasis in, 480–481
  hormones, and behavioural abnormalities, 484–486
  low testosterone, and homeostatic dysfunction, 483
  testosterone, action of, 481–482
  toxics, effects of, 481
Testosterone, 480–481
  action of, 481–482
  pollutants effect on, 483
Tetrahydrocannabinol (THC), 479
TGF, see Transforming growth factors (TGF)
THC, see Tetrahydrocannabinol (THC)
Thermodynamic fluctuations, 21
Thermogram, 21
Thermoreceptors, 188
Thoracic duct (TD)
  intrathoracic organs, lymph of, 344

Thyrocalcitonin, 456
Thyroidal peroxidase, 97, 496
Thyroid gland, 493
  agents acting directly on
    class I, 495–496
    class II, 496
    class III, 496
    class IV, 496
    class V, 496
  agents acting indirectly on, 497–498
  contaminants, and homeostatic dysfunction, 498
    aliphatic disulfides, 499
    2,4-DNP, 501
    PCBs and PBBs, 502–504
    phthalates and phthalic acids, 501–502
    polyaromatic hydrocarbons (PAHs), 504
    polyphenols, 499–500
    resorcinol, 500–501
    thiocyanate and isothiocyanates, 498–499
  physiology and pathophysiology of, 493, 495
Thyroid gland C-cells, 444
Thyroid hormones, 422
Thyroid stimulating hormone (TSH), 247
TIMPs, see Tissue inhibitors of metalloproteinases
    (TIMPs)
Tissue
  breakdown and healing process, 22
  proteins, reversible equilibrium among, 83
  remodeling after injury, 74
  response phases, 135
Tissue inhibitors of metalloproteinases (TIMPs), 152
T Lymphocytes, 243, 362–364
TNF, see Tumor necrosis factors (TNF)
Tolerance threshold of circulatory sympathetic system, 47
Toluene, case study on exposure, 430
Toroid effect of Smith, 27
Total body pollutant load, 9, 98–99
  avoidance, fasting and rotary diet, 107
  barrel, representation, 118
  dose exposure and, 38
  enzyme and nonenzyme detoxification, 143
Total environmental pollutant load, 85
  analysis of, 90
  body's homeostatic mechanism and, 93
  exposed to toxic chemicals, 86
  food pollutants, 90
  indoor air pollution, 90
  pesticide and herbicides use, 90
  process of injury, 97
  specific and nonspecific, 90
  water pollutants, 90
Toxic chemicals
  explosure, 86
  neurological effects of, 286–288
Toxic exposure
  principles of response after, 278
Toxic neuropathy, 288
Toxic Release Inventory, 87–89, 291
Toxics phenomena
  leukocytes and macrophages, effect on, 133
  in muscle and fascia, sequestration, 16
  xenobiotic catabolism, 140
Toxins in foods, 97
Transforming growth factors (TGF), 147

Traumatic injury and CTM, 13
Trichloroethylene, 293
Triglycerides role in body, 84
TRI on-site and off-site disposal or other releases by state, 87–89
Triple Camera SPECT brain scans, 249–250
Tris fire retardant (2,3-dibromopropylphosphate), 481
Trophedema, 218
TSH, see Thyroid stimulating hormone (TSH)
Tumor
    cell lysis and signal transduction, 382
    extracellular matrix regulation
        by matrix vesicles, 46
Tumor necrosis factors (TNF), 164

U

Ubretid insecticides, 237
UPK, see Uroplakins (UPK)
Urinary bladder paragangliomas, 446
Uroplakins (UPK), 33
U.S. EPA tracking program
    contaminants study, 86
Uterus and fallopian tubes, pollutant injury to, 469, 472–473; see also Endometriosis; Premenstrual syndrome (PMS)
    influence of toxics on hormonal homeostasis, 479
        PCBs, 479
        pesticides/herbicides, 479
    menopause dysfunction, 479–480
    spontaneous abortion, 478
    and uterine bleeding, 472–473

V

Vagal paraganglioma, 446
Vagina
    effect of homeostasis alteration on, 480
    vaginal pain, 480
Vaginodynia and vulvodynia, 445
Vagus nerve, 192
Vascular endothelium
    in homeostasis, 56
    swelling, 53
Vascular factors, 57
Vascular homeostasis
    cell type, 67
        basophils and mast cells, 73–75
        eosinophils, 72–73
        erythrocyte homeostasis, 80–81
        fibroblast and macrophage homeostasis, 76–77
        leukocyte homeostasis, 67–70
        lymphocytes, 76
        platelet homeostasis, 81–85
        polymorphonuclear leukocytes, 70–72
        reticuloendothelial system, 77–80
    diffusion, 48–49
    intravascular content, 62–67
    lymphatics, 56–59, 62
    microcirculation, 52–55
    vascular sieve, 49–52
    vascular tone, 55–56
Vascular optic neuropathy, 232
Vascular physiology, 55
Vascular wall sieve, 49–50
Vasculitis
    environmentally triggered, 54
Vasoactive intestinal polypeptide (VIP), 246, 420
Vasoconstriction, 221
    of skin blood vessels, 197
Vasodilatation, 147
Vessel wall, 56
Vicious downward cycle, fixed-named disease, 24
VIP, see Vasoactive intestinal polypeptide (VIP)
Virchow-Robin spaces in central nervous system (CNS), 16
Visceral-autonomic paraganglion cells, 444–445
Viscero-sensitive axons, 166
Vitamin C, 63
Vitamin E, 398
Volatile organic screening test (VOST), 96
Voltage-gated calcium and sodium channels, 208
Voluntary central nervous system
    noxious injury to blood-brain barrier, 260
VOST, see Volatile organic screening test (VOST)
Vulvitis, 480

W

Wallerian type degeneration of axon, 302
Warfarin, 478
Water diffusion, leak and fluid interchange, 49
Water pollutants, 90
    contamination areas, 95
Water retention in hypersensitive patient, 429
Water-soluble substances, 49
Well-worn mental pathways, 273
White blood cells
    and stress on bone marrow, 68
Whole-integrated homeostatic mechanism, 23
Withdrawal/overcompensation phase, 120
Wound
    healing, 144
        contraction, 151
        cytokines role in, 149–150, 152–153
        phases of, 146
        physiology, 146
        proliferative phase and collagen synthesis, 151
        scar remodeling, 152

X

X-Linked inhibitor of apoptosis, 281

Z

Zinc dependent detoxification systems
    function, 140